Soil Ecology

Soil Ecology

Editor: Rick Thomas

Callisto Reference

www.callistoreference.com

Callisto Reference,
118-35 Queens Blvd., Suite 400,
Forest Hills, NY 11375, USA

Visit us on the World Wide Web at:
www.callistoreference.com

ISBN: 978-1-63239-798-0 (Hardback)

Printed in China.

Cataloging-in-publication Data

Soil ecology / edited by Rick Thomas.
p. cm.
Includes bibliographical references and index.
ISBN 978-1-63239-798-0
1. Soil ecology. 2. Soil science. 3. Soils--Environmental aspects. 4. Agricultural ecology. I. Thomas, Rick.
QH541.5.S6 S65 2017
577.57--dc23

Table of Contents

Preface

Soil Ecology encompasses the study of the numerous interactions taking place in the soil. It includes the interactions between organisms as well as that with environmental factors. Soil Ecology directly affects the process of the plant growth. This book aims to shed light on some of the unexplored aspects of this field. Coherent flow of topics, student - friendly language and extensive use of examples make this book a valuable source of knowledge. Researchers, students and professionals engaged in the field of agriculture, botany, soil science and horticulture will find this book extremely useful.

This book aims to highlight the current researches and provides a platform to further the scope of innovations in this area. This book is a product of the combined efforts of many researchers and scientists from different parts of the world. The objective of this book is to provide the readers with the latest information in the field.

I would like to express my sincere thanks to the authors for their dedicated efforts in the completion of this book. I acknowledge the efforts of the publisher for providing constant support. Lastly, I would like to thank my family for their support in all academic endeavors.

Editor

1

Influence of Vegetation Restoration on Topsoil Organic Carbon in a Small Catchment of the Loess Hilly Region, China

Yunbin Qin, Zhongbao Xin*, Xinxiao Yu, Yuling Xiao

Institute of Soil and Water Conservation, Beijing Forestry University, Beijing, China

Abstract

Understanding effects of land-use changes driven by the implementation of the "Grain for Green" project and the corresponding changes in soil organic carbon (SOC) storage is important in evaluating the environmental benefits of this ecological restoration project. The goals of this study were to quantify the current soil organic carbon density (SOCD) in different land-use types [cultivated land, abandoned land (cessation of farming), woodland, wild grassland and orchards] in a catchment of the loess hilly and gully region of China to evaluate the benefits of SOC sequestration achieved by vegetation restoration in the past 10 years as well as to discuss uncertain factors affecting future SOC sequestration. Based on soil surveys (N = 83) and laboratory analyses, the results show that the topsoil (0–20 cm) SOCD was 20.44 Mg/ha in this catchment. Using the SOCD in cultivated lands (19.08 Mg/ha) as a reference, the SOCD in woodlands and abandoned lands was significantly higher by 33.81% and 8.49%, respectively, whereas in orchards, it was lower by 10.80%. The correlation analysis showed that SOC and total nitrogen (TN) were strongly correlated ($R^2 = 0.98$) and that the average C:N (SOC:TN) ratio was 9.69. With increasing years since planting, the SOCD in woodlands showed a tendency to increase; however, no obvious difference was observed in orchards. A high positive correlation was found between SOCD and elevation ($R^2 = 0.395$), but a low positive correlation was found between slope and SOCD ($R^2 = 0.170$, $P = 0.127$). In the past 10 years of restoration, SOC storage did not increase significantly (2.74% or 3706.46 t) in the catchment where the conversion of cultivated land to orchards was the primary restoration pattern. However, the potential contribution of vegetation restoration to SOC sequestration in the next several decades would be massive if the woodland converted from the cropland is well managed and maintained.

Editor: Manuel Reigosa, University of Vigo, Spain

Funding: This study was supported by the Fundamental Research Funds for the Central Universities (NoTD2011-2), the Open Foundation of Key Laboratory of Soil and Water Loss Process and Control on the Loess Plateau of Ministry of Water Resources (201301), the National Natural Science Foundation of China (No. 41001362) and the College Student Scientific Research Training Project of Beijing Forest University (No. 201210022013). The funders had no role in study design, data collection and analysis, decision to publish, or preparation of the manuscript.

Competing Interests: The authors have declared that no competing interests exist.

* Email: xinzhongbao@126.com

Introduction

Afforestation and other vegetation restoration techniques have been considered effective practices for the sequestration of carbon (C) to mitigate carbon dioxide (CO_2) concentrations in the atmosphere [1–4]. Soil plays an important role in the global carbon cycle. The soil C pool to a one metre depth has been estimated to sequester approximately three times the amount of carbon sequestered by the atmospheric pool and about four times that in the biotic/vegetation pool [5]. Therefore, a relatively small change in the soil C pool can significantly mitigate or enhance CO_2 concentrations in the atmosphere [6–7].

Land-use change can significantly influence the accumulation and release of SOC. When an ecosystem is disturbed by land-use change, the original equilibrium of the soil carbon pool is broken and a new equilibrium is created. During this process, soil may act as either a source or a sink of carbon depending on the ratio between inflows and outflows [8]. Many studies have reviewed the effect of land-use change on SOC. For instance, deforestation for agricultural purposes is the primary reason for the SOC loss. It was reported that approximately 25% of SOC was lost by conversion of primary forest into cropland [9]. Houghton (1999) estimated that 105 PgC was released into the atmosphere due to the conversion of forests to agricultural lands between 1850 and 1990 [10]. However, afforestation and reforestation on agricultural lands have been cited as effective methods for increasing SOC pool and reducing the atmospheric CO_2 concentration [3,11]. Morris et al. (2007) observed that placing agricultural soils in deciduous and conifer forests resulted in soil carbon accumulations of 0.35 and 0.26 $MgCha^{-1}$ yr^{-1}, respectively. Based on a meta analysis of 33 recent publications, Laganière et al. (2010) reported that afforestation increased SOC stocks by 26% for croplands. Some studies found that croplands that were converted to abandoned lands or grasslands could also increase the SOC storage [9,13–14]. Therefore, understanding the influence of land-use changes on soil organic carbon is an important step in predicting climatic change and developing potential future CO_2 mitigation strategies.

To improve the carbon sink status of afforeseation or other vegetation restoration methods on agricultural land, it is necessary to understand the control mechanisms of SOC dynamics to allow more carbon storage in soils [1]. A variety of factors will affect the quantity and quality of SOC after land-use changes. For example, climate variations have a significant effect on SOC, including temperature, precipitation, and potential evapotranspiration changes [11,15]. Laganière et al. (2010) reported that SOC restoration after afforestation was found to vary with the climate zone, with the temperate maritime zone having a higher SOC increase than others by approximately 17%. In addition, landscape and elevation also have a pronounced effect on SOC change at the catchment scale. Slope is an important topographical factor that affects soil erosion and also has an important influence on the soil nutrient loss of the slope surface [16]. Wang et al. (2012) reported that the SOCD was higher in shady slopes than in sunny slopes, and gentle slopes would generally sequester more SOC than steep slopes. Elevation differences can cause climatic and biological changes in the soil-forming environment and can influence the vertical distribution of SOC [18]. In natural ecosystems, it has been extensively documented that carbon and nitrogen are closely related [4,19]. Increased nitrogen retention may increase the carbon sequestration potential [1]. In Panama, Batterman et al. (2013) found that symbiotic N_2 fixation has potentially important implications for the ability of tropical forests to sequester CO_2 [20]. N_2-fixing tree species accumulated carbon up to nine times faster per individual than neighbouring non-fixing trees. The soil C/N ratio also has a significant effect on the rate of decomposition of organic compounds by soil microorganisms [21].

Soil erosion, as the most widespread form of soil degradation, has a large impact on the global C cycle, causing a severe depletion of SOC pools in the soil [22–24]. The total amount of C released by soil erosion each year has been estimated at approximately 4.0–6.0 Pg/year [23]. China's Loess Plateau, covering approximately 6.2×10^5 km^2, has the world's most severe soil erosion because of its unusual geographic landscape, soil and climatic conditions, and long history (over 5000 years) of human activity. Over 60–80% of the land in the Loess Plateau has been affected by soil erosion, with an average annual soil erosion of 2000 to 20000 t km^{-2} yr^{-1} [25–26]. Severe soil erosion has resulted in land degradation, which was manifested primarily in the thinning of the soil layer, nutrient loss and fertility reduction, which has directly caused decreases in the local farmers' income and has economically and socially hindered sustainable development [27–29]. Unreasonable human activities, such as deforestation and tillage on slopes, have further intensified soil erosion and land degradation in the Loess Plateau [30].

Since the 1950s, the Chinese government has launched many large-scale projects to attempt to control soil erosion and restore vegetation in the Loess Plateau, including large-scale afforestation in the 1970s and comprehensive control of soil erosion on the watershed scale in the 1980s and 1990s. Despite these efforts, there have not been significant increases in ecological benefits, which is largely due to the limitations and influences of bad natural conditions and unreasonable ecological restoration techniques. To control soil erosion and improve the quality of the local environment, "Grain for Green" was initiated by the government in the Loess Plateau in 1999. The government demanded that the agricultural lands with a slope of over 25 degrees be converted to forest, terrace orchards or grassland. To compensate famers for their economic loss, they will be given grain, cash and planting stocks by the government as subsidies and incentives for converting cultivated land back to forest, orchards or grassland. Currently, the Loess Plateau environment appears to be experiencing a recovery following more than 10 years of vegetation restoration [31–32].

Over the past decade, large-scale vegetation restoration efforts have brought obvious land-use changes to the Loess Plateau, and also have significantly influence on SOC sequestration. Therefore, understanding the effects of these dynamic changes and the corresponding changes in SOC storage caused by land-use change driven by the implementation of the Grain for Green project is important in evaluating the environmental benefits of this ecological restoration project. Recently, many studies have focused on SOC changes induced by the Grain for Green project in the Loess Plateau. These studies have mainly focused on the SOCD of different land-use types, the SOC pool, the rate of SOC change, the variation in SOC among different land-use conversions and factors which have influenced SOC sequestration in the Loess Plateau after vegetation restoration efforts [32–34]. However, for most of the researches, the study time period was shorter than 10 years as vegetation restoration was driven by the Grain for Green project. The study areas have been mostly located at the gully region of the Loess Plateau [2,17,35–36]. Little is known about the gully area of the Loess hilly region. To gain a complete understanding of the SOC change after restoration in this region, we selected the Luoyugou catchment of the Loess Plateau as the study area, which has the typical geomorphologic characteristics of the gully area of the Loess hilly region.

In this study, we hypothesised that the SOCD varied with land-use types in this catchment, and there was an increase in SOC storage of the total catchment after 10 years of restoration. Therefore, the objectives of this study were to (i) quantify the SOCD of different land-use types differ significantly in the study catchment, (ii) analyze those factors affecting SOCD, (iii) estimate the contribution of land-use conversions on SOC sequestration in the study area, in the past 10 years of restoration, and (iv) discuss the uncertainties in potential SOC sequestration during future ecological restoration efforts in the Loess Plateau.

Materials and Methods

Ethics statement

The administration of the Tianshui Experiment Station of Soil and Water Conservation, Yellow River Conservancy Commission and local farmers which are the owners of study lands gave permission for this research at each study site. We confirm that the field studies did not involve endangered or protected species.

Study area

Tianshui city (104°35′–106°44′E, 34°05′–35°10′N) of Gansu province, China, is located in the western side of the Qinling Mountains, which is the transitional zone between the Qinling Mountains and the Loess Plateau and also belongs to the third sub-region of the Loess hilly region in the middle portion of the Yellow River and the second class tributary of the Wei River. This region has the typical geomorphology of the gully area of the Loess hilly region, which includes earth-rocky mountainous areas, Loess ridges and hilly areas. In this area, the catchment landscape is an agroforestry landscape, and the climate has obvious transitional characteristics. Because of these special geomorphologic characteristics, Walter Lowermilk, deputy director of the US Department of Soil Conservation Service, chose this region in which to build an experimental station for soil and water conservation in 1941. Tianshui station, Suide station and Xifeng station now are three the best-known experimental stations for soil and water conservation in the Loess Plateau.

The Luoyugou catchment (105°30′–105°45′E, 34°34′–34°40′N), located in northern Tianshui, is a part of the observation area of the Tianshui soil and water conservation station. Its total area is 72.79 km^2, with a range from 1165–1895 m above sea level. The main topographic type is loess ridge landform, and the average slope is 19°. It has a typical continental monsoon climate with a mean annual precipitation of 548.9 mm (1986–2004). Approximately 78% of the rainfall is concentrated between July and September. The average annual temperature is 11.4°C (1986–2004),and the annual evaporation is 1293.3 mm. The main soil type in this catchment is mountain grey cinnamon soil, which, according to the Food and Agriculture Organization of the United Nations Educational, Scientific and Cultural Organization (FAO-UNESCO), belongs to the Cambisol soil group and accounts for 91.7% of the soil in the region [37–38] (Figure 1).

The main land-use types for the Luoyugou catchment include woodland, wild grassland, orchards, cultivated land, and abandoned land. In the woodland, the major tree species is black locust (*Robinia pseudoacacia L.*), most of which was artificially planted. Woodlands that are more than 30 years old have good water condition and low density because planting is located mainly beside the place beside water ditches and shady slope. They are the retained trees of the planting projects in the late 1950s to 1970s, without human management now. And most of woodland at the age of 10 to 30 years since planting is found in Fenghuang forest farm located in the gully head of this catchment, which has good management and protecting. Since 1999, the government of Tianshui city has widely implemented the "Grain for Green" project. A lot of cultivated land in the ridge top and slope top has been converted into woodland. Most of the woodland is less than 10 years since planting, and has a high density and bad management. Wild grassland is usually found on stone mountain of the northern catchment where the soil layer is thin and slope is steep. Human activities are restricted in there, only sometimes sheep may reach. The major species are dahurian bushclover (*Lespedeza dahurica (Laxm.) Schindl.*), russian wormwood (*Artemsia sacrorum Ledeb.*), digitate goldenbeard (*Bothriochloa ischaemun (L.) Keng*) and bunge needlegrass (*Stipa bungeana Trin.*), etc. The main fruit species under orchards are cherry (*Cerasus pseudocerasus (Lindl.) G. Don*), apple (*Malus pumila Mill.*) and apricot trees (*Armeniaca vulgaris Lam.*). The management system of orchards is the conventional tillage which is that the weeds beneath the trees are eliminated with herbicides and dead leaves, dried fruit and twigs are removed by manual blowers. Besides, there are not irrigation facilities and the tree water demand all comes from the rain. Fertiliser and pesticide are also applied in the orchards. This catchment is a rainfed agriculture region with a long cultivated history. The major species in the cultivated land are wheat (*Triticum aestivum L.*), maize (*Zea mays L.*) and edible rape (*Brassica campestris L.*), etc. Abandoned land is the cultivated land has been the ceased farming activities, due to the demand of "Grain for Green" project, the bad environment conditions, the shortage of rural labour and the damage by wild animals, etc. After cultivated land abandoned, old field are spontaneously colonised by various plants, while a secondary succession process will gradually develop during different plant communities. Because of short time since abandoned, the main species in the abandoned land are annual and biennial herb and a bit of perennial herb, such as virgate wormwood (*Artemisia scoparia Waldst. et Kit.*) and green bristlegrass herb (*Utricularia australis R. Br.*).

Soil sampling and laboratory analysis

Based on the main land-use types and the percentage of each land-use-type area in the total catchment area (2008), we randomly selected 83 land-use blocks that included cultivated land (34 samples), abandoned cropland (9 samples), orchards (13 samples), wild grassland (8 samples) and woodland (19 samples) as investigation plots. Their latitudes and longitudes are shown in Table 1. Spatially separated plots were located at least 3 km from one another to help avoid pseudo-replication. In July 2012, the selected plots were surveyed and collected in the field using the Global Positioning System (GPS). The information recorded from the sampling plots in the field included geographic coordinates, slope, elevation, plant species, and water conservation measures. The years since planting or abandoned of the sampling plots were evaluated by recording tree diameters and heights, plant composition, the degree of decomposition of the topsoil crust and the crop's residual body, and by interviewing local farmers. In each sampling plot — woodland and orchards (10×10 m), cultivated land, abandoned land and wild grass land (5×5 m) — we randomly collected three soil samples using a stainless steel cutting ring 5.0 cm high and 5.0 cm in diameter to measure the soil bulk density (BD) of the topsoil layer (0–20 cm) and by taking five random soil samples between 0 and 20 cm depth with a 20 cm long soil auger. The five soil samples were manually homogenised to form a composite sample for each sampling plot, and a quarter of this sample was taken to the laboratory. These samples were air-dried and passed through a 2 mm sieve, while gravel and roots were removed from each soil sample. A quarter of each sub-sample was completely passed through a 0.25 mm sieve to determine SOC and TN. SOC was determined by the $K_2Cr_2O_7$-H_2SO_4 Walkey-Black oxidation method [39]. TN was measured using the micro-Kjeldahl procedure [40].

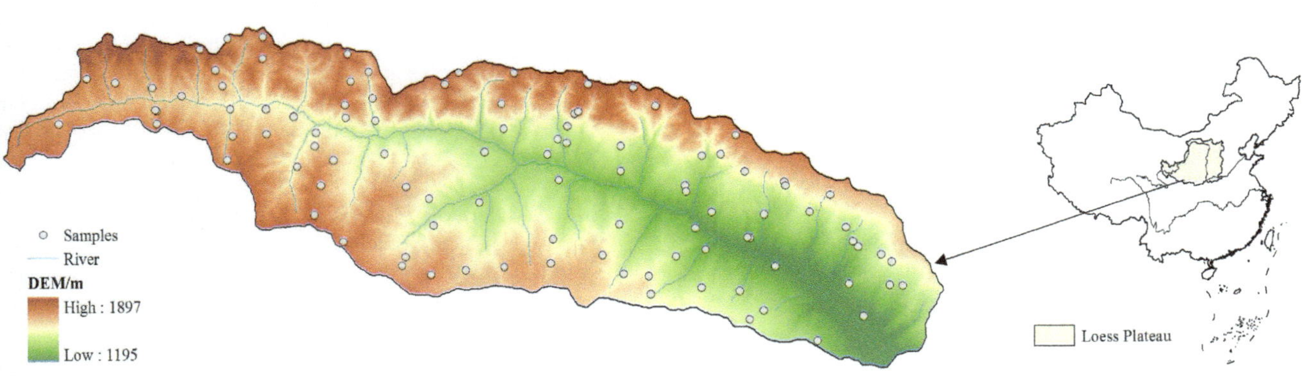

Figure 1. Location of the study area and distribution of the sampling points.

Table 1. Attributes of the studied sites.

Land-use	Sample sites	Years since planting (abandoned)/a	Elevation/m	Slope/(°)	Bulk density/(g/cm^3)	Plants
Cultivated land	34	—	1536.94	4.85	1.29	Wheat, maize, rape
Abandoned cropland	9	8–10a	1559.40	4.22	1.35	*Artemisia scoparia, Leymus secalinus*
Orchards	13	<5a, 5–10a, >10a	1446.45	1.92	1.32	Cherry, apple, pear
Woodland	24	<10a, 10a–30a, >30a	1639.14	12.13	1.39	Robinia
Wild grassland	8	—	1526.55	13.00	1.44	*Artemisia sacrorum var. incana mattf, Stipa bungeana, Bothriochloa ischaemun*

Statistical analysis

The SOCD was calculated using the following equation:

$$SOCD = SOC \times D \times BD \quad (1)$$

where SOCD was the soil organic carbon density (Mg ha^{-1}), SOC was the soil organic carbon content (%), D was the soil layer depth (cm), BD was the soil bulk density (g cm^{-3}).

The soil organic carbon storage (SOCS) was calculated using the following equation:

$$SOCS = SOCD_i \times A_i \times 100 \quad (2)$$

where SOCS was the soil organic carbon storage (t), $SOCD_i$ was the soil organic carbon density in the i land-use (Mg ha^{-1}), and A_i was the area of the i land-use (km^2).

The increase in SOCS caused by land-use conversion was calculated using the following equation:

$$\text{Increase of SOCD (t)} = SOCD_i \times (A_{i2}\text{-}A_{i1}) \times 100 \quad (3)$$

$$\text{Sequestration amount of SOCS (t)} = (SOCD_i - SOCD_c) \times (A_{i2} - A_{i1}) \times 100 \quad (4)$$

where $SOCD_i$ was the SOCD in the i land-use (Mg ha^{-1}), A_{i1} and A_{i2} were the i land-use area in 2002 and 2008, respectively (km^2), and $SOCD_c$ was the SOCD of the cultivated land (Mg ha^{-1}). All SOCD values of the different land-use types in 2002 and 2008 were the average SOCD values of the soil data collected in 2012.

All data were analysed using Excel 2007 and SPSS (Statistical Package for the Social Sciences) 18.0 software. Analysis of variance (ANOVA) was used to determine the significance of the mean difference. Fisher's LSD test was used to compare the mean values of soil variables when the results of ANOVA were significant at $p < 0.05$. Pearson's test was used to analyse the correlation between SOCD and the soil variables. For all analyses, a $p < 0.05$ was used to test statistical significance. Finally, SOCD, land-use type, elevation and slope values were used to establish the multiple regression model. SOCD was the dependent variable, and others were the independent variables.

Satellite data acquisition and processing

In our study, two types of SPOT5 multispectral images were acquired on August 25, 2002 and May 5, 2008, which were multi-spectral images (resolution: 10 m) and panchromatic images (resolution: 2.5 m). To validate the final classification result, we selected 17 points located in the Luoyugou catchment to use as the observed data based on the land-use map and some thematic maps. In July 2012, these monitoring points were collected in the field to validate the panchromatic map, while we recorded 83 points with information of land-use type and longitude and latitude to validate the accuracy of the classification results. Using ArcGIS and ERDAS software, we chose the common pre-classification, which is the supervised classification with a Maximum Likelihood Decision (MLD), to classify the preprocessed images and acquire the data of different land-use types in 2002 and 2008. The overall accuracy and Kappa coefficient of the initial classification images were 87.50% and 0.85 in 2002, respectively, and 90.10% and 0.87 in 2008, respectively. The accuracy of the 88 typical types of land-use information used to validate the classification results was 95.74%, which implied that this classification result could meet the analysis demands.

Results

Topsoil organic carbon density under different land-uses

The one-way ANOVA indicated that land-use type had a significant effect on the SOCD ($F = 8.34$, $P < 0.01$). However, only the woodland was significantly different from the other land-uses ($P < 0.05$), while no significant differences were found between other land-use types. The average SOCD of the topsoil layer (0–20 cm) in the Luoyugou catchment was 20.44 Mg/ha, and was highest in the woodland and lowest in the orchards (Table 2). Using the SOCD in the cultivated land (19.08 Mg/ha) as a reference, the SOCD in the woodland and abandoned cropland significantly increased by 33.81% and 8.49%, respectively, while the SOCD in the orchards decreased by 10.80%. The SOCD in the wild grassland was close to that of cultivated land. In this area, the coefficient of variation of the SOCD was 26.46%, and was highest (27.08%) in the woodland and smallest (16.75%) in the orchards.

Change of topsoil SOC storage in the catchment

Based on the interpretation of the remote sensing data (2000), the results showed that the total area of the four main land-use types including woodland, orchards, cultivated land and wild grassland was 67.5 km^2, accounted for 92.09% of the total region of the Luoyugou catchment. The difference of 7.02% included river, residential points, and roads that were not taken into account in the calculation of the topsoil SOC storage. From 2002 to 2008, the net decrease in the cultivated land area was

Table 2. Topsoil (0–20 cm) SOCD of different land-uses.

Land-use	Total	Woodland	Orchards	Abandoned cropland	Wild grassland	Cultivated land
Sample sites/N	83	19	13	9	8	34
SOCD Mg/ha	20.44	25.53 a	17.02 b	20.70 b	19.35 b	19.08 b
Standard deviation	5.41	6.91	2.85	4.59	3.99	3.66
Minimum	10.47	16.45	11.03	16.55	12.59	10.47
Maximum	40.71	40.71	20.72	31.14	26.02	28.86
Coefficient of variation (%)	26.46	27.08	16.75	22.17	20.62	19.16

* A different letter means a difference significant at 0.05 level.

15.11 km^2, where 7.95 km^2 of this land was converted into the woodland, and 6.51 km^2 was converted into the orchards. The increased areas of land were accounted for 52.65% and 43.12% of the total conversion area, respectively. In 2008, the SOC storage in this catchment was 1.39×10^5 t and was higher than the SOC storage in 2002 (1.35×10^5 t) by 3706.46 t. When the cultivated land was converted into the woodland, the SOC sequestration contribution was 4484.83 t; however, when the cultivated land was converted into the orchards, the contribution was −723.02 t (Table 3).

Relationship of soil organic carbon and total nitrogen content

In this catchment, soil organic carbon and total nitrogen revealed a significant positive correlation ($R^2 = 0.978$, $p<0.01$) that increased as total nitrogen content and soil organic carbon increased (Figure 2). A one-way ANOVA indicated that land-use had significant effect on the TN ($F = 3.07$. $P = 0.021$), and that had no significant effect on the soil C/N ratio ($F = 0.86$, $P = 0.48$). The average soil C/N was 9.69, with a range of 8.73–10.77. The soil C/N ratio in the woodland was the largest at 9.80 and that in the orchards was the smallest at 9.57. Other land-uses followed the order of abandoned cropland (9.76)>cultivated land (9.66)>wild grassland (9.59).

Relationship of years since planting and topsoil SOCD

For the woodland, years since planting had a significant effect on SOCD ($F = 13.00$, $P<0.001$). SOCD increased as years since planting increased. However, SOC was lost initially after afforestation in the cultivated land. The SOCD of the woodland that was older than 30 years was higher than that of the woodland under 10 years old (17.45 Mg/ha) and the cultivated land (19.08 Mg/ha), improving by 74.44% and 59.54% respectively (Figure 3a). However, years since planting had no significant effect on SOCD in the orchards ($F = 2.01$, $P = 0.146$), where the SOCD at >10a was 18.04 Mg/ha and was 10.13% higher at <10a. But, they all were below that of the cultivated land (Figure 3b).

Relationship of elevation and topsoil SOCD

The correlation analysis showed that there was a significant positive correlation between elevation and SOCD ($R^2 = 0.395$ $P<0.01$), using the land-use type as the covariate. The SOCD was calculated as the average value with each 100 m used as an elevation gradient. The results showed that with increasing elevation, SOCD had an increasing trend (Figure 4). At an elevation gradient of ≥1700 m, the SOCD was 26.57 Mg/ha, which was higher than that at the elevation gradient of <1300 m by approximately 58.10%.

Relationship of slope and topsoil SOCD

A one-way ANOVA indicated that land-use had a significant effect on slope ($F = 2.87$, $P = 0.028$) where the average slopes of the woodland and wild grassland all were more than 11.00°, and the average slopes of the cultivated land, abandoned cropland and orchards all were less than 5.00° (Table 4). The correlation analysis showed a low positive correlation between slope and SOCD ($R^2 = 0.170$, $P = 0.127$). A negative correlation was found between the slope and SOCD in the cultivated land, abandoned cropland and orchards ($R^2 = -0.210$, $F = 0.123$), and a positive correlation was found in the woodland and wild grassland ($R^2 = 0.250$, $F = 0.209$). For all correlations, the land-use type was used as the covariate.

Influence of multivariates on topsoil SOCD

Through normalisation processing of SOCD, elevation and slope and setting land-use dummy variables, the multiple regression model was established using SPSS software. The results were as follows:

$$y = 0.150 + 0.233x_1 + 0.054x_2 + 0.004s_1 - 0.025s_2 + 0.044s_3 + 0.168s_4 (R^2 = 0.329,\ p < 0.01) \quad (5)$$

where y was the topsoil SOCD, x_1 was elevation, x_2 was slope, s_1 was wild grassland, s_2 was orchards, s_3 was abandoned land, and s_4 was woodland. The cultivated land was used as the reference when the value of s_{1-4} was 0.

According to the multiple model, elevation, slope and land-use accounted for 32.9% of the SOCD variation using the cultivated land as the reference (Eqn. 5).

Discussion

SOC sequestration in different ecological restoration types

The different types of conversion result in different trends in SOC [8]. In this study, the highest SOCD in the woodland implied that the conversion from the cultivated land to the woodland may lead to the accumulation of more SOC than when cultivated land is converted into the orchards or abandoned cropland in this area (Table 2). Some previous research has also found that the conversion from cultivated land to woodland can increase SOC levels [3,17]. This is mainly because the conversion from cultivated land to woodland can increase topsoil SOCD

Table 3. Change in SOC storage from 2002 to 2008 in the Luoyugou catchment.

Land-use	2002			2008			SOC	
	Area/km^2	SOC		Area/km^2	SOC			
		Storage/t	Percentage/%		Storage/t	Percentage/%	Increased/t	Sequestration amount/t
Immature woodland	8.16	19764.05	14.62	15.48	37480.99	26.98	17716.94	3759.91
Mature woodland	2.35	7145.31	5.28	2.99	9087.78	6.54	1942.48	724.92
Orchards	4.11	7377.46	5.46	10.62	19082.52	13.74	11705.06	−723.02
Wild grassland	0.71	1374.26	1.02	1.31	2536.62	1.83	1162.36	16.22
Cultivated land	52.18	99557.47	76.63	37.07	70737.09	50.92	−28820.38	–
Total	67.50	135218.55	100.00	67.46	138925.00	100.00	3706.46	3778.03

* Because the area of this catchment in 2008 was less than that in 2002, by 0.04 km^2, so that about 71.58 t of SOC wasn't taken into the amount of SOC storage. Immature woodland refers to trees under 30 years of age since planting (SOCD = 24.22 Mg/ha), and mature woodland refers to trees more than 30 years of age since planting (SOCD = 30.44 Mg/ha).

through increasing biomass inputs into the soil, and reducing soil erosion [9,12]. In addition, compared to frequent human activity such as tillage and grazing, in the cultivated land and abandoned cropland, less human disturbance in the woodland also enhanced SOC accumulation.

Our research found that the SOCD in the immature woodland (<10 a) was lower than that in the cultivated land. However, the SOCD of the mature woodland (>30 a) was higher than that of the cultivated land, increasing by 59.54% (Figure 3a). Therefore, in the initial conversion from the cultivated land to the woodland, SOCD may decrease, and from 10a after afforestation SOCD increase rapidly and significantly. Other studies have found similar trends [4,15]. Degryze et al. (2004) in Michigan found no difference in topsoil (0–25 cm) soil C during the first ten years after afforesting in a cropland; however, soil C of the native forest (48.6 t/ha) was significantly greater than the cropland (31.9 t/ha) [41]. Lu et al. (2013), studying afforestation in the Loess Plateau, found that the time of the SOC source to sink transition was 3 to 8 years after afforestation. The decrease of SOC in the first few years following afforestation has been attributed to low net primary productivity of plants, decreased litter inputs and increased decomposition rates [12,29].

In this study, the topsoil SOCD in the orchards was less than that in the cultivated land, which implied that a decrease in SOC may occur when the cultivated land is converted into the orchards (Figure 3b). This result is similar to the results of Yang et al. in the gully region of the Loess Plateau [42]. However, Xue et al. (2011) reported that when the slope farmland was converted into the orchards, the SOC content increased slowly with increasing years since planting and reached its peak between 20 to 30 years [43]. Thirty years later, the SOC content of the converted orchards was 4.96 g/kg and was higher than that in the slope farmland by 97%. One difference between our study and that by Xue et al. (2011) is the different reference used. The reference in their study was slope cropland, whereas we used the average value of the slope and terrace cropland as the reference. Moreover, management measures in their study were better than ours. For example, they interplanted crops into the fruit trees at the seedling stage and used more fertiliser.

Management measures are one of the main factors affecting the loss of SOC in orchards. In this catchment, fruit, trimmed branches, and litter are removed from orchards, which can cause less organic compounds to enter the soil. Clear tillage in the orchards also causes more soil erosion. Therefore, in some farmland regions that are returning to their historic uses and thus need to develop fruit trees for economic purposes, methods for reducing the loss of SOC and enhancing SOC sequestration have become important. To manage this problem, some researchers have found that orchards managed with conservation practices such as growing grass and leguminous cover crop, mulching the ground, and frequently using organic and inorganic fertilisers could reduce soil erosion and improve the SOC content [44–45]. In addition, establishing new orchard management models oriented to SOC sequestration is also very necessary.

Many studies have found that the land-use change from the cultivated land to the abandoned land can increase the SOC stored in soil [13–14]. In our study, our results also showed that the SOCD may improve when the cultivated land is converted into the abandoned land (Table 2). With the termination of human disturbance, vegetation in the abandoned land may begin the process of self-succession [14], gradually forming original vegetation communities of their region, which are secondary wild grassland or secondary forest communities. However, this process of succession is slow, as the process that turns the abandoned land into the top secondary bunge needlegrass (*Stipa bungeana Trin.*) takes 40 to 50 years [46]. Further studies are needed to understand the methods needed to accelerate community succession and better sequester SOC in the abandoned land.

The research of Li et al. (2007) in the northern region of the Loess Plateau found that the SOCD of abandoned land at approximately 10 years of age was slightly lower than that in the secondary wild land where planted grassland (*Medicago sativa*) changed in the same number of years. After 6 to 10 years of restoration, degraded artificial grassland may form secondary bunge needlegrass (*Stipa bungeana Trin.*) [47–48]. Therefore, planting artificial grass in the abandoned land can shorten the time of succession, which ensures ecological benefits and increase economic benefits at the same time. However, in this study, the SOCD of the abandoned land at 8–10 years of age was higher

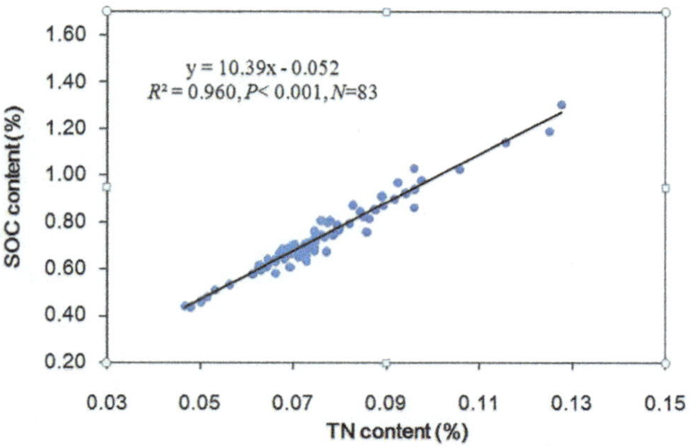

Figure 2. Relationship of topsoil soil organic carbon and total nitrogen content.

than in the wild grassland, which was different to the results of previous studies [47]. The main reason for this may be that these grasslands were distributed in the barren stony mountainous area. The average slope of these regions was 13°, and soil erosion was relatively severe. These factors led to low SOCD.

Benefits of SOC sequestration in the entire region

The SOC storage is determined by the dynamic equilibrium of SOC input and output. After approximately 10 years of ecological restoration, the area of cultivated land greatly decreased, while the area of woodland and orchards increased. From 2002 to 2008, the SOC storage of the entire area only improved by 2.74% (3706.46 t) (Table 3). Therefore, the ecological restoration has an increase in SOC storage, but not significant in the short-term. The main reason for this is that an increase in the SOC resulting from the woodland accumulation was offset by a loss in the SOC resulting from the orchards and the initial conversion stage. So, the SOC storage of the total area is in a relatively stable state. If all of the woodland values are calculated using the SOCD value of the woodland more than 30 years old (30.44 Mg/ha), then considering the loss of SOC by the orchards, the SOC storage of the total area can improve by 7.12% (9625.59 t). Therefore, SOC storage has great potential to increase in the future if the woodland converted from the cropland is well manage and maintain. Although SOC storage has not increased in the past 10 years of restoration, with increasing ecological restoration, soil erosion has been controlled and the quality of the ecological environment, such as the air, water, vegetable coverage has gradually improved in this catchment.

Influencing factors on the SOC

Results from this study showed a strong correlation between SOC and TN (Table 2), which was consistent with earlier findings [33,36]. The trend of C:N ratio varying with the land-use types was similar to that of SOC. The woodland had a higher C:N ratio than other land-use types due to increased above and below ground biomass. The most prevalent tree in the woodland is the black locust, which has N-fixing capabilities, increased TN can promote tree growth, which results in an increase in SOC; it was found that the SOC increase was greater than the TN increase [33]. In catchments, elevation is a governing variable because of its effect on various environmental factors, such as temperature and precipitation. Our study showed a significant positive correlation between elevation and the SOCD (Figure 4). With rising elevation, topsoil SOC increased. This phenomenon was similar to results of other studies [18,49], which can be explained by the lower temperatures and increased precipitation with higher elevations. Increased precipitation can promote vegetation growth, which increases the accumulation of humus, and lower temperatures can limit the decomposition and turnover of SOC, leading to enhanced SOC storage.

The slope is one of the main topographical features affecting the soil erosion intensity as well as SOC loss and enrichment [16]. In the present study, land-use types significantly affected the correlation between the slope and SOCD, where there was a positive correlation in woodland and wild grassland and a negative correlation in cultivated land, abandoned land and orchards (Table 4). These outcomes were due mainly to the influence of vegetation type and coverage. Vegetation restoration on slopes such as planting trees and grass can increase surface vegetation coverage and effectively control soil erosion, further reducing the loss of SOC. These results have been found in other regions [17,30]. Thus, the slope land should be converted to woodland and grassland, or changed to terrace to maintain and increase SOC levels. The multiple regression analysis showed that land-use, slope and elevation accounted for 32.9% of the SOCD variation. The unexplained variation may be caused by the soil particle size,

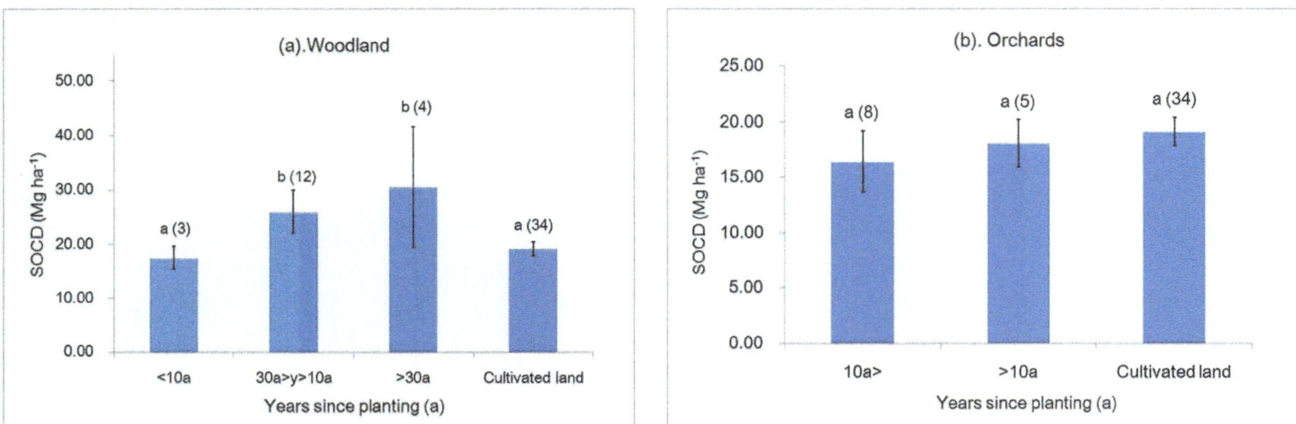

Figure 3. Relationship of years since planting and topsoil SOCD. * The error bars are the standard errors of the mean of SOCD and values above the bars is the number of observations (in parentheses). A different letter means a difference significant at P<0.05.

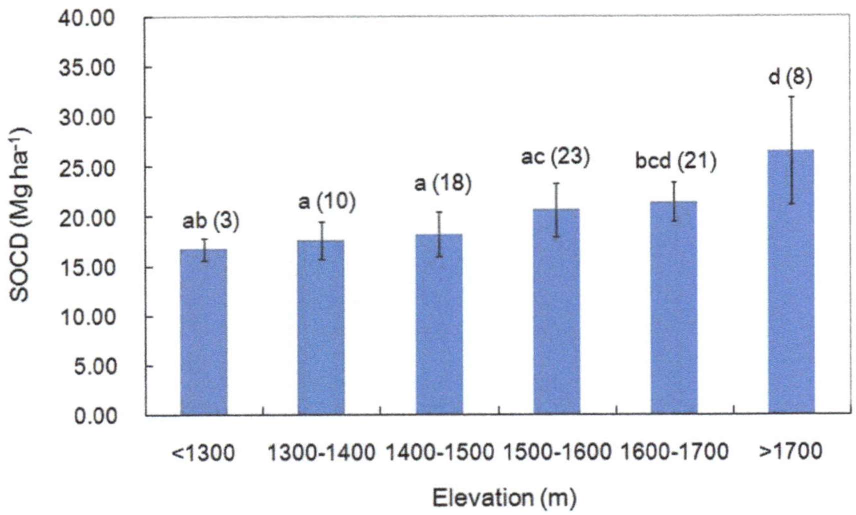

Figure 4. Relationship of elevation and topsoil SOCD. * The error bars are the standard errors of the mean of SOCD and values above the bars is the number of observations (in parentheses). A different letter means a difference significant at $P<0.05$.

soil aggregates, the aspect and the years of utilisation. Therefore, a benefits evaluation of SOC sequestration achieved through ecological restoration efforts should fully consider those influencing factors.

Uncertainties over SOC sequestration from future ecological restoration efforts

With ecological restoration efforts increasing, soil erosion in the Loess Plateau has significantly decreased, and ecological benefits such as vegetation coverage have gradually increased. However, in the long run, there are still uncertainties that will persist if we use the present SOC sequestration achievements from ecological restoration to evaluate possible future SOC sequestration benefits.

(I) Influence of national policies: with the economy developing, the present subsidy standard for returning farmland to forests in the Loess Plateau has been increasingly unable to meet the basic living demands of local farmers. The competitiveness of this subsidy is also growing increasingly weak due to the influence of factors which are raising grain prices and creating policies which benefit farmers, including the abolition of the agricultural tax and the enhancement of subsidies for farming. Based on the results of 2000 farmers' surveys, Cao et al. (2009) found that approximately 37.2% of farmers planned to return to cultivating forested areas and grasslands once the project's subsidies end in 2018. The smaller the subsidy amount to farmers, the fewer the farmers who are willing to participate in them. If the government is unable to improve farmers' income, much of the vegetation restored during the "Grain for Green" project is at risk of being re-converted into farmland and rangeland at the end of this project [50]. Therefore, it is very important and urgent to determine how to effectively consolidate the achievements of the "Grain for Green" project, especially once the project ends, while continually completing the task of returning farmland to forest. However, the government does not do well in this respect. Therefore, uncertainties exist concerning the benefits from SOC sequestration by ecological restoration in the future.

(II) Influence of the environment: Most regions converted from cultivated land to forest are experiencing drought and soil impoverishment. There are no rich water resources in the Loess Plateau, and large-scale vegetation construction will certainly consume massive amounts of water. Lü et al. (2012) found that over half of the Loess

Table 4. Slope of different land-uses.

Land-use	Total	Woodland	Orchards	Abandoned cropland	Wild grassland	Cultivated land
Sample sites/N	83	19	13	9	8	32
Slope/°	6.65	11.58 a	1.92 b	4.22 ab	13.00 a	4.85 b
Standard deviation	10.85	14.76	4.80	8.39	12.10	8.93
Minimum	0	0	0	0	0	0
Maximum	45.00	45.00	15.00	20.00	30.00	30.00
Coefficient of variation %	163.16	127.46	250.00	198.82	93.08	184.12
Correlation between slope and SOCD/R^2	0.170	0.269	−0.169	−0.459	0.480	−0.229

* A different letter means a difference significant at 0.05 level. The signification of correlation between slope and SOCD all were more than 0.05 ($P>0.05$).

Plateau experienced a decrease in runoff (2–37 mm/year) with an average of 10.3 mm/year after the implementation of the "Grain for Green" project [51]. Therefore, with water consumption increases resulting from vegetation, some regions are likely to face a deficit of water resources or even a possible depletion. Drought, decreased rainfall and the excessive depletion of deep soil water by planted vegetation causes the formation of the dried soil layer [52]. The dried soil layer can lead to land degradation and heavily influences vegetation growth, which forms large-scale, low-efficiency production forests [53–54]. Once the dried soil layer appears, it is very difficult for recovery to occur. In the process of implementing "Grain for Green" some improper phenomena have occurred, including the improper selection of vegetation restoration types and excessive close planting conducted in the blind pursuit of economic benefits and political achievement [55], and this has accelerated the formation of the dried soil layer. Therefore, in the long run, ecological restoration in the Loess Plateau is likely to face water storage restrictions. Thus, the benefits of SOC sequestration may be uncertain in the future.

(III) Influence of private and local government interests: Many regions would prefer that cultivated land be converted into orchards when the cultivated land needs to be converted due to this conversion type may generate a higher land value. However, the SOC sequestration benefits resulting from this type of conversion are lower than in the conversion from cultivated land to woodland, and this conversion type may even cause SOC loss. Over time, farmers may remove old trees in orchards converted from cultivated land, or abandoned land. In addition, according to changes in market demand, farmers may replace other tree species before the original trees have matured. The non-ecological woodland converted from cultivated land also can be harvested under reasonable conditions. However, the SOC sequestration effects of these changes have not been studied. In the future, many woodlands and orchards converted from cultivated land will almost certainly undergo cutting and regeneration. The benefits of SOC sequestration are uncertain in the future unless we can understand the SOC variation under these changes.

(IV) The influence of urbanization and economic interests: since the reforms and open policies were established, labour exports from rural areas have increased year after year. Lack of labour has caused the abandonment of large areas of rural cultivated lands [56]. Recent research has shown that the SOC content could increase when cultivated land is converted into abandoned land. However, given the increasing speed of abandonment, it is unclear how long this phenomenon can continue. In addition, these labours may choose to go home to engage in farming in the future, affecting by the economic crisis of the original working regions and other reasons such as older age and homesickness. This transform may bring new pressure on the local ecological environment, which may cause the loss of SOC. Therefore, the future benefits of SOC sequestration are uncertain.

(V) Influence of climate change: Here, climate change mainly refers to precipitation decline, temperature increase and elevated carbon dioxide levels. According to the results of Xin et al. (2011), the drying trend of the Loess Plateau was highly significant, and annual rainfall showed an obvious decreasing trend over the past five decades (1956–2008) by approximately −1.4 mm/a [57]. The reduced precipitation aggravated the water resource shortage situation in the Loess Plateau, which led to the poor tree growth. This outcome may affect SOC sequestration by vegetation restoration, because the temperature increased and elevated carbon dioxide levels have opposite effects on SOC sequestration. Increasing temperatures may promote SOC decomposition; however, elevated carbon dioxide levels may also improve the net primary productivity of plants and increase both the residual body of vegetation and SOC storage [58–59]. Over the last 50 years (1961–2010), the annual mean temperature has significantly increased by 1.91°C in the Loess Plateau [60]. The atmospheric concentration of CO_2 has increased from 280 p.p.m. to 385 p.p.m. between the pre-industrial era and 2008 [61–62]. There is no clear research concerning what functions of the increasing temperatures and elevated carbon dioxide levels for SOC storage are more pronounced. Therefore, some uncertainties exist regarding the possible benefits of SOC sequestration in the future.

Generally speaking, there are many factors that may affect SOC sequestration in the future and may increase or decrease SOC storage. Factors that may be detrimental for SOC storage include: (i) the low national subsidy standard for returning farmland into forest or grasslands and the low income of local farmers; (ii) the persistent, deteriorating environment situation and; (iii) land-use changes caused by private and local government interests. Because of these disadvantageous factors, it is very important to determine how to effectively increase SOC storage. Creating new agriculture products by the implementation of more modern agriculture techniques and offering more work opportunities in urban areas for farmers to increase their income could partially solve these problems. Additionally, the large-scale afforestation that may be exacerbating the soil water shortage in the Loess Plateau should be controlled, especially in vulnerable arid and semi-arid regions, and should fully consider the affordability of such environmental efforts before vegetation restoration is conducted.

Conclusions

In this study, land-use type had a significant influence on the topsoil SOCD. Compared with the average SOCD of cultivated land, the average SOCD in woodland was significantly larger by 33.81%, while the average SOCD in the orchards decreased by 10.80%. Therefore, the SOCD may improve when cultivated land is converted into woodland and may decline when cultivated land is converted into orchards. Over time, there was an increasing trend of SOCD in woodland, but only a small change of SOCD in the orchards. Based on the remote sensing data on land-use change, we found that the "Grain for Green" project did not significantly increase SOC storage in the Luoyugou catchment in the past 10 years of restoration. This is mainly because the increase of SOC storage by woodland was offset by a SOC storage loss in the orchards and decreases in SOC storage in the initial stage of land-use conversion. If all of the woodland area was calculated as 30.44 t/km^2 of SOCD, the potential contribution to SOC sequestration in this catchment would increase by 9625.59 t, improving the SOC storage ratio by 7.12%. Therefore, it has a huge potential for future environmental restoration efforts if we could manage and maintain the woodland converted from the cropland well. However, we cannot ignore other factors that affect SOC sequestration, including climate change and national policies.

Author Contributions

Conceived and designed the experiments: YBQ ZBX XXY YLX. Performed the experiments: YBQ ZBX. Analyzed the data: YBQ ZBX. Contributed reagents/materials/analysis tools: YBQ ZBX XXY YLX. Wrote the paper: YBQ ZBX XXY YLX.

References

1. Morris SJ, Bohm S, Haile-Mariam S, Paul EA (2007) Evaluation of carbon accrual in afforested agricultural soils. Global Change Biology, 13: 1145–1156.
2. Wang YF, Fu BJ, Lü YH, Chen LD (2011) Effects of vegetation restoration on soil organic carbon sequestration at multiple scales in semi-arid Loess Plateau, China. Catena, 85: 58–66.
3. Sauer TJ, James DE, Cambardella CA, Hernandez-Ramirez G (2012) Soil properties following reforestation or afforestation of marginal cropland. Plant Soil, 360: 375–390.
4. Li DJ, Niu SL, Luo YQ (2012) Global patterns of the dynamics of soil carbon and nitrogen stocks following afforestation: a meta-analysis. New Phytologist, 195: 172–181.
5. Lal R (2004) Agricultural activities and the global carbon cycle. Nutrient Cycling in Agroecosystems, 70: 103–116.
6. Powlson D (2005) Will soil amplify climate change? Nature, 433: 204–205.
7. Smith P, Martino D, Cai ZC, Gwary D, Janzen H, et al (2008) Greenhouse gas mitigation in agriculture. Philosophical Transactions of the Royal Society, Series B, 363: 789–813.
8. Guo LB, Gifford RM (2002) Soil carbon stocks and land-use change: a meta analysis. Global Change Biology, 8: 345–360.
9. Don A, Schumacher J, Freibauer A (2011) Impact of tropical land-use change on soil organic carbon stocks- a meta-analysis. Global Change Biology, 17: 1658–1670.
10. Houghton RA (1999) The annual net flux of carbon to the atmosphere from changes in land-use 1850–1990. Tellus, 51B: 298–313.
11. Lal R (2005) Forest soils and carbon sequestration. Forest Ecology and Management, 220: 242–258.
12. Laganière J, Angers AD, Paré D (2010) Carbon accumulation in agricultural soils after afforestation: a meta-analysis. Global Change Biology, 16: 439–453.
13. Raiesi F (2012) Soil properties and C dynamics in abandoned and cultivated farmlands in a semi-arid ecosystem. Plant Soil, 351: 161–175.
14. Novara A, Gristina L, Mantia TL, Rühl J (2013) Carbon dynamics of soil organic matter in bulk soil and aggregate fraction during secondary succession in a Mediterranean environment. Geoderma, 193–194: 213–221.
15. Paul KI, Ploglase PJ, Nyakuengama JG, Khanna PK (2002) Change in soil carbon following afforestation. Forest Ecology and Management, 168: 241–257.
16. Wang BQ, Liu GB (1999) Effects of relief on soil nutrient losses in sloping fields in hilly region of Loess Plateau. Soil Erosion and Soil and Water Conservation, 5: 18–22 (in Chinese).
17. Wang Z, Liu GB, Xu MX, Zhang J, Wang Y, et al (2012) Temporal and spatial variations in soil organic carbon sequestration following revegetation in the hilly Loess Plateau, China. Catena, 99: 26–33.
18. Zhou Y, Xu XG, Ruan HH, Wang JS, Fang YH, et al (2008) Mineralization rates of soil organic carbon along an elevation gradient in Wuyi Mountain of Southeast China. Ecology, 27: 1901–1907 (in Chinese).
19. Luo YQ, Hui DF, Zhang DQ (2006) Elevated CO_2 stimulates net accumulations of carbon and nitrogen in land ecosystems: a meta-analysis, Ecology, 87: 53–63.
20. Batterman SA, Hedin LO, Breugel Mv, Ransijn J, Craven DJ, et al (2013) Key role of symbiotic dinitrogen fixation in tropical forest secondary succession. Nature, 502: 224–227.
21. Huang CY (2000) Soil Science. Beijing: China Agriculture Press, 311p.
22. Gregorich EG, Greer KJ, Anderson DW, Liang BC (1998) Carbon distribution and losses: erosion and deposition effects. Soil & Tillage Research, 47: 291–302.
23. Lal R (2003) Soil erosion and the global carbon budget. Environment International, 29: 437–450.
24. Chartier MP, Rostagno CM, Videla LS (2013) Selective erosion of clay, organic carbon and total nitrogen in grazed semiarid rangelands of northeastern Patagonia, Argentina. Journal of Arid Environments, 88: 43–49.
25. Fu BJ (1989) Soil erosion and its control in the Loess Plateau of China. Soil Use and Management, 5: 76–81.
26. Shi H, Shao MA (2000) Soil and water loss from the Loess Plateau in China. Journal of Arid Environments, 45: 9–20.
27. Fu BJ, Chen LX, Qiu Y (2002) Land-use structure and ecological processes in the Loess Plateau. Beijing: *The Commercial Press*, pp: 1–12 (in Chinese).
28. Feng XM, Wang YF, Chen LD, Fu BJ, Bai GS (2010) Modeling soil erosion and its response to land-use change in hilly catchments of the Chinese Loess Plateau. Geomorphology, 118: 239–248.
29. Lu N, Liski J, Chang RY, Akujärvi A, Wu X, et al (2013) Soil organic carbon dynamics following afforestation in the Loess Plateau of China. Biogeosciences Discuss, 10: 11181–11211.
30. Zheng FL (2006) Effect of Vegetation Changes on soil erosion on the Loess Plateau. Pedosphere, 16: 420–427.
31. Gong J, Chen LD, Fu BJ, Huang Y, Huang Z, et al (2006) Effect of land-use on soil nutrients in the loess hilly area of the Loess Plateau, China. Land Degradation & Development, 17: 453–465.
32. Chen LD, Gong J, Fu BJ, Huang ZL, Huang YL, et al (2007) Effect of land-use conversion on soil organic carbon sequestration in the loess hilly area, loess plateau of China. Ecological Research, 22: 641–648.
33. Fu XL, Shao MA, Wei XR, Horton R (2010) Soil organic carbon and total nitrogen as affected by vegetation types in Northern Loess Plateau of China. Geoderma, 155: 31–35.
34. Chang RY, Fu BJ, Liu GH, Liu SG (2011) Soil carbon sequestration potential for "Grain for Green" project in Loess Plateau, China. Environment Management, 48: 1158–1172.
35. Wei J, Cheng JM, Li WJ, Liu WG (2012) Comparing the effect of naturally restored forest and grassland on carbon sequestration and its vertical distribution in the Chinese Loess Plateau. PLoS ONE, 7: e40123. doi:10.1371/journal.-pone.0040123
36. Lei D, ShangGuan ZP, Sweeney S (2013) Changes in soil carbon and nitrogen following land abandonment of farmland on the Loess Plateau, China. PLoS ONE 8: e71923. doi:10.1371/journal.pone.0071923
37. Yu XX, Zhang XM, Niu LL, Yue YJ, Wu SH, et al (2009) Dynamic evolution and driving force analysis of land-use/cover change on loess plateau catchment. Transaction of the CSAE, 25: 219–225 (in Chinese).
38. Zhao Y, Yu XX (2013) Effects of climate variation and land-use change on runoff-sediment yield in typical watershed of loess hilly-gully region. Journal of Beijing Forestry University, 35: 39–45 (in Chinese).
39. Nelson DW, Sommers LE (1982) Total carbon, organic carbon, and organic matter. In: Page AL, Miller RH, Keeney DR (eds) Methods of soil analysis, Part 2, Chemical and microbial properties. Agronomy Society of America, Agronomy Monograph 9, Madison, Wisconsin, pp 539–552.
40. Institute of Soil Sciences, Chinese Academy of Sciences (ISSCAS) (1978) Physical and chemical analysis methods of soil. Shanghai: Shanghai Science Technology Press, pp 7–15.
41. Degryze S, Six J, Paustian K, Morris SJ, Paul EA, et al (2004) Soil organic carbon pool changes following land-use conversions. Global Change Biology, 10: 1120–1132.
42. Yang YL, Guo SL, Ma YH, Chen SG, Sun WY (2008) Changes of orchard soil carbon, nitrogen and phosphorus in gully region of Loess Plateau. Plant Nutrition and Fertilizer Science, 14: 685–691 (in Chinese).
43. Xue S, Liu GB, Zhang C, Zhang CS (2011) Analysis of effect of soil quality after orchard established in hilly Loess Plateau. Scientia Agricultura Sinica, 44: 3154–3161 (in Chinese).
44. Umali BP, Oliver DP, Forrester S, Chittleborough DJ, Hutson JL, et al (2012) The effect of terrain and management on the spatial variability of soil properties in an apple orchard. Catena, 93: 38–48.
45. Guimarães DV, Gonzaga MIS, Silva TOd, Silva TLd, Dias NdS, et al (2013) Soil organic matter pools and carbon fractions in soil under different land-uses. Soil & Tillage Research, 126: 177–182.
46. Zou HY, Cheng JM, Zhou L (1998) Natural recoverage succession and regulation of the prairie vegetation on the Loess Plateau. Research of Soil and Water Conservation, 5: 126–138 *(in Chinese)*.
47. Li YY, Shao MA, Shang Guan ZP, Fan J, Wang LM (2006) Study on the degrading process and vegetation succession of *Medicago sativa* grassland in North Loess Plateau, China. Acta Prataculturae Sinica, 15: 85–92 (in Chinese).
48. Li YY, Shao MA, Zhang JY, Li QF (2007) Impact of grassland recovery and reconstruction on soil organic carbon in the northern Loess Plateau. Acta Ecologica Sinica, 27: 2279–2287 (in Chinese).
49. Leifeld J, Bassin S, Fuhrer J (2005) Carbon stocks in Swiss agricultural soils predicted by land-use, soil characteristics, and elevation. Agriculture, Ecosystems and Environment, 105: 255–266.
50. Cao SX, Xu CG, Chen L, Wang XQ (2009) Attitudes of farmers in China's northern Shaanxi Province towards the land-use changes required under the Grain for Green Project, and implications for the project's success. Land-use Policy, 26: 1182–1194.
51. Lü YH, Fu BJ, Feng XM, Zeng Y, Liu Y, et al (2012) A policy-driven large scale ecological restoration: quantifying ecosystem services changes in the Loess Plateau of China. PLoS ONE, 7: e31782, doi:10.1371/journal.pone.0031782
52. Pan ZB, Zhang L, Yang R, Li SB, Dong LG, et al (2012) Overview on research progress of soil drought in semiarid regions of the Loess Plateau. Research of Soil and Water Conservation, 19: 287–291, 298 (in Chinese).
53. Wang L, Shao MA (2004) Soil desiccation under the returning farms to forest on the Loess Plateau. World Forestry Research, 17: 57–60 (in Chinese).
54. Chen HS, Shao MA, Li YY (2008) Soil desiccation in the Loess Plateau of China. Geoderma, 143: 91–100.
55. Niu JJ, Zhao JB, Wang SY (2007) A study on plantation soil desiccation in the upper reaches of the Fenhe River basin based on deep soil experiments. Geographical Research, 26: 773–781 (in Chinese).

56. Duan FL, Lin Z, Xiong YQ (2007) Analysis on the phenomenon of farmland abandoned by the reason of rural laborers moving out for work. Rural Economy, 16–19 (in Chinese).
57. Xin ZB, Yu XX, Li QY, Lu XX (2011) Spatiotemporal variation in rainfall erosivity on the Chinese Loess Plateau during the period 1956–2008. Regional Environmental Change, 11: 149–159.
58. William HS (1999) Carbon and agriculture: carbon sequestration in soils. Science, 284: 2095.
59. Guo GF, Zhang CY, Xu Y (2006) Effects of climate change on soil organic carbon storage in terrestrial ecosystem. Ecology, 25: 435–442 (in Chinese).
60. Wang QX, Fan XH, Qin ZD, Wang MB (2012) Change trends of temperature and precipitation in the Loess Plateau Region of China, 1961–2010. Global and Planetary Change, 92–93: 138–147.
61. Intergovernmental Panel on Climate Change (2007) Climate Change 2007: The Science of Climate Change. Cambridge University Press, Cambridge.
62. Lal R (2009) Challenges and opportunities in soil organic matter research. European Journal of Soil Science, 60: 158–169.

2

Effects of Nitrogen Addition and Fire on Plant Nitrogen Use in a Temperate Steppe

Hai-Wei Wei[1,2], Xiao-Tao Lü[1]*, Fu-Mei Lü[3], Xing-Guo Han[1,3]

1 State Key Laboratory of Forest and Soil Ecology, Institute of Applied Ecology, Chinese Academy of Sciences, Shenyang, China, **2** Graduate University of Chinese Academy of Sciences, Beijing, China, **3** State Key Laboratory of Vegetation and Environmental Change, Institute of Botany, Chinese Academy of Sciences, Beijing, China

Abstract

Plant nitrogen (N) use strategies have great implications for primary production and ecosystem nutrient cycling. Given the increasing atmospheric N deposition received by most of the terrestrial ecosystems, understanding the responses of plant N use would facilitate the projection of plant-mediated N cycling under global change scenarios. The effects of N deposition on plant N use would be affected by both natural and anthropogenic disturbances, such as prescribed fire in the grassland. We examined the effects of N addition (5.25 g N m^{-2} yr^{-1}) and prescribed fire (annual burning) on plant N concentrations and N use characters at both species and community levels in a temperate steppe of northern China. We found that N addition and fire independently affected soil N availability and plant N use traits. Nitrogen addition increased aboveground net primary productivity (ANPP), inorganic N, and N uptake, decreased N response efficiency (NRE), but did not affect biomass-weighed N concentrations at community level. Prescribed fire did not change the community level N concentrations, but largely decreased N uptake efficiency and NRE. At the species level, the effects of N addition and fire on plant N use were species-specific. The divergent responses of plant N use at community and species levels to N addition and fire highlight the importance of the hierarchical responses of plant N use at diverse biological organization levels to the alteration of soil N availability. This study will improve our understanding of the responses of plant-mediated N cycling to global change factors and ecosystem management strategies in the semiarid grasslands.

Editor: Fei-Hai Yu, Beijing Forestry University, China

Funding: This research was supported by the National Natural Science Foundation of China (41273094, 41173086, and 31170433) and the State Key Laboratory of Forest and Soil Ecology (LFSE2013-13). The funders had no role in study design, data collection and analysis, decision to publish, or preparation of the manuscript.

Competing Interests: The authors have declared that no competing interests exist.

* E-mail: lvxiaotao@iae.ac.cn

Introduction

Nitrogen (N) availability is the dominant limiting factor for primary productivity in most terrestrial ecosystems [1–4]. Consequently, the responses of plants to the variation in soil N availability would have great implications for plant growth and species competition. It is well established that plant N economy plays an important role in affecting plant productivity and ecosystem N cycling [5,6]. Both plant N concentration and N response efficiency (NRE) are widely used to evaluate the responses of plant N economy to changes of soil N availability. Changes of plant N concentration have great implications for their N use efficiency (NUE), which is generally defined as the dry mass productivity per unit N taken up from the soil [7]. N response efficiency is the amount of biomass produced per unit of plant available N in the soil [8–10]. While plant N concentration is a indicator that integrates plant physiological and morphological responses to soil N availability, NRE is an index that integrates ecological and biogeochemical responses of plants. There is increasing empirical evidence that both plant N use indices should be simultaneously examined as they involve different processes linking the plant-soil system [11].

Both increasing atmospheric N deposition and global warming would enhance soil N availability in terrestrial ecosystems [12,13], which exerts positive effects on plant growth and primary productivity [1,14]. It should be noted that, however, the relationship between increased N availability and plant N use characters may not be straightforward in natural ecosystems. For instance, the changes of NUE in response to N availability would be the balance between the responses of its two components, NPP and the amount of N taken up by plants. Similarly, the responses of NRE would be determined by the relative changes of NPP and soil N availability. A decrease of NRE in response to N addition means that N addition shows relatively stronger impacts on soil N availability than NPP. Previous studies showed that N addition generally resulted in an increase of plant N concentration and a decrease of NRE at species level [8,11,15,16]. It has been suggested that NRE would be a more representative measurement of plant responses to increasing soil N availability [11].

Most of previous studies mainly focused on changes of foliar N characters along nutrient availability gradients [6], which may not well mirror what is happening at higher biological organization levels. For example, there is significant variation of N concentrations between leaf and stem and also between aboveground and belowground parts at the individual plant level [17–18]. Furthermore, natural communities are always composed of plant species with various nutrient use strategies [19] and different nutrient limitation status [20]. There is increasing evidence that the response of plant N use to changes of nutrient availability would vary significantly among diverse biological scales, from foliar to community level [11].

Environmental disturbances and ecosystem management strategies may also affect plant N use through their impacts on soil N availability [21–23] and community composition [24–26]. Moreover, they have the potential to alter the magnitude and direction of the impacts of N addition on plant N use [27]. Fire, as an anthropogenic or natural disturbance, exerts great control on community composition and ecosystem functioning in several terrestrial ecosystems, including grassland [25,28,29]. Reduction in fire frequency is considered as one of the major reasons for the increases of woody plant abundance in grasslands all over the world [26,30]. Prescribed burning is used as a strategy to prevent shrub encroachment in grasslands [31]. Results from a meta-analysis showed that short-term fire would enhance soil N availability while long-term repeated fire would lead to a decrease of soil N [21]. Similarly, the effects of fire on plant N status and use characters would also depend on its frequency. In the short-term, fire would increase plant NUE due to its stimulation on photosynthesis [32], whereas plant NUE would decrease under long-term repeated burning [22]. Given the interactive effects of fire and N addition on soil nutrient availability and plant production as observed in other studies [27], it is reasonable to expect that they may interact to affect plant N use. However, empirical evidence for their interactive effects on plant N use is still scarce.

The main objective of this study was to determine the effects of N addition, fire, and their interaction on plant N concentrations and N use characters at both species and community levels. We conducted a manipulative experiment in a temperate steppe in northern China, which is a typical vegetation type in the Eurasian continent. Fire was an important natural disturbance factor before the 1980s in northern China, with a frequency of once very 3–5 years [33]. In previous studies, we found that N addition and fire exert influence on plant nutrient resorption [34] and C:N:P stoichiometry [35,36]. We tested the following hypotheses: (1) N addition would enhance plant N concentrations, but reduce plant N uptake efficiency and NRE. (2) Fire would also increase plant N concentrations, but decrease plant N uptake efficiency and NRE, as the short-term fire (annual burning for two years) usually enhance soil N availability. Furthermore, we expected that the responses of plant N use at species level would be different from that at community level.

Materials and Methods

Site description

This study was conducted near the Inner Mongolia Grassland Ecosystem Research Station (IMGERS, 116°42′E, 43°38′N), which is located in the middle reach of the Xilin River, northern China. Long-term (1970–2007) meteorological data indicated that the mean annual precipitation was approximately 345 mm and mean annual temperature was 0.3°C. The length of growing season is about 150 days. The soil of this site is classified as Haplic Calcisols according to the Food and Agriculture Organization classification. Mean bulk density (top 10 cm) is 1.3 g cm^{-3} and pH is 7.4. The vegetation is dominated by *Leymus chinensis* (Trin.) Tzvel., *Stipa grandis* P. Smirn., *Cleistogenes squarrosa* (Trin.) Keng., *Caragana microphylia* Lam., *Carex korshinskyi* Kom., and *Potentilla bifurca* L.

Experimental design

This study was conducted as part of the GFE (Grassland Fire Experiment) experimental setup near IMGERS. All necessary permits were gained from Institute of Botany, Chinese Academy of Sciences and local government before the beginning of this study. The design consists of nine blocks distributed across a grassland field, with each block containing a set of 10 m×10 m plots representing fully crossed treatments of fire frequency, N addition, and mowing frequency. Plots were separated by 1 m buffers. The experiment setup was established in 2005, with all burning, N addition, and mowing treatments were started since 2006. Burning was carried out in early or late April each year before the start of growing season. Nitrogen in the form of NH_4NO_3 was added in the rainy days of late June each year. Only the treatments of two burning frequencies (never burned vs annual burning) and two N addition levels (0 vs 5.25 g N m^{-2} yr^{-1}) that without mowing in 9 blocks were used in this study, with 4 treatments ×9 replications = 36 plots in all.

Sample collection and chemical analysis

Aboveground net primary production (ANPP) was determined by clipping all plants in randomly located 1 m×1 m quadrats in each plot during the peak of growing season in August 2007. All plant samples were oven-dried at 70°C for 48 h, and then weighed. Samples of each species per plot were combined and ground to pass through a 40-mesh sieve using a mechanical mill (Retsch MM 400, Retsch GmbH & Co KG, Haan, Germany). Total N concentrations were determined colorimetrically by the Kjeldahl acid- digestion method with an Alpkem auto-analyzer (Kjektec System 1026 distilling unit, Sweden) after extraction with sulfuric acid.

Three soil cores (topsoil, 0–10 cm) were sampled in each of those quadrats following plant sampling with a metal tube (5 cm in diameter) and pooled to obtain one composite soil sample per quadrat. To measure soil water content, a soil subsample from each quadrat was weighed before and after being oven dried at 105°C for 48 h. Soil inorganic N availability was determined by extracting 10 g of fresh, root-free soil with 50 mL 2M KCl. The soil-extractant mixture was shaken for 1 h and then filtered (Whatman No. 1 filter paper). NH_4^+-N was measured with the salicylate method while NO_3^--N was measured using the cadmium reduction method on a FIAstar 5000 Analyzer (Foss Tecator, Denmark). Total soil inorganic N was the sum of NH_4^+-N and NO_3^-N.

Calculations and statistical analysis

The community level biomass-weighed N concentration was calculated as:

$$N\ concentration_{community} = \frac{\sum N\ concentration_i \times ANPP_i}{ANPP} \quad (1)$$

Where *i* is the *i*th species of *n* species in the community.

N uptake efficiency was calculated as the proportion of available N in the soil that was taken up by the plants [8]

$$N\ uptake\ efficiency = \frac{N\ uptake}{N\ available} \quad (2)$$

NRE was calculated ANPP relative to the pool of available soil N [8–10]:

$$NRE = \frac{ANPP}{N\ available} \quad (3)$$

The Kolmogorov-Smirnov test was used to test for data normality.

Two-way ANOVAs with fire and N addition as the main factors were performed to examine main and interactive effects on soil inorganic N, soil water content, ANPP, N concentrations, N uptake, N uptake efficiency, and NRE at community level. Two-way ANOVAs were used to examine the effects of N addition and fire on the plant N use characters of five dominant species at species level. All statistical analyses were performed using SPSS software (version 13.0 for windows, SPSS Inc., Chicago, IL, USA).

Results

Community level responses

Nitrogen addition and fire significantly enhanced soil inorganic N by 94.2% ($P<0.001$) and 25.8% ($P<0.05$), respectively (Fig. 1a, Table 1). Fire significantly reduced soil water content ($P<0.001$), whereas N addition did not affect soil water content (Fig. 1b, Table 1). Nitrogen addition significantly increased ANPP by 36.3% ($P<0.001$; Fig. 1c, Table 1), whereas fire reduced ANPP by 14.3% ($P<0.05$; Fig. 1c; Table 1).

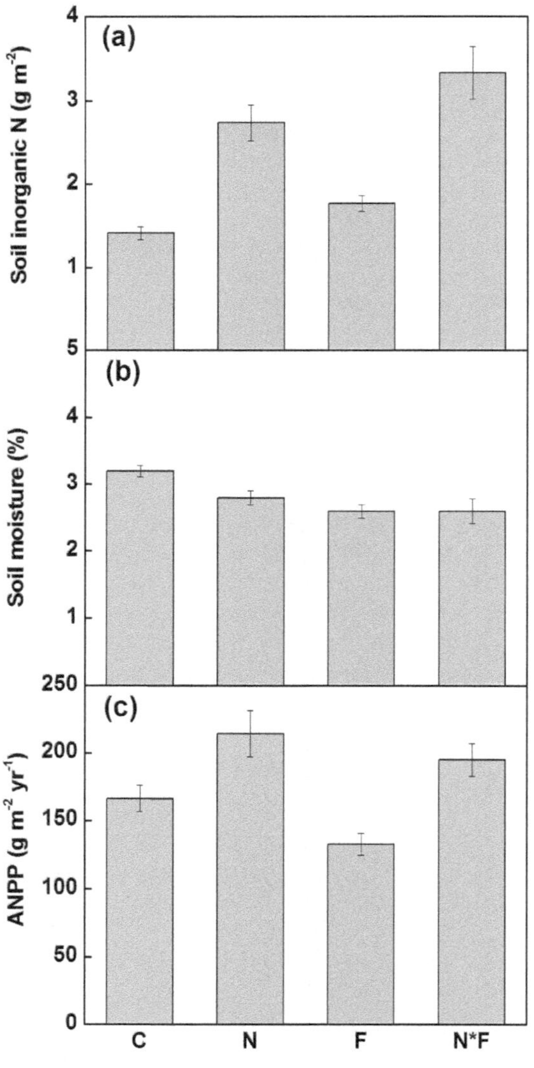

Figure 1. The effects of nitrogen addition and fire on (a) soil inorganic N concentrations, (b) soil water content, and (c) ANPP of plant community in a temperate steppe. C, control; N, nitrogen addition; F, fire; N*F, both N addition and fire. Data are adjusted means ± 1SE, where $n=9$ for all treatments.

Both N addition and fire had no impacts on biomass-weighed N concentrations at community level (Fig. 2a; Table 1). N addition enhanced plant N uptake by 41.6% ($P<0.01$; Fig. 2b; Table 1), while fire had no effects on plant N uptake (Fig. 2b; Table 1). Both N addition and fire decreased N uptake efficiency by 22.9% ($P<0.01$) and 18.2% ($P<0.05$; Fig. 2c; Table 1). NRE was significantly reduced by both N addition (31.5%, $P<0.001$, Fig. 2c; Table 1) and fire (36.6%, $P<0.001$, Table 1). There was no interaction between N addition and fire for all variables (Table 1).

Species level responses

Nitrogen addition significantly enhanced the ANPP of *C. korshinskyi* ($P<0.01$) and *L. chinensis* ($P<0.05$; Table 1; Fig. 3), but showed no impacts on that of *A. cristatum*, *C. squarrosa*, and *S. grandis*. When the species were analyzed individually, plant N use varied greatly among five dominant species (Fig. 3). N addition had no impacts on N concentration for all species, while fire significantly increased N concentration of all species except for *C. squarrosa* (Table 1; Fig. 3). N addition and fire interacted to affect N concentration of *A. cristatum* ($P<0.05$) and *S. grandis* ($P<0.01$; Table 1; Fig. 3). While N addition significantly enhanced N uptake in two out of all the five species (*A. cristatum* and *C. korshinskyi*), fire significantly reduced N uptake of *C. korshinskyi* but did not affect that of the other species (Table 1; Fig. 3).

Nitrogen addition significantly decreased N uptake efficiency of all species ($P<0.05$) except for *C. korshinskyi* (Table 1, Fig. 3), while fire only decreased the N uptake efficiency of *C. korshinskyi* ($P<0.001$). N addition significantly decreased NRE for all species ($P<0.05$), whereas fire only decreased that of *A. cristatum* ($P<0.05$) and *C. korshinskyi* ($P<0.001$). For all the species, there was no interactive effect between N addition and fire on NRE (Table 1).

Discussion

Results from this study showed that N addition and fire directly affected soil N availability, ANPP, plant N uptake and use traits at both species and community levels. We found no interaction between N addition and fire in affecting plant N use. Generally, N addition increased ANPP and plant N uptake, decreased N uptake efficiency and NRE, with no significant effect on plant N concentrations. In contrast, fire showed positive effects on plant N concentrations (especially at species level), negative effects on NRE, and neutral effects on plant N uptake.

We found that both N addition and fire increased soil inorganic N concentrations. A meta-analysis showed that the soil inorganic N would increase immediately after fire and then gradually declined to the level before fire within one year [21]. All the burning plots had been burned for two years before soils were sampled in this study. Our results suggest that short-term fire would stimulate soil N turnover in the temperate steppe. While N is assumed to be the primary limiting nutrient in this ecosystem [14], we found negative effects of fire on ANPP even it stimulated soil N availability. We suspect that water availability would play an importantly in mediating the effects of fire on ANPP in this semiarid grassland. Fire significantly reduced soil water content in this study. In the same experimental setup, Zhou et al. [33] also found that the soil moisture in unburned plots was significantly higher than that in the burned plots during the growing season. Furthermore, it has been shown that water addition instead of N addition significantly increased ANPP, indicting water availability would be the dominant limiting factor for plant growth in this ecosystem [37]. Consequently, it is reasonable to speculate that reduction of soil water availability after burning would account for

Table 1. Results (F-values) of two-way ANOVAs on the effects of nitrogen addition (N), fire (F), and their interaction on soil inorganic N, soil water content (SWC), ANPP, N concentration, N uptake, N uptake efficiency, and NRE at both community and species level in a temperate steppe.

Factor	Inorganic N	SWC	ANPP	N concentration	N uptake	N uptake efficiency	NRE
Community-level							
N	53.5***	2.5	24.2***	0	12.4**	7.9**	13.6***
F	6.0*	10.3**	6.2*	3.5	0.2	4.7*	20.7***
N*F	0.4	2.6	0.3	1.5	0.3	0.1	3
Species-level							
Agropyron cristatum							
N	-	-	3.6	0	4.6*	11.9**	10.9**
F	-	-	0.7	6.3*	0.1	1.5	5.8*
N*F	-	-	0.2	6.5*	0.5	0.2	2.5
Carex korshinskyi							
N	-	-	13.1**	3.5	13.5***	2.4	5.1*
F	-	-	16.3***	7.9**	8.4**	15.4***	27.6***
N*F	-	-	2.2	0.1	1.5	0.1	0.3
Cleistogenes squarrosa							
N	-	-	0.6	0.5	1	11.2**	12.5**
F	-	-	0.3	2.9	0	0.7	2.1
N*F	-	-	5.2*	2	5.4*	1.3	0.4
Leymus chinensis							
N	-	-	5.2*	2.3	0.9	7.0*	6.4*
F	-	-	0	6.7*	2.3	0.2	2.3
N*F	-	-	0.4	0	0	0	0
Stipa grandis							
N	-	-	0.1	1.9	0	5.1*	5.0*
F	-	-	3.8	18.2***	1.4	2.3	4.5
N*F	-	-	2	12.3**	3.7	0.7	0.1

The F-values are presented, together with their level of significance.
*$P<0.05$;
**$P<0.01$;
***$P<0.001$.
For all variables error $df = 36$.

the negative effects of fire on community ANPP as observed in the present study.

Nitrogen addition would elevate soil inorganic N availability and thus increase ANPP, which had been generally tested and confirmed by previous studies [1,14,16]. Similarly, our results showed that N addition (5.25 g N m^{-2} yr^{-1}) significantly enhanced soil inorganic N by 94.2%, and increased ANPP at both species and community levels. In the same area, Bai et al. [14] reported that N addition, even as low as 1.75 g N m^{-2} yr^{-1}, would significantly increase ANPP. It is notable that, however, the effects of N addition on ANPP would be largely dependent on soil water availability [37]. The positive effects of N addition on the growth of dominant species and the ANPP of whole community as observed in this study indicated that primary production was limited by N availability in this area. Given the projected increasing atmospheric N deposition, the primary productivity would be facilitated following the increasing N availability. We found no interactive effect of N addition and fire on primary production at community level, implying that the positive effects of N deposition on primary production would not be affect by natural fire or prescribed burning in the temperate steppe of northern China.

Our results showed that N addition significantly increased plant N uptake by 41.6%, whereas fire had no effects on plant N uptake. The positive effects of N addition on plant N uptake were resulted from the enhancement of ANPP, as we found no impacts of N addition on plant N concentrations at both species and community levels. Previous studies showed that fire would increase N concentrations in plant tissues [27]. In this study, we found two-year annual fire significantly enhanced biomass-weighed plant N concentrations at community level, which is consistent with the results of our previous study which was carried out in the same ecosystem [35]. Due to the offset between the increases of plant N concentration and the decreases of ANPP following fire, annual burning showed no significant effects on plant N uptake. Fire showed negative influence on N uptake efficiency as it had no effects on plant N uptake but enhanced soil inorganic N. It has been reported that N addition would decrease N uptake efficiency, as the increases of N uptake cannot commensurate with the increases of soil inorganic N in a peatland and a managed

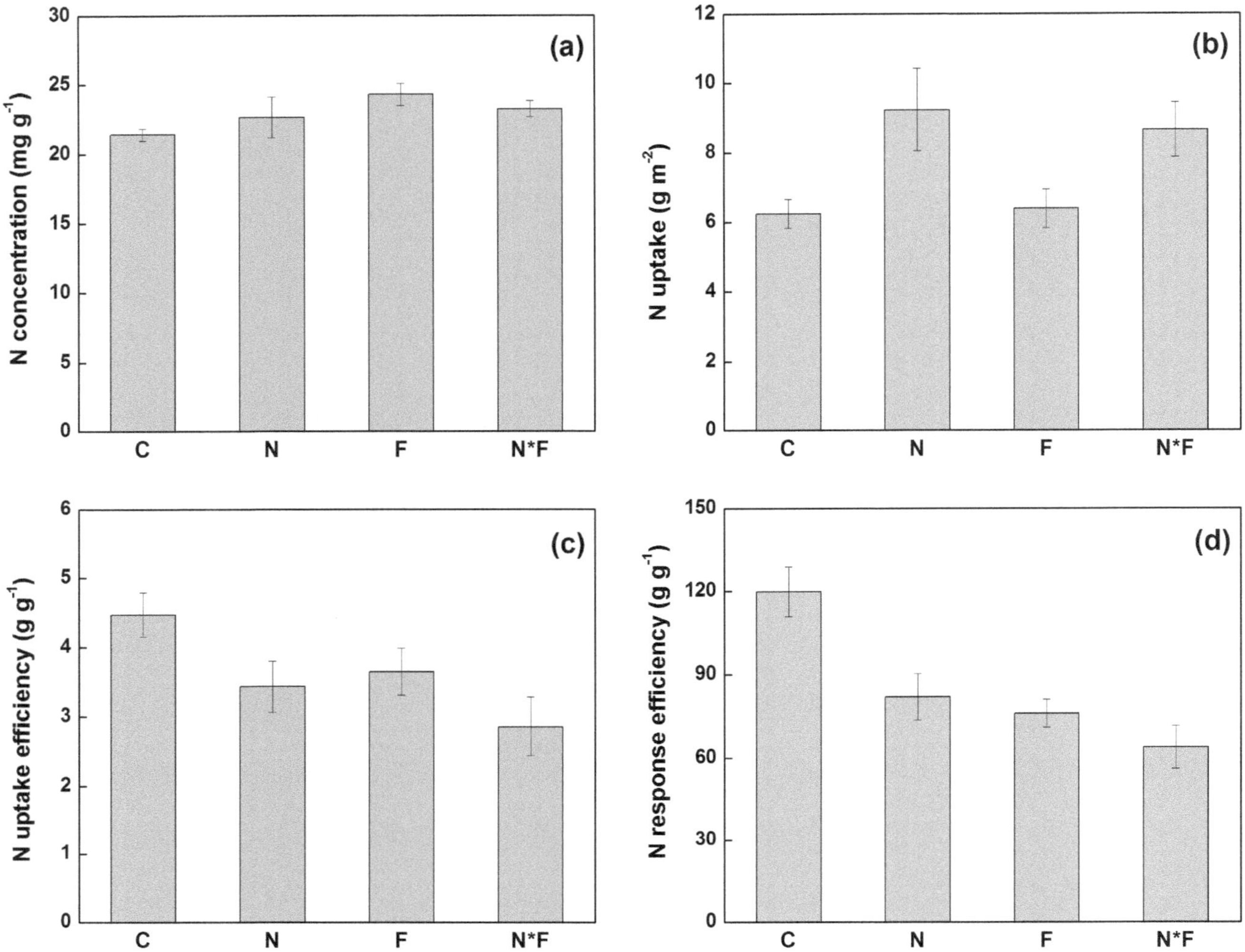

Figure 2. The effects of nitrogen addition and fire on (a) biomass-weighed N concentration, (b) N uptake, (c) N uptake efficiency, and (d) N response efficiency of plant community in a temperate steppe. C, control; N, nitrogen addition; F, fire; N*F, both N addition and fire. Data are adjusted means ± 1SE, where $n = 9$ for all treatments.

grassland [11,15]. Similarly, we found negative effects of N addition on plant N uptake efficiency at community level in this temperate steppe.

Previous studies reported that NUE often decreased with the increase of soil N availability [6,9,19], but not in other studies [38–40]. Contrary to our hypothesis, results from this study showed that N addition did not affect plant N concentrations at both species and community levels, implying that two-year N addition with the rate of 5.25 g N m^{-2} yr^{-1} did not affect NUE of plants in this temperate steppe. Our results were consistent with Reich et al. [41], who reported that increasing N availability did not change NUE in the forest ecosystem, but were different from Iversen et al. [11], who found that N addition (6 g N m^{-2} yr^{-1}) resulted in a decrease of NUE in peatlands. In temperate grassland, Yuan et al. [40] found that the responses of NUE to N addition were dependent on N addition rates. On the one hand, in our study, the N addition rate was 5.25 g N m^{-2} yr^{-1}, which might be not enough to induce a decrease of NUE. On the other hand, our results indicate that the efficiency of temperate steppe to use N would be consistent in the face of increasing N deposition, at least at the rate of 5.25 g N m^{-2} yr^{-1}. Our results that two years annual burning significantly enhanced biomass-weighed N concentrations at community level, indicating that fire would have negative effects of NUE in this community. The enhancement of plant N concentrations would have great means for litter quality and decomposition, as the litter decomposition rate and nutrient release are usually positive related to the N concentration in the litter [5,24,42].

Consistent with our hypothesis and findings from other studies [11,15], our results showed that N addition reduced NRE. Based on a natural soil N availability gradient in the semiarid grassland, Yuan et al. [43] found that NRE decreased monotonically with increasing N availability. The lower magnitude of the responses of ANPP to N addition than that of soil inorganic N availability resulted in a negative effect of N addition on NRE. Fire also led to a negative effect on plant NRE at community level, due to its negative effect on ANPP and positive effect on soil inorganic N. The negative effects of N addition and fire on community NRE suggests that part of soil inorganic N could not be taken up into plant biomass. Thus, more soil inorganic N would be lost by leaching and volatilization after N addition or burning. There is evidence that annually burning induce N loss through volatilization during fire [44], erosion, leaching, and denitrification [45]. It is notable that, the negative effects of N addition and fire on plant

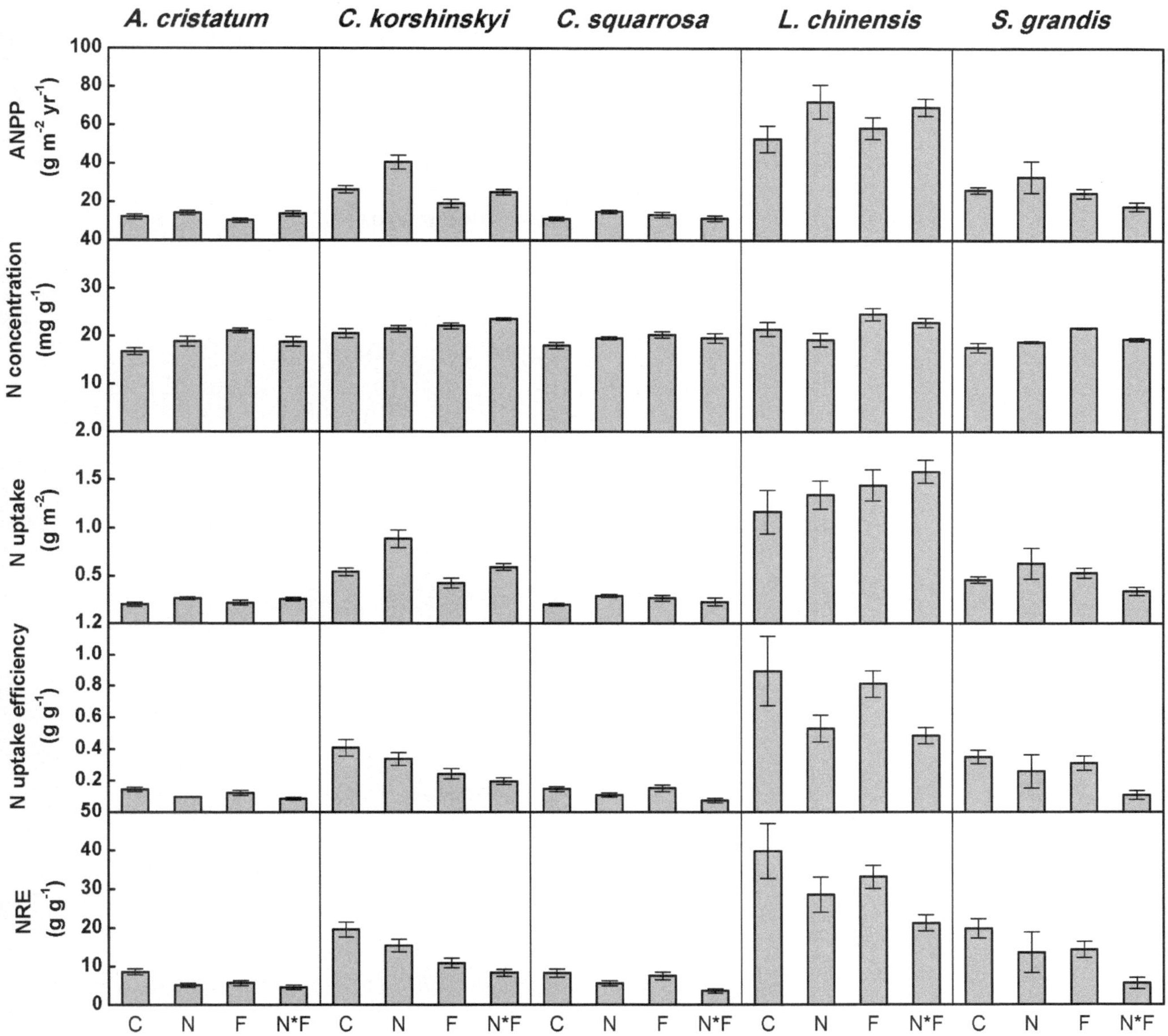

Figure 3. The effects of N addition and fire on plant N use of dominant species in a temperate stepp. C, control; N, nitrogen addition; F, fire; N*F, both N addition and fire. Data are adjusted means ±1SE, where $n = 9$ for all treatments.

community NRE were independent, rather than additive. Together, our results indicate that the temperate steppe would be less efficient in absorbing the increased soil inorganic N into plant biomass after N addition and burning.

The dominant species examined in this study greatly varied in their N use traits. In the control plots, the NRE of *L. chinensis* was almost five times higher than that of *A. cristatum* and *C. squarrosa.* The higher NRE of *L. chinensis* would help explain its dominant status in this N-limited grassland ecosystem. More soil N would be transferred to plant biomass if this ecosystem was mono-dominated by *L. chinensis*, as indicated by its higher NRE status. Consequently, any changes of community composition following N addition and fire would influence plant N use at community level. Moreover, the responses of plant N economy to N addition and fire were species-specific. While N addition significantly enhanced the N uptake of *A. cristatum* and *C. korshinskyi*, it showed no impacts on that of other species. Similarly, fire reduced NRE of *A. cristatum* and *C. korshinskyi*, but had no effects on other species. The idiosyncratic responses of different species to N addition and fire suggest that *A. cristatum* and *C. korshinskyi* were more sensitive to changes of soil N availability resulting from N addition and annual burning. Our results have great implications for the plant-mediated changes of biogeochemical cycling in response to global changes and ecosystem management strategies.

We hypothesized that the responses of plant N use traits to N addition and fire would be different at species and community levels. When all the dominant species were analyzed together, we found that the responses of plant N concentrations and NRE to N addition and fire did not different ranging from species to community level. Nitrogen addition did not alter plant N concentrations at both species and community levels, whereas it did reduce plant NRE at both levels. Fire decreased plant N concentrations and NRE at community level, while its effects on species-level plant N concentrations and NRE varied among different species. However, when those species were analyzed individually, we did found divergent responses of plant N use traits at species and community levels, which would be mainly resulted from the species-specific responses of plants to N addition and fire. In a recent study, Iversen et al. [11] reported that plant N use traits differed at different biological organizations ranging from individual plant tissue to the entire community in peatland ecosystems. Our study presented strong evidence for the hierarchical responses of plant N use to N addition and fire at different biological scales, especially for species and community levels in a grassland ecosystem.

Conclusions

Our results showed that both N addition and fire directly affected soil N availability and plant N use traits, with modest interaction between them. N addition increased ANPP, inorganic N, and N uptake, decreased NRE, and did not affect plant N concentrations. Prescribed fire, as an important management strategy in grasslands, enhanced plant N concentrations and decreased N uptake efficiency and NRE. The responses of plant N use to both N addition and fire would be different at species and community levels, mainly due to the species-specific responses. The divergence of plant N use traits among different plant species indicates that any factors leading to changes of community composition would indirectly alter plant N use at community level.

Acknowledgments

We thank Wu Wei-Jun, Li Qiang, and Li Li for assistance with field and laboratory work, two anonymous reviewers for their helpful comments on the manuscript, and Matthew Simmons for language improvement.

Author Contributions

Conceived and designed the experiments: XH XL FL. Performed the experiments: FL XL HW. Analyzed the data: XL HW. Contributed reagents/materials/analysis tools: FL XH XL. Wrote the paper: HW XL.

References

1. LeBauer DS, Treseder KK (2008) Nitrogen limitation of net primary productivity in terrestrial ecosystems is globally distributed. Ecology 89: 371–379.
2. Perakis SS, Kellogg CH (2007) Imprint of oaks on nitrogen availability and delta N-15 in California grassland-savanna: a case of enhanced N inputs? Plant Ecol 191: 209–220.
3. Cole L, Buckland SM, Bardgett RD (2008) Influence of disturbance and nitrogen addition on plant and soil animal diversity in grassland. Soil Biol Biochem 40: 505–514.
4. Vitousek PM, Howarth RW (1991) Nitrogen limitation on land and in the sea: How can it occur? Biogeochemistry 13: 87–115.
5. Vitousek P (1982) Nutrient cycling and nutrient use efficiency. Am Nat 119: 553–572.
6. Aerts R, Chapin FS (2000) The mineral nutrition of wild plants revisited: A re-evaluation of processes and patterns. Adv Ecol Res 30: 1–47.
7. Hirose T (2011) Nitrogen use efficiency revisited. Oecologia 166: 863–867.
8. Pastor J, Bridgham SD (1999) Nutrient efficiency along nutrient availability gradients. Oecologia 118: 50–58.
9. Shaver GR, Melillo JM (1984) Nutrient budgets of marsh plants: efficiency concepts and relation to availability. Ecology 65: 1491–1510.
10. Bridgham SD, Pastor J, McClaugherty CA, Richardson CJ (1995) Nutrient-Use-Efficiency: A litterfall index, a model, and a test along a nutrient availability gradient in North-Carolina peatland. Am Nat 145: 1–21.
11. Iversen CM, Bridgham SD, Kellogg LE (2010) Scaling plant nitrogen use and uptake efficiencies in response to nutrient addition in peatlands. Ecology 91: 693–707.
12. Melillo JM, Steudler PA, Aber JD, Newkirk K, Lux H, et al. (2002) Soil warming and carbon-cycle feedbacks to the climate system. Science 298: 2173–2176.
13. Wallenda T, Kottke I (1998) Nitrogen deposition and ectomycorrhizas. New Phytol 139: 169–187.
14. Bai Y, Wu J, Clark CM, Naeem S, Pan Q, et al. (2010) Tradeoffs and thresholds in the effects of nitrogen addition on biodiversity and ecosystem functioning: evidence from Inner Mongolia grasslands. Glob Change Biol 16: 358–372.
15. Keuter A, Hoeft I, Veldkamp E, Corre MD (2012) Nitrogen response efficiency of a managed and phytodiverse temperate grassland. Plant Soil 364: 193–206.
16. Chapin FS, Vitousek PM, Vancleve K (1986) The nature of nutrient limitation in plant-communities. Am Nat 127: 48–58.
17. Lu XT, Freschet GT, Flynn DFB, Han XG (2012) Plasticity in leaf and stem nutrient resorption proficiency potentially reinforces plant-soil feedbacks and microscale heterogeneity in a semi-arid grassland. J Ecol 100: 144–150.
18. Freschet GT, Cornelissen JHC, van Logtestijn RSP, Aerts R (2010) Substantial nutrient resorption from leaves, stems and roots in a subarctic flora: what is the link with other resource economics traits? New Phytol 186: 879–889.
19. Chapin FS (1980) The mineral nutrition of wild plants. Ann Rev Ecol Syst 11: 233–260.
20. Koerselman W, Meuleman AFM (1996) The vegetation N:P ratio: A new tool to detect the nature of nutrient limitation. J Appl Ecol 33: 1441–1450.
21. Wan SQ, Hui DF, Luo YQ (2001) Fire effects on nitrogen pools and dynamics in terrestrial ecosystems: A meta-analysis. Ecol Appl 11: 1349–1365.
22. Reich PB, Peterson DW, Wedin DA, Wrage K (2001) Fire and vegetation effects on productivity and nitrogen cycling across a forest-grassland continuum. Ecology 82: 1703–1719.
23. Pilkington MG, Caporn SJM, Carroll JA, Cresswell N, Phoenix GK, et al. (2007) Impacts of burning and increased nitrogen deposition on nitrogen pools and leaching in an upland moor. J Ecol 95: 1195–1207.
24. Watson PJ, Bradstock RA, Morris EC (2009) Fire frequency influences composition and structure of the shrub layer in an Australian subcoastal temperate grassy woodland. Aust Ecol 34: 218–232.
25. Bowman DM, Balch JK, Artaxo P, Bond WJ, Carlson JM, et al. (2009) Fire in the Earth system. Science 324: 481–484.
26. Bond WJ, Keeley JE (2005) Fire as a global 'herbivore': the ecology and evolution of flammable ecosystems. Trends Ecol Evol 20: 387–394.
27. Britton AJ, Helliwell RC, Fisher JM, Gibbs S (2008) Interactive effects of nitrogen deposition and fire on plant and soil chemistry in an alpine heathland. Environ Poll 156: 409–416.
28. Keeley JE, Pausas JG, Rundel PW, Bond WJ, Bradstock RA (2011) Fire as an evolutionary pressure shaping plant traits. Trends Plant Sci 16: 406–411.
29. Bradshaw SD, Dixon KW, Hopper SD, Lambers H, Turner SR (2011) Response to Keeley et al.: Fire as an evolutionary pressure shaping plant traits. Trends Plant Sci 16: 405–405.
30. Heisler JL, Briggs JM, Knapp AK, Blair JM, Seery A (2004) Direct and indirect effects of fire on shrub density and aboveground productivity in a mesic grassland. Ecology 85: 2245–2257.
31. Bowles ML, Jones MD (2013) Repeated burning of eastern tallgrass prairie increases richness and diversity, stabilizing late successional vegetation. Ecol Appl 23: 464–478.
32. Ojima DS, Schimel DS, Parton WJ, Owensby CE (1994) Long-term and short-term effects of fire on nitrogen cycling in tallgrass prairie. Biogeochemistry 24: 67–84.
33. Zhou L, Huang J, Lü F, Han X (2009) Effects of prescribed burning and seasonal and interannual climate variation on nitrogen mineralization in a typical steppe in Inner Mongolia. Soil Biol Biochem 41: 796–803.
34. Lü XT, Cui Q, Wang QB, Han XG (2011) Nutrient resorption response to fire and nitrogen addition in a semi-arid grassland. Ecol Engin 37: 534–538.
35. Cui Q, Lü XT, Wang QB, Han XG (2010) Nitrogen fertilization and fire act independently on foliar stoichiometry in a temperate steppe. Plant Soil 334: 209–219.
36. Lü XT, Lu FM, Zhou LS, Han X, Han XG (2012) Stoichiometric response of dominant grasses to fire and mowing in a semi-arid grassland. J Arid Environ 78: 154–160.
37. Lü XT, Kong DL, Pan QM, Simmons ME, Han XG (2012) Nitrogen and water availability interact to affect leaf stoichiometry in a semi-arid grassland. Oecologia 168: 301–310.
38. Aerts R, Decaluwe H (1994) Nitrogen use efficiency of *Carex* species in relation to nitrogen suplly. Ecology 75: 2362–2372.
39. Yasumura Y, Hikosaka K, Matsui K, Hirose T (2002) Leaf-level nitrogen-use efficiency of canopy and understorey species in a beech forest. Fun Ecol 16: 826–834.
40. Yuan ZY, Li LH, Huang JH, Han XG, Wan SQ (2005) Effect of nitrogen supply on the nitrogen use efficiency of an annual herb, Helianthus annuus L. J Integrat Plant Biol 47: 539–548.
41. Reich PB, Grigal DF, Aber JD, Gower ST (1997) Nitrogen mineralization and productivity in 50 hardwood and conifer stands on diverse soils. Ecology 78: 335–347.
42. Blair JM, Parmelee RW, Beare MH (1990) Decay-rates, nitrogen fluxes, and decomposer communities of single-species and mixed-species foliar litter. Ecology 71: 1976–1985.
43. Yuan ZY, Li LH, Han XG, Chen SP, Wang ZW, et al. (2006) Nitrogen response efficiency increased monotonically with decreasing soil resource availability: a case study from a semiarid grassland in northern China. Oecologia 148: 564–572.

44. Olear HA, Seastedt TR, Briggs JM, Blair JM, Ramundo RA (1996) Fire and topographic effects on decomposition rates and N dynamics of buried wood in tallgrass prairie. Soil Biol Biochem 28: 323–329.

45. Neary DG, Klopatek CC, DeBano LF, Ffolliott PF (1999) Fire effects on belowground sustainability: a review and synthesis. For Ecol Manag 122: 51–71.

3

Fractal Scaling of Particle Size Distribution and Relationships with Topsoil Properties Affected by Biological Soil Crusts

Guang-Lei Gao[1], Guo-Dong Ding[1,2]*, Bin Wu[1,2], Yu-Qing Zhang[1,2], Shu-Gao Qin[1,2], Yuan-Yuan Zhao[1,2], Yan-Feng Bao[1], Yun-Dong Liu[1], Li Wan[1], Ji-Feng Deng[1]

1 College of Soil & Water Conservation, Beijing Forestry University, Beijing, China, **2** Yanchi Research Station, Yanchi, China

Abstract

Background: Biological soil crusts are common components of desert ecosystem; they cover ground surface and interact with topsoil that contribute to desertification control and degraded land restoration in arid and semiarid regions.

Methodology/Principal Findings: To distinguish the changes in topsoil affected by biological soil crusts, we compared topsoil properties across three types of successional biological soil crusts (algae, lichens, and mosses crust), as well as the referenced sandland in the Mu Us Desert, Northern China. Relationships between fractal dimensions of soil particle size distribution and selected soil properties were discussed as well. The results indicated that biological soil crusts had significant positive effects on soil physical structure ($P<0.05$); and soil organic carbon and nutrients showed an upward trend across the successional stages of biological soil crusts. Fractal dimensions ranged from 2.1477 to 2.3032, and significantly linear correlated with selected soil properties ($R^2 = 0.494\sim0.955$, $P<0.01$).

Conclusions/Significance: Biological soil crusts cause an important increase in soil fertility, and are beneficial to sand fixation, although the process is rather slow. Fractal dimension proves to be a sensitive and useful index for quantifying changes in soil properties that additionally implies desertification. This study will be essential to provide a firm basis for future policy-making on optimal solutions regarding desertification control and assessment, as well as degraded ecosystem restoration in arid and semiarid regions.

Editor: Francisco J. Esteban, University of Jaén, Spain

Funding: This work was finacial supported by the National Technology & Science Support Program of P. R. China (2012BAD16B02)and the National Basic Research Program of P. R. China (973 Program, 2013CB429900). The funders had no role in study design, data collection and analysis, decision to publish, or preparation of the manuscript.

Competing Interests: The authors have declared that no competing interests exist.

* E-mail: dingguodong@bjfu.edu.cn

Introduction

Biological soil crusts (BSCs) refer to the cohesive, thin, horizontal layer of soil surface created by soil crust organisms composed of algae, lichens, and mosses, which are widespread throughout the world [1,2]. In the last few decades, scholars and wildlife managers have become well aware of their existences and roles in ecological processes. Generally, there have been several comprehensive reviews on BSCs including the classic papers by Evans and Johansen [3], Belnap [1] and Bowker [4], as well as the monographs by Belnap and Lange [5], which synthesized large scale and long-term studies. Although there was large variability in types and numbers of BSCs due to the differences in climate, physiognomy, soil, and vegetations, these studies clearly presented that BSCs play an increasingly important role in ecosystem processes and biogeochemical cycles [2,3,6].

Unique among all ecosystems, BSCs in deserts not only engage in multitudinous ecological processes, but also serve as the impetus of the habitation of plants, animals, and fragile physical environments [7–9]. These biological communities cover the open space between higher plants [5], and dominate the harshest habitats while enduring drought, cold, soil salinity, strong wind, high temperature and radiation in arid and semiarid regions [10]. In some extreme arid deserts, BSCs are the only plants that can survive [11].

Structures and functions of BSCs in deserts have been more thoroughly investigated recently. The first efforts mostly focused on assortment, morphological characteristics and distribution of BSCs under different environmental conditions. Phylogenetic and morphological diversity and microstructure were commonly distinguished [11–14], and spatial distribution was mapped [15–17]. BSCs potentially contribute to variability in the soil hydrological cycles in arid and semiarid land and should be considered in the development of hydrologic models [18–21]. In general, BSCs can determine the amount, position, and time of water infiltration into desert soil, and their characteristics affect the water distribution of soil profiles [22], especially in significantly increasing topsoil water [23–26]. Currently, the interactions between BSCs, soil erosion, and desertification were commonly investigated [27]. BSCs protect sandland surfaces against wind erosion and stabilize flowing sand dunes [28–30]. Physical, algae and moss crusts were found in different developmental stages of sand dunes [29]. BSCs perform several functions in carbon sequestration and nitrogen fixation, which has drawn increasing attention from scientists, especially under the conditions of global climate change. [31]. There is also a wealth of literature regarding the effects of BSCs on vegetation succession [32–34], and physiological responses of BSCs to variations in environmental conditions [35–37].

Studies regarding BSCs are numerous and have led to a much greater understanding of their place in ecosystems; however, BSCs have not been explored in full as of yet. More information is needed regarding details of the mutual relationships of BSCs and soil properties. Especially, little is known about an effective index to quantify the effects on soil properties affected by BSCs in the desertification processes. In recent decades, the possibility of characterizing soil particle size distribution (PSD) using fractal theory has been explored [38]. Significant linear correlations have been observed between fractal dimensions and various soil properties. This offers the possibility of quantifying and integrating information on soil biological, chemical, and physical characteristics measured on different spatial scales [27]. In this study, we hypothesized that the variations in topsoil properties are largely consequences of BSCs establishment and succession. Therefore, changes in topsoil properties were documented across three types of successional BSCs (algae, lichens, and mosses crust), as well as in the sandland in the Mu Us Desert, Northern China. The objectives of this study were: (1) to examine the changes in topsoil properties affected by BSCs; (2) to explore the possibility that fractal dimension (*D*) of soil PSD can be used as a practical index for quantifying variations in soil properties and the implications of desertification.

Materials and Methods

Study Area

The study area is located in Yanchi County (37° 04′ −38° 10′ N, 106° 30′−107° 47′ E; 1400–1800 m a.s.l.) in the central part of southern rim area of the Mu Us Desert in the eastern part of Ningxia Hui Autonomous Region, P. R. China (Figure 1). Yanchi County is a typical transitional zone; the terrain changes from the Loess Plateau (South) to the Ordos Plateau (North) [39]. The specific physical environments make the land quite usable and there are diverse natural resources, but it is a fragile ecological environment.

The study area has a continent-typical warm monsoon climate in a temperate zone. The average annual temperature is 7.8°C. The winter-summer and diurnal-nocturnal differences in air temperature are significantly high, with an annual range in temperature of greater than 28 and 20°C, respectively. The average annual precipitation is 292 mm with 62% occurring between July and September. The average annual evaporation is 2024 mm, which is dramatically higher than annual precipitation. The study area is perennially windy with average annual wind velocity of 2.6 m/s and a prevailing northwest wind direction, which results in frequent and strong wind erosion [40]. The landscape of this area is characterized by sand dunes. The dominated soil type in this area is the arenosols type of quartisamment (U.S. Soil Taxonomy), commonly found in arid and semiarid regions. The typical vegetation of study area is Eurasian Grassland Zone species. The natural vegetation in the study area consists largely of *Salix psammophila*, *Caragana korshinskii*, *Hedysarum scoparium*, *Artemisia ordosica*, *Glycyrrhiza uralensis*, *Sophora alopecuroides*, and *Cynanchum komanovii*. Additionally, the dominant plantation species include *Salix matsudana*, and *Populus alba*.

The field sampling sites are located in an enclosed area within Zhouzhuangzi Village, Yanchi County, which belongs to the Yanchi Research Station of Beijing Forestry University of the Chinese Terrestrial Ecosystem Research Network (CTERN).

Study Approach and Experimental Design

Soil sampling and crusts identification. The study was conducted from April to October, 2012. The field sampling sites were typical areas of desertification control implemented by fencing after the mid-1990s. In this area, algae, lichens, and mosses crusts exist collectively. To ensure the same parent rock material and similar topography, three random 2 m×2 m sites covering successional BSCs stages were chosen in the open and flat space of sandland with sporadic *Artemisia ordosica* communities in the 1 km range (Table 1). For each BSCs type, three soil sampling profiles were selected at random. Soil samples were collected for two thin layers: the top layer (BSCs layer, the thickness was consistent with BSCs) and the underlying layer (layer under BSCs, 1 cm). Meanwhile, topsoil samples were collected in the referenced uncovered sandland for two thin layers as well: 0–1 cm for the top layer, and 1–2 cm for the underlying layer. Crusts species were identified both macroscopically and microscopically.

Soil properties analysis. All the soil profile samples were naturally dried in a shaded area in the laboratory; plant stems, small stones, and large animals such as worms were carefully removed. Then, parts of the air-dried soil samples were hand-sieved through 2 and 0.25 mm screens prior to laboratory analysis. To meet the experimental needs, parts of soil samples were ground. The topsoil physicochemical properties were analyzed in the following methods (1) soil organic carbon (SOC) was determined by the potassium dichromate wet combustion method; (2) soil total nitrogen (N_T) was measured using the micro-Kjeldahl's method; (3) soil total phosphorus (P_T) was determined by the Mo-Sb colorimetry method; (4) soil total potassium (K_T) was measured using the hydrofluoric & perchloric acid (HF-HCLO acid)-flame photometer method; (5) available nitrogen (N_{Avi}) was determined by the alkali diffusion method; (6) soil available phosphorus (P_{Avi}) was measured using the sodium bicarbonate ($NaHCO_3$) digestion-Mo-Sb colorimetry method; (7) soil rapid available potassium (K_{Avi}) was determined by the ammonium acetate digestion-flame photometer method [41,42].

According to the Archimedes Principle, soil bulk density and total porosity were measured using the wax seal method [41,43–44]. Soil bulk density was calculated using Eq. (1); total porosity was subsequently calculated with Eq. (2).

$$\gamma_s = \frac{100g_1}{[(g_4 - g_3)/\rho_1 - (g_2 - g_1)/\rho_2] \times (100 + W)} \quad (1)$$

Where γ_s is soil bulk density (g·cm^{-3}), same below; g_1 is the simple weight (g); g_2 is simple weight when completely wrapped by wax (g); g_3 is the original reading of electronic balance (g); g_4 is reading of electronic balance with the sample (g); ρ_1 is specific gravity of water, $\rho_1 = 1.0$ g·cm^{-3}; ρ_2 is specific gravity of wax, $\rho_2 = 0.9$ g·cm^{-3}; W is water content of sample.

$$f = (1 - \frac{\gamma_s}{\rho_s}) \times 100 \quad (2)$$

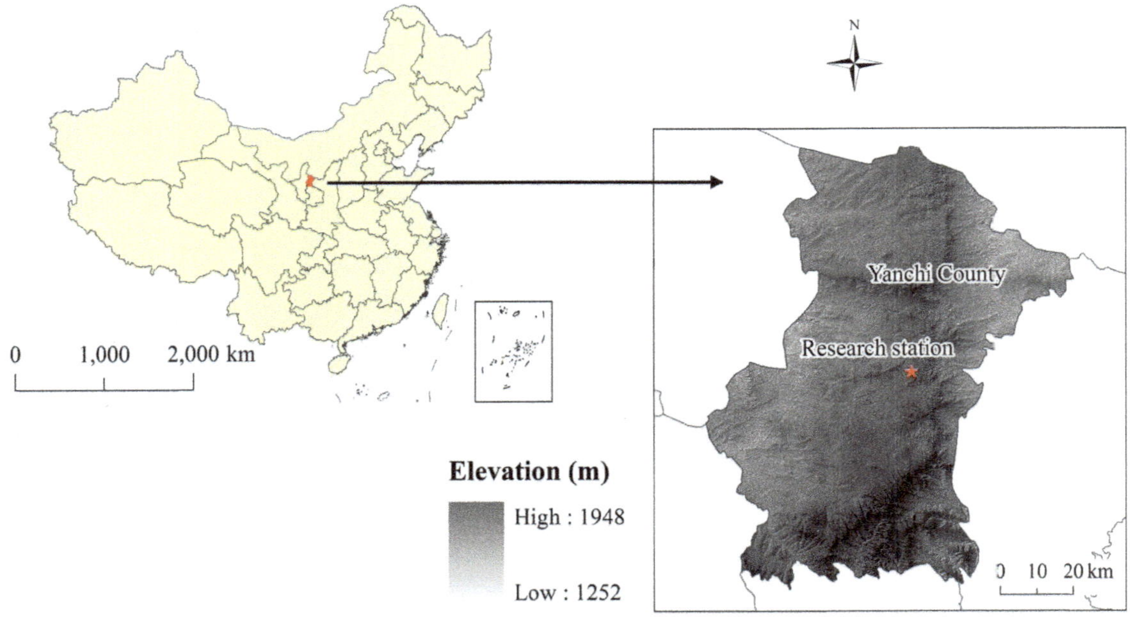

Figure 1. Location of the study area.

Where f is soil total porosity (%); ρ_s is soil particle density (g·cm^{-3}), $\rho_s = 2.65$ g·cm^{-3}.

Laser diffraction analysis and soil fractal model. To measure the topsoil particles and fractal characteristics, the unscreened air-dried soil samples were pretreated in an H_2O_2 solution (30%, w/w) to destroy any organic matter. And then, the soil aggregates were dispersed by adding sodium hexametaphosphate (NaHMP) and sonicating the samples for 30 seconds [45,46]. The pretreated soil samples were then analyzed by a laser diffraction technique using Malvern MS 2000 manufactured by Malvern Instruments in Malvern, England with measurement range and margin of error of 0.02–2000 μm and <2%, respectively. Each sample was measured 5 times and the mean value was used. The analysis results of soil PSD were output by U.S. Soil Taxonomy as follow: 0–2, 2–50, 50–100, 100–250, 250–500, 500–1000 and 1000–2000 μm.

Fractal dimension of soil PSD was calculated based on the volume distribution of soil particle size [46,47]. The equation is expressed as:

$$\frac{V(r < R_i)}{V_T} = \left(\frac{R_i}{R_{\max}}\right)^{3-D} \quad (3)$$

Where r is the soil particle size, R_i is the soil particle size of grade i, R_{max} is the maximum value of soil particle size, $V(r<R_i)$ is the volume of soil particle size less than R_i, V_T is the total volume of soil particles, D is the volume-based fractal dimension.

Data Analysis

All statistical analyses were conducted using SPSS software (Version 17.0). The differences in soil properties and D values among the BSCs sites were compared using multiple comparison and one-way analysis of variance (ANOVA) procedures. The LSD test (at $p<0.05$) was used to compare means of soil variables; the results of ANOVA were significant at the level of $p<0.05$. Simple linear regression and correlation analysis were performed to identify the relationships between D and the selected soil properties. Pearson correlation coefficient and 2-tailed test were used to distinguish the correlation and significant differences.

Results

Soil Bulk Density and Total Porosity under Different BSCs Sites

Variations in soil bulk density were remarkable (Table 2), and significant differences occurred among different BSCs types ($P<0.05$). Uncovered sandland soil had the greatest bulk density (1.71

Table 1. General information of the different BSCs sites.

No.	Type	Color	Thickness (cm)	Coverage (%)	Dominant species
1	Algae	Gray	0.37(0.06)	36.7(2.9)	*Microcoleus vaginatus, Oscillatoria chlorine*
2	Lichens	Brown	0.70(0.12)	53.3(2.9)	*Collema tenax, Collema coccophorum*
3	Mosses	Green	1.65(0.32)	71.7(5.8)	*Byum argenteum*
4	Sandland	-	-	-	-

Values in the parentheses indicates standard error (n = 3).

and 1.69 $g \cdot cm^{-3}$) compared to the BSCs sites, due to severe wind erosion. With the development of BSCs, soil bulk density decreased from algae (1.64 and 1.68 $g \cdot cm^{-3}$) to mosses crust (1.25 and 1.48 $g \cdot cm^{-3}$). Furthermore, soil bulk density within the same type of BSCs increased by 2.4%~21.2% with increasing soil depth in all sampling sites.

Total porosity showed a clear and significant difference in the top layer among different BSCs types; in the underlying layer, significant difference only occurred between mosses crust and other sites ($P<0.05$). However, as shown in Table 2, there was a clear tendency to increase from algae to mosses crust, which was the complete reverse of the contrast to soil bulk density under the different types of BSCs. The lowest total porosity of 35.58% and 32.40% were found in uncovered sandland, and soil total porosity increased successively from algae (38.29% and 34.64%) to mosses crust (52.79% and 42.58%). With the increase in soil depth, total porosity decreased by 3.65%, 13.95% and 10.21% in the same site.

SOC and Soil Nutrients under Different BSCs Sites

Accompanying the positive changes in soil physical structure, SOC and soil nutrients increased alike (Figure 2). Comparing with the sandland, SOC was higher in the BSCs sampling sites. Among them, mosses had the highest SOC in both layers (5.72 and 3.98 $g \cdot kg^{-1}$) which differed significantly with the sandland (0.94 and 0.69 $g \cdot kg^{-1}$) ($P<0.05$), but not significantly different from lichens (4.54 and 3.11 $g \cdot kg^{-1}$) and algae (3.15 and 2.45 $g \cdot kg^{-1}$) crusts ($P>0.05$). Further, SOC within the same type of BSCs deceased from the top to underlying layer.

Changes in soil nutrients varied significantly, especially the levels of K_T and N_{Avi}. However, there was a clear increasing tendency across the three types of successional BSCs, and soil nutrients under BSCs sites were greater than the uncovered sandland. For the content of K_T, N_{Avi} and P_{Avi}, no significant differences were found among the different BSCs types ($P>0.05$). However, there were significant differences in the content of N_T, P_T, and K_{Avi} at different sites, especially between the uncovered sandland and mosses crust ($P<0.05$). The spatial variations in soil nutrients in the soil profile deceased from the top to underlying layer as well.

Particles Size Distribution and Fractal Characteristics under Different BSCs

Table 3 shows the soil PSD in the different soil sampling sites, and sand particles (50–2000 μm) could be considered the dominant soil particle class accounting for more than 80% of the total, approximately. Meanwhile, there was a clear tendency to decrease of sand particle content from algae to mosses crust. Compared with the uncovered sandland, sand particles content decreased by 11.2~16.4% and 4.9~12.6%, and sand particles of mosses crust had significant differences with other crust types ($P<0.05$). Clay (0–2 μm) and silt (2–50 μm) contents were much lower than sand particles, especially the clay contents (0.4~0.7%). However, with the development of BSCs, clay and silt contents increased sharply. Compared with the uncovered sandland, from algae to mosses crust, the clay and silt contents increased by as much as 90.9~112.1%, 163.9~241.2% and 27.2~71.9%, 65.4~167.2% for the top and underlying layers, respectively. As a result, significant differences occurred between mosses crust and other sites for clay and silt contents ($P<0.05$). Furthermore, sand contents within the same type of BSCs increased from the top to underlying layer, in addition to the decrease in clay and silt contents. In contrast, clay and silt contents of the uncovered sandland were cumulative with the increase in soil depth.

Volume-based soil fractal dimensions (D) were subsequently calculated with Eq. (3) based on the soil PSD data, and the variations in D values for the different BSCs are shown in Table 4. Generally, fractal dimensions of soil PSD ranged from 2.1477 to 2.3032; with the development of BSCs, D values increased progressively from algae to mosses. Meanwhile, the D values of algae, lichens and mosses crust were higher than uncovered sandland in both layers. However, significant differences among the different BSCs types only occurred in the top layer ($P<0.05$). In the underlying layer only mosses significant differed with uncovered sandland ($P<0.05$). Additionally, D values observed in the underlying layer were lower than the top layer because the wind-induced fine particles remove surface soil.

Relationship between Soil Fractal Dimensions and Selected Soil Properties

Linear regression and correlation analysis were applied to identify the relationships between fractal dimensions (D) and selected soil properties including soil bulk density, total porosity, SOC, soil nutrients as well as soil PSD (Figure 3–5, Table 5). Results indicated that a significant negative linear correlation between soil bulk density and D values was observed with $R^2 = 0.650$, $p<0.01$ (Figure 3A). By contrast, Figure 3B showed a positive linear correlation between total porosity and D values ($R^2 = 0.626$, $p<0.01$). This reverse correlation and the different variations in bulk density and total porosity mutually verified.

Figure 4 showed the relationship between D values and SOC and soil nutrients. The recovery and accumulation of SOC and soil nutrients resulted in increased D values. Fractal dimension had considerable significant positive linear correlation with SOC, N_T, P_T, N_{Avi}, P_{Avi} and K_{Avi} ($R^2 = 0.494$~0.787, $p<0.01$). However, no significant correlation occurred between D values and K_T (Figure 4D, $R^2 = 0.060$, $p<0.01$). Therefore, soils with more

Table 2. Variations of bulk density and total porosity for different BSCs.

Item	Layer	BSCs			Sandland
		Algae	Lichens	Mosses	
Bulk density ($g \cdot cm^{-3}$)	top	1.64(0.04)a	1.32(0.05)b	1.25(0.01)c	1.71(0.08)d
	underlying	1.68(0.01)ab	1.60(0.02)b	1.48(0.08)c	1.69(0.02)a
Total porosity (%)	top	38.29(1.60)a	50.21(1.82)b	52.79(0.52)c	35.58(3.06)d
	underlying	34.64(1.40)a	36.26(1.10)a	42.58(5.57)b	32.40(0.36)a

Values in the parentheses indicates standard error (n = 3). Means with different letter in the same row are significantly different at the 0.05 level (LSD).

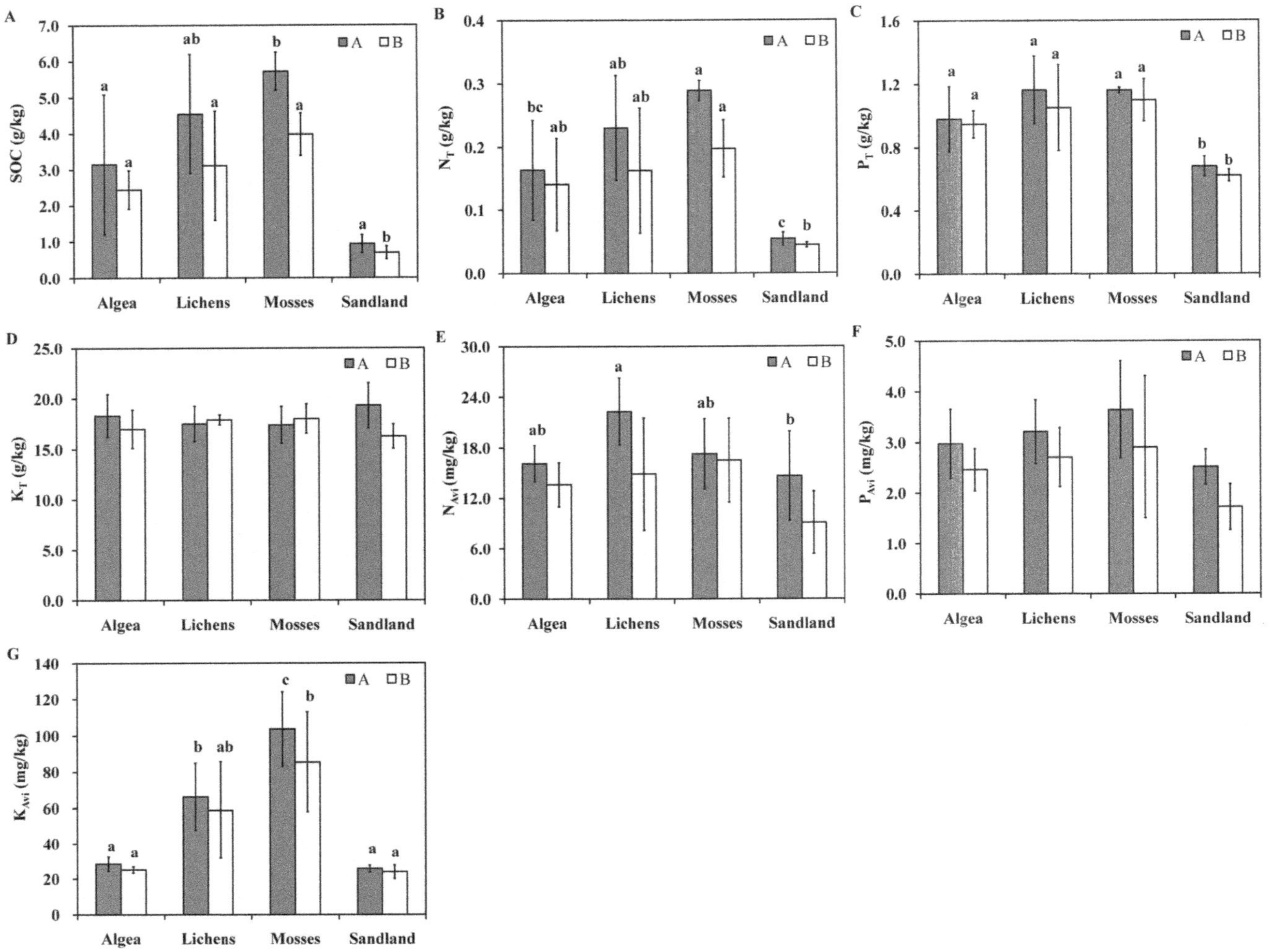

Figure 2. Variations in SOC (A) and soil nutrients (B~G) under different BSCs. Vertical bars indicate standard errors of means (n = 3). Means with the different letter in the same layers are significantly different at the 0.05 level (LSD).

fertilizer have higher fractal dimension values; values of D do not appear to be related to K_T.

Table 3. Variations of soil PSDs for different BSCs.

BSCs	Layer	Soil PSD (%)		
		Clay (0–2 μm)	Silt (2–50 μm)	Sand (50–2000 μm)
Algae	top	0.6(0.1)a	16.4(3.7)a	83.0(3.8)a
	underlying	0.5(0.1)ab	11.3(2.0)ab	88.2(2.1)ab
Lichens	top	0.7(0.1)a	20.5(7.7)a	78.9(7.8)a
	underlying	0.5(0.2)ab	15.9(7.5)a	83.6(7.6)b
Mosses	top	0.7(0.1)a	21.2(0.8)a	78.1(0.9)b
	underlying	0.7(0.1)a	18.3(2.1)a	81.1(2.2)a
Sandland	top	0.3(0.1)b	6.2(0.5)b	93.5(0.6)a
	underlying	0.4(0.1)b	6.8(0.9)b	92.8(1.1)b

Values in the parentheses indicates standard error (n = 3). Means with different letter in the same line of the same layer are significantly different at the 0.05 level (LSD).

In Figure 5, soils with greater clay and silt contents had higher soil fractal dimensions (D); soils with a greater amount of sand particles had lower D values. D values had significant positive linear correlations with clay and silt contents ($R^2 = 0.955$ and 0.814, $p<0.01$), and a negative linear correlation with sand particles contents ($R^2 = 0.821$, $p<0.01$). Therefore, wind-induced removal of fine particles resulted in decreased D values, and D values appeared to be closely related to clay particles rather than silt and sand contents.

Discussion

Effects of BSCs on Topsoil Properties

In the Mu Us Desert, soil properties can be improved through the addition of plants including the lower plants that introduce BSCs. This information is greatly important for the countries in arid and semiarid regions with a high susceptibility to wind erosion and desertification. BSCs protect sand surface from wind erosion. Erosive force and carriage ability are consumed by BSCs through increasing aerodynamic roughness length and threshold wind velocity [30]. Topsoil loss with airflow under BSCs is much lower than in the uncovered sandland; and the dusts in airstreams are captured by cyanobacteria filaments as well. These processes have positive effects on bulk density decrease and total porosity increase.

Table 4. Variations of soil fractal dimensions for different BSCs.

(cm)	BSCs			Sandland
	Algae	Lichens	Mosses	
top	2.2786(0.0281)a	2.2904(0.0262)a	2.3032(0.0214)a	2.1477(0.0403)b
underlying	2.2291(0.0224)ab	2.2409(0.0575)ab	2.2892(0.0144)a	2.1687(0.0511)b

Values in the parentheses indicates standard error (n = 3). Means with different letter in the same row are significantly different at the 0.05 level (LSD).

Moreover, most of the soil photosynthetic productivity and nitrogen fixation in sandland is concentrated within 3 mm of the surface [48]. Therefore, BSCs are meaningful to combat land degradation.

With the presence of BSCs, a significant amount of CO_2 exchange is observed between BSCs and atmospheric content. CO_2 is sequestrated and accumulated through a combination of BSCs photosynthesis and respiration. Free chlorophyll in algae and chloroplast in mosses capture CO_2 through photosynthesis and store carbon in the form of carbohydrates in their tissues. The photosynthetic rate of BSCs has been commonly reported under natural conditions or in the laboratory. However, the values are of great uncertainty, ranging from 0.1~11.5 μ mol CO^2 $m^{-2} \cdot s^{-1}$ due to the difference in various physical environments and the ability to carbon capture of different crusts species [5]. Synthesizing all these study achievements, Elbert et al [49] obtained a median flux of 16 $g \cdot m^{-2} \cdot a^{-1}$ for the net uptake of carbon by BSCs, accordingly, the total carbon sequestration by BCSs in arid and semiarid regions is approximately 1.0 $Pg \cdot a^{-1}$. Although, it is much lower in comparison with other processes of the global carbon cycle, millions tons of carbon was sequestrated by BSCs and stored in soil ecosystem, which helps soils prosper. It is of significant importance for soil environmental restoration, and can't be ignored of understated.

Nitrogen turnover is as complex as the carbon cycle; it combines nitrogen mineralization, ammonia volatilization, nitrification and denitrification. The presence of dinitrogenase in alga, and bacterium of BSCs play a significant role in this process; these biochemical processes that lead to nutrients fixation [50,51]. Photosynthesis and azotification are closely interconnected and interdependent. As an important precondition, carbohydrate produced by the BSCs' photosynthesis drives the azotification of dinitrogenase in alga and bacterium. In exchange, azotification provides much-needed organic nitrogen to sustain BSCs for photosynthesis. Like the photosynthetic rate, the average fluxes of nitrogen fixation are reported in the range of 0.1–10 $g \cdot m^{-2} \cdot a^{-1}$; the total nitrogen fixation by BCSs in arid and semiarid region is approximately 30 $Tg \cdot a^{-1}$ [49].

Finally, carbohydrates and organic nitrogen introduced by BSCs are input into soils in various ways. The most important way is through the decomposition of BSCs. At the end of BCS lifecycle, remains decompose through enzymatic activities by the combined action of soil bacterium and enzymes. Furthermore, some algal crusts can secrete polysaccharide into soil to increase carbon pool; similarly, nitrogen micromolecules are released into soil by BSCs as well. Meanwhile, SOC and soil nitrogen accumulation is beneficial to the growth of tracheophyte; successions of vascular plants help soil properties prosper. Additionally, some beetles used BSCs as habitat; these BSCs-feeding beetles are also considered as an important nutrients resource in arid and semiarid regions due to the decomposition of their bodies.

With carbon sequestration and transport at its core, BSCs increase nutrient contents and make soil more fertile. However, unique among other soil nutrients, soil K_T is an expectation. In this study, a poor relationship between D and K_T was observed which corresponded with a lack of differences in K_T among the different types of BSCs and differing degrees of desertified soils. In fact, although potassium is one of the most essential macronutrients for plant growth and development, scholars pay little attention on the conversion due to its substantial deposit in soil system of northern China. In nature, potassium of different forms can affect and transform to each other; however the rate of this conversion is

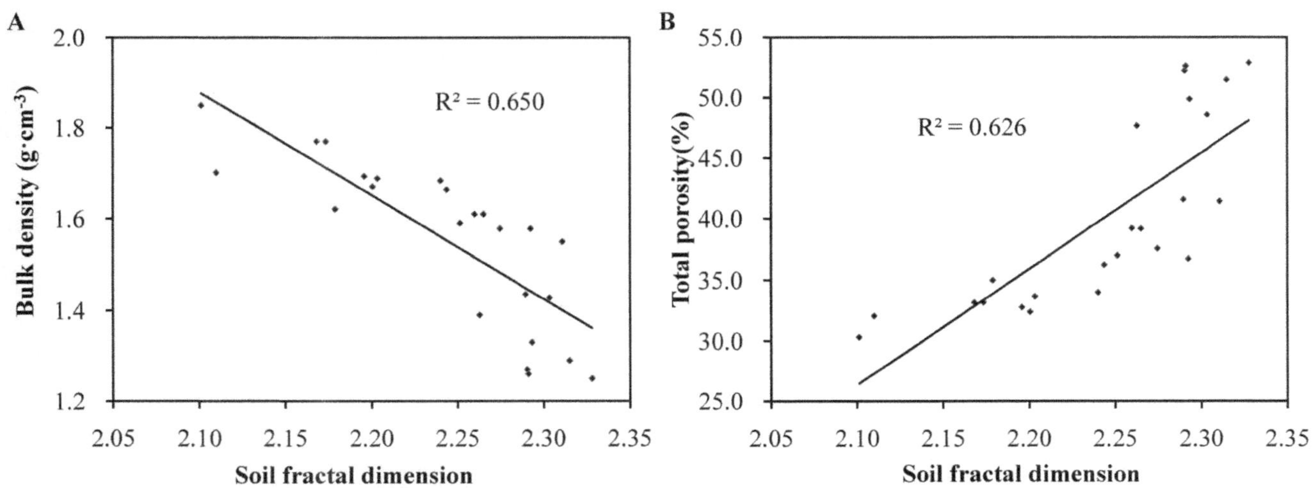

Figure 3. Relationships between *D* values and soil bulk density (A), total porosity (B).

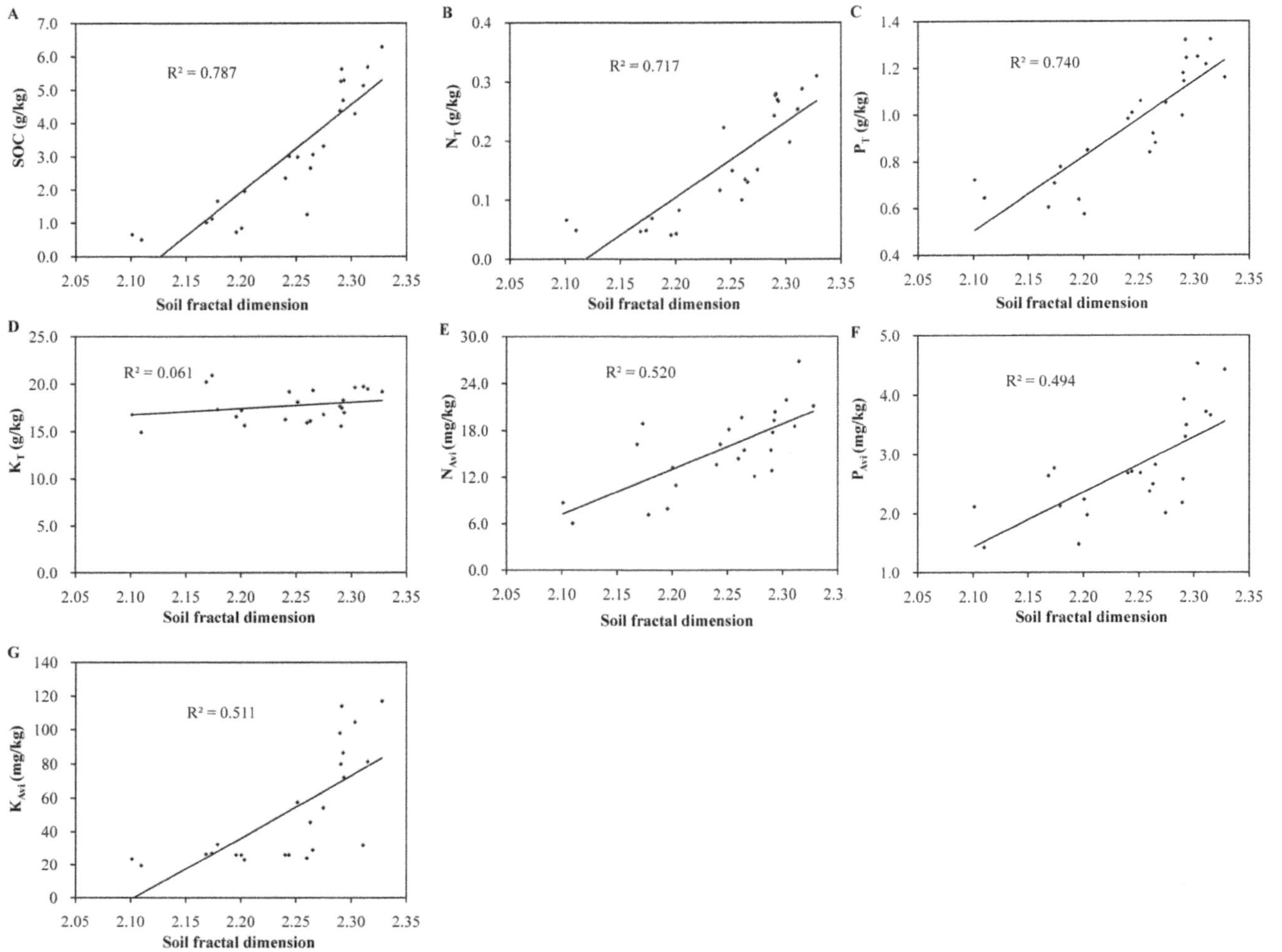

Figure 4. Relationships between D values and selected soil properties (A~G).

rather slow. Among all the forms, potassium mainly exists in a number of inorganic substances, especially the inorganic mineral [52]. These mineral is stable, ion-exchange reactions barely occur with cations in soil solutions, and are instead soaked up by plants. Therefore, soil K_T is related to soil mineralogy, and unlikely to be affected by BSCs. Besides, we suspect that K_T may lose due to the decrease of pH (from alkalescence to neutral) induced by combination efforts of vegetation recovery and wind erosion.

Variations in soil properties differ between successional stages of BSCs. Complex and evolutional BSCs achieve greater activity and ability as results of diverse community composition in numbers, species, and functions [53]. In uncovered sandland, SOC and soil nutrients are lower than at later successional stages. When algeas crusts develop into lichens, microalgal biomass increases rapidly along with carbon and nitrogen deposition. In mosses crusts, photosynthesis and azotification require a higher environmental threshold; whereas in return mosses crust have stronger resistance

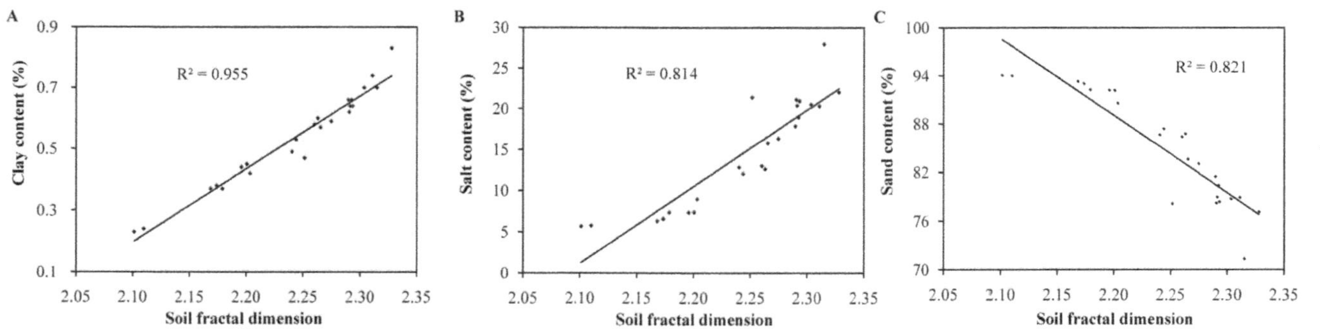

Figure 5. Relationships between *D* values and soil PSD (A~C).

Table 5. Correlations between soil fractal dimensions (*D*) and selected soil properties.

Soil properties	Correlation coefficients	Soil properties	Correlation coefficients
Bulk density	−0.806**	N_{Avi}	0.721**
Total porosity	0.791**	P_{Avi}	0.703**
SOC	0.887**	K_{Avi}	0.715**
N_T	0.847**	Clay	0.977**
P_T	0.860**	Silt	0.902**
K_T	0.247	Sand	−0.906**

**Correlation is significant at the 0.01 level (2-tailed).

to and more efficient abilities of carbon and nitrogen fixation. Therefore, although BSCs create an important increase in soil fertility, there is, at times, no significant difference occurred between sandland and algae, lichens crust. Contrarily, mosses crust significantly differs from sandland, even algae and lichens crusts.

Soil Fractal Dimension as a Practical Indicator for Desertification

Characterizing soil organization and functions with a single parameter is a question of great interest when monitoring soil degradation and desertification. In former studies, individual fractions (e.g., clay and fine fractions) textual analysis and soil organic matter were the commonly used methods to characterize soil quality. However, soil is an irregular and complex system in which many biological and physical components interact across all space and time scales [54]. Among these approaches, individual fractions always ignore coarse fractions, but emphasize fine particles which are rich in nutrients. Therefore, it cannot provide complete information and cause the waste of soil data. Moreover, it is not ideal to the real soil system, especially the soil in the deserts which coarse particles account for a greater proportion. In addition, although textual analysis provides integrated information, the precision and comparability of data are adversely affected by the descriptive indicator. Furthermore, soil organic carbon is one of the most important indexes in soil quality assessment, but it is not sensitive to environmental change in short-time scale [55]. Therefore, traditional methodologies cannot provide complete information and quantitatively represent fundamental attributes using a practical index. By contrast, fractal measure can make full use of the soil PSD information including clay, silt, and coarse particles data. Fractal dimensions significantly correlated with soil dynamics for long and short time scale. Soil degradation causes an increase in one or more intervals of fractal behavior [56]. Modeling the process of fragmentation both in rocks and soils [57], fractal geometry in soil science have shown that soil exhibits fractal characteristics of irregular shape and self-similar structure [58]. The methodology is of considerable intrinsic interest because it is practical and descriptive, rather than a simplification and simulation [59].

Soil PSD is closely related with soil functions and they are interdependent. It is generally believed that the rich fine fractions (clay and silt) mean fertile, hydrophilic, and biodiversity-rich soil system. Therefore, when wind-induced erosion occurs and soils are lost, we lose far more than just the fractions that they contain, but immeasurable ecological services. Wind erosion not only causes the transport of soil particles, but also nutrient and function losses, and further resulting in the decrease of water-holding capacity, depletion of soil structure and biological properties [27,60]. Finally, these losses in fine particles affecting by wind-induced erosion cause land degradation, and desertification. In this study, linear regression and correlation analysis indicates that soil fractal dimensions have a highly significant correlation with soil properties ($p<0.01$). Estimation of soil fractal dimension under different BSCs is helpful in determining changes in soil properties and vulnerability of desertification. More specifically, lower values indicate fine particles removal in soil coarsening process due to wind erosion in desertification-vulnerable regions. With the loss of fine particles, SOC and soil nutrients decreased concomitantly. Accordingly, higher *D* values suggest sand particles disintegration, fine particles accumulation and soil properties recovery as affected by BSCs. Furthermore, *D* values indicate the potential of desertification as well. The higher *D* values implies the greater possibility of wind erosion due to the bigger proportion of fine particles. Once the sandland loses the protection of BSCs or the actual wind velocity exceeds the threshold, accumulative fine particles will be eroded by destructive wind in a short term. Therefore, fractal dimension is sensitive and useful index for quantifying soil properties variations in the evolution of a desert, and significantly implies for soil degradation and desertification induced by wind erosion.

Conclusions

In the Mu Us desert, the establishment and succession of BSCs caused positive changes in topsoil properties. There was a clear tendency to increase from sandland to BSCs sites in soil total porosity, SOC, soil nutrients, clay and silt particle contents, accompanying decrease in soil bulk density, and sand particle contents. With the succession of BSCs, significant difference occurred between mosses crust and uncovered sandland. Linear regression and correlation analysis showed that *D* values had significant linear relationships with soil bulk density, total porosity, SOC, soil nutrients (except K_T) as well as soil PSD with R^2 values ranging from 0.494 to 0.955 ($p<0.01$). Fractal dimension is sensitive to soil coarsening and nutrients losses process in the evolution of desertification, and it is now proven to be a sensitive and practical index for quantifying changes in soil properties and additionally implies desertification and vulnerability to desertification.

Acknowledgments

We thank Prof. Hai-Yan Wang, Ph. D. student Wei Feng and Chen-Xi Shao (Beijing Forestry University) for their contribution on an earlier version of this manuscript as well as the great cooperation with the Environmental Protection and Forestry Bureau of Yanchi County, Ningxia.

Author Contributions

Conceived and designed the experiments: G-LG G-DD BW Y-QZ. Performed the experiments: G-LG LW Y-DL J-FD. Analyzed the data: G-LG Y-YZ. Contributed reagents/materials/analysis tools: S-GQ Y-FB. Wrote the paper: G-LG Y-YZ.

References

1. Belnap J (2003) The world at your feet: desert biological soil crusts. Frontiers in Ecology and Environment 1: 181–189.
2. Castillo-Monroy AP, Maestre FT (2011) Biological soil crust: recent advances in our knowledge of their structure and ecological function. Revista Chilena De Historia Natura 84: 1–21.
3. Evans RD, Johansen JR (1999) Microbiotic crusts and ecosystem processes. Critical Reviews in Plant Science 18: 183–225.
4. Bowker MA (2007) Biological soil crust rehabilitation in theory and practice: an underexploited opportunity. Restoration Ecology 15: 13–23.
5. Belnap J, Lange OL (2001) Biological soil crusts: structure, function, and management. Springer-verlag, New York, U.S. 504.
6. Maestre FT, Bowker MA, Canton Y, Castillo-Monroy AP, Cortina J, et al. (2011) Ecology and functional roles of biological crusts in semi-arid ecosystems of Spain. Journal of Arid Environments 75: 1282–1291.
7. Lalley JS, Viles HA, Henschel JR, Lalley V (2006) Lichen-dominated soil crusts as arthropod habitat in warm deserts. Journal of Arid Environments 67: 579–593.
8. Li XR, Jia RL, Chen YW, Huang L, Zhang P (2011) Association of ant nests with successional stages of biological soil crusts in the Tengger Desert, Northern China. Applied Soil Ecology 47: 59–66.
9. Mager DM, Thomas AD (2011) Extracellular polysaccharides from cyanobacterial soil crusts: A review of their role in dryland soil processes. Journal of Arid Environments 75: 91–97.
10. Schlesinger WH, Pippen JS, Wallenstein MD, Hofmockel KS, Klepeis DM, et al. (2003) Community composition and photosynthesis by photoautotrophs under quartz pebbles, southern Mojave Desert. Ecology 84: 3222–3231.
11. Bhatnagar A, Bhatnagar MB, Garg MK, Bhatnagar K (2008) Community structure and diversity of cyanobacteria and green algae in the soil of Thar Desert (India). Journal of Arid Environments 72: 73–83.
12. Graef F, Stahr K (2000) Incidence of soil surface crust types in semi-arid Niger. Soil & Tillage Research 55: 213–218.
13. Garcia-Pichel F, Lopez-Cortes A, Nubel U (2011) Phylogenetic and morphological diversity of cyanobacteria in soil desert crusts from the Colorado plateau. Applied Environmental Microbiology 67: 1902–1910.
14. Tirkey J, Adhikary SP (2005) Coanobacteria in biological soil crusts of India. Current Science 89: 515–521.
15. Zhang YM, Chen J, Wang L, Wang XQ, Gu ZH (2007) The spatial distribution patterns of biological soil crusts in the Gurbantunggut Desert, Northern Xinjiang, China. Journal of Arid Environments 68: 599–610.
16. Kidron GJ, Vonshak A, Dor I, Barinova S, Abeliovich A (2010) Properties and spatial distribution of microbiotic crusts in the negev desert Israel. Catena 82: 92–101.
17. Pietrasiak N, Johansen JR, Drenovsky RE (2011) Geologic composition influences distribution of microbiotic crusts in the Mojave and Colorado Deserts at the regional scale. Soil Biology & Biochemistry 43: 967–974.
18. Williams JD, Dobrowolski JP, West NE (1999) Microbiotic crust influence on unsaturated hydraulic conductivity. Arid Soil Research and Rehabilitation 13: 145–154.
19. Belnap J, Phillips SL, Miller ME (2004) Response of desert biological crusts to alterations in precipitation frequency. Oecologica 141: 306–316.
20. Belnap J (2006) The potential roles of biological soil crusts in dryland hydrological cycles. Hydrological Processes 20: 3159–3178.
21. Bowker MA, Belnap J, Chaudhary VB, Jonhnson NC (2007) Revisiting classic water erosion models in drylands: The strong impact of biological soil crusts. Soil Biology & Biochemistry 40: 2309–2316.
22. Belnap J, Welter JR, Grimm NB, Barger N, Ludwing JA (2005) Linkages between microbial and hydrologic processes in arid and semiarid watersheds. Ecology 86: 298–307.
23. Verrecchia E, Yair A, Kidron GJ, Verrecchia K (1995) Physical properties of the psammophile cryptogamic crust and their consequences to the water regime of sandy soils, north-western Negev Desert, Israel. Journal of Arid Environments 29: 427–437.
24. Georage DB, Roundy BA, St Clair LL, Johanen JR, Schaalje GB, et al. (2003) The effects of microbiotic soil crusts on soil water loss. Arid Land Research and Management 17: 113–125.
25. Liu LC, Li SZ, Duan ZH, Wang T, Zhang ZS, et al. (2006) Effects of microbiotic crusts on dew deposition in the restored vegetation area at Shapotou, northwest China. Journal of Hydrology 328: 331–337.
26. Gao SQ, Ye XH, Chu Y, Dong M (2010) Effects of biological soil crusts on profile distribution of soil water, organic carbon and total nitrogen in Mu Us Sandland, China. Journal of Plant Ecology 3: 279–284.
27. Su YZ, Zhao HL, Zhang WZ, Zhang TH (2004) Fractal features of soil particle size distribution and the implication for indicating desertification. Geoderma 122: 43–49.
28. Rajot JL, Alfaro SC, Gomes L, Gaudichet A (2003) Soil crusting on sandy soils and its influence on wind erosion. Catena 53: 1–16.
29. Guo YR, Zhao HL, Zuo XA, Drake S, Zhao XY (2008) Biological soil crust development and its topsoil properties in the process of dune stabilization Inner Mongolia China. Environmental Geology 54: 653–662.
30. Zhang Z, Dong Z, Zhao A, Yuan W, Han L (2008) The effect of restored microbiotic crusts on erosion of soil from a desert area in China. Journal of Arid Environments 72: 710–721.
31. Burgheimer J, Wilske B, Maseyk K, Karnieli A, Zaady E, et al. (2006) Relashionship between normalized difference vegetation index (NDVI) and carbon fluxes of biologic soil crusts assessed by ground measurements. Journal of Arid Environments 64: 651–669.
32. Li XR, Jia XH, Long LQ, Zerbe S (2005) Effects of biological soil crusts on seed bank, germination and establishment of two annual plant species in the Tengger Desert (N China). Plant and Soil 277: 375–385.
33. Su YG, Li XR, Cheng YW, Tan HJ, Jia RL (2007) Effects of biological soil crusts on emergence of desert vascular plants in North China. Plant Ecology 191: 11–19.
34. Li XR, Tian F, Jia RL, Zhang ZS, Liu LC (2010) Do biological soil crusts determine vegetation changes in sandy deserts? Implication for managing artificial vegetation. Hydrological Processes 24: 3621–3630.
35. Chen L, Xie Z, Hu C, Li D, Wang G, et al. (2006) Man-made desert algal crusts as affected by environmental factors in Inner Mongolia, China. Journal of Arid Environments 67: 521–527.
36. Chen YN, Wang Q, Li WH, Ruan X (2007) Microbiotic crusts and their interrelations with environmental factors in the Gurbantonggut Desert, Western China. Environmental Geology 52: 691–700.
37. Jia RL, Li XR, Liu LC, Gao YH, Li XJ (2008) Responses of biological soil crusts to sand burial in a revegetated area of the Tengger Desert, Northern China. Soil Biology & Biochemistry 40: 2827–2834.
38. Mandelbrot BB (1983) The fractal geometry of nature. Freeman, New York, U.S.
39. Chen XH, Duan ZH (2009) Changes in soil physical and chemical properties during reversal of desertification in Yanchi County of Ningxia Hui autonomous region, China. Environmental Geology 57: 975–985.
40. Institute of Soil Science, Chinese Academy of Sciences (CAS) (1978) Soil physical and chemical analysis. Shanghai Science Technology Press, Shanghai, China, 524–525.
41. Bao YF, Ding GD, Wu B, Zhang YQ, Liang WJ, et al. (2013). Study on the wind-sand flow structure of windward slope in the Mu Us Desert, China. Journal of Food, Agriculture & Environment, 11(2): 1449–1454.
42. Bao SD (2005) Soil agricultural chemical analysis. China Agriculture Press, Beijing, China, 1–495.
43. Baraher BR, Franzmerier DP, Valassis V, Davidson SE (1966) Use of saran resin to coat natural soil clods for bulk-density and water retention measurements. Soil Science 101: 108–109.
44. Peng XH, Zhang B, Li JT, Zhao QG (2003) Investigation on improving porous medium's porosity in wax-coated method. Journal of Soil Science 34: 19–20.
45. Wang XD, Li MH, Liu SZ, Liu GC (2006) Fractal characteristics of soils under different land use patterns in the arid and semiarid regions of the Tibetan Plateau, China. Geoderma 134: 56–61.
46. Gui DW, Lei JQ, Zeng FJ, Mu GJ, Zhu JT, et al. (2010) Characterizing variations in soil particle size distribution in oasis farmlands-A case study of the Cele Oasis. Mathematical and Computer Modelling 51: 1306–1311.
47. Wang GL, Zhou SL, Zhao QG (2005) Fractal dimension of soil particle volume distribution and its application in land use. Acta Pedologica Sinica 42: 545–550.
48. Garcia-Pichel F, Belnap J (1996) Microenvironments and microscale productivity of cyanobacterial desert crusts. Journal of Phycology 32: 774–782.
49. Elbert W, Weber B, Büdel B, Andreae MO, Pöschl U (2009) Microbiotic crusts on soil rock and plants: neglected major players in the global cycles of carbon and nitrogen? Biogeosciences 6: 6983–7015.
50. Billings SA, Schaeffer SM, Evans RD (2003) Nitrogen fixation by biological soil crusts and heterotrophic bacteria in an intact Mojave Desert ecosystem with elevated CO2 and added soil carbon. Soil Biology & Biochemistry 35: 643–649.
51. Wu N, Zhang YM, Downing A (2009) Comparative study of nitrogenase activity in different types of biological soil crusts in the Gurbantunggut Desert, Northwestern China. Journal of Arid Environments 73: 828–833.
52. Dhaliwal AK, Gupta RK, Yadvinder-Singh Bijay-Singh (2006) Potassium fixation and release characteristics of some benchmark soil series under rice-wheat cropping system in the Indo-gangetic plains of Northwestern India. Communications in Soil Science & Plant Analysis 37: 827–845.
53. Housman DC, Powers HH, Collins AD, Belnap J (2006) Carbon and nitrogen fixation between sucsessional stages of biological soil crusts in the Colorado Plateau and Chihuahuan Desert. Journal of Arid Environments 66: 620–634.
54. Watkins NW, Freeman MP (2008) Natural complexity. Science 320: 323–324.
55. Dumanski J, Pieri C (2000) Land quality indicators: Research plan. Agriculture Ecosystems & Environment 81: 93–102.

56. Pachepsky YA, Polubesova TA, Hajnos M, Sokolowska Z, Jozefaciuk G (1995) Fractal parameters of pore surface area as influenced by simulated soil degradation. Soil Science Society of American Journal 59: 68–75.
57. Perfect E (1997) Fractal models for the fragmentation of rocks and soils: a review. Engineering Geology 48: 185–198.
58. Liu X, Zhang GC, Heathman GC, Wang YQ, Huang CH (2009) Fractal features of soil particle-size distribution as affected by plant communities in the forested region of Mountain Yimeng, China. Geoderma 154: 123–130.
59. Falconer K (2003) Fractal geometry: Mathematical foundations and applications (2nd Edition), Wiley, Chichester, England.
60. Delgado-Baquerizo M, Maestre F, Gallardo A, Quero J, Ochoa V, et al. (2013) Aridity modulates N availability in arid and semiarid mediterranean grasslands. PLOS ONE 8(4): e59807.

4

Comparison of Soil Quality Index Using Three Methods

Atanu Mukherjee*, Rattan Lal

Carbon Management and Sequestration Center, School of Environment and Natural Resources, The Ohio State University, Columbus, Ohio, United States of America

Abstract

Assessment of management-induced changes in soil quality is important to sustaining high crop yield. A large diversity of cultivated soils necessitate identification development of an appropriate soil quality index (SQI) based on relative soil properties and crop yield. Whereas numerous attempts have been made to estimate SQI for major soils across the World, there is no standard method established and thus, a strong need exists for developing a user-friendly and credible SQI through comparison of various available methods. Therefore, the objective of this article is to compare three widely used methods to estimate SQI using the data collected from 72 soil samples from three on-farm study sites in Ohio. Additionally, challenge lies in establishing a correlation between crop yield versus SQI calculated either depth wise or in combination of soil layers as standard methodology is not yet available and was not given much attention to date. Predominant soils of the study included one organic (Mc), and two mineral (CrB, Ko) soils. Three methods used to estimate SQI were: (i) simple additive SQI (SQI-1), (ii) weighted additive SQI (SQI-2), and (iii) statistically modeled SQI (SQI-3) based on principal component analysis (PCA). The SQI varied between treatments and soil types and ranged between 0–0.9 (1 being the maximum SQI). In general, SQIs did not significantly differ at depths under any method suggesting that soil quality did not significantly differ for different depths at the studied sites. Additionally, data indicate that SQI-3 was most strongly correlated with crop yield, the correlation coefficient ranged between 0.74–0.78. All three SQIs were significantly correlated (r = 0.92–0.97) to each other and with crop yield (r = 0.65–0.79). Separate analyses by crop variety revealed that correlation was low indicating that some key aspects of soil quality related to crop response are important requirements for estimating SQI.

Editor: Upendra M. Sainju, Agricultural Research Service, United States of America

Funding: This research was part of a collaborative project supported by the USDA-NIFA, award number: 2011-68002-30190, "Cropping Systems Coordinated Agricultural Project (CAP): Climate Change, Mitigation, and Adaptation in Corn-based Cropping Systems" (sustainablecorn.org). The funders had no role in study design, data collection and analysis, decision to publish, or preparation of the manuscript.

Competing Interests: The authors have declared that no competing interests exist.

* Email: mukherjee.70@osu.edu

Introduction

A wide range of agricultural soils represents diversely managed arable lands while the main goal to improve soil quality, crop yield, and reduce the ecological foot print. Soil quality is defined as the soil's capacity to function within natural or managed ecosystem boundaries and to sustain plant productivity while reducing soil degradation [1–4]. As soil quality is a complex functional concept and cannot be measured directly in the field or laboratory [5] but can only be inferred from soil characteristics [6], a range of soil parameters or indicators has been identified to estimate soil quality. However, soil quality is often related to the management goal and practices as well to soil characteristics. Thus, a mathematical or statistical framework was put forward in early 1990s to estimate soil quality index (SQI) [1,3,4]. The SQI was assessed so that the management goals are not only focused on productivity per se, which may result in soil degradation [7], but also on environmental issues. Thus, an appropriate SQI may have three component goals: environmental quality, agronomic sustainability, and socio-economic viability [8].

Estimation of SQI is a complex process and difficult task [9], especially when linked with several functional goals. Yet a considerable progress has been made towards estimating SQI across a number of soil types and management practices [8–14]. Most studies indexed soil quality employing only one method, with a few exceptions [8,15,16]. As computation of SQI is difficult, a strong need exists for developing a user-friendly and credible SQI through comparison of various available methods. Thus, the objective of this study was to compare SQIs computed by three methods which are conceptually different from each other. The study is based on the hypothesis that SQIs computed from three methods have similar relationship with crop yields. Data are scarce on validation of SQI against crop yield as most studies focused on the environmental aspects of the soil as end point variable and that SQI of various soil layers has not been computed [13,16–18]. Nevertheless, there is a challenge in validating SQI against crop yield as SQI computed only from surface soils (0–10 and/or 0–20 cm) may not evaluate realistic relationship between soil quality and crop yield because root system can extend to deeper layers [19]. Thus, the other objective of this study was to compute SQI from multiple depths of the soil to examine its relationship with crop yield.

Materials and Methods

Soil sampling and analyses

Soil samples from the field (on-farm) were collected from Logan county, Ohio that included an organic (Martisco Variant silt loam: Mc, organic parent material, >26% organic matter, 40°25′12.4″N, 83°40′55.9″W), and a mineral soil (Crosby silt loam: CrB: *Fine, mixed, active, mesic Aeric Epiaqualfs*, sand: 23%, silt: 35%, clay: 42%, 40°24′52.4″N, 83°39′21.9″W), and from Franklin county, Ohio that included a Kokomo silty clay loam (Ko: *Fine, mixed, superactive, mesic Typic Argiaquolls*, sand: 9%, silt: 62%, clay: 29%, 40°00′41.2″N, 83°12′23.1″W) in May, 2013. These locations were farmer-owned field sites and before collecting the soils from specific sites permissions of the farmers were obtained from the land-owners or farmers. Initial permission was granted through The Ohio State University's extension managers and thereafter communication was established directly by the primary author. Additionally, no endangered or protected species were involved in the current study sites, and thus no such permission was required from any other regulatory agencies. The properties of the soils and other management details were presented elsewhere [20,21] and briefly presented in the discussion section. Soil samples were obtained from 0–10, 10–20, 20–40, and 40–60 cm depths under no-tillage (NT) and conventional tillage (CT) practices for Mc and CrB soil series and under NT for one year cover crop (CC) and no-cover crop (NCC) for Ko series. The organic soil (Mc) has been under practice of CT corn (*Zea mays*) for 10 years which was compared with the adjacent grassland soil (termed as NT). The mineral soil (CrB) was under NT and recent introduction (one year) of CT corn practices. Both soils received fertilizer (N:P:K as 5.5:26:30) with the rates of 336 (30, 142 and 164 kg ha^{-1} of N, P, and K, respectively) and 280 (25, 118 and 137 kg ha^{-1} of N, P, and K, respectively) kg ha^{-1} for corn, for Mc and CrB soils, respectively. The other site (Ko) was under NT corn and soybean (*Glycin max L.*) annual rotation. A mixture of pea (*Pisum sativum L.*) and turnip (*Brassica rapa L.*) cover crops with the seeding rate of 140 kg ha^{-1} was seeded prior to growing soybean in 2013. Roundup-ready soybean was seeded on 20 April, 2013 and harvested on 30 September, 2013. A row spacing of 38 cm was used during planting for corn and soybean. Soil core (inside diameter: 5.3 cm, height: 5.9 cm) samples were collected manually using soil sampler before planting (7 May, 2013) from four depths (0–10, 10–20, 20–40, and 40–60) for determination of bulk density (BD) by dividing dry mass of soil in the core by core volume [22] and hydrologic properties. Undisturbed soil cores were used to determine the water retention at field capacity (0.033 MPa), while sieved samples (<2-mm size) were used to determine the permanent wilting point (1.5 MPa). The potential available water capacity (AWC) of the soil was calculated as the difference in volumetric water content at 0.033 and 1.5 MPa moisture potentials. The water content of a soil layer was calculated by multiplying thickness of the layer with the volumetric water capacity. A minimum of three field penetration resistance (PR) measurements for soil's mechanical strength were made for each depth using a CP40II cone penetrometer [23]. Bulk samples were obtained to measure the aggregate size distributions and fraction of water stable aggregates (WSA) using the wet sieving method [24]. Five sieves of diameter sizes 4.75, 2, 1, 0.5, and 0.25 mm openings were placed into a Yoder apparatus. Air-dried soils (5–8 mm, 51 g) were slowly wetted by capillarity action by adjusting the water level in the container so that the base of the top sieve just touched the water. The sieve combination was oscillated mechanically in the water at 60 oscillations per minute for 30 min. Aggregates retained in the sieves were transferred to glass beakers and the weight of each of five fractions was measured after drying at 60°C overnight. The data were used to compute WSA, mean weight diameter (MWD), and geometric mean diameter (GMD) [25]. Soil water retention at matrix potentials of 0.033 and 1.5 MPa were measured using a pressure plate apparatus [26]. The pH and electrical conductivity (EC) of soil were determined in 1:2 soil:water ratio slurry using a Thermo-scientific Orion Star Series pH/Conductivity Meter. Concentrations of soil organic C and total N were determined using an Elemental analyzer (Vario Max, Elementar Americas, Inc., Germany) by the dry combustion method (900°C) after grinding subsamples to 0.25 mm. Total C and N stocks were calculated by multiplying the respective elemental concentrations by BD and the thickness of the soil layer [27].

Soil quality index (SQI) calculations

All the SQI methods involved a set of 72 soil samples and a number of soil quality indicators as parameters. The 13 parameters used for developing SQIs were pH, EC, BD, WSA, GMD, MWD, PR, SOC concentration, N concentration, C-Stock, N-Stock, AWC, and soil water content. As some of these parameters were synthesized and redundant, only nine parameters were chosen omitting MWD, C and N concentrations, and water content from the SQI-1 and SQI-2 calculations to avoid redundancy. However, SQI-3 was not a primarily additive approach and redundancy of the parameters was eliminated through the processes of elimination as prescribed before. Thus, all 13 parameters were initially included in SQI-3 model and only four non-redundant soil parameters with maximum variations in the dataset were finally retained in the model as described in the following section. The average and standard deviation values of these parameters grouped by soil types are presented in Table 1. Under the proposed framework an ideal soil would have SQI value of 1 for the highest quality soil and 0 for the severely degraded soil [1–4].

Simple additive SQI (SQI-1)

Simple additive SQI was estimated following the method outlined by Amacher et al. [28]. In this method, soil parameters were given threshold values based primarily on the literature review and expert opinion of the authors. The threshold levels, interpretations, and associated unitless soil index score values are listed in Table 2. The individual index values were then summed up to obtain a total SQI:

$$\sum SQI = \sum \textit{Individual soil parameter index values} \quad (1)$$

The scaled SQI (SQI-1) of individual soil, was computed by Eq. 2:

$$SQI-1 = (\sum SQI - SQI_{Min})/(SQI_{Max} - SQI_{Min}), \quad (2)$$

whereas, SQI_{Min} = Minimum value of SQI, and SQI_{Max} = Maximum value of SQI from the total dataset.

Weighted additive SQI (SQI-2)

In this approach, each soil parameter was first assigned unitless score ranging from o to 1 by employing linear scoring functions [8]. Non-linear scoring functions were avoided because of their lower capacity of predicting the end point variable or crop yield [8]. Soil parameters were divided into groups based on three mathematical algorithm functions: (a) 'more is better' (e.g., WSA, GMD, C-Stock, N-Stock, and AWC) (b) 'less is better' (e.g., BD, PR), and (c) 'optimum' (e.g., pH and EC). 'Optimum' properties are those which have positive influence up to a certain level beyond which the influence could be considered detrimental [13]. For 'more is better' parameters, each observation was divided by the highest observed value of the entire dataset so that the highest observed value would have a score of 1; for 'less is better'

Table 1. Descriptive statistics of all the soil indicators collected from four depths under three soil types used to estimate SQI.

Ko	0–10 cm		10–20 cm		20–40 cm		40–60 cm	
	Mean	SD	Mean	SD	Mean	SD	Mean	SD
pH	6.0	0.5	6.2	0.6	6.4	0.6	6.4	0.6
EC (μs cm^{-1})	190.6	151.1	165.6	84.2	151.7	71.8	175.9	92.2
BD (Mg m^{-3})	1.4	0.2	1.5	0.1	1.6	0.1	1.6	0.1
WSA %	73.1	14.8	79.5	12.0	73.3	22.5	84.7	5.6
GMD (mm)	1.4	0.4	1.6	0.4	1.4	0.3	1.5	0.3
MWD (mm)	2.6	1.7	3.1	1.8	2.3	1.7	3.4	1.1
PR (Mpa)	2.2	0.3	2.4	0.2	2.3	0.2	2.2	0.4
N (%)	0.2	0.0	0.2	0.0	0.2	0.0	0.2	0.0
SOC (%)	2.4	0.2	2.1	0.2	2.0	0.2	2.0	0.3
N-Stock (Mg/ha)	3.5	0.5	3.0	0.3	6.4	0.6	6.4	1.1
C-Stock (Mg/ha)	33.7	3.7	30.6	2.8	62.5	6.8	63.7	7.2
AWC (%)	38.5	4.1	36.7	2.1	35.8	6.4	39.9	5.8
Water content (cm)	3.8	0.4	3.7	0.2	7.2	1.3	7.9	1.0

CrB	0–10 cm		10–20 cm		20–40 cm		40–60 cm	
	Mean	SD	Mean	SD	Mean	SD	Mean	SD
pH	7.4	0.2	7.4	0.2	7.4	0.2	7.4	0.1
EC (μs cm^{-1})	250.3	44.6	217.8	26.8	212.3	31.8	223.5	45.0
BD (Mg m^{-3})	1.5	0.2	1.5	0.1	1.5	0.3	1.5	0.3
WSA %	82.4	4.4	76.4	9.7	76.5	9.5	76.5	17.7
GMD (mm)	1.2	0.1	1.1	0.2	1.1	0.2	1.2	0.3
MWD (mm)	2.0	0.5	1.7	0.8	1.5	0.7	2.0	1.1
PR (Mpa)	2.4	0.2	2.5	0.5	2.6	0.2	2.9	0.7
N (%)	0.2	0.0	0.2	0.0	0.2	0.0	0.2	0.0
SOC (%)	2.4	0.3	2.1	0.4	1.8	0.4	1.8	0.5
N-Stock (Mg/ha)	3.2	0.2	3.1	0.3	5.2	1.6	5.1	2.2
C-Stock (Mg/ha)	34.3	3.2	32.2	4.1	53.8	16.8	54.6	24.4
AWC (%)	17.8	9.0	10.1	6.8	10.7	8.2	9.0	3.2
Water content (cm)	1.8	0.9	1.0	0.7	2.1	1.6	1.8	0.6

Mc

Table 1. Cont.

Ko								
	0–10 cm		10–20 cm		20–40 cm		40–60 cm	
	Mean	SD	Mean	SD	Mean	SD	Mean	SD
	0–10 cm		10–20 cm		20–40 cm		40–60 cm	
	Mean	SD	Mean	SD	Mean	SD	Mean	SD
pH	7.4	0.1	7.4	0.1	7.4	0.1	7.5	0.1
EC (µs cm^{-1})	616.3	40.1	570.0	92.8	633.2	126.4	599.2	129.2
BD (Mg m^{-3})	0.6	0.0	0.7	0.0	0.6	0.1	0.6	0.1
WSA %	91.2	0.5	91.0	2.5	89.7	2.5	82.5	8.7
GMD (mm)	2.1	0.0	2.2	0.0	2.2	0.0	2.2	0.1
MWD (mm)	5.5	0.1	5.5	0.2	5.7	0.4	5.0	0.5
PR (Mpa)	1.3	0.1	1.5	0.3	1.6	0.3	1.9	0.1
N (%)	1.0	0.1	0.9	0.1	0.9	0.1	1.0	0.1
SOC (%)	15.1	1.1	15.0	0.8	14.9	0.9	15.9	0.6
N-Stock (Mg/ha)	6.1	0.8	6.3	0.7	11.2	1.7	12.6	1.9
C-Stock (Mg/ha)	95.7	11.7	102.8	8.8	186.6	36.2	196.6	28.0
AWC (%)	248.3	151.3	84.8	42.4	48.2	13.8	36.1	22.1
Water content (cm)	24.8	15.1	8.5	4.2	9.6	2.8	7.2	4.4

Abbreviations: SD: standard deviation, EC: electrical conductivity, BD: bulk density, WSA: water stable aggregates, GMD: geometrical mean diameter, MWD: mean weight diameter, PR: penetration resistance, N: nitrogen, SOC: soil organic carbon, AWC: available water capacity.

Table 2. Soil indicators, threshold values, interpretations and scores.

Indicators	Range	Interpretation	Score	Reference
pH	5.5–7.2	Slightly acidic to neutral: Optimum for plant growth	2	[28]
	>7.2<8.0	Slightly to moderately alkaline: Preferred by some plants, possible P and some metal deficiencies	1	
EC (us/cm)	<200	Low salt level	0	[42]
	200–500	Optimum salt level for plants	1	
	>500	High salt level, adverse effect likely	0	
BD (Mg/m^3)	<1.0	High organic soil, supports plant roots	2	[28,45–47]
	1.0–1.5	Adverse effects unlikely	1	
	>1.5	Adverse effects likely	0	
WSA (%)	<50	Infiltration and soil erosion problems likely	0	[30,42]
	50–70	Moderate constraints	1	
	70–90	Good soil	2	
	>90	Excellent soil	3	
GMD (mm)	<1.0	Infiltration and soil erosion problems likely	0	[30,42]
	1–2	Moderate limitations	1	
	>2	No limitation	2	
MWD (mm)	<1.0	Infiltration and soil erosion problems likely	0	[30,42]
	1–2	Moderate limitations	1	
	2–5	Slight limitations	2	
	>5.0	No limitation	3	
PR (Mpa)	1–2	Adverse effect on plant root unlikely	2	[42,48]
	2–3	Moderate adverse effect on plant root	1	
	>3.0	Severe adverse effect on plant roots	0	
N (%)	0.2–0.3	Moderate limitation	1	[41,42]
	>0.3	Slight to no limitation	2	
SOC (%)	2–3	Moderate limitation	1	[41,42]
	>3.0	Slight to no limitation	2	
N-Stock (Mg/ha)	<5.0	N deficient	1	Authors' opinion
	5–10	Moderate to optimum N level	2	
	>10.0	N-rich soil	3	
C-Stock (Mg/ha)	<50.0	C deficient	1	Authors' opinion
	50–100	Moderate to optimum C level	2	
	>100	C-rich soil	3	
AWC (%)	<20	Water-stress to plants	0	[31,42]
	20–50	Moderate water availability	1	
	>50	Good water capacity for plants	2	
Water content(cm)	<5.0	Water-stress to plants	0	Authors' opinion
	5–10	Moderate water availability	1	
	>10	Good water capacity for plants	2	

Abbreviations are same as Table 1.

parameters, the lowest observed value in the entire dataset was divided by each observation so that the lowest observed value received a score of 1; and 'optimum' parameters were scored up to a threshold value as 'more is better', and thereafter above the threshold values were scored as 'less is better' [8,29]. For example, pHs up to 5.5–7.2 were scored as 'more is better', and pHs >7.2 were scored as 'less is better'.

After normalizing soil parameters, the scores were integrated into a single index value for each soil using a weighted additive approach initially suggested by Karlen and Stott [4], but modified later by Fernandes et al. [13]. The following weighted additive function was used for development of SQI-2 (Eq. 3):

$$SQI-2 = [(\text{Weight1}) * \text{RDC}] + [(\text{Weight2}) * \text{WSC}] + [(\text{Weight3}) * \text{NSC}], \tag{3}$$

where, RDC (root development capacity) is the rating for the soil's ability to allow plant root development, WSC (water storage capacity) is the rating for the soil's ability to store water, NSC

Table 3. Model of SQI-2.

Soil function	Weight	Soil Indicators	Sub-weight	Scaled score	B×C	∑ B×C	D×A	%	SQI
	A		B	C		D			
RDC	0.4	BD	0.35	0.41	0.14	0.46	0.18	52.94	0.34
		PR	0.35	0.52	0.18				
		WSA	0.20	0.54	0.11				
		GMD	0.10	0.34	0.03				
WSC	0.2	AWC	1.00	0.06	0.06	0.06	0.01	2.94	
NSC	0.4	pH	0.30	0.79	0.24	0.38	0.15	44.12	
		EC	0.30	0.25	0.08				
		C-stock	0.20	0.11	0.02				
		N-stock	0.20	0.18	0.04				

Abbreviations: RDC: root development capacity, WSC: water storage capacity, NSC: nutrient storage capacity; all other abbreviations are same as Table 1.

(nutrient supply capacity) is the rating for the soil's ability to supply nutrients. Weight 1, 2 and 3 are the respective numerical weights for each soil function (RDC, WSC, and NSC). The numerical weights were assigned to each soil function according to their importance in fulfilling the management goal(s) of maintaining soil quality. While the summation of all these numerical weights (Weight 1, 2 and 3) are supposed to be 1 and distributed evenly as 0.33, 0.33, and 0.34 [4], however, a modified approach was taken for the current study following Fernandes et al. [13], where lower weightage was given to the functional attribute which had lower number of representative indicators in the model. Similar to Fernandes et al. [13] weight values were arbitrarily chosen for RDC, WSC, and NSC as 0.4, 0.2, and 0.4 in the current study. The WSC received lower weight value than others because of low number of representative indicators (only AWC) [13] among all included parameters.

The soil parameters or indicators selected for (i) RDC were BD, PR, WSA, and GMD, (ii) WSC was AWC and (iii) NSC were pH, EC, C-Stock, and N-Stock. Within this network, the sub-weight values were given to each indicator based on their importance under the particular soil functional property, field versus laboratory measurements and scope of redundancy. The sub-weight values of different soil indicators or parameters were summed up to 1 under each soil functional property [13]. The rank of the subweight values were as follows: field measurement indicators (BD, PR) > laboratory measurement indicators (WSA, AWC, pH, EC) >> synthesized parameters (GMD, C-Stock, and N-Stock). Field indicators were given the maximum weightage as these were the most representative of the soil's natural conditions and synthesized parameters received the lowest weightage to avoid data redundancy in the model. Thereafter, scaled SQI (SQI-2) of individual soil was computed by Eq. 2. An example of this model of development of SQI is shown in Table 3.

Statistically modeled SQI (SQI-3)

A statistics-based model was used to estimate SQI using principal component analysis (PCA) [8,12,14,18,30,31]. The PCA-model is used to create a minimum data set (MDS) to reduce the indicator load in the model and avoid data redundancy [8]. The main difference between the first two versus the PCA method is that the first two rely mainly on subjective expert opinion and literature review, while the PCA method is more objective of using a number of statistical tools (multiple correlation, factor and cluster analyses) which could avoid any biasness and data redundancy by choosing an MDS using mathematical formulae [8,32]. A relative advantage and disadvantage of each SQI method is explained in the later section. The preliminary function of PCA is to reduce the dimensionality of the entire data set consisting of a large number of interrelated variables, while retaining as much as possible of the variations present in the data set. This is achieved by transformation to a new set of variables, the principal components (PCs), which are uncorrelated, and ordered so that the first few retain most of the variation present in all of the original variables [33,34]. In other words, the PCA method was chosen as a data reduction tool to select the most appropriate indicator(s) to represent and estimate SQI [18].

All the original observations (untransformed data) of each soil were included in the PCA model using SPSS, version 21.0 [35]. The PCs with high eigenvalues represented the maximum variation in the dataset [8,12,14,18,30,31]. While most studies have assumed to examine PCs with eigenvalue >1.0 following Kaiser [36], the present study had the third and fourth components with eigenvalues of 0.98 and 0.89 and variances >5% (Table 4, Fig. 1). These were examined with first two PCs (eigenvalues 8.23, and 1.67) as

Table 4. Results of principal component analyses (PCA).

	PC-1	PC-2	PC-3	PC-4
Eigenvalues	8.23	1.67	0.98	0.89
% Variance	63.34	12.87	7.56	6.82
Cumulative variance	63.3	76.2	83.8	90.6
Eigen vectors or factor loading				
pH	0.423	0.351	−0.378	**0.711**
EC	**0.925**	0.113	−0.153	0.147
BD	**−0.881**	−0.034	0.212	0.002
WSA	0.590	−0.020	**0.669**	0.343
GMD	**0.927**	0.024	0.276	−0.024
MWD	**0.904**	0.001	0.342	0.046
PR	−0.770	0.268	−0.131	0.092
SOC	**0.975**	0.073	−0.109	−0.025
N	**0.972**	0.050	−0.103	−0.053
C-Stock	0.831	0.412	−0.107	−0.268
N-Stock	0.722	0.440	−0.067	−0.385
AWC	0.555	**−0.782**	−0.189	0.057
Water content	0.624	−0.693	−0.199	−0.066

Abbreviations are same as Table 1; PC: principal component; bold values under each component are highly weighted and underlined bold values are selected in minimum data set.

prescribed for the cases of fewer than three components with eigenvalue >1.0 [8,37]. Under a given PC, each variable had corresponding eigenvector weight value or factor loading (Table 2). Only the 'highly weighted' variables were retained to include in the MDS (Table 2). The 'highly weighted' variables were defined as the highest weighted variable under a certain PC and absolute factor loading value within 10% of the highest values under the same PC [37]. Thus, the bold-face values (Table 4: EC, BD, GMD, MWD, SOC, and N for PC-1, AWC for PC-2, WSA for PC-3, and pH for PC-4) were considered highly weighted eigenvectors and therefore were initially selected in the MDS. However, when more than one variable was retained under a particular PC, multivariate correlation matrix (Table 5) was used to determine the correlation coefficients between the parameters [8,12]. If the parameters were significantly correlated ($r>0.60$, $p<0.05$), then the one with the highest loading factor was retained in the MDS and all others were eliminated from the MDS to avoid redundancy. Following this procedure, except SOC, all other eigenvectors from PC-1 were eliminated from MDS due to high and significant correlation between each other (Table 5). The non-correlated parameters under a particular PC were considered important and retained in the MDS [10,12]. All the bold-faced and underlined soil parameters in Table 4 were selected in the final MDS.

After selection of parameters for the MDS, all selected observations were transformed using linear scoring functions (less is better, more is better and optimum) as described in SQI-2 method. Once the selected observations were transformed in numerical scores (ranged 0–1), a weighted additive approach was used to integrate them into indices for each soil [8,12]. Each PC explained a certain amount of variation in the dataset (Table 4), which was divided by the maximum total variation of the all PCs selected for the MDS to get a certain weightage value under a particular PC [8]. For example, the % variance (63.3) was divided by total cumulative variance (90.6) to obtain the weight value of 0.7 for PC-1 (Table 4). Thereafter, the weighted additive SQI was computed using Eq. 4:

$$SQI-3\,(PCA)=\sum Weight*Individual\,oil\,parameter\,score \quad (4)$$

Validation of SQI

The SQIs estimated from three different methods were validated against soybean (*Glycin max*) and corn (*Zea mays*) yield data of that particular year of collection of soil and crops by computing correlation coefficients [8]. The SQI values were also compared through pearson correlation to understand the effectiveness of each other.

Statistical analyses

All values are presented as means ± standard deviations of three field or laboratory measurements. Significant differences between treatments were analyzed using Tukey's test in PROC GLM in SAS version 9.2 [38]. Treatment differences were deemed significant at $p<0.05$. The PCA was performed in SPSS version 21 [35]. Descriptive statistics and linear regressions were computed in Microsoft Excel [39] and all the figures were obtained using Sigmaplot Version 12.0 [40].

Results and Discussions

Most data of three farm sites have been presented and the detailed discussions of soil type, crop and management effects on soil properties are discussed elsewhere [20,21]. The mean and

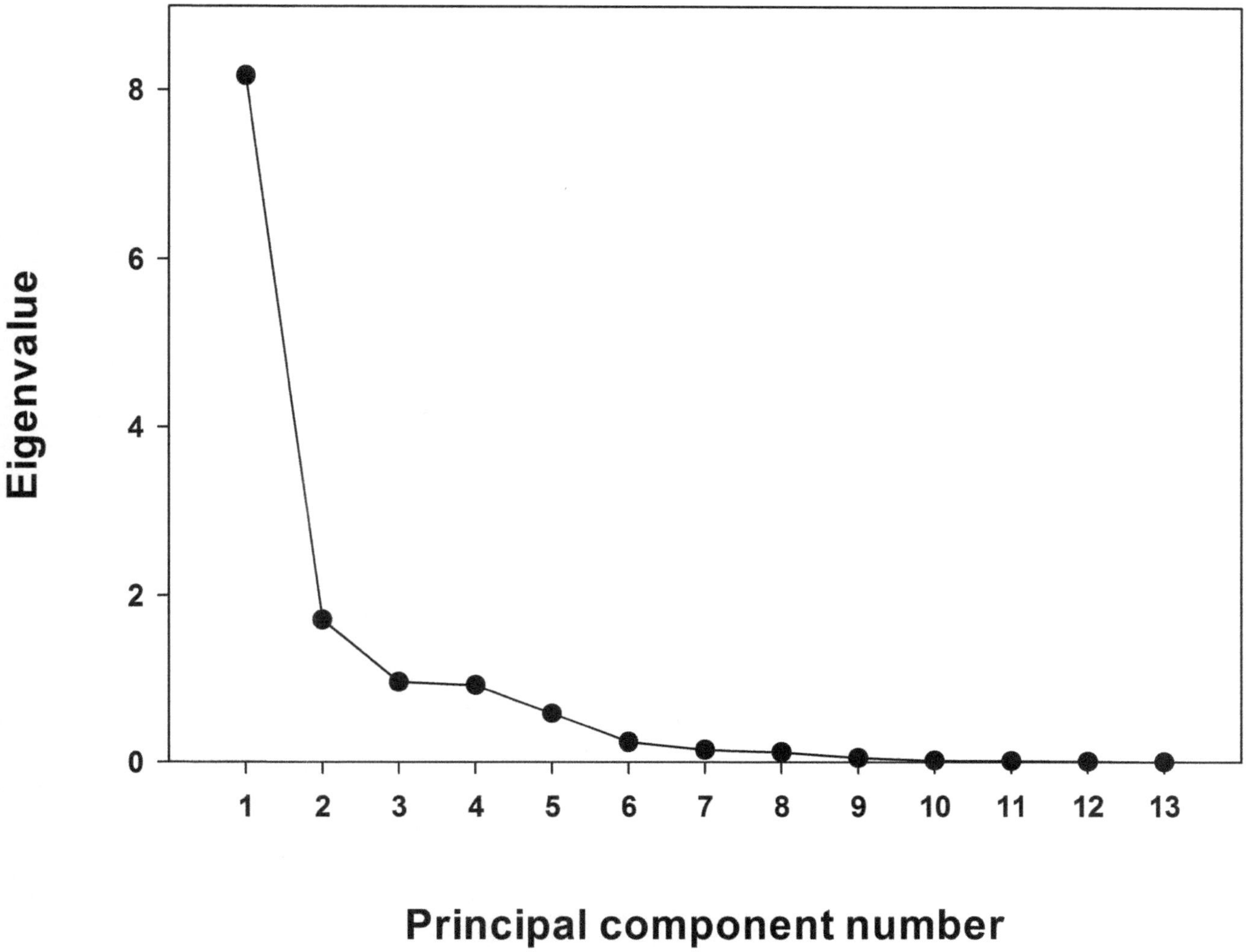

Figure 1. Scree plot of principal component analyses.

standard deviation values of the soil parameters by depth under three soil types are listed in Table 1. Most properties of mineral soils (Ko and CrB) were significantly different compared to those of the organic soil (Mc) and except for AWC and water content two mineral soils' characteristics were similar across soil profile up to 60 cm depth (Table 1). The water characteristics of Ko soil were significantly higher than those of CrB. On the other hand, organic soil (Mc) had much improved (lower BD and PR, and higher WSA, SOC, N concentrations and stock, AWC and water content) soil properties compared to those of the mineral counterparts in all four different depths (Ko and CrB). Impacts of all these parameters on overall soil quality in four soil depths calculated by three indexing methods are discussed in the following sections.

Under the framework of SQI-1, nine soil quality indicators were integrated numerically after scoring them primarily using the information from literature review. However, scoring data on some physicochemical indicators (e.g., C and N stocks) is scarce in the literature and thus, authors' opinion was used in those cases. Further, different studies used different scoring on the same indicator based on soil type, and management goals. For an example, while Amacher and Perry [28] had eight different range classes and scores for pH, Feiza et al. [41] had only four classes and scores available for pH. Thus, scoring on indicators in the current study (Table 2) had importance on experts' opinion although the knowledge was primarily based on available literature and authors' experience [28,42]. The high variability in the observations (Table 1) was reflected in the SQI-1 values. There was no statistical difference (p>0.05) across treatments (NCC versus CC, and CT versus NT) and depth under specific soil type (Table 6). Overall performance of indexing methods was determined from averaging SQI values obtained from calculations for each depth and presented in Fig. 2. The, SQI-1 was significantly higher in the following order: Mc>Ko>CrB (Fig. 2), indicating that soil quality was influenced more by soil type than management and sampling depth under the SQI-1 approach.

Overall percentage of various soil functional influences in SQI-2 is presented in Fig. 3 and depthwise percentage of influence in SQI-2 is presented in Fig. S1. Generally, mineral soils (Ko and CrB) did not have any depthwise significant changes for WSC but 0–10 cm layer of organic (Mc) soil had significantly higher influence of WSC in SQI-2 compared to lower depths. On the other hand, while both Ko and Mc had significantly higher influence of RDC in SQI-2 in the upper layers than the lower depths, the opposite trend was found for the NSC (Fig. S1). In the

Table 5. Pearson correlation coefficients (r) for all soil indicators.

Correlation (r) matrix	pH	EC	BD	WSA	GMD	MWD	PR	N	SOC	N-Stock	C-Stock	AWC
EC	0.547											
BD	−0.394	−0.88										
WSA	0.248	0.483	−0.338									
GMD	0.279	0.8	−0.762	0.673								
MWD	0.286	0.782	−0.725	0.734	0.975							
PR	−0.129	−0.655	0.684	−0.456	−0.71	−0.687						
N	0.405	0.922	−0.908	0.47	0.877	0.835	−0.737					
SOC	0.438	0.93	−0.914	0.473	0.881	0.842	−0.731	0.995				
N-Stock	0.271	0.632	−0.551	0.32	0.642	0.595	−0.396	0.709	0.707			
C-Stock	0.382	0.764	−0.702	0.367	0.742	0.695	−0.489	0.822	0.833	0.965		
AWC	0.091	0.447	−0.449	0.275	0.439	0.441	−0.525	0.504	0.489	0.106	0.181	
Water content (cm)	0.092	0.487	−0.485	0.282	0.501	0.496	−0.557	0.558	0.546	0.28	0.333	0.967
Significance level												
EC	<0.001											
BD	<0.001	<0.001										
WSA	0.018	<0.001	0.002									
GMD	0.009	<0.001	<0.001	<0.001								
MWD	0.007	<0.001	<0.001	<0.001	<0.001							
PR	0.141	<0.001	<0.001	<0.001	<0.001	<0.001						
N	<0.001	<0.001	<0.001	<0.001	<0.001	<0.001	<0.001					
SOC	<0.001	<0.001	<0.001	<0.001	<0.001	<0.001	<0.001	<0.001				
N-Stock	0.011	<0.001	<0.001	0.004	<0.001	<0.001	<0.001	<0.001	<0.001			
C-Stock	<0.001	<0.001	<0.001	0.001	<0.001	<0.001	<0.001	<0.001	<0.001	<0.001		
AWC	0.224	<0.001	<0.001	0.011	<0.001	<0.001	<0.001	<0.001	<0.001	0.195	0.07	
Water content (cm)	0.22	<0.001	<0.001	0.009	<0.001	<0.001	<0.001	<0.001	<0.001	0.01	0.003	<0.001

Abbreviations are same as Table 1.

Table 6. Treatment effects on soil quality index (SQI) under three soil types (Ko: Kokomo, CrB: Crosby silty loam, Mc: Muck) and four soil depths.

Treatments									
Ko		NCC	CC	NCC	CC	NCC	CC	NCC	CC
		0–10 cm		10–20 cm		20–40 cm		40–60 cm	
	SQI-1	0.36a	0.36a	0.40a	0.40a	0.45a	0.48a	0.48a	0.50a
	SQI-2	0.40a	0.37a	0.39a	0.36a	0.40a	0.36a	0.43a	0.38a
	SQI-3	0.21a	0.23a	0.20a	0.22a	0.19a	0.20a	0.21a	0.21a
CrB		CT	NT	CT	NT	CT	NT	CT	NT
		0–10 cm		10–20 cm		20–40 cm		40–60 cm	
	SQI-1	0.36a	0.38a	0.36a	0.17a	0.45a	0.29a	0.38a	0.26a
	SQI-2	0.42a	0.40b	0.40a	0.37a	0.41a	0.40a	0.41a	0.40a
	SQI-3	0.26a	0.24a	0.24a	0.22a	0.22a	0.22a	0.22a	0.21a
Mc		CT	NT	CT	NT	CT	NT	CT	NT
		0–10 cm		10–20 cm		20–40 cm		40–60 cm	
	SQI-1	0.88a	0.88a	0.90a	0.90a	0.90a	0.90a	0.90a	0.81a
	SQI-2	0.70a	0.72a	0.65a	0.61a	0.65a	0.66a	0.67a	0.65a
	SQI-3	0.85a	0.85a	0.82a	0.78a	0.78a	0.77a	0.82a	0.81a

Abbreviations: NCC: no cover crop, CC: cover crop, CT: conventional tillage, NT: no tillage.
Different letters under specific row and depth indicate significant differences at $p<0.05$ level, underline value is significant at $p<0.1$ level.

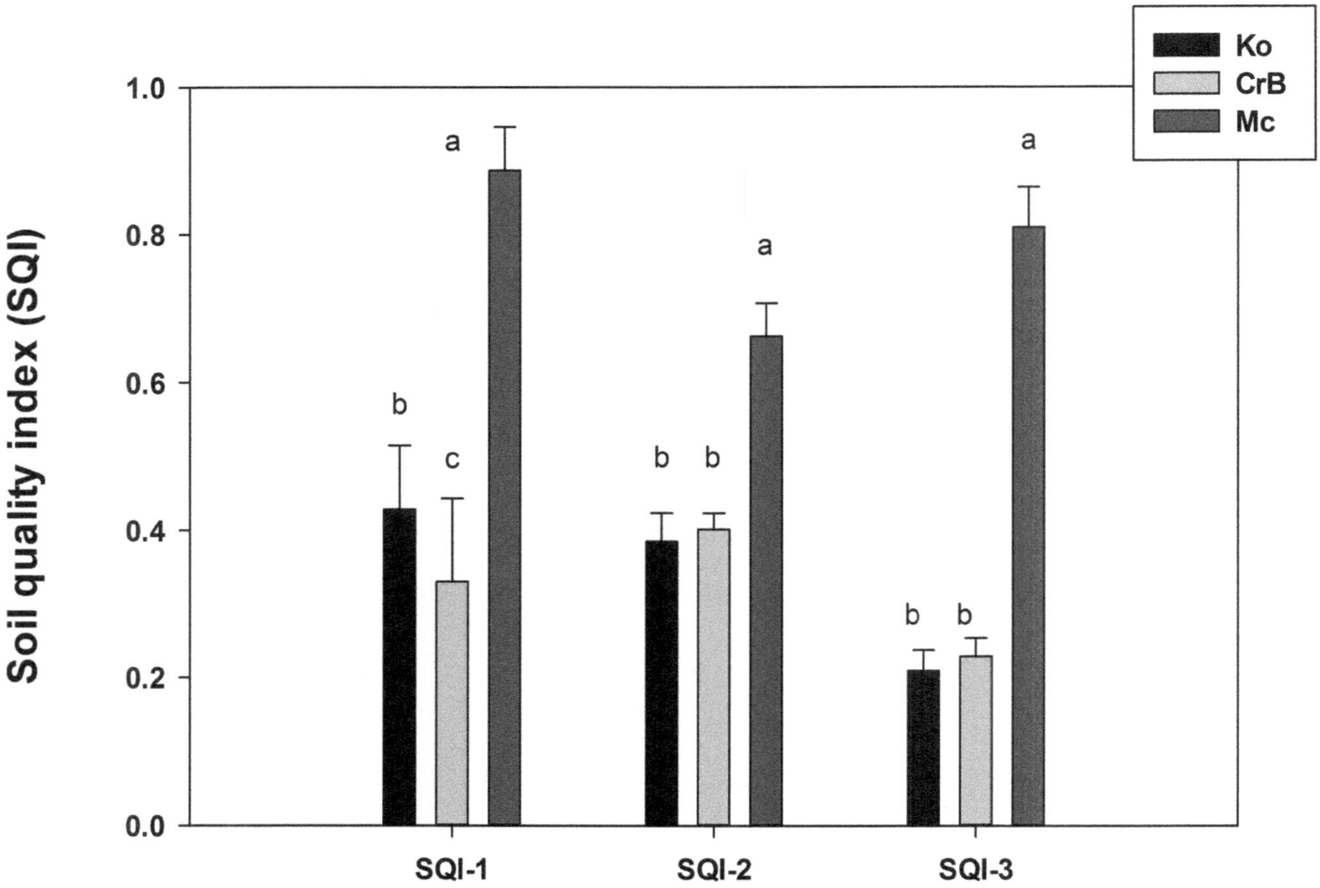

Figure 2. Overall soil quality index (SQI) values under different soil types. Different letters indicate significant differences at p<0.05 level for particular indexing method.

general composition of SQI-2, the main contributors were RDC, which ranged from 39–57% and NSC ranged from 35–59%. However, the WSC was relatively small and ranged merely from 0.1–12% (Fig. 3). The relative influence of root development functions in SQI-2 has been significantly higher in Ko than CrB, and Mc by 8 and 9%, respectively and the same for nutrient storage functions in SQI-2 has been significantly higher in CrB than Ko and Mc by 13 and 8%, respectively (Fig. 3). On the other hand, Mc soil had significantly higher water storage functional contribution to SQI-2 by 53 and 500%, respectively, compared to that of mineral (Ko and CrB) soils (Figs. 3). Thus, while the mineral soil quality was more sensitive to the functional attributes dedicated to root development and nutrient storage, muck soil quality was more influenced by soil attributes related to water storage under the scheme of SQI-2. Further, there was no significant effect on management practices on %RDC and %NSC to SQI-2 in different soil/management/crop combinations (Fig. 4). Note that CC/NCC management was used on field cultivated to soybean but CT/NT was used to soil under corn. However, contribution of WSC to SQI-2 for the case of CrB soil under NT management was significantly higher (178%) than that of CT practice (Fig. 4).

Correlation coefficients of SQIs to each other and depthwise SQIs and dry grain and straw are given in Table 7. Crop yield is obviously related to soil quality but soil quality in different layers may not be same and can influence crop yield accordingly. Data on SQI variation for different soil layers are scarce and thus no comparison to the data presented in the current study (Table 6) was possible. However, while numerical values of SQI were almost unchangeable for different depths at Mc site, SQI generally variable with depth in Ko and CrB sites (Table 6). One goal of this study was to validate SQI against crop yield as only a few studies have done so with some exceptions [8,14]. Challenge lies in establishing a correlation between crop yield versus SQI calculated either depthwise or in combination of soil layers as standard methodology is not yet available and was not given much attention to date. For example, one can have only one crop yield value for multiple soil layers and thus to date crop yield has been correlated only against SQI calculated from surface layer. Low (ranged from 0.11–0.32) and high (ranged from 0.83–0.96) correlation coefficients of SQI and crop yield across a number of soil/crop combinations were reported [8,31]. The reasons attributed to low correlation between yield and SQI were: (i) the assumptions that the indicators considered good for a soil quality may not always lead to desired outcomes and flexibility in scoring functions needed [8], (ii) indicators which may not directly relate to crop performance were given more weightage by the method employed, or (iii) some key soil indicators were not included in the MDS due to the design of the study [9,43]. Nevertheless, in order to evaluate the index performance in the current study, SQIs were correlated against end point variable (i.e., yield) at variable depths (Table 7). In the present study, depth of soil sampling had

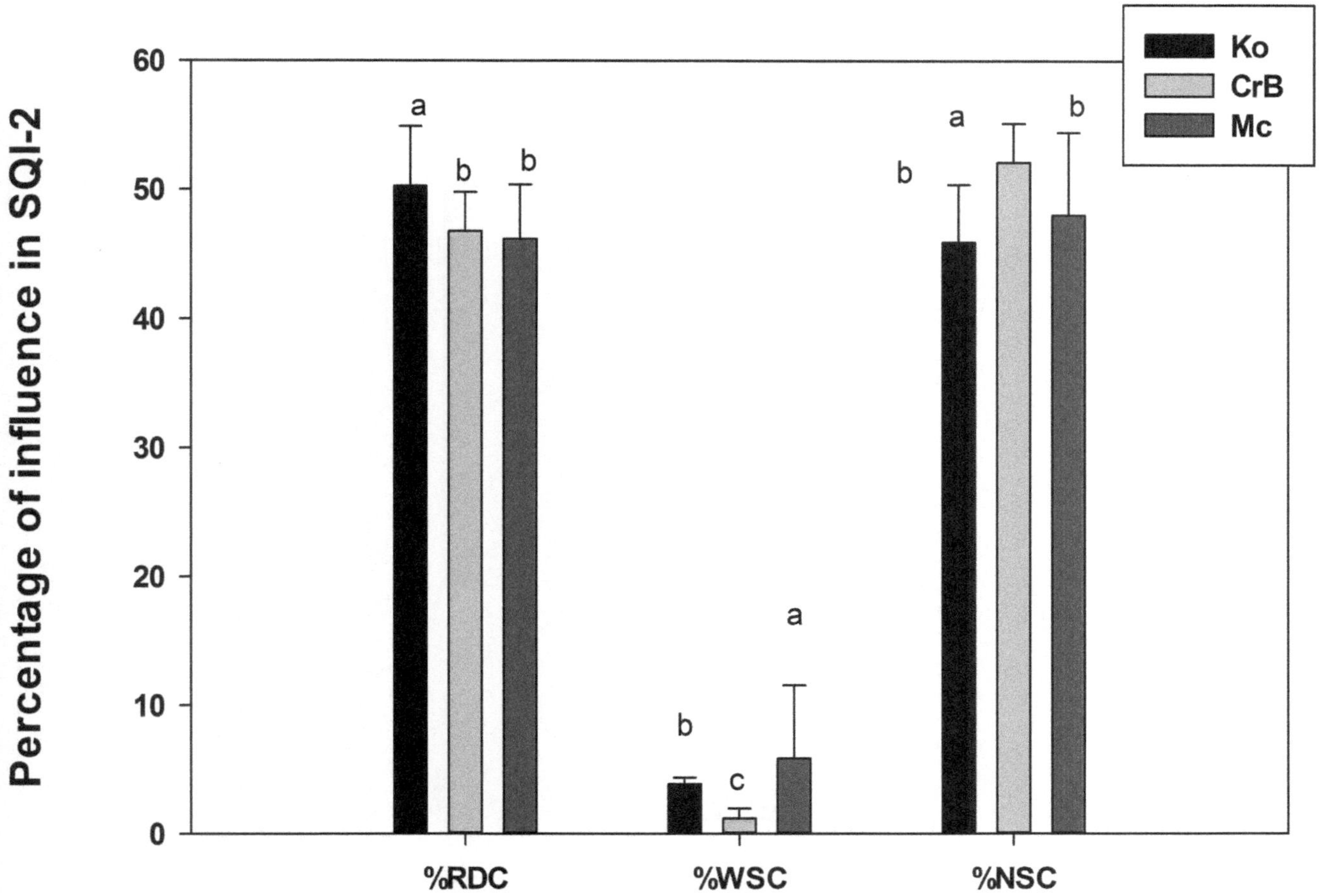

Figure 3. Percentage contribution of each soil function in SQI-2 under different soils. Different letters indicate significant differences at p<0.05 level for particular soil function. Abbreviations: RDC: root development capacity, WSC: water storage capacity, NSC: nutrient storage capacity.

no significant effect in SQI values (Table 6) and thus the individual SQI values were integrated by combining 0–20 cm and 0–60 cm profiles by averaging 0–10, and 10–20 cm for obtaining values for 0–20 cm depth and 0–10, 10–20, 20–40 and 40–60 cm for obtaining values for 0–60 cm depths and correlated the combined values with crop yield (Table 7). In the present study, grain and biomass yields were all significantly ($p<0.05$, $r = 0.65$–0.79) correlated under any SQI methods employed indicating that either of these methods was successful in predicting crop yield (Table 7). However, correlation coefficients in Table 7 are derived from the complete datasets including both corn and soybean yield but the correlation was considerably low ($r<0.20$ for soybean grain and $r<0.60$ for corn grain; data not shown), when dataset was further divided into corn and soybean. As discussed above, low correlation of SQI and crop yield was observed before and one of the listed reasons was some key soil indicators were not included in the dataset as per the experimental design and objective [8]. Crop yield is invariably related to soil fertility status and thus SQI may not always lead to high correlation with yield if some of the related characterizations are not observed as identified before [9,43] which is supported by the data of the current study (Table 7). Thus, the low correlation coefficient values may be because soil fertility (ion exchange capacities, macro and micro nutrients) and microbial (microbial biomass, microbial C, and N, and soil respiration) aspects were not monitored in the current studied sites, especially in the case of soybean grain yield. Nevertheless, relatively higher correlation coefficient was observed in the 0–10 cm soil layer than the deeper soil profile (Table 7) probably due to higher plant root density in the top layer of the soil [44].

Additionally, SQI computed from different methods were also highly correlated with each other (Table 7) suggesting that (i) a relatively easy and user-friendly SQI (SQI-1) can be computed to evaluate and compare soil quality, which is similarly useful to other approaches, (ii) giving certain appropriate weightage on scores (SQI-2) could be similarly useful in predicting particular soil quality and (ii) a minimum group of carefully selected soil indicators (SQI-3) may be used to evaluate soil quality across various soil types and management practices. As all three SQI methods were significantly correlated to each other, it is difficult to conclude which one is the best approach, however, it highly depends on factors such as design of the study, choice of soil parameters included in the model to compute SQI and end point variable (environmental aspects versus crop yield). The advantage of using SQI-1 is that the soil quality could be assessed after measuring any number (low to high) of soil parameters and this procedure is relatively easier compared to other methods as the scoring requires literature review and expert opinions only. The disadvantage of SQI-1 is that it is subjective and relies mainly on researcher's point of view. On the other hand, advantage of SQI-2 is that it includes weightage based on the design of the study, system or the dataset to offset the subjectivity of the approach

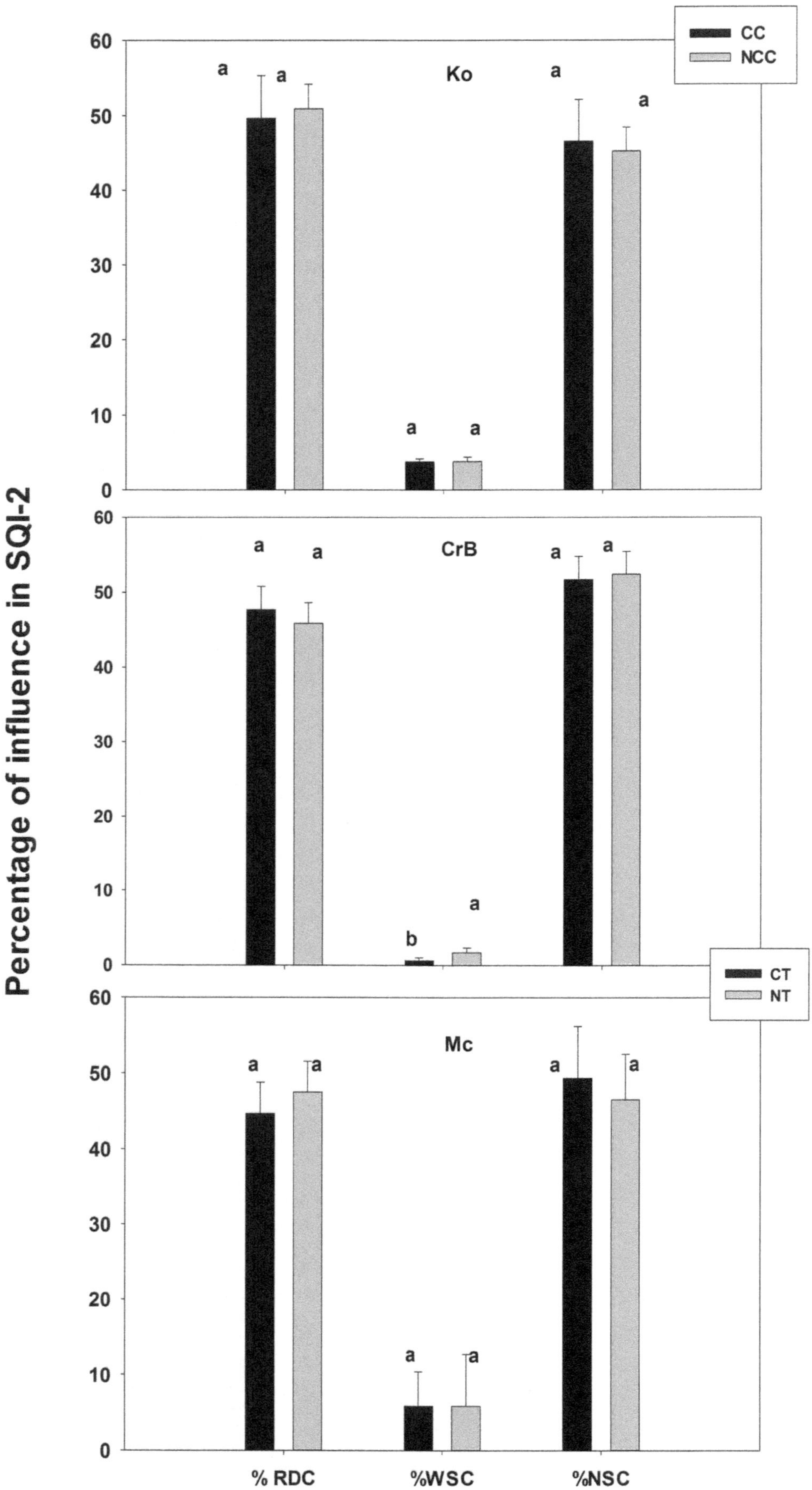
Percentage of influence in SQI-2
CC
NCC
Ko
CrB
CT
NT
Mc
0
10
20
30
40
50
60
a
b
% RDC
%WSC
%NSC

Figure 4. Percentage contribution of each soil function in SQI-2 under various management practices in different soil. Different letters indicate significant differences at p<0.05 level for particular soil function. Abbreviations: RDC: root development capacity, WSC: water storage capacity, NSC: nutrient storage capacity; NCC: no cover crop, CC: cover crop, CT: conventional tillage, NT: no tillage.

present in SQI-2. However, the disadvantage of SQI-2 is that it requires multiple numbers of soil parameters under different soil functional systems which may be expensive and time consuming in practical case. The SQI-3 is advantageous in the aspect of its ability to predict soil quality based on a reduced dataset with low number of soil parameters. Additionally, it is mostly objective approach as the statistical procedure would select a low number of soil parameters needed to calculate SQI based on the variances present in the whole dataset. So, in a long-term aspect within a particular soil/crop system, SQI-3 can be used effectively once it evaluated the most influential soil parameters required to assess soil quality of a particular soil/crop/management scenario. However, current data suggest that relative higher SQI values were obtained for Ko and Mc by SQI-1 than SQI-2 or SQI-3 and for CrB by SQI-2 than SQI-1 or SQI-3 (Fig. 2). Although SQI-1 and SQI-2 had higher values than SQI-3 in different scenarios, however, based on the current experimental design, and selected soil parameters SQI-3 successfully predicted crop yield relatively higher [Table 7 and also when grouped by crops (data not shown)] than SQI-1 and 2. Depth had little influence in any SQIs in the present study and additional chemical and biological parameters are also needed to strengthen the validity of SQI. Thus, SQI-3 appears to be the best method among the three under long-term scenario, especially due to its objective approach, relative higher correlations with crop yield and lower number of indicator selection which is more cost and time-effective over time than other approaches.

Conclusions

The data presented support the following conclusions: The SQI calculated by three different methods indicated that studied muck soil has significantly higher soil quality than that of mineral soils under on-farm conditions. The SQI was affected more by management and soil type than by depths in the studied on-farm sites. The SQI computed using three established methods were all significantly correlated to each other indicating that relatively easy and user-friendly SQI (SQI-1) can be similarly useful to evaluate soil quality, appropriate weightage on scores can predict soil's quality (SQI-2) with high performance but requires a number of soil parameters under different soil functional components and carefully chosen MDS with small numbers of soil variables (SQI-3) may adequately evaluate its quality. All three SQIs were highly and significantly correlated with crop yield. In general, however, analyses by crop type (corn and soybean) revealed that correlation was low ($r<0.20$ or 0.60 for soybean or corn grain yield, respectively; data not shown) suggesting choice of crop-specific and key soil parameters used in computing SQI. Under the current experimental on-farm conditions SQI-1 (for Ko and Mc) and SQI-2 (for CrB) values were higher than SQI-3, however, SQI-3 can be regarded as the best and easiest model given its relatively higher success to predict crop yield and objectivity approach with lower number of indicator selection ability which should be regarded as a relatively less expensive procedure over time compared to SQI-1 and 2. In addition, in order to effectively predict particular crop yield one must include soil fertility and microbial parameters in the model of SQI.

Table 7. Pearson correlation coefficients (r) of soil quality index (SQI) versus crop yield and correlations between different SQI values which were averaged up to a certain depth; all numbers are significant at $p<0.05$ level.

SQI-1	**Dry Grain**	**Dry Straw**
0–60 cm	0.65	0.67
0–20 cm	0.71	0.71
0–10 cm	0.73	0.71
SQI-2	**Dry Grain**	**Dry Straw**
0–60 cm	0.75	0.73
0–20 cm	0.75	0.72
0–10 cm	0.79	0.75
SQI-3	**Dry Grain**	**Dry Straw**
0–60 cm	0.76	0.74
0–20 cm	0.76	0.74
0–10 cm	0.78	0.76
	SQI-1	**SQI-2**
SQI-2	0.92	
SQI-3	0.93	0.97

Supporting Information

Figure S1 Percentage contribution of each soil function in SQI-2 under different soils in four soil layers; different letters indicate significant differences at p<0.05 level for particular soil function. Abbreviations: RDC: root development capacity, WSC: water storage capacity, NSC: nutrient storage capacity.

Acknowledgments

Authors would like to thank Dr. Shiguo Jiang for his help with PCA and Mr. Basant Rimal for his assistance in the laboratory. Authors would also like to thank three anonymous reviewers and an academic editor for their constructive suggestions to improve the manuscript.

Author Contributions

Conceived and designed the experiments: AM RL. Performed the experiments: AM. Analyzed the data: AM. Contributed reagents/materials/analysis tools: AM. Contributed to the writing of the manuscript: AM RL.

References

1. Doran J, Coleman D, Bezdicek D, Stewart B (1994) A framework for evaluating physical and chemical indicators of soil quality.
2. Doran JW, Parkin TB (1994) Defining and assessing soil quality. SSSA special publication 35: 3–3.
3. Karlen DL, Mausbach MJ, Doran JW, Cline RG, Harris RF, et al. (1997) Soil Quality: A Concept, Definition, and Framework for Evaluation (A Guest Editorial). Soil Sci Soc Am J 61: 4–10.
4. Karlen DL, Stott DE (1994) A framework for evaluating physical and chemical indicators of soil quality. In: Doran JW, Coleman DC, Bezdicek DF, Stewart BA, editors. Defining soil quality for a sustainable environment. Madison, WI: Soil Science Society of America. pp. 53–72.
5. Stocking MA (2003) Tropical Soils and Food Security: The Next 50 Years. Science 302: 1356–1359.
6. Diack M, Stott D (2001) Development of a soil quality index for the Chalmers Silty Clay Loam from the Midwest USA. Purdue University: USDA-ARS National Soil Erosion Research Laboratory. pp. 550–555.
7. Larson WE, Pierce FJ (1991) Conservation and enhancement of soil quality. Evaluation of Sustainable Land Management in the Developing World. International Board for Soil Research and Management, Bangkok, Thailand.
8. Andrews S, Karlen D, Mitchell J (2002) A comparison of soil quality indexing methods for vegetable production systems in Northern California. Agriculture, Ecosystems & Environment 90: 25–45.
9. Bhardwaj AK, Jasrotia P, Hamilton SK, Robertson GP (2011) Ecological management of intensively cropped agro-ecosystems improves soil quality with sustained productivity. Agriculture Ecosystems & Environment 140: 419–429.
10. Andrews SS, Carroll CR (2001) Designing a soil quality assessment tool for sustainable agroecosystem management. Ecological Applications 11: 1573–1585.
11. Andrews SS, Karlen DL, Cambardella CA (2004) The soil management assessment framework: A quantitative soil quality evaluation method. Soil Science Society of America Journal 68: 1945–1962.
12. Andrews SS, Mitchell JP, Mancinelli R, Karlen DL, Hartz TK, et al. (2002) On-Farm Assessment of Soil Quality in California's Central Valley. Agron J 94: 12–23.
13. Fernandes JC, Gamero CA, Rodrigues JGL, Mirás-Avalos JM (2011) Determination of the quality index of a Paleudult under sunflower culture and different management systems. Soil and Tillage Research 112: 167–174.
14. Mandal UK, Ramachandran K, Sharma K, Satyam B, Venkanna K, et al. (2011) Assessing Soil Quality in a Semiarid Tropical Watershed Using a Geographic Information System. Soil Science Society of America Journal 75: 1144–1160.
15. Zobeck TM, Halvorson AD, Wienhold B, Acosta-Martinez V, Karlen DL (2008) Comparison of two soil quality indexes to evaluate cropping systems in northern Colorado. Journal of Soil and Water Conservation 63: 329–338.
16. Zornoza R, Mataix-Solera J, Guerrero C, Arcenegui V, Mataix-Beneyto J, et al. (2008) Validating the effectiveness and sensitivity of two soil quality indices based on natural forest soils under Mediterranean conditions. Soil Biology and Biochemistry 40: 2079–2087.
17. Jokela W, Posner J, Hedtcke J, Balser T, Read H (2011) Midwest Cropping System Effects on Soil Properties and on a Soil Quality Index. Agronomy Journal 103: 1552–1562.
18. Navas M, Benito M, Rodriguez I, Masaguer A (2011) Effect of five forage legume covers on soil quality at the Eastern plains of Venezuela. Applied Soil Ecology 49: 242–249.
19. Brady NC, Weil RR (1984) The nature and properties of soils New York: Macmillan.
20. Mukherjee A, Lal R (2014) Long and short-term tillage effects on quality of organic and mineral soils under on-farm conditions in Ohio, Submitted. Soil and Tillage Research.
21. Mukherjee A, Lal R (2014) Short-term effects of cover cropping on quality of a Typic Argiaquolls in Central Ohio, Submitted.
22. Grossman RB, Reinsch TG (2002) Bulk density and linear extensibility. In: Dane JH, Topp GC, editors. Methods of soil analysis, Part 4: Physical methods. Madison, WI: SSSA. pp. 201–254.
23. Herrick JE, Jones TL (2002) A dynamic cone penetrometer for measuring soil penetration resistance. Soil Science Society of America Journal 66: 1320–1324.
24. Yoder RE (1936) A direct method of aggregate analysis of soils and a study of the physical nature of erosion losses. J Am Soc Agron 28: 337–351.
25. Youker RE, McGuinness JL (1957) A short method of obtaining mean weight-diameter values of aggregate analyses of soils. Soil Science 83: 291–294.
26. Dane JH, Hopmans JH (2002) Water retention and storage. In: Dane JH, Topp GC, editors. Methods of soil analysis Part 4, SSSA Book Ser 5. Madison, WI.: SSSA. pp. 671–717.
27. Chhabra A, Palria S, Dadhwal VK (2003) Soil organic carbon pool in Indian forests. Forest Ecology and Management 173: 187–199.
28. Amacher MC, Perry CH (2007) Soil vital signs: A new Soil Quality Index (SQI) for assessing forest soil health.
29. Liebig MA, Varvel G, Doran J (2001) A simple performance-based index for assessing multiple agroecosystem functions. Agronomy Journal 93: 313–318.
30. Li Q, Xu MX, Liu GB, Zhao YG, Tuo DF (2013) Cumulative effects of a 17-year chemical fertilization on the soil quality of cropping system in the Loess Hilly Region, China. Journal of Plant Nutrition and Soil Science 176: 249–259.
31. Masto RE, Chhonkar PK, Singh D, Patra AK (2008) Alternative soil quality indices for evaluating the effect of intensive cropping, fertilisation and manuring for 31 years in the semi-arid soils of India. Environmental monitoring and assessment 136: 419–435.
32. Doran JW, Parkin TB (1996) Quantitative indicators of soil quality: a minimum data set. SSSA special publication 49: 25–38.
33. Dunteman GH (1989) Principal components analysis: Sage.
34. Jolliffe I (2005) Principal component analysis: Wiley Online Library.
35. SPSS (2014) IBM SPSS, Version 21.0, Chicago, USA.
36. Kaiser HF (1960) The application of electronic computers to factor analysis. Educational and Psychological Measurement 20: 141–151.
37. Wander MM, Bollero GA (1999) Soil quality assessment of tillage impacts in Illinois. Soil Science Society of America Journal 63: 961–971.
38. SAS (2012) SAS Institute Inc.; SAS version 9.2.
39. Microsoft (2007) Microsoft Excel. Redmond, Washington.
40. Sigmaplot (2012) Scientific Software Solutions Internationals; Sigmaplot version 12.
41. Feiza V, Feiziene D, Kadziene G, Lazauskas S, Deveikyte I, et al. (2011) Soil state in the 11th year of three tillage systems application on a cambisol. Journal of Food, Agriculture & Environment 9: 1088–1095.
42. Lal R (1994) Methods and guidelines for assessing sustainable use of soil and water resources in the tropics; Washington D.C.: USDA/SMSS Technical Monograph 21.
43. Zornoza R, Mataix-Solera J, Guerrero C, Arcenegui V, García-Orenes F, et al. (2007) Evaluation of soil quality using multiple lineal regression based on physical, chemical and biochemical properties. Science of The Total Environment 378: 233–237.
44. Maurya PR, Lal R (1980) Effects of No-Tillage and Ploughing on Roots of Maize and Leguminous Crops. Experimental Agriculture 16: 185–193.
45. Brzezinska M, Sokolowska Z, Alekseeva T, Alekseev A, Hajnos M, et al. (2011) Some characteristics of organic soils irigated with municipal wastewater. Land Degradation & Development 22: 586–595.
46. Sokoowska Z (2011) Effect of phosphates on dissolved organic matter release from peat-muck soils.
47. Walczak R, Rovdan E, Witkowska-Walczak B (2002) Water retention characteristics of peat and sand mixtures. International Agrophysics 16: 161–166.
48. Carter MR (2006) Quality critical limits and standardization. In: Lal R, editor. Encyclopedia of Soil Science. New York: Marcel Dekker. pp. 1412–1415.

5

Spatial Variation in Soil Properties among North American Ecosystems and Guidelines for Sampling Designs

Henry Loescher[1,2]*, Edward Ayres[1,2,3], Paul Duffy[4], Hongyan Luo[1,2], Max Brunke[1]

1 Science Team, National Ecological Observatory Network (NEON), Boulder, Colorado, United States of America, **2** Institute of Arctic and Alpine Research (INSTAAR), University of Colorado, Boulder, Colorado, United States of America, **3** Natural Resource Ecology Laboratory, Colorado State University, Fort Collins, Colorado, United States of America, **4** Neptune and Company Inc., Lakewood, Colorado, United States of America

Abstract

Soils are highly variable at many spatial scales, which makes designing studies to accurately estimate the mean value of soil properties across space challenging. The spatial correlation structure is critical to develop robust sampling strategies (*e.g.*, sample size and sample spacing). Current guidelines for designing studies recommend conducting preliminary investigation(s) to characterize this structure, but are rarely followed and sampling designs are often defined by logistics rather than quantitative considerations. The spatial variability of soils was assessed across ~1 ha at 60 sites. Sites were chosen to represent key US ecosystems as part of a scaling strategy deployed by the National Ecological Observatory Network. We measured soil temperature (T_s) and water content (SWC) because these properties mediate biological/ biogeochemical processes below- and above-ground, and quantified spatial variability using semivariograms to estimate spatial correlation. We developed quantitative guidelines to inform sample size and sample spacing for future soil studies, *e.g.*, 20 samples were sufficient to measure T_s to within 10% of the mean with 90% confidence at every temperate and sub-tropical site during the growing season, whereas an order of magnitude more samples were needed to meet this accuracy at some high-latitude sites. SWC was significantly more variable than T_s at most sites, resulting in at least 10× more SWC samples needed to meet the same accuracy requirement. Previous studies investigated the relationship between the mean and variability (*i.e.*, sill) of SWC across space at individual sites across time and have often (but not always) observed the variance or standard deviation peaking at intermediate values of SWC and decreasing at low and high SWC. Finally, we quantified how far apart samples must be spaced to be statistically independent. Semivariance structures from 10 of the 12-dominant soil orders across the US were estimated, advancing our continental-scale understanding of soil behavior.

Editor: Ben Bond-Lamberty, DOE Pacific Northwest National Laboratory, United States of America

Funding: The National Ecological Observatory Network (NEON) is a project sponsored by the National Science Foundation (NSF) and managed under cooperative agreement by NEON, Inc. Any opinions, findings, and conclusions or recommendations expressed in this material are those of the author(s) and do not necessarily reflect the views of the NSF.' be formally acknowledged. The funders had no role in study design, data collection and analysis, decision to publish, or preparation of the manuscript.

Competing Interests: The National Ecological Observatory Network (NEON) is solely supported by the National Science Foundation (NSF). NEON is incorporated as a 501(c)3 non-profit entity, and needs to do this to engage in a cooperative agreement with the NSF. Loescher, Ayres, and Luo are employed by NEON. Brunke was employed by NEON at the time of data collection. The Neptune Company Inc. provides superlative statistical and experimental design services on contract with NEON. Duffy is employed by Neptune as a professional statistical consultant, and acted in that capacity here. Both NEON and Neptune adhere to the highest ethical standards and scientific integrity, and consider this a scholarly work with no competing interests as defined by PLOS ONE, i.e, there are no competing interests can be financial or non-financial, professional, or personal that would limit the scientific creativity and scholarly activities presented in this manuscript, nor limit any capability to provide free open-source of our data. These affiliations do (does) not alter the authors' adherence to all the PLOS ONE policies on sharing data and materials.

* E-mail: hloescher@neoninc.org

Introduction

Researchers interested in measuring the mean value for a soil property across space must make decisions about the number of samples required and how far apart they should be spaced. These decisions are inherently dependent on the variance structure and spatial independence of the data, but quantitative guidelines to inform these decisions are unavailable at most sites. For instance, intuitively researchers know that a large number of widely-spaced samples would be more likely to better measure the spatial mean of a soil property than a few samples all located close to one another. But it is typically not clear what "a large number" or "widely-spaced" means. Moreover, since increasing the sample size consumes resources (*e.g.*, time, money, and analytical capacity), it is important that sufficient samples are collected to meet the accuracy requirement to answer a specific question without wasteful over-sampling. Likewise, choosing an appropriate distance between samples is important for two reasons; 1) samples should ideally be spaced sufficiently far apart to avoid correlation and pseudoreplication in measurements at a given scale [1], thereby maximizing the amount of information provided by a limited number of samples, and 2) spacing between samples should be minimized to constrain costs, which typically increase with distance.

Here, we focus on sampling strategies designed to estimate the spatial mean value of a soil property with the smallest possible

Table 1. Site characteristics, where Lat = Latitude, Long = Longitude, El = Elevation, Temp = Mean Annual Air Temperature, Precip = Mean annual Bulk Precipitation, CH = Canopy Height, CS = Canopy Structure, CL = closed, SS = short stature, O = open, SO = semi-open.

Site	Lat (°)	Long (°)	El (m.a.s.l)	Temp (°C)	Precip (mm)	CH (m)	Ecosystem type	CS	Soil order	Owner
Harvard Forest	42.54	−72.17	348	7	1066	26.0	Temperate deciduous forest	CL	Spodosol	Harvard U
Burlington	42.52	−71.18	38	9	1095	24.0	Temperate deciduous forest	CL	Inceptisol	Town of Burlington MA
Bartlett Exp. Forest	44.06	−71.29	273	7	1219	23.0	Temperate deciduous forest	CL	Spodosol	USDA FS
Blandy Farm	39.06	−78.07	182	12	972	1.0	Shrubland	SS	Ultisol-Alfisol	U Virginia
Smithsonian CRC (SCBI)	38.89	−78.14	355	13	1057	35.0	Temperate deciduous forest	CL	Alfisol	Smithsonian Institution
Smithsonian SERC	38.89	−76.56	10	13	1097	38.0	Temperate deciduous forest	CL	Ultisol	Smithsonian Institution
Ordway-Swisher BS	29.69	−81.99	48	20	750	23.0	Temperate coniferous forest	O	Entisol	U Florida Foundation
J. Jones Ecological Research station	31.19	−84.47	47	19	1344	27.0	Temperate coniferous forest	CL	Ultisol	Robert W. Woodruff Foundation
Disney Wilderness Preserve	28.13	−81.44	20	22	1219	1.5	Grassland	SS	Spodosol	Nature Conservancy
Guanica State Forest	17.97	−66.87	126	23	840	10.0	Tropical deciduous forest	CL	Aridisol	State Forest, Dept. Natural and Environmental Resources, Puerto Rico
Lajas Ag. Exp. St.	18.02	−67.08	16	25	830	0.4	Agricultural	SS	Vertisol	U Puerto Rico
Ponce	17.99	−66.64	1	27	901	5.0	Savanna	O	Mollisol	Municipality of Ponce, PR
Steigerwaldt	45.51	−89.58	477	5	809	5.5	Temperate deciduous forest	SO	Spodosol	Steigerwaldt Land Services
Tree Haven	45.49	−89.59	461	5	809	23.0	Temperate deciduous forest	CL	Spodosol	University of Wisconsin Stevens Point
UNDERC	46.23	−89.54	520	5	751	24.0	Temperate deciduous forest	CL	Spodosol	U Notre Dame
Konza - Core	39.10	−96.56	415	13	835	1.5	Grassland	SS	Mollisol	Nature Conservancy/Kansas State U
Konza - Relocatable	39.11	−96.61	323	13	835	1.5	Agricultural	SS	Mollisol	Nature Conservancy/Kansas State U
U Kansas Bio Station	39.04	−95.19	321	14	940	19.0	Temperate deciduous forest	CL	Inceptisol	U Kansas
Oak Ridge Nat'l Lab	35.96	−84.28	342	15	1352	28.0	Temperate deciduous forest	CL	Ultisol	US DOE
Great Smokey Mtns NP	35.69	−83.50	661	13	1453	30.0	Temperate deciduous forest	CL	Ultisol	*US NPS
Mountain Lake Res. Stn.	37.38	−80.52	1170	8	1131	18.0	Temperate deciduous forest	CL	Entisol	U Virginia
Talladega Nat'l Forest	32.95	−87.39	164	15	1384	25.0	Temperate coniferous forest	SO	Ultisol	USDA FS
Choctaw NWA	31.85	−88.17	11	18	1524	35.0	Temperate deciduous forest	CL	Inceptisol	DOI NWA
Dead Lake WMA	32.54	−87.80	24	17	1415	30.0	Temperate deciduous forest	CL	Ultisol	US Army Corps of Engineers
Dakota-Coteau	47.16	−99.11	575	4	478	0.4	Grassland	SS	Mollisol	State of North Dakota Land Trust
Northern Great Plains Res. Lab	46.77	−100.92	589	5	434	0.4	Agricultural	SS	Mollisol	USGS/DOI FWS
Woodworth	47.13	−99.24	590	4	478	1.0	Agricultural	SS	Mollisol	USGS/DOI FWS
Central Plains Exp. Range	40.82	−104.75	1653	9	322	0.4	Grassland	SS	Mollisol	USDA ARS
Sterling	40.46	−103.03	1365	10	422	1.0	Agricultural	SS	Mollisol	Private,
Rocky Mtn. NP (Castnet)	40.28	−105.55	2742	7	998	19.0	Temperate coniferous forest	SO	-	DOI NPS
LBJ Nat'l grasslands	33.40	−97.57	272	18	864	13.0	Temperate deciduous forest	SO	Alfisol	USDA FS
M. Klemme Range Res. Station	35.41	−99.06	520	15	780	1.0	Agricultural	SS	Inceptisol	Oklahoma State U, Oklahoma Agricultural Experimental Station

Table 1. Cont.

Site	Lat (°)	Long (°)	El (m.a.s.l)	Temp (°C)	Precip (mm)	CH (m)	Ecosystem type	CS	Soil order	Owner
Northcutt	33.89	−96.84	212	17	970	4.0	Savanna	O	Inceptisol	Private,
Bozeman	45.66	−111.05	1503	7	491	1.0	Agricultural	SS	Mollisol	Montana State U
Loch Leven	45.46	−110.62	1448	6	432	0.4	Grassland	SS	Mollisol	Montana Dept of Fish, Wildlife and Parks
Yellowstone NP	44.95	−110.54	2129	7	249	14.0	Temperate coniferous forest	SO	-	*DOI NPS
Niwot Ridge	40.05	−105.58	3478	−4	930	0.2	Tundra	SS	Inceptisol	USDA FS
Fraser Exp. Forest	39.86	−105.86	3526	1	584	14.0	Boreal/montane coniferous forest	SO	Inceptisol	USDA FS
Moab	38.25	−109.39	1800	14	239	0.2	Grassland	SS	Aridisol	BLM
Santa Rita Exp. Range	31.91	−110.84	999	22	303	2.0	Shrubland	SS	Entisol	U Arizona
Jornada Exp. Range	32.59	−106.84	1321	18	241	0.4	Grassland	SS	Aridisol	USDA ARS
Onaqui-Benmore	40.18	−112.45	1654	9	274	1.2	Shrubland	SS	Aridisol	BLM/USDA FS
Murray	40.65	−111.92	1302	12	450	1.0	Grassland	SS	- (anthropodisol)	City of Murray, UT
Red Butte	40.78	−111.80	1676	12	450	10.0	Temperate deciduous forest	CL	Mollisol	USDA FS/U Utah
Wind River RNA	45.82	−121.95	368	9	2223	50.0	Temperate coniferous forest	CL	Andisol	USDA FS
Abby Road	45.76	−122.33	367	9	2249	1.0	Temperate coniferous forest	SS	Andisol	State of Washington Dept. Natural Resources
Thayer	45.71	−122.34	557	9	2249	12.0	Temperate coniferous forest	CL	Inceptisol	State of Washington Dept. Natural Resources
Thyme	45.71	−122.38	301	9	2249	1.0	Temperate coniferous forest	SS	Ultisol	State of Washington Dept. Natural Resources
Good Seed	45.78	−122.30	559	9	2249	1.0	Temperate coniferous forest	SS	Ultisol	State of Washington Dept. Natural Resources
Soaproot Saddle	37.03	−119.26	1210	13	957	32.0	Temperate coniferous forest	SO	Alfisol	USDA FS
Lower Teakettle	37.01	−119.01	2149	8	957	35.0	Temperate coniferous forest	SO	Entisol-Inceptisol	USDA FS
San Joaquin Exp. Range	37.11	−119.73	397	17	375	21.0	Savanna	O	Alfisol-Inceptisol	USDA FS
Toolik Lake LTER	68.66	−149.37	827	−9	316	0.3	Tundra	SS	Gelisol	BLM
Pump Station 2	69.60	−148.67	142	−9	316	0.3	Tundra	v	Gelisol	BLM
Barrow Exp. Observatory	71.28	−156.62	7	−12	105	0.3	Tundra	SS	Gelisol	*Ukpeaġvik Iṇupiat Corporation*
Caribou-Poker	65.15	−147.50	239	−3	262	8.0	Boreal/montane coniferous forest	SO	Inceptisol	UAF
Poker Flats	65.11	−147.42	472	−3	262	8.0	Boreal/montane coniferous forest	SS	Inceptisol	Alaska Dept Natural Resource/UAF
Eight Mile Lake	63.88	−149.22	684	−2	366	0.3	Tundra	SS	Gelisol	Alaska Dept Natural Resource
Delta Junction	63.88	−145.75	504	−3	305	10.0	Boreal/montane coniferous forest	SO	Inceptisol	BLM
Kenai NWA	60.53	−150.58	110	1	483	12.0	Boreal/montane coniferous forest	CL	Spodosol	DOI NWA

All sites were given access to measure soil properties by the owner, no soil samples were taken, and (*) denotes that a research permit was needed to gain access to the site.
Note: - = soil order unknown, USDA = US Department of Agriculture, FS = Forest Service, ARS = Agricultural Research Service, DOI = US Department of Interior, BLM = US Bureau of Land Management, DOE = US Department of Energy, UAF = University of Alaska, Fairbanks, FWS = US Fish and Wildlife Service, USGA = US Geological Survey, NPS = National Park Service, LTER = long term ecological research site, NWA = National Wildlife Area, RNA = Research Natural Area, WMA = Wildlife Management Area, NP = National Park, BS = Biological Station.

sample size and sample spacing. However, it is also important to note that other goals also exist that require different sampling designs, *e.g.*, redundant sampling (*i.e.*, field duplicates) involving two or more measurements at the same location to aid error detection (*e.g.*, sensor malfunction) and increase confidence in the data, while varying the distance between adjacent samples can be useful when characterizing the spatial variance structures.

The number of samples and sample spacing is rarely justified in research papers or presentations. In many, if not most studies, decisions relating to the sampling strategy appear to be based on a combination of subjective opinion, repeating previously used sampling designs (often from a different ecosystem or in relation to a different soil property), and available resources. There are only a few methodological articles that discuss sampling designs recognize the importance of choosing an appropriate sample size and sample spacing, and provide only vague- or no recommendations at all [2–4]. The advice often given is to conduct a preliminary spatial variability study (of soil properties) at the research site to determine the number of samples and sample size required to achieve a given accuracy [4,5]. While this is good advice, it is rarely followed, presumably due to a lack of time, money, and/or an appreciation of its importance.

Soils are highly variable and embody systematic (*e.g.*, soil forming processes) and random sources of variability across space [6,7]. Some studies have used the coefficient of variation to assess this variability at a given site, and used in sample size analyses [8–14]. Geostatistical techniques, in particular semivariograms, have increasingly been used over the last few decades to characterize spatial variation in soil properties [15,16]. Semivariograms provide two useful components for designing a robust sampling strategy: 1) an estimate of the variance (*i.e.*, the semivariogram sill), which can be used to inform sample size in future studies; and 2) an estimate of the minimum distance required for samples to be considered spatially independent (*i.e.*, the semivariogram range), which can be used to inform sample spacing (Figure S1). We emphasize that the use of sample size analyses in this paper is to inform decisions relating to future studies, not to assess the power of a sampling design retrospectively, which may be fundamentally flawed [17].

Site-specific studies that have assessed spatial variation in soil properties have shown that spatially structured variability is ubiquitous. For instance, Robertson et al. [11] studied variation in soil physical, chemical, and biological properties across an agricultural field in Michigan, USA. Almost every soil property exhibited spatially structured variation, which in most cases accounted for the majority of the variability that was observed Robertson et al. [11], and indicated that measurements of soil properties at one location could be used to predict values for those properties up to ~60 m away (*i.e.*, the range). Similar findings have been reported elsewhere, across a range of ecosystems and soils [9,12,18–24], although there are a few exceptions, with some properties exhibiting little or no spatial dependence at the scale studied [9,11,19,20]. The identification of spatial patterns in soil properties among sites, in particular at larger continental scales has been hampered, because most studies have investigated these quantities at a single site.

The few cross-site comparisons studies have focused on soil moisture variability located only within a small region with relatively narrow ecoclimatic variability [13,14,25] or, if they spanned larger scales, involved only a few sites [26] or combined data from different studies with widely differing designs *e.g.*, spatial extent, physical quantities, sampling depth, seasonal timing, measurement technique, etc. [12,27–30]. Moreover, previously published data are often unavailable for additional analyses. A single study that employs a uniform sampling and analysis approach, *i.e.*, the above mentioned quantities, as well as being representative of site characteristics (*e.g.*, ecosystem structure, soil type, etc), made under similar environmental conditions for a wide range of ecosystems is needed for meaningful cross-site comparisons and to develop quantitative sampling guidelines that can inform sampling strategies in future soil studies.

Here, we studied the spatial variation in soil properties using a consistent approach at 60 sites throughout the USA and developed quantitative relationships to provide: 1) an ecological understanding of spatial variability in soils, and 2) guidance on sampling designs for future soil studies. We studied local scale (~1 ha) variation in soil temperature (T_s) and soil water content (SWC) at sites that included representatives from all major terrestrial ecosystem types, soil types, and climates found in the USA. Soil temperature and moisture were used as proxies for other soil properties because they can be measured quickly and accurately in

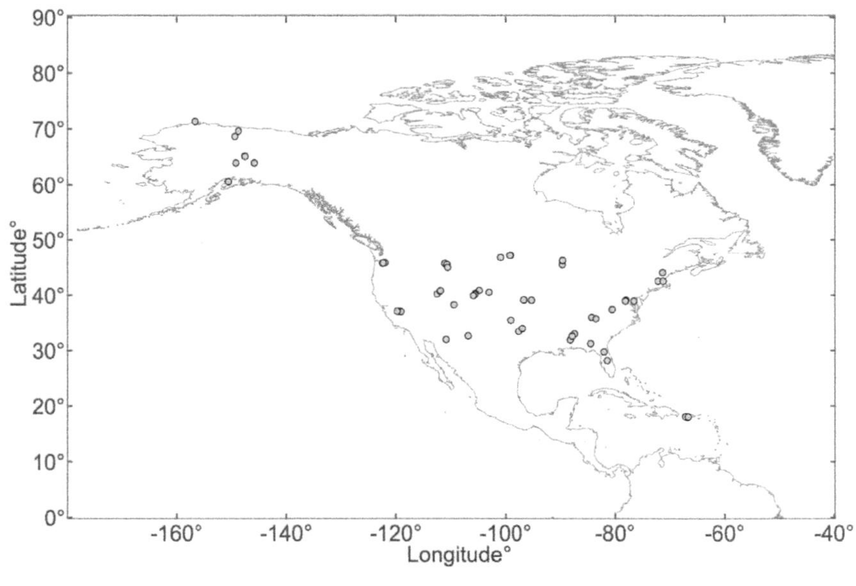

Figure 1. The sampling sites (open circles) were widely distributed throughout the US.

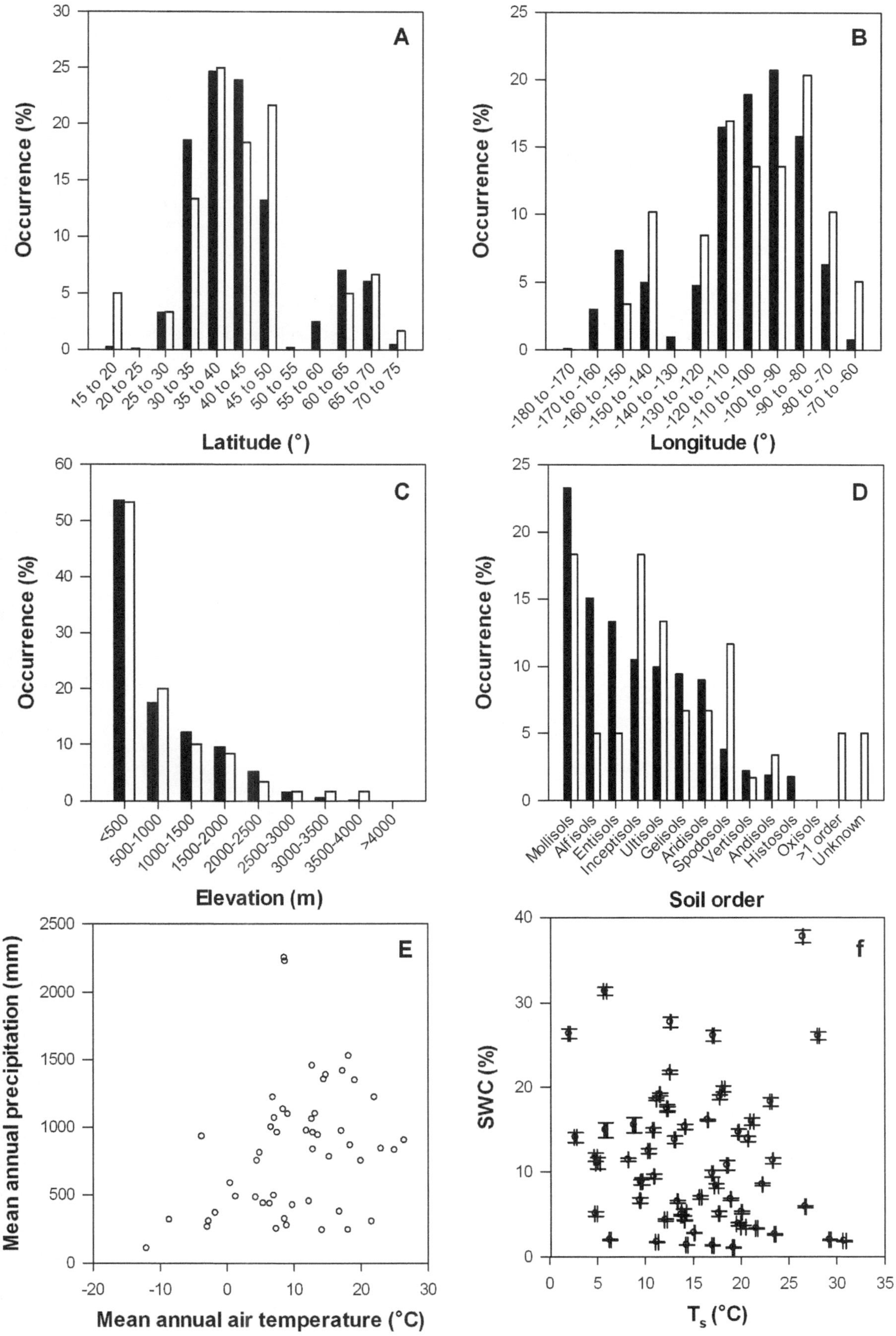
A
Occurrence (%)
0
5
10
15
20
25
30
15 to 20
20 to 25
25 to 30
30 to 35
35 to 40
40 to 45
45 to 50
50 to 55
55 to 60
60 to 65
65 to 70
70 to 75
Latitude (°)
B
Occurrence (%)
0
5
10
15
20
25
-180 to -170
-170 to -160
-160 to -150
-150 to -140
-140 to -130
-130 to -120
-120 to -110
-110 to -100
-100 to -90
-90 to -80
-80 to -70
-70 to -60
Longitude (°)
C
Occurrence (%)
0
10
20
30
40
50
60
<500
500-1000
1000-1500
1500-2000
2000-2500
2500-3000
3000-3500
3500-4000
>4000
Elevation (m)
D
Occurrence (%)
0
5
10
15
20
25
Mollisols
Alfisols
Entisols
Inceptisols
Ultisols
Gelisols
Aridisols
Spodosols
Vertisols
Andisols
Histosols
Oxisols
>1 order
Unknown
Soil order
E
Mean annual precipitation (mm)
0
500
1000
1500
2000
2500
-20
-10
0
10
20
30
Mean annual air temperature (°C)
f
SWC (%)
0
10
20
30
40
0
5
10
15
20
25
30
35
T_s (°C)

Figure 2. The sites were broadly representative of the US. Percent occurrence of land in latitudinal (a), longitudinal (b), elevational (c), and soil order (d) categories in the US (black bars), and our study sites (white bars). The study sites also occupied a wide range of climates (e) and T_s and SWC conditions at the time of samplings (f). Error bars in (f) represent ±1 SE.

the field, they are major drivers of soil biotic and biogeochemical properties and processes, their spatial pattern integrates the spatial pattern of ecosystem canopy structure and has been shown to correlate with spatial patterning of other soil properties at the local scale, and they are increasingly recognized as important variables in understanding land-atmosphere interactions *e.g.*, soil moisture is recognized as a Global Climate Observing System Essential Climate Variable [9,11,31–36]. The sampling scale (~1 ha) corresponds to many ecosystem-level soil studies and is the basis of many large emergent observatories (National Ecological Observatory Network, NEON, www.neoninc.org; Integrated Carbon Observing System, www.icos-infrastructure.eu). We used these data to further examine the relationship between the spatial structure of variability in soil properties and site characteristics at broader landscape- to continental scales (*e.g.*, latitude, climate, ecosystem type, soil type, and vegetation structure). Lastly, we used analyses to estimate the sample size required to accurately estimate the spatial mean T_s and SWC at each site.

This study was made to inform the sensor-based soil plot designs at NEON sites. In addition, several hypotheses were tested. We hypothesized that;

1) the spatial variability in soil properties would be associated with variability in ecosystem structure, *e.g.*, we expect open canopy ecosystems, which exhibit large spatial variability in vegetation height (*e.g.*, savannas) to have larger variability in soil properties than closed canopy ecosystems,
2) agricultural ecosystems would exhibit lower spatial variability in soil properties than other ecosystems, since management activities (*e.g.*, tillage and irrigation) aim to homogenize properties,
3) younger soils (*e.g.*, inceptisols) would exhibit both lower spatial variability and have smaller range values than older soils (*e.g.*, ultisols). We expected that some soil properties (*e.g.*, soil texture) that influence the spatial structure of variability are controlled by processes that operate over long temporal scales (*e.g.*, weathering),
4) the spatially structured variability (*e.g.*, semivariogram range, nugget, and sill), as well as the sample size required to accurately estimate the mean, would be positively correlated for soil temperature (T_s) and soil water content (SWC), becuase some underlying site characteristics (*e.g.*, canopy gaps, etc.) are likely to influence both variables,
5) some forms of semivariograms may be more common than other forms, *e.g.*, , that small semivariogram range values would be more commonly associated with small semivariogram sills than large sills, and
6) variability would be larger for SWC than T_s since water is more mobile than heat and sensitive to microtopography, as well as being actively transported by plants.

We also aimed to develop empirically-based rules that could be used by other researchers to guide sampling designs to accurately estimate the spatial mean of soil properties for future studies.

Methods

Site descriptions and data collection

The study was conducted at 60 NEON sites (Table 1). These sites were chosen for landscape level representativeness using the approaches of Hargrove and Hoffman [37,38], which divided the USA into 20 eco-climatic domains. Within each domain, 3 representative site were chosen; 1 wildland (*i.e.*, relatively undisturbed) site and typically 2 sites that address key ecological issues for that domain (*e.g.*, urbanization, land management, advance of invasive species, or climate change) [39,40]. Measurements at these sites included ecosystem scale (*i.e.*, 10^3 to 10^6 m^2) tower-based estimates of eddy covariance [41–44], and the long term sensor-based soil measurements are required to be made in the flux footprint of the tower, *i.e.*, at the flux scale [45–47] and occur on locally (co-) dominant soil series. No sampling occurred requiring any sampling permits (re soil sampling), no human sampling occurred, nor the use or impact to any endangered or protected animals were involved in this study. The sampling area for this study corresponded to the expected location of NEON's long term sensor-based soil measurements at each site. We received permission to access and to measure soil properties by the owner at all sites, and no soil samples were taken (hence not requiring a sampling permit of any kind). Two National Park Service sites required a research permit to gain access to the site (see Table 1).

Soil taxonomic information at a site was gathered from the USDA Natural Resource Conservation Service's Web Soil Survey, except for Wind River [48], Bartlett Experimental Forest (www.fs.fed.us/ne/durham/4155/bartlett.htm#SOI), and Eight Mile Lake [49]. Soil taxonomic information was not available for Yellowstone National Park, WY, or Rocky Mountain National Park, CO, while the soil at Murray, UT, was classified by the Web Soil Survey as *dumps* (*i.e.*, anthropogenosol), which cannot be assigned to a specific soil order. Mean annual temperature and mean annual total precipitation data were gathered from local site sources when available, and from climate data for nearby towns (www.usclimatedata.com) when site-specific information was unavailable.

Field-based soil temperature and moisture data were collected over 1 to 2 days at each site between August 2009 and April 2011 (with most sites visited between March and October 2010) and were timed to coincide with peak growing season. The only exception was sampling at Mountain Lake, VA, which occurred prior to budburst due to logistical constraints. Our measurements corresponded to the growing season because this is the time of year when most biological activity occurs. It should be noted that spatial variability of soil properties can change seasonally [44], *i.e.*, there is often intra-annual nonstationarity in both functional linkages and spatial correlation), although this does not always occur [12,21,23,26,50]. As such, the spatial variability that we observed may not reflect spatial variation among other seasons or years.

Soil temperature, T_s (°C), was measured at 0–12 cm depth with platinum resistance temperature sensors (RTD 810, Omega Engineering Inc., Stamford CT) and soil water content, SWC (vol H_2O vol^{-1} soil expressed as a percentage) was measured at a 0–15 cm depth with time domain dielectric sensors (CS616, Campbell Scientific Inc., Logan UT). Due to the large number of sites and soil types it was not possible to calibrate the soil moisture sensor to the soil type at each site; instead a manufacturer recommended equation was used to convert the sensor measurement of the dielectric constant to SWC [51,52]. The sensors were inserted into the soil and allowed to stabilize prior to data

Table 2. Summary statistics of T_s and SWC measurements, as well as range, sill, and nugget estimates based on the semivariogram model, at each site.

	T_s										SWC										Sampling date‡
Site	Mean (°C)	SD	$CV_{Traditional}$	Maximum lag	Range (m)*	Nugget (% of sill)*	Sill*	CV_{Sill}	Sample size†	n	Mean (%)	SD	$CV_{Traditional}$	Maximum lag	Range (m)*	Nugget (% of sill)*	Sill*	CV_{Sill}	Sample size†	n	
Harvard Forest	17.7	0.6	0.04	100	74	55	0.46	0.05	0.8	144	5.0	3.5	0.69	100	17	62	9.60	0.87	203.8	144	7/8/2010
Burlington	19.2	0.8	0.04	100	9	56	0.67	0.06	1.0	119	1.1	0.6	0.52	100	10	31	0.44	0.88	209.5	120	7/7/2010
Bartlett Exp. Forest	17.0	0.9	0.05	100	40	67	0.98	0.08	1.8	126	9.8	4.6	0.47	100	68	27	29.41	0.78	166.1	126	7/6/2010
Blandy Farm	22.2	0.4	0.02	100	-	-	-	-	-	129	8.5	1.8	0.21	100	47	53	3.56	0.31	26.8	129	8/18/2010
Smithsonian CRC (SCBI)	20.7	0.4	0.02	80	5	16	0.05	0.02	0.1	126	13.9	3.6	0.26	80	-	-	-	-	-	126	8/19/2010
Smithsonian ERC	23.3	0.4	0.02	80	34	15	0.16	0.02	0.2	123	11.3	3.9	0.34	80	32	41	13.40	0.46	56.7	123	8/17/2010
Ordway-Swisher BS	14.3	1.2	0.08	100	6	5	0.58	0.08	1.5	144	1.4	0.3	0.22	100	23	64	0.09	0.31	25.7	144	3/24/2010
J. Jones Ecological Research station	14.1	0.3	0.02	100	35	31	0.08	0.03	0.2	129	4.3	0.7	0.17	100	8	96	0.54	0.24	15.8	129	3/21/2010
Disney Wilderness Preserve	16.5	0.6	0.04	100	4	25	0.31	0.05	0.6	144	16.1	1.0	0.06	95	78	77	1.00	0.09	2.1	144	3/23/2010
Guanica State Forest	26.7	0.7	0.03	84	-	-	-	-	-	129	5.8	1.4	0.24	84	15	32	2.35	0.37	37.3	129	3/18–19/2010
Lajas Ag. Exp. St.	28.1	1.2	0.04	65	12	22	0.63	0.04	0.4	120	26.0	5.3	0.20	65	24	57	27.17	0.28	21.7	120	3/17–18/2010
Ponce	26.5	0.4	0.02	100	25	18	0.21	0.02	0.2	126	37.8	7.9	0.21	100	25	9	73.66	0.32	28.0	121	10/12/2010
Steigerwaldt	19.8	0.8	0.04	100	-	-	-	-	-	144	14.6	4.9	0.33	100	-	-	-	-	-	144	8/12/2010
Tree Haven	18.6	0.9	0.05	80	-	-	-	-	-	129	10.8	6.5	0.60	80	12	40	15.87	0.52	74.2	129	8/11/2010
UNDERC	18.9	0.4	0.02	100	70	68	0.13	0.03	0.2	141	6.7	1.6	0.23	100	20	75	2.70	0.35	32.4	141	8/10/2010
Konza - Core	14.2	0.2	0.02	100	18	58	0.04	0.02	0.1	129	15.3	3.6	0.23	100	52	75	13.70	0.34	31.7	129	5/12/2010
Konza - Relocatable	12.6	0.2	0.02	100	35	34	0.04	0.02	0.1	123	21.8	2.6	0.12	100	67	38	9.02	0.20	10.3	123	5/13/2010
U Kansas Bio Station	12.7	0.5	0.04	84	7	18	0.06	0.03	0.2	129	27.7	6.7	0.24	84	-	-	-	-	-	129	5/11/2010
Oak Ridge Nat'l Lab	13.4	0.4	0.03	84	30	16	0.18	0.04	0.5	129	6.4	1.7	0.26	84	-	-	-	-	-	129	4/19/2010
Great Smokey Mtns NP	11.2	0.3	0.02	80	70	54	0.05	0.03	0.2	129	18.6	1.9	0.10	80	-	-	-	-	-	129	4/21/2010
Mountain Lake Res. Stn.	8.2	0.4	0.05	100	79	42	0.11	0.06	0.9	129	11.4	2.2	0.19	100	70	70	5.55	0.29	23.1	129	4/22/2010

Table 2. Cont.

	T_s										SWC										Sampling date‡
Site	Mean (°C)	SD	$CV_{Traditional}$	Maximum lag	Range (m)*	Nugget (% of sill)*	Sill*	CV_{Sill}	Sample size†	n	Mean (%)	SD	$CV_{Traditional}$	Maximum lag	Range (m)*	Nugget (% of sill)*	Sill*	CV_{Sill}	Sample size†	n	
Talladega Nat'l Forest	20.2	3.0	0.15	100	94	36	4.78	0.15	6.3	129	3.5	2.5	0.72	100	20	20	5.63	0.97	252.9	129	4/16/2010
Choctaw NWA	17.1	1.1	0.06	84	16	8	0.63	0.07	1.2	129	26.1	7.4	0.28	84	45	42	51.93	0.39	41.3	129	4/14/2010
Dead Lake WMA	17.8	0.3	0.02	80	-	-	-	-	-	132	18.8	3.1	0.17	80	-	-	-	-		132	4/19/2011
Dakota-Coteau	12.3	0.8	0.07	100	10	19	0.43	0.08	1.5	144	17.4	3.2	0.19	100	19	60	8.06	0.23	14.4	144	5/26/2010
Northern Great Plains Res. Lab	11.6	0.5	0.04	100	8	23	0.12	0.04	0.5	144	19.1	2.0	0.11	100	4	63	4.37	0.15	6.5	144	5/25/2010
Woodworth	15.7	1.3	0.08	100	14	65	0.89	0.09	2.0	144	6.9	2.4	0.35	100	8	53	2.89	0.35	32.5	144	5/27/2010
Central Plains Exp. Range	23.5	1.0	0.04	100	95	32	0.54	0.04	0.5	144	2.6	0.8	0.32	100	29	40	0.72	0.46	57.8	144	8/13/2009
Sterling	23.1	1.3	0.06	100	50	57	0.60	0.05	0.6	156	18.2	6.5	0.36	100	2	0	49.25	0.54	80.1	156	8/10/2009
Rocky Mtn. NP (Castnet)	11.2	1.7	0.15	84	-	-	-	-	-	132	1.7	0.5	0.32	84	12	42	0.31	0.46	57.3	132	8/18–19/2009
LBJ Nat'l grasslands	20.1	0.6	0.03	100	9	2	0.48	0.05	0.6	129	5.2	1.5	0.29	100	3	17	2.49	0.43	50.0	129	9/28/2010
M. Klemme Range Res. Station	18.1	1.6	0.09	100	22	11	2.31	0.12	3.8	129	19.8	4.1	0.21	100	72	23	21.98	0.33	30.4	128	4/30/2010
Northcutt	19.7	2.8	0.14	100	11	27	2.49	0.11	3.5	141	3.8	2.3	0.59	100	78	15	7.59	1.02	280.8	141	4/28/2010
Bozeman	21.1	2.3	0.11	100	25	7	3.47	0.12	4.2	123	15.9	4.8	0.30	100	23	49	25.99	0.45	55.6	123	7/20/2010
Loch Leven	21.6	1.4	0.06	100	15	44	0.82	0.06	0.9	138	3.3	0.6	0.19	100	70	36	0.57	0.32	27.9	138	7/21/2010
Yellowstone NP	12.1	1.7	0.14	100	31	13	3.02	0.20	11.1	132	4.3	1.4	0.33	100	15	28	2.36	0.51	69.0	132	7/22/2010
Niwot Ridge	8.8	1.7	0.20	80	5	3	2.60	0.26	18.0	123	15.5	10.0	0.65	84	27	9	83.27	0.83	187.9	123	7/30/2010
Fraser Exp. Forest	6.3	1.2	0.19	84	21	23	1.00	0.22	13.5	132	2.0	0.7	0.33	80	15	94	0.44	0.47	60.3	132	9/15/2010
Moab	15.1	0.5	0.04	100	16	64	0.12	0.03	0.3	144	2.8	0.4	0.16	100	37	58	0.22	0.24	15.6	144	10/6/2010
Santa Rita Exp. Range	30.8	1.9	0.06	100	2	0	0.79	0.04	0.5	141	1.8	0.5	0.30	100	17	38	0.32	0.44	52.7	141	8/25/2010
Jornada Exp. Range	29.3	1.2	0.04	100	46	73	0.57	0.04	0.4	141	2.0	0.5	0.26	100	-	-	-	-	-	141	8/23/2010
Onaqui-Benmore	13.8	0.8	0.06	84	14	12	0.61	0.08	1.7	135	4.9	0.8	0.16	84	-	-	-	-	-	135	5/18/2010
Murray	17.4	2.6	0.15	84	-	-	-	-	-	135	8.3	3.1	0.37	84	2	0	6.86	0.45	54.0	135	5/19/2010

Table 2. Cont.

	T_s										SWC										Sampling date‡
Site	Mean (°C)	SD	$CV_{Traditional}$	Maximum lag	Range (m)*	Nugget (% of sill)*	Sill*	CV_{Sill}	Sample size†	n	Mean (%)	SD	$CV_{Traditional}$	Maximum lag	Range (m)*	Nugget (% of sill)*	Sill*	CV_{Sill}	Sample size†	n	
Red Butte	10.8	1.0	0.09	65	22	42	0.81	0.12	3.7	132	14.9	3.2	0.21	65	15	79	10.44	0.31	25.4	132	5/20/2010
Wind River RNA	10.9	0.4	0.03	100	4	31	0.09	0.04	0.4	138	9.5	3.1	0.33	100	3	63	10.27	0.48	62.1	138	6/8/2010
Abby Road	9.6	1.0	0.11	100	3	30	1.02	0.15	6.0	141	8.8	3.7	0.42	100	21	44	16.15	0.65	114.1	141	10/20/2010
Thayer	9.5	0.6	0.06	70	40	58	0.42	0.10	2.5	111	6.6	3.4	0.51	70	-	-	-	-	-	111	10/21/2010
Thyme	12.3	0.6	0.05	100	17	32	0.40	0.07	1.4	129	17.5	5.1	0.29	100	27	53	27.81	0.43	49.3	129	6/10/2010
Good Seed	13.1	0.8	0.06	100	20	13	0.79	0.10	2.5	126	13.8	5.0	0.36	100	43	46	25.47	0.52	72.7	126	6/9/2010
Soaproot Saddle	17.0	1.1	0.07	84	9	13	1.18	0.09	2.2	129	1.3	0.5	0.36	84	3	18	0.24	0.52	73.5	129	7/15/2010
Lower Teakettle	14.0	2.9	0.20	80	11	2	6.75	0.26	18.6	129	5.2	4.4	0.86	80	13	2	20.06	1.22	404.6	129	7/15/2010
San Joaquin Exp. Range	9.7	1.2	0.12	100	6	0	0.96	0.14	5.6	144	8.8	5.0	0.57	100	8	6	19.91	0.71	137.8	144	2/22/2011
Toolik Lake LTER	2.7	1.6	0.59	100	3	40	2.65	0.85	194.6	144	14.0	6.9	0.49	100	5	58	52.77	0.73	145.1	144	6/19/2010
Pump Station 2	2.1	1.2	0.59	100	31	42	1.63	0.88	207.6	128	26.3	7.1	0.27	100	17	47	55.17	0.40	43.1	144	6/20/2010
Barrow Exp. Observatory	5.8	2.2	0.39	100	14	15	4.31	0.51	69.1	141	31.4	5.4	0.17	100	-	-	-	-	-	141	7/26/2010
Caribou-Poker	5.0	3.1	0.61	100	6	44	9.90	0.88	211.2	144	11.0	8.1	0.73	100	7	49	67.34	1.05	300.9	144	6/22/2010
Poker Flats	10.4	1.4	0.14	100	3	3	1.58	0.17	8.0	144	12.5	3.6	0.29	100	8	59	13.29	0.41	46.4	144	6/28/2010
Eight Mile Lake	5.9	2.1	0.36	80	72	81	4.65	0.52	73.5	138	14.9	10.6	0.71	80	14	74	94.74	0.92	230.4	138	6/29/2010
Delta Junction	4.8	1.9	0.40	100	44	31	4.28	0.61	100.4	138	11.7	5.8	0.49	100	6	34	33.23	0.70	131.2	138	6/23/2010
Kenai NWA	4.8	1.5	0.32	100	84	80	2.66	0.48	61.9	144	5.0	3.2	0.63	100	44	55	11.47	0.95	245.9	144	7/6/2010

The standard deviation (SD) is reported here because it was required to calculate $CV_{Traditional}$.
*- indicates that the semivariogram model did not reach its sill, which prevented sill, range, and nugget values from being accurately estimated.
†Sample size required to estimate to with 10% of the spatial mean with 90% confidence.
‡Month/day/year format.

Table 3. Correlation coefficients indicating covariation among many site characteristics.

	Latitude	Longitude	Mean annual air temperature	Mean annual precipitation	Elevation	Canopy height	Mean T_s	Mean SWC
Latitude	1	−0.808	−0.921	−0.288	−0.001	−0.211	−0.700	0.084
Longitude	−0.808	1	0.693	0.277	−0.173	0.309	0.654	−0.018
Mean annual air temperature	−0.921	0.693	1	0.284	−0.252	0.196	0.701	−0.061
Mean annual precipitation	−0.288	0.277	0.284	1	−0.271	0.412	0.005	0.025
Elevation	−0.001	−0.173	−0.252	−0.271	1	−0.154	−0.129	−0.325
Canopy height	−0.211	0.309	0.196	0.412	−0.154	1	0.012	−0.199
Mean T_s	−0.700	0.654	0.701	0.005	−0.129	0.012	1	−0.115
Mean SWC	0.084	−0.018	−0.061	0.025	−0.325	−0.199	−0.115	1

acquisition. Data at each measurement point were acquired at 1 s (execution interval), and descriptive statistics were calculated over 30 s averaging periods with a datalogger (CR3000, Campbell Scientific Inc.). The sampling design was similar among the sites, but it was not identical due to site-specific space constraints (*e.g.*, property boundaries) and obstacles (*e.g.*, rock outcrops, roads, streams, tree trunks, and stone walls). The measurements were taken along 2 to 4 intersecting transects at each site, with the number and length of transects depending on site topography and property size. The intersection point of the transects was typically at the center of each transect. The length of transects ranged from 42 m to 210 m. The starting, center and ending points of each transect were recorded by a GPS unit, which was used to calculate distances during semivariogram analysis.

The determination of sampling locations followed the cyclic sampling logic used by Bond-Lamberty *et al.* [23], with measurements taken at 0, 2, 8, 28, 38, and 42 m (noted as transect points) and with this spacing scheme repeated until the end of the transect was reached (*i.e.*, the next sampling points would occur at 44, 50, 70, 80, 84, 86, 92 … m). In addition, two more T_s and SWC measurements were made at −0.3 m and +0.3 m from each transect point along the axis of the transect, thus resulting in a total of 3 T_s and 3 SWC measurements made at each transect point (Figure S1). The minimum and maximum distance between measurements ranged from 0.3 m to at least 84 m at each site, and up to 210 m at some sites, as a result, we were able to capture both small spatial scale (~0.3 m) and the plot-to-footprint scale (~100's m^{-2}) variability. Measurements were taken at 134±1 (mean ±1 standard error, convention used throughout the text, unless otherwise noted) locations per site (maximum = 156, minimum = 111). The number of measurement points at each site exceeded the recommendation

Table 4. Statistical significance (p values) of site characteristics on mean T_s and SWC, as well as semivariogram properties and derived values.

	T_s					SWC				
	Mean	Range	Sill	Nugget	Sample size	Mean	Range	Sill	Nugget	Sample size
Latitude	<0.001	NS	NS	NS	NS	NS	NS	NS	NS	NS
Longitude	NS	NS	0.013	NS	0.003	NS	0.008	0.023	0.002	0.004
Elevation	NS	0.091	NS	NS	NS	0.032	NS	NS	NS	NS
Mean annual air temperature	NS	NS	0.003	NS	0.004	NS	NS	0.004	0.082	0.009
Mean annual precipitation	NS	NS	NS	NS	NS	0.007	NS	0.050	NS	0.002
Canopy height	NS	0.065	NS	0.024	0.087	NS	0.037	NS	NS	NS
Ecosystem type	<0.001	NS	NS	NS	NS	<0.001	NS	NS	0.015	0.071
Canopy structure	NS	0.019	<0.001	0.006	<0.001	NS	NS	0.053	<0.001	<0.001
Soil type	NS	0.009	NS	0.046	0.021	NS	NS	NS	0.026	NS
Mean T_s	NA	NS	0.012	NS	NS	0.032	NS	0.046	0.001	0.004
Mean SWC	NS	NS	NS	0.005	NS	NA	NS	<0.001	0.072	0.037
Model r^2	0.73	0.40	0.57	0.48	0.83	0.51	0.18	0.70	0.62	0.66

The r^2 of the statistical model (observed versus predicted) is also shown.
NS: not significant (*i.e.*, p>0.1 and factor was removed from the statistical model), NA: not applicable.

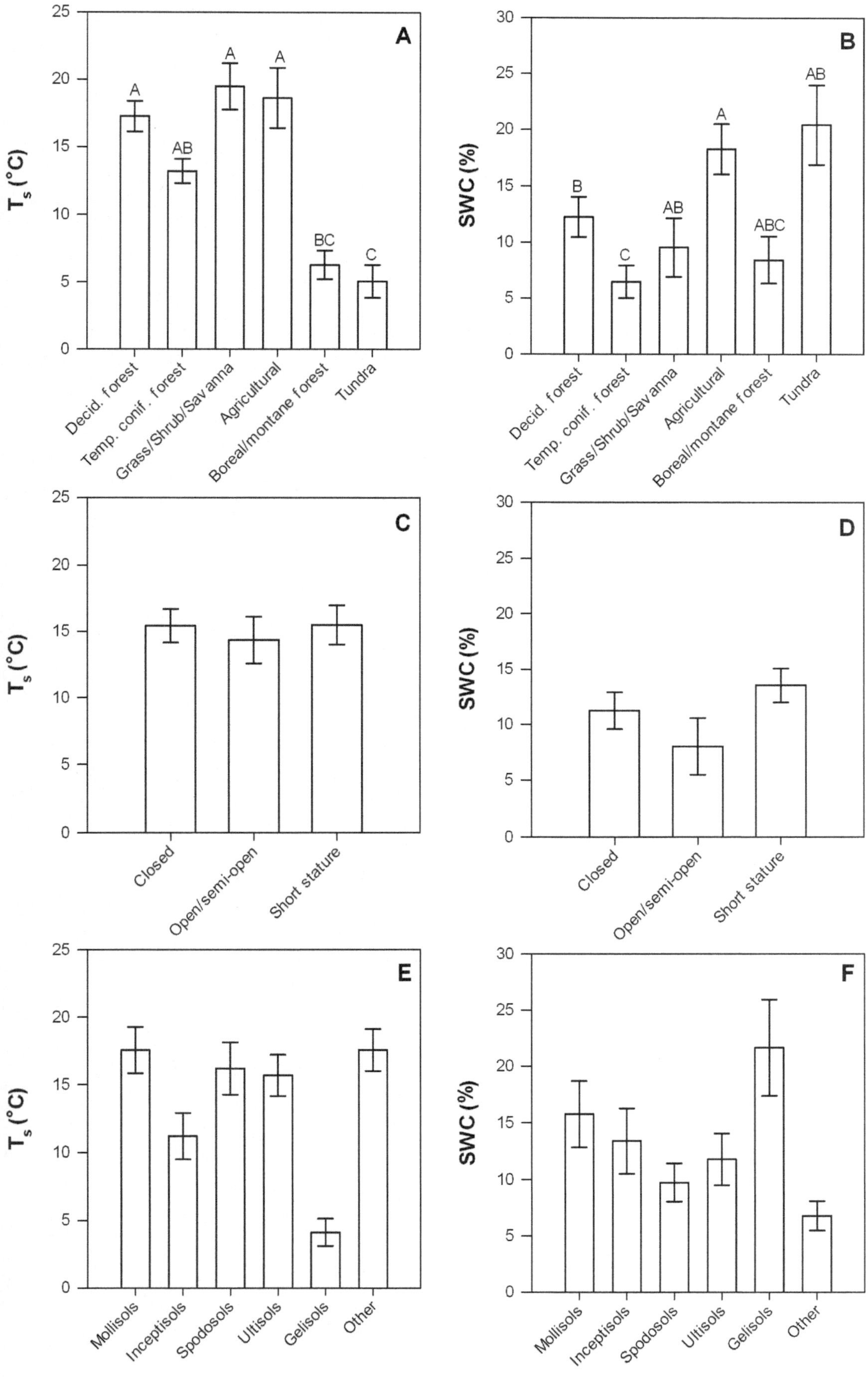
A
B
C
D
E
F
T_s (°C)
SWC (%)
0
5
10
15
20
25
30
A
AB
A
A
BC
C
B
C
AB
A
ABC
AB
Decid. forest
Temp. conif. forest
Grass/Shrub/Savanna
Agricultural
Boreal/montane forest
Tundra
Closed
Open/semi-open
Short stature
Mollisols
Inceptisols
Spodosols
Ultisols
Gelisols
Other

Figure 3. Mean T_s (left column) and SWC (right column) at each site in relation to ecosystem type (a, b), canopy structure (c, d), and soil type (e, f). Error bars represent ±1 SE. Bars with different letter were significantly different (p<0.05) based on Tukey HSD tests. The letter "A" corresponds to the largest least square mean(s) and subsequent letters (B, C, ...) correspond to progressively smaller least square means.

of at least 100 points to construct representative isotropic (*i.e.*, non-directional) semivariograms with soil data [53,54].

It typically took ~2.5 to 5 hours to collect the data at each site, during which time natural changes to T_s and SWC could occur, *e.g.*, solar heating of the soil, evaporation and/or transpiration. Thus, we also continuously collected T_s and SWC with a second, stationary suite of 3 pairs of identical T_s and SWC sensors (spaced 0.3 m apart) adjacent to the point where all the transects intersected to estimate diurnal and other weather-related influences on the measurements. Data at were acquired at 1 s (execution interval), and descriptive statistics were calculated over 30 sec averaging periods with a datalogger (Models CR1000, Campbell Scientific Inc.). T_s and SWC data from the transects and the stationary system, as well as measurement coordinates, that were used to create the semivariograms are available available by completing the request form at the bottom of the webpage: http://neoninc.org/pds/FIU007.php [55].

Semivariograms and model fitting

Two primary approaches were used to de-trend temporal changes from the data collected along the transects, such that the remaining variability in the data can be attributed to its spatial sources and quantified with a semivariogram. First, mean T_s and SWC values from the stationary location were subtracted from transect data at each corresponding time of day to produce *corrected* data. Second, a linear regression based on the relationship between time of day and the *corrected* T_s and SWC data was fitted, and the residuals calculated. These residuals were then used to construct the semivariogram. In the context of geostatistics, de-trending is driven by the need to satisfy the second-order stationarity requirement, which needs to be met in order to ensure that the resulting covariance function is valid, *e.g.*, [24]. At 5 sites (Sterling, Rocky Mountain NP, Central Plains Exp. Range, Caribou Poker, and Northcutt) the stationary system was either unavailable or malfunctioned. Therefore, de-trending was done using the linear regression approach only. If additional patterns (*i.e.*, gradients in T_s or SWC across the sampling site or non-normal distribution) were still visible in the data after removing temporal trends, we used a third method to de-trend the data based on topographic relief (elevation, aspect, and slope calculated from a digital elevation map) to meet the assumptions of the semivariogram model. This third method consisted of adding elevation, aspect, and slope in all combinations (*i.e.*, elevation×aspect×slope, elevation×aspect, elevation×slope, etc.) to the time of day linear regression described above. Linear regressions that were not significant (p>0.05) were excluded, and the residuals were recalculated. De-trending using elevation, slope, and/or aspect was necessary at 18 sites (Barrow Environmental Observatory, Poker Flats, Eight Mile Lake, Woodworth, Northern Great Plains Research Lab, Tree Haven, Steigerwaldt, Upper Teakettle, Konza – Core, UK Biological Station, Harvard Forest, Niwot Ridge, Smithsonian Conservation Biology Institute, Ordway-Swisher BS, Great Smoky NP, Talladega Nat'l Forest, Choctaw WMA, and Murray).

Data collected were used for geospatial analyses in the R statistical computing language with the geoR package [56,57]. At each site the empirical semivariogram, $\hat{\gamma}$(h), which is half the average squared difference between data pairs, was calculated using the following equation,

$$\hat{\gamma}(h) = \frac{1}{2N(h)} \sum_{a=1}^{N(h)} [z(u_a) - z(u_a + h)]^2 \qquad (1)$$

where, $z(u_\alpha)$, $\alpha = 1, 2, \ldots, n$ denotes the set of T_s or SWC data, u_α is the vector of spatial coordinates of the α_{th} observation, h represents a distance separating pairs of data, and N(h) is the number of data pairs separated by a given distance. In general, the covariance function for any pair of data that is h units apart can be represented as,

$$C(h) = s\gamma(h) \qquad (2)$$

where, s is a variance parameter and $\gamma(h)$ is any positive definite correlation function. The correlation function used in this work was a spherical model with the following form,

$$\hat{\gamma}(h) = \begin{cases} (1 - 1.5\left(\frac{h}{a}\right) + 0.5\left(\frac{h}{a}\right)^3, & if \quad h < a \\ 0 & , otherwise \end{cases} \qquad (3)$$

where, a is the semivariogram range. In addition, we allowed for a nugget term, which is simply added to the covariance function. Approximate values for the range, sill, and nugget were estimated from the experimental semivariogram to seed the model and best-fit values were subsequently estimated using weighted least squares. It is important to note that a key goal of this study is to provide meaningful comparisons of the results across a number of sites, and this comparability was weighed as more important than a potentially negligible difference in the fit of a model for varying functional forms of the variogram, i.e., same sources of uncertainty in the model fit for all sites. Based on these two constraints, we selected the functional form of the variograms (spherical) that provided the best fit at most of the sites and applied it to all sites for the sake of consistency.

Lag spacing was set to 1 m for every empirical semivariogram. Preliminary tests with T_s and SWC data from Eight Mile Lake, AK, and Klemme, OK, showed that changing the lag spacing to 1, 2, or 5 m did not alter the estimated range value in any systematic manner, and the change was ±9% on average, which corresponded to ±4 m. Similarly, the coefficient of variation estimated from the semivariogram model was strongly positively correlated with the coefficient of variation calculated via the traditional method (see *Coefficient of variation and sample size* subsection in the *Results* section), indicating that 1 m lag spacing allowed the model to accurately describe the variability in the dataset.

Calculating CVs and sample size

The coefficient of variation for T_s and SWC was calculated for each site in two ways: 1) the traditional method of dividing the standard deviation by the mean (referred to as $CV_{Traditional}$); and 2) using the sill to estimate standard deviation and dividing that value by the mean (referred to as CV_{Sill}) as follows,

$$CV_{sill} = \frac{\sqrt{2 \times \hat{S}}}{\bar{x}} \qquad (4)$$

Table 5. Site characteristics where the range value was larger than or less than the maximum lag distance covered by the semivariogram model, (mean ±1 SE). Range>maximum lag indicates that the semivariogram did not reach an asymptot.

	Latitude (°)	Longitude (°)	Elevation (m.a.s.l.)	Mean annual temperature (°C)	Mean annual precipitation (mm)	Canopy height (m)	Mean T_s (°C)	Mean SWC (%)	Maximum lag of semivariogram (m)	n
T_s										
Range<maximum lag	43±2	−106±3	741±111	9±1	861±80	12±2	15±1	12±1	94±1	53
Range>maximum lag	37±4	−90±6	759±367	12±3	899±109	13±4	19±2	10±2	87±3	7
P value	NS	NS	NS	NS	NS	NS	0.098	NS	NS	
SWC										
Range<maximum lag	42±2	−105±3	778±122	9±1	841±76	11±2	15±1	11±1	95±1	50
Range>maximum lag	42±4	−102±8	572±168	10±3	990±211	16±4	15±2	14±3	86±3	10
P value	NS	NS	NS	NS	NS	NS	NS	NS	0.024	

NS: not significant (*i.e.,* p>0.1 and factor was removed from the statistical model).

where, $\bar{x}$ represents the sample mean, and $\hat{S}$ represent the semivariogram sill.

We calculated the sample size (n) needed to estimate the mean T_s and SWC to within 10% of the spatial mean and with 90% confidence using [4],

$$n = \frac{(1.645)^2 \left(\frac{\sqrt{2 \times \hat{S}}}{\overline{X}} \right)}{(0.1)^2} \quad (5)$$

where, $\bar{x}$ represents the sample mean, and $\hat{S}$ represents the semivariogram sill. The constants 0.1 and 1.645 represent the accuracy requirement (*i.e.,* 10%) expressed as a proportion and the value of t statistic that corresponds to the 90% confidence interval, respectively. These accuracy and confidence thresholds were chosen because they are conservative enough to meet the goals of most studies. An assumption of the hypothesis test underlying this analysis is that the data are normally distributed, which we ensured prior to constructing the semivariograms.

Statistical analyses. Several of the hypotheses that we tested involved relating site characteristics to dependent variables. Some site characteristics (*i.e.,* factors in the statistical tests) were continuous variables (latitude, longitude, mean annual air temperature, mean annual precipitation, elevation, canopy height, mean T_s, and mean SWC), while others were categorical variables (ecosystem type, canopy structure, and soil type). Some levels of the categorical site characteristics contained too few sites to allow statistical analyses to be conducted. For example, Andisols were only found at 2 sites (Wind River and Abby Road, WA) and a tropical deciduous forest ecosystem only occurred at 1 site (Guanica, PR). As a result, levels were combined to increase the number of sites within each level prior to conducting the statistical tests. Ecosystem types were grouped into 6 levels: deciduous forests (both tropical and temperate), temperate coniferous forests, grasslands (incl. shrublands, and savannas), agricultural, boreal (including montane forests), and tundra. While we recognize that short stature ecosystems (*e.g.,* grasslands) can have closed canopies with only small gaps between neighboring plants or open canopies with substantial amounts of bare ground between plants, they clearly differ in structure from taller stature ecosystems. The ecosystem and canopy structure classifications were somewhat subjective and some sites could have been reasonably placed in two or more different groups, but the chosen category represents our best judgment. Canopy structure was grouped into 3 levels: closed canopy ecosystems, open and semi-open canopy ecosystems, and short stature ecosystems. Vegetation canopy heights were measured in the field and reflect the average height of the

Table 6. Correlation coefficients and statistical significance for T_s and SWC semivariogram properties and derived values.

	Spearman correlation coefficient	p value
T_s range – SWC range	0.26	0.089
T_s sill – SWC sill	0.36	0.016
T_s nugget – SWC nugget	0.34	0.021
T_s CV_{Sill} – SWC CV_{Sill}	0.59	<0.001
T_s sample size – SWC sample size	0.59	<0.001

tallest component of the plant community (*e.g.*, in a savanna ecosystem the canopy height reflects the average height of the trees, rather than the grasses). Soils were grouped according to soil orders: Inceptisols, Spodosols, Ultisols, Mollisols, Gelisols, and other (incl. Vertisols, Andisols, Alfisols, Entisols, Aridisols, as well as sites with >1 soil order and sites where the soil order was unknown).

In all statistical models described below, factors in the model that had a p value of >0.1 were removed in a stepwise manner and the model was re-run until all remaining factors had a p value of <0.1. Statistical significance was assumed at p<0.05 for all tests. Post-hoc Tukey HSD tests were performed when ANOVAs identified a significant effect of the categorical factor(s).

We expected that many of the site characteristics would covary (*e.g.*, latitude and mean annual air temperature) and it is necessary to understand this covariation when interpreting the statistical tests that we performed. To assess covariation in the explanatory variables used in the statistical tests, we calculated correlation coefficients for each combination of site characteristics (latitude, longitude, mean annual air temperature, mean annual precipitation, elevation, canopy height, mean T_s, and mean SWC). Categorical site characteristic factors (*i.e.*, ecosystem type, canopy structure, and soil type) were excluded from this analysis. Stepwise removal of non-significant factors from statistical tests tends to remove factors that covary with one-another, therefore, interpretation of a significant effect caused by a site characteristic should also consider other covarying site characteristics.

At some sites it was not possible to estimate the range or sill because the semivariogram did not reach an asymptote (see *Semivariograms* subsection in the *Results* section); therefore these sites were excluded from further statistical analysis. We tested whether these sites had different characteristics than sites where the semivariograms did reach an asymptote to determine whether this could influence the interpretation of our results. To achieve this, nominal logistic models were performed in JMP (SAS Institute Inc., Cary, NC) to determine whether significant differences in latitude, longitude, elevation, mean annual air temperature, mean annual precipitation, canopy height, mean T_s, mean SWC, and maximum semivariogram lag distance existed between sites where semivariograms did and did not reach an asymptote. Categorical site characteristic (*i.e.*, ecosystem type, canopy structure, and soil type) were excluded from this analysis due to the limited number of sites within each level of the categories.

ANOVAs were performed in JMP to determine the relationship between site characteristics and T_s and SWC. The factors in the ANOVAs were latitude, longitude, elevation, mean annual air temperature, mean annual precipitation, canopy height, ecosystem type, canopy structure, and soil type and the dependent variables were mean T_s and mean SWC. Both mean T_s and mean SWC were $\log_{10}$ transformed to meet assumptions of normality and homogeneity.

ANOVAs were also used to test hypotheses 1–4. ANOVAs were performed in JMP to assess the relationship between site characteristics and semivariogram range, nugget, and sill, as well as estimated sample size required to meet the accuracy requirement for T_s and SWC. Semivariogram nugget data for T_s and SWC met the assumptions of normality and homogeneity; however, range, sill, and sample size data required $\log_{10}$ transformations to meet these assumptions.

Because some factors often explained the majority of variation in an ANOVA, relatively weak relationships that were statistically significant were sometimes masked by the dominant factors when plotted as graphs. This is because the ANOVA accounts for variability in the data caused by others factors, whereas the graphs,

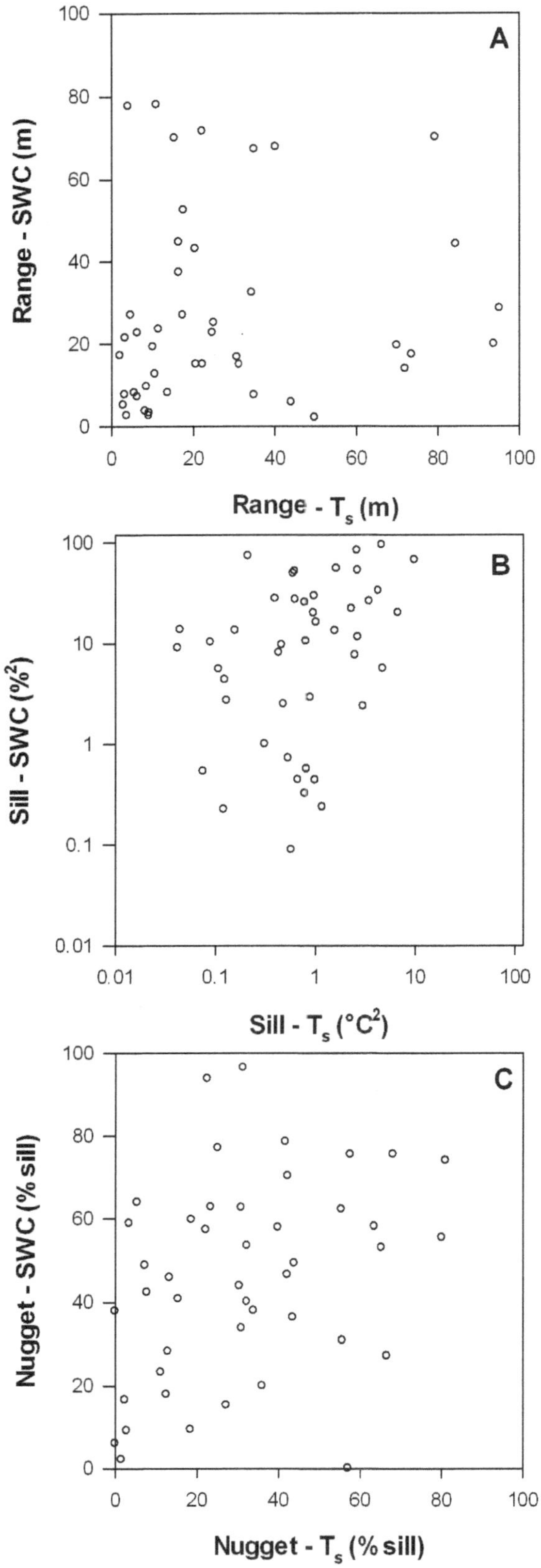

Figure 4. Semivariogram range (a), sill (b), and nugget (c) for T_s versus SWC at the sites. Sill values are plotted on a log scale due to large differences in magnitude across the sites.

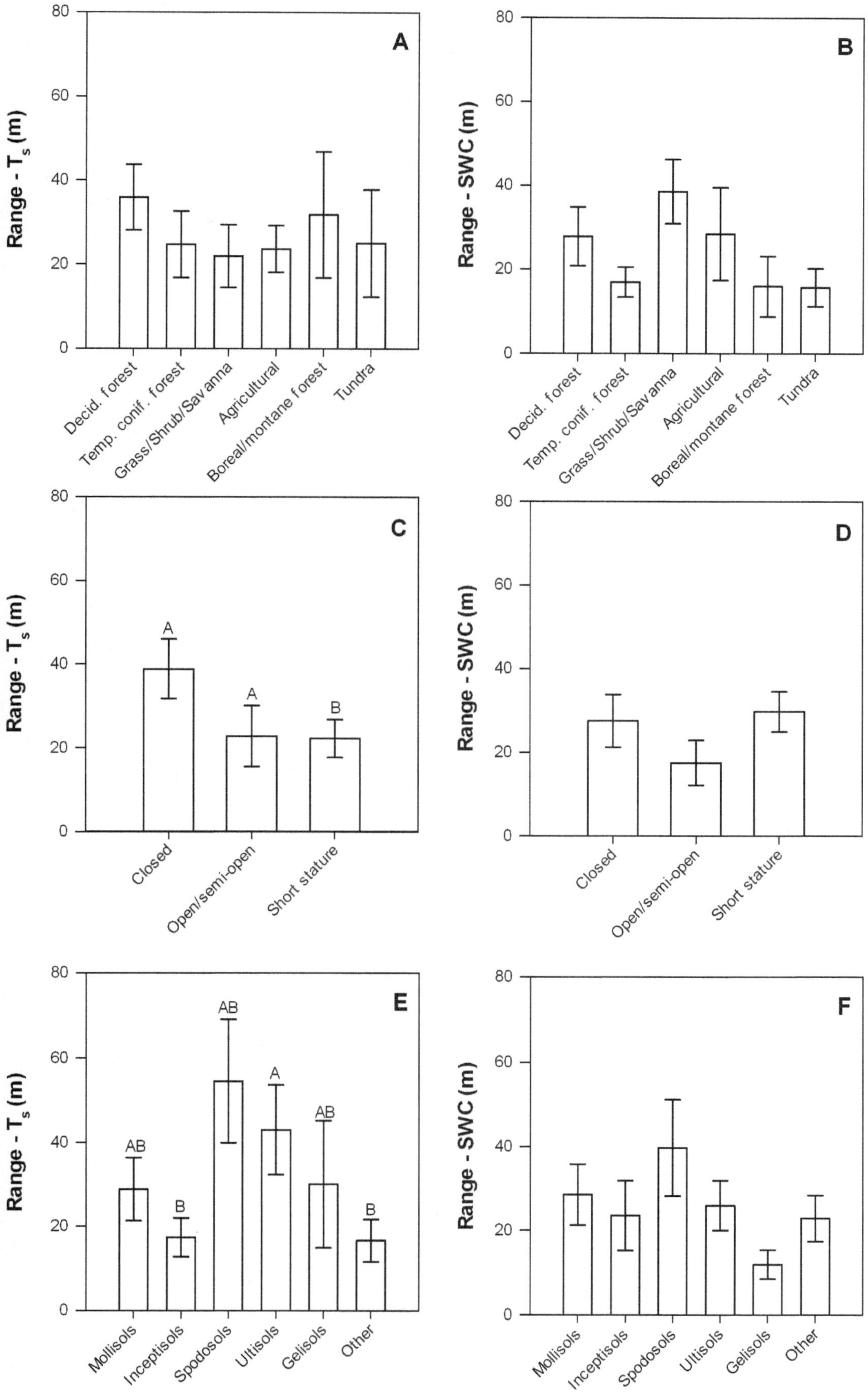
A
B
C
D
E
F
Range - T_s (m)
Range - SWC (m)
0
20
40
60
80
Decid. forest
Temp. conif. forest
Grass/Shrub/Savanna
Agricultural
Boreal/montane forest
Tundra
Closed
Open/semi-open
Short stature
Mollisols
Inceptisols
Spodosols
Ultisols
Gelisols
Other
A
A
B
AB
B
AB
A
AB
B

Figure 5. Semivariogram range for T_s (left column) and SWC (right column) at each site in relation to ecosystem type (a, b), canopy structure (c, d), and soil type (e, f). Error bars represent ±1 SE. Bars with different letter were significantly different ($p<0.05$) based on Tukey HSD tests. The letter "A" corresponds to the largest least square mean(s) and subsequent letters (B, C, …) correspond to progressively smaller least square means.

which are based on raw data, do not. This should be taken into account when comparing the statistical results to the graphs.

A Spearman's rank correlation was performed in JMP to assess the relationships between mean T_s and mean SWC to determine whether they were significantly correlated. Additional Spearman's rank correlations were performed in JMP to assess the relationships among T_s range and SWC range, T_s nugget and SWC nugget, T_s sill and SWC sill, and T_s sample size and SWC sample size (Hypothesis 5); and T_s range, nugget, and sill, as well as SWC range, nugget, and sill (Hypothesis 6).

Paired one sample *t*-tests (JMP) were used to assess whether range, nugget, sill, and sample size values differed between T_s and SWC among the sites to test Hypothesis 6. In addition, to test whether the semivariograms accurately described the variability in T_s and SWC at each site, linear regressions (JMP) were used to assess the relationship between $CV_{Traditional}$ and CV_{Sill} for both T_s and SWC.

The range and sample size for T_s and SWC were plotted against the cumulative proportion of sites to investigate the frequency distribution. These relationships were modeled with a 3-parameter logistic curve using SigmaPlot (Systat Software Inc., Chicago, Illinois).

In the interest of brevity, we only present graphs relating to a subset of site characteristics in this paper, namely ecosystem type, canopy structure, and soil order. The relationship between dependent variables and other site characteristics, including latitude, longitude, elevation, mean annual air temperature, mean annual precipitation, mean T_s, and mean SWC, are presented in the Appendices.

Results

Site characteristics

The sites encompassed a continental-scale range of eco-climatic properties (Table 1, Figure 1), and were broadly representative of the proportion of US land in different latitudinal, longitudinal, and elevational categories, as well as spanning a wide range of soil types, climate space, and soil temperature and moisture conditions at the time of sampling (Figure 2). T_s varied from 2.1±0.1°C (mean ±1SE) at Pump Station 2, AK, to 30.8±0.2°C at Santa Rita Exp. Range, AZ, while SWC varied from 1.1±0.1% at Burlington, MA, to 37.8±0.7% at Ponce, PR (Table 2). T_s and SWC estimates were not correlated with each other among the sites ($p=0.143$), resulting from several sites with contrasting T_s and SWC combinations (*i.e.*, warm and wet, warm and dry, cool and wet, and cool and dry, Figure 2f).

As expected, correlations demonstrated a substantial amount of covariation in the site characteristics (Table 3). For example, mean annual air temperature, mean T_s, and longitude were all strongly negatively correlated with latitude. Weaker correlations were observed between mean annual precipitation and canopy height, and elevation and mean SWC. These correlations should be considered when interpreting the results, since the stepwise removal of non-significant factors from the ANOVA models tends to remove factors that covary from the model. For example, a significant relationship between longitude and a dependent variable would likely also be significant for latitude, mean annual air temperature, or mean T_s if longitude had not been included as a factor.

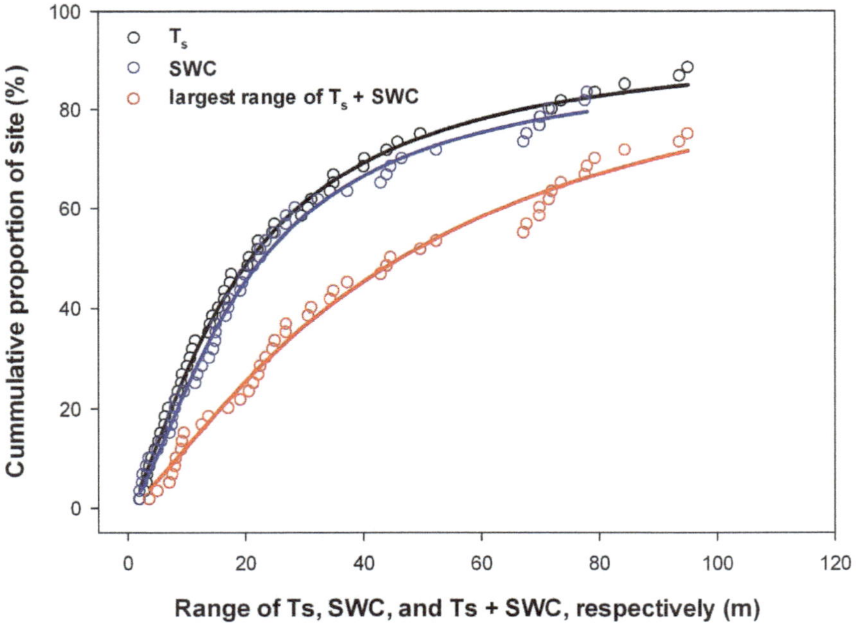

Figure 6. Semivariogram range for T_s, SWC, and the largest range for T_s or SWC at each site versus the cumulative proportion of sites. A logistic curve was fitted to the data with R^2 of 1.00, 0.99, and 0.99 for T_s, SWC, and T_s and SWC, respectively. Equations for the curves can be found in the results section (Eq. 6a–c).

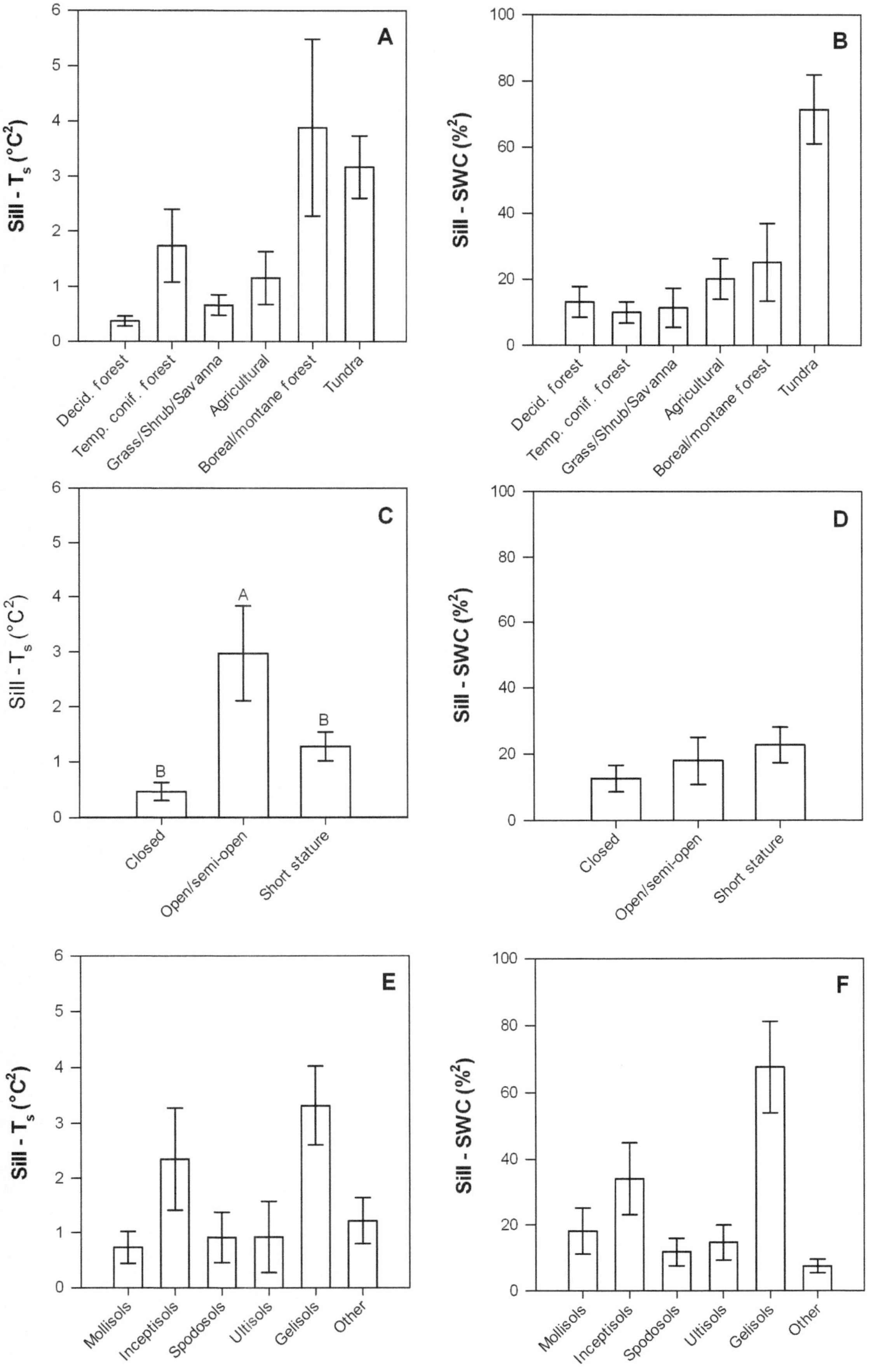
A
Sill - T_s (°C²)
Decid. forest
Temp. conif. forest
Grass/Shrub/Savanna
Agricultural
Boreal/montane forest
Tundra
B
Sill - SWC (%²)
C
A
B
B
Closed
Open/semi-open
Short stature
D
E
Mollisols
Inceptisols
Spodosols
Ultisols
Gelisols
Other
F

Figure 7. Semivariogram sill for T_s (left column) and SWC (right column) at each site in relation to ecosystem type (a, b), canopy structure (c, d), and soil type (e, f). Error bars represent ±1 SE. Bars with different letter were significantly different ($p<0.05$) based on Tukey HSD tests. The letter "A" corresponds to the largest least square mean(s) and subsequent letters (B, C, ...) correspond to progressively smaller least square means.

Mean T_s among sites was negatively related to latitude and differed among ecosystems, *i.e.*, not surprisingly being higher in temperate ecosystems, and lower in montane and high-latitude ecosystems (Table 4, Figure 3). Mean T_s among sites followed expected relationships to mean annual air temperature, elevation, and soil type (*e.g.*, being lower in Gelisols, which are associated with cold climates; Figure 3), but the effects were not significant due to covariation among these factors and the factors that were statistically significant in the model (*i.e.*, latitude and ecosystem type). Mean SWC was positively related to mean total annual precipitation, decreased with increasing elevation and mean T_s, and was significantly related to ecosystem type, being largest in tundra ecosystems, which have low evapotranspiration rates, and in agricultural ecosystems, several of which were irrigated, and the lowest in temperate coniferous forests (Table 4, Figure 3).

Semivariograms

In 12% and 17% of cases the semivariograms for T_s and SWC, respectively, had not reached an asymptote at the maximum lag distance used to construct the modeled semivariogram (Table 2). As a result, sill, range, and nugget (expressed as a percent of sill) values for these semivariograms were based on extrapolation, which may not accurately represent the *true* values. At 75% of sites both the T_s and SWC semivariogram reached an asymptote, at 13% of sites only the T_s semivariogram reached an asymptote, at 8% of sites only the SWC semivariogram reached an asymptote, and at the remaining 3% of sites no asymptote was reached for T_s or SWC. Sites with semivariograms that did not reach an asymptote were not different from other sites in terms of their environmental characteristics (*i.e.*, latitude, longitude, elevation, mean annual air temperature, mean annual precipitation, canopy height, mean soil temperature, or mean soil moisture). However, the SWC semivariograms that did not reach an asymptote were associated with shorter maximum lag distances (Table 5). Unless stated otherwise, all subsequent results are based solely on estimated sill, range, and nugget values from semivariograms that reached an asymptote. Since there were no significant differences in environmental characteristics between sites where the semivariogram did and did not reach an asymptote, there is no evidence that excluding sites where the semivariogram did not reach an asymptote introduced a systematic bias into the interpretation of our results.

Semivariogram range

The mean range value was 27±4 m and 26±3 m for T_s and SWC, respectively. A paired *t*-test demonstrated that T_s range values did not significantly differ from SWC range values ($p=0.773$). Despite the similarity in the mean range, range values for T_s were not correlated with range values for SWC (Table 6; Figure 4a). At some sites, the T_s and SWC range values were similar, but at many sites one range value was substantially larger than the other.

Range values for T_s were significantly influenced by canopy structure and soil type, while those for SWC were significantly related to longitude and canopy height (Table 4, Figure 5). T_s range values were lower in short stature ecosystems than those in closed canopy or open and semi-open canopy ecosystems, and also lower in Inceptisols and "other" soils than in Ultisols (Figure 5). SWC range values increased with longitude (*i.e.*, increased from west to east) and decreased with canopy height (Table 4).

The relationship between the T_s range and the cumulative proportion of sites exhibited a logistic pattern (Figure 6). For example, 88%, 75%, and 55% of sites had a range value for T_s of less than 100, 50, and 25 m, respectively. Similar patterns were observed for SWC range values, as well as the largest range value for T_s or SWC (*i.e.*, whichever was largest at a site; Figure 6). The equations that described these relationships were,

$$\text{Soil temperature}: \quad y=\frac{94.3705}{\left[1+(x/19.1852)^{-1.3735}\right]} \tag{6a}$$

$$\text{Soil moisture}: \quad y=\frac{90.0096}{\left[1+(x/19.801)^{-1.4759}\right]} \tag{6b}$$

$$\text{Soil temperature/moisture}: \quad y=\frac{100.4066}{\left[1+(x/46.6724)^{-1.2824}\right]} \tag{6c}$$

where, x is the range distance expressed in meters, and y is the percent of sites that had smaller ranges than x.

Semivariogram sill

Sill values for T_s and SWC were positively related to one another, suggesting that underlying site properties, in part, mediate the spatial variability in both variables (Table 6, Figure 4b). The sill values for T_s were negatively related to longitude (*i.e.*, decreasing from west to east), and mean annual air temperature, positively related to mean T_s, and higher for open and semi-open ecosystems than for closed canopy or short stature ecosystems (Table 4, Figure 7). The sill values for SWC were also negatively related to longitude and mean annual air temperature, and positively related to mean T_s as well as mean SWC (Table 4). Notably, neither ecosystem type nor soil type were significantly related to sill values for T_s or SWC (Table 4).

Semivariogram nugget

The nugget was expressed as a percent of the sill and indicates the proportion of variation that occurred at scales of <1 m. The nugget ranged from 0–81% for T_s and 0–96% for SWC, and a paired *t*-test demonstrated that the mean T_s nugget (32±3%) was significantly different from the mean SWC nugget (44±3%, $p<0.001$). Similar to the sill values, T_s nugget values were positively correlated with SWC nugget values, indicating that underlying site characteristics mediate sub-meter variability of both parameters (Table 6, Figure 4c).

Nugget values for T_s were negatively related to canopy height and mean SWC, and were higher in closed canopy ecosystems than short stature or open/semi-open ecosystems (Table 4, Figure 8). There was also a significant effect of soil type on T_s nugget values, however, a subsequent post-hoc test did not reveal significant differences among soil types (Figure 8). Nugget values for SWC were positively related to longitude (*i.e.*, increasing from

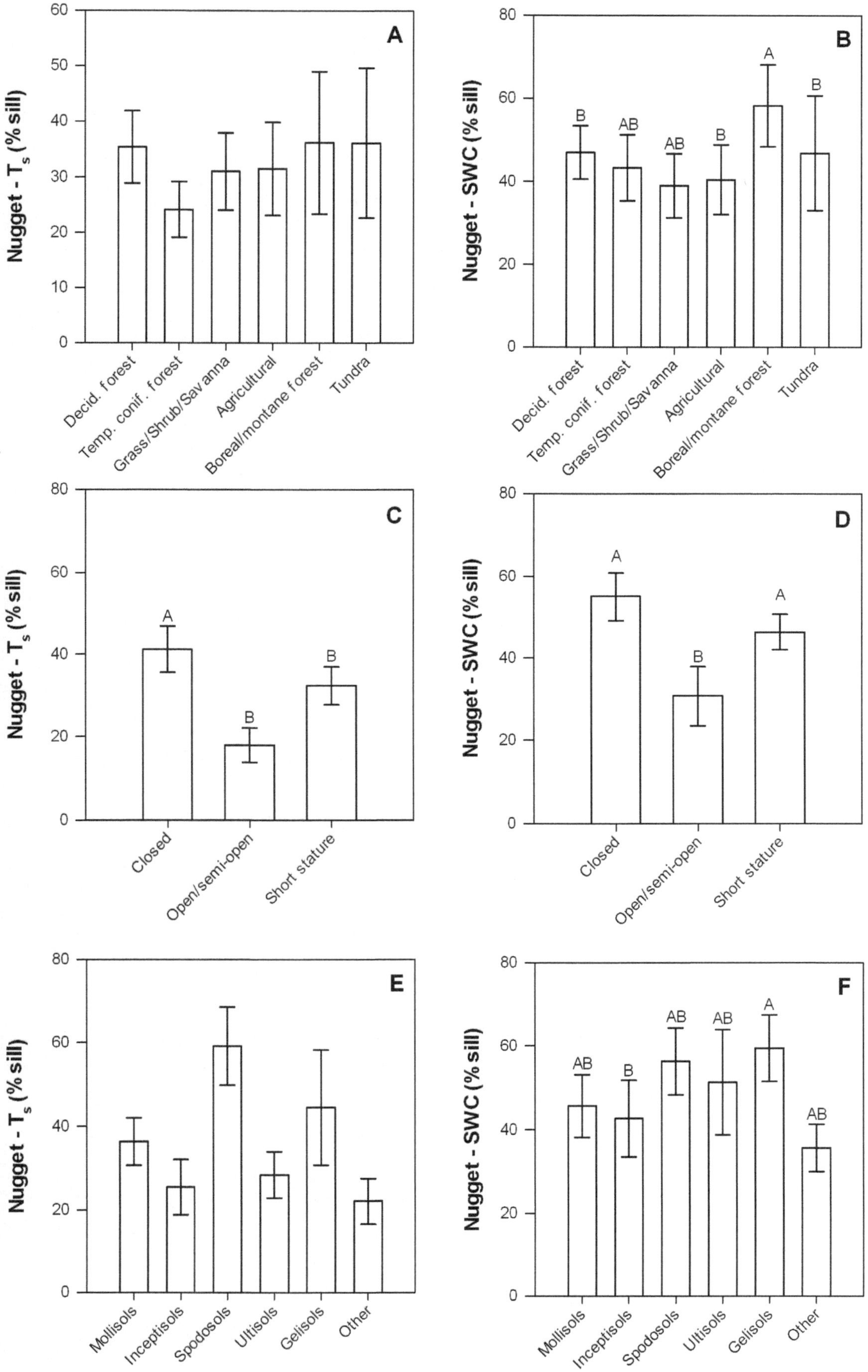
A
Nugget - T_s (% sill)
0
10
20
30
40
50
60
Decid. forest
Temp. conif. forest
Grass/Shrub/Savanna
Agricultural
Boreal/montane forest
Tundra
B
Nugget - SWC (% sill)
0
20
40
60
80
B
AB
AB
B
A
B
Decid. forest
Temp. conif. forest
Grass/Shrub/Savanna
Agricultural
Boreal/montane forest
Tundra
C
Nugget - T_s (% sill)
0
20
40
60
80
A
B
B
Closed
Open/semi-open
Short stature
D
Nugget - SWC (% sill)
0
20
40
60
80
A
B
A
Closed
Open/semi-open
Short stature
E
Nugget - T_s (% sill)
0
20
40
60
80
Mollisols
Inceptisols
Spodosols
Ultisols
Gelisols
Other
F
Nugget - SWC (% sill)
0
20
40
60
80
AB
B
AB
AB
A
AB
Mollisols
Inceptisols
Spodosols
Ultisols
Gelisols
Other

Figure 8. Semivariogram nugget expressed as a percent of the sill for T_s (left column) and SWC (right column) at each site in relation to ecosystem type (a, b), canopy structure (c, d), and soil type (e, f). Error bars represent ±1 SE. Bars with different letter were significantly different (p<0.05) based on Tukey HSD tests. The letter "A" corresponds to the largest least square mean(s) and subsequent letters (B, C, ...) correspond to progressively smaller least square means.

west to east), negatively related to mean T_s, as well as being significantly related to ecosystem type, canopy structure, and soil type (Table 4, Figure 8). SWC nugget values were larger in forest ecosystems (particularly boreal/montane forests), and lower in tundra ecosystems, as well as being larger in closed canopy and short stature ecosystems and lower in open/semi-open canopies ecosystems (Figure 8). SWC nugget values were larger for Gelisols than for Inceptisols (Figure 8).

Relationships among the semivariogram sill, range, and nugget

Range values were positively related to nugget values (expressed as a percent of the sill) for the T_s semivariograms (Table 7, Figure 9a). However, there was no relationship between range and sill values or nugget and sill values for T_s (Table 7). Likewise, range, sill, and nugget values were all unrelated to one another for the SWC semivariograms (Table 7, Figure 9b).

The relative abundance of different semivariogram shapes was assessed by assigning them to four categories: small nugget (<50% sill) and small range (<50 m); large nugget (>50% sill) and small range (<50 m); small nugget (<50% sill) and large range (>50 m); and large nugget (>50% sill) and large range (>50 m). For T_s the most common combination was a small nugget and small range, which accounted for 37 of the 60 sites (62%; Figure 9c). Small nugget and range values were also the most common combination for SWC, accounting for 24 of the 60 sites (40%). In addition, 18 of the sites (30%) had SWC semivariograms with a large nugget combined with a small range value. For both T_s and SWC, it was relatively rare to observe a large range value with either a small or large nugget value (Figure 9c).

Coefficient of variation and sample size

The two estimates of the coefficient of variation ($CV_{Traditional}$ and CV_{Sill}) were strongly and positively correlated with one another (T_s, $r^2=0.96$, $p<0.001$; SWC, $r^2=0.95$, $p<0.001$; Figure 10), demonstrating that the semivariograms accurately described the pattern of spatial variability among the sites. However, $CV_{Tradtional}$ values were typically smaller than corresponding CV_{Sill} values for both T_s and SWC ($p<0.001$). This occurred because the traditional method of calculating CV does not account for spatial correlation in data, yet spatial correlation was ubiquitous at our sampling sites (data not shown). Across all sites, CV_{Sill} values were 24±3% and 44±2% larger than $CV_{Tradtional}$ values for T_s and SWC, respectively.

The CV_{Sill} values ranged from 0.02 to 0.88, with a mean of 0.16±0.03, and 0.09 to 1.22 with a mean of 0.51±0.04 for T_s and SWC, respectively (Table 2). A paired *t*-test demonstrated that T_s CV_{Sill} values significantly differed from SWC CV_{range} values ($p<0.001$), demonstrating that SWC was more variable across space. Similar to the sill values, there was a positive correlation between CV_{Sill} values for T_s and SWC (Table 6, Figure 10), again suggesting that variability in these parameters may be controlled by the same site characteristics.

We used a sample size analysis based on CV_{Sill} (Eq. 5) to calculate the number of samples required to estimate T_s and SWC to within 10% of the mean with 90% confidence (Table 2). We used CV_{Sill} rather than $CV_{Traditional}$ to calculate sample size because $CV_{Traditional}$ underestimated variability (Figure 10), which in turn would cause the sample size to be underestimated. For example, at Harvard Forest the sample size required to estimate SWC to within 10% of the mean with 90% confidence was 127 when calculated with $CV_{Traditional}$ (0.69), but was 204 when calculated with CV_{Sill} (0.87). Across all the sites, the sample size calculated using $CV_{Traditional}$ underestimated the sample size calculated with CV_{Sill} by a factor of 1.6±0.1 (maximum: 2.6) and 2.1±0.1 (maximum: 3.0) for T_s and SWC, respectively. Hereafter, we focus on sample sizes calculated using CV_{Sill}.

Since we used the sill to represent the variance in the sample size analysis, the number of samples required assumes that the sample spacing will be larger than or equal to the range value from the semivariogram (*i.e.*, the distance at which the sill is reached). In addition, because the sample size was based on CV_{Sill}, both parameters (CV_{Sill} and sample size) exhibited similar patterns. For example, the sample size analyses indicated that the number of samples necessary to meet the accuracy requirement varied from 1 to 211 for T_s (mean of 20±7) and from 2 to 405 for SWC (90±13), and a paired *t*-test demonstrated that sample size was significantly larger for T_s than SWC ($p<0.001$), which is similar to findings for CV_{Sill}. In addition, the sample size required for T_s was positively correlated to the sample size required for SWC (Table 6, Figure 11).

The number of samples required to estimate T_s to within 10% of the mean with 90% confidence was negatively related to longitude (*i.e.*, decreasing from west to east) and mean annual air temperature, and was also influenced by canopy structure and soil types (Table 4, Figure 12). More samples were required in open/semi-open canopy ecosystems than other ecosystems, and in Gelisols than in Mollisols (Figure 12). At 87% of sites, 20 or fewer samples were sufficient to meet the accuracy requirement for T_s. The remaining 13% of sites that required more samples were all Alaskan sites that shared a number of similarities, including high latitudes, westerly longitudes, low mean annual air temperature (<2°C), low mean T_s at the time of sampling (<6°C), low mean annual precipitation (<500 mm), tundra or boreal forest ecosystem types, primarily Gelisols or Inceptisols, low to mid elevations (<1000 m.a.s.l.), and low to mid-vegetation canopy heights (<15 m; Figure 12).

Table 7. Spearman correlation coefficients and statistical significance for T_s and SWC semivariogram properties.

	Correlation coefficient	P value
T_s		
Range - nugget	0.53	<0.001
Range - sill	−0.07	0.628
Nugget - sill	−0.15	0.278
SWC		
Range - nugget	0.10	0.492
Range - sill	0.03	0.863
Nugget - sill	−0.17	0.237

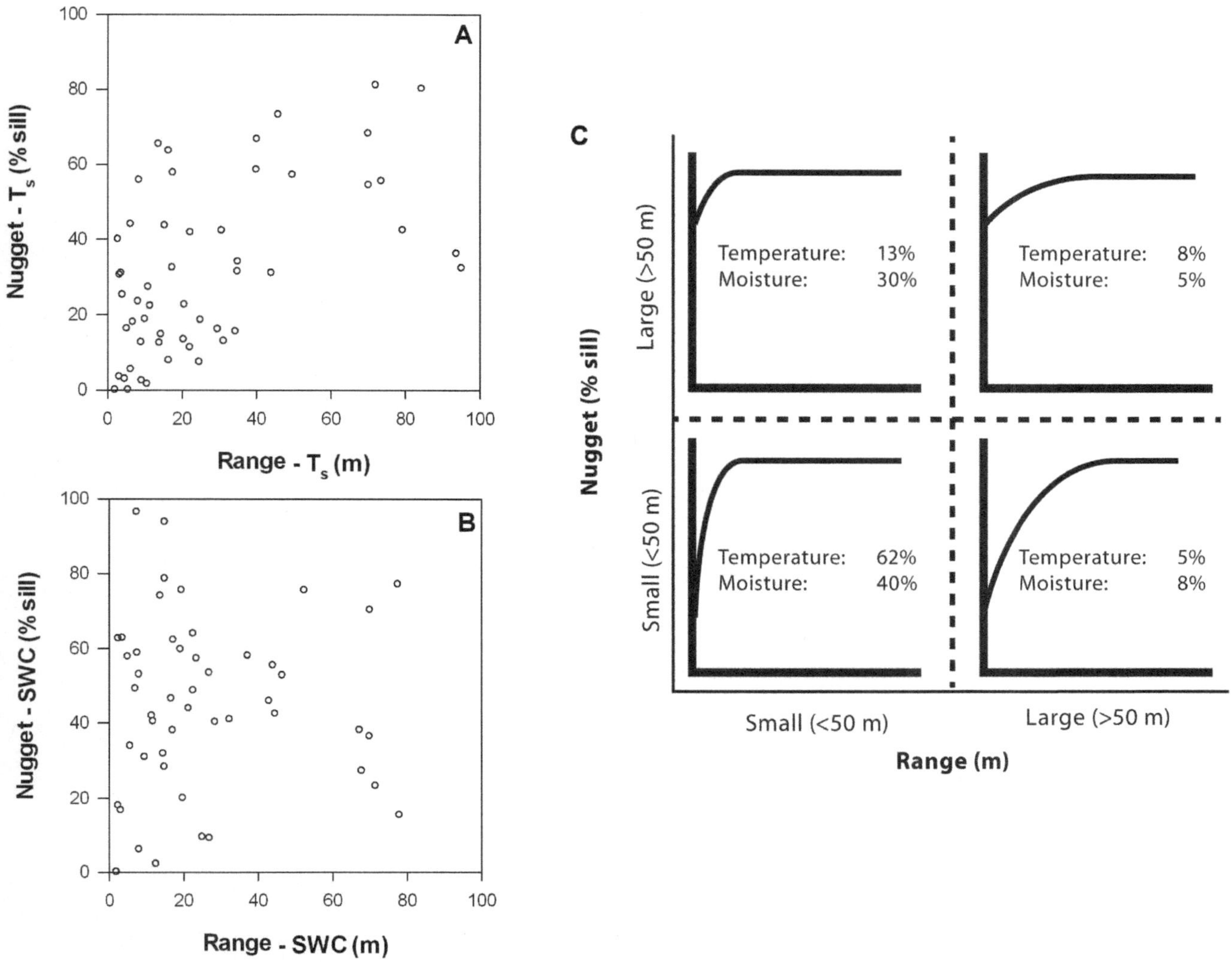

Figure 9. Semivariogram range versus nugget for T_s (a) and SWC (b), and the relative occurrence of semivariograms with small (>50 m) or large (>50 m) range values and small (<50% sill) or large (>50% sill) nugget values. The percentages in (c) are based on all 60 sites, including the sites where range values exceeded the maximum distance of the semivariogram model (*i.e.*, 7 sites (12%) for T_s and 10 sites (17%) for SWC). As a result, the percentages do not add up to 100%. Since the semivariograms that did not reach an asymptote all had range values that exceeded 50 m, the percentages in the right two quadrants are underestimated by a total of 12 and 17 percentage points for T_s and SWC, respectively.

The number of samples required to estimate SWC to within 10% of the mean with 90% confidence was positively related to mean annual precipitation and mean T_s, negatively related to longitude, mean annual air temperature, mean SWC, and was also influenced by canopy structure (Table 4, Figure 12). More samples were required to meet the accuracy requirement in closed and open/semi-open canopy ecosystems than short stature ecosystems (Figure 12). Unlike T_s, sites that required a large number of samples (*i.e.*, >100) to accurately measure SWC did not share many similar characteristics, although none had a mean soil moisture of >16% (data not shown).

When the required sample size for T_s was plotted against the cumulative proportion of sites the relationship could be accurately described by a logistic equation (Figure 13). For example, based on the sample size analyses, one sample was sufficient to estimate T_s to within 10% of the spatial mean with 90% confidence at 45% of sites, while 10 samples was sufficient to meet this requirement at 79% of sites. A similar pattern emerged for SWC sample size plotted against the cumulative proportion of sites, although at least an order of magnitude more samples were required than for T_s (Figure 13). In addition, the pattern was similar for the number of samples required for T_s or SWC (whichever was largest at a site; Figure 13). The equations that described these relationships were,

$$\text{Soil temperature}: \quad y = \frac{94.3653}{\left[1+(x/1.2306)^{-0.9435}\right]} \qquad (7a)$$

$$\text{Soil moisture}: \quad y = \frac{95.9175}{\left[1+(x/49.799)^{-1.8399}\right]} \qquad (7a)$$

$$\text{Soil temperature/moisture}: \quad y = \frac{99.5316}{\left[1+(x/56.6125)^{-1.4619}\right]} \qquad (7c)$$

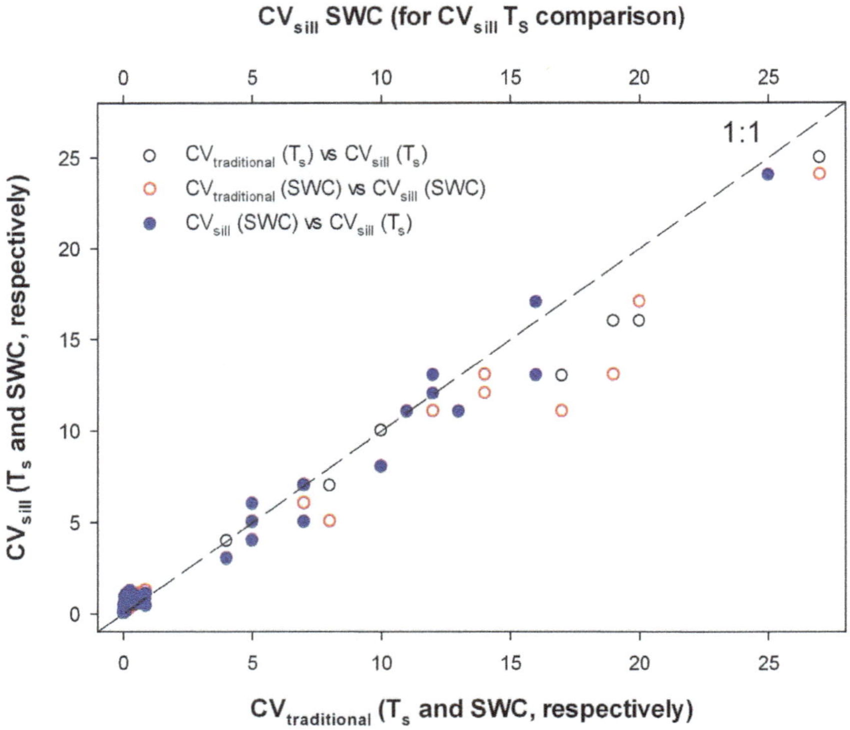

Figure 10. Coefficient of variation calculated by dividing the standard deviation by the mean ($CV_{Traditional}$) versus coefficient of variation calculated using the semivariogram sill (CV_{Sill}) for T_s, SWC, and CV_{Sill} for T_s versus SWC. Dotted line is 1:1.

where, x represents a number of samples and y is the percent of sites that require x (or fewer) samples to estimate the spatial mean to within 10% of the mean with 90% confidence.

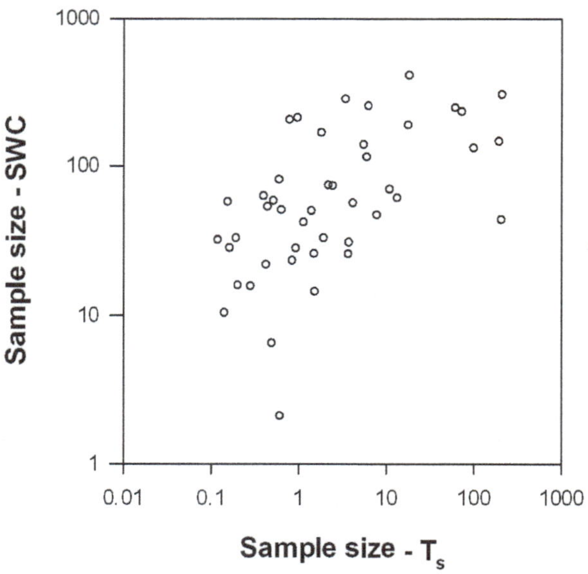

Figure 11. Number of samples required to measure T_s to within 10% of the spatial mean with 90% confidence versus the number of samples required to meet the same accuracy requirement for SWC. Values are plotted on a semi-log scale due to large differences in magnitude across the sites.

Efficacy of sampling strategies

To estimate the mean T_s across the area we sampled at a particular site with the fewest samples, sample spacing should equal or exceed the range value estimated from the T_s semiovariogram at that site. This is because samples spaced this distance apart are effectively independent, thereby providing the maximum amount of information from each individual sample. In contrast, if the sample spacing was less than the range value then adjacent samples would be correlated (*i.e.*, each sample would provide less than the maximum amount of information), which meant that more samples would be needed to accurately estimate the spatial mean of T_s. We emphasize that using the semivariogram range to inform sample spacing is appropriate here because we are only constraining the sampling to accurately estimate the mean. However, this may not be appropriate for studies with different goals.

Since Eq. 6a describes the proportion of sites with a soil temperature range less than a given value and Equation 7a describes the proportion of sites where a given sample size would be required to estimate soil temperature to within 10% of the mean with 90% confidence, we created a matrix that outlines the efficacy of different sampling strategies based on these two equations (Figures 14–16). For example in Figure 14, two temperature measurements spaced 5 m apart would only be sufficient to estimate mean T_s to within 10% of the mean with 90% confidence over ~1 ha (the approximate size of the area we sampled) at 7% of our sites. Interestingly, increasing the number of samples while keeping sample spacing at 5 m barely improved the efficacy of the sampling strategy and this occurs for two reasons, i) because spatial variation in T_s is often low (Figure 10), so few samples are needed at most sites (Figure 13), and ii) because only 12% of sites had a range value of 5 m or less (Figure 6). However, if the sample spacing were increased to 100 m, just two samples

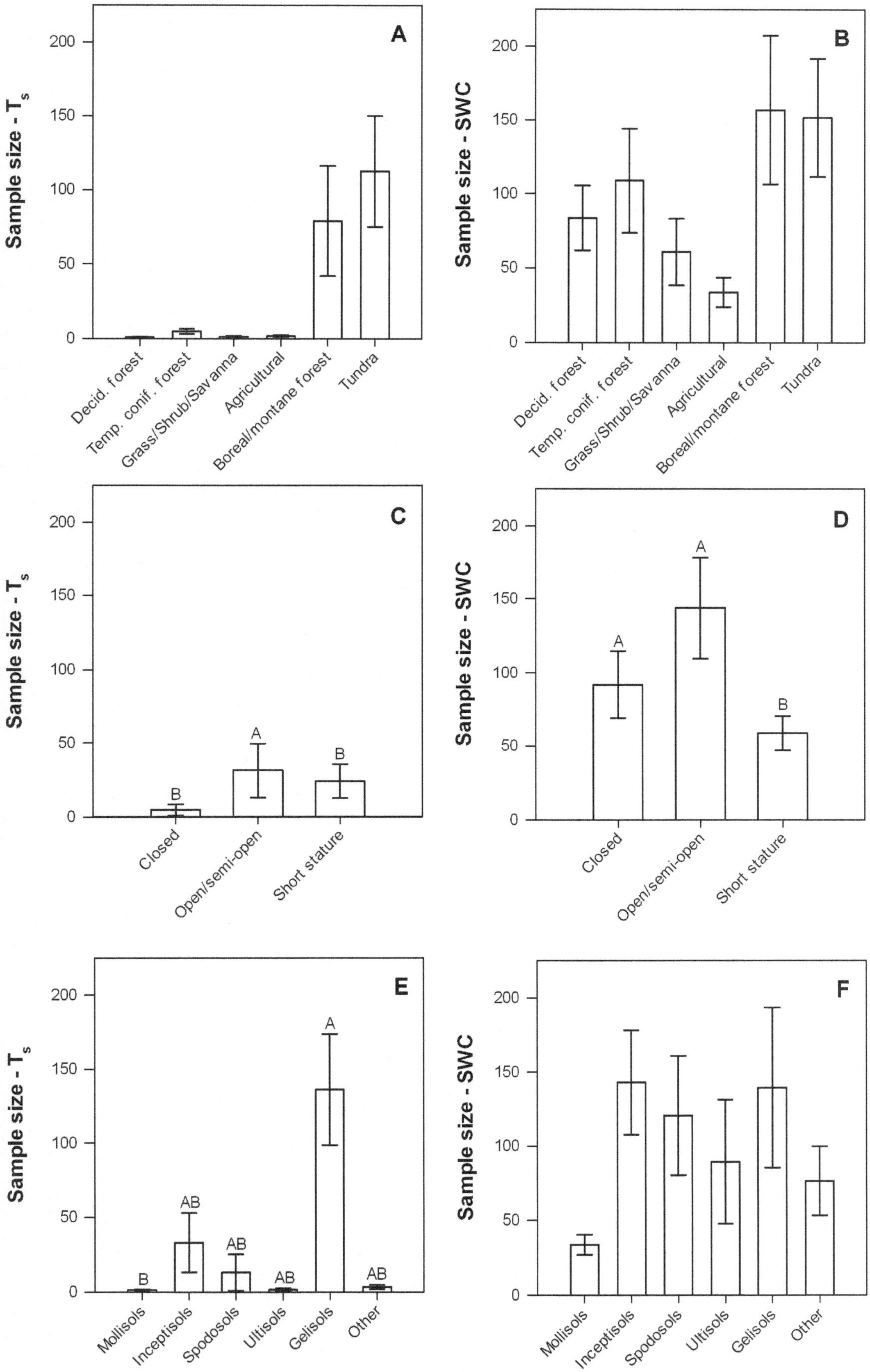
A
Sample size - T_s
0
50
100
150
200
Decid. forest
Temp. conif. forest
Grass/Shrub/Savanna
Agricultural
Boreal/montane forest
Tundra
B
Sample size - SWC
0
50
100
150
200
Decid. forest
Temp. conif. forest
Grass/Shrub/Savanna
Agricultural
Boreal/montane forest
Tundra
C
Sample size - T_s
0
50
100
150
200
B
A
B
Closed
Open/semi-open
Short stature
D
Sample size - SWC
0
50
100
150
200
A
A
B
Closed
Open/semi-open
Short stature
E
Sample size - T_s
0
50
100
150
200
B
AB
AB
AB
A
AB
Mollisols
Inceptisols
Spodosols
Ultisols
Gelisols
Other
F
Sample size - SWC
0
50
100
150
200
Mollisols
Inceptisols
Spodosols
Ultisols
Gelisols
Other

Figure 12. Number of samples to measure T_s (left column) and SWC (right column) to within 10% of the spatial mean with 90% confidence at each site in relation to ecosystem type (a, b), canopy structure (c, d), and soil type (e, f). Error bars represent ±1 SE. Bars with different letter were significantly different ($p<0.05$) based on Tukey HSD tests. The letter "A" corresponds to the largest least square mean(s) and subsequent letters (B, C, ...) correspond to progressively smaller least square means.

would be needed to meet the requirement of measuring T_s to within 10% of the spatial mean with 90% confidence at half the sites, although common sense dictates that more samples would be preferable (*e.g.*, 5 samples). This matrix can be used as a guide to quantitatively evaluate the efficacy of increasing sample size and/or sample spacing to measure the spatial mean of T_s at scales of ~1 ha at our sampling sites.

Using Equation 6b and 7b, we created a similar matrix to evaluate the efficacy of different sampling strategies for measuring SWC at spatial scales of ~1 ha (Figure 15). As with T_s, 2 samples spaced 5 m apart would not be an effective sampling strategy to measure SWC to within 10% of the mean with 90% confidence. Increasing the sample size or sample spacing only slightly improves the efficacy, however, by increasing both the sample size and sample spacing the sampling strategy can become substantially more effective at meeting the accuracy requirement. Figure 15 also demonstrates how a sampling strategy that is effective at a given proportion of sites can be achieved in many different ways. For instance, 150 samples spaced 10 m apart would meet the accuracy requirement for SWC at the same proportion of sites as 60 samples 15 m apart or 30 samples 55 m apart.

Lastly, using Equation 6c and 7c we created a matrix to evaluate the efficacy of measuring both T_s and SWC to within 10% of the mean with 90% confidence at spatial scales of ~1 ha (Figure 16). I n this case, more samples spaced further apart were always required to meet the accuracy requirement at a given proportion of sites than for T_s or SWC alone.

Discussion

We characterized the spatial structure of variability in soil properties over ~1 ha at 60 sites throughout the US. Our study included nearly an order of magnitude more sites than any previous study that examined the spatial variation in soil properties, and as a result, allowed us to derive some empirical rules to describe the observed variability in soils. Moreover, we were able to evaluate the efficacy of different sampling strategies (*i.e.*, sample size and sample spacing) at measuring the spatial mean of T_s and SWC, which can guide future soil studies. We emphasize that our guidelines for sampling strategies are designed to accurately estimate the spatial mean of a soil property, and as a result, our guidelines may not be suitable for studies with different goals. Because our sites were broadly representative of US geography and ecology (*i.e.*, including every major terrestrial ecosystem type and soil type), our findings are broadly applicable to US soils, and to a lesser extent other regions of the world.

Variability

Generally, variability was larger for SWC than T_s, which resulted in a larger sample size required to measure SWC to within 10% of the spatial mean with 90% confidence than for T_s. This is consistent with other studies that have also observed larger CV values for SWC than T_s [23,24,58–60]. We did not specifically set out to determine the sources of variability, or why variability was larger for SWC than T_s, but a likely explanation is that water is more mobile than heat, and as a result more sensitive to

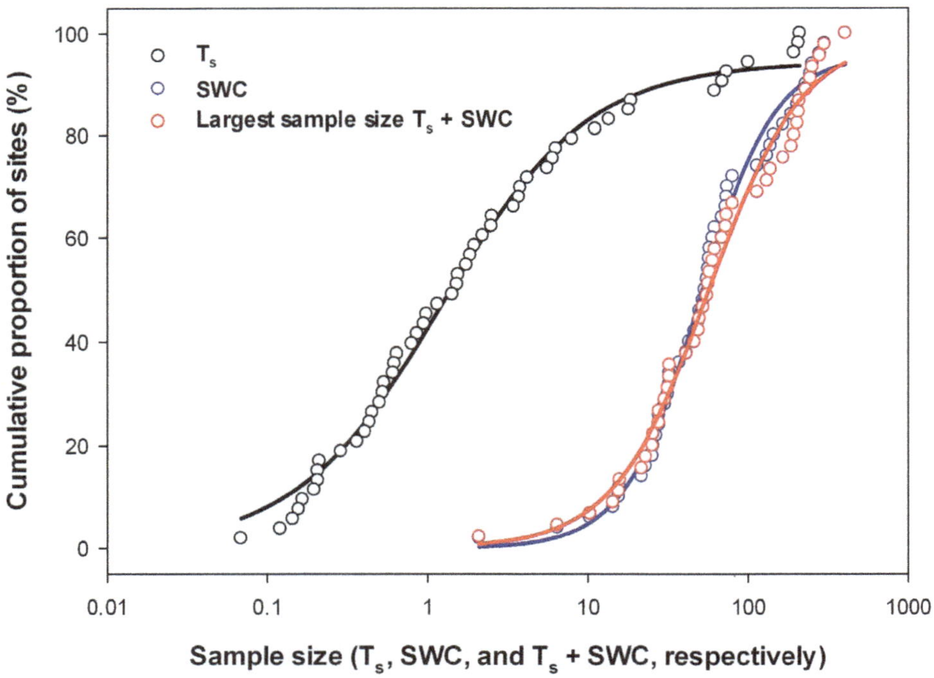

Figure 13. Number of samples required to measure T_s, SWC, or T_s and SWC to within 10% of the spatial mean with 90% confidence versus the cumulative proportion of sites. A logistic curve was fitted to the data with $R^2=0.99$ for each equation (Eqs. 7a–c). Values are plotted in a semi-log scale due to large differences in magnitude across the sites.

Legend

0	10	20	30	40	50	60	70	80

Sample spacing (m)

Sample size	5	10	15	20	25	30	35	40	45	50	55	60	65	70	75	80	85	90	95	100
2	7	16	23	28	32	35	38	40	42	43	44	45	46	47	47	48	48	49	49	49
3	8	18	26	32	37	40	43	46	47	49	50	51	52	53	54	55	55	56	56	56
4	9	19	28	34	40	43	47	49	51	53	54	55	56	57	58	59	59	60	60	61
5	10	20	29	36	41	46	49	52	54	55	57	58	59	60	61	62	62	63	63	64
6	10	21	30	37	43	47	51	53	56	57	59	60	61	62	63	64	64	65	65	66
8	10	22	32	39	45	49	53	56	58	60	62	63	64	65	66	67	67	68	68	69
10	11	23	33	40	46	51	54	57	60	62	63	65	66	67	68	69	69	70	70	71
15	11	24	34	42	48	53	57	60	62	64	66	67	69	70	71	71	72	73	73	74
20	11	24	35	43	49	54	58	61	63	65	67	69	70	71	72	73	74	74	75	75
25	11	24	35	43	50	55	59	62	64	66	68	70	71	72	73	74	74	75	76	76
30	12	25	35	44	50	55	59	62	65	67	69	70	72	73	74	74	75	76	76	77
35	12	25	36	44	50	55	59	63	65	67	69	71	72	73	74	75	76	76	77	77
40	12	25	36	44	51	56	60	63	66	68	69	71	72	73	74	75	76	77	77	78
45	12	25	36	44	51	56	60	63	66	68	70	71	73	74	75	76	76	77	78	78
50	12	25	36	44	51	56	60	63	66	68	70	71	73	74	75	76	77	77	78	78
60	12	25	36	45	51	56	60	64	66	68	70	72	73	74	75	76	77	78	78	79
70	12	25	36	45	51	57	61	64	67	69	71	72	73	75	76	76	77	78	78	79
80	12	25	36	45	52	57	61	64	67	69	71	72	74	75	76	77	77	78	79	79
90	12	25	36	45	52	57	61	64	67	69	71	72	74	75	76	77	77	78	79	79
100	12	25	36	45	52	57	61	64	67	69	71	73	74	75	76	77	78	78	79	79
125	12	26	37	45	52	57	61	64	67	69	71	73	74	75	76	77	78	79	79	80
150	12	26	37	45	52	57	61	65	67	69	71	73	74	75	76	77	78	79	79	80
175	12	26	37	45	52	57	61	65	67	70	71	73	74	75	76	77	78	79	79	80
200	12	26	37	45	52	57	61	65	67	70	71	73	74	76	77	77	78	79	79	80
250	12	26	37	45	52	57	62	65	68	70	72	73	75	76	77	78	78	79	80	80

Figure 14. Efficacy of different sampling strategies at measuring T_s to within 10% of the spatial mean with 90% confidence at scales of ~1 ha. Values represent the percent of sites where the corresponding sampling strategy would achieve the accuracy requirement. The x-axis is based on Equation 6a and the y-axis is based on Equation 7a.

microtopography (*i.e.*, water drains from hummocks to depressions, whereas heat does not), and is actively transported by plants (*i.e.*, transpiration induced uptake and hydrologic lift), resulting in larger variability at this spatial scale.

Despite generally larger spatial variability in SWC than T_s, (*i.e.*, sill and CV_{Sill}), T_s and SWC were positively correlated suggesting that they were likely controlled by similar site properties at the scale we sampled. Indeed, ANOVAs also indicated that sill values for both measures decreased with increasing longitude (*i.e.*, from west to east) and mean annual air temperature, while they increased with mean T_s. In addition, among the sites sill values for temperature and moisture increased with mean T_s and SWC, respectively, exhibiting the commonly observed pattern of increased variability with increasing mean values, *i.e.*, heteroscedastic [61]. Similarly, data presented by Brocca et al. [12] shows that the standard deviation for SWC increased from the driest site (MON, mean ± SD: 22.8±2.0) to the wettest site (LEC, 38.5±3.0) for seven Italian sites. In contrast, Western et al. [26] found no clear relationship between SWC and variance at locations in Australia and New Zealand, while data reported by Famiglietti et al. [25] show a negative relationship among six sites in Oklahoma. These studies however, differ in their design (*e.g.*, spatial scale, sampling frequency, geographic region), making it difficult to determine why different patterns were observed among the studies. But one explanation could be that the driest sites in the other studies had a mean SWC of between 10% and 23%, whereas half of our sites had a mean SWC of <10% (all of which had relatively low semivariogram sill values) and previous studies that have monitored SWC over time at individual sites often observed that low mean SWC coincided with relatively low variability, *e.g.*, [13,14]. We studied the relationship between the mean and variability (*i.e.*, sill) across space, whereas most previous studies have investigated this relationship at individual sites across time and have often (but not always) observed the variance or standard deviation increasing with intermediate SWC and decreasing at low and high SWC [12–14,20,26].

Ecosystem type did not significantly influence sill values for either T_s or SWC, despite the common assumption that agricultural soils were less variable than soils in other ecosystems [4,11]. Similarly, Robertson et al. [11] were surprised by the large variability in a wide range of soil properties from a soybean field that visibly appeared homogeneous, supporting the notion that agricultural ecosystems are typically more variable than many researchers appreciate.

The ANOVA models left 43% and 30% of the variability in T_s and SWC sill values unexplained for, respectively, indicating that other site properties (*i.e.*, besides the ones we measured) also are sources of variability. In part this may relate to vegetation canopy structure at the sites. For example, soil beneath canopy gaps would receive both more radiation and precipitation inputs than soil

Sample spacing (m)

Sample size	5	10	15	20	25	30	35	40	45	50	55	60	65	70	75	80	85	90	95	100
2	0	0	0	0	0	0	0	0	0	0	0	0	0	0	0	0	0	0	0	0
3	0	0	0	0	0	0	0	0	0	0	0	0	0	0	0	0	0	0	0	0
4	0	0	0	0	0	1	1	1	1	1	1	1	1	1	1	1	1	1	1	1
5	0	0	0	1	1	1	1	1	1	1	1	1	1	1	1	1	1	1	1	1
6	0	0	1	1	1	1	1	1	1	1	1	1	1	1	2	2	2	2	2	2
8	0	1	1	1	2	2	2	2	2	2	2	2	2	2	3	3	3	3	3	3
10	0	1	2	2	3	3	3	3	3	3	4	4	4	4	4	4	4	4	4	4
15	1	2	3	4	5	6	6	6	7	7	7	7	7	7	8	8	8	8	8	8
20	2	4	5	7	8	9	9	10	10	11	11	11	12	12	12	12	12	12	12	12
25	2	5	8	10	11	12	13	14	15	15	16	16	16	16	17	17	17	17	17	17
30	3	7	10	12	14	16	17	18	19	19	20	20	21	21	21	22	22	22	22	22
35	3	8	12	15	17	19	21	22	23	24	24	25	25	26	26	26	27	27	27	27
40	4	9	14	17	20	22	24	26	27	28	28	29	29	30	30	31	31	31	31	32
45	5	10	16	20	23	25	27	29	30	31	32	33	33	34	34	35	35	35	36	36
50	5	12	17	22	25	28	30	32	33	35	35	36	37	38	38	38	39	39	39	40
60	6	13	20	25	30	33	35	37	39	40	41	42	43	44	44	45	45	46	46	46
70	7	15	22	28	33	36	39	42	43	45	46	47	48	49	49	50	50	51	51	52
80	7	16	24	31	36	39	43	45	47	49	50	51	52	53	53	54	55	55	55	56
90	7	17	26	33	38	42	45	48	50	51	53	54	55	56	57	57	58	58	59	59
100	8	18	27	34	40	44	47	50	52	54	55	57	58	59	59	60	61	61	62	62
125	8	19	29	37	43	47	51	54	56	58	60	61	62	63	64	65	65	66	66	67
150	9	20	30	38	45	49	53	56	59	61	62	64	65	66	67	68	68	69	69	70
175	9	21	31	40	46	51	55	58	61	63	64	66	67	68	69	70	70	71	71	72
200	9	21	32	40	47	52	56	59	62	64	66	67	68	69	70	71	72	72	73	73
250	10	22	33	41	48	53	57	61	63	65	67	69	70	71	72	73	74	74	75	75

Figure 15. Efficacy of different sampling strategies at measuring SWC to within 10% of the spatial mean with 90% confidence at scales of ~1 ha. Values represent the percent of sites where the corresponding sampling strategy would achieve the accuracy requirement. The x-axis is based on Equation 6b and the y-axis is based on Equation 7b. Legend in Figure 14.

beneath a closed-canopy, which would increase spatial variability in both measures at a site [62]. Indeed, T_s sill values were significantly larger in open/semi-open ecosystems (*i.e.*, sites with high variability in vegetation cover) than in closed-canopy or short stature ecosystems, but this pattern did not occur for SWC. In a rangeland ecosystem, diurnal maximum T_s at 5 cm were found to be much lower beneath sagebrush plants (10°C) than in interspaces between plants (17°C), indicating that canopy structure strongly influenced spatial variability in T_s [63]. Similar results have been found with SWC [64,65]. In addition, previous studies have shown that topography influences the spatial variability of SWC [12,66,67], which may also account for some of the unexplained variation found here.

Sample size

We estimated the number of samples required to measure T_s and/or SWC to within 10% of the spatial mean with 90% confidence at each site for prospective purposes (*i.e.*, to inform future study designs). This represents a conservative number of samples for most studies, as this level of accuracy is often deemed more than sufficient and some have supported even lower accuracy requirements [4,5]. Of course each study should evaluate their required accuracy, and if these criteria were relaxed, design elements like sample size could be reduced.

Previous estimates of the sample size required to measure soil properties to a given accuracy have often assumed that their data was not spatially correlated [12,13,14,24]. However, our data, and data from many other studies [9,18,23,24,26,59,68–72], show that spatial correlation is ubiquitous for T_s and SWC. Since $CV_{Traditional}$ did not account for spatial correlation (*i.e.*, it assumes that all measurements are independent regardless of how closely they are spaced) it underestimated the true variability in the data, causing the required sample size to be underestimated [73]. In contrast, CV_{Sill} did account for spatial correlation, therefore, it reliably estimated the variability in the data and is appropriate for calculating the required sample size. Had we used $CV_{Traditional}$ rather than CV_{Sill} to calculate sample size, we would have substantially underestimated the required sample size (*i.e.*, by 60% for T_s and 109% for SWC when averaged across all sites). As a result, caution should be used when relying on sample size estimates that assume no spatial correlation in T_s and SWC data.

Other methods can be used to estimate the number of samples required to accurately estimate the spatial mean of a soil property. For example, previous research has shown that fewer samples are sufficient to meet an accuracy requirement if the spatial mean is calculated by kriging, rather than by averaging the data [74–77]. However, the kriging method requires knowledge of the semivariogram at the sampling site, which is frequently unknown. Moreover, our data shows that it is not possible to reliably estimate the key features of T_s and SWC semivariograms (*i.e.*, sill, nugget, and range) based on the site characteristics we studied. Similarly, the temporal stability (or rank stability) approach and the random combination method can also be used to reduce the required sample size, but these approaches require detailed knowledge of spatial variation of soil properties at the sampling site, which is also often unknown [14,24,77,78]. It should be noted that these approaches address a slightly different goal (*i.e.*, estimating the number of samples required to estimate mean soil

Sample spacing (m)

Sample size	5	10	15	20	25	30	35	40	45	50	55	60	65	70	75	80	85	90	95	100
2	0	0	0	0	0	0	0	0	0	0	0	0	0	0	0	0	1	1	1	1
3	0	0	0	0	0	0	1	1	1	1	1	1	1	1	1	1	1	1	1	1
4	0	0	0	1	1	1	1	1	1	1	1	1	1	1	1	1	1	1	1	1
5	0	0	1	1	1	1	1	1	1	1	2	2	2	2	2	2	2	2	2	2
6	0	0	1	1	1	1	1	2	2	2	2	2	2	2	2	2	2	3	3	3
8	0	1	1	1	2	2	2	2	3	3	3	3	3	3	4	4	4	4	4	4
10	0	1	1	2	2	3	3	3	4	4	4	4	4	5	5	5	5	5	5	5
15	1	2	2	3	4	5	5	6	6	7	7	7	8	8	8	8	9	9	9	9
20	1	2	3	5	6	6	7	8	9	9	10	10	11	11	12	12	12	13	13	13
25	1	3	4	6	7	8	9	10	11	12	13	13	14	15	15	15	16	16	17	17
30	2	3	5	7	9	10	12	13	14	15	16	16	17	18	18	19	19	20	20	21
35	2	4	6	8	10	12	14	15	16	17	18	19	20	21	21	22	23	23	24	24
40	2	5	7	9	12	14	15	17	18	20	21	22	23	24	24	25	26	26	27	27
45	2	5	8	11	13	15	17	19	20	22	23	24	25	26	27	28	28	29	30	30
50	2	6	9	11	14	16	19	20	22	24	25	26	27	28	29	30	31	32	32	33
60	3	6	10	13	16	19	21	23	25	27	29	30	31	33	34	35	36	36	37	38
70	3	7	11	15	18	21	24	26	28	30	32	33	35	36	37	38	39	40	41	42
80	3	8	12	16	19	23	25	28	30	33	34	36	38	39	40	42	43	44	44	45
90	4	8	13	17	21	24	27	30	32	35	37	38	40	42	43	44	45	46	47	48
100	4	8	13	18	22	25	28	31	34	36	38	40	42	44	45	46	48	49	50	51
125	4	9	14	19	24	28	31	34	37	40	42	44	46	48	49	51	52	53	54	55
150	4	10	15	20	25	29	33	36	39	42	45	47	49	51	52	54	55	56	57	59
175	5	10	16	21	26	30	34	38	41	44	46	49	51	53	54	56	57	59	60	61
200	5	11	16	22	27	31	35	39	42	45	48	50	52	54	56	57	59	60	62	63
250	5	11	17	23	28	32	37	40	44	47	50	52	54	56	58	60	61	63	64	65

Figure 16. Efficacy of different sampling strategies at measuring T_s and SWC to within 10% of the spatial mean with 90% confidence at scales of ~1 ha. Values represent the percent of sites where the corresponding sampling strategy would achieve the accuracy requirement. The x-axis is based on Equation 6c and the y-axis is based on Equation 7c. Legend in Figure 14.

moisture over a number of time steps) than the statistical method to calculate sample size, which focuses on a single time step. Because of these limitations, we calculated the required sample size using the classical statistical approach (Eq. 5) to allow our findings to be more easily applied to sampling designs at sites where the spatial structure of variation may be unknown.

Spatial variability in T_s was so low at many sites that the sample size analyses indicated that even just one sample was sufficient to meet our accuracy requirement (although common sense dictates that more samples would be preferable), while 10 samples were sufficient at 79% of sites, and 100 samples were sufficient at all but 3 sites. In contrast, SWC often required at least an order of magnitude more samples to meet the accuracy requirement at a similar proportion of sites (*e.g.*, 100 samples were sufficient at 72% of sites). Because more samples were required to meet our accuracy requirement for SWC than T_s at almost every site, the sample size needed when examining both T_s and SWC was very similar to that found for SWC alone.

The large sample sizes needed to accurately measure the spatial mean SWC based on our findings differ substantially from previous studies that assumed SWC was not spatially correlated [12–14]. For example, Brocca et al. [12] reported that 15 or fewer samples would be needed at 4 out of 8 sites to measure SWC to within ±2% volumetric with 95% confidence, and at most 40 samples would be needed to meet this accuracy requirement at any of the sites. We largely attribute this difference to accounting for spatial correlation in our estimates of sample size, but other factors may also play a role including that many previous studies have focused on agricultural and grassland sites, which tend to require smaller sample sizes (Figure 13). In addition, at site with a mean SWC of <20%, our relative error threshold (10% of the mean) is more stringent than the 2% absolute error that has often been used in previous SWC studies. We are not aware of previous studies that have estimated the sample size required to measure T_s to a given accuracy, but our data suggests these estimates will only be reliable if they account for spatial correlation.

Based on our finding, 20 samples would be more than sufficient to meet the accuracy requirement for T_s at most sites, although Alaskan sites typically required many more samples to meet the same requirement. As a result, measuring T_s across space in a temperate or sub-tropical ecosystem one might reasonably assume that 20 samples would be sufficient (assuming appropriate sample spacing), while working at high latitude (>50°N) one would likely need an order of magnitude more samples to have confidence in meeting the same accuracy requirement. Sites where a large number of samples were required were more closely associated with high latitudes, westerly longitudes, and cold mean T_s at the time of sampling (<6°C).

There are plausible reasons why sites with both low mean T_s and high latitude would result in the need for large sample sizes. First, sites with low mean T_s statistically result in high CV values because the equation divides by the mean T_s (see Eq. 4), which in turn results in large sample sizes (see Eq. 5). Second, the low angle of the sun at high latitude sites, which results in large differences in shading (*i.e.*, radiation inputs) of locations north or south of an obstacle (*e.g.*, tree trunk, tussock, or rock) will spend a large amount of time shaded by the obstacle and when unshaded, the sun will be at a low angle (*i.e.*, low radiation inputs), and collectively cause large spatial variability in T_s. The converse is true at lower latitude sites where shading by an obstacle will be much less than at high latitude sites creating less spatial variation in T_s and in turn results in fewer samples necessary to meet the accuracy requirement. Indeed, the positive relationship that we observed between T_s sill values and latitude lends additional support to the first hypothesis, suggesting that high latitudes, as well as low mean T_s values, both result in large sample sizes. In contrast, we are not aware of a plausible reason why westerly longitudes should be expected to increase the number of samples required, instead it seems likely that relationship between longitude and the number of samples is an artifact of US geography, since all high latitude sites in the US are located in Alaska, which is located further west than most of the temperate and sub-tropical US. Thus, it appears that the low mean T_s and high latitudes, not westerly longitudes, of the Alaskan sites resulted in the large sample sizes for T_s required at those sites.

Unlike T_s, the sample sizes required to accurately measure spatial SWC were only weakly related to site characteristics making it difficult to estimate a sample size based solely these variables. However, it is notable that a sample size of 100 was sufficient at most sites (72%), while sites that needed >100 samples all had mean SWC of <16%, and sites that needed >250 samples had mean SWC of ≤11%. Similar to T_s, the reduction in the required sample size at high mean SWC is presumably partly due to the fact that low mean SWC results in large CVs because this is calculated by dividing by the mean SWC (see Eq. 4), which in turn causes large sample sizes to be needed to accurately estimate the spatial mean (see Eq. 5).

The large sample size that was required to accurately estimate the spatial mean SWC across ~1 ha at most sites is beyond the scope of many research projects. However, the recent development of sensors that aim to measure SWC at scales of hundreds and thousands of m^2 may hold promise for estimating mean SWC at this scale with just 1 or a handful of sensors [79,80] and may represent a more cost-effective way of meeting the accuracy requirement than deploying tens or hundreds of sensors that each measure SWC in a small volume of soil.

Sample spacing

The number of samples required to meet our accuracy requirement assumes that the samples are spaced at least at the distance of the range in order to avoid correlation and pseudoreplication [1]. The mean range for both T_s and SWC was ~27 m. Similar range values have been observed for T_s and SWC in other studies that sampled at a similar scale [8,18,20,59,71,81], demonstrating that range values of tens of meters are the norm for temperature and moisture at this scale.

Unlike the sill values, the range values for T_s and SWC were not correlated, indicating that they are controlled by different site properties. The ANOVAs supported this conclusion, showing that range values for T_s were related to canopy structure and soil type, whereas for SWC they were related to longitude and canopy height. However, the ANOVA models left 60% and 82% of the variability in range values unexplained for T_s and SWC, respectively, indicating that we did not measure the most important factor(s) that control the spatial structure of variability. As a result it is difficult to determine what controls the spatial structure of variability at our sites, but potential factors may include microtopography, canopy gap spacing, and variation in soil texture.

Since we cannot reliably predict the semivariogram range for T_s or SWC based on the site characteristics that we measured, a researcher wishing to measure T_s and/or SWC at scales of ~1 ha could use Equations 6a–c to estimate if a given sampling spacing would likely be sufficient (*i.e.*, larger than the range value). For instance, a sample spacing of 5 m would likely be deemed insufficient since only ~12% of sites had a smaller T_s or SWC range, whereas a sample spacing of 40 m may be considered more appropriate since ~2/3 of sites had a smaller T_s or SWC range than this. This sample spacing could then be coupled with the estimated number of samples required based on latitude and mean T_s (for spatial measurements of temperature) and/or mean SWC (for spatial measurements of moisture) to develop a site-specific sampling strategy.

Sampling strategies

Figures 14–16 can guide a sampling design. These figures clearly demonstrate the potential benefit of increasing sample number and/or sample spacing at the spatial scale we sampled. In particular, there are many different ways of achieving a sampling strategy that is likely to meet our accuracy requirement at a given proportion of sites (Figures 14, 15, 16). This demonstrates a particularly pertinent finding that increasing sample spacing is as important as increasing sample number to achieve a given sampling efficacy. This is significant because doubling the samples size often creates much more work and/or expense than simply doubling sample spacing.

Several existing networks that monitor T_s and SWC (*e.g.*, USDA Soil Climate Analysis Network (SCAN) and the US NOAA Climate Reference Network (USCRN)) use sampling designs that would likely require supplemental sampling effort in order to accurately estimate the spatial (~1 ha) mean of these properties at most sites (rf. Figures 14, 15, 16). To illustrate this point, USCRN deploys 3 T_s and SWC sensors 5.2 m apart in a range of depths at >100 sites throughout the US (M Palecki, pers. comm.), while SCAN deploys 1 T_s and SWC sensor at several depths at sites throughout the US [33]. Both the number of sensors and their spatial distribution would be inadequate to estimate these quantities within 10% of the mean with 90% confidence. To highlight the utility of our data, an effective way to increase the local scale (~1 ha) spatial accuracy in this case would be to increase sample spacing and add more SWC sensors (rather than T_s sensors).

Using the results found here, we have constructed a simplified matrix to inform researchers on the efficacy of different sampling designs (Figures 14, 15, 16), given the caveat that they do not account for information provided by samples spaced at distances less than the range; 2) recognize that one T_s sample (*i.e.*, 0 m sample spacing) is sufficient at several sites; and 3) recognize that in practice a sampling design operates at multiple spatial scales, *e.g.*, 9 samples spaced 5 m apart and arranged in a 3×3 grid also includes 4 samples spaced 10 m apart (Figure 14). Similarly, 9 samples spaced 5 m apart along a transect also includes 5 samples spaced 10 m apart, 3 samples spaced 20 m apart, and 2 samples spaced 40 m apart (Figure 17). Nonetheless, these figures provide an empirically-derived quantitative starting point for developing robust soil sampling strategies.

In addition to the limitations associated with Figures 14, 15, 16, researchers interested in using our findings to guide their sampling strategy should also bear in mind the conditions under which our data were generated. For example: 1) the sites were typically dominated by a single vegetation type, therefore our findings may not adequately capture variability at locations that cross major vegetation ecotones; 2) data were collected during growing season, so may not represent other seasons; 3) since our sampling typically occurred on a single day, which may have represented anomalous conditions (*e.g.*, usually hot, cold, wet, or dry), we expect that the general patterns we observed across multiple sites are more

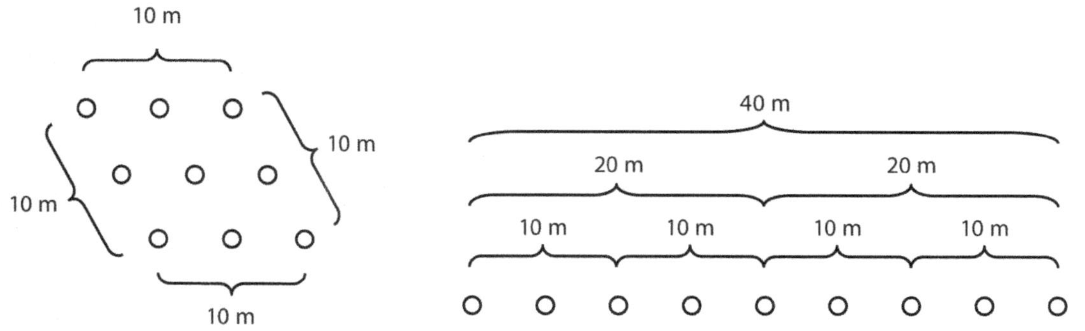

Figure 17. Nine sampling points (open circles) spaced 5 m apart and arranged (left panel) in a 3×3 grid and (right panel) along a transect. Besides 9 samples spaced 5 m apart, the 3×3 grid also includes 4 samples spaced 10 m apart, while the transect also includes 5 samples spaced 10 m apart, 3 samples spaced 20 m apart, and 2 samples spaced 40 m apart.

reliable than data from an individual site; 4) our data are representative of the scale that we sampled (~1 ha) and may not be applicable at substantially smaller or larger scales; and 5) our data are based on surface soil measurements, which typically exhibits larger spatial variability than deeper soils [63,82].

Emergent continental scale properties

Soil forming processes, parental material, time, climate, weathering type and rates, and biological activity have been recognized in soil classification schemes [83,84]. But because soil genesis is difficult to measure directly, the current US soil taxonomy, rf. [84] relies on soil properties (pedology) that only implies genesis. This classification system provides an underlying framework of physical mechanisms that contribute towards the spatial correlation that we attempt to quantify. Here, we have shown new evidence of emergent continental scale soil properties, *i.e.*, how ecosystem scale spatial variability is patterned across the US. Granted, this variability can be either interpreted directly as functions of soil temperature and moisture, or by proxy for other soil properties. We also fully recognize that there are a variety of local scale controls on this variability, *i.e.*, canopy structure, canopy height, etc. (that mostly control the temperature and/or water microclimate), but this does not preclude the existence of continental scale patterns that we identified, as these influences manifest themselves at smaller spatial resolutions.

Continental scale patterns included: 1) SWC range values increased with longitude (*i.e.*, increased from west to east), 2) sill values for T_s and SWC were negatively related to longitude (*i.e.*, decreasing from west to east), 3) nugget values for SWC were positively related to longitude (*i.e.*, increasing from west to east), and 4) range values for T_s increased from inceptisols (younger soils) to spodosols (intermediate age) to ultisols (older soils), suggesting that the spatial structure of variability is in part related to soil development. Such patterns could not have been identified from previous studies since they involved too few sites (≤6), and a meta-analysis approach was not possible due to large methodological inconsistencies among the studies (*i.e.*, different sampling scales, measurement methods, sampling depths, etc).

Conclusions

We characterized the spatial variability of soil properties at 60 US sites to inform NEON's sensor-based soil sampling strategy. In addition, we developed quantitative guidelines that can be used to inform soil sampling decisions throughout the US and elsewhere. While our data can be used as a guide, the most reliable way to generate a robust sampling design would be to conduct a site specific assessment of variability in soil properties on several occasions during the relevant season and at the relevant scale, as others have previously suggested [4,5]. Indeed, as Klironomos *et al.* [5] highlighted, although a preliminary study of spatial variation may seem like a large amount of extra work, it significantly contributes towards avoiding a sampling strategy that does not provide data that are both necessary and sufficient to meet the design constraints. However, since soil researchers have only rarely followed these recommendations, we also present our findings as a useful tool to help guide future soil sampling designs.

Supporting Information

Figure S1 Data collected across (a) space can be used to construct a (b) semivariogram, which describes the relationship between semivariance (*i.e.*, half the variance) and distance. A typical sampling layout used in this study is shown in (a). In addition to collecting data at each point shown on the graph, data were also collected at −0.3 m and −0.3 m from each point along the axis of each transect. The three components that describe the shape of the (b) semivariogram are the sill, nugget, and range. The sill represents the maximum semivariance that is encountered at a site and is equivalent to half the variance in the data set used to create the semivariogram. The nugget represents the variance that exists at spatial scales smaller than the minimum sampling distance, as well as sampling error. The range represents the distance beyond which samples are effectively independent at the scale sampled.

Acknowledgments

Staff associated with each site provided invaluable assistance and logistical support. Authors wish to thank M. Slater and L. Goldman for graphical assistance. We thank Don Julio for inspiring many late night scientific discussions.

Author Contributions

Conceived and designed the experiments: HWL EA PD. Performed the experiments: HWL EA HL MB. Analyzed the data: HWL EA PD HL MB. Contributed reagents/materials/analysis tools: HWL EA PD. Wrote the paper: HWL EA PD MB. Secured funding and grantsmanship: HWL.

References

1. Hurlbert SH (1984) Pseudoreplication and the design of ecological field experiments. Ecological Monographs 54: 187–211.
2. Peck TR, Melsted SW (1973) Field sampling for soil testing. In: Walsh LM, Beaton JD, editors. Soil Testing and Plant Analysis. Madison, Wisconsin: Soil Science Society of America, Inc., pp 67–75.
3. Petersen RG, Calvin LD (1996) Sampling. In: Sparks DL, Page AL, Helmke PA, Loeppert RH, Soltanpour PN, Tabatabai MA, Johnston CT, Summer ME, editors. Methods of Soil Analysis Part 3: Chemical Methods. Madison, Wisconsin: Soil Science Society of America, Inc., pp 1–18.
4. Boone RD, Grigal DF, Sollins P, Ahrens RJ, Armstrong DE (1999) Soil sampling, preparation, archiving, and quality control. In: Robertson GP, Coleman DC, Bledsoe CS, Sollins P, editors. Standard Soil Methods for Long-Term Ecological Research. Oxford: Oxford University Press. pp 3–28.
5. Klironomos JN, Rillig MC, Allen MF (1999) Designing belowground field experiments with the help of semi-variance and power analyses. Applied Soil Ecology 12: 227–238.
6. Jenny H (1941). *Factors of Soil Formation*. New York: McGraw-Hill Press. 281 p.
7. Burgess TM, Webster R (1980) Optimal interpolation and isarithmic mapping of soil properties. I. The variogram and punctual kriging. European Journal of Soil Science 31: 315–331.
8. Beckett PHT, Webster R (1971) Soil variability: A review. Soils and Fertilizers 34: 1–15.
9. Robertson GP, Huston MA, Evans FC, J M Tiedje JM (1988) Spatial variability in a successional plant community - Patterns of nitrogen availability. Ecology 69: 1517–1524.
10. Grigal DF, McRoberts RE, Ohmann LF (1991) Spatial variation in chemical-properties of forest floor and surface mineral soil in the North Central United-States. Soil Science 151: 282–290.
11. Robertson GP, Klingensmith KM, Klug MJ, Paul EA, Crum JR, et al. (1997) Soil resources, microbial activity, and primary production across an agricultural ecosystem. Ecological Applications 7: 158–170.
12. Brocca L, Morbidelli R, Melone F, Moramarco T (2007) Soil moisture spatial variability in experimental areas of central Italy. Journal of Hydrology 333: 356–373.
13. Famiglietti JS, Ryu DR, Berg AA, Rodell M, Jackson TJ (2008) Field observations of soil moisture variability across scales. Water Resources Research 44: W01423.
14. Brocca L, Melone F, Moramarco T, Morbidelli R (2010) Spatial-temporal variability of soil moisture and its estimation across scales. Water Resources Research 46: W02516.
15. Burgess TM, R. Webster R (1980) Optimal interpolation and isarithmic mapping of soil properties. I. The variogram and punctual kriging. European Journal of Soil Science 31: 315–331.
16. Goovaerts P (1998) Geostatistical tools for characterizing the spatial variability of microbiological and physico-chemical soil properties. Biology and Fertility of Soils 27: 315–334.
17. Hoenig J, Heisey DM (2001) The abuse of power: The pervasive fallacy of power calculations for data analysis. The American Statistician 55: 1–6.
18. Nyberg L (1996) Spatial variability of soil water content in the covered catchment at Gardsjon, Sweden. Hydrological Processes 10: 89–103.
19. Dobermann A, Goovaerts P, Neue HU (1997) Scale-dependent correlations among soil properties in two tropical lowland rice fields. Soil Science Society of America Journal 61: 1483–1496.
20. Ettema CH, Coleman DC, Vellidis G, Lowrance R, Rathbun SL (1998) Spatiotemporal distributions of bacterivorous nematodes and soil resources in a restored riparian wetland. Ecology 79: 2721–2734.
21. Western AW, Bloschl G, Grayson RB (1998) Geostatistical characterisation of soil moisture patterns in the Tarrawarra a catchment. Journal of Hydrology 205: 20–37.
22. Castrignano A, Giugliarini L, Risaliti R, Martinelli N (2000) Study of spatial relationships among some soil physico-chemical properties of a field in central Italy using multivariate geostatistics. Geoderma 97: 39–60.
23. Bond-Lamberty B, Brown KM, Goranson C, Gower ST (2006) Spatial dynamics of soil moisture and temperature in a black spruce boreal chronosequence. Canadian Journal of Forest Research-Revue Canadienne De Recherche Forestiere 36: 2794–2802.
24. Herbst M, Prolingheuer N, Graf A, Huisman JA, Weihermuller L, et al. (2009) Characterization and understanding of bare soil respiration spatial variability at plot scale. Vadose Zone Journal 8: 762–771.
25. Famiglietti JS, Devereaux JA, Laymon CA, Tsegaye T, Houser PR, et al. (1999) Ground-based investigation of soil moisture variability within remote sensing footprints during the Southern Great Plains 1997 (SGP97) Hydrology Experiment. Water Resources Research 35: 1839–1851.
26. Western AW, Zhou SL, Grayson RB, McMahon TA, Bloschl G, et al. (2004) Spatial correlation of soil moisture in small catchments and its relationship to dominant spatial hydrological processes. Journal of Hydrology 286: 113–134.
27. Choi M, Jacobs JM, Cosh MH (2007) Scaled spatial variability of soil moisture fields. Geophysical Research Letters 34: L01401.
28. Brocca L, Tullo T, Melone F, Moramarco T, Morbidelli R (2012). Catchment scale soil moisture spatial-temporal variability. Journal of Hydrology 422–423: 63–75.
29. Li B, Rodell M (2012) Spatial variability and its scale dependency of observed and modeled soil moisture under different climate conditions. Hydrology and Earth System Sciences 17:1177–1188.
30. Mittelbach H, Seneviratne SI (2012) A new perspective on the spatio-temporal variability of soil moisture: Temporal dynamics versus time invariant contributions. Hydrology and Earth System Sciences 16: 2169–2179.
31. Kaspar TC, Bland WL (1992) Soil temperature and root growth. Soil Science 154: 290–299.
32. Sitaula BK, Bakken LR (1993) Nitrous-oxide release from spruce forest soil - Relationships with nitrification, methane uptake, temperature, moisture and fertilization. Soil Biology and Biochemistry 25: 1415–1421.
33. Kirschbaum MUF (1995) The temperature-dependence of soil organic-matter decomposition, and the effect of global warming on soil organic-C storage. Soil Biology and Biochemistry 27: 753–760.
34. Rey A, Petsikos C, Jarvis PG, Grace J (2005) Effect of temperature and moisture on rates of carbon mineralization in a Mediterranean oak forest soil under controlled and field conditions. European Journal of Soil Science 56: 589–599.
35. Global Carbon Observing System GCOS (2007) Global climate observing system: Ensuring the availability of global observations for climate. Available: http://www.wmo.int/pages/prog/gcos/Publications/gcos-138.pdf. Accessed: 1 July 2013
36. Schaefer GL, Cosh MH, Jackson TJ (2007) The USDA Natural Resources Conservation Service Soil Climate Analysis Network (SCAN). Journal of Atmospheric and Oceanic Technology 24: 2073–2077.
37. Hargrove WW, Hoffman FM (1999) Using multivariate clustering to characterize ecoregion borders. Computing in Science and Engineering 1: 18–25.
38. Hargrove WW, Hoffman FM (2004) Potential of multivariate quantitative methods for delineation and visualization of ecoregions. Environmental Management 34: S39–S60.
39. Keller M, Schimel DS, Hargrove WW, Hoffman FM (2008) A continental strategy for the National Ecological Observatory Network. Frontiers in Ecology and the Environment 6: 282–284.
40. Schimel D, Keller M, Berukoff S, Kao R, Loescher HW, et al. (2011) NEON 2011 Science Strategy; Enabling continental-scale ecological forecasting. Boulder CO: NEON Inc. 55 p.
41. Lemon ER (1960) Photosynthesis under field conditions 2. An aerodynamic method for determining the turbulent carbon dioxide exchange between the atmosphere and a corn field. Agronomy Journal 52: 697–703.
42. Monteith JL, Szeicz G (1960) The carbon-dioxide flux over a field of sugar beet. Quarterly Journal of the Royal Meteorological Society 86: 205–214.
43. Kaimal JC, Wyngaard JC (1990) The Kansas and Minnesota experiments. Boundary-Layer Meteorology 50: 31–47.
44. Loescher HW, Law BE, Mahrt L, Hollinger DY, Campbell J, et al. (2006) Uncertainties in, and interpretation of, carbon flux estimates using the eddy covariance technique. Journal of Geophysical Research-Atmospheres 111. D21S90, doi:10.1029/2005JD006932.
45. Schmid HP (1994) Source areas for scalars and scalar fluxes. Boundary-Layer Meteorology 67: 293–318.
46. Schmid HP (1997) Experimental design for flux measurements: matching scales of observations and fluxes. Agricultural and Forest Meteorology 87: 179–200.
47. Horst TW, Weil JC (1995) How far is far enough - the fetch requirements for micrometeorological measurement of surface fluxes. Journal of Atmospheric and Oceanic Technology 12: 447–447.
48. Shaw DC, Franklin JF, Bible K, Klopatek J, Freeman E, et al. (2004) Ecological setting of the wind river old-growth forest. Ecosystems 7: 427–439.
49. Lee H, Schuur EAG, Vogel JG (2010) Soil CO2 production in upland tundra where permafrost is thawing. Journal of Geophysical Research-Biogeosciences 115: G01009.
50. Gorres JH, Dichiaro MJ, Lyons JB, Amador JA (1998) Spatial and temporal patterns of soil biological activity in a forest and an old field. Soil Biology & Biochemistry 30: 219–230.
51. Giese K, Tiemann R (1975) Determination of the complex permittivity from thin-sample time domain reflectometry, Improved analysis of the step response waveform. Adv Molec Relax Processes 7: 45–49.
52. Topp GC, Davis JL, Annan AP (1980) Electromagnetic determination of soil water content: measurements in coaxial transmission lines. Water Resources Research, 16: 574–582.
53. McBratney AB, Webster R (1986) Choosing functions for semi-variograms of soil properties and fitting them to sampling estimates. Journal of Soil Science 37: 617–639.
54. Webster R, Oliver MA (1992) Sample adequately to estimate variograms of soil properties. Journal of Soil Science 43: 177–192.
55. Ayres E, Loescher HW, Brunke M, Luo H (2011) Soil temperature and moisture data from NEON candidate sites. National Ecological Observatory Network pub., Boulder, Colorado. http://neoninc.org/pds/FIU007.php.
56. Trangmar BB, Yost RS, Uehara G (1986) Application of geostatistics to spatial studies of soil properties. Advances in Agronomy: 45–94.
57. Riberiro JR, Diggle PJ (2001) geoR: A package for geostatical analysis. R-NEWS 1: 15–18. ISSN 1609-3631.

58. Kosugi Y, Mitani T, Ltoh M, Noguchi S, Tan M, et al. (2007) Spatial and temporal variation in soil respiration in a Southeast Asian tropical rainforest. Agricultural and Forest Meteorology 147: 35–47.
59. Huang XD, Grace P, Mengersen K, Weier K (2011) Spatio-temporal variation in soil derived nitrous oxide emissions under sugarcane. Science of the Total Environment 409: 4572–4578.
60. Panosso AR, Marques J, Pereira GT, La Scala N (2009) Spatial and temporal variability of soil CO2 emission in a sugarcane area under green and slash-and-burn managements. Soil and Tillage Research 105: 275–282.
61. Sokal RR, Rohlf FJ (1995) Biometry. 3rd edition. New YorkNY: Freeman Press, 937 p.
62. Loescher HW, Powers JS, Oberbauer SF (2002) Spatial variation of throughfall volume in an old growth tropical rain forest. Journal of Tropical Ecology 18: 397–407.
63. Pierson FB, Wight JR (1991) Variability of near-surface soil-temperature on sagebrush rangeland. Journal of Range Management 44: 491–497.
64. Elkins NZ, Sabol GV, Ward TJ, Whitford WG (1986) The influence of subterranean termites on the hydrological characteristics of a Chihuahuan desert ecosystem. Oecologia 68: 521–528.
65. Schlesinger WH, Reynolds JF, Cunningham GL, Huenneke LF, Jarrell WM, et al. (1990) Biological feedbacks in global desertification. Science 247: 1043–1048.
66. Famiglietti JS, Rudnicki JW, Rodell M (1998) Variability in surface moisture content along a hillslope transect: Rattlesnake Hill, Texas. Journal of Hydrology 210: 259–281.
67. Chaplot V, Walter C (2003) Subsurface topography to enhance the prediction of the spatial distribution of soil wetness. Hydrological Processes 17: 2567–2580.
68. Mohanty BP, Klittich WM, Horton R, Vangenuchten MT (1995) Spatio-temporal variability of soil temperature within three land areas exposed to different tillage systems. Soil Science Society of America Journal 59: 752–759.
69. Fitzjohn C, Ternan JL, Williams AG (1998) Soil moisture variability in a semi-arid gully catchment: implications for runoff and erosion control. Catena 32: 55–70.
70. Anctil F, Mathieu R, Parent LE, Viau AA, Sbih M, et al. (2002) Geostatistics of near-surface moisture in bare cultivated organic soils. Journal of Hydrology 260: 30–37.
71. Gallardo A (2003) Spatial variability of soil properties in a floodplain forest in northwest Spain. Ecosystems 6: 564–576.
72. Ryu D, Famiglietti JS (2006) Multi-scale spatial correlation and scaling behavior of surface soil moisture. Geophysical Research Letters 33: L08408, doi: 10.1029/2006GL025831.
73. Legendre P (1993) Spatial autocorrelation - Trouble or new paradigm? Ecology 74: 1659–1673.
74. McBratney AB, Webster R (1983) How many observations are needed for regional estimation of soil properties. Soil Science 135: 177–183.
75. Di HJ, Trangmar BB, Kemp RA (1989) Use of geostatistics in designing sampling strategies for soil survey. Soil Science Society of America Journal 53: 1163–1167.
76. Chung CK, Chong SK, Varsa EC (1995) Sampling strategies for fertility on a stony silt loam soil. Communications in Soil Science and Plant Analysis 26: 741–763.
77. Wang CM, Zuo Q, Zhang RD (2008) Estimating the necessary sampling size of surface soil moisture at different scales using a random combination method. Journal of Hydrology 352: 309–321.
78. Teuling AJ, Uijlenhoet R, Hupet F, van Loon EE, Troch PA (2006) Estimating spatial mean root-zone soil moisture from point-scale observations. Hydrology and Earth System Sciences 10: 755–767.
79. Larson KM, Small EE, Gutmann ED, Bilich AL, Braun JJ, et al. (2008) Use of GPS receivers as a soil moisture network for water cycle studies. Geophysical Research Letters 35: L2445, doi:10.1029/2008GL03613.
80. Zreda M, Desilets D, Ferre TPA, Scott RL (2008) Measuring soil moisture content non-invasively at intermediate spatial scale using cosmic-ray neutrons. Geophysical Research Letters 35: L21402, doi: 10.1029/2008GL035655.
81. Loague K (1992) Soil-water content at R-5.1. Spatial and temporal variability. Journal of Hydrology 139: 233–251.
82. Scharringa M (1976) Representativeness of soil temperature measurements. Agricultural Meteorology 16: 263–276.
83. Dokuchaev VV (1883) Russian Chernozems (Russkii Chernozems). Israel Prog. Sci. Trans., Jerusalem, 1967. Transl. from Russian by N. Kaner. Springfield, VA: U.S. Dept. of Commerce.
84. Coffey GN (1912) A Study of the Soils of the United States. U. S. Dept. of Agriculture Bur., Soils Bull. 85, US Govt. Printing Office, Washington, D.C. 114 p.

6

Structure, Composition and Metagenomic Profile of Soil Microbiomes Associated to Agricultural Land Use and Tillage Systems in Argentine Pampas

Belén Carbonetto[1]*, Nicolás Rascovan[1], Roberto Álvarez[2], Alejandro Mentaberry[3], Martin P. Vázquez[1]*

1 Instituto de Agrobiotecnología de Rosario (INDEAR), Predio CCT Rosario, Santa Fe, Argentina, **2** Facultad de Agronomía, Universidad de Buenos Aires, Buenos Aires, Argentina, **3** Departamento de Fisiología y Biología Molecular y Celular, Facultad de Ciencias Exactas y Naturales, Universidad de Buenos Aires, Buenos Aires, Argentina

Abstract

Agriculture is facing a major challenge nowadays: to increase crop production for food and energy while preserving ecosystem functioning and soil quality. Argentine Pampas is one of the main world producers of crops and one of the main adopters of conservation agriculture. Changes in soil chemical and physical properties of Pampas soils due to different tillage systems have been deeply studied. Still, not much evidence has been reported on the effects of agricultural practices on Pampas soil microbiomes. The aim of our study was to investigate the effects of agricultural land use on community structure, composition and metabolic profiles on soil microbiomes of Argentine Pampas. We also compared the effects associated to conventional practices with the effects of no-tillage systems. Our results confirmed the impact on microbiome structure and composition due to agricultural practices. The phyla *Verrucomicrobia, Plactomycetes, Actinobacteria*, and *Chloroflexi* were more abundant in non cultivated soils while *Gemmatimonadetes, Nitrospirae* and WS3 were more abundant in cultivated soils. Effects on metabolic metagenomic profiles were also observed. The relative abundance of genes assigned to transcription, protein modification, nucleotide transport and metabolism, wall and membrane biogenesis and intracellular trafficking and secretion were higher in cultivated fertilized soils than in non cultivated soils. We also observed significant differences in microbiome structure and taxonomic composition between soils under conventional and no-tillage systems. Overall, our results suggest that agronomical land use and the type of tillage system have induced microbiomes to shift their life-history strategies. Microbiomes of cultivated fertilized soils (i.e. higher nutrient amendment) presented tendencies to copiotrophy while microbiomes of non cultivated homogenous soils appeared to have a more oligotrophic life-style. Additionally, we propose that conventional tillage systems may promote copiotrophy more than no-tillage systems by decreasing soil organic matter stability and therefore increasing nutrient availability.

Editor: Kathleen Treseder, UC Irvine, United States of America

Funding: Funding for this work was provided by Agencia Nacional de Promoción Científica y Tecnológica, Argentina-PAE 37164. The funders had no role in study design, data collection and analysis, decision to publish, or preparation of the manuscript.

Competing Interests: The authors have declared that no competing interests exist.

* E-mail: martin.vazquez@indear.com (MPV); belen.carbonetto@indear.com (BC)

Introduction

Agriculture is facing major challenges nowadays. Production will have to double in the next 50 years in order to face growing food demand and bioenergy needs [1,2]. This must be done without increasing environmental threats such as climate change, biodiversity loss and degradation of land and freshwater. Achieving such a goal represents one of the greatest scientific challenges ever. This is in part because of the trade-offs among economic and environmental goals and because of the insufficient knowledge about the biological, biogeochemical and ecological processes that are relevant for sustainable ecosystem functioning [3,4]. Much has been done during the last decade to gain sufficient information on agricultural ecosystem biology, still, more work needs to be done to gain deeper comprehension and to be able to reduce the negative environmental impacts of agriculture [2,5,6]. The main focus should be oriented to soil degradation. Soil fertility, as the capacity to sustain abundant crop production, needs to be preserved. Nowadays soil fertility is maintained by dependence on external inputs; with increasing water contamination [7]. In this context, the key to understand the behavior of life-supporting elements in soil, such as carbon, nitrogen, and phosphorus lies in the fluxes between their various forms in the environment, which are modulated by biology [8]. Comprehension of soil microorganism dynamics is then essential to understand soil processes that affect fertility. Ecological approaches are being taken into account in soil microbial studies trying to address these questions. These approaches involve diversity and functional analyses of soil communities [9,10]. Scholes & Scholes point out that this complex view is necessary for the comprehension of soil systems and that soil restoration of biological processes is the key to achieving lasting food and environmental security [8].

Argentine Pampas is an important player in this scenario. With a plain area of 50 million ha., nearly 50% of the whole Pampas area is devoted to crop production [11]. Cultivation began in the 19th Century in the central humid portion of the region, in soils of high fertility, and spread in last decades to the south and the semiarid west [12]. Soil degradation (i.e. intense erosion and net loss of nutrients and organic carbon) caused by the use of conventional tillage systems were reported in the Pampas [13–17].

Nowadays between 60 and 80% of production is conducted under conservational no-till practices [18]. Extensive research was done to evaluate the effects of reduced tillage and no-tillage systems on soil physical properties, water content, fertility and crop yields[19]. The main outcome of these analyses points that the adoption of limited tillage systems led to soil improvement, by augmenting organic matter content and soil structure. Still, external fertilization is needed in order to restore nutrient levels and fertility regardless the tillage system employed.

Even though the effects of different tillage practices on soil physical and chemical characteristics have been deeply studied, changes in microbial biodiversity and functioning have been poorly reported in Argentinean Pampas. Most works have studied tillage effects on microbial biomass or specific microbial activities (i.e. utilization of specific substrates, extracellular enzyme production, mineralization, etc.) rather than on full microbiome [13,20,21]. Other studies have focused on the behavior of specific bacterial taxa [22,23]. Reports with an ecological approach (i.e. microbial community analysis) have usually focused on individual effects of land use such as the application of herbicides [24,25]. In these cases, biodiversity variability has been assessed using classical fingerprinting techniques (such as RFLP and DGGE) that lack information about microbial taxonomic identity and only capture the most dominant species in the environment [26,27]. In the last few years, 16S amplicon pyrosequencing has been largely implemented to determine microbial diversity and structure of many different ecosystems worldwide [28,29]. This strategy allows a more exhaustive characterization of community patterns and composition. Moreover, some works have incorporated the use of shotgun metagenomics to study the metabolic potential of soil microbiomes [10,30]. The shotgun approach generates a massive amount of data using random high-trhoughput sequencing of soil isolated DNA. This allows the identification of functional capabilities by gene annotation and the comparison of metabolic profiles between samples. To our knowledge, Figuerola et al. [31] were the only authors studying microbial communities in agronomical soils of Argentine Pampas using high throughput sequencing approaches. They observed differences in microbial community composition of soils under no-tillage systems using 16S pyrosequencing. As a novelty, our efforts focused on assessing the impact of long-term agriculture on Pampas soil microbiomes using both shotgun metagenomics and deep 16S amplicon sequencing approaches. We evaluated the effect of more than a hundred years of agronomical land use on both community features and metabolic profiles of soil microbiomes in comparison with nearby control soils with no agricultural records. We also addressed the differences between the effects of two tillage systems: conventional tillage vs. no-tillage on microbial communities.

Several previous studies of soil microbiomes from different parts of the world showed the effects of agronomical land use on soil microbial communities [10,32–35]. Some of these studies showed differences in trophic strategies between microbial communities related to tillage; and most of them were done in experimental plots. As a novelty, we tested the impact of long term agriculture in soils sampled in production fields in the Argentine Pampas, allowing a deeper insight to the effects of intense land use on soil ecosystems functioning. We confirmed the hypothesis that agronomical practices affected Pampas soil microbiomes by promoting a shift of life-history and trophic strategies. We also showed differences in the effects of contrasting tillage systems (i.e. conventional vs. no- tillage) on community taxonomic and metabolic composition on a long term experiment.

Materials and Methods

Sites description and sampling

Soil samples were taken in production and experimental fields between June and August 2010. To address the effects of agricultural land use on soil microbial communities, three different production farms were sampled in the Rolling Pampas area: "La Estrella", "La Negrita" and "Criadero Klein" (See Rascovan et al and Table S1 for details). Rolling Pampas soils are classified as Typic Argiudolls [36] and mean annual rainfall and temperature were1002 mm and 16.8°C respectively. Two treatments were defined: *cultivated* for production plots, and *no cultivated* for farm-houses parks. Production plots were under cultivation for at least one century under conventional tillage systems, with a mixed rotation of pastures and annual grain crops. During the last 15 years before sampling plots were subjected to continuous crop cultivation under no-tillage systems (i.e. minimal soil disturbance, permanent soil cover, rotations and fertilization). The last crop rotation before sampling was wheat-soybean. Nitrogen and phosphorus fertilizers were applied. Samples were collected one month after soybean harvest. Soil samples were also collected nearby the farmers' houses where no agricultural land use (no tillage nor cultivation) was recorded for the last 30 years except from grass mowing. Parks around farmers' houses are usually considered as undisturbed environments in Argentine Pampas [37]. Soils under no land use were covered with grass and other herbaceous (non-woody) plants common in the region such as *Cirsium sp, Trifolium sp, Micropsis sp, Festuca sp, Dichondra sp, Cyperus sp* and *Taraxacum officinale*. For numerical analyses purposes the three farms are treated as experimental replicates. Four soil samples were taken with an auger from the upper 20 cm soil layer in each farm and treatment. A total of 24 samples were collected in Rolling Pampa soils.

In order to compare effects of contrasting tillage systems, samples were also collected in a 34-year-old experiment located in Balcarce in the Southern Pampas (See Rascovan et al and Table S1 for more details). Samples were taken in experimental plots because no production fields are using conventional tillage for crop production nowadays in the Pampas. Soils in Balcarce are a complex of Typic Argiudolls and Petrocalcic Paleudols and mean annual rainfall and temperatures were 875 mm and 13.8°C respectively. The experiment was carried out in three (175 m^2) experimental plots ($n = 3$). Treatments were defined as: *no tillage* (NT) and *conventional tillage* (CT). NT plots had minimal soil disturbance and permanent soil cover combined with rotations; which have included pastures and grain crops (soybean, corn, wheat) during the last 16 years. CT plots were managed with moldboard plough. Nitrogen fertilization was performed in NT and CT plots (60 kg N ha-1). Last rotation before sampling was corn-soybean. Two sub-samples were collected from all treatment and replicate plots a month after soybean harvest. A total of 12 samples were collected.

Samples were immediately sent to the lab after collection. Samples used for DNA purification were air dried and sieved through 1 mm mesh to thoroughly homogenize, break aggregates and remove roots and plant detritus, then stored at −80°C. DNA purification and library preparation was previously described in Rascovan. et al.[38].

None of the sampling sites is located in protected areas. Permissions were obtained directly from each farm owner or manager: Alejandro Cattaneo at La Negrita and La Estrella, Roberto Klein at Criadero Klein and Guillermo Studdert at Balcarce experiment.

Soil chemical and physical measurements

Soil organic carbon was determined by wet digestion and organic matter was estimated [39]. Nitrate-nitrogen was analyzed by 2 M KCL extraction and the phenoldisulfonic acid method [40]. Extractable phosphorus was determined by the Bray method [41]. The pH was measured in a soil:water ratio 1:2.5. Salinity was estimated by the determination of electrical conductivity [42]. Texture analysis was performed by the hydrometer method [43] and nitrogen was determined by Kjeldahl method [40].

16S amplicon sequencing and shotgun metagenomic datasets

To analyze the effect of agronomical practices on soil microbiomes, sequence data from the previously reported Pampa dataset [38] was used. In order to evaluate microbial community structure and taxonomic composition, a total of 112,800 high-quality filtered 16S rRNA gene amplicon sequences, obtained from the 42 soil DNA samples (replicates and subsamples were included). DNA shotgun metagenomic data was used to analyze metabolic profiles. Shotgun metagenomic data completed a total of 10,445,170 sequences. In this case sequences were obtained from one subsample per sampling replicated plot.

In brief, libraries were prepared as follows: DNA was isolated from 10 g of soil of each of the 42 soil samples using the Power MaxSoil DNA Isolation Kit following the manufacturer's instructions (MO BIO Laboratories, Inc.). For amplicon libraries the V4 hyper variable region of the 16s rRNA gene was amplified. Duplicated reactions were performed using barcoded bacterial universal primers containing Roche- 454 sequencing A and B adaptors and a nucleotide multiple identifier (MID) to sort samples: 563F: 5′-CGTATCGCCTCCCTCGCGCCATCA-GACGAGTGCGTAYTGGGYDTAAAGNG -3′ (where AC-GAGTGCGT is an example, different MIDs for each sample were used) and 802R (5′-CTATGCGCCTTGCCAGCCCGCT-CAGTACCRGGGTHTCTAATCC, 5′-CTATGCGCCTTGC-CAGCCCGCTCAGTACCAGAGTATCTAATTC, 5′-CTAT-GCGCCTTGCCAGCCCGCTCAGCTACDSRGGTMTCTA-ATC, 5'-CTATGCGCCTTGCCAGCCCGCTCAGTACNVG-GGTATCTAATCC) [44]. All amplicons were cleaned using Ampure DNA capture beads (Agencourt- Beckman Coulter, Inc.) and pooled in equimolar concentrations before sequencing on a Genome Sequencer FLX (454-Roche Applied Sciences) using Titanium Chemistry according to the manufacturer's instructions.

Shotgun metagenomic libraries were prepared by nebulization, followed by tagging with GS-FLX-Titanium Rapid Library MID Adapters Kit (454-Roche Applied Sciences) and sequenced with a Genome Sequencer FLX (454-Roche Applied Sciences) using Titanium Chemistry according to the manufacturer's instructions. Sequencing runs were performed in INDEAR sequencing facility.

All the sequences used in the present study are available in The Sequence Read Archive (SRA) under accession number SRA058523 and SRA056866. See Rascovan et al. [38] for more information.

Amplicon sequence processing, OTU classification and taxonomic assignment

Sequence data were quality controlled and denoised with the ampliconnoise.py script of QIIME [45].This script also eliminated chimeras. Sequences obtained from Rolling Pampa soil libraries and Balcarce soil libraries were processed separately. Sequences were clustered into Operational Taxonomic Units (OTUs) using the pick_otus.py script with the Uclust method [46] at 97% sequence similarity. Rolling Pampas samples yielded 2,591sequences on average (ranging from 1,455 to 3,991sequences). Balcarce samples yielded 2,329 reads on average (ranging from 1,211 to 4,755 reads).OTU representative sequences were aligned using PyNast algorithm [47] with QIIME default parameters. Phylogenetic trees containing the aligned sequences were then produced using FastTree [48]. All downstream analyses were determined after each sample was randomly rarefied to 70% the number of reads of the smallest sample (i.e. 1,080 reads for Rolling Pampa libraries and 850 reads for Balcarce libraries). Phylogenetic distances between OTUs were calculated using unweighted and weighted Unifrac [49]. Taxonomic classification of sequences was done with Ribosomal Database Project (RDP) Classifier using Greengenes database using a 50% confidence threshold [50,51].

Microbial community analyses

Unifrac phylogenetic pairwise distances among samples were visualized with principal coordinates analysis (PCoA). Analysis of similarity statistics (ANOSIM) was calculated to test a-priori sampling groups. BIOENV analysis was performed to elucidate which soil properties correlated with community patterns. All calculations were carried out with R packages 'BiodiversityR' and 'Vegan' [52,53]. T- tests were performed with QIIME script otu_category_significance.py, and R scripts in order to elucidate differences in read abundances.

Shotgun metagenomic sequence processing and analysis

SSF files obtained from shotgun sequencing runs were uploaded to the MG-RAST webserver [54] for sequence filter and analyses. Reads more than two standard deviations away from the mean read length were discarded. For dereplication removal MG-RAST used a simple k-mer approach to rapidly identify all 20 character prefix identical sequences. This step is required in order to remove artificial duplicate reads. We obtained an average of 1.28×10^6 filtered reads per sample for Rolling Pampa shotgun libraries, and an average of 304,258 filtered reads per sample for Balcarce libraries.

Filtered high quality sequences were assigned to Cluster of orthologous groups (COG) by the MG-RAST sever pipeline using a similarity-based approach. COGs were assigned with a maximum E value of 10^{-20}, an average alignment of 80 amino acids length and 70% average identity.

Relative abundances were calculated by dividing the number of hits for each COG or COG-category by the total number of filtered reads in each sample. Euclidean distances based on relative abundances were calculated between sample pairs. PCoA visualizations and ANOSIM calculations were performed. All calculations were carried out with R packages 'BiodiversityR' and 'Vegan'. T- tests were performed with QIIME script otu_category_significance.py, and R scripts in order to elucidate differences in COG relative abundances between samples.

Results

Microbiome community changes related to agricultural land use

The PCoA visualization revealed clear differences between cultivated and noncultivated soils (ANOSIM $R = 0.8406$, $p \leq 0.001$; Figure 1A).Similar results were obtained when using Bray Curtis distance matrices (Figure S1).

The soil properties (Table S1) that best explained the phylogenetic variation observed in microbial communities were determined using Clarke and Ainsworth's BIOENV analysis. Our results showed that variables that best correlated with community

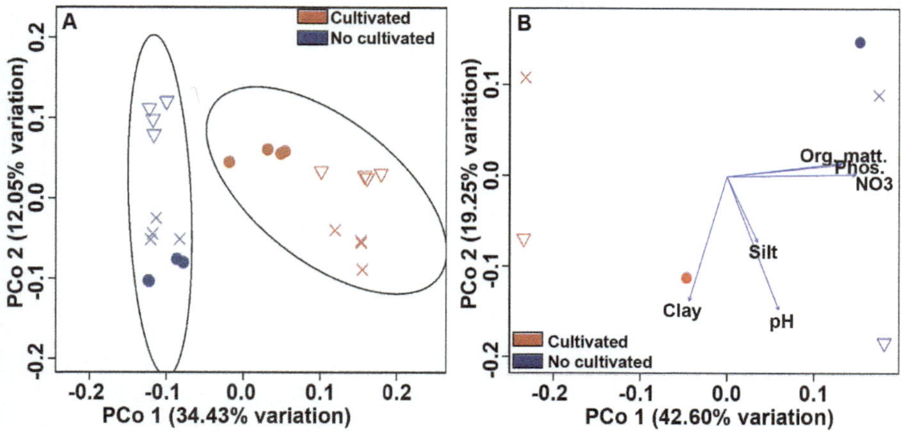

Figure 1. PCoA plots of Pampa production field soil microbiomes based on 97% similarity Weighted Unifrac distance matrices. A) PCoA of cultivated and non-cultivated soil microbiomes. All sub-samples are plotted. Standard error ellipses show 95% confidence areas. B) PCoA biplot of soil properties that best explained variation in community structure. Correlations were calculated using BIOENV on average data of each sampled site (Mantel r = 0.6107, p≤0.05). Circles represent samples from "La Estrella", crosses represent "Criadero Klein" samples and triangles represent "La Negrita" samples.

differences were organic matter, clay and silt content, nitrates, phosphorus and pH (Mantel r = 0.6107, p≤0.05).The PCoA biplot (Figure 1B) showed that organic matter, phosphorus and nitrate levels correlated with the first ordination axis that discriminates between cultivated and non cultivated soils. These three properties were higher in soils under no land use.

Regarding the taxonomic analyses, we observed that members of phyla *Verrucomicrobia, Planctomycetes, Actinobacteria* and *Chloroflexi* were more abundant in non cultivated soils (p≤0.05) (Figure 2). On the other hand, we found that sequences related to *Gemmatimonadetes*, candidate division WS3 and *Nitrospirae* were enriched in cultivated soils (p≤0.05) (Figure 2). No significant differences were found for *Proteobacteria* (for non of the Clases), *Acidobacteria* and *Bacteroidetes* phyla. The ten mentioned taxa represent on average 95% of total sequences of each sample.

Microbiome metagenomic profile changes related to agricultural land use

We found that cultivated and non cultivated soils also clustered apart when metagenomic functional categories were used for the analysis. The first two components of the PCoA explained over 60% of the variability between samples (Figure 3). Standard deviation ellipses overlapped in the ordination plot, indicating that some features are shared between metagenomes. Still, a positive correlation was observed between metabolic and weighted-Unifrac distance matrices (Mantel r: 0.5036 p≤0.05). The analyses of individual COG categories revealed that the relative abundances of COG categories associated with transcription, protein modification, nucleotide transport and metabolism, wall and membrane biogenesis and intracellular trafficking and secretion were higher in cultivated soils (Figure 4, p≤0.05). A deeper analysis inside COG categories revealed that COGs related to Coenzyme A and acetyl-Coa metabolism, energy storage and starvation or quiescence such as, pantothenate kinase, phospho-transacetylase, and trehalose utilization protein were more abundant in non cultivated than in cultivated soils (Figure S2, p≤0.05). On the other hand, COGs related to rapid regulation systems, tricarboxylic acid cycle and nitrogen assimilation such as urease, citrate synthase, glutamate synthase, fumarate hydratase, S-adenosyl- homocysteine hydrolase, S-adenosyl-methionine synthetase, cobalamin biosynthesis protein and ABC transporters, were more abundant in cultivated soils (Figure S2, p≤0.05).

Microbiome community structure and composition related to conventional tillage and no-tillage systems.

To compare the structure of microbiomes under different tillage systems, we collected samples from an experimental field located in Balcarce in the Southern Pampas. The 34-year-old experiment compared two tillage systems: no-tillage (NT) and conventional tillage (CT). Weighted Unifrac analysis showed differences in community structure associated to the tillage system employed

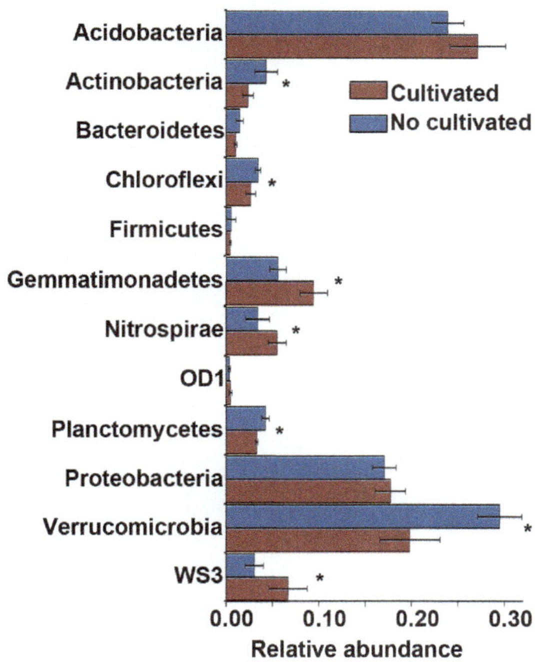

Figure 2. Relative abundances of taxonomic groups in Pampa production field soil microbiomes. Bars represent ± 1 standard error. (*) indicate significant differences (p≤0.05).

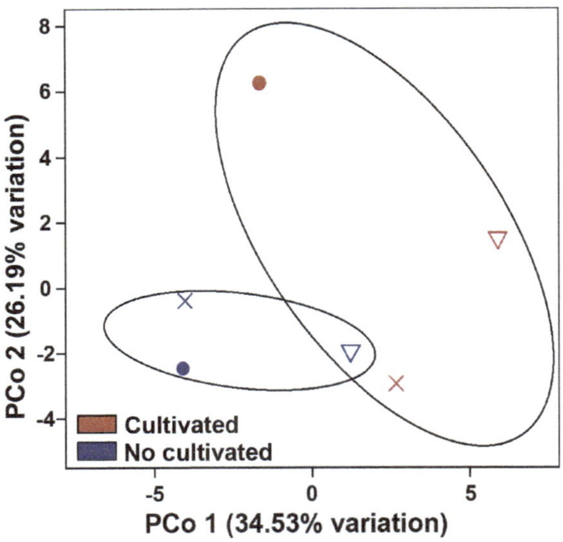

Figure 3. PCoA plot of metagenomic data based on Euclidean distance matrices of COG categories of Pampa production field soil microbiomes. Standard error ellipses show 95% confidence areas. Circles represent samples from "La Estrella", crosses represent "Criadero Klein" samples and triangles represent "La Negrita" samples.

(ANOSIM R = 0.9009, p<0.05, Figure 5A). The first axis explained 31.56% of total variation and separated NT from CT. Additionally, we showed that nitrates were the only soil variable (Table S1) that significantly correlated with community structure (BIOENV analysis, Mantel r = 0.7721, p≤0.01, Figure 5B).

Moreover, microbiomes of CT and NT soils also differed in taxonomic composition. Members of *Acidobacteria*, *Gemmatimonadetes*, candidate division TM7 and class *Gammaproteobacteria* were more abundant in CT soils, while *Nitrospirae*, candidate divisionWS3 and *Deltaproteobacteria* were more represented in NT soils (Figure 6, p≤0.05).

Microbiome metabolic profiles related to conventional tillage and no- tillage systems

Variation in metagenomic profiles between CT and NT microbiomes was analyzed with PCoA based on Euclidean distance matrices of COG abundances. We could not find significant differences in overall profile metabolic structure between tillage systems (Figure S3). Additionally, we did not find significant correlation between Euclidean metabolic matrices and phylogenetic matrices. However, categories related to intracellular trafficking and secretion, amino acid transport and metabolism, and energy production and conversion were shown to be more abundant in soil under CT (Figure 7).

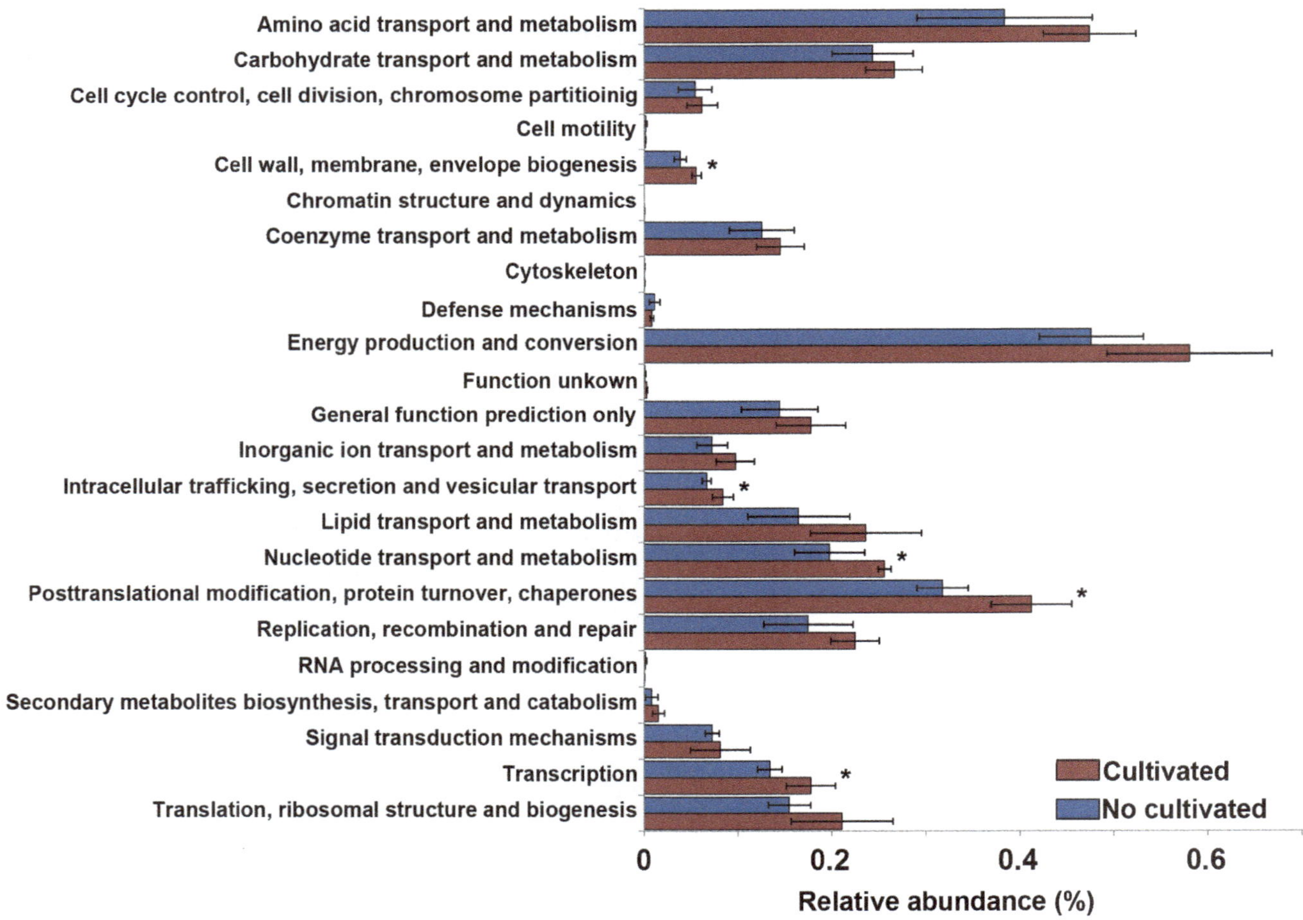

Figure 4. Relative abundances of COG categories in Pampas production field soil microbiomes. Bars represent ±1 standard error. (*) indicate significant differences (p≤0.05).

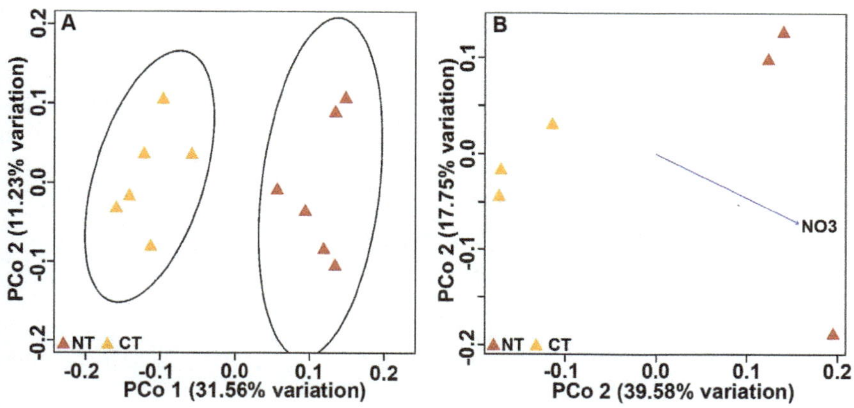

Figure 5. PCoA plots of Balcarce experimental field soil microbiomes based on 97% similarity Weighted Unifrac distance matrices. A) CA: conventional tillage; NT: no-tillage. All sub-samples are plotted. Standard error ellipses show 95% confidence areas. B) PCoA biplot, nitrate was the variable that best explained variation in community structure. Correlations were calculated using BIONEV on average data of each experimental plot (Mantel r = 0.7721, p≤0.01).

Discussion

Much work still needs to be done to get a comprehensive view of the soil microbiomes. Our work is one of the firsts done in the Argentine Pampas at this resolution level, with a combination of metagenomic and phylogentic approaches; and it is aimed to contribute to the comprehension of soil microbiomes function and dynamics. In that context, our results are in agreement with previous works that showed differences in soil microbial community structure and taxonomic composition due to the presence of agricultural land use [9,10,32,33].

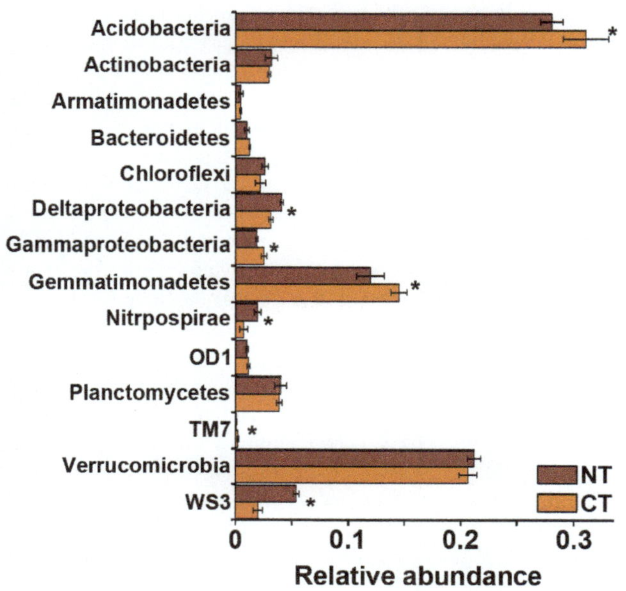

Figure 6. Comparison of relative abundances of taxonomic groups in NT and CT soils. CA: conventional tillage; NT: no-tillage. Bars represent ±1 standard error. (*) indicate significant differences (p≤ 0.05).

Effects of agricultural land use on community composition of soil microbiomes in Argentine Pampas

We observed effects on taxonomic composition at the phylum level. *Verrucomicrobia, Plactomycetes, Actinobacteria* and *Chloroflexi* were more abundant in soils that were never cultivated; while *Gemmatimonadetes, Nitrospirae* and WS3 were more abundant in crop cultivated soils. These results are in agreement with the copiotroph/oligotroph hypothesis [55], that propose that high number of oligotrophic prokaryotes may be found in bogs and soils with high amounts of recalcitrant organic matter [56]. On the other hand, copiotrophic organisms are able to use labile nutrient fractions and to grow at higher rates as a consequnece. In our study, non-cultivated soils are considered to be oligotrophic since they present high levels of organic matter highly rich in humic acids [12] and cultivated soils as copiotrophic environments due to fertilization and the seasonal presence of crop residues, which increases organic matter and nitrogen accessibility [57]. Under this assumption, bacteria in non-cultivated soils are expected to be K selected and to present low growth rates and very efficient nutrient uptake systems with higher substrate affinities. In contrast, bacteria in cultivated soils are expected to be r-selected and to have higher rates of activity per biomass unit, higher turnover rates and faster growth rates. Our results showed a trend toward these statements since a reduction in the abundance of taxa with oligotrophic characteristics, such as *Verrucomicrobia* [34], and *Planctomycetes* [58] were detected in cultivated fertilized soils. Moreover, this is in agreement with recent findings that confirm a correlation between *Verrucomicrobia* abundance patterns and conditions of limited nutrient availability in Prairie Soils in the United States [9]. On the other hand, the relative abundance of phylum *Gemmatimonadetes* was increased in fertilized cultivated soils as previously described for nitrogen-fertilized forest soils [59]. Consistently with our results, these authors observed that nitrogen fertilization was related to a higher abundance of *Gemmatimonadetes* and detected no presence of *Verrucomicrobia*. Little is known about *Gemmatimonadetes* ecology and metabolism since only one representative from this phylum has been isolated and characterized [60]. Even though, their presence in environments with a wide range of nutrient concentrations and redox states suggests versatile metabolisms [61].

We can also say that cultivated soils are more heterogeneous environments than non cultivated soils. Crop rotation, periodic fertilization and pesticide application generate temporal and

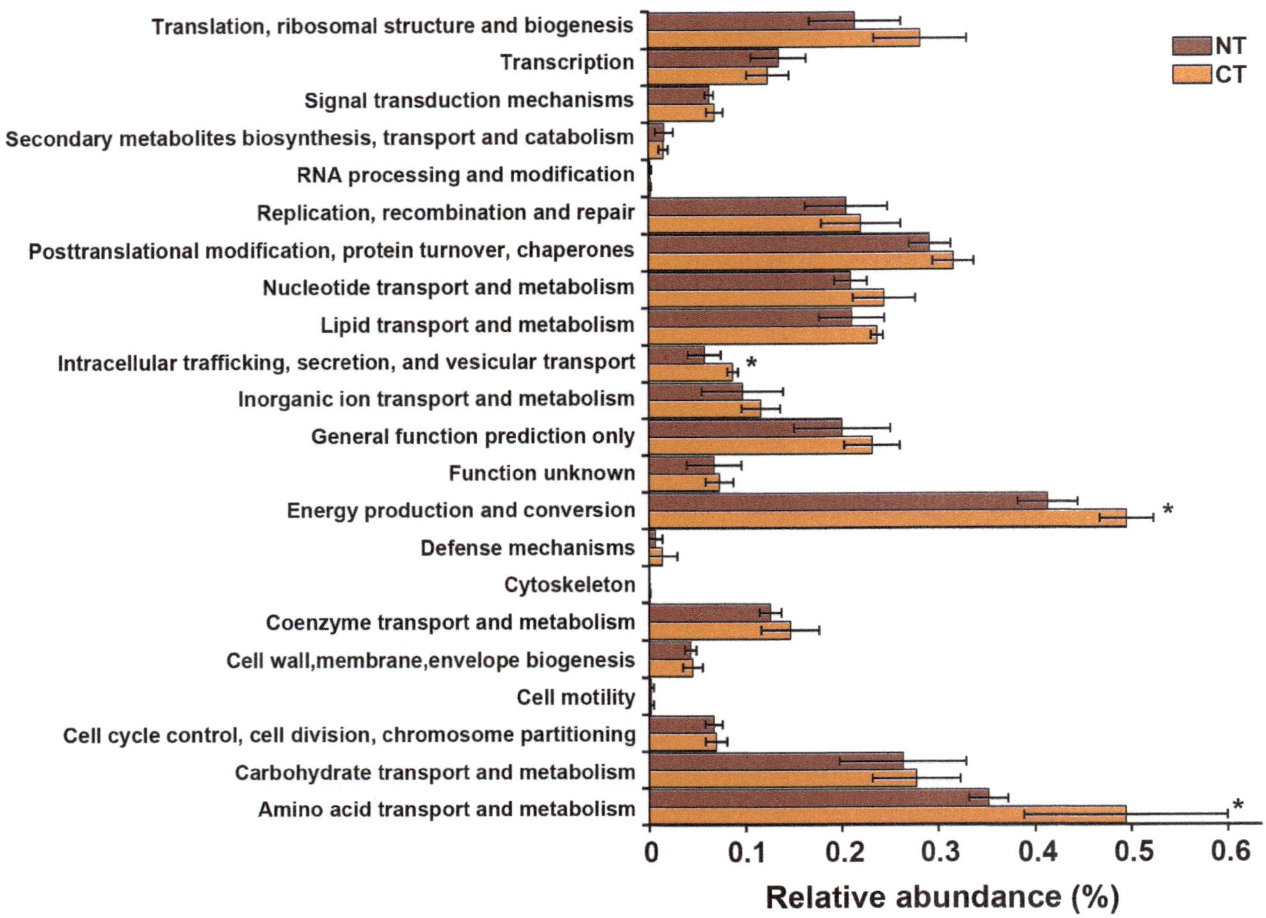

Figure 7. Relative abundances of COG categories in soil microbiomes of tillage systems comparison experiment in Balcarce. CA: conventional tillage; NT: no-tillage. Bars represent ±1 standard erro. (*)indicate significant differences ($p \leq 0.05$).

spatial changes in soil chemical properties, and therefore in nutrient availability and approachability for microorganisms. Bacteria dominating in this type of soils should be adapted to heterogeneity. These microorganisms should be able to fine-tune carbon and nitrogen intakes according to their metabolic needs under frequent external changes. This seems to be the case for *Gemmatimonadetes* suggesting a generalist ecological strategy. The high abundance of *Nitrospirae* may also respond to heterogeneous conditions. It has been proposed that *Nitrospirae* lineages occupy different positions on an imaginary scale reaching from K- to r-strategies [62]. We hypothesize that in cultivated soils, *Nitrospirae* K-strategists would be exploiting nitrite in high N microenvironments, while r-strategist would be mining low concentration areas in a nitrogen gradient enhanced by fertilization [63]. This competition would be less fierce in non cultivated soils due to the highest homogeneity and a less marked nitrogen gradient due to the lack of fertilization and more recalcitrant organic matter forms.

Effects of agricultural land use on metabolic profiles of soil microbiomes in Argentine Pampas

Our results from shotgun metagenomic data also indicated a tendency for the microbiomes of cultivated soils towards adaptation to nutrient heterogeneity. The highest relative abundances of sequences assigned to COGs related to transcription, protein modifications, nucleotide transport and metabolism, wall and membrane biogenesis and intracellular trafficking and secretion in cultivated fertilized soils are consistent with a copiotrophic strategy (i.e. rapid tight metabolism regulation and fast grow rates).Moreover, some of these COG categories were previously shown to be up-represented in copiotrophic marine microorganism genomes [58]. In a deeper look we detected that the relative abundance of sequences assigned to riboswitch regulated genes was higher in cultivated soils (i.e. cobalamin biosynthesis protein, S-adenosylmethionine synthetase, S-adenosyl- homocysteine hydrolase) [64–66]. These ancient regulators may be playing a central role in a life-style strategy adapted to nutrient heterogeneity since they were described to be the most 'economical' and fast-reacting regulatory systems (no intermediate factors involved) [67]. Moreover, the abundant riboswitch- COGs in cultivated soils were related to synthesis of B vitamins. The levels of these vitamins have already been linked to differences in community composition in marine ecosystems [68]. More studies are needed in order to address this relationship in soil environments; still, our results suggest an important role of B-vitamins in fertilized heterogeneous soils.

Another interesting observation was the highest abundance of glutamate synthase (GOGAT) related COGs in cultivated soils metagenomic profiles. The combined role of GOGAT with glutamine synthetase and glutamate dehydrogenase allows the cells to sense ammonia external levels [69]. The high abundance of this regulation system detected in cultivated soil metagenomes suggests its importance in the detection of N fluxes related with fertilization. Moreover COGs related to the tricarboxylic acid cycle (TCA), such as citrate synthase and succinate dehydrogenase, were more abundant in cultivated soil microbiomes. GOGAT nitrogen assimilation pathway and TCA are related. It has been shown that the concentration of compounds of both pathways changes considerably and rapidly upon nitrogen up shift; in contrast, the concentrations of glycolytic intermediates remains homeostatic [70]. In addition, abundance of urease related COGs were also higher in cultivated soils. The soils sampled in this study have a long history of agricultural land use and have been long fertilized with both type of nitrogen (i.e. ammonia-based and urea-based fertilizers). This kind of environmental pressure finally selected a microbiome adapted to changing N sources and availability.

It is important to mention that these results do not infer that expression or activity of these metabolisms will be necessarily increased in cultivated soil microbiomes, still, the highest abundance of these COGs could be reflected in a highest diversification and specialization. The presence of a highest copy number of strategic genes have already been linked to copiotrophic or oligotrophic life-styles [58].

Microbiomes of non cultivated soils showed a higher abundance of sequences related to Coenzyme A and acetyl-Coa metabolism than microbiomes of cultivated soils. It is known that acetyl-CoA is a fundamental building block and energy source [71]. High acetyl-CoA levels would indicate a "proliferative" or "fed" state, while low acetyl-CoA levels (and high CoA levels) would be indicative of a "quiescent" or "starved" state. Pantothenate kinase (PanK), the key enzyme in CoA syntehsis, was also highest in non cultivated soil microbiomes [72]. In addition, it was stated that some oligotrophs preferentially use lipids as immediate and stored sources of carbon and energy in marine environments [58]. The observed abundance of Pank genes may ensure a correct CoA intracellular level in fasted moments, allowing proper lipid utilization. In addition, non cultivated metagenomic profiles showed higher abundance of sequences related to trehalose utilization. Trehalose is known to serve as energy source in many microorganisms [73]. Members of *Actinobacteria* genera *Mycobacterium* and family *Frankiaceae* are known to produce and/or utilize trehalose [74–76]. As mentioned above, our results showed higher abundance of *Actinobacteria*-related sequences in non cultivated soils. Moreover, sequences classified within family *Frankiaceae* were only present in these soils and the abundance of sequences assigned as *Mycobacterium* was higher than in cultivated soils (not shown). These observations are congruent with an oligotrophic strategy based on the use of storage components for carbon and energy sources.

The responses of the phylogenetic structure and metagenomic profile to agronomic land use were significantly correlated, suggesting some degree of correspondence between these different microbiome features. These results agree with previous observations done in soils from different biomes [9,77]. Moreover, Fierer et al. found similar a similar correlation between metagenomic and phylogenetic data for microbial communities of agricultural soils under different nitrogen gradients in experimental plots [10].

Correlation with soil properties

In addition, soil properties such as organic matter, phosphorus and nitrate levels explained most of the variability observed between cultivated and non cultivated soils. As mentioned, the highest amount of organic matter is found mostly in a recalcitrant form in these soils. Moreover, it was already established that nitrate accumulation exhibits a negative correlation with organic carbon availability [78]. In addition it has been proved that organic matter source and quality played an important role in regulating the magnitude of carbon metabolism and could be as important as nutrient abundance in water environments [79]. Our results are congruent with these observations since non cultivated soils, with highest levels of organic carbon, presented metagenomic profiles with tendencies to oligotrophy and the best explanation for this scenario is the low lability of the recalcitrant forms of carbon and nitrogen.

Assessing the effects of different tillage systems on Pampas soil microbiomes

Our results also showed differences in the structure and composition of soil microbiomes between no-till and conventional tillage soils. Sequences related to phyla *Gemmatimonadetes*, candidate division TM7 and *Acidobacteria*, were highest in CT soils, while the abundances of *Nitrospirae* and candidate division WS3 were highest in NT soils. Tillage is the principal agent producing soil disturbance and subsequent soil structure modification [80,81]. The negative effects of CT in soil stabilization and macroaggregate losses were previously registered in Pampas soils [82]. It has also been shown that NT increases macroaggregate abundance and organic matter content [83].These increments are related with a more recalcitrant organic matter, with increased humic acid contents and nutrient retention [84–86]. The higher abundance of *Nitrospirae* in NT soils reinforces the idea of a community adapted to a better N mining in these environments. Moreover, we observed very low abundance of *Nitrobacter* related sequences in Balcarce soils and no significant difference was observed between CT and NT soils (not shown). Changes in ammonia oxidation due to a decrease in ammonia availability by humic substances have already been proposed in microcosm experiments [87]. If this is the case of Balcarce soils, higher humic acids in NT soils would be decreasing ammonia availability for oxidation into nitrite and therefore decreasing nitrate availability, compared to CT soils. The predicted highest stability of organic matter and abundance of macroaggregates in NT soils are probably generating more marked nitrite gradients than in CT soils. The highest abundance of *Nitrospirae* in NT may be reflecting a highest lineage diversity that would be better adapted to these gradients.

In addition, communities under NT showed higher abundance of sequences related to the order *Syntrophobacterales* (*Deltaproteobacteria*, Figure S4). These are known anaerobic and syntrophic organisms [88,89]. These characteristics may be an advantage in stable soils with higher number of highly humic macro aggregates since syntrophy is known to be important for community functioning in micro-environments with low nutrient levels [90]. On the other hand, CT soils presented higher relative abundance of *Gammaproteobacteria* related sequences. The order *Xanthomonadales* was the main responsible for these differences (Figure S4). This taxon has already been associated to CT practices [32].

Even though we could not find significant differences associated to tillage systems at the community structure level for metabolic profiles, COG categories related to intracellular trafficking and secretion, amino acid transport and metabolism and energy production and conversion were more abundant in CT mirobiomes. These results suggest a tendency of CT microbiomes to a more copiotrophic life-style strategy than NT microbiomes.

Conclusion

Our results are consistent with the hypothesis that microbiomes exhibit different life- history and trophic strategies in Pampean soils under different land uses and tillage systems. Our data suggest that microbiomes of fertilized cultivated soils have more flexible metabolisms adapted to nutrient fluxes with tendencies to copiothropy while microorganisms in non cultivated soils are better adapted to lowest external nutrient availability and homogenous environment. The lowest nutrient accessibility in non cultivated soils may be explained by the higher amount of humic substances, recalcitrant organic matter and the lack of fertilizer amendments. Moreover, NT soils, with most stable structure and highest macroaggregate abundance, presented microbiomes better adapted to recalcitrant environments; while CT microbiomes presented a higher tendency to copiotrophy.

This work is of major contribution to understand how historical changes in soil properties due to agronomical land use have altered the diversity and function of below-ground communities. The importance of high-throughput characterization for the reconstruction of pre- agricultural microbiomes is being reinforced nowadays [9]. Following this direction, our findings will be very useful in future restoration and monitoring programs of Argentine Pampas ecosystems.

Supporting Information

Figure S1 PCoA plots of Pampa production field soil microbiomes based on average Bray Curtis distance matrices. A) PCoA of cultivated and non cultivated soil microbiomes. Standard error ellipses show 95% confidence areas. B) PCoA biplot of soil properties that best explained variation in community structure. Correlations were calculated using BIOENV on average data of each sampled site (Mantel $r = 0.6214$, $p \leq 0.05$). Circles represent samples from "La Estrella", crosses represent "Criadero Klein" samples and triangles represent "La Negrita" samples.

Figure S2 Relative abundances of Cluster of Orthologous groups (COGs) in Pampa production field soil microbiomes. Bars represent ± 1 standard error. Only significant COGs are showed.

Figure S3 Comparison of tillage systems effects on the structure of metabolic profiles. PCoA plot based on Euclidean distance matrices. CA: conventional tillage; NT: no-tillage. Standard error ellipses show 95% confidence areas.

Figure S4 Relative abundances of reads assigned to orders within classes *Gammaproteobacteria* and *Deltaporteobacteria* in NT and CT soils.

Table S1 Soil chemical and physical properties.

Acknowledgments

We acknowledge Germán F. Domínguez and Guillermo A. Studdert from the "Manejo Sustentable del suelo" research group (from Facultad de Ciencias Agrarias, Universidad Nacional de Mar del Plata, Unidad Integrada Balcarce) for providing samples of the tillage systems experiment. We thank Gonzalo Berhongaray, Josefina de Paepe, and María Rosa Mendoza for his help during sampling in the Rolling Pampas. We also acknowledge Roxana Colombo and Marcelo Soria for helpful discussion and advice.

Author Contributions

Conceived and designed the experiments: BC NR RÁ AM MV. Performed the experiments: BC NR. Analyzed the data: BC. Contributed reagents/materials/analysis tools: BC NR RÁ AM MV. Wrote the paper: BC NR MV. Obtained permission for sampling: RÁ.

References

1. Foley JA, Ramankutty N, Brauman KA, Cassidy ES, Gerber JS, et al. (2011) Solutions for a cultivated planet. Nature 478: 337–342.
2. Tilman D, Balzer C, Hill J, Befort BL (2011) Global food demand and the sustainable intensification of agriculture. Proc Natl Acad Sci U S A 108: 20260–20264.
3. Tilman D, Cassman KG, Matson PA, Naylor R, Polasky S (2002) Agricultural sustainability and intensive production practices. Nature 418: 671–677.
4. Balmford A, Green RE, Scharlemann JPW (2005) Sparing land for nature: exploring the potential impact of changes in agricultural yield on the area needed for crop production. Global Change Biology 11: 1594–1605.
5. Power AG (2010) Ecosystem services and agriculture: tradeoffs and synergies. Philos Trans R Soc Lond B Biol Sci 365: 2959–2971.
6. Foley JA, Defries R, Asner GP, Barford C, Bonan G, et al. (2005) Global consequences of land use. Science 309: 570–574.
7. Bennett EM, Carpenter SR, Caraco NF (2001) Human Impact on Erodable Phosphorus and Eutrophication: A Global Perspective: Increasing accumulation of phosphorus in soil threatens rivers, lakes, and coastal oceans with eutrophication. BioScience 51: 227–234.
8. Scholes MC, Scholes RJ (2013) Dust Unto Dust. Science 342: 565–566.
9. Fierer N, Ladau J, Clemente JC, Leff JW, Owens SM, et al. (2013) Reconstructing the Microbial Diversity and Function of Pre-Agricultural Tallgrass Prairie Soils in the United States. Science 342: 621–624.
10. Fierer N, Lauber CL, Ramirez KS, Zaneveld J, Bradford MA, et al. (2012) Comparative metagenomic, phylogenetic and physiological analyses of soil microbial communities across nitrogen gradients. ISME J 6: 1007–1017.
11. Satorre EH, Slafer GA (1999) Wheat: Ecology and Physiology of Yield Determination: Taylor & Francis.
12. Hall AJ, Rebella CM, Ghersa CM, Culot JP (1992) Field crop systems of the Pampas. In: Pearson CJ, editor. Field Crop Ecosystems of the World. Amsterdam: Elsevier Science & Technology Books. pp. 413–450.
13. Alvarez R, Díaz RA, Barbero N, Santanatoglia OJ, Blotta L (1995) Soil organic carbon, microbial biomass and CO2-C production from three tillage systems. Soil and Tillage Research 33: 17–28.
14. Bernardos JN, Viglizzo EF, Jouvet V, Lértora FA, Pordomingo AJ, et al. (2001) The use of EPIC model to study the agroecological change during 93 years of farming transformation in the Argentine pampas. Agricultural Systems 69: 215–234.
15. Alvarez R (2001) Estimation of carbon losses by cultivation from soils of the Argentine Pampa using the Century Model. Soil Use and Management 17: 62–66.
16. Hevia GG, Buschiazzo DE, Hepper EN, Urioste AM, Antón EL (2003) Organic matter in size fractions of soils of the semiarid Argentina. Effects of climate, soil texture and management. Geoderma 116: 265–277.
17. Quiroga AR, Buschiazzo DE, Peinemann N (1996) Soil Organic Matter Particle Size Fractions in Soils of the Semiarid Argentinian Pampas. Soil Science 161.
18. Kassam A, Friedrich T, Shaxson F, Pretty J (2009) The spread of Conservation Agriculture: justification, sustainability and uptake. International Journal of Agricultural Sustainability 7: 292–320.
19. Alvarez R, Steinbach HS (2009) A review of the effects of tillage systems on some soil physical properties, water content, nitrate availability and crops yield in the Argentine Pampas. Soil and Tillage Research 104: 1–15.
20. Gomez E, Bisaro V, Conti M (2000) Potential C-source utilization patterns of bacterial communities as influenced by clearing and land use in a vertic soil of Argentina. Applied Soil Ecology 15: 273–281.
21. Aon MA, Cabello MN, Sarena DE, Colaneri AC, Franco MG, et al. (2001) I. Spatio-temporal patterns of soil microbial and enzymatic activities in an agricultural soil. Applied Soil Ecology 18: 239–254.
22. Agaras B, Wall LG, Valverde C (2012) Specific enumeration and analysis of the community structure of culturable pseudomonads in agricultural soils under no-till management in Argentina. Applied Soil Ecology 61: 305–319.
23. Nievas F, Bogino P, Nocelli N, Giordano W (2012) Genotypic analysis of isolated peanut-nodulating rhizobial strains reveals differences among populations obtained from soils with different cropping histories. Applied Soil Ecology 53: 74–82.
24. Zabaloy MC, Garland JL, Gómez MA (2008) An integrated approach to evaluate the impacts of the herbicides glyphosate, 2,4-D and metsulfuron-methyl on soil microbial communities in the Pampas region, Argentina. Applied Soil Ecology 40: 1–12.
25. Zabaloy MC, Gómez E, Garland JL, Gómez MA (2012) Assessment of microbial community function and structure in soil microcosms exposed to glyphosate. Applied Soil Ecology 61: 333–339.

26. Deng W, Xi D, Mao H, Wanapat M (2008) The use of molecular techniques based on ribosomal RNA and DNA for rumen microbial ecosystem studies: a review. Molecular Biology Reports 35: 265–274.
27. Pontes D, Lima-Bittencourt C, Chartone-Souza E, Amaral Nascimento A (2007) Molecular approaches: advantages and artifacts in assessing bacterial diversity. Journal of Industrial Microbiology & Biotechnology 34: 463–473.
28. Sogin ML, Morrison HG, Huber JA, Welch DM, Huse SM, et al. (2006) Microbial diversity in the deep sea and the underexplored "rare biosphere". Proceedings of the National Academy of Sciences 103: 12115–12120.
29. Fortunato CS, Herfort L, Zuber P, Baptista AM, Crump BC (2012) Spatial variability overwhelms seasonal patterns in bacterioplankton communities across a river to ocean gradient. ISME J 6: 554–563.
30. Delmont TO, Prestat E, Keegan KP, Faubladier M, Robe P, et al. (2012) Structure, fluctuation and magnitude of a natural grassland soil metagenome. ISME J 6: 1677–1687.
31. Figuerola ELM, Guerrero LD, Rosa SM, Simonetti L, Duval ME, et al. (2012) Bacterial Indicator of Agricultural Management for Soil under No-Till Crop Production. PLoS ONE 7: e51075.
32. Souza RC, Cantão ME, Vasconcelos ATR, Nogueira MA, Hungria M (2013) Soil metagenomics reveals differences under conventional and no-tillage with crop rotation or succession. Applied Soil Ecology 72: 49–61.
33. Lauber CL, Ramirez KS, Aanderud Z, Lennon J, Fierer N (2013) Temporal variability in soil microbial communities across land-use types. ISME J 7: 1641–1650.
34. Ramirez KS, Craine JM, Fierer N (2012) Consistent effects of nitrogen amendments on soil microbial communities and processes across biomes. Global Change Biology 18: 1918–1927.
35. Ramirez K, Lauber CL, Knight R, Bradford MF (2010) Consistent effects of nitrogen fertilization on soil bacterial communities in contrasting systems. Ecology: 3463–3470
36. Lavado RS (2008) La Región Pampeana: historia, características y uso de sus suelos. In: Alvarez R, editor. Materia Orgánica Valor agronómico y dinámica en suelos pampeanos. Buenos Aires: Editorial Facultad de Ingenieria. pp. 1–11.
37. Berhongaray G, Alvarez R, De Paepe J, Caride C, Cantet R (2013) Land use effects on soil carbon in the Argentine Pampas. Geoderma 192: 97–110.
38. Rascovan N, Carbonetto B, Revale S, Reinert M, Alvarez R, et al. (2013) The PAMPA datasets: a metagenomic survey of microbial communities in Argentinean pampean soils. Microbiome 1: 21.
39. Nelson DW, Sommers LE, Sparks DLE, Page ALE, Helmke PAE, et al. (1996) Total Carbon, Organic Carbon, and Organic Matter. Methods of Soil Analysis Part 3-Chemical Methods: Soil Science Society of America, American Society of Agronomy. pp. 961–1010.
40. Bremner JME, Sparks DLE, Page ALE, Helmke PAE, Loeppert RH (1996) Nitrogen-Total. Methods of Soil Analysis Part 3-Chemical Methods: Soil Science Society of America, American Society of Agronomy. pp. 1085–1121.
41. Kuo SE, Sparks DLE, Page ALE, Helmke PAE, H LR (1996) Phosphorus. Methods of Soil Analysis Part 3-Chemical Methods: Soil Science Society of America, American Society of Agronomy. pp. 869–919.
42. Rhoades JDE, Sparks DLE, Page ALE, Helmke PAE, H. LR (1996) Salinity: Electrical Conductivity and Total Dissolved Solids. Methods of Soil Analysis Part 3-Chemical Methods: Soil Science Society of America, American Society of Agronomy. pp. 417–435.
43. Gee GW, Bauder JWEKA (1986) Particle-size Analysis. Methods of Soil Analysis: Part 1—Physical and Mineralogical Methods: Soil Science Society of America, American Society of Agronomy. pp. 383–411.
44. Cole J, Wang Q, Cardenas E, Fish J, Chai B, et al. (2009) The Ribosomal Database Project: improved alignments and new tools for rRNA analysis. Nucleic acids research: 141–145.
45. Caporaso JG, Kuczynski J, Stombaugh J, Bittinger K, Bushman FD, et al. (2010) QIIME allows analysis of high-throughput community sequencing data. Nat Meth 7: 335–336.
46. Edgar RC (2010) Search and clustering orders of magnitude faster than BLAST. Bioinformatics 26: 2460–2461.
47. Caporaso JG, Bittinger K, Bushman FD, DeSantis TZ, Andersen GL, et al. (2010) PyNAST: a flexible tool for aligning sequences to a template alignment. Bioinformatics 26: 266–267.
48. Price MN, Dehal PS, Arkin AP (2009) FastTree: Computing Large Minimum Evolution Trees with Profiles instead of a Distance Matrix. Molecular Biology and Evolution 26: 1641–1650.
49. Lozupone C, Knight R (2005) UniFrac: a New Phylogenetic Method for Comparing Microbial Communities. Applied and Environmental Microbiology 71: 8228–8235.
50. DeSantis TZ, Hugenholtz P, Larsen N, Rojas M, Brodie EL, et al. (2006) Greengenes, a Chimera-Checked 16S rRNA Gene Database and Workbench Compatible with ARB. Applied and Environmental Microbiology 72: 5069–5072.
51. Wang Q, Garrity GM, Tiedje JM, Cole JR (2007) Naïve Bayesian Classifier for Rapid Assignment of rRNA Sequences into the New Bacterial Taxonomy. Applied and Environmental Microbiology 73: 5261–5267.
52. Kindt R, Coe R (2005) Tree Diversity Analysis: A Manual and Software for Common Statistical Methods for Ecological and Biodiversity Studies: World Agrofirestry Centre.
53. Dixon P (2003) VEGAN, a package of R functions for community ecology. Journal of Vegetation Science 14: 927–930.
54. Meyer F, Paarmann D, D'Souza M, Olson R, Glass EM, et al. (2008) The metagenomics RAST server - a public resource for the automatic phylogenetic and functional analysis of metagenomes. BMC Bioinformatics 9: 386.
55. Fierer N, Bradford MA, Jackson RB (2007) Toward an ecological classification of soil bacteria. Ecology 88: 1354–1364.
56. Dion P, Nautiyal CS (2008) Microbiology of Extreme Soils: Springer.
57. Galantini J, Rosell R (2006) Long-term fertilization effects on soil organic matter quality and dynamics under different production systems in semiarid Pampean soils. Soil and Tillage Research 87: 72–79.
58. Lauro FM, McDougald D, Thomas T, Williams TJ, Egan S, et al. (2009) The genomic basis of trophic strategy in marine bacteria. Proceedings of the National Academy of Sciences 106: 15527–15533.
59. Nemergut DR, Townsend AR, Sattin SR, Freeman KR, Fierer N, et al. (2008) The effects of chronic nitrogen fertilization on alpine tundra soil microbial communities: implications for carbon and nitrogen cycling. Environmental Microbiology 10: 3093–3105.
60. Zhang H, Sekiguchi Y, Hanada S, Hugenholtz P, Kim H, et al. (2003) Gemmatimonas aurantiaca gen. nov., sp. nov., a Gram-negative, aerobic, polyphosphate-accumulating micro-organism, the first cultured representative of the new bacterial phylum Gemmatimonadetes phyl. nov. International Journal of Systematic and Evolutionary Microbiology 53: 1155–1163.
61. DeBruyn JM, Nixon LT, Fawaz MN, Johnson AM, Radosevich M (2011) Global Biogeography and Quantitative Seasonal Dynamics of Gemmatimonadetes in Soil. Applied and Environmental Microbiology 77: 6295–6300.
62. Maixner F, Noguera DR, Anneser B, Stoecker K, Wegl G, et al. (2006) Nitrite concentration influences the population structure of Nitrospira-like bacteria. Environmental Microbiology 8: 1487–1495.
63. Attard E, Poly F, Commeaux C, Laurent F, Terada A, et al. (2010) Shifts between Nitrospira- and Nitrobacter-like nitrite oxidizers underlie the response of soil potential nitrite oxidation to changes in tillage practices. Environmental Microbiology 12: 315–326.
64. Edwards AL, Reyes FE, Héroux A, Batey RT (2010) Structural basis for recognition of S-adenosylhomocysteine by riboswitches. RNA 16: 2144–2155.
65. Loenen WA (2006) S-adenosylmethionine: jack of all trades and master of everything? Biochem Soc Trans 34: 330–333.
66. Winkler WC, Breaker RR (2005) Regulation of bacterial gene expression by riboswitches. Annual Review of Microbiology 59: 487–517.
67. Nudler E, Mironov AS (2004) The riboswitch control of bacterial metabolism. Trends in biochemical sciences 29: 11–17.
68. Sañudo-Wilhelmy SA, Cutter LS, Durazo R, Smail EA, Gómez-Consarnau L, et al. (2012) Multiple B-vitamin depletion in large areas of the coastal ocean. Proceedings of the National Academy of Sciences.
69. Yan D (2007) Protection of the glutamate pool concentration in enteric bacteria. Proceedings of the National Academy of Sciences 104: 9475–9480.
70. Doucette CD, Schwab DJ, Wingreen NS, Rabinowitz JD (2011) α-ketoglutarate coordinates carbon and nitrogen utilization via enzyme I inhibition. Nat Chem Biol 7: 894–901.
71. Cai L, Tu BP (2011) On Acetyl-CoA as a Gauge of Cellular Metabolic State. Cold Spring Harbor Symposia on Quantitative Biology 76: 195–202.
72. Leonardi R, Rehg JE, Rock CO, Jackowski S (2010) Pantothenate Kinase 1 Is Required to Support the Metabolic Transition from the Fed to the Fasted State. PLoS ONE 5: e11107.
73. Elbein AD, Pan YT, Pastuszak I, Carroll D (2003) New insights on trehalose: a multifunctional molecule. Glycobiology 13: 17R–27R.
74. Barabote RD, Xie G, Leu DH, Normand P, Necsulea A, et al. (2009) Complete genome of the cellulolytic thermophile Acidothermus cellulolyticus 11B provides insights into its ecophysiological and evolutionary adaptations. Genome Research 19: 1033–1043.
75. Lopez MF, Fontaine MS, Torrey JG (1984) Levels of trehalose and glycogen in Frankia sp. HFPArI3 (Actinomycetales). Canadian Journal of Microbiology 30: 746–752.
76. Tropis M, Meniche X, Wolf A, Gebhardt H, Strelkov S, et al. (2005) The Crucial Role of Trehalose and Structurally Related Oligosaccharides in the Biosynthesis and Transfer of Mycolic Acids in Corynebacterineae. Journal of Biological Chemistry 280: 26573–26585.
77. Fierer N, Leff JW, Adams BJ, Nielsen UN, Bates ST, et al. (2012) Cross-biome metagenomic analyses of soil microbial communities and their functional attributes. Proceedings of the National Academy of Sciences 109: 21390–21395.
78. Taylor PG, Townsend AR (2010) Stoichiometric control of organic carbon-nitrate relationships from soils to the sea. Nature 464: 1178–1181.
79. Apple JK, del Giorgio PA (2007) Organic substrate quality as the link between bacterioplankton carbon demand and growth efficiency in a temperate salt-marsh estuary. ISME J 1: 729–742.
80. Bayer C, Mielniczuk J, Amado TJC, Martin-Neto L, Fernandes SV (2000) Organic matter storage in a sandy clay loam Acrisol affected by tillage and cropping systems in southern Brazil. Soil and Tillage Research 54: 101–109.
81. Langdale GW, West LT, Bruce RR, Miller WP, Thomas AW (1992) Restoration of eroded soil with conservation tillage. Soil Technology 5: 81–90.
82. Bongiovanni MD, Lobartini JC (2006) Particulate organic matter, carbohydrate, humic acid contents in soil macro- and microaggregates as affected by cultivation. Geoderma 136: 660–665.
83. Plaza-Bonilla D, Cantero-Martínez C, Viñas P, Álvaro-Fuentes J (2013) Soil aggregation and organic carbon protection in a no-tillage chronosequence under Mediterranean conditions. Geoderma 193–194: 76–82.

84. Jiao Y, Whalen JK, Hendershot WH (2006) No-tillage and manure applications increase aggregation and improve nutrient retention in a sandy-loam soil. Geoderma 134: 24–33.
85. Tivet F, de Moraes Sá JC, Lal R, Borszowskei PR, Briedis C, et al. (2013) Soil organic carbon fraction losses upon continuous plow-based tillage and its restoration by diverse biomass-C inputs under no-till in sub-tropical and tropical regions of Brazil. Geoderma 209–210: 214–225.
86. Slepetiene A, Slepetys J (2005) Status of humus in soil under various long-term tillage systems. Geoderma 127: 207–215.
87. Dong L, Córdova-Kreylos AL, Yang J, Yuan H, Scow KM (2009) Humic acids buffer the effects of urea on soil ammonia oxidizers and potential nitrification. Soil Biology and Biochemistry 41: 1612–1621.
88. Sieber JR, McInerney MJ, Gunsalus RP (2012) Genomic Insights into Syntrophy: The Paradigm for Anaerobic Metabolic Cooperation. Annual Review of Microbiology 66: 429–452.
89. McInerney MJ, Sieber JR, Gunsalus RP (2009) Syntrophy in anaerobic global carbon cycles. Current Opinion in Biotechnology 20: 623–632.
90. Kim HJ, Boedicker JQ, Choi JW, Ismagilov RF (2008) Defined spatial structure stabilizes a synthetic multispecies bacterial community. Proceedings of the National Academy of Sciences 105: 18188–18193.

7

Inorganic and Organic Nitrogen Acquisition by a Fern *Dicranopteris dichotoma* in a Subtropical Forest in South China

Xingliang Xu[1], Qingkang Li[1]*, Jingyuan Wang[1], Leiming Zhang[1], Shengni Tian[3], Lin Zhi[3,4], Qianru Li[1,2], Yue Sun[1,2]

1 Key Laboratory of Ecosystem Network Observation and Modeling, Institute of Geographic Sciences and Natural Resources Research, Chinese Academy of Sciences, Beijing, China, **2** Graduate School of the Chinese Academy of Sciences, Beijing, China, **3** College of Life Science, Anhui Agriculture University, Hefei, Anhui Province, China, **4** Xikou Junior Middle School, Xikou City, Zhejiang Province, China

Abstract

The fern *Dicranopteris dichotoma* is an important pioneer species of the understory in Masson pine (*Pinus massoniana*) forests growing on acidic soils in the subtropical and tropical China. To improve our understanding of the role of *D. dichotoma* in nitrogen (N) uptake of these forests, a short-term ^{15}N experiment was conducted at mountain ridge (MR, with low N level) and mountain foot (MF, with high N level). We injected ^{15}N tracers as $^{15}NH_4$, $^{15}NO_3$ or ^{15}N-glycine into the soil surrounding each plant at both MR and MF sites. Three hours after tracer injection, the fern *D. dichotoma* took up $^{15}NH_4^+$ significantly faster at MF than at MR, but it showed significantly slower uptake of $^{15}NO_3^-$ at MF than at MR. Consequently, $^{15}NO_3^-$ made greater contribution to the total N uptake (50% to the total N uptake) at MR than at MF, but ^{15}N-glycine only contributed around 11% at both sites. Twenty-four hours after tracer injection, *D. dichotoma* preferred $^{15}NH_4^+$ (63%) at MR, whereas it preferred $^{15}NO_3^-$ (47%) at MF. We concluded that the *D. dichotoma* responds distinctly in its uptake pattern for three available N species over temporal and spatial scales, but mainly relies on inorganic N species in the subtropical forest. This suggests that the fern employs different strategies to acquire available N which depends on N levels and time.

Editor: Xiujun Wang, University of Maryland, United States of America

Funding: This study was supported by National Basic Research Program of China (2012CB416903) and the National Natural Science Foundation of China (41071209 and 30970484) and National Key Technologies R & D Program (2012BAC11B05-3). The funders had no role in study design, data collection and analysis, decision to publish, or preparation of the manuscript.

Competing Interests: The authors have declared that no competing interests exist.

* E-mail: liqk@igsnrr.ac.cn

Introduction

Nitrogen (N) is a major limiting element in many terrestrial ecosystems [15,32]. In the past three decades a large number of plants have been identified to have the capacity to directly take up organic N, mainly in the form of amino acids from soil solution [3,11,21,22]. Therefore, uptake pattern of different N species may be an important mechanism responsible for species coexistence in plant communities[19]. Numerous studies on plant N acquisition of organic and inorganic N species have been conducted in subtropical and tropical forests, most focusing on tree species [1,27,28,29,36,37,38], bryophytes and lichens [6,34] as well as some epiphytes [8,35]. While there was one study investigating N acquisition by a tropical fern [39], little is done to study organic and inorganic N uptake by subtropical ferns.

Dicranopteris dichotoma is an important terrestrial fern and is widely distributed in southern China as an early-stage colonizer of acidic and oligotrophic soils [9,40]. *D. dichotoma* is characterized by rapid clonal growth and often forms a dense understory layer in humid subtropical and tropical forests (Figure 1). Numerous studies have suggested that *Dicranopteris* species can influence many ecological processes in these forests, such as soil erosion, nutrient cycling, tree regeneration, and plant community succession [5,25,26,41]. Recently, understory removal experiments showed that the *Dicranopteris*-dominated understory can form favorable soil microclimates and acts as a major driver of soil biota and ecological processes in forest ecosystems [18,40]. However, N acquisition mechanisms of *D. dichotoma* remain unknown in these tropical forests, and clarification of *D. dichotoma* N uptake patterns could improve the mechanistic understanding of its role in these subtropical and tropical forests.

To investigate organic and inorganic N acquisition by *D. dichotoma*, we selected two different habitats in a subtropical forest (Figure 1): one located at the mountain ridge (MR) and the other located at the mountain foot (MF). Compared to the MF site, the MR site is characterized by heavy soil erosion and relatively lower N availability. Numerous studies have demonstrated that plants under high available N levels show higher N uptake rates [7,30]. Therefore, we hypothesized that *D. dichotoma* would have greater N uptake capacity at high N level than at low N level. Additionally, we hypothesized that *D. dichotoma* could acquire more nitrate than ammonium and organic N because it is more mobile in soil solution [31]. To test these hypotheses, we conducted a short-term ^{15}N labeling experiment in both *D. dichotoma* communities with different available N levels.

a. mountain ridge (MR)

b. mountain foot (MF)

Figure 1. The two sites located along the mountain slope: (a) mountain ridge (MR), (b) mountain foot (MF).

Materials and Methods

Study Site

This study was carried out at Xingguo Soil Erosion Observation Station (which belongs to the Institute of Geographic Sciences and Natural Resources, Chinese Academy of Sciences) in Xingguo County (26°30′N, 115° 28′E, 80 m above sea level) of Jiangxi Province, southern China, where no additional special permission for the research site was needed given that the site is owned by the institute and a long-term research permission from the local government exists since 1993. Moreover, our study did not involve any endangered or protected species. Two sites were selected along a mountain slope (Figure 1): one was located at the mountain ridge (MR) and the other was located at the foot of the mountain (MF). The understory was dominated by *D. dichotoma* at both sites. The *Pinus massoniana* trees at both sites had been planted more than 20 years before the study. In the MR sites, the trees are very small because of heavily eroded soil. In contrast, the trees grow better in the MF site. Soil was classified as loamy Lixisol and some properties are presented in Table 1, showing higher nutrient concentrations at the MF site than at the MR site.

Thirty-six similar sized clusters of *D. dichotoma* were randomly selected at both sites. They were divided into three groups with each group including 12 individual plants. Each of the groups was labeled either with $^{15}NH_4^+$ ($^{15}NH_4NO_3$, 98.4 atom% ^{15}N), $^{15}NO_3^-$ ($NH_4{}^{15}NO_3$, 98.2 atom% ^{15}N), and ^{15}N-glycine ($C_2H_5{}^{15}NO_2$, 95.0 atom% ^{15}N), by injecting 10 µg N g^{-1} dw soil. The tracers were injected into the soil at 3 cm depth in a pattern representing the three points of a triangle, with 5 cm length between points and the plant at the centre of the triangle. Six clusters were harvested 3 h after the tracer was injected. The remaining 6 clusters for each treatment were harvested 24 h after the tracer was injected. These harvested plants were classified into roots and shoots. An additional 6 clusters were injected with the same amount of water following the same pattern and harvested as the control after 24 h. The roots were put into 0.5 mM $CaCl_2$ solution for 30 min. Then, they were rinsed with purified H_2O and dried at 75°C. These plant materials were weighed for biomass. Dried roots and shoots were ground to a fine powder using a ball mill (MM200, Retsch, Germany) for the measurements of N content and $^{15}N/^{14}N$ ratios. Fresh soil (upper 10 cm) was collected for nutrient analysis.

Soil NO_3^--N and NH_4^+-N were determined in 0.5 M K_2SO_4 extracts on an auto-analyzer (AA3, Bran-Luebbe, Germany). Soil glycine concentrations were measured by high performance liquid chromatography (Waters 515, Waters Inc., USA) in the same extracts [23]. Aliquots of ground plant material (about 2 mg) and soil (about 40 mg) were weighed into tin capsules for analysing organic C, total N and $^{15}N/^{14}N$ ratios using isotope ratio mass spectrometry (IRMS, MAT 253, Finnigan MAT, Germany), with a Flash EA1112 interfaced by ConFlo III to the IRMS. Soil pH was measured using a glass electrode on a 1:2 soil-to-water ratio by weight.

N Uptake Calculation

Atom% excess ^{15}N (APE) was calculated as the atom% ^{15}N difference between plants from ^{15}N treated and from control plants. ^{15}N uptake by plants was estimated by calculating the ^{15}N excess of each plant part (biomass×%N/100×APE/100 for shoot and root individually) and then summing them up and dividing this number by root biomass and expressed as µg ^{15}N g^{-1} dw root h^{-1}.

The standard errors of means are presented in figures and tables as a variability parameter. T-test was used to compare the

Table 1. Characteristics of topsoil (0–10 cm) of the mountain ridge (MR) and the mountain foot (MF) sites where *Dicranopteris dichotoma* grows.

	Mountain ridge	Mountain foot
Total N (%)	0.037±0.003	0.075±0.019
Organic carbon (%)	0.65±0.10	1.54±0.45
C/N	16.85±1.57	19.47±1.33
pH (H_2O)	4.54±0.04	4.45±0.04
Soil moisture (%)	8.93±0.50*	18.30±1.00*
Ammonium (µg N g^{-1} dw soil)	5.07±0.61*	14.52±0.79*
Nitrate (µg N g^{-1} dw soil)	0.09±0.03	0.05±0.01
Glycine (µg N g^{-1} dw soil)	0.05±0.01*	0.12±0.03*

Means (±1SE) of six replicates are presented (n = 6). Asterisks indicate significant differences between two sites at $P<0.05$.

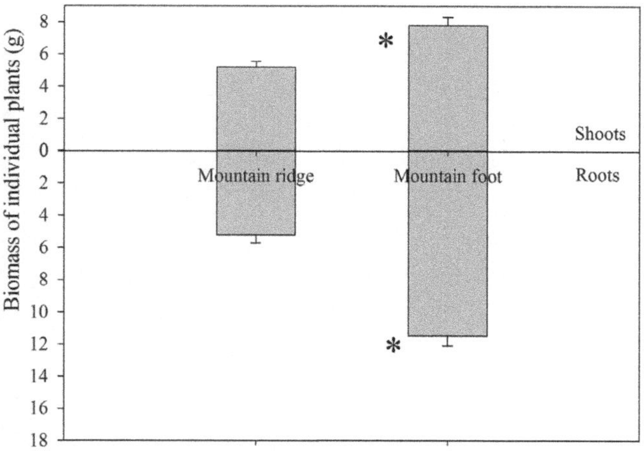

Figure 2. Above-ground and below-ground biomass of the fern *D. dichotoma* located at the mountain ridge (MR) and the mountain foot (MF). Values are means (±1 SE) of 18 replicates. Asterisks indicate significant differences between MR and MF sites at $P<0.05$.

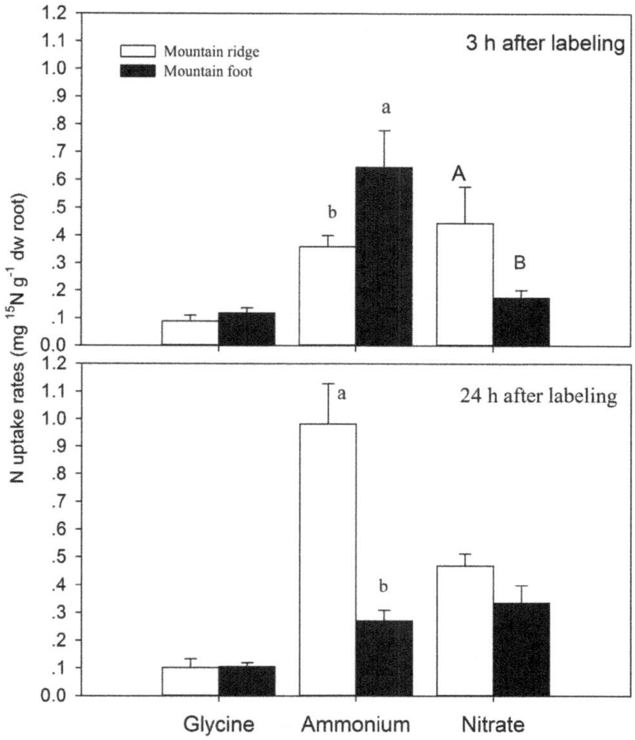

Figure 3. N uptake rates of different N species by the fern *D. dichotoma* at the mountain ridge and mountain foot. Different letters indicate significant differences between the two sites at $P<0.05$.

difference in N uptake rates and soil characteristics between sites. Tukey HSD test was used to compare the contribution of three N species to total N uptake between MR and MF. Three-way ANOVA was performed to test the effects of site, N species and time and their interactions on N uptake rates. Data transformation of Ln (data) was applied to meet preconditions of variance homogeneity and normal distribution before ANOVA analysis. All differences were tested for significance at $P<0.05$ and all statistical analysis were performed on SPSS 17.0 software package (SPSS Inc., Chicago, IL, USA).

Results

Concentrations of soil organic C and total N were consistently but not significantly higher at MF than MR (Table 1). Compared to MR, concentrations of NH_4^+ and glycine were significantly higher at MF (Table 1). NH_4^+ was the dominant N species among the three N species, while nitrate and glycine-N concentrations were comparably low at both sites. Total biomass of *D. dichotoma*, including shoots and roots was significantly lower at MR than at MF ($P<0.05$) (Fig. 2). *D. dichotoma* at MR also had significantly lower ratios of root to shoot than at MF (MR vs MF: 1.15±0.11 vs 1.76±0.17) (Fig. 2).

Results of three-way ANOVA showed significant effects of site, N species and their interactions on N uptake rates but no significant effect of time (Table 2). Three hours after tracer injection, the fern *D. dichotoma* took up $^{15}NH_4^+$ significantly faster (MF vs MR: 0.65±0.13 vs 0.36±0.04 μg ^{15}N g^{-1} dw root h^{-1}), while it took up less $^{15}NO_3^-$ (MF vs MR: 0.17±0.03 vs 0.44±0.13 μg ^{15}N g^{-1} dw root h^{-1}) at MF than at MR. Compared to $^{15}NH_4^+$ and $^{15}NO_3^-$, *D. dichotoma* showed considerably lower uptake rates for ^{15}N-glycine (MF vs MR: 0.10±0.03 vs 0.09±0.02 μg ^{15}N g^{-1} dw root h^{-1}) without significant differences between sites (Fig. 3).

Twenty-four hours after tracer injection, uptake rates of ^{15}N-glycine by *D. dichotoma* did not change at both sites (Fig. 3). Surprisingly, uptake rates of $^{15}NH_4^+$ significantly increased at MR (0.98±0.15 μg ^{15}N g^{-1} dw root h^{-1}) while they decreased at MF

Table 2. Results of multifactorial ANOVA for the effects of site, N species, time, and their interactions on N uptake rates of *D. dichotoma* in a subtropical forest.

Source of variation	*F*	*P*
Site	5.10	0.03
N species	69.54	0.00
Time	2.16	0.15
Site * N species	4.71	0.01
Site * Time	5.16	0.03
N species * Time	1.12	0.33
Site * N species * Time	7.72	0.00

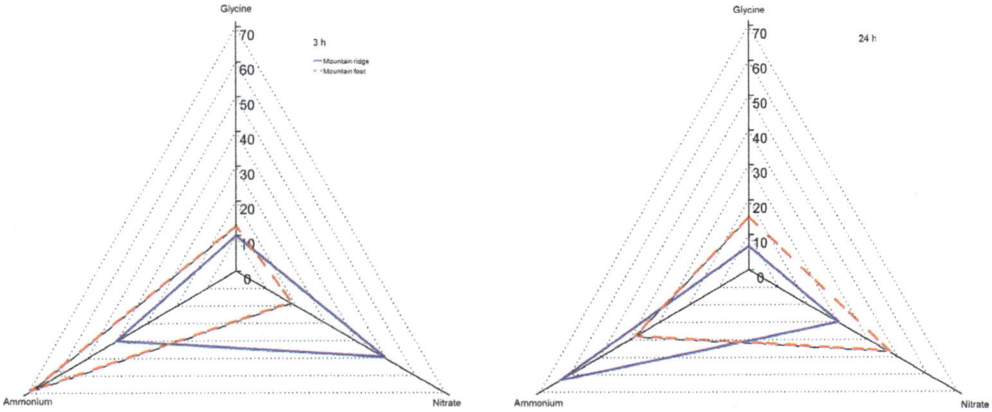

Figure 4. Chemical niche shifts in terms of N acquisition by the fern *D. dichotoma* under two available N levels. The area enclosed by the solid line refers to MR site (low N availability), whereas the dashed line encloses MF site data (high N availability). Values are means (±1SE) of 6 replicates. Asterisks indicate significant differences between mountain ridge and mountain foot MR and MF at $P<0.05$.

(0.27 ± 0.04 μg ^{15}N g^{-1} dw root h^{-1}). By comparison, uptake rates of $^{15}NO_3^-$ remained unchanged at MR, but they significantly increased at MF (Fig. 3, T-test, P = 0.034).

Three hours after tracer injection, the fern *D. dichotoma* preferentially took up $^{15}NO_3^-$ at MR (50% of total N uptake) while it preferentially took up $^{15}NH_4^+$ at MF (69% of total N uptake). By comparison, $^{15}NH_4^+$ contributed 40% and 69% to the total N uptake at MR and at MF, respectively. The contribution of $^{15}NO_3^-$ was significantly higher at MR than at MF. The contribution of ^{15}N-glycine was around 11% to the total N uptake at either of the sites (Fig. 4).

Distinct N uptake pattern of *D. dichotoma* was observed at 24 h after tracer injection. *D. dichotoma* preferred $^{15}NH_4^+$ at MR, which contributed 63% to the total N uptake, while the fern preferred $^{15}NO_3^-$ at MF, which contributed about 47% (Fig. 4). The contribution of ^{15}N-glycine uptake was significantly higher at MF than at MR (14.8% vs 6.6%) (Fig. 4).

Discussion

We investigated inorganic and organic N acquisition patterns by the terrestrial fern *D. dichotoma* at two habitats in a subtropical forest using a short-term ^{15}N labeling experiment. A previous study in a tropical forest showed that amino acid uptake of a low-light understory terrestrial fern (*Danaea wendlandii*) reached up to about 330 μg N g^{-1} dw root h^{-1} [39]. In this study, the terrestrial fern *D. dichotoma* showed a much lower uptake rate for glycine-N, only around 0.1 μg ^{15}N g^{-1} dw root h^{-1}. One possible explanation is that Watkins [39] tested uptake of amino acids using excised roots in solution while this study was performed in the field. Roots likely had more opportunity to take up glycine-N in solution than in soil. Available soil N concentrations will be changed through microbial competition and due to microbial mineralization of this organic tracer. Low glycine-N concentrations in the soil (Table 1) also reflect this situation. Besides, the injected soil volume was certainly smaller as the volume of soil extracted to collect the roots afterwards, i.e. many of the collected roots never came into contact with ^{15}N. Additionally, excised roots may have been more efficient compared to the roots used in this study, which included more inefficient roots not responsible for nutrient uptake. Nonetheless, it has to be noted that excised roots will loose lose uptake potential over time due to C starvation, because the phloem sugar import has been cut off after root excision.

Our hypothesis that *D. dichotoma* would have greater N uptake capacity at high N level than at low N level was partly supported by our data. Three hours after tracer injection, the fern *D. dichotoma* demonstrated faster uptake rates of $^{15}NH_4^+$ at MF than at MR. One possible explanation could be ascribed to good water status at MF (Table 1), which may enhance the diffusion rate of NH_4^+ from soil solution to root surface[2, 10]. However, 24 h after tracer injection, the fern *D. dichotoma* growing at MR showed higher uptake rates for the dominant N form NH_4^+. On the basis of the difference in root biomass between MR and MF (Fig. 2), the efficiency of $^{15}NH_4^+$ acquisition by roots at MR is higher than at MF. This indicates that the optimal performance of NH_4^+ transporters with high affinity on the root surface [12,24,33] could be more important than soil moisture in these soils where NH_4^+ is the dominant N species. Besides, strong microbial competition for $^{15}NH_4^+$ could be responsible for low uptake rates by roots at MF [13].

Although soil NO_3^- concentrations were considerably lower compared to NH_4^+ (Table 1), the fern *D. dichotoma* showed higher uptake rates for $^{15}NO_3^-$ and its uptake was similar to $^{15}NH_4^+$ uptake rates in some cases (e.g., 3 h after tracer injection at MR and 24 h after tracer injection at MF). One possible explanation is that NO_3^- is more mobile in soil solution [31]. These results indicate that our second hypothesis was partly supported by our observations. This could be explained by the fact that NH_4^+ is the dominant N species in both habitats (Table 1), and therefore the fern *D. dichotoma* prefers NH_4^+. The preference for NH_4^+ was also observed in the tropical terrestrial fern (*D. wendlandii*) despite high NO_3^- concentration [39].

The contribution of the three N species to the total N uptake strongly relies on time and site (Table 2, Fig. 4). Under low N levels (MR), *D. dichotoma* preferentially take up NO_3^- 3 h after tracer injection, but shifted to the dominant N species (NH_4^+) 24 h after tracer injection. Under high N levels (MF), *D. dichotoma* preferred the dominant N species (NH_4^+) at 3 h after tracer injection and shifted to NO_3^- at 24 h after tracer injection. The distinct uptake pattern for these available N species observed over temporal and spatial scales suggests that the *D. dichotoma* employs different strategies to acquire available N, depending on N levels and time. This could be ascribed to rapid regulation of N uptake modulated by a variety of biotic and abiotic factors in addition to tracer (^{15}N) dilution of the available soil N pool, such as the carriers of nitrate, ammonium and amino acids located at the surface of the roots [16,17,20]; soil supply rates of available N [4,14]; delivery of N to the rhizosphere through mass flow and diffusion [10]; as well as competition with soil microorganisms[13]. Further investigations should be focused on how interactions between these biotic and abiotic factors affect N uptake by *D. dichotoma* for a better understanding of the underlying mechanisms.

Supporting Information

Table S1 T-test for individual biomass allocation of *Dicranopteris dichotoma* growing at Mountain Ridge (MR) and Mountain Foot(MF) sites. Supplementary data, including the characteristics of top soil (0–10 cm), individual biomass allocation of *Dicranopteris dichotoma*, uptake rate of different N species and their contribution to total N uptake rate. And statistical analysis results are also included inside.

Acknowledgments

We would like to thank Christine Verhille at the University of British Columbia and Dr. Luke McCormack at the Institute of Geographic Sciences and Natural Resources Research, Chinese Academy of Sciences for their assistance with English language and grammatical editing of the manuscript and thank Ms Jinrong Yang for sample analysis and Mr. Shanyuan Yin for field sample collection.

Author Contributions

Conceived and designed the experiments: XX Qingkang Li. Performed the experiments: XX Qingkang Li JW ST L. Zhi Qianru Li YS. Analyzed the data: XX Qingkang Li YS. Contributed reagents/materials/analysis tools: ST L. Zhang Qingkang Li YS. Wrote the paper: XX Qingkang Li.

References

1. Andersen KM, Turner BL (2013) Preferences or plasticity in nitrogen acquisition by understorey palms in a tropical montane forest. J Ecol 101: 819–825.
2. Barber SA, Walker JM, Vasey EH (1963) Mechanisms for the movement of plant nutrients from the soil and fertilizer to the plant root. J Agr Food Chem 11: 204–207.
3. Chapin FS, Moilanen L, Kielland K (1993) Preferential use of organic nitrogen for growth by a nonmycorrhizal arctic sedge. Nature 361: 150–153.
4. Clarkson DT, Hanson JB (1980) The mineral nutrition of higher plants. Ann Rev Plant Physiol Plant Mol Biol 31: 239–298.
5. Cohen AL, Singhakumara BMP, Ashtoni PMS (1995) Releasing rain forest succession: a case study in the *Dicranopteris linearis* fernlands of Sri Lanka. Restor Ecol 3: 261–270.
6. Dahlman L, Persson J, Palmqvist K, Näsholm T (2004) Organic and inorganic nitrogen uptake in lichens. Planta 219: 459–467.
7. Gray JT, Schlesinger WH (1983) Nutrient use by evergreen and deciduous shrubs in southern California: ii. Experimental investigations of the relationship between growth, nitrogen uptake and nitrogen availability. J Ecol 71: 43–56.
8. Hietz P, Wanek W, Pop M (1999) Stable isotopic composition of carbon and nitrogen and nitrogen content in vascular epiphytes along an altitudinal transect. Plant Cell Environ 22: 1435–1443.
9. Hou X (1983) Vegetation of China with reference to its geographical distribution. Ann Missouri Bot Gard 70: 509–549.
10. Inselsbacher E, Näsholm T (2012) The below-ground perspective of forest plants: soil provides mainly organic nitrogen for plants and mycorrhizal fungi. New Phytol 195: 329–334.
11. Jones DL, Healey JR, Willett VB, Farrar JF, Hodge A (2005) Dissolved organic nitrogen uptake by plants-an important N uptake pathway? Soil Biol Biochem 37: 413–423.
12. Kronzucker HJ, Siddiqi MY, Glass ADM (1996) Kinetics of NH_4^+ influx in spruce. Plant Physiol 110: 773–779.
13. Kuzyakov Y, Xu XL (2013) Competition between roots and microorganisms for N: Mechanisms and ecological relevance. New Phytol 198: 656–669 doi:10.1111/nph.12235.
14. Leadley PW, Reynolds JF, Chapin FSA (1997) A model of nitrogen uptake by *Eriophorum vaginatum* roots in the field: ecological implications. Ecol Monogr 67: 1–22.
15. LeBauer DS, Treseder KK (2008) Nitrogen limitation of net primary productivity in terrestrial ecosystems is globally distributed. Ecology 89: 371–379.
16. Li Z-C, Bush DR (1990) ΔpH-dependent amino acid transport into plasma membrane vesicles isolated from sugar beet leaves. Plant Physiol 94: 268–277.
17. Li Z-C, Bush DR (1991) ΔpH-dependent amino acid transport into plasma membrane vesicles isolated from sugar beet (*Beta vulgaris* L.) leaves. Plant Physiol 96: 1338–1344.
18. Liu Z, Wu J, Zhou L, Lin Y, Fu S (2012) Effect of understory fern (*Dicranopteris dichotoma*) removal on substrate utilization patterns of culturable soil bacterial communities in subtropical *Eucalyptus* plantations. Pedobiologia 55: 7–13.
19. McKane RB, Johnson LC, Shaver GR, Nadelhoffer KJ, Rastetter EB, et al. (2002) Resource- based niches provide a basis for plant species diversity and dominance in arctic tundra. Nature 415: 68–71.
20. Nacry P, Bouguyon E, Gojon A (2013) Nitrogen acquisition by roots: physiological and developmental mechanisms ensuring plant adaptation to a fluctuating resource. Plant and Soil 370: 1–29 doi:10.1007/s11104-013-1645-9.
21. Näsholm T, Ekblad A, Nordin A, Giesler R, Högberg M, et al. (1998). Boreal forest plants take up organic nitrogen. Nature 392: 914–916.
22. Näsholm T, Kielland K, Ganeteg U (2009) Uptake of organic nitrogen by plants. New Phytol 182: 31–48.
23. Näsholm T, Sandberg G, Ericsson A (1987) Quantitative-analysis of amino-acids in conifer tissues by high-performance liquid-chromatography and fluorescence detection of their 9-fluorenylmethyl chloroformate derivatives. J Chromatogr 396: 225–236.
24. Ninnemann O, Jauninaux JC, Frommer WB (1994) Identification of a high affinity NH_4^+ transporter from plants. EMBO J 13: 3464–3471.
25. Russell AE, Raich JW, Vitousek PM (1998) The ecology of the climbing fern *Dicranopteris linearis* on windward Mauna Loa, Hawaii. J Ecol 86: 765–779.
26. Russell AE, Vitousek PM (1997) Decomposition and potential nitrogen fixation in *Dicranopteris linearis* litter on Mauna Loa, Hawaii. J Trop Ecol 13: 579–594.
27. Schmidt S, Stewart GR (1997) Waterlogging and fire impacts on nitrogen availability and utilization in a subtropical wet heathland (wallum). Plant, Cell Environ 20: 1231–1241.
28. Schmidt S, Stewart GR (1999) Glycine metabolism by plant roots and its occurrence in Australian plant communities. Funct Plant Biol 26(3): 253–264.
29. Schmidt S, Mason M, SangtieanT, Stewart GR (2003) Do cluster roots of *Hakea actities* (Proteaceae) acquire complex organic nitrogen? Plant and Soil 248(1): 157–165.
30. Stoelken G, Simon J, Ehlting B, Rennenberg H (2010) The presence of amino acids affects inorganic N uptake in non-mycorrhizal seedlings of European beech (*Fagus sylvatica*). Tree Physiol 30: 1118–1128.
31. Tinker PB, Nye P (2000) *Solute Movement in the Rhizosphere*. Oxford, UK: Oxford University Press.
32. Vitousek PM, Howarth RW (1991) Nitrogen limitation on land and in the sea: how can it occur? Biogeochemistry 13: 87–115.
33. von Wirén N, Gazzarrini S, Frommer WB (1997) Regulation of mineral nitrogen uptake in plants. Plant Soil 196: 191–199.
34. Wanek W, Pörtl K (2008) Short-term ^{15}N uptake kinetics and nitrogen nutrition of bryophytes in a lowland rainforest, Costa Rica. Funct Plant Biol 35(1): 51–62.
35. Wanek W, Arndt SK, Huber W, Popp M (2002) Nitrogen nutrition during ontogeny of hemiepiphytic *Clusia* species. Funct Plant Biol 29(6): 733–740.
36. Warren CR (2009) Uptake of inorganic and amino acid nitrogen from soil by *Eucalyptus regnans* and *Eucalyptus pauciflora* seedlings. Tree Physiol 29(3): 401–409.
37. Warren CR (2006) Potential organic and inorganic N uptake by six *Eucalyptus* species. Functional Plant Biol 33(7): 653–660.
38. Warren CR, Adams PR (2007) Uptake of nitrate, ammonium and glycine by plants of Tasmanian wet eucalypt forests. Tree Physiol 27(3): 413–419.
39. Watkins JE (2006) Functional Ecology of the Gametophytes and Sporophtyes of Tropical Terns. Ph D dissertation.
40. Zhao J, Wan S, Li Z, Shao Y, Xu G, et al. (2012) *Dicranopteris*-dominated understory as major driver of intensive forest ecosystem in humid subtropical and tropical region. Soil Biol Biochem 49: 78–87.
41. Zheng H, Chen F, Ouyang Z, Tu N, Xu W, et al. (2008) Impacts of reforestation approaches on runoff control in the hilly red soil region of southern China. J Hydrol 356: 174–184.

8

Ecomorphological Variation of the Wireworm Cephalic Capsule: Studying the Interaction of Environment and Geometric Shape

Hugo A. Benítez[1,2]*, Thomas Püschel[1], Darija Lemic[3], Maja Čačija[3], Antonela Kozina[3], Renata Bažok[3]

1 Faculty of Life Sciences, University of Manchester, Michael Smith Building, Oxford Road, Manchester, United Kingdom, **2** Instituto de Alta Investigación, Universidad de Tarapacá, Casilla, Arica, Chile, **3** University of Zagreb, Faculty of Agriculture, Department for Agricultural Zoology, Svetošimunska, Zagreb, Croatia

Abstract

Studying the association between organismal morphology and environmental conditions has been very useful to test hypothesis regarding the influence of climate on shape. It has been long recognized that different environments produce dissimilar stress levels in insects, which can be reflected on the ability of an individual to overcome these pressures and spread further. *Agriotes* (Coleoptera: Elateridae) species infest agricultural fields in different parts of Croatia, inhabiting different climatic conditions. Previous biological studies have indicated that there is a relationship between some *Agriotes* biological parameters such as density and climatic conditions such as soil moisture and temperature. However, it is still unknown how these environmental properties influence the wireworm morphological structure. This is highly relevant because the head of this species is directly involved in the mobility in the soil, thus affecting the invasive capacity of this insect. Therefore the aim of this study was to assess the association between different climatic conditions and the morphological variation of *Agriotes* cephalic capsule. Advanced multivariate analysis and geometric morphometric tool were applied to study the covariation between shape and environmental variables. Partial Least Squares methods were used in order to analyse the association between the wireworm head shape and three different climatic conditions: soil type, temperature and rainfall. Our results showed that there is a high covariation between the wireworm head shape and the climatic conditions. It was suggested that the observed shape–environment association could be result of the high plasticity of this species in relation to its invasive capacity.

Editor: Hector Escriva, Laboratoire Arago, France

Funding: This research was supported by the Ministry of Science, Education and Sports of the Republic of Croatia by project: "The spatial distribution of economically important pests with the use of GIS" (178-1782066-2065). The funders had no role in study design, data collection and analysis, decision to publish, or preparation of the manuscript.

Competing Interests: The authors have declared that no competing interests exist.

* Email: hugo.benitez@postgrad.manchester.ac.uk

Introduction

Taxonomic classification and biological diversity analyses have been traditionally based on morphological descriptions [1]. Driven by a mathematical quantitative revolution, morphological studies have experienced a significant revitalization due to the development of shape analysis based on statistical multivariate techniques and novel visualisation methods. Broadly, morphometrics refer to the quantitative analysis of form (i.e. shape and size) and how it covaries with respect to other factors (e.g. ecology, development, biomechanics, and genetics among others) [2–5]. Geometric morphometrics (GMM) is a coordinate-based method, which means that their primary data are 2D or 3D Cartesian coordinates of anatomically distinguishable landmarks (i.e. discrete anatomical loci that are arguably homologous among all the individuals under analysis). Coordinates are better when compared to linear data, because they preserve the spatial information of a structure, providing a relatively complete description of an organism's shape [2,3,6]. One of the great characteristics of GMM is that it allows studying the association between shape and other kinds of data, such as ecological, genetic, biomechanical, or other relevant factors. This is really useful because one of the traditional interests of ecologists is to associate character states or different phenotypic values with environmental data. There are several matters for which morphological analyses play an important role in ecology. For instance, ecomorphological studies have revealed constraints and selective factors affecting the phenotypic response to certain environments [7–9], how morphology influences the ecological distribution of a particular phenotype [10–13] and evolutionary trends such as phylogenetically conserved morphologies [11,14–16]. In all these cases, morphology represents certain organismal aspects that relate and individual to its environment, hence its importance. Indeed, the association between morphology and ecology could provide useful insights about the expression of the phenotype-environment interaction and the related evolutionary history. There is a plethora of available methods to study association between morphological and ecological variables (e.g. regression analysis; canonical correlation analysis; Mantel test; principal coordinate analysis; etc.). However, Partial Least Squares (PLS) method is probably one of the mostly applied when assessing the covariation between shape and other factors [17]. For example, it has been used to relate morphometric and ecological variables [5,13,18–20], morphometric and allele frequency data [21], shape and behavior [18], different parts of the same configuration of landmarks [21–23], and even data from different parts or different views of the same specimens [24,25].

Wireworms are click beetle larvae from the genus *Agriotes* (Coleoptera: Elateridae) that considerably damage field crops, especially potatoes [26–28]. They are long-lived soil insects; most species spend 3–5 years in the larval stage [26,28]. Their body are elongated and hard, characterized by a reddish-brown coloration [29]. They are polyphagous and usually inhabit most kinds of soil. Some wireworm species are serious agricultural pests [28,30,31]. Larvae feed on germinating seeds and plant roots, affecting negatively seedlings and young plants, often causing their death [32,33]. Among over 10,000 species known worldwide [29], there are just around 150 considered dangerous enough to cause significant damages to agricultural cultigens. The five most important *Agriotes* species in Croatia include *A. lineatus* L., *A. sputator* L., *A. obscurus* L., *A. brevis* Cand. and *A. ustulatus* Schall. They can produce significant economic losses in agriculture, due to their feeding behaviour [34]. *Agriotes* "species infest agricultural fields and cause serious economic damages in the continental part of Croatia. Therefore, studying these species is relevant due to the impact that they have on agricultural production.

East continental Croatia is characterized by chernozemic soils, while the central region is defined by ground water gley and alluvial muds soils, and the west part is characterized by luvic and pseudoglay soils [35]. As a result of the dissimilar soils found in each one of these regions, the prevailing microhabitat parameters are different. The most important parameters are soil moisture and temperature, both of which are known to impact upon oviposition, egg survival and larval development [27,30,34,36]. Unlike the other four species, *Agriotes ustulatus* overwinters only as larvae and develops for two (three calendar) years [27], and the higher temperatures positively affect its abundance [27,37]. According to previous literature, *A. ustulatus* was the most frequent species only in the eastern regions of Croatia [32,38,39] but recent studies have shown that this species has also spread to other regions of continental Croatia [37,40].

It is well known that the adaptation over time to a specific environment is the result of both environmental pressures and geographic distance [41–43]. Moreover, it is well documented that adverse temperatures, nutritional stresses, presence of chemicals, population density and many other factors that affect development can lead to an increase in morphological asymmetry as a result of high intraspecific variation (e.g. [42,44,45]). Therefore, it is expected that when environmental conditions change, organisms and populations should adapt to the new conditions [45]. In this context, adaptive variation plays a key role since it reflects historical evolution and determines the population's phenotypic response [46–48]. Moreover, Bouyer et al. [49] have suggested that the influence of environment on an organism's genotype takes more time to manifest than on the phenotype, and as such the study of environmental influence on populations and individuals should be made on phenotypic, rather than genotypic, characters. Hence, we preferred the application of geometric morphometric methods instead of molecular analyses, because our goal was to analyse the association between environmental factors and morphological structure, rather than determining the genetic response to different climatic conditions. Previous studies have analyzed the effect of different ecological factors on insect morphology from diverse perspectives, for instance anti-predator defences [50–52], behaviour and sexual dimorphism [53–57], physiology [58–60] and environmental adaptations [61–64]. However our approach is innovative, because it is a novel attempt to characterize the relationship between a larval morphology with closely related environmental conditions. The aim of this study was to assess the association between different climatic conditions (i.e. temperature, rainfall and soil type) on the morphological differentiation of *Agriotes ustulatus* cephalic capsule. Furthermore, this study applied for the first time geometric morphometric techniques to analyze beetle larvae shape variation.

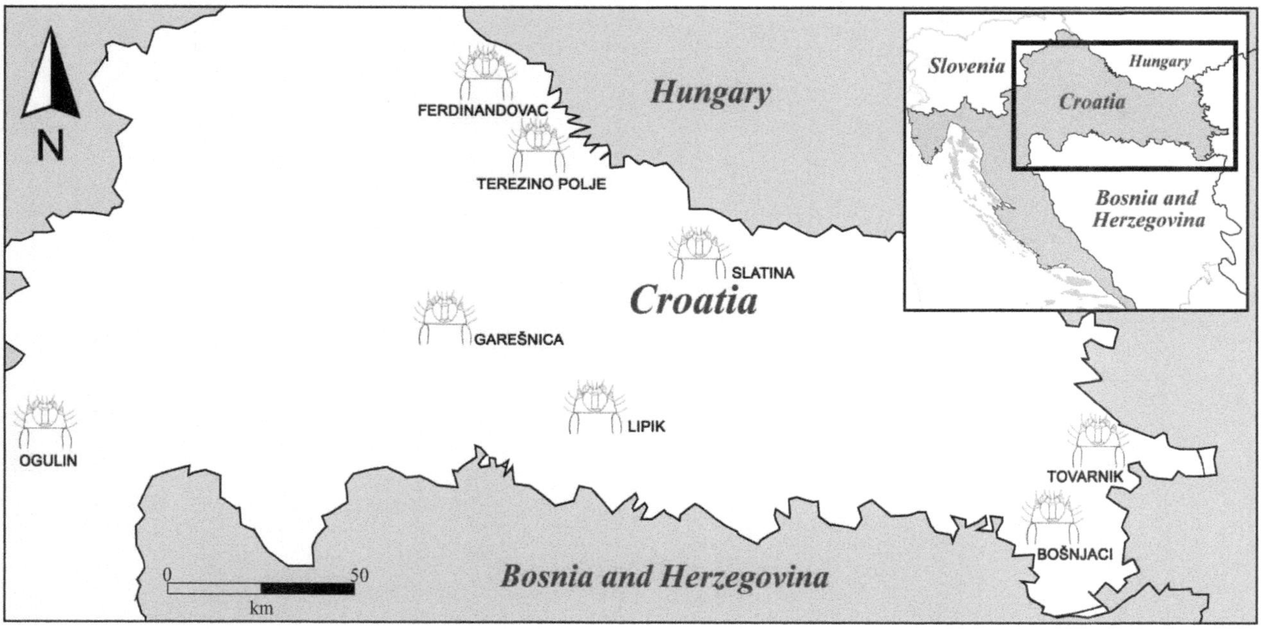

Figure 1. Geographic distribution of the eight Croatian localities of *A. ustulatus* examined in this study.

Table 1. Localities with their corresponding geographic coordinates and environmental variables: n, number of specimens; T, mean annual between 2009–2013 temperature (°C); Rain, Rainfall (mm); Type of soil (group 1: chernozemic soil, group 2: alluvial soil, group 3: ground water gley soil and group 4: luvic and pseudoglay soil).

Location	Samples (n)	Coordinates	Temperature (Average 2009–2013)	Rainfall (Average 2009–2013)	Type of Soil
a. GAREŠNICA	40	45°34′15.39″N; 16°55′56.02″E	11.87	58.83	ground water gley soil
b. LIPIK	35	45°24′59.61″N; 17°9′24.10″E	11.90	70.6	ground water gley soil
c. TOVARNIK	37	45°9′24.43″N; 19°9′14.51″E	12.35	40.99	chernozemic soil
d. BOŠNJACI	39	45°2′54.64″N; 18°44′59.59″E	12.49	60.97	chernozemic soil
e. SLATINA	40	45°42′51.70″N; 17°41′59.18″E	11.26	61.57	alluvial soil
f. TEREZINO POLJE	2	45°56′5.47″N; 17°27′27.82″E	11.513	71.69	alluvial soil
g. FERDINANDOVAC	13	46°3′42.85″N; 17°10′59.92″E	11.078	66.71	alluvial soil
h. OGULIN	46	45°15′27.46″N; 15°12′38.78″E	10.96	129.171	luvic and pseudoglay soil

Materials and Methods

Ethics Statement

"N/A".

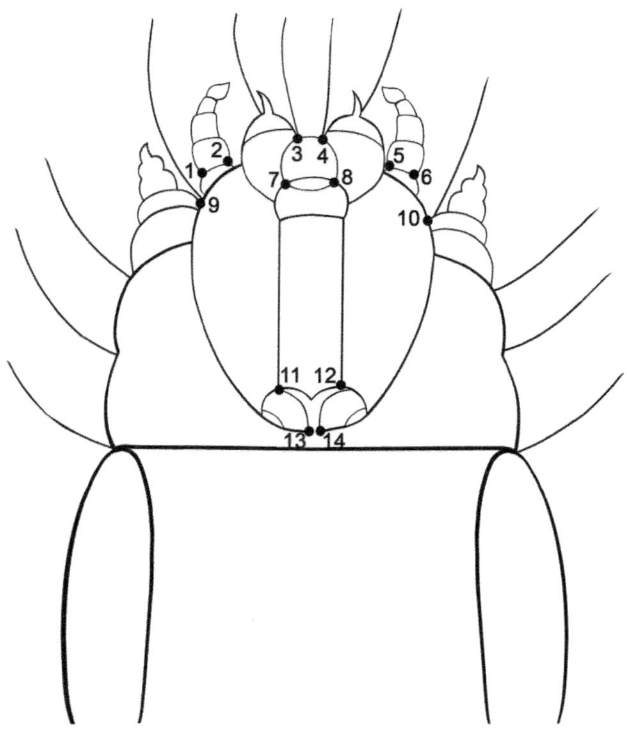

Figure 2. Representation of the 14 morphological landmarks identified on the cephalic capsule of *A. ustulatus*.

Our samples not needed any specific permission to collect (locations/activities) and this article is not involved in any endangered or protected species.

Data collection

In this study we analysed morphological differences in 258 wireworms of *Agriotes ustulatus*. The wireworm larvae were sampled during spring (May) and autumn (September) from 2009 until 2013. Although the taxonomic identification of the *Agriotes* species at the larval stage is extremely difficult, the analyzed larvae were positively classified as member of *A. ustulatus* based on the distinctive presence of specific marks on the last abdominal segment. This particular character is an exclusive trait of this species and none of the other members of this genus shows it [65]. They came from eight different locations of continental Croatia. Locations were grouped according to climatic conditions and soil type. Group 1 corresponds to the eastern locations of Tovarnik and Bošnjaci, which had drier weather conditions and chernozemic soils. Group 2 includes Slatina, Terezino Polje and Ferdinandovac, which had lower temperatures as compared to the eastern location but higher moisture and alluvial soils. Group 3 are Garešnica and Lipik locations, which had climate conditions in between the two previous groups and ground water gley soils. Group 4 consisted of one location on the west, Ogulin, which is on higher altitude and had cooler climate (perhumid climate), with luvic and pseudoglay soils (Fig. 1, Table 1). All sampled areas were crop fields, cultivated with oilseed rape, maize, barley, soybean, sugar beet, tobacco, alfalfa and potato. The sampling method consisted in digging a series of ca. five holes (dimensions 25×25×25 cm) on each field [38]. All the excavated soil was manually crumbled to small pieces on a black foil and examined thoroughly to detect any wireworm presence. The collected specimens were preserved in 70% ethanol. Species determination was done by examining the ninth abdominal segment [65]. Only larvae of the approximately same developmental stage (7th–13th instar, larger than 10 mm) of *Agriotes ustulatus* species were selected

Table 2. Measurement error Procrustes ANOVA for both centroid size and shape of *A. ustulatus*, characterised by matching symmetry.

Centroid size					
Effect	**SS**	**MS**	**df**	**F**	**p-value**
Individual	4.682	0.142	33	12.1	<0.0001
Error 1	**0.374**	**0.012**	**32**		
Shape					
Effect					
Individual	0.156	3.95E-04	396	3.09	<0.0001
Side	0.0049	0.0004	12	3.23	0.0002
Ind×Side	0.051	0.0001	396	2.47	<0.0001
Error 1	**0.039**	**5.17E-05**	**768**		

Sums of squares (SS) and mean squares (MS) are in units of Procrustes distances (ie. dimensionless).

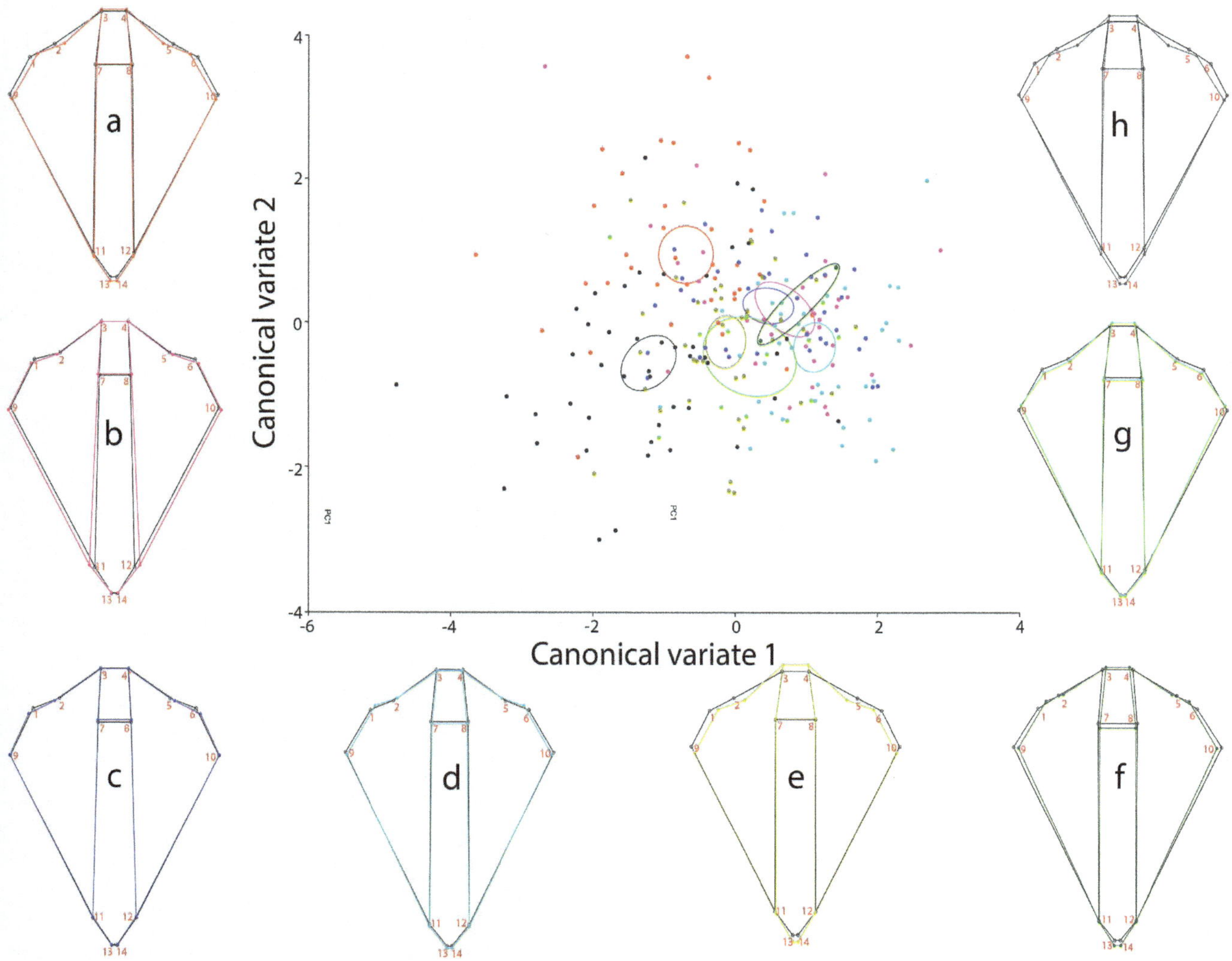

Figure 3. Canonical Variate analysis of the eight *A. ustulatus* populations. The figure shows the firsts two CV axes and the wireframe visualization of the average shape for all populations. a: GAREŠNICA (red) b: LIPIK (pink) c: TOVARNIK (blue) d: BOŠNJACI (light blue) e: SLATINA (dark yellow) f: TEREZINO POLJE (green) g: FERDINANDOVAC (light green) h: OGULIN (black).

Table 3. Procrustes ANOVA for both centroid size and shape of *A. ustulatus*, characterized by object symmetry.

Centroid size					
Effect	**SS**	**MS**	**df**	**F**	**p-value**
Location	7.09	1.013	7	17.31	<0.0001
Individual	14.28	0.058	244		
Shape					
Effect					
Location	0.233	0.0028	84	7.17	<0.0001
Individual	1.136	0.0004	2928	3.78	<0.0001
Side	0.009	0.0008	12	8.1	<0.0001
Ind×Side	0.309	0.0001	3012		

Sums of squares (SS) and mean squares (MS) are in units of Procrustes distances (dimensionless) SH: Shape, CS: centroid size.

for further analysis [27]. All the specimens are housed in the Department for Agricultural Zoology, University of Zagreb.

Landmark Acquisition

Cephalic capsules were photographed using a Leica DFC295 digital camera (3M Pixel) on a trinocular mount of a Leica MZ16a stereo-microscope and saved in JPEG format using the Leica Application Suite v3.8.0 (Leica Microsystems Limited, Switzerland). Fourteen homologue type 1 landmarks (Fig. 2), (Table S1, Table S2) were marked on the wireworm mandible, so that they could be accurately and repeatedly identified. Each landmark was digitised using tpsDIG V2.17 [66] and imported into MorphoJ v1.04d for further statistical analyses [67]. The dataset containing the raw coordinates used in this study is available as supplementary material. Measurement error (ME) has a critical importance when analyzing shape. In order to assess the ME level, the cephalic capsules of 35 individual beetles were digitized twice [68].

Shape Analysis

Once obtained the Cartesian coordinates for all landmarks, the shape information was extracted by performing a full Procrustes fit [69,70] taking into account the object symmetry of this anatomical structure. Procrustes superimposition is a mathematical procedure that removes the variation due to rotation, position, orientation and size [69]. Due to the existence of object symmetry in the analysed structure, reflection was removed by including the original and mirror image of all configurations in the analysis and simultaneously superimposing all of them [71].

Allometry is a key factor influencing shape variation [72,73], therefore allometric effects were assessed by performing a multivariate regression of shape on centroid size, pooling the dataset by location. Then the covariance matrix of the residuals pooled by location was computed to perform the following analyses.

Multivariate Analysis

The main patterns of variation in the shape space were visualized by carrying out a principal component analysis (PCA), calculated from the covariance matrix of the averaged population symmetric component of shape [71]. In order to statistically test if there were differences between the locations, a canonical variate analysis (CVA) of the symmetric component of shape variation was applied [74]. CVA maximizes the differences between groups

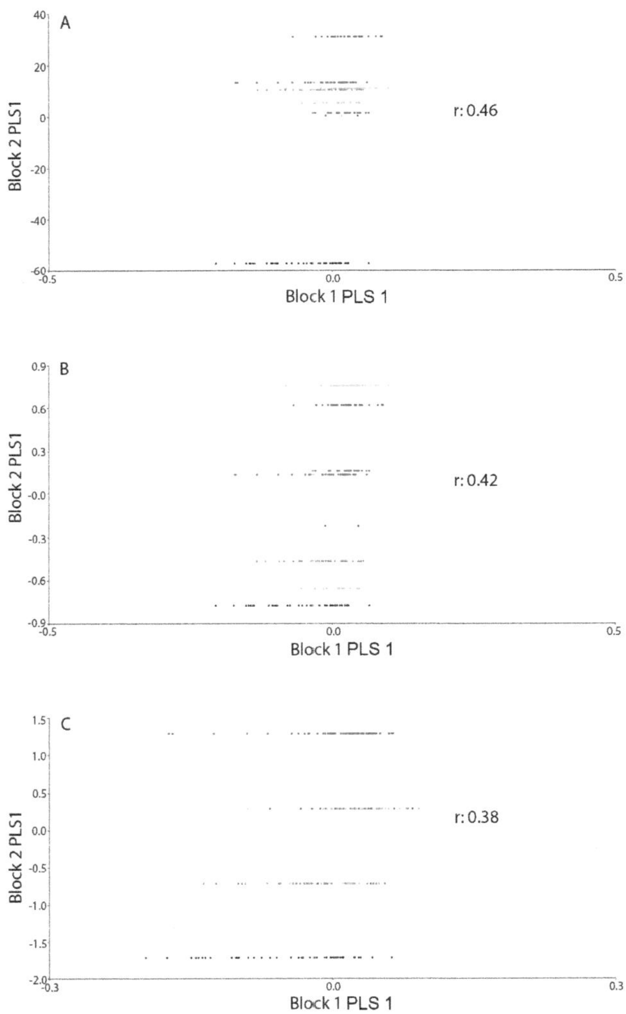

Figure 4. Partial least squares analysis of head shape and environmental condition in *A. ustulatus*. The horizontal axis of the scatter plot is the first PLS axis for head shape and the vertical axis is the first PLS axis for environmental condition for the eight localities. A: temperature, B: Rainfall and C: Soil Type.

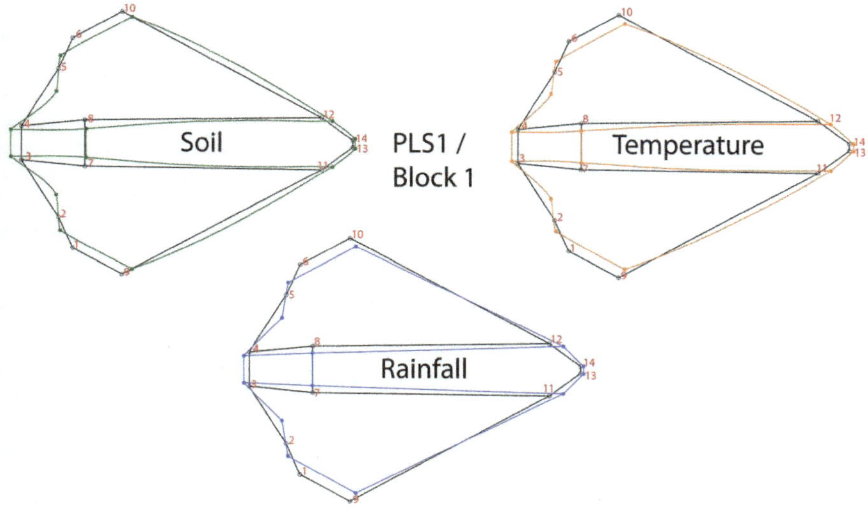

Figure 5. Wireframe visualization of the average shape extracted from the PLS1 associated with the different environmental conditions. The figure colors shows the particular displacements of landmarks relative to each one of the environmental conditions, green: soil, orange: temperature and blue: rainfall.

relative to the variation within groups and it is therefore one of the most used methods to distinguish among groups [74]. The statistical significance of the pairwise differences in mean shapes was assessed with performing a permutation test of the Procustes distance (10,000 permutations per test).

The amount of symmetric variation was assessed by computing a Procrustes ANOVA as applied in other studies to analyse object symmetry [71,75,76]. This test was used to compare the individual-reflection interactions as compared to measurement error, being this latter estimated from the total variation of the entire landmark configuration.

Climatic Correlation

In order to assess the degree of association between shape and climatic variables, a two block partial least square (PLS) was performed. The geographic coordinates and climatic condition (temperature, rainfall and soil type) (Table 1) were extracted from Croatian Meteorological and Hydrological Service (for complete values since 2009 to 2013 see Table S3), for each one of the analysed sites. PLS was preferred due to the fact that climatic variables are strongly collinear [17]. The two-block PLS analysis is a widely used morphometric method, and it is based on the singular value decomposition of the common covariance matrix of the two variables sets, which in our case consist of the shape and climatic conditions matrices [19,20]. PLS analysis estimates the correlated pairs of lineal combinations (singular vectors) between these two sets. These singular vectors are generated as pairs of latent variables (one per block), explaining most of the covariance between the original variable blocks, thus maximizing the covariation between the two sets of variables [17,19,20]. In order to test the significance of the correlation between the pairs of singular vectors (i.e. climatic variable singular vectors and shape singular vectors), a permutation test with 10,000 rounds randomizations was performed.

Mantel Test

In order to statistically test whether there was relationship between shape differences and geographic location, a Mantel test was performed. The correlation level was assessed comparing Procrustes (i.e. the morphometric distances between all the wireworms at each location), and geographic (i.e. the geodesic distances between all the locations where the wireworms were sampled) distance matrices by computing their product moment correlation and the Mantel test statistic (observed Z values compared to their permutation distribution after 10,000 rounds).

Results

The Procrustes ANOVA applied to assess the measurement error showed that the mean square for individual variation exceeded the measurement error; therefore it was negligible (Table 2). The same analysis applied to assess the population differences showed significant differences among localities for both size and shape ($P<0.0001$) (Table 3). The PCA of the residuals of the regression (used to excluded the allometric effects) of the head shape variation showed that the first three PCs accounted for 79.2% (PC1 = 52.8%; PC2 = 17.04%; PC3 = 9.3%) of the total shape variation, hence providing a reasonable approximation of the total amount of variation. The CVA showed a significant differentiation among localities based on Procrustes distances (Fig. 3). The Mantel test indicated that in spite of a relatively high (0.727) correlation value between the morphometric (i.e. Procrustes) and geographic distances (i.e. km), it was not significant (p-value = 0.076; 10,000 permutation rounds). The two-block, partial least-squares analyses of the head shape indicated a moderate association between the shape variables and the environmental ones (Rainfall: $r = 0.422$, p-value<0.001; Temperature: $r = 0.46$, p-value<0.001; Type of Soil: r: 0.38, p-value<0.001) (Fig. 4). A strong and significant amount of covariation between the head shape and the environmental variables was graphically visualised by the first block and PLS1 (Fig. 5).

Discussion

The eight populations were distinguished according to the relative displacement of the symmetric landmarks 1, 2–5, 6 and 9–10. *A. ustulatus* presents a relatively long larval stage as compared to most insects (ca. 3–5 years) [27]. This interesting life history trait has led to several studies focusing on the effect of soil type on population density or invasive capacity [26,36,37,77]. Nonetheless, there have been no studies trying to assess the influence of soil type (and other climatic conditions) on the cephalic capsule shape. It was expected to contribute to the understanding of the influence of environmental factors on the morphological development of wireworms.

The first dimensions of the PCA accounted for most of the morphological variation, showing that the wireworm head shape has a relative regional ordination (wireframe Fig. 3). This means that each population has relatively different head shape with respect to the others, in such a manner that the majority of the observed shape differences are due to regional differences. The CVA results confirm this interpretation, although there was some overlap between the confidence interval ellipses (Fig. 3). The Mantel test was performed in order to assess whether this regional ordination was expressed in a morphoclinal way (i.e. in accordance with a gradual differentiation relative to distance). Our results did not support the latter perspective, but on the contrary showed that there was no lineal relationship between the shape of the wireworms and the distance between the collection sites. Hence, it was possible to discard the existence of a morphoclinal variation in the larvae head shape. Although, this conclusions must be cautiously considered, since it has been shown that the Mantel test is not always the best method to test for linear relationships [78].

The PLS results showed that the shape variation of the head observed in *A. ustulatus* covaries with the environmental factors. Our study has confirmed a high covariation between the wireworm head shape and the analyzed climatic conditions, although the correlation values between PLS axes were moderate. It is important to keep in mind that these correlations are not the decisive factor that has been maximized by the PLS analysis. In the present study, the shape changes associated to the climatic variables were clearly visible in PLS1, showing that the head shape had an expansion of landmarks 3 and 4 (base of the hair on the mentum) due to soil type differences resulting in a morphotype defined by an elongated head (Fig. 5). The average effect of temperature showed a head shape expansion of landmarks on the maxillary palpi (1–6 and 2–5) and a contraction of the mentum, showing a narrowed head that increasingly expands towards the cardo and alocardo (Fig. 5). Finally the rainfall effect was characterized by a contraction of the landmarks of the maxillary palpi and a moderate elongation of the juncture between the submentum and the cardo (Fig. 5). Nevertheless the shape changes associated with the PLS1 axes showed fairly complex local shifts of landmarks (Fig. 5). Both temperature and rainfall showed similar covariation patterns with respect to shape, exhibiting relative displacements of individual head traits rather than large-scale deformations. However, because the landmark configuration does not include all the traits of the head, this finding should be interpreted cautiously.

It is important to always take into account the possible functional consequences of the observed shape differences when an association between morphology and environment is found [20]. Therefore, the observed shape–environment association noticeable in the first pair of PLS vectors could be interpreted in functional terms. The observed covariation between climatic conditions and shape changes could be a result of the high plasticity of this species. Actually, those populations inhabiting ground water gley soil (i.e. soils were drainage is poor due to a high phreatic surface), have more narrow cephalic capsules (Fig. 3b), suggesting a better penetrating capacity. On the other hand, those individuals occupying wetter muds have more expanded heads (Fig. 3h), allowing them to possibly use it as a shovel in order to dig these kinds of soils in an enhanced way. This plasticity may also explain the capacity of this species as invasive pest, occupying a broad range of host plants both as larva and adult [26,36]. Previous studies in a diverse range of organisms have shown that phenotypic plasticity can explain the observed morphological differences, especially when considering the ecological distribution of the analysed species [79,80]. A similar study carried out on the Chrysomelid pest beetle (*Diabrotica virgifera virgifera*), have also established the existence of a relationship between an anatomical structure and soil type [81]. This species exhibit an elongated hind wing morphotype in environments dominated by water gley soil type. By contrast, where a chernozemic or alluvial soil type dominated, a narrow hind wing morphotype was found. Even though the precise mechanism underlying the relationship between *A. ustulatus* and *D. virgifera* phenotypes and soil type is probably different, it is highly suggestive that both are pest species with wide ecological ranges. Perhaps this observed phenotypic plasticity is related to the invasive capacity of this kind of species, although more analyses are needed.

Further geostatistical analyses of spatial patterning and clustering could adjust sampling methods, in order to properly understand the key factors determining population size and pest dynamics [82]. Toepfer et al. [83] have studied the clustering of this group, concluding that this species are environmentally mediated by mortality factors that affect pest populations, which might vary spatially as well as seasonally and due to biological factors such as limited movement of larvae [83]. Hence, it is highly relevant to continue studying the relationship between shape and climatic conditions, but including as well some specific kinematic measurements of the different larvae population displacing through dissimilar soil types. Additionally, future studies should assess the relationship between vegetational structure and climatic conditions (i.e. host plant differences due to climate) and their effect on the morphology of *A. ustulatus* and how this influences its pest dynamics.

Supporting Information

Table S1 Anatomical Description of the 14th Landmarks in *Agriotes* Schall.

Table S2 Raw Landmark dataset.

Table S3 Average monthly temperatures and Rainfall.

Acknowledgments

We thank the two anonymous reviewers that provided us with valuable comments that clearly improved our manuscript. We give great gratitude to Ivana Marušić and Marina Buketa for the technical assistance. We give great gratitude to the Ministry of Science, Education and Sports of the Republic of Croatia by project: "The spatial distribution of economically important pests with the use of GIS" (178-1782066-2065). HB and TP are grateful with the Becas Chile scholarship program, CONICYT Chile.

Author Contributions

Conceived and designed the experiments: HAB TP DL MC AK. Performed the experiments: DL MC AK. Analyzed the data: HAB TP. Contributed reagents/materials/analysis tools: HAB TP DL MC AK RB. Contributed to the writing of the manuscript: HAB TP DL.

References

1. Daly HV (1985) Insect Morphometrics. Annual Review of Entomology 30: 415–438.
2. O'Higgins P (2000) The study of morphological variation in the hominid fossil record: biology, landmarks and geometry. Journal of Anatomy 197: 103–120.
3. Rohlf FJ, Marcus LF (1993) A revolution in morphometrics. Trends in Ecology & Evolution 8: 129–132.
4. Henderson A (2006) Traditional morphometrics in plant systematics and its role in palm systematics. Botanical Journal of the Linnean Society 151: 103–111.
5. Adams DC, Rohlf FJ (2000) Ecological character displacement in *Plethodon*: Biomechanical differences found from a geometric morphometric study. Proceedings of the National Academy of Sciences of the United States of America 97: 4106–4111.
6. Adams DC, Rohlf FJ, Slice DE (2013) A field comes of age: geometric morphometrics in the 21st century. Hystrix-Italian Journal of Mammalogy 24: 7–14.
7. Lauder GV (1982) Historical biology and the problem of design. Journal of Theoretical Biology 97: 57–67.
8. Lande R, Arnold SJ (1985) Evolution of mating preference and sexual dimorphism. Journal of Theoretical Biology 117: 651–664.
9. De Esteban-Trivigno S (2011) Ecomorphology of extinct Xenarthrans: analysis of the mandible using geometric morphometrics methods. Ameghiniana 48: 381–398.
10. De Esteban-Trivigno S (2011) Searching ecomorphological patterns shared by extant ungulates and extinct Xenarthrans. Ameghiniana 48: 189–209.
11. Schluter D (1988) Character Displacement and the Adaptive Divergence of Finches on Islands and Continents. The American Naturalist 131: 799–824.
12. Ricklefs Re (1987) Community Diversity: Relative Roles of Local and Regional Processes. Science 235: 167–171.
13. Klingenberg CP, Ekau W (1996) A combined morphometric and phylogenetic analysis of an ecomorphological trend: Pelagization in Antarctic fishes (Perciformes: Nototheniidae). Biological Journal of the Linnean Society 59: 143–177.
14. Paton DC, Collins BG (1989) Bills and tongues of nectar-feeding birds: A review of morphology, function and performance, with intercontinental comparisons. Australian Journal of Ecology 14: 473–506.
15. Monteiro LR (2013) Morphometrics and the comparative method: studying the evolution of biological shape. Hystrix-Italian Journal of Mammalogy 24: 25–32.
16. Monteiro LR, Nogueira MR (2010) Adaptive radiations, ecological specialization, and the evolutionary integration of complex morphological structures. Evolution 64: 724–743.
17. Rohlf FJ, Corti M (2000) Use of two-block partial least-squares to study covariation in shape. Systematic Biology 49: 740–753.
18. Corti M, Rohlf FJ (2001) Chromosomal speciation and phenotypic evolution in the house mouse. Biological Journal of the Linnean Society 73: 99–112.
19. Monteiro LR, Duarte LC, dos Reis SF (2003) Environmental correlates of geographical variation in skull and mandible shape of the punaré rat *Thrichomys apereoides* (Rodentia: Echimyidae). Journal of Zoology 261: 47–57.
20. Piras P, Marcolini F, Raia P, Curcio M, Kotsakis T (2010) Ecophenotypic variation and phylogenetic inheritance in first lower molar shape of extant Italian populations of *Microtus* (*Terricola*) *savii* (Rodentia). Biological Journal of the Linnean Society 99: 632–647.
21. Klingenberg CP, Zaklan SD (2000) Morphological integration between developmental compartments in the Drosophila wing. Evolution 54: 1273–1285.
22. Bookstein FL, Gunz P, Mitteroecker P, Prossinger H, Schaefer K, et al. (2003) Cranial integration in *Homo*: singular warps analysis of the midsagittal plane in ontogeny and evolution. Journal of Human Evolution 44: 167–187.
23. Drake AG, Klingenberg CP (2008) The pace of morphological change: historical transformation of skull shape in St Bernard dogs. Proceedings of the Royal Society B-Biological Sciences 275: 71–76.
24. Klingenberg CP, Badyaev AV, Sowry SM, Beckwith NJ (2001) Inferring developmental modularity from morphological integration: Analysis of individual variation and asymmetry in bumblebee wings. American Naturalist 157: 11–23.
25. Tabachnick RE, Bookstein FL (1990) The Structure of Individual Variation in *Miocene globorotalia*. Evolution 44: 416–434.
26. Traugott M, Schallhart N, Kaufmann R, Juen A (2008) The feeding ecology of elaterid larvae in central European arable land: New perspectives based on naturally occurring stable isotopes. Soil Biology and Biochemistry 40: 342–349.
27. Furlan L (1998) The biology of *Agriotes ustulatus* Schaller (Col., Elateridae). II. Larval development, pupation, whole cycle description and practical implications. Journal of Applied Entomology-Zeitschrift Fur Angewandte Entomologie 122: 71–78.
28. Tóth Z (1984) Click beetles (Elateridae) in the soils of central Europe - their distribution and description. Part I. (Gen.: Agriotes). Acta Phytopathologica Academiae Scientiarum Hungaricae 19: 13–29.
29. Laibner S (2000) Elateridae of the Czech and Slovak Republics. Zlin: Kabourek.
30. Parker WE, Howard JJ (2001) The biology and management of wireworms (*Agriotes spp.*) on potato with particular reference to the U.K. Agricultural and Forest Entomology 3: 85–98.
31. Tóth M (2013) Pheromones and attractants of click beetles: an overview. Journal of Pest Science 86: 3–17.
32. Kovačević Ž (1960) Problematika zemljišnih štetnika u istočnoj Slavoniji. Sav Polj: 567–580.
33. Brunner N, Kromp B, Meindl P, Pázmándi C, Traugott M (2005) Evaluation of different sampling techniques for wireworms (Coleoptera, Elateridae) in arable land. IOBC/wprs Bulletin 28: 117–122.
34. Benefer C, Herk WG, Ellis JS, Blackshaw RP, Vernon RS, et al. (2013) The molecular identification and genetic diversity of economically important wireworm species (Coleoptera: Elateridae) in Canada. Journal of Pest Science 86: 19–27.
35. Bogunović M, Vidaček Ž, Racz Z, Husnjak S, Sraka M (1996) Namjenska pedološka karta 1:300000 Faculty of Agriculture: University of Zagreb.
36. Benefer CM, Knight ME, Ellis JS, Hicks H, Blackshaw RP (2012) Understanding the relationship between adult and larval *Agriotes* distributions: The effect of sampling method, species identification and abiotic variables. Applied Soil Ecology 53: 39–48.
37. Kozina A (2012) The factors for the temporal and spatial distribution of the economically important maize pests: University of Zagreb.
38. Maceljski M (2002) Poljoprivrednaentomologija. Čakovec: Zrinski.
39. Bažok R (2007) Žičnjaci. Glasilo biljne zaštite 5: 339–344.
40. Bažok R, Igrc-Barčić J (2010) Pheromone applications in maize pest control. New York: Nova Science.
41. Bravi R, Benitez HA (2013) Left-right asymmetries and shape analysis on *Ceroglossus chilensis* (Coleoptera: Carabidae). Acta Oecologica-International Journal of Ecology 52: 57–62.
42. Benitez HA, Sanzana M-J, Jerez V, Parra LE, Hernandez CE, et al. (2013) Sexual Shape and Size Dimorphism in Carabid Beetles of the Genus *Ceroglossus*: Is Geometric Body Size Similar Between Sexes Due to Sex Ratio? Zoological Science 30: 289–295.
43. Alibert P, Moureau B, Dommergues JL, David B (2001) Differentiation at a microgeographical scale within two species of ground beetle, *Carabus auronitens* and *C-nemoralis* (Coleoptera, Carabidae): a geometrical morphometric approach. Zoologica Scripta 30: 299–311.
44. Benitez H, Briones R, Jerez V (2008) Fluctuating asymmetry in two populations of *Ceroglossus chilensis* (Eschscholtz, 1829) (Coleoptera: Carabidae) in agroecosystem of *Pinus radiata* d. Don, Bio-Bio region, Chile Gayana 72: 131–139.
45. Clarke G (1998) Developmental Stability and Fitness: The Evidence Is Not Quite So Clear. The American Naturalist 152: 762–766.
46. Ghalambor CK, McKay JK, Carroll SP, Reznick DN (2007) Adaptive versus non-adaptive phenotypic plasticity and the potential for contemporary adaptation in new environments. Functional Ecology 21: 394–407.
47. Pfennig DW, Wund MA, Snell-Rood EC, Cruickshank T, Schlichting CD, et al. (2010) Phenotypic plasticity's impacts on diversification and speciation. Trends in Ecology & Evolution 25: 459–467.
48. Parsons SA, Joern A (2014) Life history traits associated with body size covary along a latitudinal gradient in a generalist grasshopper. Oecologia 174: 379–391.
49. Bouyer J, Ravel S, Dujardin J-P, De Meeüs T, Vial L, et al. (2007) Population Structuring of *Glossina palpalis gambiensis* (Diptera: Glossinidae) According to Landscape Fragmentation in the Mouhoun River, Burkina Faso. Journal of Medical Entomology 44: 788–795.
50. Stone GN, Schönrogge K (2003) The adaptive significance of insect gall morphology. Trends in Ecology & Evolution 18: 512–522.
51. Haine ER, Moret Y, Siva-Jothy MT, Rolff J (2008) Antimicrobial Defense and Persistent Infection in Insects. Science 322: 1257–1259.
52. Vogelweith F, Thiéry D, Moret Y, Colin E, Motreuil S, et al. (2014) Defense strategies used by two sympatric vineyard moth pests. Journal of Insect Physiology 64: 54–61.
53. Gross P (1993) Insect Behavioral and Morphological Defenses Against Parasitoids. Annual Review of Entomology 38: 251–273.
54. Moczek AP, Emlen DJ (2000) Male horn dimorphism in the scarab beetle, *Onthophagus taurus*: do alternative reproductive tactics favour alternative phenotypes? Animal Behaviour 59: 459–466.
55. Roy D, Seehausen O, Nosil P (2013) Sexual dimorphism dominates divergent host plant use in stick insect trophic morphology. BMC Evolutionary Biology 13: 135.
56. McCullough EL, Emlen DJ (2013) Evaluating the costs of a sexually selected weapon: big horns at a small price. Animal Behaviour 86: 977–985.
57. Benitez HA, Avaria-Llautureo J, Canales-Aguirre CB, Jerez V, Parra LE, et al. (2013) Evolution of sexual size dimorphism and its relationship with sex ratio in carabid beetles of Genus *Ceroglossus* Solier. Current Zoology 59: 769–777.
58. Cook DF (1990) Differences in courtship, mating and postcopulatory behaviour between male morphs of the dung beetle *Onthophagus binodis* Thunberg (Coleoptera: Scarabaeidae). Animal Behaviour 40: 428–436.
59. Nijhout HF, Emlen DJ (1998) Competition among body parts in the development and evolution of insect morphology. Proceedings of the National Academy of Sciences 95: 3685–3689.
60. Castañeda LE, Nespolo RF (2013) Phenotypic and Genetic Effects of Contrasting Ethanol Environments on Physiological and Developmental Traits in *Drosophila melanogaster*. PLoS ONE 8: e58920.
61. Demirci B, Lee Y, Lanzaro GC, Alten B (2012) Altitudinal genetic and morphometric variation among populations of *Culex theileri* Theobald (Diptera: Culicidae) from northeastern Turkey. Journal of Vector Ecology 37: 197–209.

62. Benitez HA, Pizarro-Araya J, Bravi R, Sanzana M-J, M AF (2014) Morphological variation on isolated populations of *Praocis* (*Praocis*) *spinolai*. Journal of Insect Science 14: 1–12.
63. Gurgel-GonÇAlves R, Ferreira JBC, Rosa AF, Bar ME, GalvÃO C (2011) Geometric morphometrics and ecological niche modelling for delimitation of near-sibling triatomine species. Medical and Veterinary Entomology 25: 84–93.
64. Sanzana M-J, Parra LE, Sepulveda-Zuniga E, Benitez HA (2013) Latitudinal gradient effect on the wing geometry of *Auca coctei* (Guerin) (Lepidoptera, Nymphalidae). Revista Brasileira De Entomologia 57: 411–416.
65. Kaupp A, Wurst C (1997) Familie Elateridae – Nachträge und Berichtigungen. In: Klausnitzer B, editor. Die Larven der Käfer Mitteleuropas. Jena: Gustav Fischer Verlag. 330–345.
66. Rohlf FJ (2013) TPSdig, v. 2.17. NY State University at Stony Brook.
67. Klingenberg CP (2011) MorphoJ: an integrated software package for geometric morphometrics. Molecular Ecology Resources 11: 353–357.
68. Arnqvist G, Martensson T (1998) Measurement error in geometric morphometrics: empirical strategies to assess and reduce its impact on measures of shape. Acta Zoologica Academiae Scientiarum Hungaricae 44: 73–96.
69. Dryden I, Mardia K (1998) Statistical Shape Analysis. Chichester: John Wiley and Son.
70. Rohlf FJ, Slice D (1990) Extensions of the Procustes methods for the optimal superimposition of landmarks. Systematic Zoology 39: 40–59.
71. Klingenberg CP, Barluenga M, Meyer A (2002) Shape analysis of symmetric structures: Quantifying variation among individuals and asymmetry. Evolution 56: 1909–1920.
72. Gould SJ (1966) Allometry and size in ontogeny and phylogeny. Biological Reviews of the Cambridge Philosophical Society 41: 587-&.
73. Klingenberg CP (1996) Multivariate allometry. In: Marcus LF, Corti M, Loy A, Naylor GJP, Slice DE, editors. Advances in Morphometrics. 23–49.
74. Campbell NA, Atchley WR (1981) The Geometry of Canonical Variate Analysis. Systematic Zoology 30: 268–280.
75. Klingenberg CP, Monteiro LR (2005) Distances and directions in multidimensional shape spaces: Implications for morphometric applications. Systematic Biology 54: 678–688.
76. Klingenberg CP, McIntyre GS (1998) Geometric morphometrics of developmental instability: Analyzing patterns of fluctuating asymmetry with procrustes methods. Evolution 52: 1363–1375.
77. Kozina A, Čačija M, Igrc Barčić J, Bažok R (2013) Influence of climatic conditions on the distribution, abundance and activity of *Agriotes lineatus* L. adults in sex pheromone traps in Croatia. International Journal of Biometeorology 57: 509–519.
78. Legendre P, Fortin M-J (2010) Comparison of the Mantel test and alternative approaches for detecting complex multivariate relationships in the spatial analysis of genetic data. Molecular Ecology Resources 10: 831–844.
79. Villani MG, Allee LL, Díaz A, Robbins PS (1999) Adaptive strategies of edaphic arthropods. Annual Review of Entomology 44: 233–256.
80. Téllez TR, Møller AP (2006) Fluctuating asymmetry of leaves in *Digitalis thapsi* under field and common garden conditions. International Journal of Plant Sciences 167: 321–329.
81. Benitez HA, Lemic D, Bazok R, Gallardo-Araya CM, Mikac KM (2014) Evolutionary directional asymmetry and shape variation in *Diabrotica virgifera virgifera* (Coleoptera: Chrysomelidae): an example using hind wings. Biological Journal of the Linnean Society 111: 110–118.
82. Liebhold AM, Rossi RE, Kemp WP (1993) Geostatistics and Geographic Information Systems in Applied Insect Ecology. Annual Review of Entomology 38: 303–327.
83. Toepfer S, Ellsbury MM, Eschen R, Kuhlmann U (2007) Spatial clustering of *Diabrotica virgifera virgifera* and *Agriotes ustulatus* in small-scale maize fields without topographic relief drift. Entomologia Experimentalis et Applicata 124: 61–75.

9

Sequence Depth, Not PCR Replication, Improves Ecological Inference from Next Generation DNA Sequencing

Dylan P. Smith, Kabir G. Peay*

Department of Biology, Stanford University, Stanford, California, United States of America

Abstract

Recent advances in molecular approaches and DNA sequencing have greatly progressed the field of ecology and allowed for the study of complex communities in unprecedented detail. Next generation sequencing (NGS) can reveal powerful insights into the diversity, composition, and dynamics of cryptic organisms, but results may be sensitive to a number of technical factors, including molecular practices used to generate amplicons, sequencing technology, and data processing. Despite the popularity of some techniques over others, explicit tests of the relative benefits they convey in molecular ecology studies remain scarce. Here we tested the effects of PCR replication, sequencing depth, and sequencing platform on ecological inference drawn from environmental samples of soil fungi. We sequenced replicates of three soil samples taken from pine biomes in North America represented by pools of either one, two, four, eight, or sixteen PCR replicates with both 454 pyrosequencing and Illumina MiSeq. Increasing the number of pooled PCR replicates had no detectable effect on measures of α- and β-diversity. Pseudo-β-diversity – which we define as dissimilarity between re-sequenced replicates of the same sample – decreased markedly with increasing sampling depth. The total richness recovered with Illumina was significantly higher than with 454, but measures of α- and β-diversity between a larger set of fungal samples sequenced on both platforms were highly correlated. Our results suggest that molecular ecology studies will benefit more from investing in robust sequencing technologies than from replicating PCRs. This study also demonstrates the potential for continuous integration of older datasets with newer technology.

Editor: Christina A. Kellogg, U.S. Geological Survey, United States of America

Funding: Financial support for this work was provided by National Science Foundation (NSF) Dimensions of Biodiversity grant (DBI 1249341) to KGP. The funders had no role in study design, data collection and analysis, decision to publish, or preparation of the manuscript.

Competing Interests: The authors have declared that no competing interests exist.

* E-mail: kpeay@stanford.edu

Introduction

Next generation DNA sequencing (NGS) has changed the face of microbial ecology in the space of a few years. As a result, we have gained unprecedented insight into the community dynamics of morphologically cryptic organisms such as fungi, bacteria and viruses [1][2][3]. However, the outcome of NGS based ecological inquiry may be sensitive to technical practices that in many cases have not been adequately tested. For instance, the assumption that NGS read counts accurately reflect absolute abundance in ecological analyses may not be appropriate due to taxon specific PCR and sequencing biases [4]. These technical practices have important effects on our view of underlying biological reality, but also on the allocation of resources (time, money, reagents) that often define the scope of ecological inquiry.

Early optimization of NGS methods has focused on correcting platform specific sequencing issues, such as the known homopolymer error rates in 454 pyrosequencing [5], often with bioinformatic solutions. However, potential distortions may also arise prior to DNA sequencing during sample collection [6], DNA extraction [7][8], or PCR amplification [9][10]. Recognition of these problems has led to a loosely knit collection of best lab and bioinformatics practices that have emerged in the microbial ecology literature and that are aimed at increasing the robustness of whole community amplification. Among other things, such practices include the use of hot-start Taq polymerase, reducing the number of amplification cycles, and the pooling of multiple PCR replicates per sample [11]. The necessity of pooling PCR replicates is thought to arise from stochasticity in the PCR process that results in variable composition of DNA fragments across individual PCR reactions. Possible causes for this may be sampling effects that lead to variation in the initial population of DNA template used to start the reaction, slight variation in initial conditions, or priority effects of amplification in the early rounds of PCR. The few studies that have actually reported results from replicate NGS of the same sample (e.g. [8][12]) have found sample-to-sample variance in sequence composition that seem to support the importance of stochastic PCR effects. However, these studies have focused on the comparison of individual PCR replicates that were sequenced separately, and therefore the extent to which their conclusions rely on PCR or sequencing stochasticity is still unknown.

Though many studies have suggested pooling PCR replicates prior to sequencing, to our knowledge no study has directly tested whether samples comprised of multiple, pooled PCR replicates capture a more robust sample of the true diversity within the sample. Because there have been no explicit tests of PCR pooling,

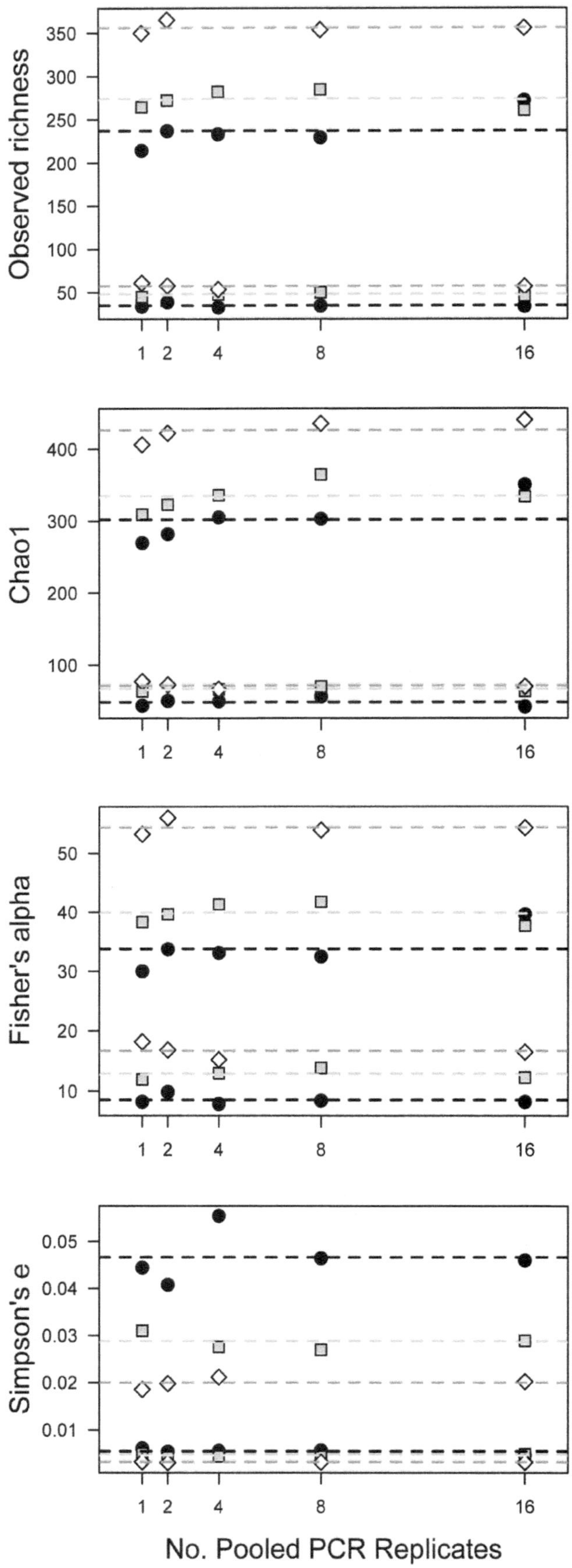

Figure 1. Estimated species richness does not depend on the number of PCR reactions pooled prior to sequencing. Plots of independent replicates representing different levels of PCR pooling for samples CT2, OR1, and OR4 against four different diversity indicators. Points are colored by sample ID. Dotted lines represent the average between different replicates of the same sample. The top three lines represent samples sequenced with Illumina MiSeq and the bottoms three lines represent the same samples sequenced with 454.

papers vary wildly in the number of pooled PCR replicates they use or recommend. While three appears to be a somewhat canonical number (e.g. it has been adopted by the Earth Microbiome Project [13]) other studies have used a single PCR replicate [14], five [15] or even ten [16].

The rapid rise and fall of NGS sequencing platforms also raises major concerns about portability of data across studies. One of the major advantages of DNA based community profiling is the collection of standardized data that can be compared or combined across studies. However, if the observed structure of a community is platform dependent it would seriously weaken the additive nature of sequencing efforts in microbial ecology [17].

In this study our primary goals were (1) to determine quantitatively the number of pooled PCR replicates that maximizes ecological inference in NGS studies and (2) to test the robustness of ecological inferences about community structure using two different NGS sequencing platforms. We did this using a two-pronged experimental approach where the same DNA sample was sequenced from a pool of 1, 2, 4, 8 or 16 separate PCR reactions using both Roche's 454 Pyrosequencing (454) and Illumina's MiSeq (MiSeq). We then compared patterns of α- and β-diversity (the primary response variables in most community ecology studies) among samples and replicates. In addition, we used both NGS platforms to sequence a larger set of soil samples taken from Pine forests in geographically distinct parts of North America where we expected to see differences in community composition.

Based on previous studies, we hypothesized that increasing the number of PCR replicates prior to sequencing would increase α-diversity, reduce β-diversity and increase reproducibility by averaging out PCR noise. Surprisingly, we found that increasing PCR replication did not meaningfully change any of our ecological response variables and may be a poor investment of resources in molecular ecology studies. By contrast, increased sequencing depth markedly improved estimates of β-diversity using both NGS platforms. In addition, we found that community sequencing results from 454 and MiSeq provide largely similar results, suggesting that data from the two platforms can be combined in a meaningful way.

Materials and Methods

Experimental design

We investigated the effects of PCR replication and sequencing platforms on ecological inference using samples from an ongoing project to characterize ectomycorrhizal fungal communities across North America pine forests. To test the hypothesis that pooling PCR replicates improves ecological inference, we selected three soil samples taken from two sites in Oregon (42.79° E –121.62° N) and one in Connecticut (41.82° E, –72.96° N). To test whether NGS sequencing platform affects ecological inference, we selected 60 samples from two sites in North Carolina (NC1 36.01° E, –78.97° N, NC2 35.99° E, –79.10° N), two sites in California (CA1 37.84° E, –119.94° N, CA2 37.81° E, –119.91° N), and two sites in

Table 1. Analysis of variance tests for an effect of PCR replicate number, sample identity, and sequencing method on fungal richness from the CT2, OR1, and OR4 samples.

	No. PCR Replicates		Sample ID		Method		Replicates × Sample ID		Replicates × Method	
	$F_{1,22}$	P	$F_{2,22}$	P	$F_{1,22}$	P	$F_{2,22}$	P	$F_{1,22}$	P
Observed	0.372	0.548	18.748	<0.001*	646.450	<0.001*	0.320	0.730	0.261	0.615
Chao1	2.380	0.137	15.480	<0.001*	717.970	<0.001*	0.171	0.844	2.428	0.134
Fisher's Alpha	0.415	0.526	38.256	<0.001*	490.635	<0.001*	0.439	0.650	0.430	0.519
Simpson	0.060	0.809	42.190	<0.001*	17.250	<0.001*	0.185	0.832	0.000	0.991
Simpson's E	0.001	0.977	13.502	<0.001*	137.701	<0.001*	0.054	0.947	0.001	0.979

Samples were sequenced with both 454 and Illumina MiSeq.

Alaska (AK1 64.77° E, −148.27° N, AK2 64.76° E, −148.25° N) (for all sites n = 10).

All sample sites in Oregon and California were located on United States Department of Agriculture (USDA) National Forest land. The Alaskan sites were located at the Bonanza Creek Long Term Ecological Research (LTER) site. Sites in North Carolina were located on private land owned by Duke University, and sites in Connecticut were located in the Gold's Pine State Forest. No permits were required for any site, and all necessary permissions were obtained prior to sampling. Soil samples were taken from either the homogenized organic or mineral layer of a core approximately 7.5 cm diameter ×14 cm deep (**File S1**). Soils were stored cool until ~0.25 g were extracted using the Powersoil DNA extraction kit (MoBio, Carlsbad CA). DNA extracts were diluted 1:20 and then 1 µl used for PCR.

Molecular methods

For sequencing using the 454 platform, PCR was carried out using modified versions of the fungal specific primer set ITS1F [18] and ITS4 [19]. The 5′ end of the ITS1F primer was modified to include the 454 Lib-L A adapter plus a 10-bp molecular identification (MID) tag to allow for sample multiplexing as in [14]. The 5′ end of the ITS4 primer was modified to include the 454 Lib-L B adapter.

For sequencing using the Illumina MiSeq platform, we designed modified versions of the primer set ITS1F and ITS2 [19]. This primer set targets a shorter section of the fungal ITS region because of the shorter read lengths possible with MiSeq. The 5′ end of the ITS1F primer was modified to include the forward Illumina Nextera adapter and a two basepair "linker" sequence designed to mismatch against all major fungal lineages immediately upstream of the gene primer (**Fig S1**). The induced mismatch is designed to decrease potential taxon-specific PCR bias from downstream matches to the adapter or barcode. The 5′ end of the ITS2 primer was modified with the appropriate reverse Illumina Nextera adapter, linker sequence, and a 12-bp error-correcting Golay barcode as in [17]. Using the program NetPrimer (Premier Biosoft, Palo Alto CA) we designed three custom sequencing primers that demonstrated low dimerization potential and high thermodynamic compatibility with each other and with the Illumina-specific PhiX sequencing primer. The Read 1 and Read 2 sequencing primers were designed to anneal to the gene priming regions of the amplicons and extend further into the conserved 18S portion of the amplified region, thereby maximizing the amount of ITS sequence returned by the reads. The Index sequencing primer was designed to sequence only the 12 bp barcode of each amplicon.

PCR was carried out in 25 µl reactions including 1 µl genomic DNA, 0.5 µl of each 10 µM primer, 5 µl of 5× OneTaq Standard Reaction Buffer (New England BioLabs, Ipswitch MA), 0.5 µl of 10 mM dNTPs (New England BioLabs, Ipswitch MA), and 0.63 units Taq polymerase. All PCR reactions were set up on ice and using Fusion hot start Taq polymerase (New England Biolabs, Ipswitch MA) to minimize non-specific amplification and primer dimerization. PCR conditions were: denaturation at 94°C for 1 min; 30 amplification cycles of 30 sec at 94°C, 30 sec at 52°C and 30 sec at 68°C; followed by a 7 min final extension at 68°C. PCR products were visualized using gel electrophoresis and successful samples cleaned using the Agencourt Ampure XP kit (Beckman Coulter, Brea CA). For the replication experiment, the three samples were each amplified 1, 2, 4, 8 or 16 times using a separate MID tag or barcode for each replication treatment (N = 3 samples ×5 replication levels = 15). Individual PCR reactions for a given sample × replication treatment were pooled and then 20 µl of each pool cleaned using the Ampure Kit as above.

Cleaned PCR products were quantified using the Qubit hs-DS-DNA kit (Invitrogen, Carlsbad CA) on a Tecan Infinite F200 Pro plate reader reading at 485 nm excitation and 530 nm emission. PCR products to be sequenced with 454 were then pooled in equimolar concentration and sent to the Duke University Institute for Genome Sciences & Policy core and sequenced on a ¼ plate partition using Titanium FLX chemistry. PCR products generated for Illumina sequencing were pooled at equimolar concentration and then multiplexed with 44 additional bacterial samples containing 16S rDNA amplicons used for an unrelated study. The final pool containing both loci was sent to the Stanford Functional Genomics Facility for 250 bp paired-end sequencing on an Illumina MiSeq. Bacterial and fungal sequencing primers were also pooled for each read before submission to the sequencing facility. A spike of 30% PhiX was included in the amplicon library in order to achieve sufficient sample heterogeneity. Raw sequence data are deposited at NCBI's Short Read Archive under study accession SRP035367. Sample metadata information is provided as File S1.

Bioinformatics

Sequence de-multiplexing and bioinformatic processing of the 454 and Illumina datasets were performed using aspects of the QIIME [20] and the UPARSE [21] pipelines. Initial quality filtering of 454 sequences excluded all sequences <350 or >1200 bp, with any primer mismatches, with a homopolymer run >10 bp, or with a mean quality score below 25. The remaining sequences were denoised using flowgram clustering [22]. Pre-filtered forward and reverse reads from the Illumina

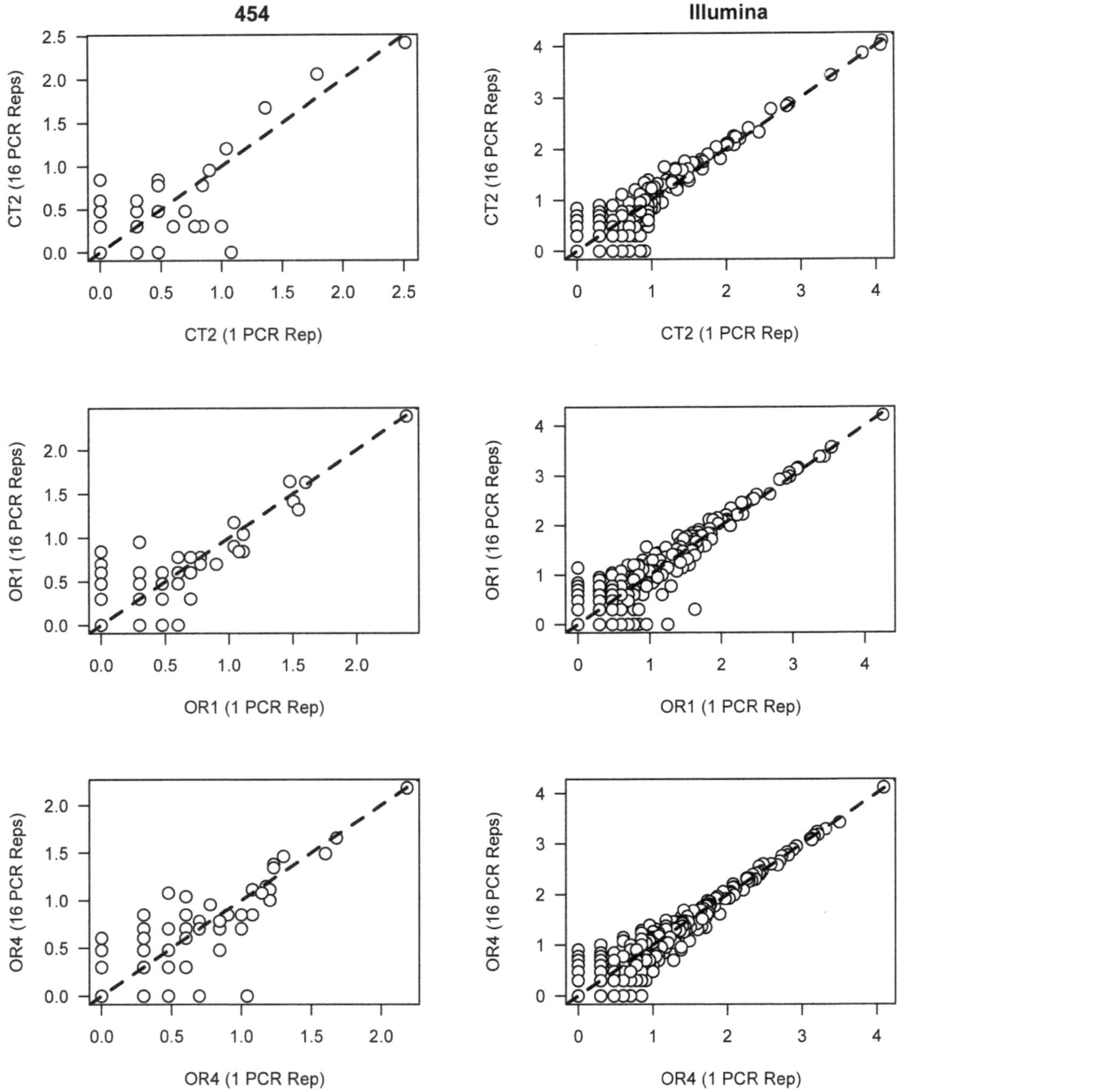

Figure 2. Taxon abundance is strongly correlated across sequencing runs with different levels of PCR replication. Circles represent individual taxa, and the relationship between the $\log_{10}$ of their abundance in samples comprised of 16 PCR replicates (y axis) and one PCR replicate (x axis) for CT2 (top two panels), OR1 (middle two panels), and OR4 (bottom two panels). Dashed lines represent a 1:1 relationship. The left three panels show samples sequenced with 454 and the right three panels show samples sequenced with Illumina MiSeq.

dataset were 238 bp long and our multiplexing strategy resulted in high quality sequences for both fungal and bacterial samples. For the fungal samples analyzed in this study, reads were trimmed with CutAdapt [23] to the point where the sequence met the distal priming site, and further trimmed using Trimmomatic [24] to remove any additional low quality end regions. After quality trimming forward reads averaged 208 bp and reverse reads averaged 185 bp. Reads were paired using USEARCH v. 7.0.1001 with a minimum Phred score sequence cutoff threshold of 3 and a minimum sequence length of 75 bp. Paired reads averaged 230 bp and were discarded if they contained >0.25 expected errors. The final fasta file containing all sequences used for analysis is available from the authors upon request.

All final, high-quality sequences from both the 454 and Illumina datasets were combined and grouped into operational taxonomic units (OTUs) in USEARCH using the UPARSE-OTU and UPARSE-OTUref algorithms (which included chimaera detection and filtering and dropped all global singleton reads) at a 97% sequence similarity cutoff. OTUs were given taxonomic assignments in QIIME based on a previously published sequence

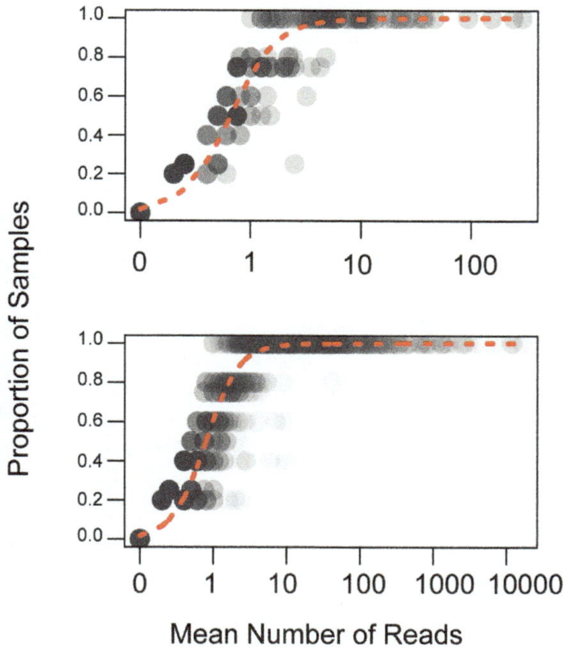

Figure 3. Taxa represented by more sequences are observed most consistently across replicates of the same sample. Plots of the proportion of samples in which an OTU is observed against the average number of reads representing that OTU. Zeros were assigned a value of 0.1 prior to log transformation. Dashed lines represent the logistic model predicting the relationship. The top panel shows results from 454 and the bottom panel shows results from Illumina.

database [8] modified for QIIME compatibility as in [25]. To compare samples on an equal basis all samples were rarefied to even sampling depths prior to statistical analysis. Rarefaction depths were determined ad-hoc to maximize the number of samples included while still maintaining a reasonable number of sequences. For the replication experiment, the 454 samples were rarefied to 500 sequences and the Illumina samples to 38,000 sequences. For the larger, cross platform-comparison dataset, 454 samples were rarefied to 1000 sequences and the Illumina samples to 40,000 sequences.

Statistical analysis

To see how replication affects ecological inference we calculated a number of common α-diversity (observed richness, Fisher's Alpha, Chao 1 and Simpson, Simpson Evenness) and β-diversity (Jaccard, Bray-Curtis, β-sim) metrics used in community ecology. We used a linear model to test whether or not the number of PCR replicates and sequencing platform affected different richness estimators (S = Replication × Sample ID × Platform). We used a similar approach to test whether average β-diversity changed in any predictable way with the number of replicates used to generate each sample. This was done by calculating β-diversity (Bray Curtis or Jaccard) for each replication level compared with all other samples sequenced from those plots.

This dataset also allowed us a unique look at data reproducibility with repeated sequencing of the same sample. To see how sequencing depth affects estimates of sample β-diversity, we calculated within sample β-diversity (that is, β-diversity between independent replicates of the same sample – hereafter termed pseudo-β-diversity) at a range of sequencing depths, from 50–1000 (454) and 100 to 80,000 (Illumina). Because there has been much debate about the handling and validity of low abundance OTUs, we tested the effect of within sample sequence abundance on the repeatability with which an OTU is detected across replicate sequencing of the same sample. We used logistic regression to model the relationship between $\log_{10}$ transformed mean within sample abundance and frequency of detection across samples (for this analysis 0 values were assigned ½ the minimum observed value prior to log transformation). We also looked at quantitative reproducibility by comparing OTU read abundance ($\log_{10}$ X+1 transformed) between the single replication treatment and the 16 replication treatment for each DNA sample.

Finally, with the larger dataset we compared the similarity of ecological inferences made with different sequencing platforms. α-diversity estimates were generated for each sample based on rarefaction to a common sequence depth within each platform. We used a Mantel test to determine whether community similarity estimates were similar across platforms. Pairwise sample similarity across samples and across platforms was visualized using non-metric dimensional scaling (NMDS), and a perMANOVA tested for the effect of sequencing platform and geographic origin on estimates of community similarity. To compare whole community overlap we generated a Venn diagram to illustrate the proportion of shared and unique taxa generated with each platform. To look for taxonomic bias we plotted relative abundance of lineages for shared and unique OTUs across platforms. Statistics were performed using the R software package [26] and the Vegan community analysis package for NMDS and perMANOVA [27].

Several data points were left out of our analyses due to either insufficient or low-quality sequences or sample mishandling (454 data: the 2 PCRs treatment for OR1, the 8 PCRs treatment for OR4, and individual points CA1.A5.OH and CA1.A5.AH from the larger dataset. MiSeq data: the 4 PCRs treatment for OR1 and individual point NC2.0.OH from the larger dataset).

Results

After quality control, denoising, and chimera removal of the smaller dataset, sequencing depth for successful samples ranged from 573–1783 sequences in the 454 dataset and from 38,423–92,189 sequences in the Illumina dataset. Observed richness at 500 sequences (454 dataset) ranged from approximately 30 to 60 OTUs/sample and at 38,000 sequences (Illumina dataset) ranged from approximately 200 to 350 OTUs/samples. As expected for Pine soils the most commonly observed taxa belonged to lineages of Basidiomycota ectomycorrhizal fungi and saprotrophs (data not shown).

Increasing the number of PCR replicates pooled prior to sequencing had no effect on the estimated α-diversity of a sample regardless of the sequencing method used (**Table 1**). This effect was consistent regardless of the richness metric chosen (**Fig 1**) and whether or not an interaction term was included in the model. Similarly, estimates of β-diversity compared with other samples in the same site did not show any trends related to the number of pooled PCR replicates (454 Jaccard ANOVA $F_{1,7} = 0.588$, $P = 0.468$; Bray-Curtis ANOVA $F_{1,7} = 1.682$, $P = 0.236$; Illumina Jaccard ANOVA $F_{1,8} = 0.118$, $P = 0.741$; Bray-Curtis ANOVA $F_{1,8} = 1.685$, $P = 0.231$; **Fig S2**). That is to say, a sample sequenced from 1 PCR did not show higher or lower β -diversity with other samples from the same plot than the same sample sequenced from 16 PCR replicates. When ordinated the different replication levels from the same sample clustered together with little variation from the centroid (**Fig S2**).

Sequence counts for individual OTUs were highly correlated across resequencing instances of the same sample. This relation-

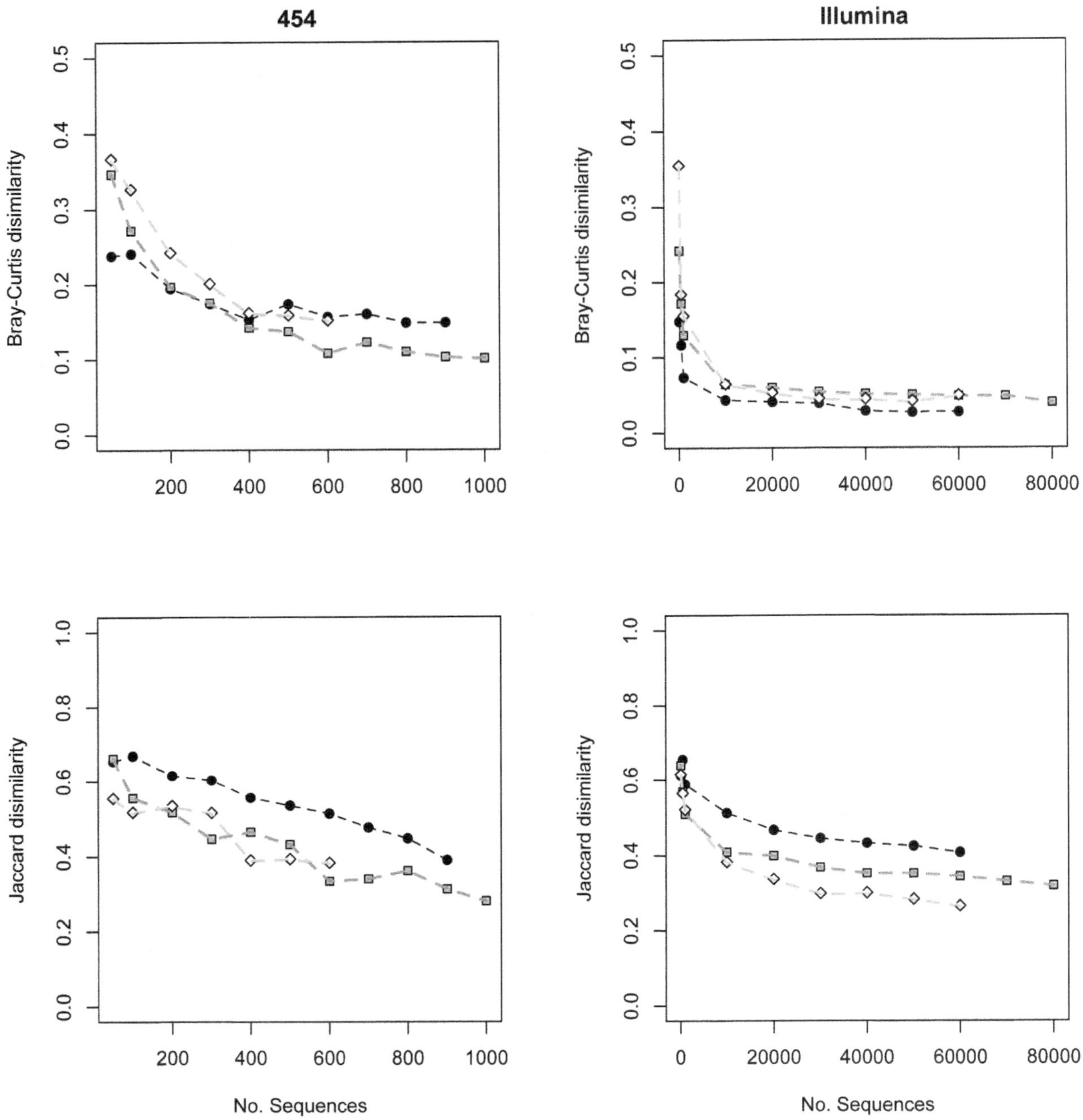

Figure 4. Increasing sequence depth reduces pseudo-β-diversity. Plots of the between-sample Bray-Curtis dissimilarity (top two panels) and Jaccard dissimilarity (bottom two panels) in CT2, OR1, and OR4 against per-sample sequencing depth. Points represent the β-diversity values between different replicates of the same sample and are colored by sample ID. Dashed lines connect each symbol within a sample. The left two panels show samples sequenced with 454 and the right two panels show samples sequenced with Illumina MiSeq.

ship was true regardless of the PCR replicate number. Plots relating the number of sequences per taxon in a sample composed of 16 PCR replicates to that of a single PCR replicate were strongly log linear and in almost all cases followed a 1:1 relationship (**Fig 2**). As a result, per OTU sequence counts were highly significantly correlated (Pearson's product moment correlation: 454 dataset CT2 $r = 0.84$, $P<0.001$; OR1 $r = 0.89$, $P<0.001$; OR4 $r = 0.87$, $P<0.001$; Illumina dataset CT2 $r = 0.92$, $P<0.001$; OR1 $r = 0.94$, $P<0.001$; OR4 $r = 0.96$, $P<0.001$).

High abundance OTUs were detected more consistently across replicate sequencing runs. The proportion of samples that an OTU was observed in increased significantly with average within sample read depth for that taxon (**Fig 3**; Overall effects tests: 454 No. reads $\chi^2_1 = 1{,}496$, $P<0.001$, Site $\chi^2_2 = 1.0$, $P = 0.60$; Illumina No. reads $\chi^2_1 = 10{,}742$, $P<0.001$, Site $\chi^2_2 = 2.3$, $P = 0.32$). There were no differences in this relationship across samples and the same patterns were seen for analyses run with median, maximum and minimum read depth (data not shown). For both 454 and Illumina OTUs with mean read abundance >10 sequences were detected nearly 100% of the time. However, many low abundance OTUs were also detected with a high degree of regularity.

Pseudo-β-diversity estimates decreased exponentially with increasing sequencing depth (**Fig 4**). In the 454 dataset,

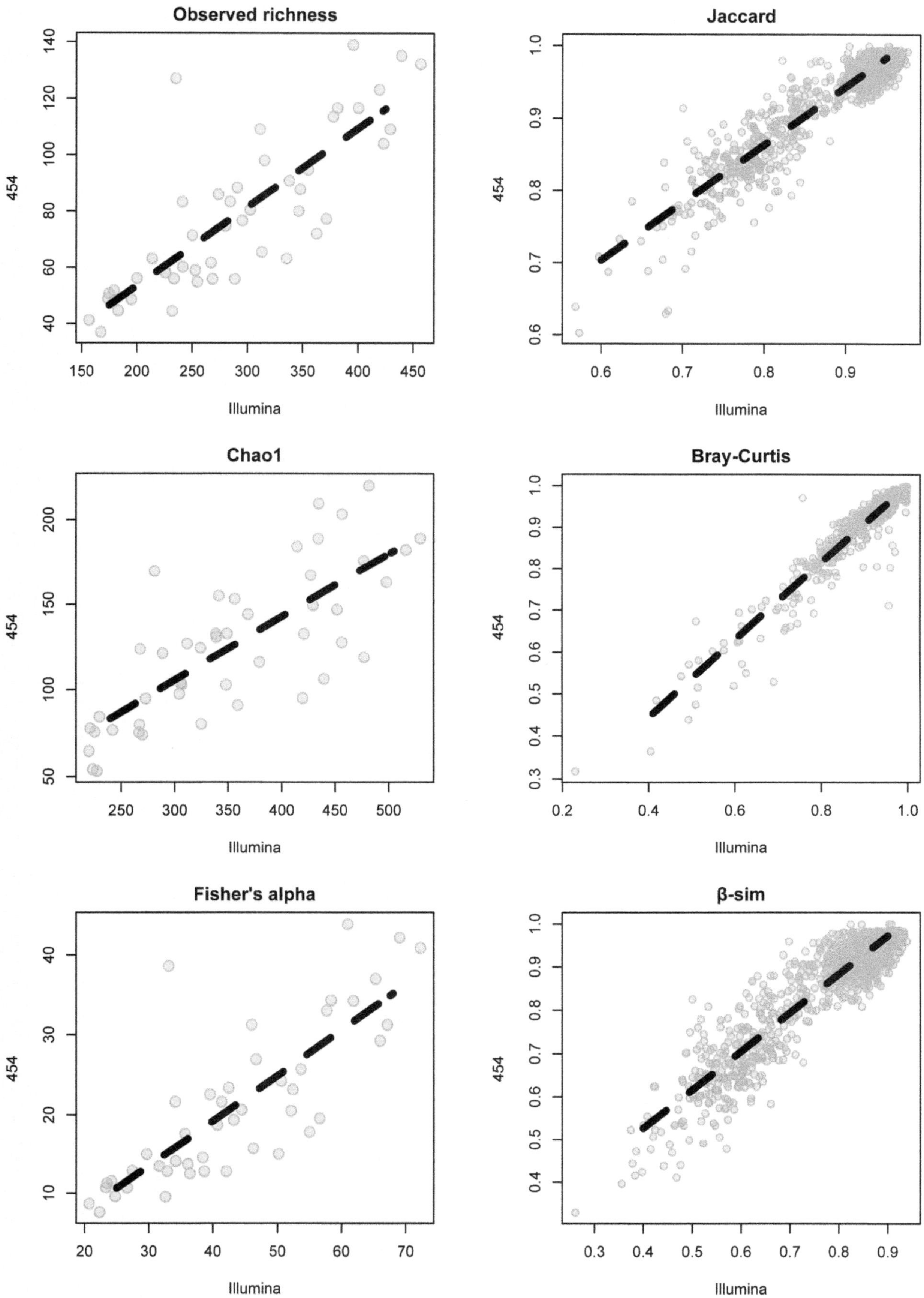
Observed richness
454
40
60
80
100
120
140
150
200
250
300
350
400
450
Illumina
Jaccard
454
0.6
0.7
0.8
0.9
1.0
0.6
0.7
0.8
0.9
Illumina
Chao1
454
50
100
150
200
250
300
350
400
450
500
Illumina
Bray-Curtis
454
0.3
0.4
0.5
0.6
0.7
0.8
0.9
1.0
0.2
0.4
0.6
0.8
1.0
Illumina
Fisher's alpha
454
10
20
30
40
20
30
40
50
60
70
Illumina
β-sim
454
0.4
0.5
0.6
0.7
0.8
0.9
1.0
0.3
0.4
0.5
0.6
0.7
0.8
0.9
Illumina

Figure 5. Patterns of α- and β-diversity are highly reproducible when samples are sequenced on different platforms. Regressions of diversity found in 55 soil samples sequenced on both platforms. Left three columns: points represent individual samples, and the relationship between the total richness per sample found when sequenced with 454 (y axis) and Illumina MiSeq (x axis). Right three columns: points represent pairwise differences in between-sample community composition and the relationship between dissimilarity found with 454 (y axis) and Illumina (x axis). Dashed lines represent the linear models predicting the relationships.

dissimilarity measures consistently remained above 0.1 (Bray-Curtis) and 0.3 (Jaccard) at the maximum amount of sequences/sample recovered. In the Illumina dataset, Bray-Curtis dissimilarity approached zero above 20,000 sequences/sample while Jaccard dissimilarity remained above 0.3 at the maximum amount of sequences/sample recovered.

From the larger set of samples, sequencing on the 454 and Illumina MiSeq platforms resulted in 3,660 total OTUs. Of these 3 (0.08%) were unique to the 454 dataset, 1,798 (49.13%) were unique to the Illumina dataset, and 1,859 (50.79%) were shared (**Fig S3**). Richness was significantly higher for samples sequenced with Illumina MiSeq than for the same samples sequenced with 454 (**Table 2**). Regressions of per-sample observed species, Chao1 estimated richness, and Fisher's Alpha index between the 454 dataset and the Illumina dataset indicated that diversity recovered with either sequencing method was highly correlated (**Fig 5**: Observed slope = 0.28±0.03, $r^2 = 0.67$, $P<0.001$; Chao1 slope = 0.37±0.05, $r^2 = 0.59$, $P<0.001$, Fisher's Alpha slope = 0.57±0.06, $r^2 = 0.65$, $P<0.001$).

β-diversity estimates of between-sample dissimilarity were somewhat affected by sequencing platform (Jaccard perMANOVA $F_{1,98} = 30.26$, $r^2 = 0.08$ $P = 0.001$; Bray-Curtis perMANOVA $F_{1,94} = 45.40$, $r^2 = 0.09$, $P = 0.001$; β-sim perMANOVA $F_{1,94} = 29.30$, $r^2 = 0.05$, $P = 0.001$). However, regressions of β-diversity between the 454 dataset and the Illumina dataset showed that between-sample dissimilarity was highly correlated between the two sequencing platforms (**Fig 5**: Jaccard slope = 0.80±0.01, Mantel r = 0.94, P = 0.001; Bray-Curtis slope = 0.93±0.01, Mantel r = 0.96, P = 0.001, β-sim slope = 0.92±0.01, Mantel r = 0.92, P = 0.001). Both sequencing methods recovered the significant differences in community structure expected between sampling bioregions (Jaccard perMANOVA $F_{2,98} = 124.40$, $r^2 = 0.62$, $P = 0.001$; Bray-Curtis perMANOVA $F_{2,98} = 145.32$, $r^2 = 0.61$, $P = 0.001$; β-sim perMANOVA $F_{2,98} = 204.45$, $r^2 = 0.74$, $P = 0.001$) which explained an order of magnitude more variation than did sequencing method. All samples from both sequencing datasets ordinated primarily by sampling bioregion (**Fig 6**).

Taxonomic assignment was highly consistent between OTUs found in the 454 dataset and OTUs found in the Illumina dataset at the phylum, class, and ordinal levels (**Fig 7, Fig S4**). Taxonomic bias between the two sequencing platforms was primarily limited to low abundance taxa (e.g. relatively more species of Zygomycota and Chytridomycota taxa found in the 454 dataset), but taxonomic composition of each dataset looked nearly identical when relative abundance of taxonomic groups was considered.

Discussion

In this study we compare the effects of different lab, sequencing and bioinformatic protocols on a number of ecological metrics of α- and β-diversity. These metrics form the basis for conclusions in most microbial community studies and so our results should have important ramifications for how this work is carried out. This is particularly true because of the limited resources and options that can be explored in a single study.

Despite the popularity of PCR replication in molecular ecology studies, we find that increasing the number of PCR replicates that are pooled prior to sequencing has no meaningful effect on ecological measures of diversity or community structure and thus likely no effect on the conclusions of a given study. We present several lines of evidence to support this conclusion. First, the effect of PCR replication is highly insignificant in all statistical models predicting α- and β-diversity of replicates of three soil samples taken from different pine biomes across North America (**Table 1, Fig 1, Fig S2**). Although visually there appears to be a slight increase in species richness, Chao1 richness, and Fisher's Alpha diversity as more PCR replicates are pooled, the trend does not hold in a linear fashion and is both sample, sequencing method, and metric-specific (e.g. only seen in the Illumina dataset, only true for OR1 and OR4, and is most apparent only for Chao1 richness estimates). On the other hand, sample ID (i.e. sampling location, a more relevant ecological factor) has a comparably large effect on α- and β-diversity metrics. In all models tested, sample ID and sequencing method are the predominant drivers between differences in diversity and community structure.

Second, the number of sequences observed for each taxon between low-replicated samples (e.g. 1 PCR replicate) and high-replicated samples (16 PCR replicates) is highly correlative in a 1:1 relationship (**Fig 2**), suggesting that pooling more PCR replicates prior to sequencing does not affect the relative abundances of taxa found in each sample, and that sequence abundance per taxon in one PCR can accurately predict sequence abundance per taxon in a pool of 16 replicates. It is important to note that the relationship between sequence abundance per taxon in high vs. low replicated samples is weakest with low-abundance taxa, highlighting the

Table 2. Average per-sample fungal richness for 55 soil samples from pine forests.

	454 Avg. Richness/Sample	Illumina Avg. Richness/Sample	$F_{1,102}$	*P*
Observed	79.907	289.000	322.800	<0.001*
Chao1	125.835	349.610	269.600	<0.001*
Fisher's Alpha	21.190	42.522	82.110	<0.001*
Simpson	0.814	0.819	0.032	0.858
Simpson's E	0.019	0.005	83.970	<0.001*

Samples were sequenced with both 454 and Illumina MiSeq.

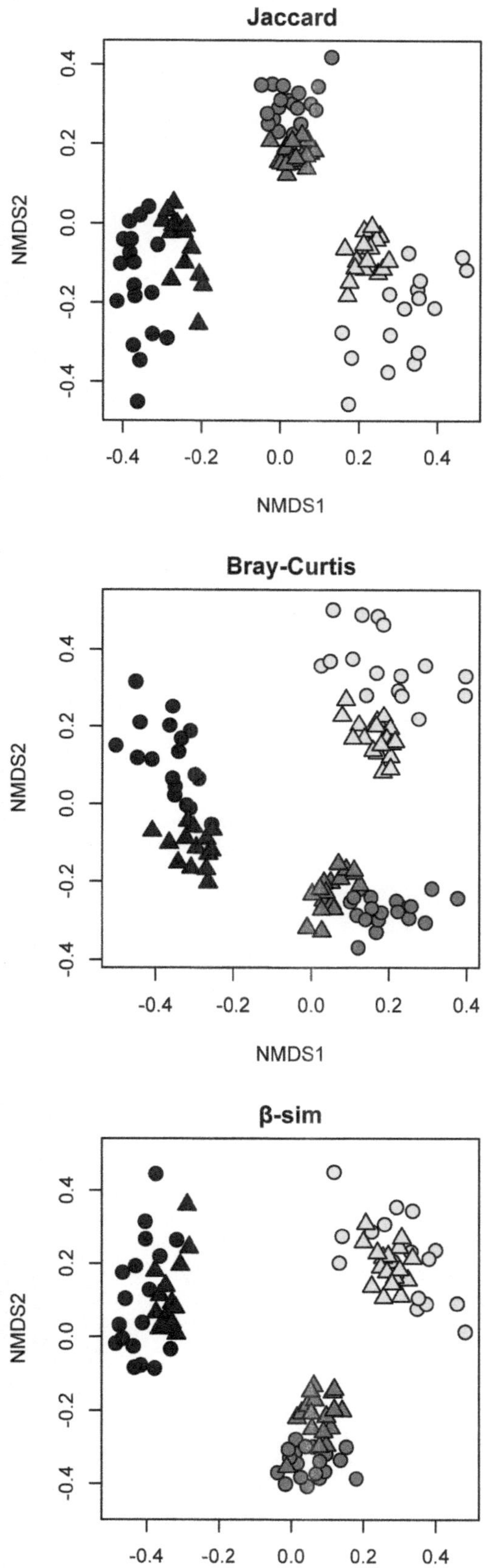

Figure 6. Broad ecological patterns of β-diversity are recovered equally well with each sequencing platform. Non-metric multidimensional scaling of fungal communities from 55 soil samples sequenced with both 454 (circles) and Illumina MiSeq (triangles). Points are colored by the three regions of sample collection. Ordinations are based on between-sample dissimilarity calculated with Jaccard (top panel), Bray-Curtis (middle panel), and β-sim (bottom panel).

importance of adequate sequencing depth for obtaining an accurate depiction of diversity within samples. Together, these results suggest that ecological studies focused on comparing diversity levels and differences between multiple samples across gradients or treatments varying in space or environmental conditions will not be improved significantly by using multiple PCR replicates.

In searching through the literature the origins of this practice are actually somewhat unclear. In general, most evidence in favor of PCR replication is based on differences in OTUs detected in repeated sequencing of single PCR replicates of the same sample [8][11][12]. However, we would argue that sampling error during PCR is a small problem compared with sampling error in the actual sequencing process, in which a few thousand molecules are down sampled from an overall population of billions. Our data show that replicates of the same sample sequenced multiple times at low sequencing depth can lead to higher levels of between-replicate dissimilarity – i.e. pseudo-β-diversity - than should ideally be the case. This is most likely due to inadequate sequencing depth or suboptimal levels of rarefaction making low-abundance taxa unlikely to be appear in all samples. We find that the proportion of replicates of the same sample in which an OTU is present increases logarithmically with the average number of reads representing that OTU (**Fig 3**). This is to say that the reproducibility of taxon coverage and composition of a sample will improve with increasing sequencing depth per taxon. For both 454 and Illumina, OTUs represented by an average of >10 sequences are detected nearly 100% of the time. While many low abundance OTUs are detected repeatedly across samples and thus likely to be real, restricting analyses to these core OTUs may restrict the influence of pseudo-β-diversity due to limited sampling when making ecological conclusions about similarity of microbial communities.

In addition, we find that pseudo-β-diversity between sequencing replicates of the same sample decreases with an increasing number of sequences per sample (**Fig 4**). Pseudo-β-diversity of abundance sensitive metrics like Bray-Curtis dissimilarity decreases exponentially as more sequences are added, approaching zero (indicating little to no difference between replicates of the same sample) at a depth of >10,000 sequences/sample. In our study, sequencing on the 454 platform was unable to capture this sequencing depth and thus dissimilarity between replicates in this dataset remain higher (>0.1) at the maximum sequencing depth recovered. Sequencing on the Illumina platform recovered approximately 70× more sequences per sample and thus easily reaches the lowest Bray-Curtis dissimilarity values possible between replicates within the recovered sequencing depth. Interestingly, values of a binary metric like Jaccard dissimilarity remain higher than might be desired for multiple replicates of the same sample even at the maximum sequencing depth recovered for both platforms. This indicates that rare taxa continue to be detected in low abundance as more sequences are recovered on either platform, regardless of the total amount of additional sequences.

The implications of this result are several. First, it could suggest that extremely low-abundance microbial taxa are always present in high diversity systems such as soils. As technology progresses to achieve orders of magnitude more sequences per sample with each new sequencing platform, microbial ecology studies will tend to detect more and more rare taxa, perhaps without ever saturating

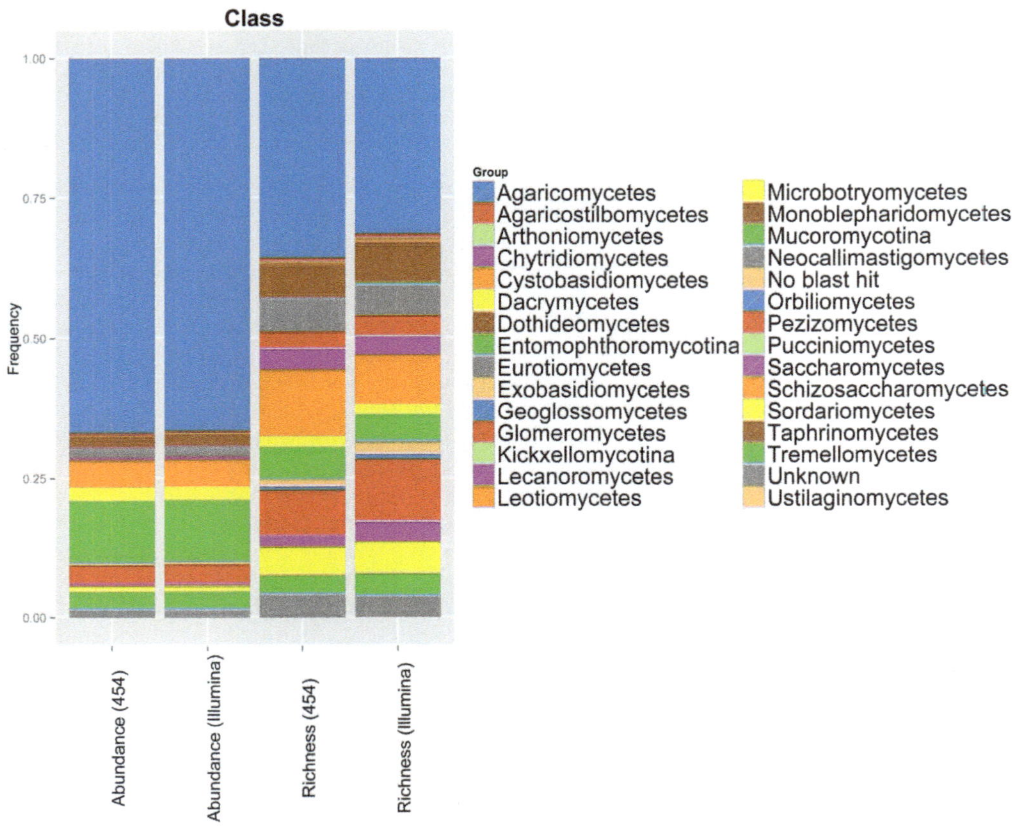

Figure 7. Taxonomic assignment to OTUs observed between both sequencing platforms is highly consistent. Bar chart indicating the proportional richness and abundance of taxa identified to the class level in 55 soil samples sequenced with both 454 and Illumina MiSeq.

taxa-accumulation curves. Using binary metrics like Jaccard could lead to artificially high estimates of between sample dissimilarity given how low-abundance and rare these new taxa are, and thus should be used with caution. Perhaps more realistically, the rare taxa that persist in detection at extremely low abundance as sequencing depth grows could also be a result of sequencing error and spurious OTU formation during sequence processing [21][28][29][30]. As a result, the ability to saturate taxa-accumulation curves by increasing sequencing depth could be somewhat confounded by subsequent increases in sequencing or processing errors. However, given that we observed many low-abundance taxa in different replicates of the same sample (**Fig 3**), it is likely that many of these taxa represent real organisms. Again, we would argue that this should encourage microbial ecologists to consider relative sequence abundance when examining β-diversity comparisons between highly diverse samples. While there is no silver bullet, the right choice of metrics will depend on the relative risk of pseudo-β-diversity vs. taxon bias in addressing the particular ecological question at hand.

Our results from the larger dataset of soil samples from three different North American pine biomes reveal interesting insights about the influence of sequencing platform on ecological conclusions and thus the adaptability of microbial ecology studies to the latest NGS platforms like Illumina MiSeq. Since MiSeq reads are at present maximum only 300 bp in either direction, adapting ecological sequencing studies to newer technology presents challenges for longer loci previously sequenced with 454. Often this means designing new primers or switching loci altogether, and the degree to which similar patterns and conclusions can be drawn from sequencing the same organisms as labs adapt their protocols for the future remains uncertain. In our study, the 454 dataset and Illumina dataset differ in the reverse PCR primers used and in the length of the amplicons.

Sequencing of the same samples on the Illumina platform vs. the 454 platform results in a ~40× increase in high quality read coverage as well as considerably more total OTUs and levels of richness per sample (**Table 2**). Encouragingly, the Illumina dataset finds nearly all OTUs that are detected with 454 but largely expands the total taxonomic coverage (**Fig S3**). The only three OTUs unique to the 454 dataset are identified as taxa in the genera Trichophaea (Ascomycota), Camarophyllopsis (Basidiomycota), and Amanita (Basidiomycota). Interestingly, eight Trichophaea taxa and 14 Amanita taxa are also present in the Illumina dataset, indicating no explicit lineage bias of the Illumina primers against members of these genera. The unique Thrichophaea and Amanita OTUs found only in the 454 dataset are thus likely due to sampling effects or error in the OTU clustering step of the sequencing processing pipeline. Additionally, since the read pairing step of Illumina processing is designed to correct sequencing errors from one read with higher-quality base calls from the other [31], it is possible that the 454-uniqe OTUs simply represent sequencing error that was not otherwise corrected.

Despite the large increase in diversity recovered with Illumina vs. 454, values of α- and β-diversity remain highly correlative between the two sequencing platforms (**Fig 5**). Sample bioregion is the predominant driving factor in the ordination of all samples from both sequencing methods (**Fig 6**), with 454 and Illumina replicates of the same samples clustering strongly by their sampling location. Within each bioregion cluster, samples cluster further by sequencing method. Variation around the centroid is greater for the 454 dataset, suggesting that β-diversity relationships between samples will depend slightly on the method used to sequence samples. This is likely due to differences in sequencing depths and species richness attainable with the two platforms. Some of the platform-specific β-diversity differences disappear in the ordination based on the β-sim metric, which controls for differences in richness between samples and thus a major difference with varied sequencing depth. It is important to note, however, that a certain degree of variation in ordination seen between the two datasets is to be expected given the different primer sets used for each platform. Still, our results strongly suggest that larger scale patterns of α- and β-diversity are as equally and consistently recoverable with newer, Illumina sequencing technology as with older 454 methods, and that ecologists should be able to transition their research with little hesitation to newer, more high-resolution sequencing technology.

Taxonomic assignment to OTUs is also consistent across the two sequencing platforms (**Fig 7, Fig S4**). At the phylum level, we observe almost complete agreement in the number and types of taxa identified. Differences between taxonomic assignment in the two datasets are primarily in the number of different Neocallimastigomycota, Glomeromycota, and Chytridiomycota taxa present. However, these groups represent a proportionally small amount of the total sequences recovered, and thus differences in the abundance of each taxonomic group recovered by the two datasets are accordingly quite small and could be due to stochasticity rather than bias. This relationship additionally holds true for the class and order groups.

Despite it's advent several years ago, amplicon sequencing with Illumina for higher order eukaryotic organisms like fungi remains scarce in the literature (but see [12] and [32]). Our results from the larger pine biome dataset present novel evidence that large scale ecological patterns of diversity, structure, and taxonomic resolution are easily attainable with an Illumina-specific fungal ITS primer set, and that ecological studies have much to gain by adopting newer NGS methods.

Supporting Information

Figure S1 Primer constructs for the amplification and sequencing of ITS1 for Illumina MiSeq. a) Sequences of PCR and sequencing primers designed to amplify and sequence ITS1, specific to the Illumina MiSeq platform. b) Partial diagram of the ITS region in fungi (not to scale), with approximate annealing locations of PCR and sequencing primers. The PCR primers are designed to generate large amplicons comprising the variable ITS1 region and conserved 18S and 5.8S regions. The Read 1 and Read 2 sequencing primers are designed to sequence a smaller region comprised mostly of ITS1, eliminating most of the conserved flanking regions. The Index sequencing primer sequences the barcode on each amplicon.

Figure S2 Pseudo-β-diversity is not significantly affected by the number of PCR replicates pooled prior to sequencing. The top four panels show the average between-replicate dissimilarity between independent replicates of CT2, OR1, and OR4 plotted against increasing PCR replication level, as determined by sequencing with 454 and Illumina MiSeq. The bottom two panels show non-metric dimensional scaling (NMDS) ordinations of the same dissimilarity values. Different colored symbols represent the different sample IDs; different shaped symbols represent the PCR replication level of each replicate.

Figure S3 Sequencing on the Illumina platform greatly expands the detectable taxonomic diversity. Venn diagram of the total fungal OTUs found from 55 soil samples sequenced with both Illumina MiSeq (light grey circle) and 454 (dark grey circle). OTUs found only when samples were sequenced with Illumina or 454 are represented by the non-overlapping regions of the circles on the left (1798 OTUs) and right (3 OTUs), respectively. OTUs present in both sequencing runs are represented by the overlapping region in the middle (1859 OTUs).

Figure S4 Taxonomic assignment to OTUs observed between both sequencing platforms is highly consistent. Bar charts indicating the proportional richness and abundance of taxa in 55 soil samples sequenced with both 454 and Illumina MiSeq at the phylum, class, and order levels.

File S1 Metadata for all samples collected and sequenced for this study.

Acknowledgments

We gratefully acknowledge Devin Leopold for helpful discussions on our sequence processing methods and for help with bioinformatic scripts.

Author Contributions

Conceived and designed the experiments: KGP. Performed the experiments: DPS. Analyzed the data: DPS KGP. Contributed reagents/materials/analysis tools: DPS KGP. Wrote the paper: DPS KGP. Performed fieldwork: DPS KGP.

References

1. Angly F, Felts B, Breitbart M, Salamon P, Edwards RA, et al. (2006) The marine viromes of four oceanic regions. PLoS Biol 4: e368.
2. Buée M, Reich M, Murat C, Morin E, Nilsson RH, et al. (2009) 454 Pyrosequencing analyses of forest soils reveals an unexpectedly high fungal diversity. New Phytol 184: 449–456.
3. Amend AS, Seifert KA, Bruns TD (2010) Quantifying microbial communities-with 454 pyrosequencing: does read abundance count? Mol Ecol 19: 5555–5565.
4. Fierer N, Strickland MS, Liptzin D, Bradford MA, Cleveland CC (2009) Global patterns in belowground communities. Ecol Lett 12: 1238–1249
5. Quince C, Lanzen A, Curtis TP, Davenport RJ, Hall N, et al. (2009) Accurate determination of microbial diversity from 454 pyrosequencing data. Nat Methods 6: 639–U627
6. Wu GD, Lewis JD, Hoffmann C, Chen Y, Knight R, et al. (2010) Sampling and pyrosequencing methods for characterizing bacterial communities in the human gut using 16S sequence tags. BMC Microbiol 10: 206.
7. Feinstein LM, Sul WJ, Blackwood CB (2009) Assessment of bias associated with incomplete extraction of microbial DNA from soil. Appl Envrion Micriobiol 75: 5428–5433.
8. Tedersoo L, Nilsson RH, Abarenkov K, Jairus T, Sadam A, et al. (2010) 454 Pyrosequencing and Sanger sequencing of tropical mycorrhizal fungi provide similar results but reveal substantial methodological biases. New Phytol 188: 291–301.
9. Bellemain E, Carlsen T, Brochmann C, Coissac E, Taberlet P, et al. (2010) ITS as an environmental DNA barcode for fungi: an in silico approach reveals potential PCR biases. BMC microbiol 10.
10. Ihrmark K, Bodeker IT, Cruz-Martinez K, Friberg H, Kubartova A, et al. (2012) New primers to amplify the fungal ITS2 region–evaluation by 454-sequencing of artificial and natural communities. FEMS Microbiol Ecol 82: 666–677.
11. Lindahl BD, Nilsson RH, Tedersoo L, Aberkenov K, Carlsen T, et al. (2013) Fungal community analysis by high-throughput sequencing of amplified markers–a user's guide. New Phytol 199: 288–299.
12. Schmidt PA, Balint M, Greshake B, Bandow C, Rombke J, et al. (2013) Illumina metabarcoding of a soil fungal community. Soil Biol Biochem 65: 128–132.
13. Gilbert JA, Meyer F, Antonopoulos D, Balaji P, Brown CT, et al. (2010) Meeting Report. The terabase Metagenomics Workshop and the Vision of an Earth Microbiome Project. Stand Genomic Sci 3:3.

14. Peay KG, Schubert MG, Nguyen NH, Bruns TD (2012) Measuring ectomycorrhizal fungal dispersal: macroecological patterns driven by microscopic propagules. Mol Ecol 16: 4122–4136.
15. Orgiazzi A, Lumini E, Nilsson RH, Girlanda M, Vizzini A, et al. (2012) Unravelling soil fungal communities from different Mediterranean land-use backgrounds. PLoS One 7: e34847.
16. Medinger R, Nolte V, Pandey RV, Jost S, Ottenwälder B, et al. (2010) Diversity in a hidden world: potential and limitation of next-generation sequencing for surveys of molecular diversity of eukaryotic microorganisms. Mol Ecol 19 Suppl 1: 32–40.
17. Caporaso JG, Lauber CL, Walters WA, Berg-Lyons D, Huntley J, et al. (2012) Ultra-high-throughput microbial community analysis on the Illumina HiSeq and MiSeq platforms. ISME J 6: 1621–1624.
18. Gardes M, Bruns T (1993) ITS primers with enhanced specificity for basidiomycetes - application to the identification of mycorrhizae and rusts. Mol Ecol 2: 113–118.
19. White TJ, Bruns TD, Lee S, Taylor J (1990) Amplification and direct sequencing of fungal ribosomal RNA genes for phylogenetics. In: Innis MA, Gefland DH, Sninsky JJ, White TJ, editors. PCR protocols: a guide to method and applications. San Diego, Academic Press. pp. 315–322.
20. Caporaso JG, Kuczynski J, Stombaugh J, Bittinger K, Bushman FD, et al. (2010) QIIME allows analysis of high-throughput community sequencing data. Nat Methods 7: 335–336.
21. Edgar RC (2013) UPARSE: highly accurate OTU sequences from microbial amplicon reads. Nat Methods 10: 996–998.
22. Knight R, Reeder J (2010) Denoising pyrosequencing reads by flowgram clustering. Nat Methods 7: 668–669.
23. Martin M (2011) Cutadapt removes adapter sequences from high-throughput sequencing reads. EMBnet.journal North America. Available: http://journal.embnet.org/index.php/embnetjournal/article/view/200/458. Accessed 2013 September 18.
24. Lohse M, Bolger AM, Nagel A, Fernie AR, Lunn JE, et al. (2012) RobiNA: a user-friendly, integrated software solution for RNA-Seq-based transcriptomics. Nucleic Acids Res 40: W622–627.
25. Peay KG, Baraloto C, Fine PV (2013) Strong coupling of plant and fungal community structure across western Amazonian rainforests. ISME J 7:1852–1861.
26. R Core Team (2013) R: A language and environment for statistical computing. R Foundation for Statistical Computing, Vienna, Austria. Available: http://www.R-project.org/. Accessed 2013 August 10.
27. Oksanen J, Kindt R, Legendre P, O'Hara B, Simpson GL, et al. (2008) vegan: Community Ecology Package. R package version 2.0-9. Available: http://CRAN.R-project.org/package = vegan. Accessed 2013 October 10.
28. Dickie IA (2010) Insidious effects of sequencing errors on perceived diversity in molecular surveys. New Phytol 188: 916–918.
29. Kunin V, Engelbrektson A, Ochman H, Hugenholtz P (2009) Wrinkles in the rare biosphere: pyrosequencing errors lead to artificial inflation of diversity estimates. Environ Microbiol 12: 118–123.
30. Quince C, Curtis TP, Sloan WT (2008) The rational exploration of microbial diversity. ISME J 2: 997–1006.
31. Masella AP, Bartram AK, Truszkowski JM, Brown DG, Neufeld JD (2012) PANDAseq: paired-end assembler for illumina sequences. BMC Bioinformatics 13: 31.
32. McGuire KL, Payne SG, Palmer MI, Gillikin CM, Keefe D, et al. (2013) Digging the New York City Skyline: soil fungal communities in green roofs and city parks. PLoS One 8: e58020.

10

Significant Increase in Ecosystem C Can Be Achieved with Sustainable Forest Management in Subtropical Plantation Forests

Xiaohua Wei[1], Juan A. Blanco[2]*

1 Department of Earth and Environmental Sciences, University of British Columbia, Kelowna, British Columbia, Canada, 2 Departmento de Ciencias del Medio Natural, Universidad Pública de Navarra, Pamplona, Navarra, Spain

Abstract

Subtropical planted forests are rapidly expanding. They are traditionally managed for intensive, short-term goals that often lead to long-term yield decline and reduced carbon sequestration capacity. Here we show how it is possible to increase and sustain carbon stored in subtropical forest plantations if management is switched towards more sustainable forestry. We first conducted a literature review to explore possible management factors that contribute to the potentials in ecosystem C in tropical and subtropical plantations. We found that broadleaves plantations have significantly higher ecosystem C than conifer plantations. In addition, ecosystem C increases with plantation age, and reaches a peak with intermediate stand densities of 1500–2500 trees ha^{-1}. We then used the FORECAST model to simulate the regional implications of switching from traditional to sustainable management regimes, using Chinese fir (*Cunninghamia lanceolata*) plantations in subtropical China as a study case. We randomly simulated 200 traditional short-rotation pure stands and 200 sustainably-managed mixed Chinese fir – *Phoebe bournei* plantations, for 120 years. Our results showed that mixed, sustainably-managed plantations have on average 67.5% more ecosystem C than traditional pure conifer plantations. If all pure plantations were gradually transformed into mixed plantations during the next 10 years, carbon stocks could rise in 2050 by 260.22 TgC in east-central China. Assuming similar differences for temperate and boreal plantations, if sustainable forestry practices were applied to all new forest plantation types in China, stored carbon could increase by 1,482.80 TgC in 2050. Such an increase would be equivalent to a yearly sequestration rate of 40.08 TgC yr^{-1}, offsetting 1.9% of China's annual emissions in 2010. More importantly, this C increase can be sustained in the long term through the maintenance of higher amounts of soil organic carbon and the production of timber products with longer life spans.

Editor: Dafeng Hui, Tennessee State University, United States of America

Funding: Funding for this research was provided by: 1) Forest Carbon Counting Method Project (XDA05060101) of the Chinese Academy of Science Strategic and Flagship Program on the Chinese Ecological and Engineering Projects and their Contributions to Carbon; and 2) BC-China Innovation and Commercialization Strategic Development Program (ICSD-2007-Wei-Kimmins). The funders had no role in study design, data collection and analysis, decision to publish, or preparation of the manuscript.

Competing Interests: The authors have declared that no competing interests exist.

* E-mail: juan.blanco@unavarra.es

Introduction

Reforestation efforts are promoted globally to meet the growing demand of forest products, especially in developing countries. Subtropical forests are therefore promoting the maintenance of forest cover in the region and the production of goods that improve the livelihoods of local communities. Subtropical forests cover 7–10% of the global land area, and store 40–50% of carbon present in terrestrial vegetation [1]. However, subtropical forests also account for most of current deforestation, with accumulated losses in the 1996–2010 period of ~100 million ha [2]. Because forests are a major carbon pool, their management is crucial to develop successful policies for climate change mitigation. Carbon storage versus production of timber and non-timber forest products are seen sometimes as incompatible objectives to be implemented at the same time in a forest, but this view is being challenged [3]. Clarifying if subtropical forest plantations can support multi-objective forestry and estimating how much carbon could be stored in working forest plantations would help to maximize the outcome of forest management plans while helping to reach a more sustainable development of the subtropical regions.

New planted forests are typically established on sites that have rapid growth and access to processing facilities and growing markets. Hence, very large areas of new forest plantations have been established in subtropical countries, particularly in Latin America and parts of Asia. In spite of such rapid growth, only 3% of the world forest land is covered with productive forest plantations. However, this area expanded by 2 million ha annually in the 1990s and by 2.8 million ha in the 2000s [4].

While much attention and discussion have been devoted on controlling and reverting deforestation, how to enhance forest carbon stocks by improving forest ecosystem condition has received less emphasis until recently [5]. Reforestation has been suggested not only as an effective way to restore degraded ecosystems, but also as a way to mitigate elevated atmospheric CO_2, hence contributing towards the reduction of climate change [3,6]. Whether and how these efforts can maximize and sustain forest C storage largely remains unexamined [7,8,9]. Recent decisions by the U.N. Framework Convention on Climate Change have encouraged consideration of other ecosystem services while implementing REDD+ (Reducing greenhouse gas Emissions from Deforestation and forest Degradation) projects for forest carbon sequestration and storage. Therefore, selecting the most appropriate approaches for implementing multipurpose forest management in planted forests is just as important as growing more forests [10]. However, few incentives and policy guidelines are in place to encourage growing better forests globally, perhaps due to the lack of a clear understanding of long-term benefits from growing better forests as well as operational difficulties.

Traditional forest management practices in forest plantations, particularly in developing countries, usually involve application of short rotations, usage of monoculture and exotic species, and high-level removal of biomass, with the main purpose of maximizing short-term economic gains [11]. The ecological problems associated with those traditional forestry practices have been well documented: declines in long-term productivity [12,13], biodiversity loss [14], increased susceptibility to insects and diseases [15], physicochemical changes in forest soils [16], and losses of other ecological services [17]. In the context of climate change mitigation, traditional forestry practices may lead to short-term carbon sequestration in wood products, but those gains are unlikely sustainable due to losses in ecosystem C linked to the above-mentioned ecological problems. Sustainable forestry practices, however, could have an important role in mitigating climate change impacts, if they are well designed and implemented [3]. For example, transformation of conifer monocultures into mixed conifer-broadleaved plantations has been considered as an efficient strategy to sustain forest productivity and restore degraded forests [18].

China has both subtropical and temperate forests, and aggressive reforestation policies have turned it into one of the five most forest-rich countries in the world, accounting for the largest gain in forested areas globally [12,19]. However, most of ecosystems reforested during the past decades in China have been by monocultures dominated by coniferous species. It has been widely reported that timber yield declines and degradation in soil fertility in coniferous plantations over several continuous rotations (see [20], and references therein). On the other hand, in subtropical China, evergreen broadleaved forests are the main native ecosystem type in the region, with a clear potential capacity for carbon sequestration [11]. Among the native species, the broadleaf *Phoebe bournei* (Hemsley) Yang is one of the most valuable tree species in this subtropical region because of its high-quality wood properties, with significant economic and ecological benefits [21]. Sharing the same distribution area, the conifer Chinese fir (*Cunninghamia lanceolata* (Lamb.) Hook. is the most important commercial forest species in subtropical China. It has been widely planted in the southeast provinces of China [20,22], becoming one of the most abundant planted species in the world [23].

The traditional management of Chinese fir plantations is representative of the type of management usually carried out in

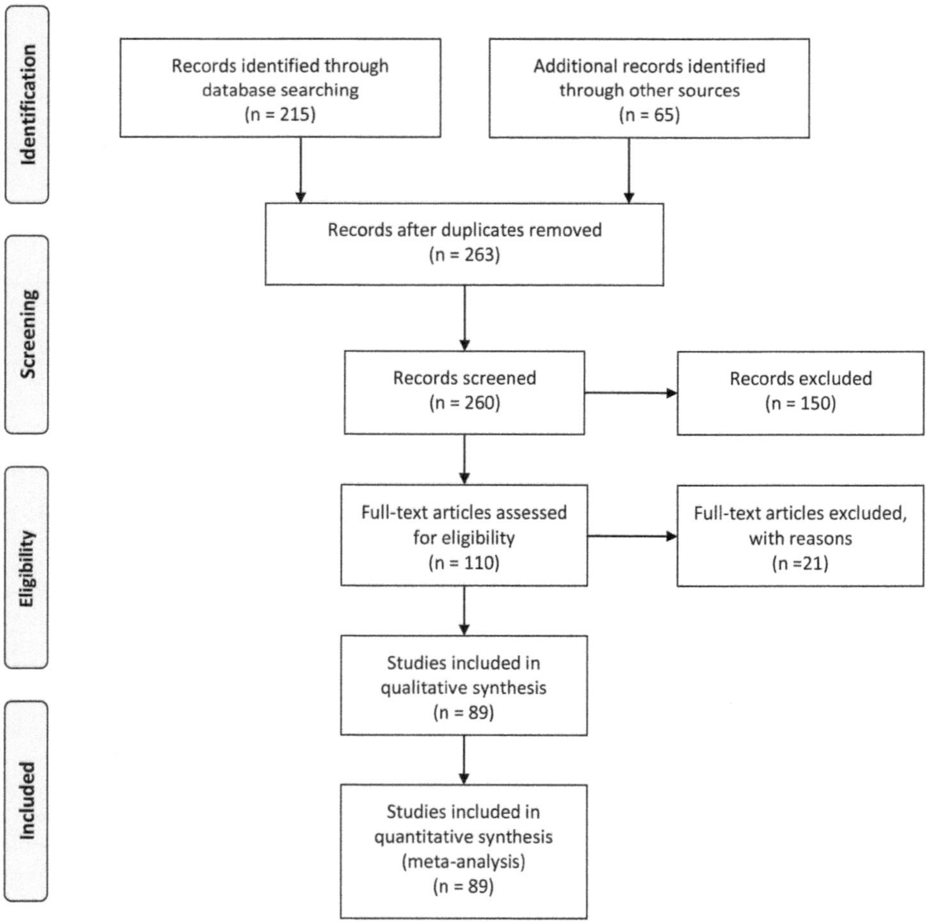

Figure 1. PRISMA flow diagram for the creation of the database of carbon pools in tropical and subtropical forest plantations.

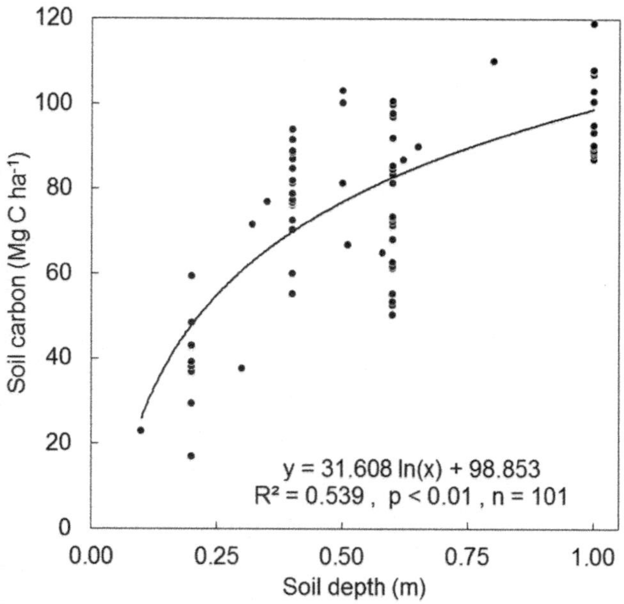

Figure 2. Values of soil carbon *vs.* soil depth in the reviewed literature.

many other tropical and subtropical forest regions around the world. After harvesting and slash-burning the native mixed evergreen broad-leaved forests, pure plantations of fast-growing conifers were established. The plantation sites used to be abandoned after one or two rotations and allowed to regenerate naturally to mixed species stands by stump sprouting and natural seeding, which then acted as a fallow period to restore the site [24]. However, since the 1950s, the plantation area of Chinese fir has been enlarged, and this species has been repeatedly planted on the same sites without intercropping or periods of fallow. Local foresters have generally used a 25-year rotation, with variation from 20 to 30 years depending on soil nutrient abundance [25]. A trend towards rotation length shortening is being observed in the latest years, driven by increasing demand for timber products fueled by economic development and population increase in China. Continuous cultivation of Chinese fir at the same site has resulted in a well-known ecological problem: yield decline over consecutive rotations [20,26]. With these conditionings, an important question has recently been raised in Chinese forestry with implications for the management of forest plantations in other subtropical regions of the world: can forest plantations be managed to produce forest products but in a way that also maintains or even increases ecosystem C? If so, can the differences in ecosystem C between traditional and sustainable management be quantified?

To answer these questions, in this work we tested the hypothesis that subtropical tree plantations accumulate less carbon under traditional exploitative management than under sustainable forest management, which is more similar to natural stand development. We also aim to quantify the potential increase in ecosystem C at a regional level. To test this hypothesis we first reviewed available information on C pools in subtropical forest plantations and related them to factors defining their management (species type, rotation length, stand density). Also, the literature review allowed us to identify the management prescriptions with the highest carbon storage, and to define realistic sustainable management scenarios. Finally, we estimated potential C sequestration gains through time at a regional level by comparing ecosystem C values under traditional and sustainable management regimes in subtropical plantations in China, simulating plantation growth with the forest ecosystem model FORECAST [27].

Materials and Methods

We first carried out a literature review to test if differences in ecosystem C among subtropical plantation types exist, and also to identify the management factors that have maximized ecosystem C in those plantations in the past. Then, we used ecological modelling to estimate the average differences between traditional forest management and sustainable forest management, with consideration of the factors identified in the literature review. As ecological models need to be calibrated for specific ecosystem types [28], we used Chinese fir plantations (*Cunninghamia lanceolata* (Lamb.) Hook.) located in the subtropical region of China as a case study. Continuous repetitions of traditional practices over several rotations (monoculture, short rotations, whole-tree harvesting, and slash burning) have resulted in severe yield decline [26]. Among various possible reasons, N limitation by nutrient exports and resource competition between tree seedlings and understory are the most important factors [20]. Therefore, there is a need to improve the management of this species to establish a more sustainable and multi-purpose forestry in China.

Literature Review of Carbon in Subtropical Plantations

We conducted a structured search of published scientific documents to define the observed range in carbon storage in the ecosystem, as well as in the belowground and aboveground fractions, and their relationships with plantation type (broadleaf or coniferous), plantation age, and stand density. We searched in the on-line databases Google Scholar (by Google Inc. http://scholar.google.com), ISI Web of Knowledge (by Thomson Reuters, http://www.isiwebofknowledge.com), China Academic Journal Database (by China National Knowledge Infrastructure, http://cnki.en.eastview.com/kns50), CAB Abstracts (by the Centre for Agricultural Bioscience International, http://www.cabdirect.org), and the databases hosted by the Canadian Forest Service Bookstore (http://bookstore.cfs.nrcan.gc.ca), the University of British Columbia's library (http://www.ubc.ca/library), and the Public University of Navarre's library (http://www1.unavarra.es/biblioteca).

We divided the population of collected documents into two groups: 1) documents reporting C pools in broadleaf or other non-coniferous forests, and 2) documents describing coniferous or mixed conifer-broadleaf plantations. The final number of documents and sites used for further analysis after a standardized selection process [29] were, for broadleaves 38 documents and 138 sites; whereas for conifers 51 documents and 160 sites (Figure 1; see the full reference list and values of carbon and management variables for each site in Material S1). For each site, we gathered data on plantation age, density, aboveground C density (MgC ha^{-1}), belowground C density, and ecosystem C density. Aboveground C was calculated as tree C plus understory C (if reported). Belowground C carbon was calculated as C in tree roots (and in plant roots if reported), litter C and mineral soil C. Values from graphs were extracted using DataThief III [30]. Original values for mineral soil C were often not directly comparable, as they were measured for different soil depths. To standardize soil carbon values, we modified the method by [31] and used bulk density and soil C concentration to calculate and plot soil C content versus soil depth, obtaining a significant correlation between both variables (Figure 2). We then estimated the

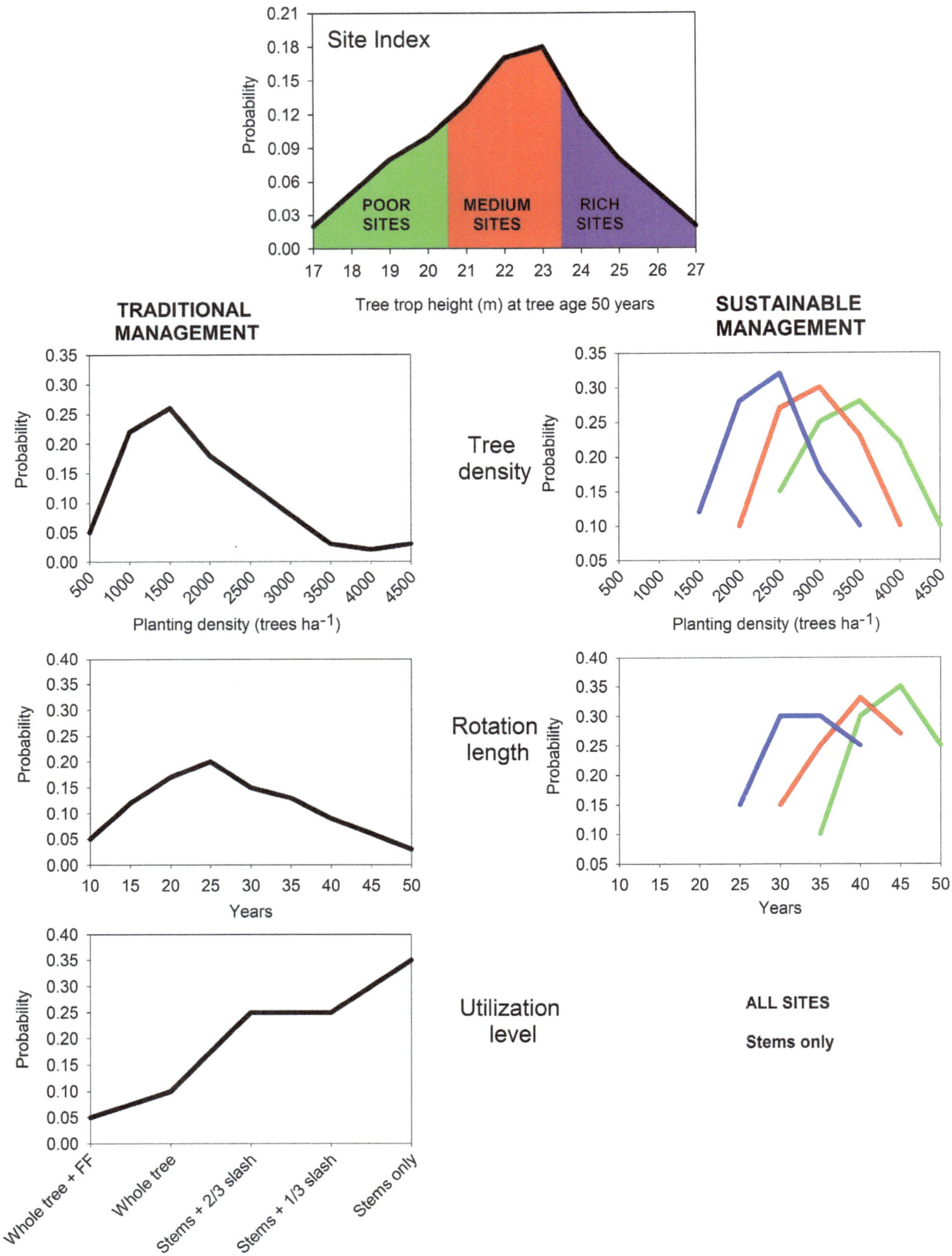

Figure 3. Probability distributions of the four variables defining the management scenarios. Different color lines correspond to the different site qualities as defined in the top graph. FF: Forest floor removal.

Table 1. Values used to calibrate FORECAST parameters related to Chinese-fir.

Parameter	Unit	Rich site	Poor site
Chinese fir parameters			
Nitrogen concentration in needles young/old/dead	%	1.53/1.36/1.13	1.21/1.11/0.93
Nitrogen concentration in stem sapwood/heartwood	%	0.14/0.03	0.12/0.03
Nitrogen concentration in bark live/dead	%	0.44/0.27	0.37/0.24
Nitrogen concentration in branches live/dead	%	0.67/0.52	0.55/0.47
Nitrogen concentration in root sapwood/heartwood	%	0.37/0.06	0.35/0.06
Nitrogen concentration in fine roots live/dead	%	1.17/0.97	0.96/0.79
Shading by maximum foliage biomass	% of full light	8	30
Soil volume occupied at maximum fine root biomass	%	100	95
Efficiency of N root capture	%	98	100
Retention time for young/old foliage/dead branches	years	1/2/40	1/2/40
Fine roots turnover	$year^{-1}$	0.95	1.35
Decomposition rates			
Sapwood (by litter age)	% $year^{-1}$	1–5 years (2.0); 6–10 years (10.0); 11–15 years (30.0); 16–20 years (20.0); > 20 years (4.0)	
Heartwood	% $year^{-1}$	1–10 years (0.4); 11–15 years (10.0); 16–25 years (15.0); 25–40 years (10.0); >40 years (2.0)	
Bark	% $year^{-1}$	1–5 years (2.0); 6–20 years (12.0); 20–40 years (20.0); >40 years (4.0)	
Branches and large roots	% $year^{-1}$	1–5 years (10.0); 6–10 years (45.0); 11–15 years (35.0); >15 years (4.0)	
Needles (poor site)	% $year^{-1}$	1–2 years (20.0); 3–5 years (30.0); 6–10 years (40.0); >10 years (2.0)	
Needles (good site)	% $year^{-1}$	1–2 years (27.0); 3–5 years (30.0); 6–10 years (40.0) >10 years (3.0)	
Fine roots	% $year^{-1}$	1–2 years (30.0); 3–4 years (50.0); >4 years (9.0)	

Decomposition rates indicate the mass loss in one year as a fraction of the initial mass at that year. See text for list of bibliographical sources for model calibration.

standardized mineral soil carbon content at 60 cm depth as the value that would correspond if soil carbon distribution in each soil followed a relationship with soil depth parallel to the regression line.

We tested the equality of means for aboveground, belowground and ecosystem C between the two types of plantations using Welch's modified *t*-test, (a robust test that does not assume equality of variances; [32]), setting significance level at 0.05. We also carried out regressions (weighted by the error associated to each measure) of carbon density vs. plantation age and stand density, lumping together data from conifer and broadleaf plantations as initial analysis did not show improvement of the regression results by splitting the data into species groups or location (data not shown). All the statistical analyses where carried out with Sigmaplot 10.0 (Systat Software Inc.), JMP v.5.0.1 (SAS Institute, NC, USA), and S-PLUS 6.1 (Insightful Corp.).

Simulating Ecosystem C in Planted Forests with FORECAST

In the second part of our research, we used the results from the literature review to define probability distributions for the two main factors defining forest plantation management (stand density and rotation length) that maximized ecosystem C, and we assumed that they are indicators of more sustainable forest management. Probability distribution of site quality was defined using the distribution map of site qualities for Chinese fir in south-eastern China [33]. Probability distributions of planting density, rotation length and utilization level for traditional management were also created for both sets of simulations after consultations with panels of local experts composed by academics, forest officers, and forest managers (Figure 3). The panel of local experts indicated the need for creating different management scenarios based on site quality for sustainable management. Hence, probability distributions of tree density and rotation length were different in each site quality for sustainable management, but they were the same in all sites under traditional management.

Using the probability functions, we constructed 200 Monte Carlo projections for each set of simulations (traditional management and sustainable forest management). Each run simulated a management scenario (a random combination of site quality, rotation length, and stand density). In traditional management, all trees planted were Chinese fir. In sustainable management, probability distributions show total tree density, which was composed by Chinese fir and *Phoebe bournei* in a 1:1 ratio, the most suitable species mix for carbon sequestration [21].

We then simulated each scenario for 120 years with the ecosystem model FORECAST, which is a stand-level model specially designed for assessing the effects of various forest management practices on forest carbon storage and sequestration. FORECAST has been successfully tested and applied in many world forests before, including tropical and subtropical forests [28].

FORECAST is a management-oriented, deterministic, non-spatial stand-level forest growth and ecosystem dynamics simulator that operates at annual time steps. The model has been described in detail before [27,28], and therefore only a summary of the main driving function used to calculate tree growth is provided in Material S1. The model uses a mass balance approach to estimate ecosystem nutrient circulation, and how nutrient availability limits tree growth together with available light in the canopy.

FORECAST has three application stages: 1) assembling calibration data and generating historical rates of key ecosystem processes; 2) model initialization by establishing the ecosystem condition for the beginning of a simulation run; and 3) simulation of tree and plant growth.

Input data came from sites covering the observed range of Chinese-fir plantation growth, qualitatively described as poor, medium and good (Table 1). The sites were scaled quantitatively as 17, 21 and 27 (from poor to good, based on canopy top height in m at age 50 years) to represent a relative index of tree growth required for extrapolation within the model. Tree data on biomass, mortality, stand density, tree height, canopy height, nutrient concentrations of live tissues and other data were based on reported values for similar Chinese-fir plantations [25,34,35,36,37,38,39,40,41,42,43].

Model calibration data for *Phoebe bournei* forests (Table 2) came from literary sources and field data [21]. They included data of biomass accumulation and stand density [44,45,46,47,48,49,50], decomposition rates [51,52,53,54], soil nutrients under different site conditions [55,56,57], and photosynthetic efficiency [58,59,60,61]. Understory was simulated as grass and shrub complexes [20]. Published data were used to characterize shrub biomass, height, tissue nutrient concentrations and other relevant data [62,63,64,65] (Table 3).

Data describing soil processes and nutrient inputs in precipitation and slope seepage, mineral soil cation and anion exchange capacities, humus mass, nutrient concentrations in litterfall, litter decomposition rates and others. were derived from literature [38,39,40,66,67,68,69,70,71] (Table 3). To establish initial site conditions the model was run in set-up mode, forcing the model to match the observed site conditions [72,73]. Initial conditions were created for each site quality, by running several cycles of tree growth ending with stand-replacing windthrow, as typhoons are the most common natural disturbance in this region [74]. A library of initial conditions was created with initial soil organic matter values (humus+litter) ranging from 90.9 to 296.3 Mg ha^{-1} for the poorest and richest sites, respectively.

Published field data on several chronosequences at different site qualities were used to evaluate FORECAST performance for Chinese fir and *Phoebe bournei* plantations (see Material S1). After obtaining a satisfactory model performance, we run the model for the 400 randomly-generated scenarios described above. A sub-set of documents gathered during the literature review dealing with Chinese fir plantations was also used to evaluate the model performance for ecosystem C.

Results

From our review, 23.9% of the studies on bradleaves used local native species, but *Populus* spp. (19.6%), *Tectionia sp.* (13.1%) and *Eucalyptus* spp. (8.10%) were the most common commercial species used. For conifers, 81.4% of the studies reported data from *Cunninghamia* sp. plantations, and 14.3% came from *Pinus* spp. plantations. Our results showed that subtropical broadleaves plantations can, on average, sustain higher carbon densities than subtropical conifer plantations (Figure 4). The differences, however, are mostly caused by differences in belowground C, whereas differences among plantations types in aboveground C are non-significant. Not surprisingly, the trend in subtropical forest plantations is to increase ecosystem carbon storage as plantation age increases. This is basically the result of increase in tree size,

Table 2. Values used to calibrate FORECAST parameters related to *Phoebe bournei.*

Parameter	Unit	Rich site	Poor site
Chinese fir parameters			
Nitrogen concentration in leaves young/old/dead	%	1.35/0.92/0.69	1.15/0.80/0.67
Nitrogen concentration in stem sapwood/heartwood	%	0.18/0.04	0.16/0.04
Nitrogen concentration in bark live/dead	%	0.55/0.47	0.49/0.33
Nitrogen concentration in branches live/dead	%	0.33/0.29	0.28/0.26
Nitrogen concentration in root sapwood/heartwood	%	0.37/0.29	0.34/0.27
Nitrogen concentration in fine roots live/dead	%	0.81/0.57	0.65/0.56
Shading by maximum foliage biomass	% of full light	8	25
Soil volume occupied at maximum fine root biomass	%	100	100
Efficiency of N root capture	%	100	100
Retention time for young/old foliage/dead branches	years	1/0.5/30	1/0.5/30
Fine roots turnover	year^{-1}	1.00	1.15
Decomposition rates			
Sapwood (by litter age)	% year^{-1}	1–3 years (0.1); 4–15 years (2.0); 15–20 years (12.0); >20 years (9.0)	
Heartwood	% year^{-1}	1–3 years (0.1); 4–15 years (2.0); 15–20 years (12.0); 20–40 years (9.0); > 40 years (2.0)	
Bark	% year^{-1}	1–10 years (20.0); 11–20 years (40.0); 21–30 years (35.0); >30 years (2.0)	
Branches and large roots	% year^{-1}	1–5 years (15.0); 6–15 years (35.0); 16–20 years (45.0); >20 years (15.0)	
Leaves (poor site)	% year^{-1}	1–2 years (45.0); 3–5 years (35.0); >6 years (20.0)	
Leaves (good site)	% year^{-1}	1–2 years (50.0); 3–5 years (40.0); >6 years (20.0)	
Fine roots	% year^{-1}	1–5 years (45.0); >6 years (35.0)	

Decomposition rates indicate the mass loss in one year as a fraction of the initial mass at that year. See text for list of bibliographical sources for model calibration.

Table 3. Values used to calibrate FORECAST parameters related to the shrub complex and soil processes.

Shrub complex parameters	**Unit**	**Rich site**	**Poor site**
Nitrogen concentration in leaves live/dead	%	1.68/1.38	1.12/0.76
Nitrogen concentration in stems live/dead	%	0.48/0.14	0.32/0.10
Nitrogen concentration in rhizomes live/dead	%	1.14/1.05	0.76/0.70
Nitrogen concentration in roots live/dead	%	1.10/0.84	0.73/0.56
Shading by maximum foliage biomass	% of full light	0.20	0.45
Soil volume occupied at maximum fine root biomass	%	60	50
Efficiency of N root capture	%	99	99
Transfer from live to dead stem/rhizomes/roots	% year^{-1}	20/20/30	20/20/40
Retention time for foliage	years	1	1
Decomposition rates		Litter age in years (decomposition rate in %)	
Foliage (poor site)	% year^{-1}	1 year (80.0); 2–3 years (60.0); 4–5 years (50.0); >5 years (2.0)	
Foliage (rich site)	% year^{-1}	1 year (95.0); 2–3 years (60.0); 4–5 years (50.0); >5 years (2.0)	
Stems & roots	% year^{-1}	1 year (20.0); 2–3 years (30.0); 4–5 years (40.0); >5 years (2.0)	
Soil parameters		**Rich site**	**Poor site**
Nitrogen concentration in slow/fast humus	%	3.00/1.40	3.0/1.40
Decomposition rate slow/fast humus	% year^{-1}	0.15/2.00	0.15/2.00
CEC soil (CEC humus)/AEC	kg N ha^{-1}	80.0 (0.2)/20.0	40.0 (0.2)/5.0
Atmospheric deposition/non-symbiotic fixation	kg N ha^{-1} year^{-1}	4.9/1.0	4.9/1.0
Non-symbiotic N fixation rate	Kg N ha^{-1}	2.5	3.0

Decomposition rates indicate the mass loss in one year as a fraction of the initial mass at that year.

and accumulation of litter and coarse woody debris biomass. This trend, however, seems to slow down over time, with little expected gains beyond plantation age ~90, as described by the significant exponential model (Figure 5). Ecosystem C initially increases as plantation density increases, reaching a peak at about 1500–2500 stems ha^{-1}. For higher densities the trend is to reduce ecosystem C (Figure 5).

During the simulation runs, the most frequent scenario for sustainable forest management was 3000 trees ha^{-1} (1500 Chinese fir, 1500 *P. bournei*), rotation length of 40 years and stem-only harvesting. For traditional management, the most frequent scenario was 1500 trees ha^{-1} (Chinese fir only), rotation length of 25 years, stem plus slash removal, and site preparation by understory burning. Comparing these two scenarios when run for 240 years, we can see how plantations managed for sustainable, long-term goals, maintain ecosystem productivity higher and during longer periods that traditional intense management.

Yearly net primary productivity under traditional management quickly drops in each consecutive rotation, and as a consequence tree biomass (and the associated merchantable volume) also decreases rapidly over time (Figure 6). In addition, there is an increasingly stronger competition with understory. Although with traditional management higher peaks of available soil N are achieved than in the sustainable management, the total amount of ecosystem N decreases over time. On the other hand, under sustainable management higher levels of soil organic matter are sustained (Figure 6).

When all the inputs and outputs are accumulated through the 240 years of simulation, it is clear how current management is depleting ecosystem N reserves. On the other hand, sustainable management can maintain and even increase N pools (Figure 7). Traditional management removes N from the ecosystems during the site preparation (directly through slash-burning and indirectly by erosion caused by heavy machinery and losses of plant cover), with the export of harvested products, and by N leaching. Under sustainable forest management, slash-burning is not used and erosion is greatly reduced as practices affecting soil stability and plant cover are minimized. N removal in harvested products is also lower as only stems (with low N concentrations) leave the plantations. On the other hand, as N reserves under sustainable management are higher, leaching losses are also higher, just because the soil N pool is bigger than in traditional management. All flows considered, N outputs with current practices are not only higher, but N inputs are also smaller, as there is less N fixation due to the reduced biological activity in sites of lower productivity. Also, because any given stand gets less N inputs by seepage as it is surrounded by stands that also have lower N capitals.

When simulating how the C pools in Chinese fir stands would evolve over the 21st century, our results showed that if sustainable forestry practices were introduced at a rate of 10% of the total planted area, the region would keep on average a higher and sustained level of carbon storage over the long term, compared to traditional management (Figure 8). At the beginning of our scenarios a large overlapping range for ecosystem C between traditional and sustainable plantations can be seen (grey area). Such fact indicates that some time is needed for the sustainable management to significatively reduce C loses compared to traditional management. This is not surprising as productivity losses are accumulative and occur mostly at the change of rotation. However, by the middle and until the end of the simulated 120 years the overlap has been greatly reduced. By that time, only the sites with the highest C contents and managed traditionally would be comparable to the poor sites under sustainable management. Such increase in carbon density could be 12% by 2020. This difference would grow over time, reaching 17%, 52%, 79%, and 67% greater carbon density under sustainable forest management

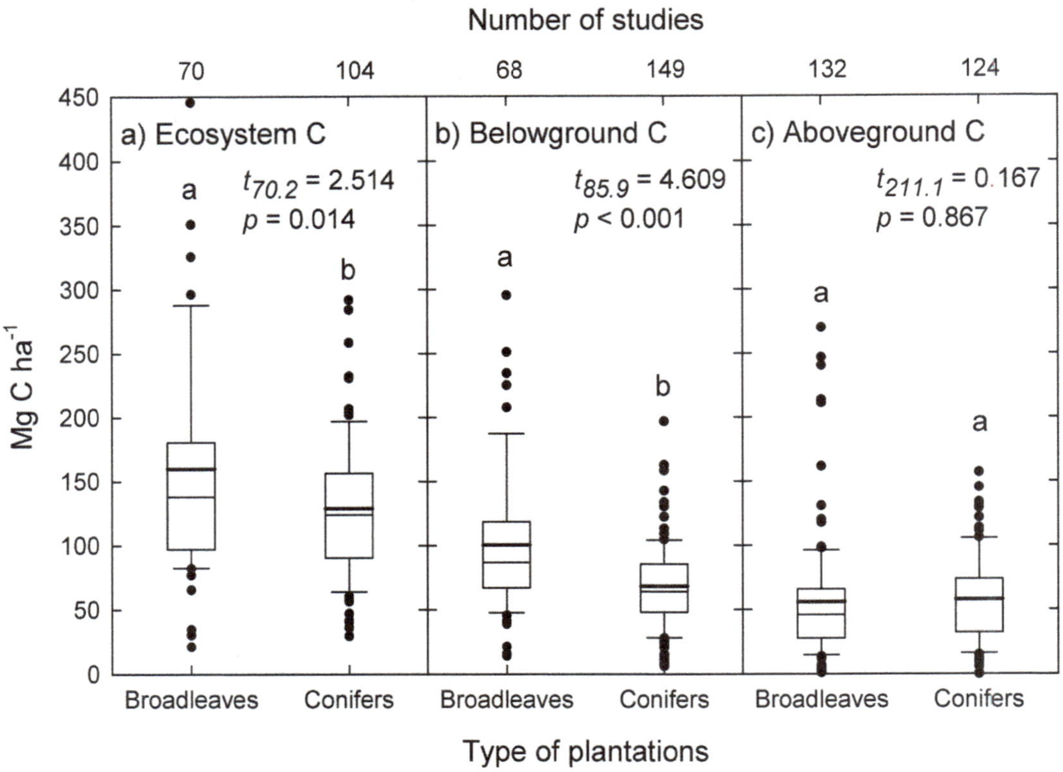

Figure 4. Distribution of C storage in published bibliography for different subtropical plantation types. The thick horizontal line indicates the average of all observations. Letters indicate significant differences with Welch's *t* at *P*<0.05. See Material S1 for the complete list of references used.

after 25, 50, 75 and 100 years, respectively. The average weighted carbon density increase over the 100-year simulation period was 67.5%.

As a way to validate our estimations at the stand level, when analyzing the average and 95% confidence intervals for the whole set of projected ecosystem C values they were always inside the range of reported values for Chinese-fir plantations (Figure 8), providing confidence on our estimations.

Stand-level differences between management regimes would have large regional implications. Extrapolated to the whole area dedicated to Chinese fir plantations (7,046,520 ha [11]), together with the plan of substituting traditional by sustainable management implemented at a ratio of 10% of the land every year, the increase in ecosystem C stored in the mixed Chinese fir plantations could reach 261.2 TgC by 2050 (Figure 8). After that year, the differences in ecosystem carbon storage between management options show a temporal decline. This is as a consequence of the

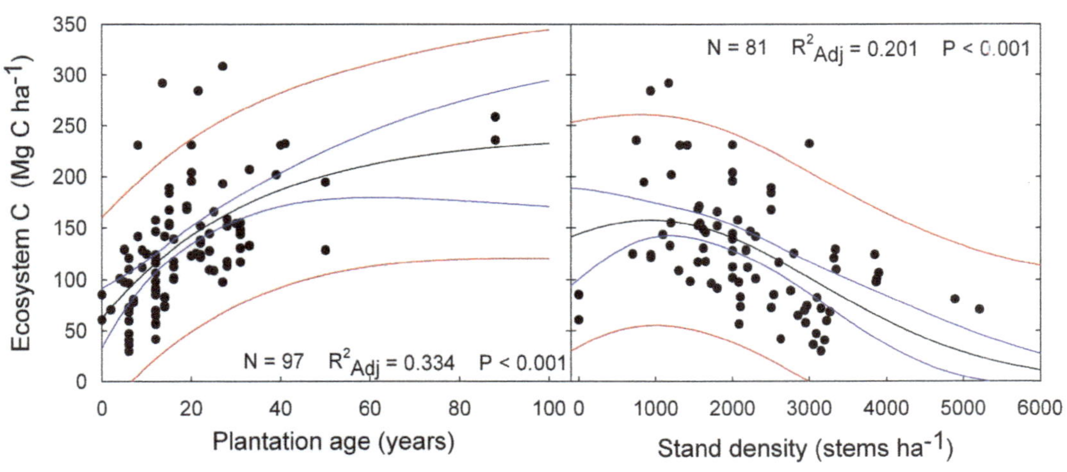

Figure 5. Carbon stocks in subtropical plantations for different A) rotation ages, and B) stand densities. Black lines indicate the best-fit model, blue lines 95% confidence interval and red lines 95% prediction intervals.

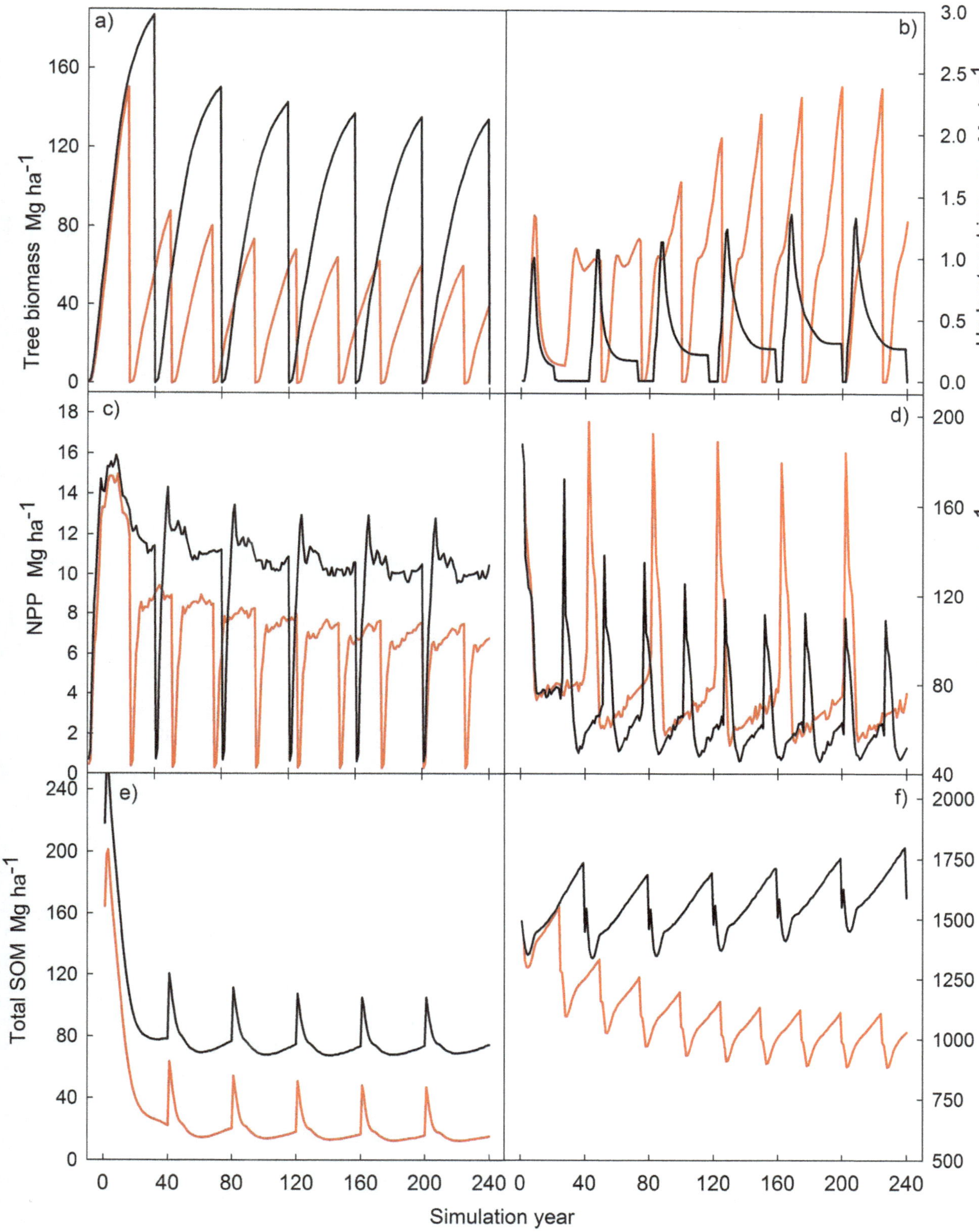

Figure 6. Change in ecological variables in Chinese fir plantations in 240 years of simulation under the average plan two types of management. Current (red line) 25-year rotation, whole-tree harvesting, 1500 trees ha^{-1} Chinese fir planted; and close-to-nature management (black line) 40-year rotations, stem-only harvesting, 1500 trees ha^{-1} Chinese fir +1500 trees ha^{-1} *Phoebe bournei*.

harvesting of sustainable forest plantations primarily clustered around the years 40 to 60. The second rotation growth would again produce an increase in stored C, which would remain above 200 TgC after 120 years of simulation.

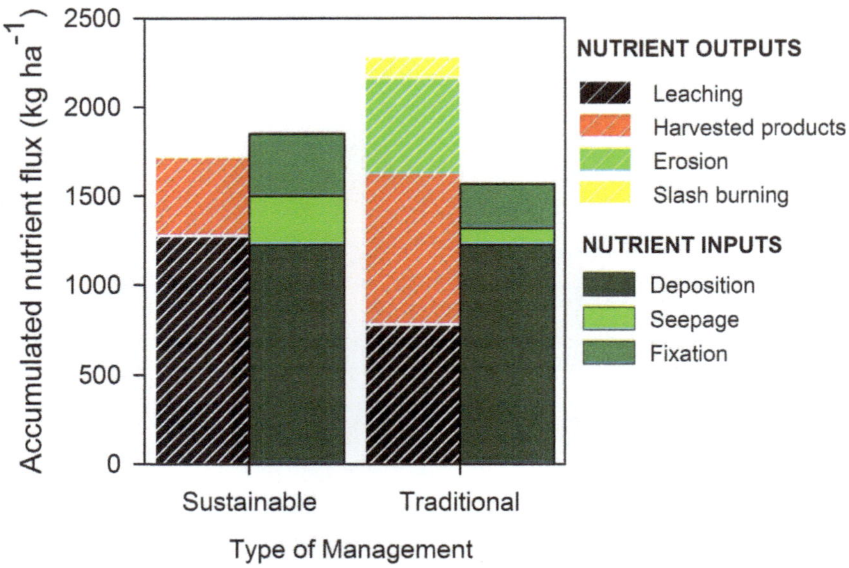

Figure 7. Accumulated N inputs (solid bars) and outputs (dashed bars) in Chinese subtropical plantations after 240 years of management under two most typical management regimes for traditional and sustainable plantations.

Discussion

Subtropical plantations have been described as the most effective for C sequestration [75,76,77]. In subtropical regions, local broadleaved species are usually planted, but the available data on C stocks in subtropical plantations came mostly from the five genuses more widely planted around the globe: *Eucalyptus, Populus, Tectonia, Pinus* and *Cunninghamia.* These are the most valuable among the fast-growing species currently used, and our results show that selecting among the species type (broadleaves or conifers) can produce significant differences in C pools depending on plantation type and management regime. Ecosystem C increases with tree age in subtropical plantations, as these systems develop ecological structures that make them closer to natural forests over time [78]. Older plantation age has also been associated to higher soil carbon levels [79]. Therefore, increasing rotation length should be encouraged when implementing sustainable forestry plans for increasing carbon stocks, as well as generating more biophysical structure for higher niche diversity and its associate biodiversity.

[80] have pointed out that density plays an important role in the productivity of mixed stands, with higher productivities in mixed stands than in monocultures at medium to high densities (>1500 trees ha^{-1}). Our results also indicate that density should be maintained at medium values, avoiding excessive inter-tree competition for resources such as light, water and nutrients that could reduce stand productivity, but also avoiding low densities that may leave some resources unused by trees. Our estimates also fit with several field experiments in mixed plantations of Chinese fir with other trees species, which have consistently shown higher biomass and C density in mixed stands with a tree density ratio of 1:1 for Chinese fir/companion species, for total densities in the 2000–3000 trees ha^{-1} range [81,82,83,84].

Our results indicate that managing plantations designed for multiple values could sustain higher carbon storage potentials if they include broadleaves, are managed for rotations lengths of 25–50 years, and have planting densities of 1500–2500 stems ha^{-1}. Our estimations corroborate previous field research that have reported maximum biomass values in mixed Chinese fir plantations for similar rotation and density ranges, reporting little gain or even biomass decrease with plantation densities above 3000 trees ha^{-1} [82,85]. In this regard, forest plantations seem to have important functional similarities with native forests, where intermediate tree densities have been reported as the ones reaching the carrying capacity of the ecosystem, without surpassing it [86].

Differences in C density among management types can be better understood by comparing the typical (most frequent in our set of simulations) management regimes for traditional and sustainable forest management practices. Depletion in ecosystem N reserves reduces ecosystem productivity, which in turn produces a long-term decrease in tree biomass [87]. This productivity reduction is much more pronounced under traditional practices, as shorter rotation lengths accumulate higher and faster nutrient exports. In addition, short rotations and lower stand densities produce an under-canopy environment with more light available and therefore more suitable for understory growth (Figure 6). In these plantations, understory grass and shrubs usually sprout from rhizomes and other underground structures, rather than from seeds [20]. This phenomenon provides a competitive advantage against planted tree seedlings, which is exacerbated by the advantage of understory's lower nutrient needs in increasingly nutrient-poor sites as a consequence of intensive organic matter extraction after harvesting [88], slash burning [89], and N leaching [90] (Figure 7).

Reductions in soil organic matter and its associated sources of nutrients, water retention and physical properties would also be more severe under traditional management [16], leading to the already reported yield decline in Chinese fir plantations [26]. On the other hand, our results show that using longer rotations, moderately higher stand densities, and mixing conifers and broadleaves would prevent site productivity decline. With more sustainable management, the ecosystem would maintain and even can increase N reserves by keeping a positive N balance, an indicator of ecological sustainability of forest management in the long-term [26,91].

Theoretical forest ecology indicates that in tree species mixtures, if each species has different growing patterns (differential shade

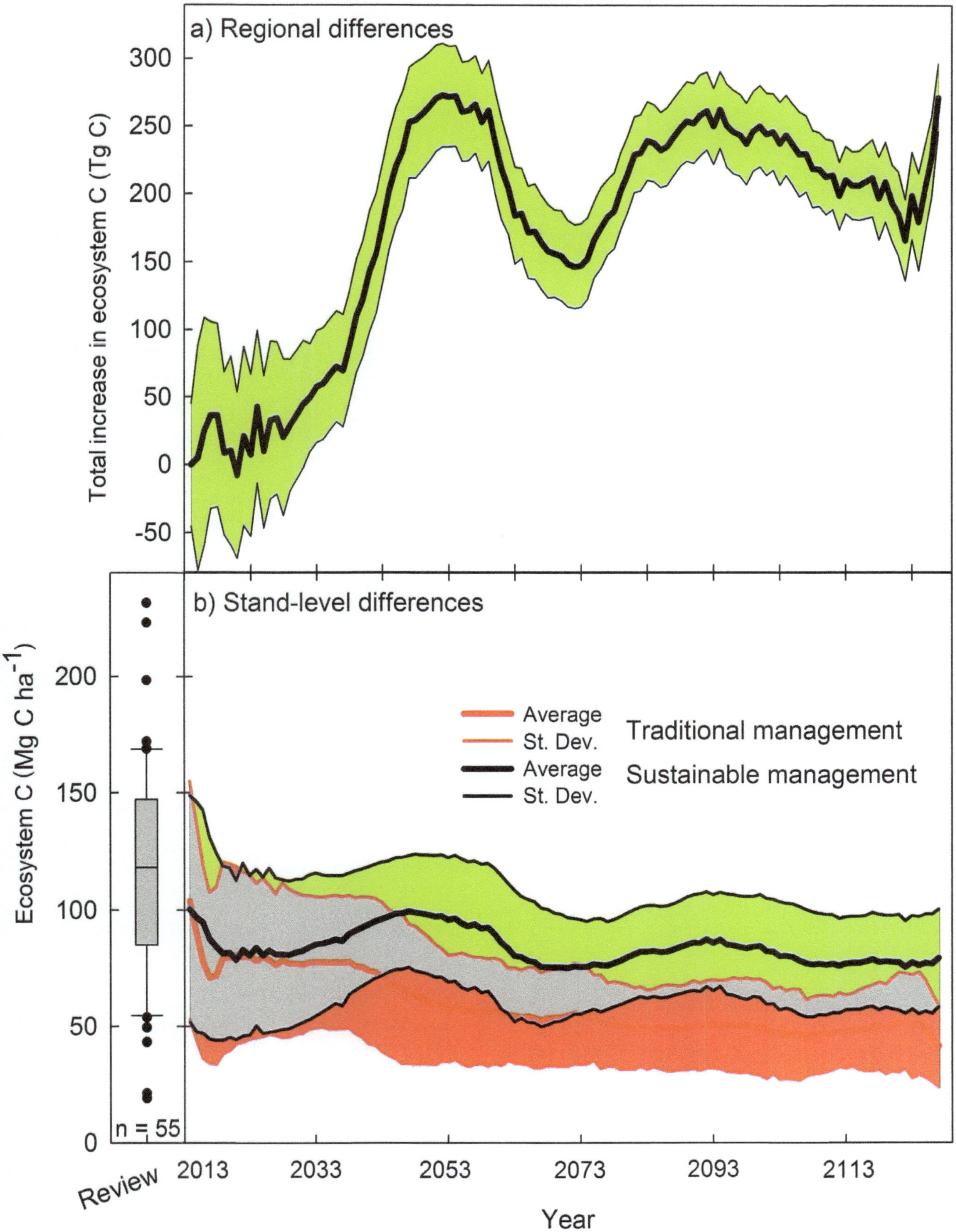

Figure 8. Total carbon gains when switching from current traditional intensive forestry to sustainable forest management at an annual rate of 10% of total area dedicated to Chinese-fir plantations (panel A). Ecosystem carbon density in Chinese fir plantations under current intensive management (red lines), and sustainable management (green lines) (panel B). Thick lines represent average and color areas

represent one standard deviation. The grey area indicates values of ecosystem C that can be achieved by either sustainable or traditional management plans. The box-plot in the bottom-left summarizes the values obtained from a review of 55 published peer-reviewed papers on carbon stocks in Chinese fir plantations (see Material S1 for the complete reference list).

tolerance, physical separation of canopies, phenological differences, successional separation, and differences in soil resource utilization), inter-specific competition may be reduced. Growing evidence indicates a generalized pattern in mixed forests, in which niche separation at both canopy and root levels leads to a more complete use of site resources [80,92,93,94]. This phenomenon has already been reported in mixed tropical plantations [95,96,97]. Therefore, resources in the mixed stand are used more efficiently, hence total stand productivity increases. This phenomenon is due to two facts: complementary resource use between species and facilitative improvement in nutrition in mixtures with a nitrogen-fixing species [93,95,98].

Our results fit with this theoretical background, showing that in Chinese fir plantations, implementing mixtures with *Phoebe bournei* will not only increase productivity, but it will also prevent SOM and N losses and their associated yield decline over time. The additional increase in sequestered ecosystem C is another positive effect of species mixtures in these sites. Similar results have also been reported when simulating boreal [99], temperate [73] and tropical [100] forests.

Forest management can greatly affect net C exchange with the atmosphere, not only because it changes ecosystem C pools, but also because the C removed from the ecosystem during harvesting is stored in wood products, such as paper, cardboard, furniture, structural timber, etc. [101]. We have shown that subtropical plantations managed under sustainable forest management can store more ecosystem carbon. However, their net effect on C sequestration depends on forest types, their management regimes, the type of wood products produced, and the efficiency of biomass conversion, as well as assumptions about how the wood and wood residues will substitute for other products with greater GHG emissions and for fossil energy [102].

In Chinese fir plantations under traditional management, most of the production goes to the market as small diameter trunks. Over time, harvested sizes become increasingly smaller as plantation sites lose productivity due to the nutrient losses and competition with understory. However, using a more sustainable forest management, harvesting rates are reduced as it takes more time to complete the harvesting turn, but on the other hand harvested tree sizes are bigger and stay bigger through several rotations. These wood products are expected to reach better prices and last longer because they will be used more as structural timber and high quality furniture [103]. The increased ecosystem C storage, combined with the longer lifespan of higher quality wood products, could compensate the reduced harvesting rates compared to traditional management. Although a life-cycle analysis is needed to test this hypothesis, previous work has shown that secondary forests allowed growing rather than being harvested and their wood products stored could result in lower net CO_2 emissions [104].

Effects of forest management actions on carbon should be examined for large areas and over long time periods [105]. In the case of China, government's commitment to increase forest cover by 40 million ha with new plantations may not have an impact on carbon storage as important as intended unless better forest management is also implemented [106]. In this context, applying a more ecologically-based forestry could make a significant difference. Assuming those new 40 million ha were evenly distributed among all Chinese forest types, there could be 17.55 million ha of new subtropical plantations by 2020 [8]. If the differences between sustainable and traditional forestry were similar to the ones we have estimated for Chinese fir plantations, a more ecologically-based forestry implemented in Chinese subtropical plantations could add 650.8 TgC by 2050.

Previous research on long-term C storage under sustainable forest management regimes in Chinese temperate forests have estimated ecosystem C densities of 600–750 MgC ha^{-1} in poplar (Dai *et al.* under review), 450–800 MgC ha^{-1} in spruce [107], and 200–325 MgC ha^{-1} larch plantations [108]. In all cases, estimates C densities for temperate plantations were higher than our estimates for subtropical plantations. Therefore, it seems reasonable to assume that differences between traditional and sustainable forest management regimes in temperate plantations are at least similar to the ones calculated for subtropical plantations. Under this assumption, C storage in Chinese plantations could potentially be increased by 1,482.8 TgC by 2050. These estimations could even be conservative, as temperate plantations have on average higher levels of ecosystem carbon stocks [109].

In terms of emissions, applying sustainable forestry to all new Chinese plantations could potentially lead to an average net increase in stored C of 0.04 PgC y^{-1} in 2050, equivalent to offsetting 1.91% of China's annual carbon emissions in 2010 [110]. More importantly, this carbon increase would be sustainable with opportunities to maintain other ecological services and functions linked to increased biodiversity and less degraded forest soil [3].

In conclusion, our results show that applying a more ecologically-based forestry would have positive impacts not only on long-term plantations productivity, but it can also be a better strategy to implement climate change mitigation programs. Carbon stored in plantations could be further increased if sustainable forestry practices were combined with other practices not examined here, such as genetic improvement of trees, landscape design, etc., although the ecological sustainability of such practices should also be assessed. Finally, socio-economic incentives for farmers to extend rotations and to market non-conifer timber and other products would also be important.

Supporting Information

Material S1 Additional information describing the FORECAST model, its validation for subtropical forests in China, and the list of references used for the review on carbon pools in subtropical planted forests.

Author Contributions

Conceived and designed the experiments: XW JAB. Performed the experiments: JAB. Analyzed the data: XW JAB. Contributed reagents/materials/analysis tools: XW JAB. Wrote the paper: XW JAB.

References

1. Lewis SL, López-González G, Sonké B, Affum-Baffoe K, Baker TR, et al. (2009) Increasing carbon storage in intact African tropical forests. Nature 457: 1003–1007.
2. Food and Agriculture Organization (2012) State of World's Forests 2012. Rome: Food and Agriculture Organization.
3. Paquette A, Messier C (2010) The role of plantations in managing world's forests in the Anthropocene. Fron Ecol Environ 8: 27–34.
4. Kirilenko AP, Sedjo RA (2007) Climate change impacts on forestry. PNAS 104, 19697–19702.
5. Rhemtulla JM, Mladenoff DJ, Clayton MK (2009) Historical forest baselines reveal potential for continue carbon sequestration. PNAS 105: 6082–6087.
6. Kimmins JP, Blanco JA, Seely B, Welham C, Scoullar K (2008) Complexity in Modeling Forest Ecosystems; How Much is Enough? For Ecol Manage 256: 1646–1658.
7. Niu XZ, Duiker SW (2006) Carbon sequestration potential by afforestation of marginal agricultural land in the Midwestern U.S. For Ecol Manage 223: 415–427.
8. State Forestry Administration People's Republic of China (2009). People's Republic of China forestry outlook study. FAO Working Paper No. APFSOS II/WP/2009/11. Bangkok.
9. Chen X, Zhang X, Zhang Y, Wan C (2009) Carbon sequestration potential of the stands under the Grain for Green program in Yunnan Province, China. For Ecol Manage 258: 199–206.
10. Nabuurs GJ et al. (2007) Forestry. In: Metz, B., Davidson, O.R., Bosch, P.R., Dave, R., Meyer, L.A. (eds) Climate Change 2007: Mitigation. Contribution of Working Group III to the Fourth Assessment Report of the Intergovernmental Panel on Climate Change. Cambridge: Cambridge University Press.
11. Wei X, Blanco JA, Jiang H, Kimmins JP (2012) Effects of nitrogen deposition on carbon sequestration in Chinese fir forests. Sci Tot Environ 416: 351–361.
12. Food and Agriculture Organization (2001) Biological Sustainability of productivity in successive rotations. Report based on the work of J. Evans. Forest Plantation Thematic Papers, Working Paper 2. Forest Resources Development Service, Forest Resources Division. Rome: Food and Agriculture Organization.
13. Carson KM, Curran LM, Asner GP, Pittman AM, Trigg SM, et al. (2013) Carbon emissions from forest conversion by Kalimantan oil palm plantations. Nature Clim Change 3: 283–287.
14. Gibson L, Lee TM, Koh LP, Brook BW, Gardner TA, et al. (2011) Primary forests are irreplaceable for sustaining tropical biodiversity. Nature 478: 378–381.
15. Jactel H, Brockerhoff EG (2007) Tree diversity reduces herbivory by forest insects. Ecol Let 10: 835–848.
16. Liao C, Luo Y, Fang C, Chen J, Li B (2012) The effects of plantation practice on soil properties based on the comparison between natural and planted forests: a meta-analysis. Global Ecol Biogeogr 21: 318–327.
17. Constanza R, D'arge R, De Groot R, Farber S, Grasso M, et al. (1997) The value of the world's ecosystem services and natural capital. Nature 387: 253–260.
18. Lo YH, Lin YC, Blanco JA, Yu CH, Guan BT (2012) Moving from ecological conservation to restoration: an example from central Taiwan, Asia. In: Blanco JA, Lo YH (Eds.) Forest Ecosystems: more than just trees. Rijeka: InTech. 339–354.
19. Lewis SL (2006) Tropical forests and the changing Earth system. Phil Trans R Soc Lond B 261: 195–210.
20. Bi J, Blanco JA, Kimmins JP, Ding Y, Seely B, et al. (2007) Yield decline in Chinese Fir plantations: A simulation investigation with implications for model complexity. Can J For Res 37: 1615–1630.
21. Wang W, Wei X, Liao W, Blanco JA, Liu Y, et al. (2013) Evaluation of the effects of forest management strategies on carbon sequestration in evergreen broad-leaved (*Phoebe bournei*) plantation forests using FORECAST ecosystem model. For Ecol Manage 300: 21–32.
22. Lu ZH, Wu G, Ma X, Bai GX (2002) Current situation of Chinese forestry tactics and strategy of sustainable development. J For Res 13: 319–22.
23. Food and Agriculture Organization (2006) Global planted forests thematic study: results and analysis, by D. Del Lungo, J. Ball, J. Carle. Planted Forests and Trees Working Paper 38. Rome: Food and Agriculture Organization.
24. Yu XT (1997) (Ed.) Silviculture of Chinese Fir. Fuzhou: Science and Technology Press of Fujian. 250 pp. (in Chinese).
25. Wu ZL (1984) Chinese-fir China Forestry. Beijing: Publishing House. (in Chinese).
26. Tian DL, Xiang WH, Chen XY, Yan WD, Fang X, et al. (2011) A long-term evaluation of biomass production in first and second rotations of Chinese fir plantations at the same site. Forestry 84: 411–418.
27. Kimmins JP, Mailly D, Seely B (1999) Modelling forest ecosystem net primary production: the hybrid simulation approach use in FORECAST. Ecol Model 122: 195–224.
28. Kimmins JP, Blanco JA, Seely B, Welham C, Scoullar K (2010) Forecasting Forest Futures: A Hybrid Modelling Approach to the Assessment of Sustainability of Forest Ecosystems and their Values. London: Earthscan. 218 p.
29. Moher D, Liberati A, Tetzlaff J, Altman DG, The PRISMA Group (2009) Preferred Reporting Items for Systematic Reviews and Meta-Analyses: The PRISMA Statement. PLoS Med 6(6): e1000097. doi:10.1371/journal.pmed1000097.
30. Tummer B (2006) DataThief III v1.6. Available: http://datathief.org. Accessed: 22 October 2013.
31. Guo LB, Gifford M (2002) Soil carbon stocks and land use change: a meta analysis. Global Chan Biol 8: 345–360.
32. Quinn GP, Keough MJ (2002) Experimental design and data analysis for biologist. Cambridge: Cambridge University Press.
33. Yu XT (1999) Chinese fir: In Chinese Forests Forests Editorial Committee (Eds) (China Forestry Publishing House, Beijing) 1–29 (in Chinese).
34. Pan WC, Tian DL, Lei ZX, Kang WX (1983) Studies on the nutrient cycling in the Chinese-fir plantations: (II) Contents, accumulation rate and biological cycling of nutrient elements in the fast-growing Chinese-fir forest in the hill regions. J Central South For Inst 3: 1–12 (in Chinese).
35. Shao L (1992) Chinese fir. In: Sheng W, Chen B, Xu H, Xu X (Eds) Research on Site Degradation of Timber Plantation. Beijing: Chinese Science & Technology Press. 87–92 (in Chinese).
36. Yang YS, He ZM, Qju RH, Yu XT, Huang BL (1999) Growth pattern of 29-year-old Chinese-fir grown from seed in different rotations. Sci Silvae Sin 35: 32–36 (in Chinese).
37. Xiao W, Nie D, Zhang J (1999) Study on biomass and energy use efficiency of the stands of *Cunninghamia lanceolata* in China. For Res 3: 253–258.
38. Liu F, Luo R, Jiang J (1991) Quantities and effects of various phosphorus compounds in soil under Chinese-fir plantation. J Nanjing For Univ 4: 7–12 (in Chinese).
39. Tian DL, Zhao K (1989) Studies on the litter in a Chinese-fir plantation ecosystem: (I) Amounts, composition and dynamics of litter. J Central South For Inst 9(Sup): 38–43 (in Chinese).
40. Tian DL (1994) Study on bio-geochemistry cycling in Chinese-fir ecosystem. In: Forestry Ministry of China (Ed.) Studies on Forest Ecosystems in China. Harbin: Publishing House of Northeast Forestry Univ. 136–145 (in Chinese).
41. Zhong AL, Hsiung WY (1993) Evaluation and diagnosis of tree nutritional status in Chinese-fir (*Cunninghamia lanceolata* (Lamb) Hook) plantations, Jiangxi, China. For Ecol Manage 62: 245–270.
42. Zhou X (1994) Long-Term Research On China's Forest Ecosystems. Harbin: Northeast Forestry Univ. Press. 320 pp. (in Chinese).
43. Zhou H (1999) Study on productivity of mixed forest planted on site of continuously planted Chinese-fir. J Fujian College For 2: 165–169. (in Chinese).
44. Du J, Lu Y (2009) Study on growth regularity of *Phoebe zhennan* plantation. J Zhejiang For Sci Technol 29: 9–12 (in Chinese).
45. Ma M, Jiang H, Liu Y (2008). Biomass, carbon content, carbon storage and their vertical distribution of *Phoebe bournei* artificial stand. Scien Silv Sin 44: 34–39 (in Chinese).
46. Peng L (2003) Biomass and productivity in 35 year old *Phoebe bournei* plantation. J Fujian College For 23: 128–131 (in Chinese).
47. Peng L (2008) Preliminary study on the biomass of *Phoebe bournei* plantation with different density and site condition. J Fujian For Sci Technol 35: 15–23 (in Chinese).
48. Wu W (2009) Study on stand structure and growth law of near mature *Phoebe bournei* plantation. Develop. For Sci Technol 23: 75–78 (in Chinese).
49. Lu M (2006) Study On Biomass And Nutrient Of Phoebe Bournei Plantation In Heterogeneity Spatiality Of Density And Site. PhD Thesis. Fujian: Fujian Agriculture and Forestry Univ (in Chinese).
50. Liao H, Di D, Zhang C, et al. (1989) Planting density management to *Cinnamoum camphora* and *Phoebe bournei*. J. Fujian College For. 9: 80–84 (in Chinese).
51. Cheng Y (2003) Study On Litter Fall Decomposing Dynamic And Nutrients Releasing Rule Of Phoebe Bournei. PhD Thesis. Fujian: Fujian Agriculture and Forestry Univ. (in Chinese).
52. Guo Y (2007) Decomposition and the nutrient dynamics of *Phoebe bournei* leaf litter. J Fujian College For 27: 199–202 (in Chinese).
53. Lin KM, Zhang ZQ, Cao GQ, He ZM, et al. (2006) Decomposition characteristics and its nutrient dynamics of leaf litter mixtures of both Chinese fir and *Phoebe bournei*. Acta Ecol Sin 26: 2732–2738 (in Chinese).
54. Lin KM, Zhang ZQ, Ye FM, Lin Y, Li QS. (2010) Dynamic analysis of decomposition characteristics and content change of nutrient elements of leaf litter of *Cunninghamia lanceolata*, *Phoebe bournei* and *Schima superba* under *C. lanceolata* artificial forest. J Plant Resour Environ 19: 34–39 (in Chinese).
55. Peng L (2008) Study on nutrient elements of Phoebe bournei plantation in the different site conditions. J Fujian For Sci Tech 35: 10–15 (in Chinese).
56. Chen S (2010) Effect of different site conditions on the growth of *Phoebe bournei*. J. Fujian College For. 30: 157–160 (in Chinese).
57. Liu Q (2009) Study of soil fertility of *Phoebe bournei* planted forest. J Jiangxi For Sci Technol 1: 31–33 (in Chinese).
58. Du J, Fan ZX, Ye DY, Lu CT (2010) The predicting models of crown volume and LAI for *Phoebe zhennan* plantation. J Zhejiang For Sci Technol 30: 37–41 (in Chinese).

59. Hu J, Liu G (2010) Preliminary investigation on photosynthetic characteristics of saplings of *Phoebe zhennan* and *Machilus leptophylla*. J Anhui Agri Univ 37: 541–546 (in Chinese).
60. Ren CH, Chen H, Zhou LH (1990) Study on L.A.I. (Leaf Area Index) of *Phoebe bournei* stand in Youxi county. J Fujian College For 10: 67–71 (in Chinese).
61. Wu Z, Chen S (2004) The effect of light on growth of *Phoebe bournei* plantation. J Fujian College For 24: 371–373 (in Chinese).
62. Fan X, Ma X, Fu R, Liu A (2001) Comparative study on underground vegetation develop of different generation plantations of Chinese-fir. For Res 14: 11–19 (in Chinese).
63. Lin KM, Yu XT, Hong W, Huang BL (2001) Effect of minor vegetation on soil fertility in Chinese-fir plantation. Scient Silv Sin 37: 94–98 (in Chinese).
64. Xiang WH, Tian DL, Yan WD, Kang WX, Fang X (2003) Biomass and nutrient content of vegetation on fallow site after clear-cutting of Chinese-fir plantation. Acta Ecol Sin 23: 695–702 (in Chinese).
65. Yan W, Tian D, Jiao X (2003) A study on biomass dynamics and distribution of undervegetation in the secondary generation of Chinese-fir plantation in Hui Tong. For Res 16: 82–86 (in Chinese).
66. Zhou X, Luo R, Ye J (1991) Effect of continuous cropping with Chinese-fir upon soil nutrients and its feedback. J Nanjing For Univ 3: 47–52 (in Chinese).
67. Tian DL, Zhu XP, Cai BY, Gao YM (1989) Studies on the litter in a Chinese-fir plantation ecosystem: (II) Nutrient contents and decomposition rate of litter. J Central South For Inst 9(Sup): 45–55 (in Chinese).
68. Liao LP (2000) Decomposition of leaf litter of Chinese-fir in mixture with major associated broad-leaved plantation species. Acta Phytoecol Sin 24: 27–33 (in Chinese).
69. Ding YX, Tian Y, Qi L (1999) A testing simulation with FORECAST on long-term productivity of Chinese-fir plantations. For Studies China 1: 34–38 (in Chinese).
70. Ding YX, Chen JL (1995) Effect of continuous plantation of Chinese-fir on soil fertility. Pedosphere 5: 57–66.
71. Yang YS, Chen GS, Huang BL (2000) Variation in the soil water and nutrients between different rotation stands of Chinese-fir. J Nanjing For Univ 24: 25–28 (in Chinese).
72. Seely B, Welham C, Kimmins JP (2002) Carbon sequestration in a boreal forest ecosystem: results from the ecosystem simulation model, FORECAST. For Ecol Manage 169: 123–135.
73. Blanco JA, Seely B, Welham C, Kimmins JP, Seebacher TM (2007) Testing the performance of FORECAST, a forest ecosystem model, against 29 years of field data in a *Pseudotsuga menziesii* plantation. Can J For Res 37: 1808–1820.
74. Guan BT, Wrigth WE, Chung CH, Chang ST (2012a) ENSO and PDO strongly influence Taiwan spruce height growth. For Ecol Manage 267: 50–57.
75. Masera OR, Garza-Caligaris JF, Kanninen M, Karjalainen T, Liski J, et al. (2003) Modeling carbon sequestration in afforestation, agroforestry and forest management projects: the CO2FIX V.2 approach. Ecol Model 164: 77–199.
76. Cannell MGR (2003) Carbon sequestration and biomass energy offset: Theoretical, potential and achievable capacities globally, in Europe and the UK. Biomass Bioener 24: 97–116.
77. Van Minnen JG, Strngers BJ, Eickhout B, Swart RJ, Leemans R (2008) Quantifying the effectiveness of climate change mitigation through forest plantations and carbon sequestration with an integrated land-use model. Carbon Balance Manage 3: 3.
78. Blanco JA (2012) Forests may need centuries to recover their original productivity after continuous intensive management: an example from Douglas-fir. Sci Total Environ 437: 91–103.
79. Marín-Spiotta E, Sharma S (2013) Carbon storage in successional and plantation forest soils: a tropical analysis. Global Ecol Biogeogr 22: 105–117.
80. Amoroso MM, Turnblom MM (2006) Comparing productivity of pure and mixed Douglas-fir and western hemlock plantations in the Pacific Northwest. Can J For Res 36: 1484–1496.
81. Wu Z (2005) Study on growth effect of the mixed forest of *Cunninghamia lanceolata* and *Phoebe bournei*. J Fujian College For 25(2): 142–146.
82. Duan W, Zhu B, Lu JL, Lu R, Yang CJ, et al. (2008) Carbon Storage Effects in Artificial Mixed Forest of *Cryptomeria fortune - Cunninghamia lanceolata* in MontaneAreas of Western Sichuan Province of China. Sichuan Environment 27(5): 57–63 (in Chinese).
83. Zhang G (2009) Growth and biomass in the Chinese fir mixtures. Anhui Agri. Sci. Bull. 15(21): 131–132.
84. Lin T (2011) Mixed Forests of *Michelia macclurei* and *Cunninghamia lanceolata* in Coastal Mountains Land of Southern Fujian. Jiangxi Forestry Sci Tech 3: 16–19.
85. Chen XY (2009) Stand biomass and its distribution of mixed forests of *Alnus crematogyne* and *Cunninghamia lanceolata*. Jiangxi Forest Science and Technology 1, 14–17.
86. Keith H, MaKey B, Berry S, Lindenmayer D, Gibbons P (2010) Estimating carbon carrying capacity in natural forest ecosystems across heterogeneous landscapes: addressing sources of error. Glob Change Biol 16: 2971–2989.
87. Berthrong ST, Jobbagy EG, Jackson RB (2009) A global-meta-analysis of soil exchangeable cation, pH, carbon and nitrogen with afforestation. Ecol Appl 19: 2228–2241.
88. Minghe L, Ritchie GA (1999) Eight hundred years of clonal forestry in China: I. Traditional afforestation with Chinese fir (*Cunninghamia lanceolata* (Lamb.) Hook.). New For 18: 131–142.
89. Zhang XQ, Kirschbaum MKU, Hou Z, Guo Z (2004) Carbon stock changes in successive rotations of Chinese-fir (*Cunninghamia lanceolata* (Lamb.) Hook) plantations. For Ecol Manage 202: 131–147.
90. Blanco JA, Wei X, Jiang H, Jie CY, Xin ZH (2012) Impacts of enhanced nitrogen deposition and soil acidification on biomass production and nitrogen leaching in Chinese fir plantations. Can J For Res 42: 437–450.
91. Blanco JA, Zavala MA, Imbert JB, Castillo FJ (2005) Sustainability of forest management practices: Evaluation through a simulation model of nutrient cycling. For Ecol Manage 213: 209–228.
92. Kelty MJ (1992) Comparative productivity of monocultures and mixed-species stands. In: Kelty, M.J., Larson, B.C., Oliver, C.D. (Eds.), The Ecology and Silviculture of Mixed-Species Forests. Kluwer Academic Publishers, Dordrecht, Boston, 125–141.
93. Pretzsch H, Bielak K, Block J, Bruchwald A, Dieler J, et al. (2013) Productivity of mixed versus pure stands of oak (*Quercus petraea* (Matt.) Liebl. and *Quercus robur* L.) and European beech (*Fagus sylvatica* L.) along an ecological gradient. Eur J For Res 132: 263–280.
94. Primicia I, Camarero JJ, Imbert JB, Castillo FJ (2013) Effects of thinning and canopy type on growth dynamics of *Pinus sylvestris*: inter-annual variations and intra-annual interactions with microclimate. Eur J Forest Res 132: 121–135.
95. Bauhus J, Khanna PK, Menden N (2000) Aboveground and belowground interactions in mixed plantations of *Eucalyptus globulus* and *Acacia mearnsii*. Can J For Res 30: 1886–1894.
96. Erskine PD, Lamb D, Bristow M (2006) Tree species diversity and ecosystem function: Can tropical multi-species plantations generate greater productivity? For Ecol Manage 233: 206–210.
97. Forrester DI, Bauhus J, Cowie AL, Vanclay JK (2006) Mixed-species plantation of *Eucalyptus* with nitrogen fixing trees: a review. For Ecol Manage 233: 211–230.
98. Kelty MJ (2006) The role of species mixtures in plantation forestry. For Ecol Manage 233: 195–204.
99. Seely B, Welham C, Blanco JA (2010) Towards the application of soil organic matter as an indicator of ecosystem productivity: Deriving thresholds, developing monitoring systems, and evaluating practices. Ecol Ind 10: 999–1008.
100. Blanco JA, González E (2010) Exploring the sustainability of current management prescriptions for Pinus caribaea plantations in Cuba: a modelling approach. J. Trop. For. Sci. 22: 139–154.
101. Fahey TJ, Woodbury PB, Battles JJ, Goodale CL, Hamburg SP, et al. (2010) Forest carbon storage: ecology, management, and policy. Front Ecol Environ 8: 245–252.
102. Eriksson E, Gillespie AR, Gustavsson L, Langvll O, Olsson M, et al. (2007) Integrated carbon analysis of forest management practices and wood substitution. Can J For Res 37: 671–681.
103. Lippke B, Oneil E, Harrison R, Skog K, Gustavsson L, et al. (2011) Life cycle impacts of forest management and wood utilization on carbon mitigation: knowns and unknowns. Carbon Manage 2: 303–333.
104. Perez-Garcia J, Lippke B, Comnick J, Manriquez C (2005) An assessment of carbon pools, storage, and wood products market substitution using life-cycle analysis results. Wood Fiber Sci 37: 140–148.
105. McKinley DC, Ryan MG, Birdsey RA, Giardina CP, Harmon ME, et al. (2011) A synthesis of current knowledge on forests and carbon storage in the United States. Ecol Appl 21: 1902–1924.
106. Liu P, Yin R (2012) Sequestering carbon in China's forest ecosystems: potential and challenges. Forests 3: 417–430.
107. Jie C, Jiang H, Zhou G, Wei X, Blanco JA, et al. (2011) Simulating the carbon storage of spruce forests based on the FORECAST model and remotely sensed data. Proceedings - 2011 19th International Conference on Geoinformatics, Geoinformatics 2011. Article number 5981581. ISBN: 978-161284848-8. DOI: 10.1109/GeoInformatics.2011.5981581.
108. Qi G (2011) Carbon pool and sequestration potential of larch plantations in Northeast China. PhD thesis. Shenyang: Chinese Academy of Sciences.
109. Keith H, Mackey BG, Lindenmayer DB (2009) Re-evaluation of forest biomass carbon stocks and lessons from the world's most carbon-dense forests. PNAS 106: 11635–11640.
110. Guan D, Liu Z, Geng Y, Lindner S, Hubacek K (2012b) The gigatonne gap in China's carbon dioxide inventories. Nature Clim Change 2: 672–675.

11

Automated Analysis of Two-Dimensional Positions and Body Lengths of Earthworms (*Oligochaeta*); MimizuTrack

Naomi Kodama[1*], Toshifumi Kimura[2], Seiichiro Yonemura[1], Satoshi Kaneda[1], Mizue Ohashi[2], Hidetoshi Ikeno[2]

1 National Institute for Agro-Environmental Sciences (NIAES), 3-1-3 Kannondai, Tsukuba, Ibaraki, Japan, **2** School of Human Science and Environment, University of Hyogo, 1-1-12 Shinzaike-honcho, Himeji, Hyogo, Japan

Abstract

Earthworms are important soil macrofauna inhabiting almost all ecosystems. Their biomass is large and their burrowing and ingestion of soils alters soil physicochemical properties. Because of their large biomass, earthworms are regarded as an indicator of "soil heath". However, primarily because the difficulties in quantifying their behavior, the extent of their impact on soil material flow dynamics and soil health is poorly understood. Image data, with the aid of image processing tools, are a powerful tool in quantifying the movements of objects. Image data sets are often very large and time-consuming to analyze, especially when continuously recorded and manually processed. We aimed to develop a system to quantify earthworm movement from video recordings. Our newly developed program successfully tracked the two-dimensional positions of three separate parts of the earthworm and simultaneously output the change in its body length. From the output data, we calculated the velocity of the earthworm's movement. Our program processed the image data three times faster than the manual tracking system. To date, there are no existing systems to quantify earthworm activity from continuously recorded image data. The system developed in this study will reduce input time by a factor of three compared with manual data entry and will reduce errors involved in quantifying large data sets. Furthermore, it will provide more reliable measured values, although the program is still a prototype that needs further testing and improvement. Combined with other techniques, such as measuring metabolic gas emissions from earthworm bodies, this program could provide continuous observations of earthworm behavior in response to environmental variables under laboratory conditions. In the future, this standardized method will be applied to other animals, and the quantified earthworm movement will be incorporated into models of soil material flow dynamics or behavior in response to chemical substances present in the soil.

Editor: Jan M. Hemmi, University of Western Australia, Australia

Funding: This work was supported by Grant for Environmental Research Projects from the Sumitomo Foundation (Contract nr.:123225.) The funders had no role in study design, data collection and analysis, decision to publish, or preparation of the manuscript.

Competing Interests: The authors have declared that no competing interests exist.

* E-mail: naomi.kodama@googlemail.com

These authors contributed equally to this work.

Introduction

Earthworms, the so-called ecosystem engineers, are important soil macrofauna belonging to the subclass *Oligochaeta*. They inhabit various types of soil ranging from 10 to 2000 individuals per meter squared, resulting in a large biomass of between 0.5 to 305 g dry mass m^{-2} [1]. They significantly change soil properties and fertility in the vicinity of the large burrowing areas [2]. Their important role in soil formation has been already emphasized by Darwin [3]. Earthworms are one of the main components of the drilosphere, one of the biological systems regulating mineralization of soil organic matter, where they activate microbes by mixing the litter and soil, breaking down the physical protection of soil organic matter, and developing soil aggregates during gut transit [2]. The changes in soil pore structures and the extent of aggregations also alter soil gas diffusion coefficients physically, which could increase emissions of global warming gases, such as CO_2, CH_4 and N_2O [4]. Because of their behaviors and features, earthworms are regarded as important biological indicators of chemical toxicity in soil ecosystems and many studies have examined the effects on population dynamics of earthworms such as fecundity, reproductive activity, and mortality [5]. Their behaviors affect not only below-ground material flow dynamics, but also indicate "soil health" conditions.

Evaluation of the functional role of earthworms on soil ecosystems has been, however, insufficient because of methodological difficulties [6]. Traditional research methods including descriptive observations of their behavior, gut content analysis, choice tests, and litter bags have increased our understanding of their ecological importance. Recent advances of molecular tools provide information on enzyme activities of microorganisms which interact with earthworms. Isotope labeling techniques have been used to quantify materials exchanged between earthworms and soils or atmosphere, and between earthworms and microorganisms. These new techniques are powerful but still limit our knowledge on the soil biogeochemical processes regulated by earthworms because they are indirect and static methods. In combination with existing methods, additional development of the tools to quantitatively and mechanistically evaluate their movements and exploring factors controlling their feeding behavior

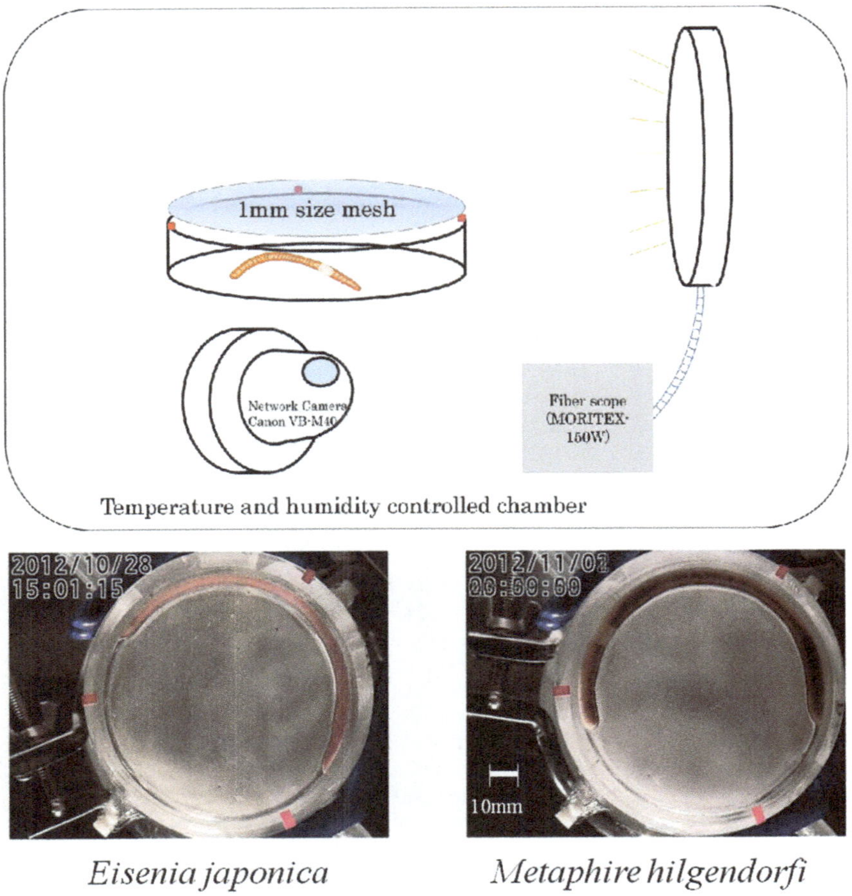

Figure 1. Webcam setup and study animals. The upper diagram shows the system for webcam recording in a controlled chamber. The earthworms used in this study are shown in the bottom panel. The diameter of the glass Petri dish was 85 mm (inside diameter) and the height of the Petri dish was 20 mm.

would be helpful to further understand their ecological roles. Such information will also enable us to establish more realistic models of soil ecosystems because many of the current models do not take the impact of soil animals into account [5]. Thus, observing earthworm feeding behavior, such as their response to some stimuli (such as light, temperature, vibration, and odor cues of food) in vitro without soil will be of importance albeit under very simplified artificial condition. The understanding of their physical activities resulting from their feeding behavior will advance quantification of the earthworm's contribution to whole below-ground biogeochemical processes. To determine earthworm activity, it is first necessary to quantify the extent to which they move and stretch their bodies and relate this to gas emissions from the soil. To confirm the activities, therefore, direct observation is a powerful tool and the combination of conventional methods with

Figure 2. Earthworm body thinning procedure and determination of the head, tail, and central points. The captured image data were used for the latter.

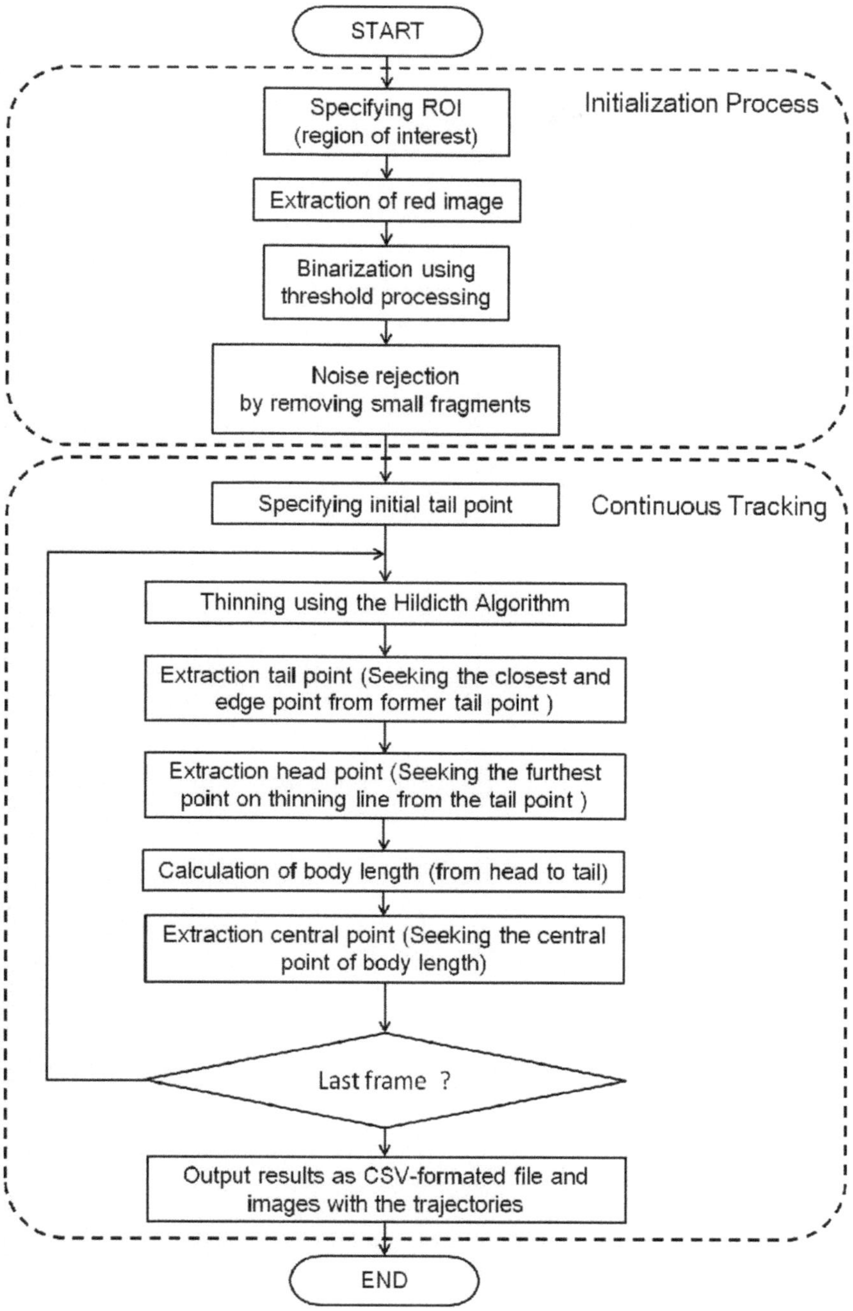

Figure 3. Workflow of the proposed method. Our method consists of two processes; 'Initialization Process' and 'Continuous Tracking'.

direct observation such as using a web camera or scanner will be of great merit in this field.

In addition to the importance of earthworms in ecosystem functions, they are also regarded as an important biological indicator of chemical toxicity in the soil. Ecotoxicology has traditionally used biological indicators to quantify or predict the changes in soil "health" or "quality" [5]. To date many studies have found serious effects of chemicals and heavy metals on the reproductivity, growth rate, and mortality of earthworms ([7] [8,9]), and experimental protocol and earthworm culture techniques in vitro have been developed [10]. However, the observation of earthworms in response to such chemical pollution has been conducted over a relatively long time span with a lower resolution time (on the scale of minutes), and less attention has been paid to the signs in early stages, such as irregular behavior or morphological alterations induced by the uptake of pollutants. However, since the sudden changes in their activity and features might be also tightly connected to the damages and types of the pollutant, observations at high time resolutions (on the scale of seconds) should also be conducted.

To date, automated systems have been used to track movement with high time resolution continuously using a web camera for nematodes, ants, honeybees, and drosophila, and to subsequently analyze the continuous image data [11] [12] [13] [14] [15] [16] [17]. Image analysis is an efficient method for quantifying the tracking data. Previous studies have attempted to track earth-

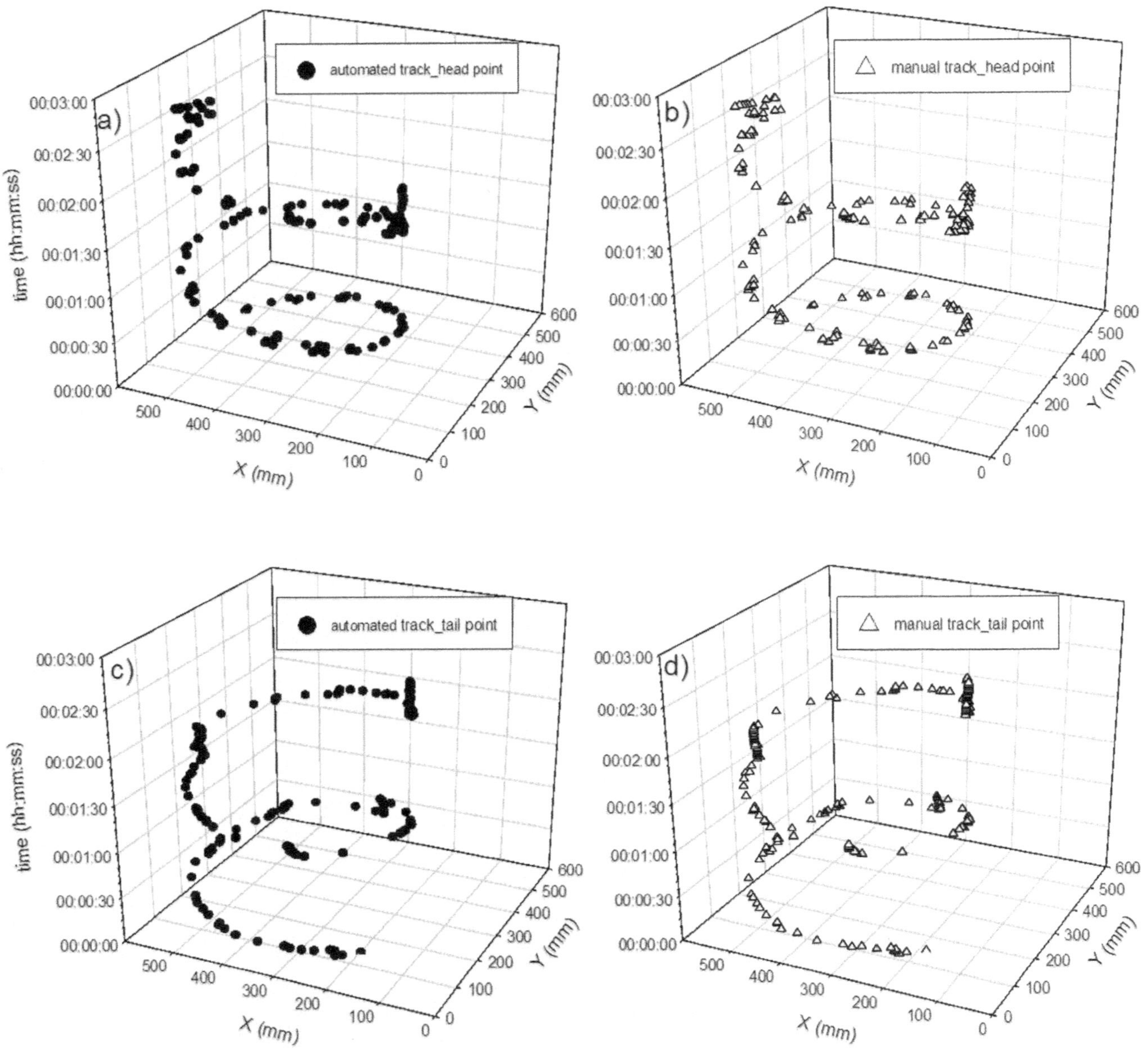

Figure 4. The automatically and manually measured two-dimensional coordinates of *Eisenia japonica* over time. Each panel shows the head point in the upper panel and the tail point in the bottom panel.

worms directly in the soil by means of X-ray or minirhizotron [18] [19]; the time resolution, however, was low to track their dynamic changes over a long time period on a time scale of seconds. Revealing their habits without soil but with a high time resolution would be a first step in investigating their behavior. However, those tools cannot be applied directly to earthworms as some of their traits are different from other animals in terms of the extent to which they can stretch their bodies (their body stretch can reach 1.5 imes that of their default body length) [20].

A number of automated tracking systems for a model worm, *Caenorhabditis elegans*, have not considered changes in the extent of body stretch [20]. The WormScan is indeed a useful tool to observe population dynamics such as fecundity, reproductive activity, or mortality along with the growth rate that occur over a long time span. However, we still lack a program that can track dynamic change in their activities to observe a fast response to chemical pollutants, environmental changes or toxic substances. WormScan's default settings are for the nematode as a target animal. The features of nematodes, such as the movement and body size, vary widely from those of earthworms, therefore it requires intensive modification of the parameters of the program to apply it to earthworms. In addition, a standardized tracking system is required so that reliable comparisons can be made between species or phenotypes when tracking is done by different users. Standardized measurements will reduce misestimation and will output more reliable quantification.

Although existing methods such as WormScan [20] are candidates that may have been applicable in this study, the current WormScan program, which was originally developed for tracking nematodes, was not capable of following the changes in morphology of earthworms because earthworms move more quickly and over a wide area compared with nematodes.

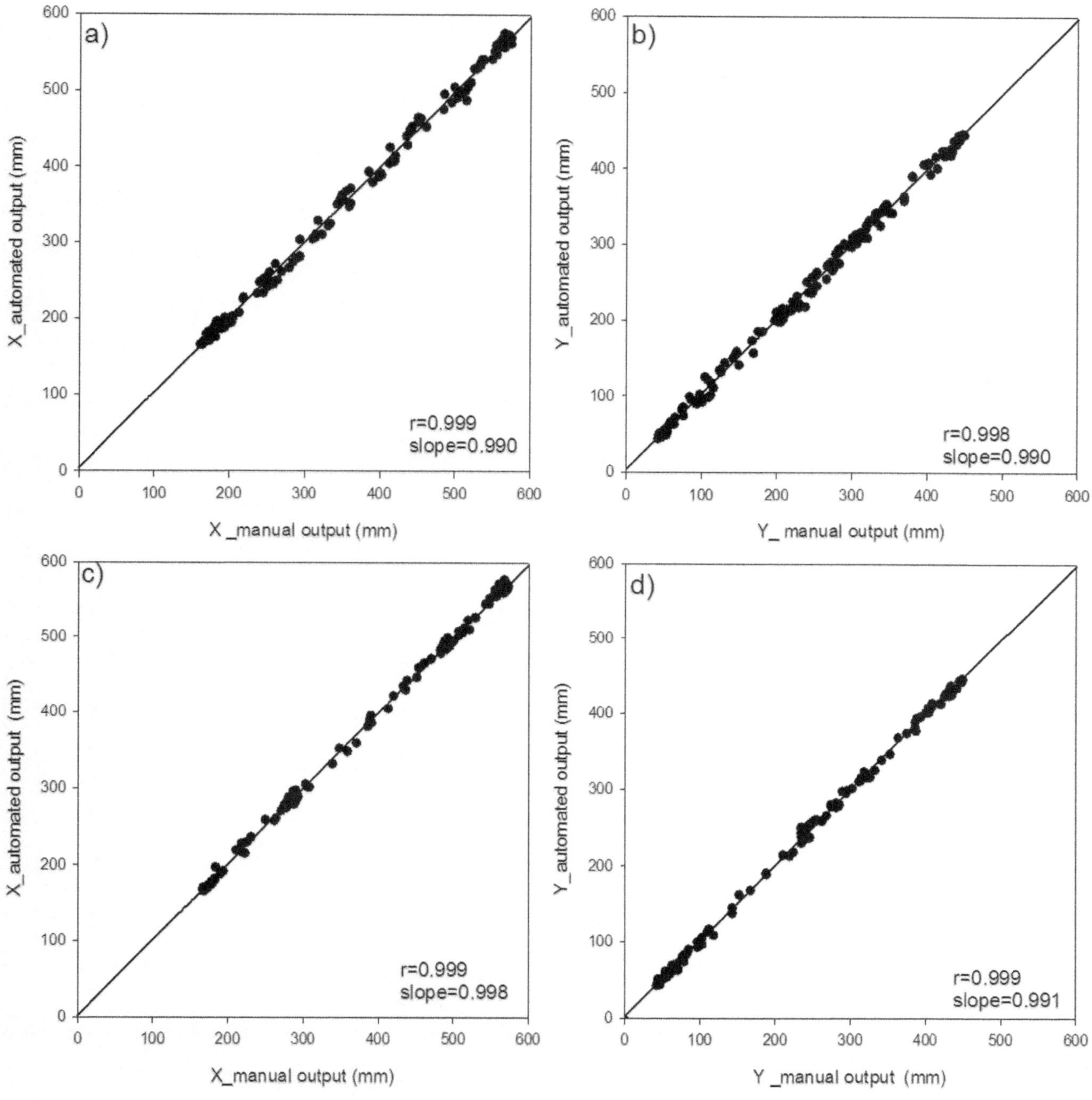

Figure 5. Correlations between the automatically and manually measured two-dimensional coordinates of *Eisenia japonica*. Each panel shows the head point in the upper panel and the tail point in the bottom panel.

WormScan has been developed for observing population dynamics such as reproductive activity, fecundity or mortality which occur over a longer term span (several days to weeks), and the behavior is also relatively static. WormScan, therefore does not suit observations at a continuous and high time resolution, which captures dynamic and rapid movements. This is a remarkable property of earthworms and is an important factor in quantifying their physical activity. We believe that WormScan can track some objects of varying sizes, however we found that earthworms' movements are too fast and dynamic, and WormScan could not cope with this. The observation of dynamic changes would help to understand the ecological background of earthworm behaviors; chemical assays in vitro without soils could also be conducted. The existing application ImageJ (Fiji) (http://rsbweb.nih.gov/ij/ version 1.46 for Windows) can provide a plug-in to manually track the two-dimensional positions and lengths of objects. Even when using this application, analyzing large data sets is labor intensive; therefore, a more reliable and automated quantification method is required for continuous image data.

In the present study, we developed a high-throughput program that tracks two-dimensional coordinates of multiple points and body lengths of objects in continuously acquired image data from earthworms; to the program was able to process approximately 1 frame per second. Following user initialization, the program tracks

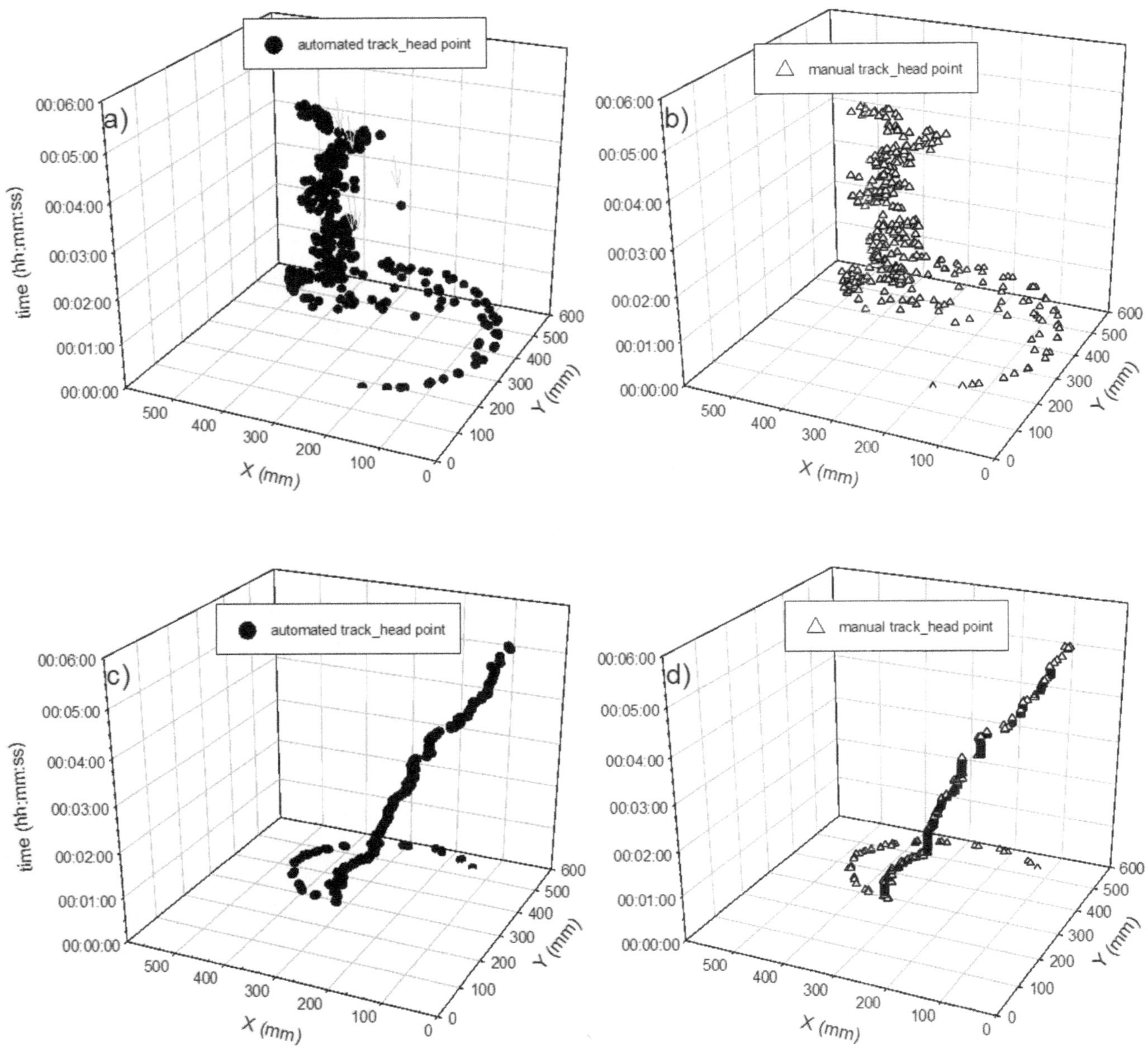

Figure 6. The automatically and manually measured two-dimensional coordinates of *Metaphire hilgendorfi* over time. Each panel shows the head point in the upper panel and the tail point in the bottom panel.

the head, central position, and tail of a single earthworm. This program provides a means to track animals that change their body lengths over time both accurately and efficiently, though still with limited performance. The output was subsequently used to calculate the velocity of those points. Finally, in the future, the quantified behavior could be incorporated into soil material flow dynamic models or applied to measure their reactions to chemical inputs in the soil.

Materials and Methods

Study animals

Two different species of earthworm were used in this study; *Eisenia japonica* [21] with a body size ranging from 4–17 cm [22], and *Metaphire hilgendorfi* [21] with a body size ranging from 9–30 cm [23]. The lifespan of both is usually 6 months from spring to fall. Both species are common in Japanese mountains and fields, and both genera have a wide global distribution [24]. *M. hilgendorfi* is epigeic because it primarily inhabits the litter layer and *E. japonica* is polyhumic-endogeic because it mainly lives in the topsoil layer and has more inorganic content in the gut [25].

Continuous recording of earthworm movement in glass containers (image capture)

Adult earthworms were captured from a field at the National Institute for Agro-Environmental Sciences, Tsukuba, Japan on 28 October, 2012. Earthworms were kept in a plastic box (inner diameter 15 cm, height 6 cm) with approximately 100 g of alluvial soil at 60% of its water holding capacity until the experiment. The weights of *E. japonica* and *M. hilgendorfi* used in the experiment were 1.45 and 4.75 g, respectively. *E. japonica* and *M. hilgendorfi* were transferred from the plastic box to glass containers for network-

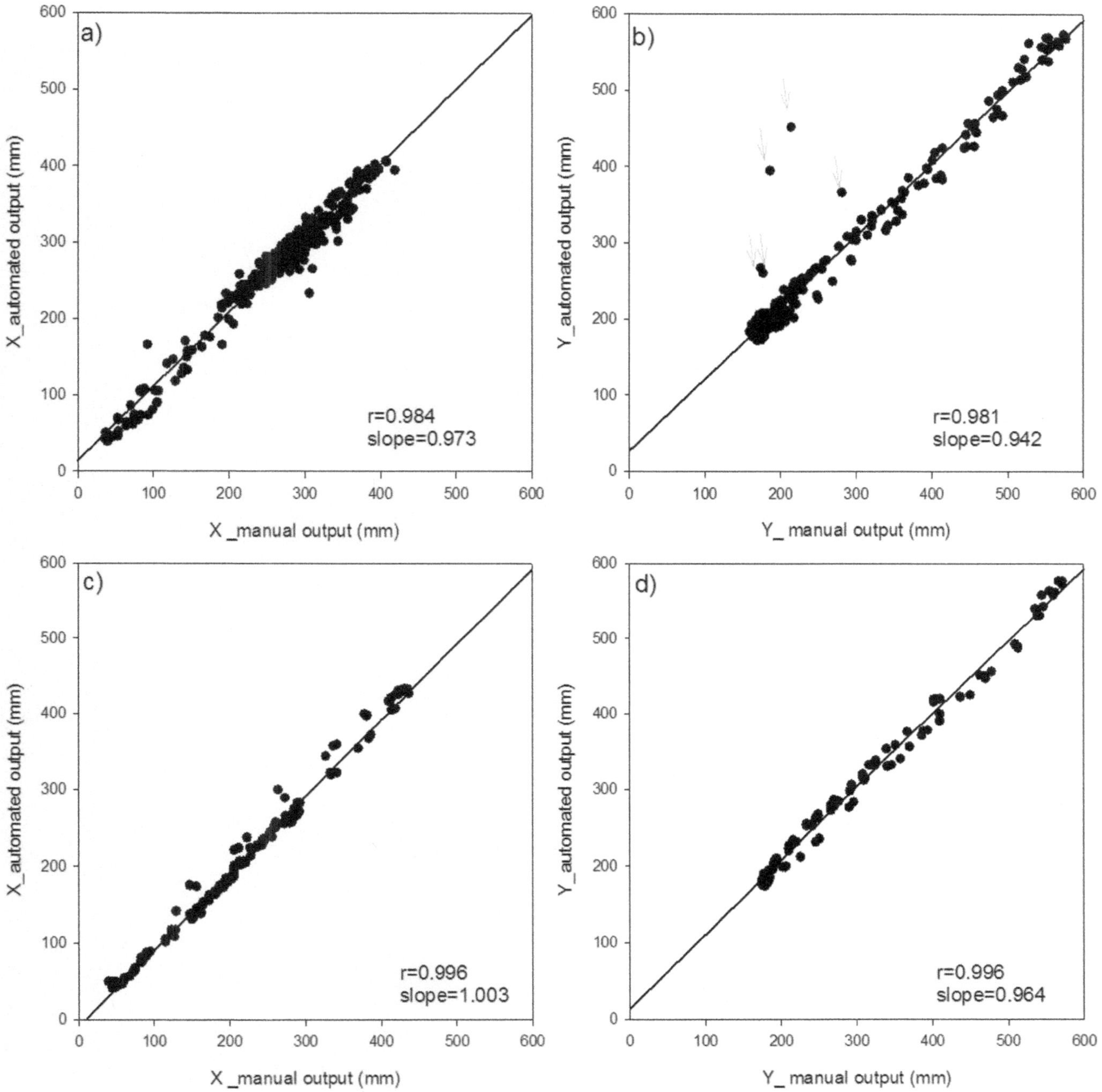

Figure 7. Correlations between the automatically and manually measured two-dimensional coordinates of *Metaphire hilgendorfi*. Each panel shows the head point the in upper panel and the tail point the in bottom panel.

camera recording inside an incubator, where the mean ambient temperature was set at 15°C and the ambient air humidity was saturated at the given temperature (Fig. 1). A fiber scope light was used to maintain constant light conditions (MHAB-150W, Moritex, Tokyo) at approximately 500 lux outside of the container and app. 250 lux inside the container; this was measured with a radiometer (T-10A, Konica Minolta Optics, Inc., Tokyo, Japan). The glass containers were covered with 1-mm mesh plastic cloths and fixed with double-sided cellophane tape to prevent the earthworms from escaping during the experimental period. A small fan (OD2510, Orion Fans, Dallas, USA) was placed on the top of the glass container to ventilate the air inside and wet tissue was put in to keep the humidity constant during the measurement period. A network camera (VB-M40, Canon Inc., Tokyo, Japan) was fixed to the bottom of the glass container with approximately 60 cm of distance between the camera and the glass container (Fig. 1). The network camera was fixed in parallel to the glass container firmly while capturing the image. The position of the camera was calibrated using the ruler to ensure the accurate pixilation of the actual length. The network camera uses a lens that is 40 mm in diameter and can cover 60.4° from the horizontal angle. The captured image data was 640×480 pixels in size. To obtain the coordinates of the glass container, marks were fixed at three positions on the edge of the containers with red tape. The

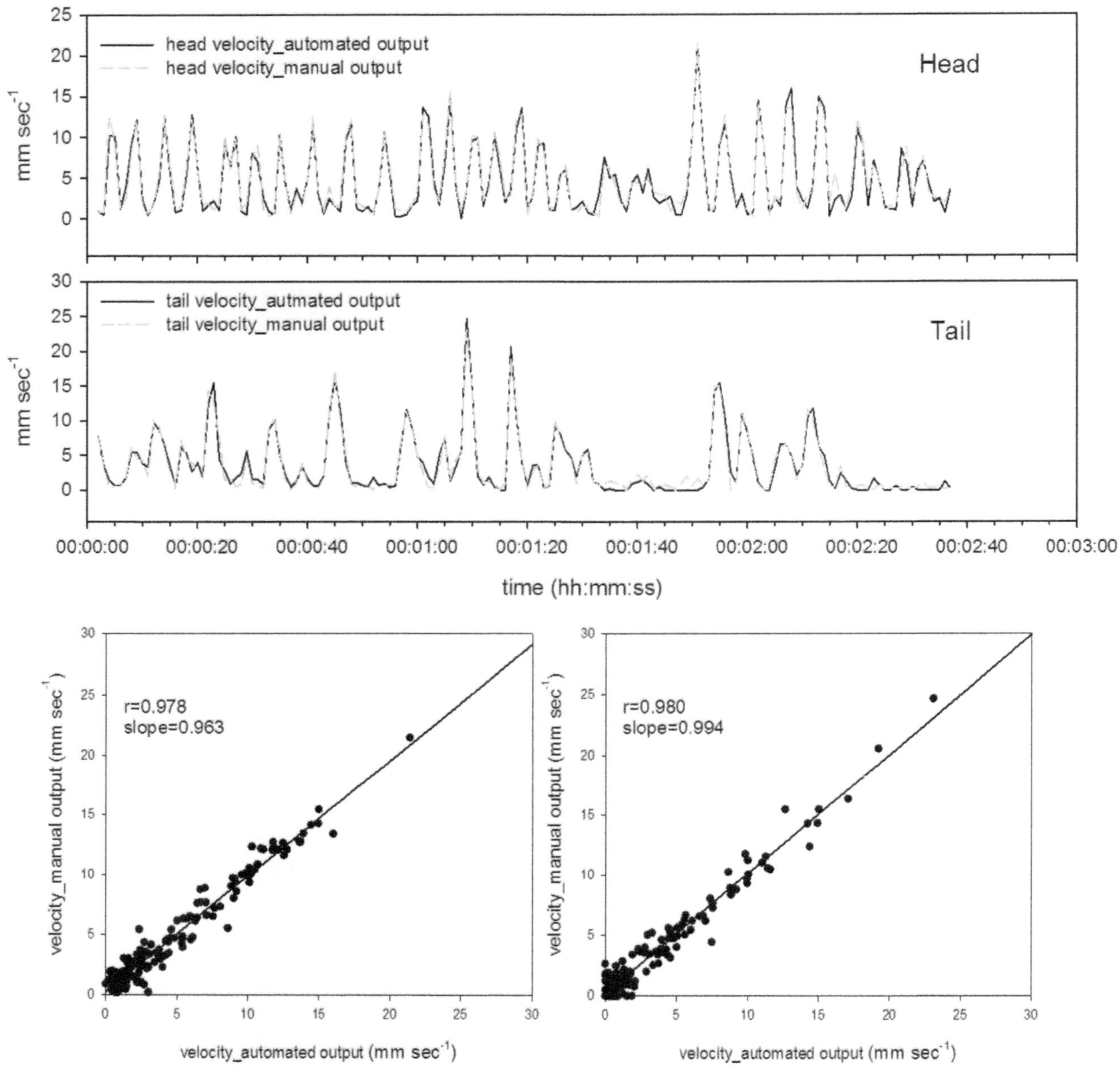

Figure 8. Comparison of the automatically and manually calculated velocity of each point of *Eisenia japonica*. The upper panel shows the calculated velocity of the head point and the central panel shows that of the tail point. Correlations between the automatically and manually calculated velocities of *Eisenia japonica*.

round glass containers were 85 mm in inside diameter and 2.5 mm in thickness. The image data were recorded as a movie using a network camera continuously recording every second and saved as JPEG files. One pixel was equivalent to 0.2 mm in the captured image data.

To estimate local image distortions, checkerboard (8 rows and columns) was located for image capturing system on the same position as the glass container. The length of the diagonal lines were measured using a vernier caliper with a 0.1 mm precision and calculated the errors. The maximum error in mm was 0.2 mm that resulted the precision less than 0.1%.

Image analysis

Two methods were implemented: a manual tracking method and a new automated tracker. First we used a manual tracking method relies on the Fiji implementation of ImageJ (http://rsbweb.nih.gov/ij/version 1.46 for Windows) with an additional plugin, Manual Tracking (http://fiji.sc/wiki/index.php/Manual_Tracking). Currently manual tracking is a standard method to track animal behavior using image data. Second, we then used a newly developed program (MimizuTrack) that was compared with Manual Tracking in Fiji.

Manual tracking

Using the Manual Tracking plug-in on the Fiji implementation of ImageJ, head and tail of the body were tracked separately. The static image data (JPEG) captured at 1 second intervals were converted to movie data, using a function in Fiji to convert from JPEG into.avi formatted data. Each coordinate per one second was tracked to obtain the velocity of the movement. We then

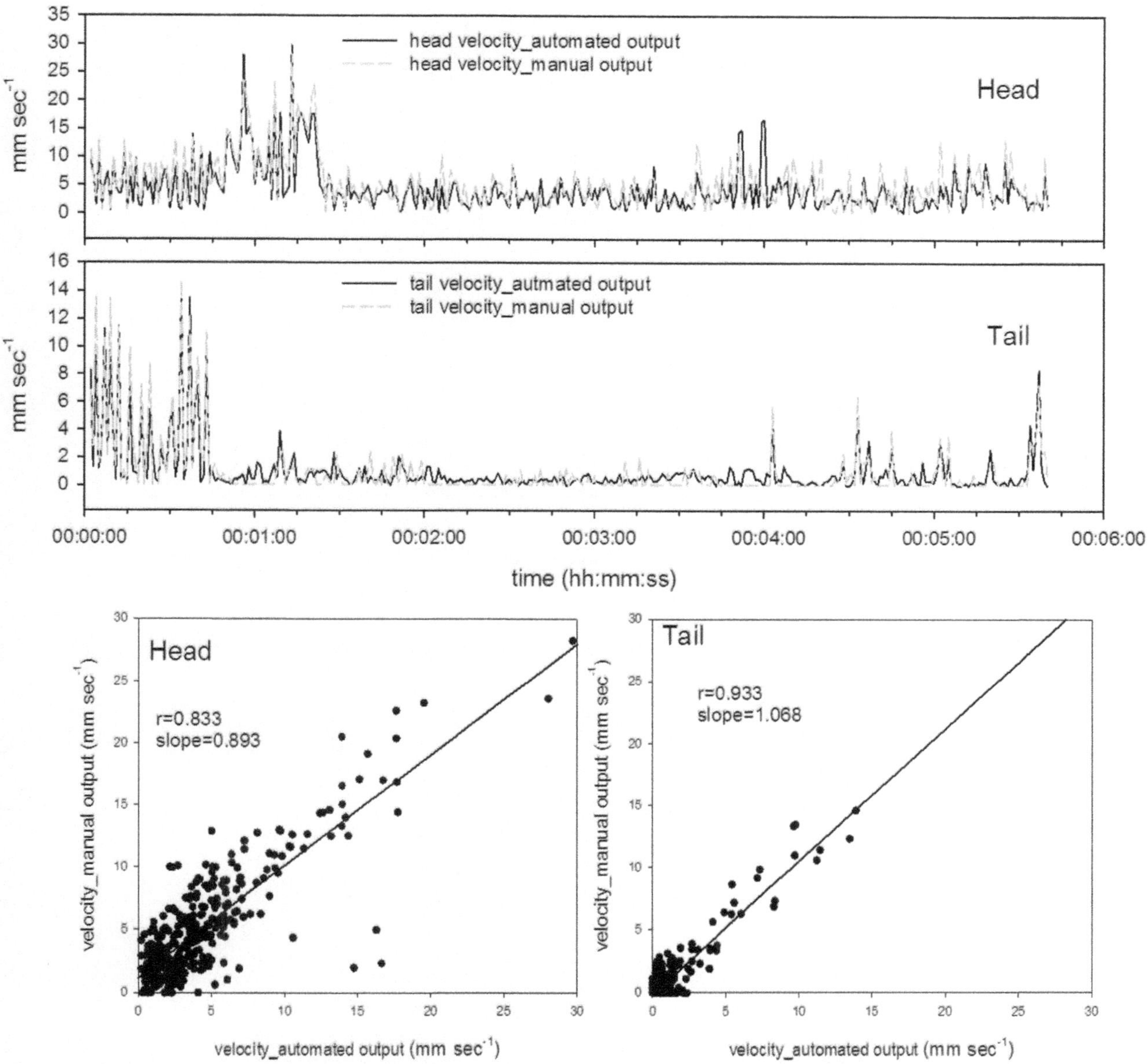

Figure 9. Comparison of the automatically and manually calculated velocity of each point of *Metaphire hilgendorfi*. The upper panel shows the calculated velocity of the head point and the central panel shows that of the tail point. Correlations between the automatically and manually calculated velocities of *Metaphire hilgendorfi*.

handled the output data as ground truth for further evaluation since the data were input by expert users. The output data were used for evaluation of the program described in the next section. Our program was benchmarked against user input only. This was because no automated tracking system currently exists that is directly comparable to our tracker in this specific setting.

Automatic tracking - Initialization Process

The image data obtained were first processed to separate the objects (earthworms from the background). In the original image, the contrast between the earthworms and the background was highest in the red channel so only this component was used for the process. The procedures are done manually as an initial step, then the image data are processed using a program (image shown in Fig. 2); We developed several programs to track moving objects over time, such as honeybees or plant roots in previous studies [12] [26] [27] the source code for which is original, and we have developed a new program (MimizuTrack) for earthworms based on the previous programs. The steps are shown in Fig. 3 and detailed below. The detailed initialization process are described as follows:

- The outline of the inside of the glass container was determined using a tool for drawing around the shape and cutting the outside of the image.
- A color threshold was determined to highlight the color of the earthworm body and split the color channels into three colors (green, blue, and red). The red image was then used for further analysis. Earthworms were assigned a value of 255 and the non-earthworm areas were eliminated by applying a value of 0.

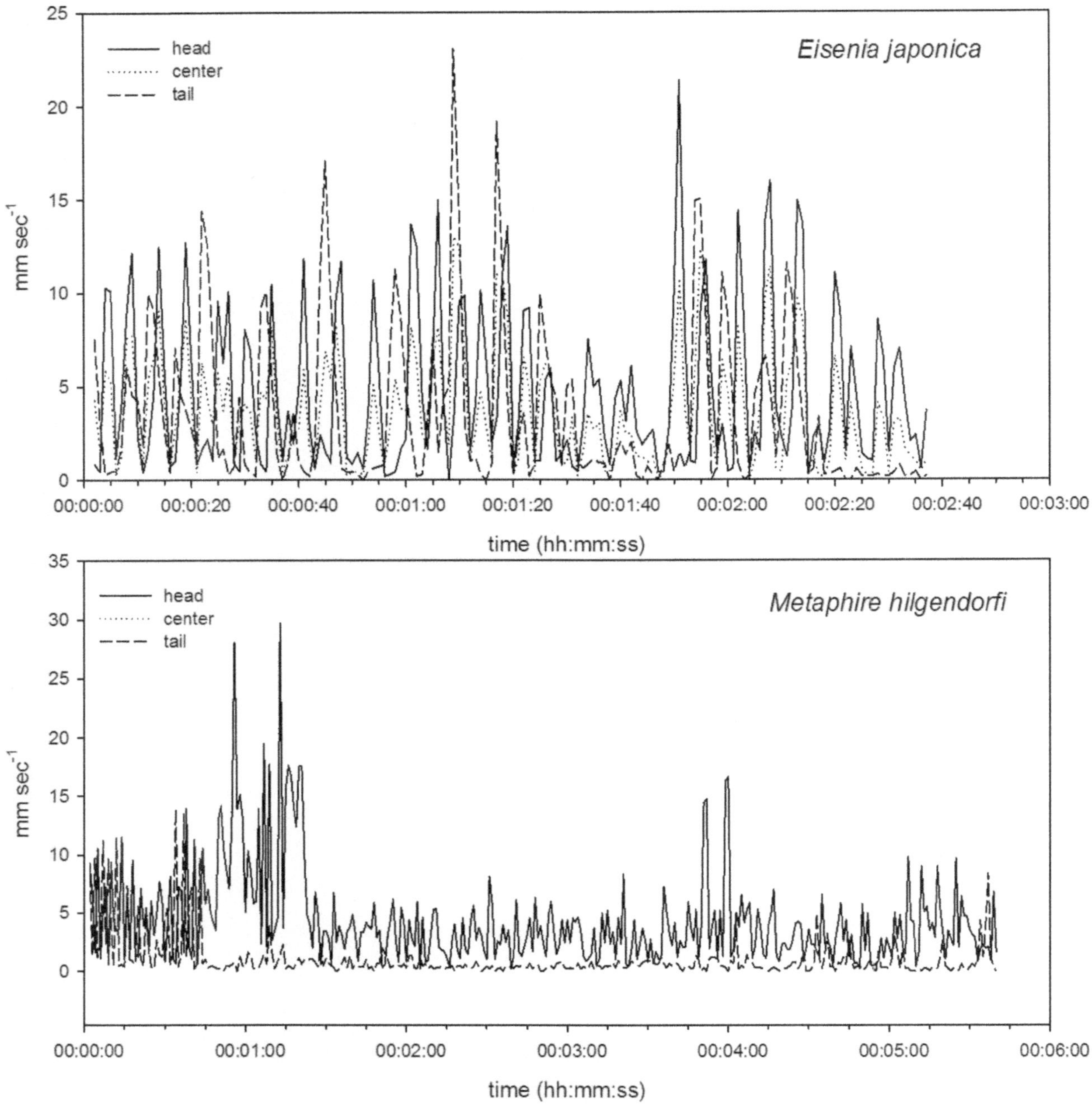

Figure 10. Comparison of velocities among different points (head, central, and tail points). The upper panel shows the velocities of each point of *Eisenia japonica* and the bottom panel shows those of *Metaphire hilgendorfi*.

- A threshold to extract the earthworm body was determined and a binary image with white for the object and black for the background was generated.
- The noise was eliminated by removing small fragments from the background.

Automatic tracking - Continuous Tracking

The rest of the image processing was carried out by our newly developed program. The procedures are shown in Fig. 2. We developed a prototype program using Microsoft Visual Studio 2010 (Visual C++2010) and OpenCV 2.31 on a computer with an Intel Core i5 2.50 GHz (CPU), 16 GB (Memory), 640 GB (HDD), and running Microsoft Windows 7 Enterprise 64-bit (OS). Our developed software is executed as a Win64 console application on Windows.

The tracking method is as follows;

- Thinning the earthworm body areas

To extract feature points, such as the head, tail, and center points, the program was tasked with identifying the end-points (tails) of the line segment corresponding to the worm body. That is as thin as possible while remaining fully connected and centered. Using binary images of the earthworm body, the program was able to obtain three feature points on each frame. First, the Hilditch algorithm is used to reduce the pixel region representing the

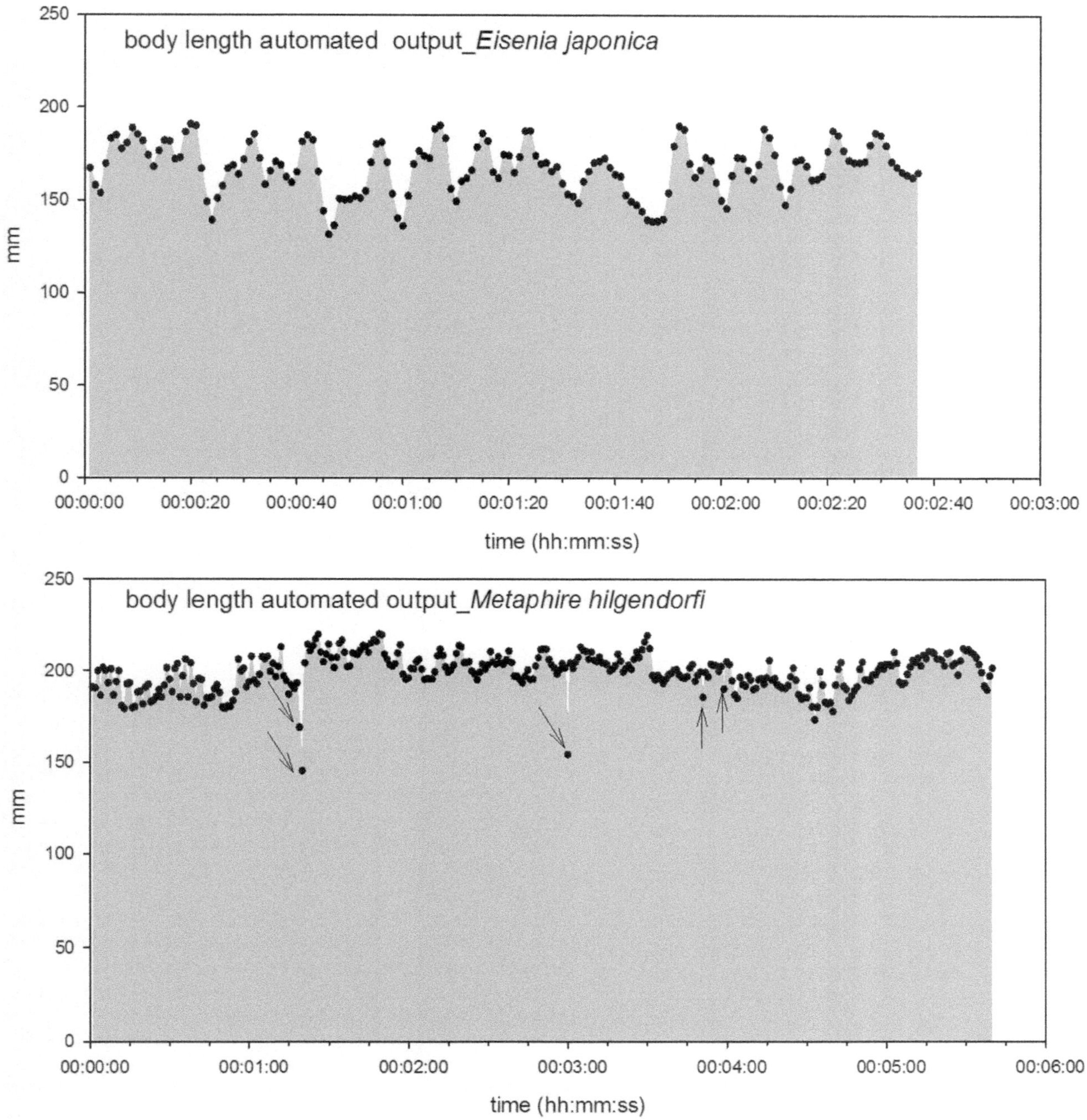

Figure 11. Change in body length over time. The upper panel shows the length of *Eisenia japonica* and the bottom panel shows that of *Metaphire hilgendorfi.*

earthworm to its skeletal form [28]. The end-points of the center line are recognized as candidates of earthworm feature points.

- Extracting the head, tail, and center point of individuals (feature extraction)

The next step was to determine the head, tail, and center point. The program extracted the feature points of the earthworm body by first identifying the head and tail using the different features between the two points (tail points always move slower than the other points), then the center point was determined. To initialize the program, the tail point was determined manually. By doing this, the program automatically calculated four-neighbor distances of the center line from the tail point and extracted the point at the furthest distance. The program defined this as the head point of the earthworm. Then, the program recognized the halfway point between the head and tail as the central point of the body. Using this process, the program could obtain the initial feature points of the earthworm (i.e. head, tail, and center points). From the second frame, the program calculated new feature points using the previous ones. First, the program extracted new tail points based on the fact that the tail points of the earthworm move slower than the other points. Using this feature, the program recognized the edge points as new tail points with the shortest distance from the previous points. Both head and center points were obtained by the same procedure, which were extracted from the initial head and

Table 1. Mean, maximum, and minimum values of automatically and manually calculated velocities of each point (head and tail) of two earthworm species (*E. japonica* and *M. hilgendorfi*).

Species	velocity	mean (mm sec^{-1})	max. (mm sec^{-1})	mini. (mm sec^{-1})
E. japonica				
Program	head	4.75±4.36	21.37	0
	tail	3.44±4.38	23.09	0
manual	head	4.81±4.29	21.45	0.21
	tail	3.55±4.44	24.69	0
M. hilgendorfi				
program	head	4.38±4.13	29.71	0.21
	tail	1.03±1.92	13.9	0
manual	head	5.09±4.43	28.29	0
	tail	0.92±2.21	14.58	0

The errors indicate standard deviations over time.

center points. These processes were carried out on all of the images. The program outputs positional data in each frame to a CSV file, allowing trajectories over time to be visualized.

Statistical Analysis

All statistical analyses were performed using SPSS 10.05 (SPSS Inc., Chicago, IL, USA). Relationships between two variables (program output and manual output) were assessed using standard bivariate correlation procedures. The correlation coefficient is shown in the text below and the figures.

Results

Evaluations of automatic output using the manually tracked values

Tracking and extraction of each point. First, we were successful in tracking the three target points, (the head, center, and tail of the body) on a time scale of seconds, continuously followed by body length calculations using the newly developed program. We used a part of the obtained data for the evaluation of the program which shows typical movement of two species of earthworms every 150 seconds and 300 seconds for *E. japonica* and *M. hilgendorfi* respectively (Figs. 4–11). The tracking of the head and tail with the automatic and manual programs over time are shown in Figs. 4 and 6 respectively. All of the regression lines between the automatically and manually tracked coordinates were close to 1 (>0.98), especially in the case of *E. japonica* (Fig. 5). In contrast, some outliers were observed in *M. hilgendorfi* as indicated by the arrows in Fig. 7. There were no significantly different measurements observed in *E. japonica* between different body parts; however, significant differences were observed between the different body parts of *M. hilgendorfi*. In general, the errors in measurement of the head point ($r = 0.981$, $r = 0.984$, X and Y coordinates, respectively) were larger than those for the tail point ($r = 0.996$, $r = 0.996$, X and Y coordinates, respectively). One reason for this was that the head moved more frequently while the tail stayed in the same position. Since the program tracks and determines the feature points by predicting the next steps, the program may fail to track the objects if the objects (in this case, the heads) move more broadly in the image-capturing interval (1 second). In the case of *M. hilgendorfi*, when they looped, coiled, or crossed their bodies, the program faced difficulties in recognizing the edges, and failed to detect the head positions.

Table 2. Mean maximum values of differences between automatically and manually calculated velocities of each point (head and tail) of two earthworm species (*E. japonica* and *M. hilgendorfi*).

Species	mean (mm sec^{-1})
E. japonica	
head	0.07±0.92
tail	0.11±0.88
M. hilgendorfi	
head	0.71±2.49
tail	0.11±0.80

The errors indicate standard deviations over time.

Calculations of velocity of each point and body length. Using their coordinates, the velocity of each point was calculated. The output data obtained with the automated program were further confirmed by comparing them with the manually acquired data (Figs. 9 and 10). The accuracy of each output was satisfactory for the application to quantify earthworm activity over time. Some outliers in head velocity in *M. hilgendorfi* were due to the misestimation of head points by the automated program, in which case the head was attached to a part of the body. Otherwise, the automated program satisfactorily detected the peaks in terms of timing and absolute values. As shown in Table1, the mean velocities of the head and tail of *E. japonica* calculated by the automated program were 4.75±4.36 and 3.44±4.38 mm sec^{-1}, respectively, while those from the manually calculated values were 4.81±4.29 and 3.56±4.44 mm sec^{-1}, respectively. The mean velocities of the head and tail of *M. hilgendorfi* calculated by the automated program were 4.38±4.13 and 1.03±1.92 mm sec^{-1}, respectively, while those from the manually calculated values were 5.09±4.43 and 0.92±2.21 mm sec^{-1}, respectively. The differences between the automated and manually calculated values were 0.07±0.92 and 0.11±0.88 mm sec^{-1} for the head and tail of *E. japonica*, respectively and 0.71±2.49 and 0.11±0.80 mm sec^{-1} for

Table 3. Mean, maximum, and minimum values of body length of two earthworm species (*E. japonica* and *M. hilgendorfi*).

body length	mean (mm)	max. (mm)	mini.(mm)
E. japonica	167.02±13.62	190.73	131.44
M. hilgendorfi	199.28±10.05	220.38	145.33

The errors indicate standard deviations over time.

those of *M. hilgendorfi*, respectively (Table2). The difference was largest in the head of *M. hilgendorfi* but the others matched quite well, with the differences being less than 0.2 mm sec^{-1}. The tail points tended to move slower than the head points according to the calculated velocities, and the head point of *M. hilgendorfi* moved faster than that of *E. japonica* by 1.65±7.07 mm sec^{-1}. The body lengths were calculated from the binary data, which were converted into a line (shown in Table3).

The mean lengths of *E. japonica* and *M. hilgendorfi* were 167.02±13.62 and 199.28±10.05 mm, respectively, while the maximum and minimum lengths of *E. japonica* were 190.73 and 131.44 mm, respectively, and those of *M. hilgendorfi* were 220.38 and 145.33 mm, respectively. Therefore, the stretch from minimum to maximum was 59.19 and 75.05 mm for *E. japonica* and *M. hilgendorfi*, respectively. The extent of the stretchiness was in agreement with previously reported values [1], which indicated that the body could reach 1.5 times longer than its default length. Moreover, our program also generated the change in length over time per second, so that we could observe the frequency of the stretchiness for further applications (Fig. 11).

Discussion

To date, there exists no method to process and extract such high time resolution image data (on a time scale of seconds) capturing three feature points along with tracking elastic objects such as earthworms, which move quickly and widely, although the program is still a prototype at this stage. A number of automated systems have been developed to track insects' movements, and our earthworm tracking system might be similar to that of *Caenorhabditis elegans* (one of the model nematodes) [20,29]. Our program can deal with a time scale of seconds, though there remain some problems with the program. The program could not cope with irregular shapes of earthworms, such as coiling or overlapping. This could be achieved by modifying the program in the future using the Knot Theory, or topological concept. The program also failed to follow the feature points when the head moved more frequently than the general movement. Furthermore, in the case of *M. hilgendorfi*, when they coiled or crossed their bodies, the program faced difficulties in recognizing the edges, and failed to detect the head positions. To overcome such problems, one solution would be to obtain images at higher time resolutions so that the program could keep tracking more frequently moving objects without losing the target points because the higher time resolution is closer to continuous data. Increasing the contrast between the background and the objects for image analysis would also achieve more precise results. Furthermore, there would be alternative methods to detect feature points such as differentiating the shapes or brightness of head or tail, which will further improve the program.

The velocity of each point of the different body parts of *E. japonica* was synchronized; in contrast, those of *M. hilgendorfi* were not. In the case of *M. hilgendorfi*, the head and center points were synchronized but the tail point was not; it tended to stay still on the same coordinate. We recognize that the velocities were under non-soil conditions; therefore, earthworms in the soil will be much slower than our calculations indicate. However, the program picked up the features of each species when observed under the same conditions. We consider our newly developed program to be sufficient to measure earthworm activity; this method can be used to compare dynamic changes in these parameters between species or in response to environmental variables such as temperature, light, and moisture, as well as chemical contaminants.

Image data is powerful, direct information that can be used to quantify physical information of changes in earthworm's activities. The intention of this study is to apply the program to quantify the physical parameters to indicate earthworm's activity from their movement speed or body stretchiness. It will be more beneficial if these data are combined with other data, such as data on gas emissions from the body, enzyme activities using molecular tools, or pathway of elements using isotopic data for mechanistic understanding. The method used in this study is artificial; a small Petri dish without soil as used. However, this study will advance our understanding of the basic behavior of earthworms depending on the alteration of their living environment. This basic behavior can be incorporated in models in terms of material flow belowground or reactions in response to chemical contaminant in soil [5].

Combining different kinds of studies will help to unravel or predict the complex processes in belowground systems [6]. Furthermore, although this system could not be applied to observe their behavior directly in the soil, culturing earthworms in transparent gels or agar medium would be a useful method to observe them three-dimensionally. Furthermore, applying other rapidly developing state of the art technologies, such as MRI or radar, to observe earthworms under soil conditions, it will be theoretically possible to directly observe their behavior in soil, though the time resolution will not be high enough. In fact, radar is already being used for root growth observations in soil.

Conclusions

We demonstrated that our newly developed program successfully tracks the movement of two species of earthworms that have different features in terms of body size, mass, and morphology as well as movement. Based on our data, the program reduced the time required for manual tracking 3-fold and the program could track the movement which is inherent in earthworm behavioral ecology studies. Large continuous image data sets can be immediately processed and the program can determine earthworm body parts along with their changes in elastic body lengths over time. In the future, quantified earthworm behaviors could be incorporated in material flow dynamic models in soil or in models of the response of earthworms to chemical inputs to soil.

The software "MimizuTrack" will be released on INCF Software Center (http://software.incf.org/) under the BSD license. Otherwise the program is open upon request to the following contact person; Assis. Prof. Toshifumi Kimura (email: kimura@shse.u-hyogo.ac.jp).

Acknowledgments

We would like to acknowledge Mrs. Sugawara for her cheerful support in the setup of the measurement systems and taking care of the earthworms.

Author Contributions

Conceived and designed the experiments: NK SY SK. Performed the experiments: NK SY SK. Analyzed the data: NK TK. Contributed reagents/materials/analysis tools: NK TK SY SK MO HI. Wrote the paper: NK TK SY SK MO HI.

References

1. Edwards CA, Bohlen PJ (1996) Biology and ecology of earthworms: Chapman & Hall.
2. Lavelle P, Spain AV (2001) Soil ecology: Kluwer Academic Pub.
3. Darwin C (1892) The formation of vegetable mould, through the action of worms, with observations on their habits: J. Murray.
4. Lubbers IM, van Groenigen KJ, Fonte SJ, Six J, Brussaard L, et al. (2013) Greenhouse-gas emissions from soils increased by earthworms. Nature Climate Change 3 187–194.
5. Bartlett MD, Briones MJI, Neilson R, Schmidt O, Spurgeon D, et al. (2010) A critical review of current methods in earthworm ecology: from individuals to populations. European Journal of Soil Biology 46: 67–73.
6. Curry JP, Schmidt O (2007) The feeding ecology of earthworms - A review. Pedobiologia 50: 463–477.
7. Helling B, Reinecke SA, Reinecke AJ (2000) Effects of the Fungicide Copper Oxychloride on the Growth and Reproduction of *Eisenia fetida* (Oligochaeta). Ecotoxicology and Environmental Safety 46: 108–116.
8. Neuhauser EF, Callahan CA (1990) Growth and reproduction of the earthworm *Eisenia fetida* exposed to sublethal concentrations of organic chemicals. Soil Biology and Biochemistry 22: 175–179.
9. Anderson GL, Boyd WA, Williams P (2001) Assessment of sublethal endpoints for toxicity testing with the nematode Caenorhabditis elegans. Environmental Toxicology and Chemistry 20: 833–838.
10. Yasmin S, D'Souza D (2010) Effects of pesticides on the growth and reproduction of earthworm: a review. Applied and Environmental Soil Science.
11. Feldman A, Balch T (2003) Automatic identification of bee movement. pp. 53–59.
12. Kimura T, Ohashi M, Okada R, Ikeno H (2011) A new approach for the simultaneous tracking of multiple honeybees for analysis of hive behavior. Apidologie 42: 607–617.
13. Mersch DP, Crespi A, Keller L (2013) Tracking Individuals Shows Spatial Fidelity Is a Key Regulator of Ant Social Organization. Science 340: 1090–1093.
14. Rizvandi NB, Pizurica A, Rooms F, Philips W (2008) Skeleton analysis of population images for detection of isolated and overlapped nematode C. elegans.
15. Straw AD, Branson K, Neumann TR, Dickinson MH (2011) Multi-camera real-time three-dimensional tracking of multiple flying animals. Journal of The Royal Society Interface 8: 395–409.
16. Colomb J, Reiter L, Blaszkiewicz J, Wessnitzer J, Brembs B (2012) Open Source Tracking and Analysis of Adult Drosophila Locomotion in Buridan's Paradigm with and without Visual Targets. PLoS ONE 7: e42247.
17. Kimura T, Ohashi M, Crailsheim K, Schmickl T, Okada R, et al. (2014) Development of a New Method to Track Multiple Honey Bees with Complex Behaviors on a Flat Laboratory Arena PLoS ONE.
18. Ligthart TN, Peek GJWC, Taber EJ (1993) A method for the three-dimensional mapping of earthworm burrow systems. Geoderma 57: 129–141.
19. Joschko M, Mueller PC, Kotzke K, Doehring W, Larink O (1993) Earthworm burrow system development assessed by means of X-ray computed tomography. Geoderma 56: 209–221.
20. Mathew MD, Mathew ND, Ebert PR (2012) WormScan: A technique for high-throughput phenotypic analysis of Caenorhabditis elegans. PLoS One 7: e33483.
21. Michaelsen W (1891) Oligochaeten des Naturhistorischen Museums in Hamburg. I [-IV].
22. Yamaguchi E (1965) Allolobophora japonica. In: Okada K editor Illustrated encyclopedia of the fauna of Japan 1 Hokuryukan co ltd: 547.
23. Ishizuka K (2001) Taxonomic Study of the Genus Pheretima s. lat (Oligochaeta, Megascolecidae) from Japan. Bulletin of Seikei University 33: 1–125.
24. Blakemore RJ, Ito MT, Kaneko N (2006) Alien earthworms in the Asia/Pacific region with a checklist of species and the first records of Eukerria saltensis (Oligochaeta: Ocnerodrilidae) and Eiseniella tetraedra (Lumbricidae) from Japan, and Pontoscolex corethrurus (Glossoscolecidae) from Okinawa. Assessment and Control of Biological Invasion Risks: 173–181.
25. Uchida T, Kaneko N, Ito MT, Futagami K, Sasaki T, et al. (2004) Analysis of the feeding ecology of earthworms (Megascolecidae) in Japanese forests using gut content fractionation and ホエ 15N and ホエ 13C stable isotope natural abundances. Applied Soil Ecology 27: 153–163.
26. Kimura T, Ohashi M, Crailsheim K, Schmickl T, Odaka R, et al. Tracking of Multiple Honey Bees on a Flat Surface; 2012. IEEE. pp. 36–39.
27. Nakano A, Ikeno H, Kimura T, Sakamoto H, Dannoura M, et al. (2012) Automated analysis of fine-root dynamics using a series of digital images. Journal of Plant Nutrition and Soil Science 175: 775–783.
28. Hilditch CJ (1969) Linear skeleton from square cupboards. Machine intelligence 6: 403–733.
29. Geng W, Cosman P, Berry CC, Feng Z, Schafer WR (2004) Automatic tracking, feature extraction and classification of C. elegans phenotypes. Biomedical Engineering, IEEE Transactions on 51: 1811–1820.

12

Microbial Community Diversities and Taxa Abundances in Soils along a Seven-Year Gradient of Potato Monoculture Using High Throughput Pyrosequencing Approach

Xing Liu[1,2], Junlian Zhang[3,4], Tianyu Gu[5], Wenming Zhang[1,2], Qirong Shen[6], Shixue Yin[5]*, Huizhen Qiu[1,2]*

1 College of Resources and Environmental Sciences, Gansu Agricultural University, Lanzhou, Gansu Province, China, **2** Gansu Provincial Key Laboratory of Aridland Crop Science, Gansu Agricultural University, Lanzhou, Gansu Province, China, **3** College of Agronomy, Gansu Agricultural University, Lanzhou, Gansu Province, China, **4** Gansu Key Laboratory of Crop Genetic and Germplasm Enhancement, Gansu Agricultural University, Lanzhou, China, **5** College of Environmental Science and Engineering, Yangzhou University, Yangzhou, Jaingsu Province, China, **6** Jiangsu Provincial Key Lab of Organic Solid Waste Utilization, Nanjing Agricultural University, Nanjing, China

Abstract

Background: Previous studies have focused on linking soil community structure, diversity, or specific taxa to disturbances. Relatively little attention has been directed to crop monoculture soils, particularly potato monoculture. Information about microbial community changes over time between monoculture and non-monoculture treatments is lacking. Furthermore, few studies have examined microbial communities in potato monoculture soils using a high throughput pyrosequencing approach.

Methodology/Principal Findings: Soils along a seven-year gradient of potato monoculture were collected and microbial communities were characterized using high throughput pyrosequencing approach. Principal findings are as follows. First, diversity ($H_{Shannon}$) and richness (S_{Chao1}) indices of bacterial community, but not of fungal community, were linearly decreased over time and corresponded to a decline of soil sustainability represented by yield decline and disease incidence increase. Second, *Fusarium*, the only soilborne pathogen-associated fungal genus substantially detected, was linearly increased over time in abundance and was closely associated with yield decline. Third, *Fusarium* abundance was negatively correlated with soil organic matter (OM) and total nitrogen (TN) but positively with electrical conductivity (EC). Fourth, *Fusarium* was correlated in abundances with 6 bacterial taxa over time.

Conclusions: Soil bacterial and fungal communities exhibited differential responses to the potato monoculture. The overall soil bacterial communities were shaped by potato monoculture. *Fusarium* was the only soilborne pathogen-associated genus associated with disease incidence increase and yield decline. The changes of soil OM, TN and EC were responsible for *Fusarium* enrichment, in addition to selections by the monoculture crop. *Acidobacteria* and *Nitrospirae* were linearly decreased over time in abundance, corresponding to the decrease of OM, suggesting their similar ecophysiologial trait. Correlations between abundance of *Fusarium* with several other bacterial taxa suggested their similar behaviors in responses to potato monoculture and/or soil variables, providing insights into the ecological behaviors of these taxa in the environment.

Editor: Kevin McCluskey, University of Missouri, United States of America

Funding: This work was supported by the Agricultural Ministry of China (201103004), the Natural Science Foundation of China (31360500) and the National Nature Science Foundation of China (41071177). The funders had no role in study design, data collection and analysis, decision to publish, or preparation of the manuscript.

Competing Interests: The authors have declared that no competing interests exist.

* E-mail: hzqiu@gsau.edu.cn (HQ); sxyin@yzu.edu.cn (SY)

Introduction

Soil microbial community is extremely important in maintaining soil quality, functions, sustainability and global nutrient cycling. Majority of past studies focused on linking the microbial community structure, diversity, or specific taxa to soil variables/disturbances, such as soil pH [1,2], soil types [3], soil texture [4], soil layers [5], plant species [6], nitrogen level [7], temperature [8], water regime [9], tillage [10,11], amendments [12], biocontrol agents [13], land uses [14], and contaminations [15]. However, relatively limited attention has been directed to the links between crop monoculture and soil microbial communities.

Crop monoculture is when the same crop is grown year after year on the same soil, without interruption crops that could potentially induce different crop-specific microbial communities [16]. Crop monoculture can be viewed as a simple and an ideal model system to study the relationship between agricultural practices and soil microbial communities. When crop monoculture is practiced, the microbial community is continuously exposed to the roots of the same crop that selects and enriches certain groups of microorganisms including yield-debilitating populations (i. e. soilborne pathogens) of that crop [17]. This consequently shapes the soil microbial community that can often be easily observed. Several studies have determined differences of soil microbial communities between rotation/mixed culture and monoculture of cotton [18], maize [19], wheat (and maize) [20], rice [21], soybean [22], oilseed rape [23] and potato [24]. With a few notable exceptions [24], these studies typically used an one time-point soil sampling strategy. Soil samples were collected at a given time-point after a certain time span, generally years, over which the monoculture was practiced. The microbial communities were then compared between monoculture and non-monoculture treatments. This sampling strategy is useful for revealing differences in communities between treatments, but it does not allow measurements of community changes over time (i. e. the number of years a crop monoculture has been maintained). This is important information to understand the assembly processes of microbial communities under continuous selection of crop monoculture practice [25,26,27,28].

An alternative strategy is to sample soils along a year-gradient of monoculture. This strategy allows observations of changes in patterns of different microbial taxa over the years that may occur in a predictable manner within the experimental time span. In addition, the change pattern of one taxon may be correlated with another in abundance, allowing us to perceive the behaviors of these taxa in response to potato monoculture. In spite of these benefits, studies using year-gradient samples are limited in literature. Some early reports followed this strategy (see review [29]), but community changes were based on culture-dependent methods only, which have been proven to have strong biases against yet-to-cultured majorities [30]. The relationship between crop monoculture practices and the soil microbial community is thus not sufficiently understood. Relevant knowledge on potato monoculture agroecosystems is particularly lacking.

In a previous study [31], we examined the fungal community in potato monoculture soils. Further re-examination of our original data revealed that the diversity (H_{Shannon}) and richness (S_{Chao1}) indices of the fungal community were nearly un-changed over the years of monoculture (time). Furthermore, *Fusarium* was the only soilborne pathogen-associated fungal genus substantially detected using a 454 pyrosequencing approach. *Fusarium* was significantly enriched over time, and closely associated with disease incidence observed in the field (see the Results section below). These findings encouraged us to further examine whether the diversity and richness indices of the bacterial community were also significantly changed. Specifically we are interested in knowing change patterns of different bacterial taxa with reference to that of *Fusarium*. Based on the above findings, we hypothesized that the overall soil bacterial community were significantly changed under the continuous monoculture pressure, and abundance of some bacterial taxa were correlated with *Fusarium* abundances over time. As soil variables play key roles in microbial community structuring[1,7,14], associations between soil variables and the microbial change patterns were also studied.

Materials and Methods

Ethics Statement

Authors confirm that no specific permissions were required for the locations/activities involved in this study, because the land accessed is publically owned and no protected species were sampled. Authors also confirm that the field studies did not involve endangered or protected species.

Field experiment, soil sampling and variable measurements

Detailed descriptions of the field experiment, soil sampling and variable measurements can be found elsewhere [31]. Only a brief description is given here. The experimental field, located at the Tiaoshan Farm, a warm terrestrial arid area with a mean annual precipitation 186 mm is located in Gansu Province, China. This field contained 21 plots, each being 9.0×6.1 m in size. Prior to potato monoculture, the field was under a corn-potato rotation. Each year, starting from year 2005, 3 plots (replicates) were selected for practicing potato ("Atlantic" cultivar) monoculture, leaving other plots to continue with the corn-potato rotation. The previous crop of selected plots or the potato year of the rotation was always corn. By 2011, all plots were used up. This experimental design provides opportunity to collect soil samples after potato monoculture from 1 to 7 years simultaneously.

Potato was typically seeded in April 25 with a few days variation among years. Tuber pieces were buried on the top of raised paths (~40 cm in height and ~135 cm in bottom width) and plants were spaced 17 cm apart along the row. Two rows were planted on each raised path with 70 cm between the two rows. This resulted in a plant density of 460 plants per plot (equivalent to 84,075 plants ha^{-1}). Blended fertilizer (15-15-15), additionally supplemented with urea and K_2SO_4, was used at the rate of 210 kg N ha^{-1}, with the ratio $N:P_2O_5:K_2O$ being 1.4:1.0:2.0. Nitrogen form in blended fertilizer was $(NH_4)_2SO_4$. Fertilizers were applied before seeding by machine. All other field management activities were performed manually. Once seeded and fertilized, the raised paths were covered manually with a plastic film. The field was irrigated by introducing water into the ditch between two raised paths through pipes. Irrigation was done three times during the growth period, typically at June 1 (seedling stage), July 1 (early flowering stage) and July 20 (tuber enlargement stage) at the rate of 2,700 t ha^{-1} each time. Potato was harvested in later August. For yield evaluation, the tubers were dug out manually. Stolons, underground stems, true roots, scale leaf rhizomes and adhered soils were removed. Tubers smaller than 2 cm in diameter were considered as developing tubers and were excluded from the yield calculations. Disease incidence was calculated by totaling the number of plants with symptoms of any recognizable disease, and converting to percent by dividing by the total number of plants and then multiplying by 100. Fusarium wilt was the major disease, covering ~90% of diseased plants. The disease incidence was recorded three times during the growth period, but only the records at harvest time are presented.

Soil samples were collected in July 20, 2011. The plots with 5 years of monoculture were not sampled. From each plot, 15 spots were randomly sampled from the 0–20 cm soil layer and the soil from these spots were placed into a single aggregate sample and then well mixed. Totally, 18 samples were obtained. Samples were put into sterile plastic bags, placed into an ice box, transported to laboratory and used as soon as possible, or stored in a refrigerator at −80°C if not immediately used. Selected soil variables measured included organic matter (OM, by dichromate oxidation), total nitrogen (TN, by total Kjeldahl N), NH_4^+–N, NO_3^-–N (by 1 M KCl extraction), pH and electrical conductivity (EC) (both at 5:1, water:soil ratio).

DNA extraction, amplification, sequencing and sequence treatment

For each soil sample, bulk DNA was extracted from three 0.5 g aliquots of soils (dry weight basis) with a FastDNA SPIN Kit for Soil (Bio 101, Carlsbad, Calif.) following the manufacturer's instructions. The triplicate DNA samples were pooled together, resulting 18 DNA samples in total, representing 18 field plots. The integrity of the extracted DNA was confirmed by running on 0.8% agarose gel with 0.5 TBE buffer (45 mM Tris-borate, 1 mM EDTA, pH 8.0). The extracted DNA (as well as PCR products) was quantified using TBS-380 Mini-Fluorometer (Turner Biosystems, CA, USA). All DNA samples were diluted to 100 ng μl^{-1} and used as PCR templates.

For the fungal community, detail description on primer pairs, amplification, sequencing and sequence treatment can be found elsewhere [31]. Operational taxonomic units (OTUs) were assigned at the 97% identity level.

For the bacterial community, the primer pair 563F (5′-AYTGGGYDTAAAGVG-3′) and 802R (5′-TACNVGGGTATCTAATCC-3′) [32] was used for amplifying the V4 hypervariable region of the 16 S rRNA gene. The PCR mixture (20 µl) contained 5×FastPfu buffer 4 µl, 2.5 mM dNTPs 2 µl, each primer(5 µM) 0.4 µl, DNA 0.5 µl, FastPfu polymerase 0.4 µl, and ddH_2O (TransGen Biotech, Beijing, China). DNA samples were amplified using the following program: 95°C 2 min, 30 cycles: 95°C 30 sec, 55°C 30 sec, 72°C 30 sec; 72°C 5 min; 10°C forever. A second round of PCR was performed under the same conditions using Amplicon Fusion Primers 5′-*A*-*x*-563F-3′ and 5′-*B*-802R-3′, where *A* and *B* represent the pyrosequencing adaptors (CCATCTCATCCCTGCGTGTCTCCGACGACT and CCTATCCCCTGTGTGCCTTGGCAGTCGACT) and *x* represents an 10 bp-tag for sample identification. For each sample, three independent PCRs were performed. The triplicate products were pooled and purified using AxyPrep PCR Clean-up Kit (Axygen Biosciences, CA, USA). The purified PCR products from different soil samples contained 1.06~3.30 ng μl^{-1} DNA and were pooled in equal quantities for pyrosequencing in one run, which was performed commercially at Shanghai Majorbio Bio-pharm Biotechnology Co., Ltd., on a Roche GS 454 FLX platform (http://www.majorbio.com).

Sequence reads were treated using MOTHUR v. 1.29.2 [33]. The forward primer (563F), tags and adaptor *A* and *B* were removed. Low quality nucleotides were filtered using qwindowaverage 27 and qwindowsize 50. Reads shorter than 120 bp were removed. Filtered sequences were aligned against Silva database (106 v.). Chimera reads were removed using both *chimera.slayer* and *chimera.uchime* scripts. A few reads belonging to *Archaea* were excluded. OTUs were assigned at the 97% identity level. On average, 11,950 original reads per sample were obtained, and after filtering 8,750 high-quality sequences per sample were left, of which 97.5% were 155 bp in length. The OTU counts were normalized into relative abundances (%).

Statistics was done with SPSS (v 19) and curve fitting with Sigmaplot (v. 12.5). Unless stated, p values <0.05 were considered significant. Raw sequence data for bacteria and fungi were submitted to DDBJ database under the accession number DRA001116 and DRA000962 respectively.

Results

Overall microbial communities

The overall bacterial communities are summarized in Figure 1. For comparison purposes, the summarized fungal community is also shown. Average relative abundances are listed in a column to the right of the pie chart. As for bacteria, *Acidobacteria*, *α-Proteobacteria* and *Gemmatimonadetes* were the top three abundant phyla with average relative abundances over 10%. As for fungi, "unidentified fungus", *Hypocreales* and *Sordariales* were the top three abundant taxa. All the fungal taxa in Figure 1A, except "unidentified fungus" and some "others", belong to phylum *Ascomycota*, which accounted for 58.2% in average relative abundance. *Basidiomycota* accounted for only 0.6%. The richness (S_{Chao1}) and diversity ($H_{Shannon}$) indices of the fungal community showed nearly flat lines over time, with slightly increasing trends by fitting to a linear model. Note that the p value of the fitted line for the fungal diversity index was not significant. In contrast, the richness and diversity indices of the bacterial community showed significant decreasing trends over time.

The change trends in relative abundances of these fungal taxa over 18 samples were shown in Figure S1. By curve fitting, the abundances of *Hypocreales, Pezizomycotina, Onygenales* and "fungi unclassified" showed significant increasing trends over time whereas *Mortierellales* and "*Ascomycota* incertae sedis" showed decreasing trends. By p values, *Hypocreales* was best fitted. *Hypocreales* and *Sordariales* were further broken down into genera and their relative abundances across soil samples were listed in Table S1. Within order *Hypocreales*, *Fusarium* was the top abundant genus with an average relative abundance 8%. All other genera were 1% or below in abundance. Within order *Sordariales*, *Chaetomium* and *Thielavia* were the top abundant genera with average relative abundances over 3%. All other genera in this order (except "unidentified fungus") were 1% or lower in abundance.

Yield trend and *Fusarium* abundance

Potato yield declined substantially after 2 years of monoculture, and was accompanied by an increase in disease incidence (Figure 2). Major diseases found in the field were fungi-associated, such as fusarium dry rots, late bright and black scurf/stem canker [31] that are caused by *Fusarium* spp., *Phytophthora infestans*, and *Rhizoctonia solani*, respectively. Because all these pathogens are soilborne, we further looked at how abundant they were in the soils. Because phytopathogens are often species-, even variety-specific, a pathogen in a strict taxonomic sense must be identified by pure culture, and can not be identified by environmental DNA-based approaches only (as in this present study). Therefore, we grouped the fungal abundances by genus that were detected in the soil samples, and checked them from a list of well-recognized potential soilborne potato pathogens [34,35] by genus. If found among the list, the genera detected were referred to as pathogen-associated genera. Obviously we are not saying all *Fusarium* are pathogens, but viewing the data in this way showed that *Fusarium* was the only pathogen-associated fungal genus substantially detected in our soil samples with a range of relative abundance from 1.9% to 14.1% (Table S2). Other pathogen-associated genera were either below the detection limit (e. g. *Phytophthora* and *Rhizoctonia*) or detected but with too low relative abundances (e. g. *Phoma* and *Pythium*). The relative abundance of *Fusarium* increased with the increase of monoculture time, with the increase of disease incidence, and with the decrease of yield (Figure 2).

Correlation between *Fusarium* abundance and soil variables

Using a linear model to fit soil variables with the year of monoculture, soil EC and pH were significantly increased over time, whereas OM, TN, and NH_4^+-N were decreased (Figure S2). NO_3^--N was also decreased but did not fit the model. Further fitting soil variables with *Fusarium* abundance, it was shown that

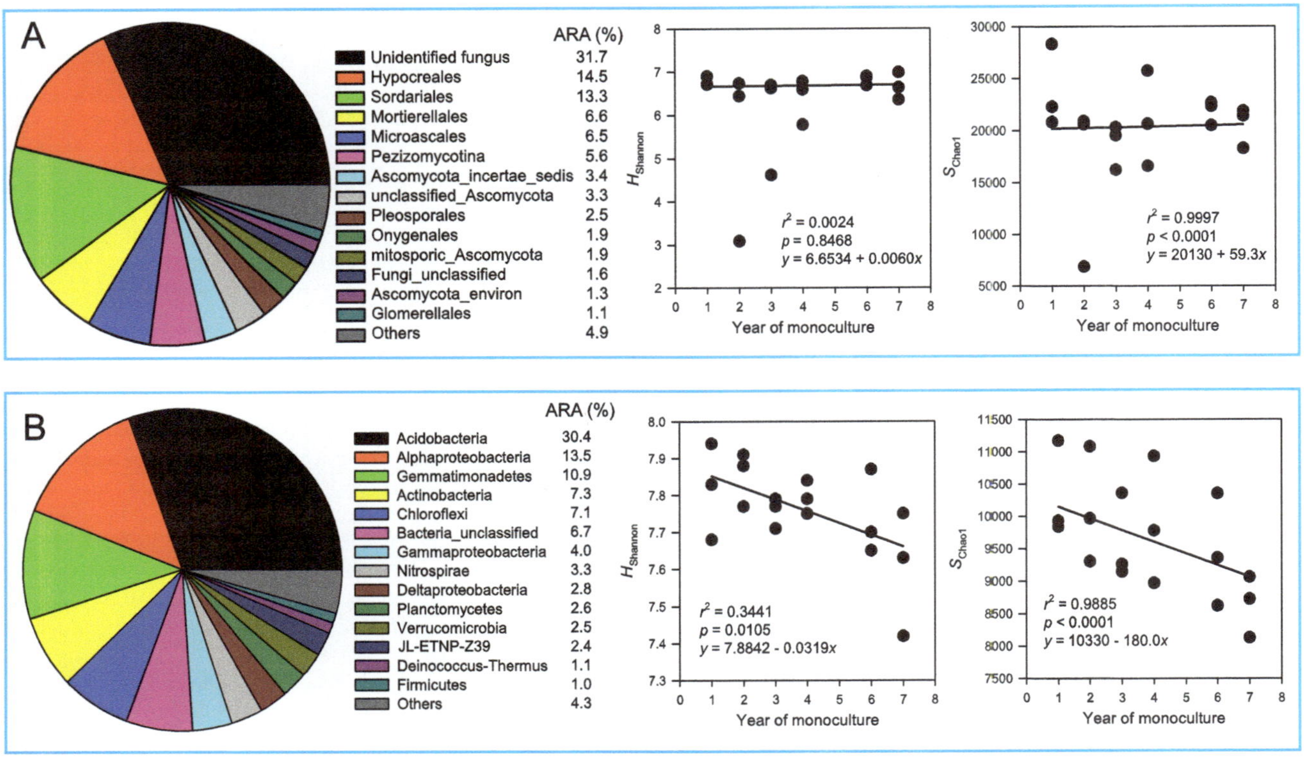

Figure 1. Overall fungal (A) and bacterial (B) communities in the soil samples along a year-gradient of potato monoculture. Pie charts show the average relative abundance over 18 samples. OTUs were assigned at the 97% identity level, based on which the $H_{Shannon}$ and S_{Chao1} indices were calculated, which are shown by graphs. Phylogenetic lineages of bacteria were assigned by phylum/subphylum and of fungi by order or higher levels. ARA stands for average relative abundance. The "unidentified fungus" is a taxon able to be assigned to kingdom Fungi but unable to any lower taxon.

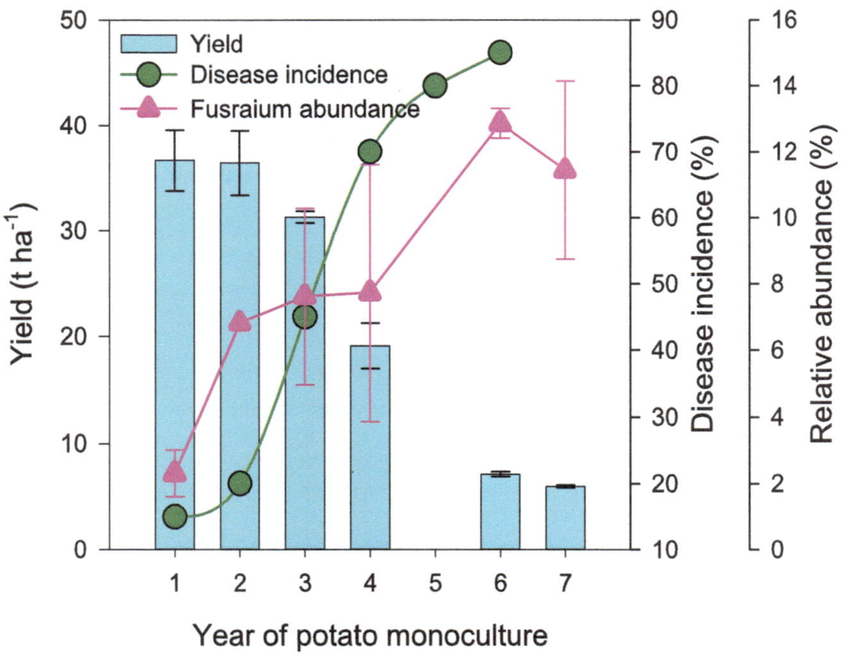

Figure 2. The change of potato yield, disease incidence and *Fusarium* abundance over the years of potato monoculture. Disease incidences shown in this figure were measured at harvest time.

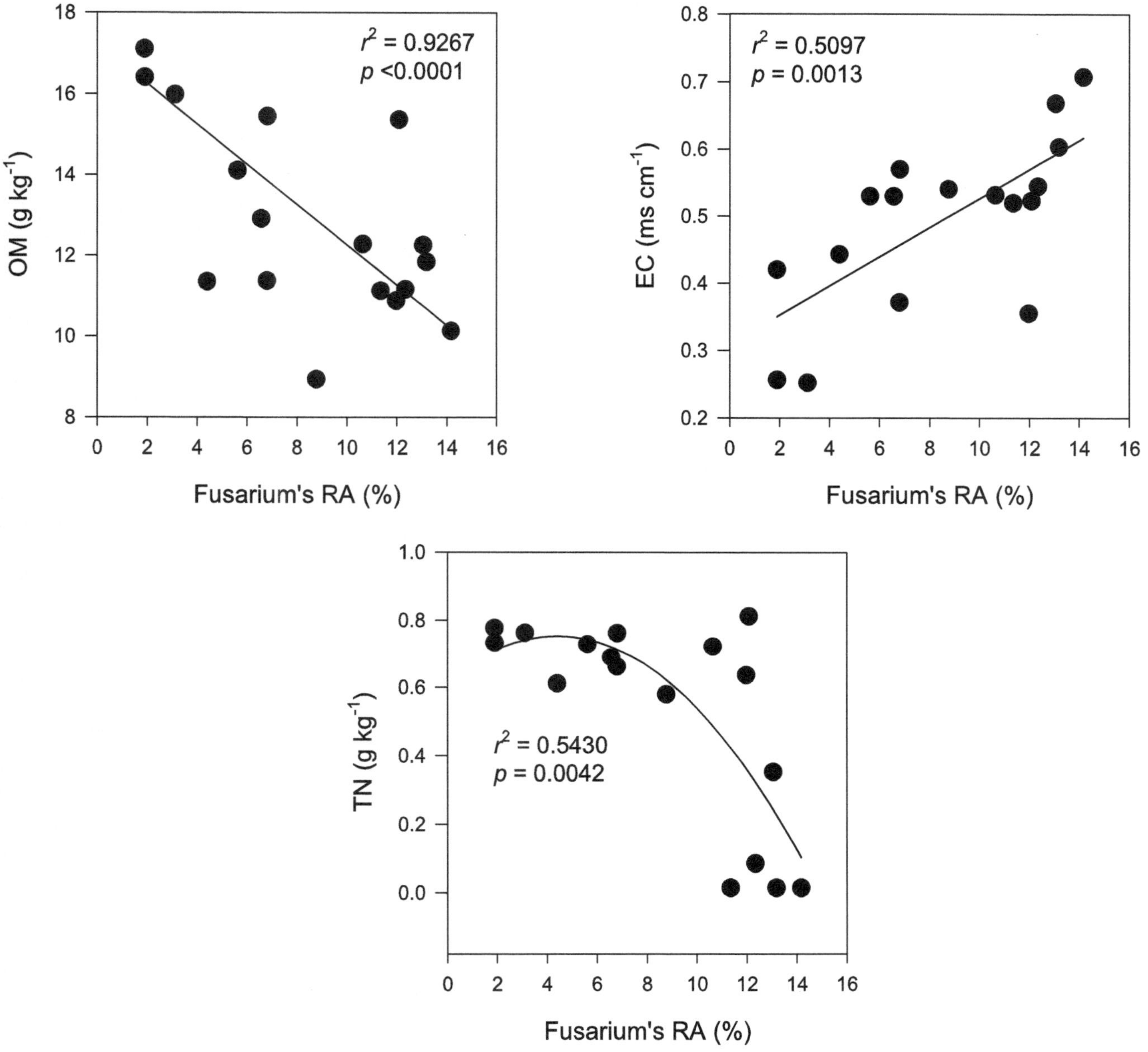

Figure 3. The correlations between *Fusarium* abundances and soil variables.

only OM, TN and EC were significantly correlated with the relative abundance of *Fusarium* (Figure 3). Other soil variables (NH_4^+-N and pH) were not significantly correlated, or not fitted with any of the models integrated in the software (NO_3^--N).

Correlation between abundance of bacterial taxa and the year of monoculture

To determine what bacterial taxa were significantly changed over time (the year of monoculture), correlations between taxa abundances and time were conducted. In total, seven taxa showed significant correlations (Table 1). The sum of average relative abundances of these taxa was 57.6%. *Acidobacteria* and *Nitrospirae* were linearly decreased in abundance (Figure 4), whereas *Gemmatimonadetes*, *Chloroflexi*, *Planctomycetes*, and JL-ETNP-Z39 were linearly increased. β-*Proteobacteria* was also decreased but was best fitted with a non-linear model.

Regression between *Fusarium* and other bacterial taxa in abundances

Regressions between *Fusarium* and the bacterial taxa abundances showed that among the seven taxa that significantly changed over time (Table 1), six showed significant regressions (Figure 5). *Acidobacteria* and *Nitrospirae* showed negative linear regressions whereas β-*Proteobacteria* showed negative non-linear regression. *Gemmatimonadetes*, *Chloroflexi* and JL-ETNP-Z39 showed positive linear regressions.

Discussion

The relative abundance rank profiles of fungal and bacterial phyla reported in this study generally agree with previous pyrosequencing surveys of soil microbial communities. As for bacteria, *Acidobacteria* and α-*Proteobacteria* were the top two abundant phyla across all samples (Figure 1), agreeing with the abundance rank profiles derived from 88 soils collected from North and South America [1], as well as from forest and grassland soils in Germany[2]. Acosta-Martínez et al [36] also reported that *Actinobacteria* and *Bacteriodetes* were the top two phyla in Texas (USA) soils with different managements. The average relative abundance of *Gemmatimonadetes* in our samples was 10.9% (Figure 1), which is much higher than the <3% reported in previous studies [1,2,36]. DeBruyn et al [37] reported that *Gemmatimonadetes* is a drier soil-adapted phylum, which is probably the reason why its abundance was so high in our soils. As for fungi, *Ascomycota* was the top abundant phylum in our samples, accounting for 58.2% in average relative abundance, generally agreeing with the results reported by Buée et al [38] and by McGuire et al [39] for forest soils. However, our results contrasted with those derieved from tallgrass prairie soils [40], in which basidiomycetes (*Basidiomycota*) accounted for 54% relative abundance and ascomycetes (*Ascomycota*) accounted for 41%. In our samples, *Basidiomycota* accounted for 0.6% only. The difference was probably due to different soils in nature. Different primers targeting different genes/regions and different databases used for phylogentical assignment also affected the results.

Using the soil samples along a year-gradient of potato monoculture, several facts about potato monoculture and soil microbial communities were revealed. First, Pearson, Kendall, and Spearman correlation methods were used simultaneously in order to catch all possible significant correlations between bacterial taxa and the years of monoculture. Different methods have different merits. Pearson method measures the degree of linear dependence between two variables, Kendall measures the similarity of the orderings of the variables when ranked by each of the quantities (rank correlation), and Spearman measures how well the relationship between two variables can be described using a monotonic function. Generally speaking, Pearson measures linear relationships, whereas Kendall and Spearman measure nonlinear relationships. Results showed that the majority of the bacterial community (57.6% by abundance) was significantly changed over time (Table 1), including the top abundant taxon *Acidobacteria*. Bacterial diversity and richness indices were significantly and linearly decreased with an increase of monoculture time (year), but the corresponding values for fungal community was nearly unaffected (Figure 1). This suggests differential responses of bacterial and fungal communities to potato monoculture. Using fatty acid methyl ester profiles to characterize soil microbial communities, several other studies [41,42,43,44,45] also reported differential responses of fungal and bacterial communities to potato monoculture. Larkin and Honeycutt [42], for example, reported that continuous potato resulted in lower proportions of polyunsaturated fatty acids, which are most closely associated with fungi, resulting in a lower fungi-to-bacteria ratio than several other rotations. By substrate utilization approach, Larkin et al [44] reported that potato monoculture produced the lowest values in soil microbial activity, richness and diversity among other rotation treatments. Although these studies did not use high throughput sequencing approach as present study did, similar conclusion was drawn, suggesting that the differential responses of fungal and bacterial communities were common in potato monoculture agroecosystems. However, whether differential responses exist in other crop monoculture agroecosystems is not clear. A few studies have reported decreased bacterial diversities in monoculture ecosystems of rice [21], cotton [36], soybean [22] and watermelon [46]. Unfortunately, these studies did not include fungal communities, or they were included but such differential responses were not looked for. Nevertheless, our results point out that the link between microbial diversity and soil quality/sustainability depends on the focus of the microbial community, i.e. bacterial or fungal. In the present case, the soil sustainability decline represented by a potato yield decline and disease incidence increase (Figure 2) corresponded to a decrease in bacterial, but not fungal, diversity.

Second, *Fusarium* was the only soilborne pathogen-associated fungal genus substantially detected by our present high-throughput pyrosequencing protocol (Table S1 and Table S2), and was continuously and significantly enriched (Figure 2). The change pattern of *Fusarium* was closely associated with disease incidence and yield decline, suggesting its reponsibility for yield decline. *Fusarium* enrichment in soils is reportedly related to disease resistance of cultivars used [46,47]. The cultivar "Atlantic" used in the present experiment was reportedly susceptible to *Fusarium roseum* [48]. Root exudates and debris left in soils from previous crops, especially a monoculture potato crop that was sampled in this study, would likely favor *Fusarium* proliferation and enrichment in soils.

Third, *Fusarium* enrichment over time was correlated with soil variables. Among the six selected variables, OM and TN were significantly and negatively correlated with *Fusarium* abundances, and EC was significantly and positively correlated (Figure 3). These three correlated soil variables are all associated with field management. The site of the experimental plots for this study was located in an arid region where the mean annual precipitation is only 186 mm. Cations introduced from fertilizers are not easily leached out of the soils, leading to an increase of EC values (Figure S2). Arid soils are also water limited. Irrigation improves the soil water regime, promotes microbial activity, thus enhancing decomposition of soil OM [49]. Without organic amendments to compensate OM loss (as in the present experiment), soil OM content will be inevitably decreased (Figure S2), instead of being preserved by dryness. TN will be decreased as well because it is mostly associated with soil OM. Therefore, soil management causes change in the soil variables of OM, TN and EC, thus causing *Fusarium* enrichment in soils.

Fourth, among the seven taxa that were significantly changed over time under continued potato monoculture pressure (Table 1), *Acidobacteria* and *Nitrospirae* were linearly decreased over time in abundance (Figure 4). This is probably relevant to their ecophysiologies. As for *Acidobacteria*, it has been shown that the abundance of *Acidobacteria* was negative correlated with soil carbon mineralization rates [50]. Although the carbon mineralization rate was not measured in this study, it was reasonable to believe that microbial activity and carbon mineralization rates in the arid soils

Table 1. Correlation coefficient between abundance of microbial taxa and the year of monoculture[a].

Taxa	Average RA (%)	Pearson	Kendall τ_b	Spearman ρ
Acidobacteria	30.4	−0.550*	−0.424*	−0.561*
Gemmatimonadetes	10.9		0.369*	0.486*
Chloroflexi	7.1	0.524*	0.466**	0.649**
Nitrospirae	3.3	−0.554*	−0.452*	−0.561*
Planctomycetes	2.6	0.618**	0.424*	0.567*
JL-ETNP-Z39	2.4	0.706**	0.522**	0.680**
β-Proteobacteria	0.9	−0.500*	−0.369*	−0.505*

[a]RA is relative abundances. * and ** represent significance level at <0.05 and < 0.01 respectively. The taxa with average relative abundance over 18 samples < 0.5% and with significance level >0.05 were not shown.

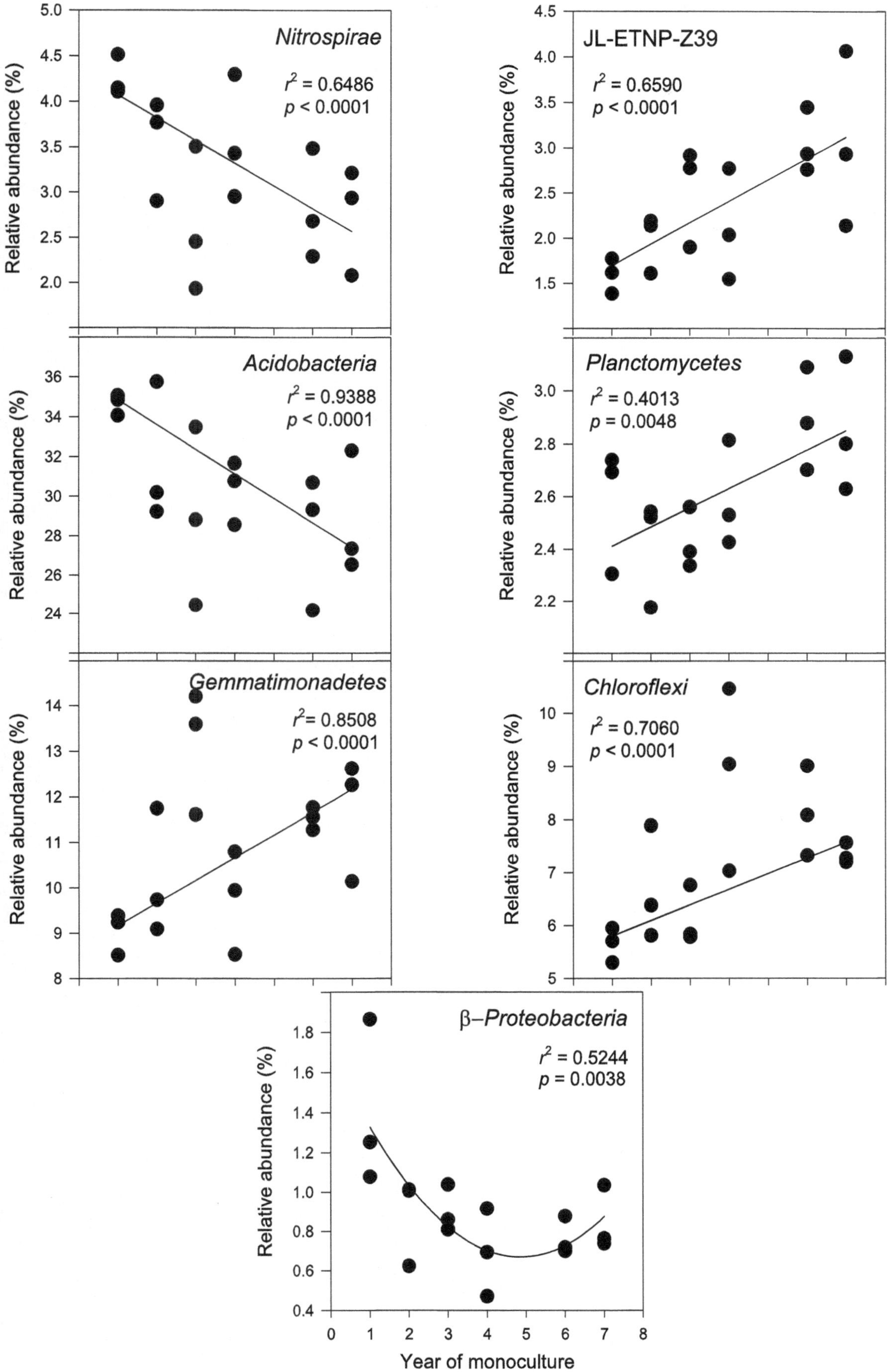

Figure 4. Regressions between bacterial taxa abundances and the year of potato monoculture.

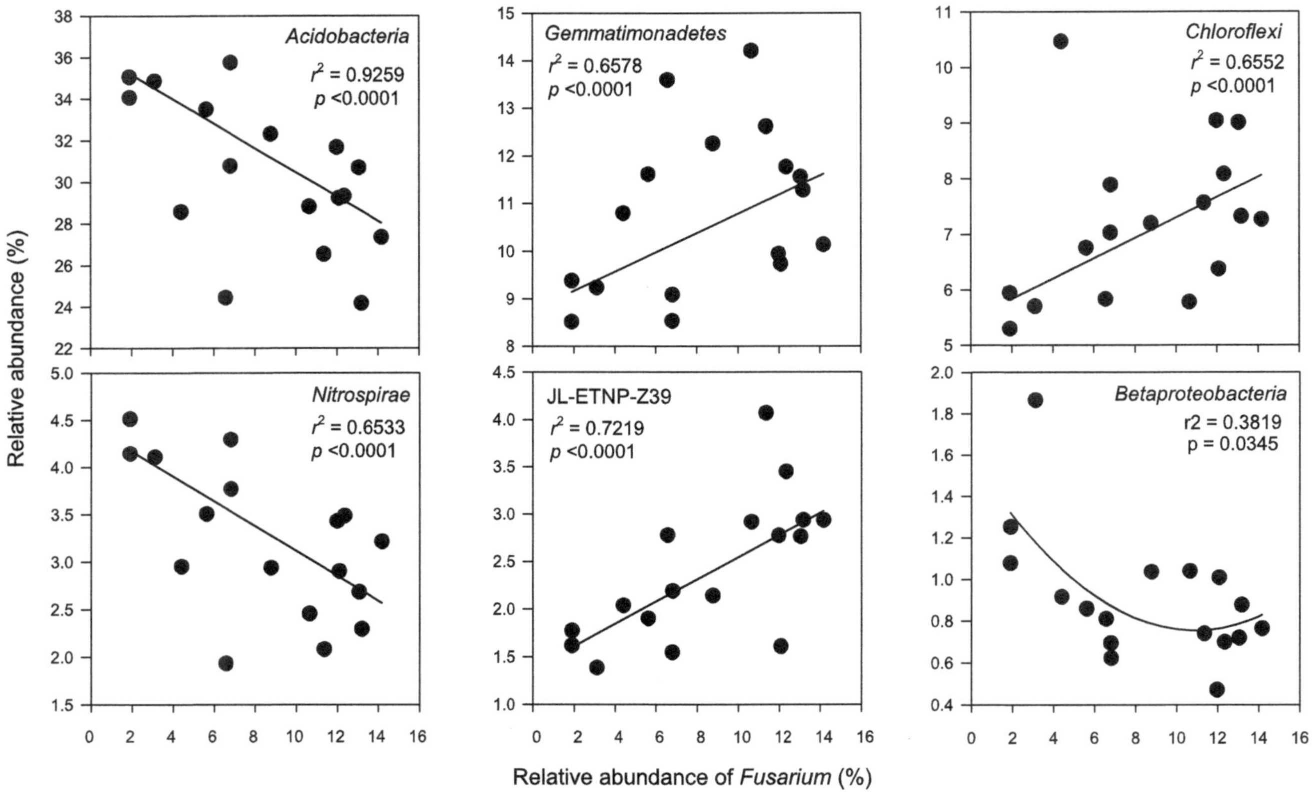

Figure 5. Regressions between relative abundance of *Fusarium* and bacterial taxa. Regressions with *p* values >0.05 are not shown.

were increased due to the improved water regime by irrigation, which consequently led to the decrease of soil OM contents (Figure S2). Therefore our data indirectly supported the ecological classification of *Acidobacteria* as an oligotrophic phylum. As for *Nitrospirae*, its ecophysiology is not clear. However, there have been evidences indicating that it tended to adapt to substrate-limited and well-aerated conditions [51,52], implying that the ecophysiology of *Nitrospirae* was similar to that of *Acidobacteria*. Furthermore, *Nitrospirae* is a phylum most likely involved in nitrite oxidation [51,52], although phylogenetic heterogeneity implies it has diverse functions. Nevertheless, the decrease of *Nitrospirae* abundance suggested negative effects on the process of soil nitrification.

Fifth, six bacterial taxa were significantly correlated with the abundance of *Fusarium* (Figure 5). Positive correlations were found between *Fusarium* and *Gemmatimonadetes*, *Chloroflexi* and JL-ETNP-Z39. Negative correlations occurred between *Fusarium* and *Acidobacteria*, *Nitrospirae* and β-*Proteobacteria*. The sum of average relative abundances of these correlated bacterial taxa was 55% (Figure 5), suggesting that potato monoculture not only enriched *Fusarium* but also shaped the overall soil bacterial communities by enriching and impoverishing certain bacterial taxa. Furthermore, the data also suggested these taxa exhibited similar behaviors in responses to potato monoculture and/or soil variables, providing additional insights into the ecological behaviors of different taxa in the environments.

Conclusion

Fungal and bacterial communities exhibited different responses to potato monoculture. Fungal diversity was nearly unaffected, whereas bacterial diversity was significantly decreased over time (years). Seven bacterial taxa were significantly changed, including the top abundant taxa *Acidobacteria* and *Gemmatimonadetes*. *Fusarium* was the only soilborne pathogen-associated fungal genus substantially detected by our high throughput pyrosequencing approach. *Fusarium* became enriched over the experimental time span, and was closely associated with potato yield decline and disease increases. *Fusarium* abundance was also significantly and negatively correlated with soil OM and TN, but significantly and positively with EC. *Fusarium* was positively correlated with abundances of *Gemmatimonadetes*, *Chloroflexi*, and JL-ETNP-Z39, and negatively correlated with *Acidobacteria*, *Nitrospirae*, and β-*Proteobacteria* abundances. These correlations suggested their similar behaviors in responses to potato monoculture and/or soil variables, providing further insights into the effects of potato monoculture on soil microbial communities.

Supporting Information

Figure S1 The changing trends in abundances of fungal taxa over time. Phylogenetic lineages were assigned by order or higher levels.

Figure S2 Regression between soil variable and the year of monoculture.

Table S1 Relative abundance of genus within the order Hypocreales and Sodariales.

Table S2 The relative abundance of pathogen-associated fungal genus detected in the soil samples.

Acknowledgments

Authors are grateful for Professor Warren A. Dick, School of Natural Resources, The Ohio State University, for his comments and language editing, and Professor Yuhua Shan, College of Environmental Science and Engineering, Yangzhou University, for his professional consultancies on statistical analyses. Many graduate students and staff members who were involved in maintaining the field plots, collecting field data, and collecting soil samples, but are not listed as co-authors are grateful. Authors would like to thank three anonymous reviewers for their comments, which significantly improve the manuscript.

Author Contributions

Conceived and designed the experiments: QRS HZQ. Performed the experiments: XL WMZ JLZ. Analyzed the data: SXY TYG. Contributed reagents/materials/analysis tools: JLZ. Wrote the paper: SXY. Pyrosequencing data and data analyses: SXY.

References

1. Lauber CL, Hamady M, Knight R, Fierer N (2009) Pyrosequencing-based assessment of soil pH as a predictor of soil bacterial community structure at the continental scale. Appl Environ Microbiol 75: 5111–5120.
2. Nacke H, Thurmer A, Wollherr A, Will C, Hodac L, et al. (2011) Pyrosequencing-based assessment of bacterial community structure along different management types in German forest and grassland soils. PLoS One 6: e17000.
3. Girvan MS, Bullimore J, Pretty JN, Osborn AM, Ball AS (2003) Soil type is the primary determinant of the composition of the total and active bacterial communities in arable soils. Appl Environ Microbiol 69: 1800–1809.
4. Sessitsch A, Weilharter A, Gerzabek MH, Kirchmann H, Kandeler E (2001) Microbial population structures in soil particle size fractions of a long-term fertilizer field experiment. Appl Environ Microbiol 67: 4215–4224.
5. Will C, Thurmer A, Wollherr A, Nacke H, Herold N, et al. (2010) Horizon-specific bacterial community composition of German grassland soils, as revealed by pyrosequencing-based analysis of 16S rRNA genes. Appl Environ Microbiol 76: 6751–6759.
6. Garbeva P, Postma J, van Veen JA, van Elsas JD (2006) Effect of above-ground plant species on soil microbial community structure and its impact on suppression of Rhizoctonia solani AG3. Environ Microbiol 8: 233–246.
7. Fierer N, Lauber CL, Ramirez KS, Zaneveld J, Bradford MA, et al. (2012) Comparative metagenomic, phylogenetic and physiological analyses of soil microbial communities across nitrogen gradients. ISME J 6: 1007–1017.
8. Norris TB, Wraith JM, Castenholz RW, McDermott TR (2002) Soil microbial community structure across a thermal gradient following a geothermal heating event. Appl Environ Microbiol 68: 6300–6309.
9. Brockett BFT, Prescott CE, Grayston SJ (2012) Soil moisture is the major factor influencing microbial community structure and enzyme activities across seven biogeoclimatic zones in western Canada. Soil Biol Biochem 44: 9–20.
10. Jansa J, Mozafar A, Kuhn G, Anken T, Ruh R, et al. (2003) Soil tillage affects the community structure of mycorrhizal fungi in maize roots. Ecol Appl 13: 1164–1176.
11. Yin C, Jones KL, Peterson DE, Garrett KA, Hulbert SH, et al. (2010) Members of soil bacterial communities sensitive to tillage and crop rotation. Soil Biol Biochem 42: 2111–2118.
12. Stark C, Condron LM, Stewart A, Di HJ, O'Callaghan M (2007) Influence of organic and mineral amendments on microbial soil properties and processes. Appl Soil Ecol 35: 79–93.
13. Luo J, Ran W, Hu J, Yang X, Xu Y, et al. (2010) Application of Bio-organic fertilizer significantly affected fungal diversity of soils. Soil Sci Soc Am J 74: 2039–2048.
14. Ederson CJ, Marsh TL, Tiedje JM, de S Moreira FM (2009) Changes in land use alter the structure of bacterial communities in Western Amazon soils. ISME J 3: 1004–1011.
15. Lee J, Lee T, Löffler F, Park J (2011) Characterization of microbial community structure and population dynamics of tetrachloroethene-dechlorinating tidal mudflat communities. Biodegradation 22: 687–698.
16. Garbeva P, van Veen JA, van Elsas JD (2004) Microbial diversity in soils: selection of microbial populations by plant and soil type and implications for disease suppressiveness. Annu Rev Phytopathol 42: 243–270.
17. Cook RJ (2006) Toward cropping systems that enhance productivity and sustainability. P Natl Acad Sci USA 103: 18389–18394.
18. Acosta-Martínez V, Dowd SE, Bell CW, Lascano R, Booker JD, et al. (2010) Microbial community composition as affected by dryland cropping systems and tillage in a semiarid sandy soil. Diversity 2: 910–931.
19. Ceja-Navarro J, Rivera F, Patiño-Zúñiga L, Govaerts B, Marsch R, et al. (2010) Molecular characterization of soil bacterial communities in contrasting zero tillage systems. Plant Soil 329: 127–137.
20. Govaerts B, Mezzalama M, Unno Y, Sayre KD, Luna-Guido M, et al. (2007) Influence of tillage, residue management, and crop rotation on soil microbial biomass and catabolic diversity. Appl Soil Ecol 37: 18–30.
21. Xuan D, Guong V, Rosling A, Alström S, Chai B, et al. (2012) Different crop rotation systems as drivers of change in soil bacterial community structure and yield of rice, *Oryza sativa*. Biol Fert Soils 48: 217–225.
22. Li C, Li X, Kong W, Wu Y, Wang J (2010) Effect of monoculture soybean on soil microbial community in the Northeast China. Plant Soil 330: 423–433.
23. Hilton S, Bennett AJ, Keane G, Bending GD, Chandler D, et al. (2013) Impact of shortened crop rotation of oilseed rape on soil and rhizosphere microbial diversity in relation to yield decline. PLoS One 8: e59859.
24. Larkin RP (2003) Characterization of soil microbial communities under different potato cropping systems by microbial population dynamics, substrate utilization, and fatty acid profiles. Soil Biol Biochem 35: 1451–1466.
25. Pereira e Silva MC, Dias AC, van Elsas JD, Salles JF (2012) Spatial and temporal variation of archaeal, bacterial and fungal communities in agricultural soils. PLoS One 7: e51554.
26. Gobet A, Boer SI, Huse SM, van Beusekom JE, Quince C, et al. (2012) Diversity and dynamics of rare and of resident bacterial populations in coastal sands. ISME J 6: 542–553.
27. Pereira e Silva MC, Poly F, Guillaumaud N, van Elsas JD, Falcão Salles J (2012) Fluctuations in ammonia oxidizing communities across agricultural soils are driven by soil structure and pH. Front Microbiol 3: 77.
28. Bruggen AC, Semenov A, Diepeningen A, Vos O, Blok W (2006) Relation between soil health, wave-like fluctuations in microbial populations,and soil-borne plant disease management. Plant disease epidemiology: facing challenges of the 21st Century: Springer Netherlands. pp. 105–122.
29. Shipton PJ (1977) Monoculture and soilborne plant pathogens. Annu Rev Phytopathol 15: 387–407.
30. Amann RI, Ludwig W, Schleifer KH (1995) Phylogenetic identification and in situ detection of individual microbial cells without cultivation. Microbiol Rev 59: 143–169.
31. Lu L, Yin S, Liu X, Zhang W, Gu T, et al. (2013) Fungal networks in yield-invigorating and -debilitating soils induced by prolonged potato monoculture. Soil Biol Biochem 65: 186–194.
32. Cardenas E, Wu WM, Leigh MB, Carley J, Carroll S, et al. (2010) Significant association between sulfate-reducing bacteria and uranium-reducing microbial communities as revealed by a combined massively parallel sequencing-indicator species approach. Appl Environ Microbiol 76: 6778–6786.
33. Schloss PD, Westcott SL, Ryabin T, Hall JR, Hartmann M, et al. (2009) Introducing mothur: open-source, platform-independent, community-supported software for describing and comparing microbial communities. Appl Environ Microbiol 75: 7537–7541.
34. Fiers M, Edel-Hermann V, Chatot C, Le Hingrat Y, Alabouvette C, et al. (2012) Potato soil-borne diseases. A review. Agron Sustain Dev 32: 93–132.
35. Jeger M, Hide G, Van Den Boogert P, Termorshuizen A, Van Baarlen P (1996) Pathology and control of soil-borne fungal pathogens of potato. Potato Res 39: 437–469.
36. Acosta-Martínez V, Dowd S, Sun Y, Allen V (2008) Tag-encoded pyrosequencing analysis of bacterial diversity in a single soil type as affected by management and land use. Soil Biol Biochem 40: 2762–2770.
37. DeBruyn JM, Nixon LT, Fawaz MN, Johnson AM, Radosevich M (2011) Global biogeography and quantitative seasonal dynamics of *Gemmatimonadetes* in soil. Appl Environ Microbiol 77: 6295–6300.
38. Buée M, Reich M, Murat C, Morin E, Nilsson RH, et al. (2009) 454 Pyrosequencing analyses of forest soils reveal an unexpectedly high fungal diversity. New Phytol 184: 449–456.
39. McGuire K, Fierer N, Bateman C, Treseder K, Turner B (2012) Fungal Community Composition in Neotropical Rain Forests: the influence of tree diversity and precipitation. Microbial Ecol 63: 804–812.
40. Jumpponen A, Jones KL, Blair J (2010) Vertical distribution of fungal communities in tallgrass prairie soil. Mycologia 102: 1027–1041.
41. Larkin RP (2008) Relative effects of biological amendments and crop rotations on soil microbial communities and soilborne diseases of potato. Soil Biol Biochem 40: 1341–1351.
42. Larkin RP, Honeycutt CW (2006) Effects of different 3-year cropping systems on soil microbial communities and rhizoctonia diseases of potato. Phytopathol 96: 68–79.
43. Larkin RP, Griffin TS, Honeycutt CW (2010) Rotation and cover crop effects on soilborne potato diseases, tuber yield, and soil microbial communities. Plant Dis 94: 1491–1502.
44. Larkin RP, Honeycutt CW, Griffin TS, Olanya OM, Halloran JM, et al. (2011) Effects of different potato cropping system approaches and water management on soilborne diseases and soil microbial communities. Phytopathol 101: 58–67.
45. Bernard E, Larkin RP, Tavantzis S, Erich MS, Alyokhin A, et al. (2012) Compost, rapeseed rotation, and biocontrol agents significantly impact soil microbial communities in organic and conventional potato production systems. Appl Soil Ecol 52: 29–41.
46. Hopkins DL, Lobinske RJ, Larkin RP (1992) Selection for *Fusarium oxysporum* f.sp. niveum race 2 in monocultures of watermelon cultivars resistant to *Fusarium* wilt. Phytopathol 82: 290–293.

47. Hopkins DL, Elmstrom GW (1984) Fusarium wilt in watermelon cultivars grown in a 4-year monoculture. Plant Dis 68: 129–131.
48. Leach SS, Webb RE (1981) Resistance of selected potato cultivars and clones to *Fusarium* dry rot. Phytopathol 71: 623–629.
49. Arroita M, Causapé J, Comín FA, Díez J, Jimenez JJ, et al. (2013) Irrigation agriculture affects organic matter decomposition in semi-arid terrestrial and aquatic ecosystems. J Hazard Mater 263, Part 1: 139–145.
50. Fierer N, Bradford MA, Jackson RB (2007) Toward an ecological classification of soil bacteria. Ecology 88: 1354–1364.
51. Lücker S, Wagner M, Maixner F, Pelletier E, Koch H, et al. (2010) A *Nitrospira* metagenome illuminates the physiology and evolution of globally important nitrite-oxidizing bacteria. P Natl Acad Sci USA 107: 13479–13484.
52. Schramm A, de Beer D, van den Heuvel JC, Ottengraf S, Amann R (1999) Microscale distribution of populations and activities of *Nitrosospira* and *Nitrospira* spp. along a macroscale gradient in a nitrifying bioreactor: quantification by in situ hybridization and the use of microsensors. Appl Environ Microbiol 65: 3690–3696.

13

Identification of Factors Influencing the Restoration of Cyanobacteria-Dominated Biological Soil Crusts

Chongfeng Bu[1,2]*, Shufang Wu[1,3], Yongsheng Yang[2], Mingguo Zheng[4]

1 Institute of Soil and Water Conservation, Northwest A & F University, Yangling, Shaanxi, China, **2** Institute of Soil and Water Conservation, Chinese Academy of Sciences and Ministry of Water Resources, Yangling, Shaanxi, China, **3** College of Water Resources and Architectural Engineering, Northwest A & F University, Yangling, Shaanxi, China, **4** Institute of Geographic Sciences and Natural Resources Research, Chinese Academy of Sciences, Beijing, China

Abstract

Biological soil crusts (BSCs) cover >35% of the Earth's land area and contribute to important ecological functions in arid and semiarid ecosystems, including erosion reduction, hydrological cycling, and nutrient cycling. Artificial rapid cultivation of BSCs can provide a novel alternative to traditional biological methods for controlling soil and water loss such as the planting of trees, shrubs, and grasses. At present, little is known regarding the cultivation of BSCs in the field due to lack of knowledge regarding the influencing factors that control BSCs growth. Thus, we determined the effects of various environmental factors (shade; watering; N, P, K, and Ca concentrations) on the growth of cyanobacteria-dominated BSCs from the Sonoran Desert in the southwestern United States. The soil surface changes and chlorophyll *a* concentrations were used as proxies of BSC growth and development. After 4 months, five factors were found to impact BSC growth with the following order of importance: $NH_4NO_3 \approx$ watering frequency>shading>$CaCO_3 \approx KH_2PO_4$. The soil water content was the primary positive factor affecting BSC growth, and BSCs that were watered every 5 days harbored greater biomass than those watered every 10 days. Groups that received NH_4NO_3 consistently exhibited poor growth, suggesting that fixed N amendment may suppress BSC growth. The effect of shading on the BSC biomass was inconsistent and depended on many factors including the soil water content and availability of nutrients. KH_2PO_4 and $CaCO_3$ had nonsignificant effects on BSC growth. Collectively, our results indicate that the rapid restoration of BSCs can be controlled and realized by artificial "broadcasting" cultivation through the optimization of environmental factors.

Editor: Brett Neilan, University of New South Wales, Australia

Funding: The National Natural Science Foundation of China (Grant Nos 41071192 & 40701096), the West Light Foundation of the Chinese Academy of Sciences (No. B2008132), and the 111 project (No. B12007) provided funding for this research. The funders had no role in study design, data collection and analysis, decision to publish, or preparation of the manuscript.

Competing Interests: The authors have declared that no competing interests exist.

* E-mail: buchongfeng@163.com

Introduction

Biological soil crusts (BSCs) are highly complex communities composed of mosses, cyanobacteria, lichens, bacteria, and fungi with soil particles [1,2]. BSCs, which are widely distributed in arid and semiarid areas of hot and cold zones, impact biogeochemical cycles and hydrological processes and function to prevent erosion [3,4,5,6]. BSCs are known to enrich soil nutrients, enhance soil stability, and reduce soil erosion by wind and water [7,8,9]. For example, in the Loess Plateau, moss-dominated BSC were shown to reduce water runoff and sediment loss from soils by >30% and >80%, respectively [10]. In a separate wind tunnel experiment, moss- and cyanobacteria-dominated BSCs reduced the wind erosion rate of soils by more than 90% as compared to uncovered desert soils [11]. Therefore, attempts have been made to establish artificial BSCs to promote soil stability and to reduce losses due to wind and water erosion. Wei [12] proposed the concept of BSC carpet engineering, i.e., the creation of artificial BSCs to control sand movement. Subsequent research on BSC inoculation has progressed rapidly, especially regarding the use of artificial cyanobacteria-dominated BSC cultures. For instance, in a 200-ha experimental soil plot established in the Hobq Desert [13,14,15], an artificial cyanobacterial crust enhanced the activity of soil communities and accelerated soil development [16]. Moreover, the average thickness of this artificial BSC ranged from 2.23 to 5.36 mm, with coverage of 70%, after 3 years of growth, suggesting that this approach may have long-term positive effects on soil stability [17].

BSCs are generally classified into moss, lichen, and cyanobacterial crusts based on their predominant compositions. Mosses with stems, leaves, and rhizoids are considered the higher plant and have relatively strong photosynthetic capacities. Cyanobacteria, which usually act as the pioneer organisms for desert soils, are considered the lower plant characterized by low photosynthesis. Lichens are complexes of algal cells enveloped within epiphyte mycelia and can endure extremely arid environments [18]. These three organisms are characterized by different physiological properties and require multifarious environments and conditions for rapid growth, which results in the complexity of artificial restoration for each type of BSC. The influences affecting the development of BSCs can be divided into internal and external factors. Internal factors include the physiological characteristics of the species present in the BSCs and the interactions between these species, whereas external factors involve soil moisture, light intensity, temperature, and the availability of nutrients. A previous study [19] showed that the production of biomass and exopolysaccharides

(EPS) by *Microcoleus vaginatus* Gom. was not linearly related to temperature, light, or the availability of nutrients. Li et al. [20] found that the colonization of cyanobacteria and algae in the earliest successional stages of BSCs was facilitated by a higher soil pH and total potassium content in the topsoil. Recent research [21] showed that BSCs amended with composted sewage sludge promoted N and C fixation and increased chlorophyll *a*. However, the frequent watering of these BSCs led to a lower diversity of cyanobacteria populations. In addition, the organic matter content, soil water content, total N, total P, bioavailable P, bioavailable K, pH, electrical conductivity, and total salt content also significantly affected the abundance of microorganisms in BSCs [22]. External factors, such as those described above, are likely to significantly influence key ecological processes such as nitrogen fixation. *Collema tenax* is the dominant N-fixing lichen in many BSCs, and the gross photosynthesis of this species was surprisingly found to be suppressed by the addition of nutrients such as P, K, and Zn [8]. It remains unclear, however, how nutrient amendment affects biogeochemical processes in the long term and how this may influence BSC stability [23].

For cyanobacterial BSCs, knowledge on the rapid cultivation through the direct broadcasting of crust samples remains scarce. The experimental study of multiple factors will be beneficial to realizing the rapid restoration of cyanobacterial crusts in the field. Based on crust sampling in the field, lab cultivation, and the investigation of biomass and surface changes, the present study examined the effects of light intensity, watering frequency, and nutrient amendment of N, P, K, and Ca on the development of cyanobacteria-dominated BSCs located in the Sonoran Desert in the southwestern United States. The results of this work improve our understanding of the factors that affect BSC development, and this information may narrow gaps between application and research to restore and engineer BSCs in arid and semiarid regions.

Materials and Methods

Soil Sampling and Preparation

Samples were collected from a plot regulated by Arizona State University and located in the Sonoran Desert near Phoenix, Arizona, USA (32°59′27″N, 111°45′38″W). Cyanobacterial crusts were collected from the soil surface (<10 mm depth), and the underlying soil was sampled from a depth >10 mm as the substrate. The crust sample and soil substrate were transported to the laboratory and sifted using a 5-mm sieve; the materials were then air dried and kept for use in subsequent experiments. This study was approved by the Department of Life Sciences, Arizona State University.

The crust sample was dominated by cyanobacteria, and no moss or lichen was observed. The dominant cyanobacteria at the end of the experiment was identified via microscopy as *Microcoleus vaginatus* Gom.

Properties of the Crust and Soil Substrate

The initial chlorophyll *a* concentration of the BSC sample, determined using an acetone extraction method [24], was 1.28±0.08 μg/cm^2, a value much lower than that of other well-developed cyanobacterial crusts (typically 10 to 50 μg/cm^2, [25,26]). The physical and chemical soil properties of both the crust and substrate samples were determined using standard methods [27], and the results are shown in Table 1.

Table 1. Selected physical and chemical properties of the soils of the study.

Sample	Particle size distribution			Total N (g kg^{-1})	Total P (g kg^{-1})	Total K (g kg^{-1})	$CaCO_3$ (g kg^{-1})	TOC (%)	pH
	Sand (g kg^{-1})	Silt (g kg^{-1})	Clay (g kg^{-1})						
Crust	60.1	34.4	5.5	0.60	0.61	0.37	1.2	0.44	7.44
Substrate	72.5	23.3	4.2	0.20	0.44	0.30	1.2	0.16	7.72

Table 2. Treatments used in the cyanobacteria-dominated crust cultivation experiments.

Treatment	W10															
	S								NS							
	N				NN				N				NN			
	KP		NKP		KP		NKP		KP		NKP		KP		NKP	
	Ca	NCa	Ca	NCa	Ca	NCa	Ca	NCa	Ca	NCa	Ca	NCa	Ca	NCa	Ca	NCa
Treatment No.	1	2	3	4	5	6	7	8	9	10	11	12	13	14	15	16
Combination	W10+S+N+KP+Ca	W10+S+N+KP	W10+S+N+Ca	W10+S+N	W10+S+KP+Ca	W10+S+KP	W10+S+Ca	W10+S	W10+N+KP+Ca	W10+ N+KP	1W10+N+Ca	W10+N	W10+KP+Ca	W10+KP	W10+Ca	W10

Treatment	W5															
	S								NS							
	N				NN				N				NN			
	KP		NKP		KP		NKP		KP		NKP		KP		NKP	
	Ca	NCa	Ca	NCa	Ca	NCa	Ca	NCa	Ca	NCa	Ca	NCa	Ca	NCa	Ca	NCa
Treatment No.	17	18	19	20	21	22	23	24	25	26	27	28	29	30	31	32
Combination	W5+S−N+KP+Ca	W5+S+N+KP	W5+S+N+Ca	W5+S+N	W5+S+KP+Ca	W5+S+KP	W5+S+Ca	W5+S	W5+N+KP+Ca	W5+N+KP	W5+N+Ca	W5+N	W5+KP+Ca	W5+KP	W5+Ca	W5

Note: W10:300 ml deionized water was added to each sample every 10 days. W5:300 ml deionized water was added to each sample every 5 days. S: each sample was covered by a 60% shade cloth at 20 cm above the pot. NS: no shading. N: 2.10 g NH_4NO_3 added to each pot. NN: no NH_4NO_3 added to any pots. KP: 1.05 g KH_2PO_4 added to each pot. NKP: no H_2PO_4 added to any pots. Ca: 2.10 g $CaCO_3$ added to each pot. NCa: no $CaCO_3$ added to any pots.

Figure 1. Visible changes in the soil surfaces after 1 and 4 months of cultivation in selected treatments.

Experimental Design

Substrate soil (approximately 1,000 g) was encased in 15 cm×15 cm pots at a depth of 4.5 cm. Nutrient amendments (described below) were stirred into the crust sample, and the crust sample (approximately 200 g) was then homogenized and spread on top of the soil substrate to a depth of 0.5 cm in each pot.

Five external factors were evaluated: frequency of watering [5 or 10 day interval (W5 and W10, respectively)]; shade (S) or no shade (NS); NH_4NO_3 amendment (N) or no NH_4NO_3 amendment (NN); KH_2PO_4 amendment (KP) or no KH_2PO_4 amendment (NKP); and $CaCO_3$ amendment (Ca) or no $CaCO_3$ amendment (NCa). Full factorial combinations of these conditions were tested as described in Table 2. Three replicate growth experiments were performed for each treatment.

The treatment groups were watered every 5 or 10 days using a dripping system until each soil pot was saturated by 300 ml water. Double-deionized water was used to avoid the introduction of additional nutrients. A knitted black shade cloth, which excluded 60% of all incident light, was placed 20 cm above each pot in the shaded treatments. Nutrient amendments were supplied at the following levels: 2.1 g of NH_4NO_3 (300 kg/ha), 1.05 g of KH_2PO_4 (150 kg/ha), and 2.1 g of $CaCO_3$ (300 kg/ha). Pots were incubated in a greenhouse with natural light for a total of 4 months, spanning April to August. The greenhouse was ventilated continuously and did not allow the incubating pots to be exposed to rain. The greenhouse temperatures fluctuated with the changing seasons.

Indices and Data Analysis

The chlorophyll *a* content of the cyanobacteria-dominated BSCs was used as a proxy for their growth as biomass. At the end of the experiment, only the upper layer (≈ 4 mm) of the crust was used for the measurement of chlorophyll *a*. Exopolysaccharides (EPS, indicating the metabolic capacity of cyanobacterial crust) and soil organic matter (SOM) were measured using a combustion method and the phenol-sulfuric acid (PSA) test, respectively [27,28]. Changes at the soil surface were recorded using a digital camera, and the crust coverage was measured using wire meshes with 100 grid squares (5×5 mm). Using the program SPSS 14.0, a paired-sample *t*-test (16 samples) was used to determine the significance of each factor with respect to the chlorophyll *a* concentration. The differences between each set of two factors were analyzed using a one-way ANOVA and LSD analysis (16 samples). The corresponding index is presented as the mean ± standard error.

Results

Changes in the Crust Surface Area

Green surfaces indicative of photosynthetic pigments (e.g., chlorophyll *a*) were not present in any of the pots at the start of the experiment. However, the cyanobacterial crusts developed rapidly, as indicated by the development of green pigment after only 1 month. The coverage increased continually, and the surface color became darker over time, as shown in the photographs of the four treatment groups (treatments 1, 16, 17, and 21; Figure 1) which are the most representative of the surface differences. Figure 1 showed as well that, the treatment groups without NH_4NO_3 generally grew much faster than the other treatment groups. The treatment groups exposed to NH_4NO_3 developed patchy crusts, although additional watering produced better growth in these treatments [e.g., treatments 1 (W10+S+N+KP+Ca) and 17 (W5+ S+N+KP+Ca)]. Watering every 5 days in combination with shade produced the best growth. The effects of $CaCO_3$ and KH_2PO_4 on the cyanobacterial crusts were not significant based on the image analysis alone.

In addition to the visual illustration described above, the corresponding coverage of the cyanobacterial crust for those 4 treatments after 1 and 4 months was consistent with the photographic evidence. For example, in treatment groups 16 (less watering) and 21 (more watering), both without NH_4NO_3, the crust coverage reached 82.2±2.19% and 93.3±6.43%, respectively, within 1 month and increased to 95.7±1.86% and 98.3±0.33%, respectively, after 4 months. For treatment groups 1 (less watering) and 17 (more watering), both with NH_4NO_3, the crust coverage was 2.8±2.84% and 12.3±9.61%, respectively, after 1 month and increased to only 4.0±1.15% and 26.7±10.7%,

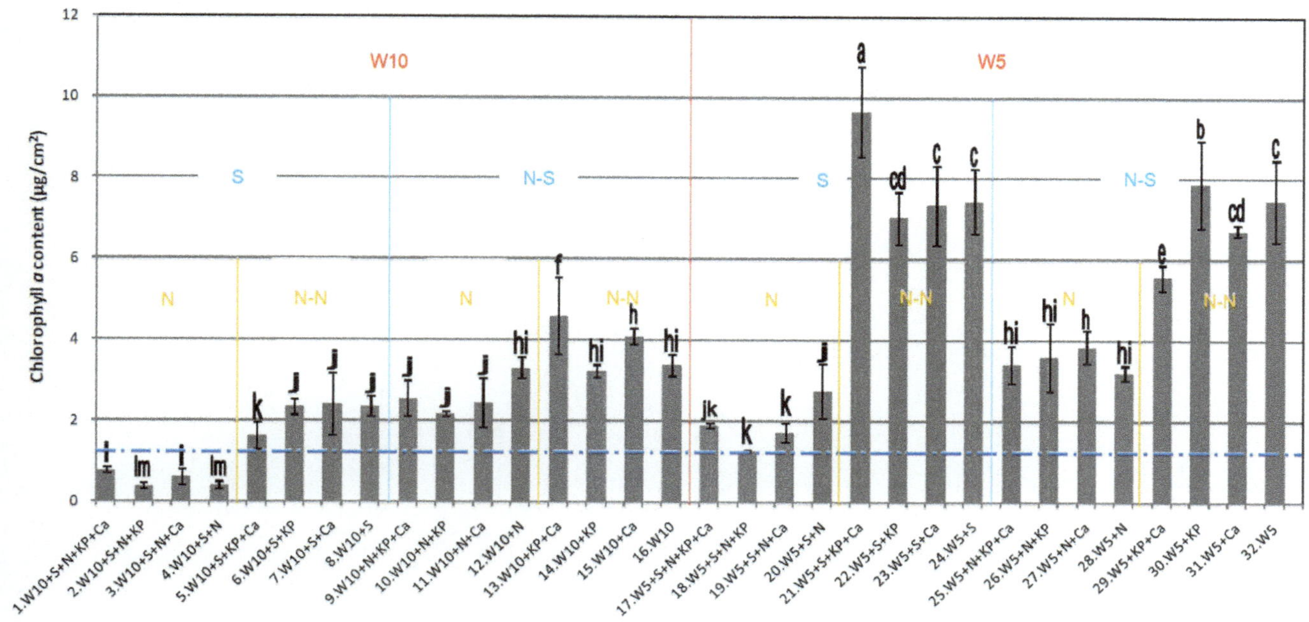

Figure 2. Chlorophyll *a* content in each of the 32 treatments after 4 months of cultivation. Note: W10:300 ml deionized water was added to each sample every 10 days. W5:300 ml deionized water was added to each sample every 5 days. S: each sample was covered by a 60% shade cloth positioned 20 cm above the pot. NS: no shading. N: 2.10 g NH_4NO_3 added to each pot. NN: no NH_4NO_3 added to any pots. KP: 1.05 g KH_2PO_4 added to each pot. NKP: no H_2PO_4 added to any pots. Ca: 2.10 g $CaCO_3$ added to each pot. NCa: no $CaCO_3$ added to any pots. The capital letters indicate the different treatments, while different lower case letters above the bars indicate significant differences between any two treatments at $P<0.01$.

respectively, after 4 months. These results indicated that the crust coverage of the treatment without NH_4NO_3 can reach high levels within a relatively short time, and additional watering enhanced this trend.

Differences of Crust Biomass

Figure 2 shows the chlorophyll *a* levels of each treatment group after 4 months. The watering frequency, light intensity, and different nutrient amendments significantly influenced the crust development. The frequency of watering significantly increased the growth of crust biomass. The comparison of the treatment

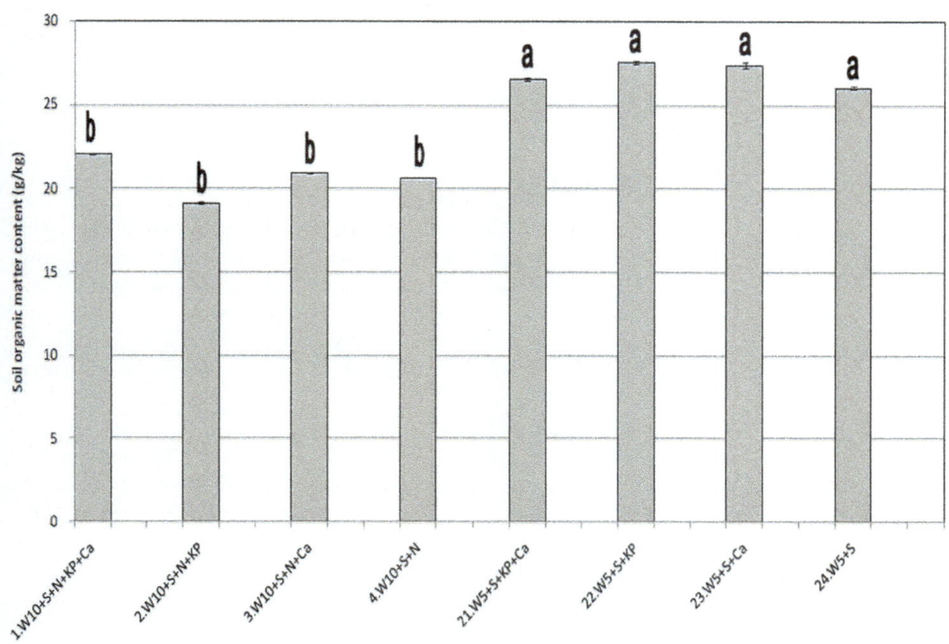

Figure 3. Soil organic matter content of the cyanobacterial crust layer in the treatments with the highest (21–24) and lowest (1–4) measured biomass. Note: Different letters above the bars indicate significant differences ($P<0.01$) between treatments.

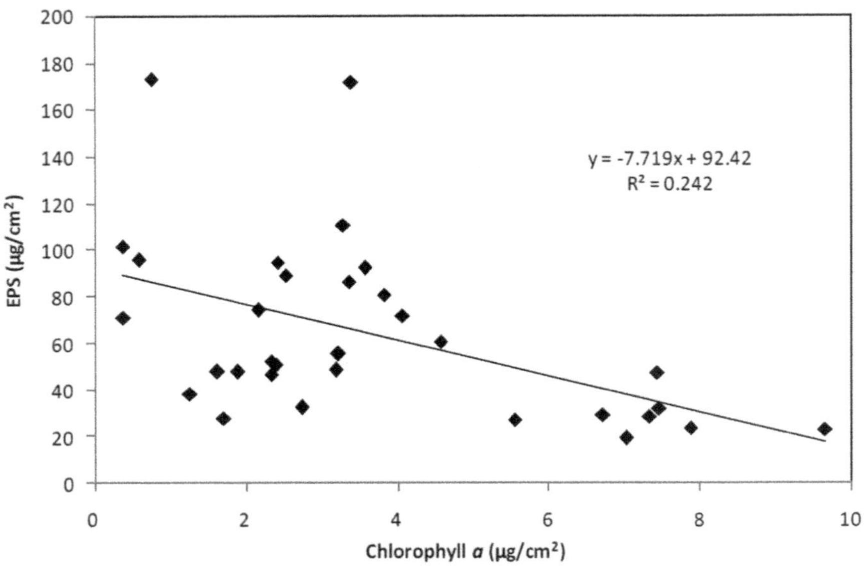

Figure 4. Relationship between chlorophyll *a* and exopolysaccharides (EPS).

groups watered every 10 days (1–16) and the same treatments watered every 5 days (17–32) clearly indicated that a greater frequency of watering produced higher biomass. The highest biomass and best crust growth were produced under sufficient watering (treatments 21–24). However, the effects of watering on cyanobacterial growth varied with different the nutrient levels. The amendment of NH_4NO_3 inhibited the development of the cyanobacterial crust. In particular, the biomass levels of the crusts treated with NH_4NO_3 were consistently lower than those of the untreated crusts in the treatment groups that were watered every 5 days. The effects of KH_2PO_4 and $CACO_3$ on the development of crusts were inconsistent; i.e., higher/lower biomass levels were observed in the presence/absence of both KH_2PO_4 and $CACO_3$.

For the 5-day watering frequency treatments, the highest biomass was observed in the two shaded treatment groups [21 (W5+S+KP+Ca) and 23 (W5+S+Ca)]. This result indicated that shade could positively affect the development of crust biomass when the soil moisture content was high. However, the effects of shading were not consistent and often inhibited biomass accumulation. For example (Figure 2), shading led to the development of less crust biomass in the treatment groups watered every 10 days (treatment groups 1–16); this effect was more pronounced in the groups treated with NH_4NO_3 (1–4, 9–12). Shade inhibited crust growth in the treatment groups that received NH_4NO_3 (17–20, 25–28) but led to a slight enhancement of growth in the groups that were not treated with NH_4NO_3 (21–24, 29–32). Thus, the effect of shade may vary according to other factors, including growth stage, season, and availability of nutrients.

Further statistical analyses were performed to quantify the effects of the five factors on crust development. The paired-sample *t*-test showed that NH_4NO_3 and watering frequency had highly significant effects ($P<0.01$), whereas the degree of shading had a lower, although significant, effect on crust biomass development ($P<0.05$). $CaCO_3$ and KH_2PO_4 had no significant effects on crust biomass development. The one-way ANOVA and LSD analysis indicated there was no significant difference between treatments that received NH_4NO_3 and the frequency of watering. Likewise, no significant difference was observed between treatments that received $CaCO_3$ and KH_2PO_4. The effects of the five factors on cyanobacteria growth (in descending order of importance) were as follows: NH_4NO_3 ≈ watering frequency>shade>$CaCO_3$ ≈ KH_2PO_4. The significant differences between all treatments (Figure 2) showed a similar trend as in the statistical analysis presented above.

Crust Organic Matter and EPS

To further examine the changes in crust development, the treatments with the lowest (treatments 1–4, watered every 10 days, not treated with NH_4NO_3) and highest (treatments 21–24, watered every 5 days, treated with NH_4NO_3) biomass levels were selected for the measurement of the soil organic matter in the cyanobacterial crust layer after 4 months of cultivation (Figure 3). This analysis indicated that the treatments with higher biomass levels contained more soil organic matter. We also measured the EPS contents as an indication of the metabolic capacity of the cyanobacterial crusts. The EPS contents of the crust layers that received the same treatments (Figure 4) were inversely proportional to their chlorophyll *a* contents and soil organic matter levels.

Discussion

For a certain cyanobacterial crust, the external factors including the soil water content, availability of nutrients, light, and temperature are the main factors for rapid crust growth and development. Our results were both consistent and inconsistent with the findings of the previous studies. The reasons or interpretations behind these conclusions are discussed as follows.

Nutrients Factor

In general, BSCs catalyze the fixation of CO_2 and N_2 and function to enrich the nutrient composition of soil [29,30,31,32]. Thus, we assumed that nutrient amendment would facilitate the growth of BSCs, however, our results surprisingly contradicted this hypothesis. Specifically, the application of NH_4NO_3 inhibited the development of cyanobacterial crusts in our study, regardless of the soil moisture content or shade level, although BSCs are known to possess a high capacity for N_2 fixation [33]. It was postulated that the disruption of the N_2 fixation component of the niche inhibited the cyanobacteria growth. This result was in contrast to previous reports that most BSC types require significant amounts of C and N nutrients to maintain vigorous development [21,34]. Furthermore, we found that higher K and Ca levels did not benefit the development of cyanobacterial crusts, which also contrasts with the results of a previous report [20]. This result may be due to the secondary roles of these nutrients in controlling crust development in comparison to the primary variables, such as light intensity and watering frequency. Clearly, these different results or abnormal phenomenon cannot be explained solely based on the current experiments. Further studies are warranted to determine how and why cyanobacterial crusts from different regions differentially respond to nutrient amendment, with the goal of developing a holistic understanding of the environmental conditions that promote optimal crust development.

Soil Moisture Factor

The soil water content was found to represent a critical and positive factor for the development of BSCs in this study, a finding that is consistent with the results of previous studies. For example, the optimum moisture levels for net and gross photosynthesis in the cyanobacteria-dominated BSCs of China's Loess Plateau were 40%–80% of the field water capacity [10]. Likewise, a study conducted in the Negev Desert indicated a strong positive relationship between the daylight wetness duration and the chlorophyll content of the crust, whereas the moisture content played an important role in stabilizing the BSC habitats [35]. As expected, our results were consistent with those of previous reports indicating that a higher soil water content or longer wetness duration significantly promotes the growth and development of cyanobacterial crust. Therefore, increasing the surface soil water and extending the wetness duration should be the most important mechanisms for rapidly restoring crusts in the field.

Light and Temperature Factor

Previously, it was shown that the production of biomass was not linearly correlated with the temperature, light, or the availability of nutrients in the culture. Instead, the optimal biomass levels of *Microcoleus vaginatus* Gom. were produced when the temperature was 30°C ±2°C, the light was 600–700 $\mu E/m^2{\cdot}s$, and the renewal rate was 35% [19]. Zhao et al. [10] showed that the light compensation point and saturation point of a cyanobacteria-dominated BSC in the hilly Loess Plateau of China were < 10 $\mu mol/m^2{\cdot}s$ and 800 $\mu mol/m^2{\cdot}s$, respectively. In our study, light intensity also impacted the development of crust biomass. However, compared with previous reports focused on single factor, our experiment was conducted under variable temperature, light intensity, and soil water content conditions similar to the natural environment. For example, the ranges of light intensity with and without shading were 1.1–401.6 $\mu mol/m^2{\cdot}s$ and 8.4–1164.3 $\mu mol/m^2{\cdot}s$, respectively on July 28th; meanwhile, the temperature fluctuated at 27.2–36.4°C and 27.8–39.4°C with and without shading, respectively, on July 28th. The total reduction of light associated with the shade cloth was 53.6%–86.9%, whereas the effect of shading on the temperature was minimal (0.6–3.0°C). Clearly, shading simultaneously changed the light intensity and temperature, which most likely resulted in actual temperature or light values that were lower than the optimal level on some days but higher on other days. For this reason, both inhibitory and promoting effects of shading were observed in the study, indicating that shading or the shading level in the field must be considered to obtain the optimal conditions.

Relationship between Biomass and EPS

Unexpectedly, the EPS content was not positively correlated with the biomass level, as reported by a previous study [19]. We hypothesized that EPS might be consumed by heterotrophic organisms in the short term (during this period between the lasting watering event and EPS measurement, EPS was measured on the 5th and 10th day after the last watering event for the W5 and W10 treatments, respectively) or to produce filamentous cells of cyanobacteria in the soil crusts. These explanations would account for the higher biomass and lower EPS levels in these crusts. In addition, compared with studies conducted in the field [36,37], our crust was cultivated in lab conditions and maintained for only a relatively short period (4 months), which may explain the negative relationship between the biomass and EPS. Further experimental studies are necessary to clearly address this question.

Feasibility of Enhanced Restoration of Cyanobacterial Crusts

Wang et al. [16] and Rao et al. [38] both independently assessed the feasibility of using cyanobacterial culture inoculation to promote BSC formation in desert areas in Inner Mongolia. They found that crust coverage could reach 48.5% with an accompanying increase in the chlorophyll *a* content during the second year. However, they used a "culture inoculation" method experiencing isolation, purification, propagation and inoculation, which has a high cost and technical complexity. In this study, we attempted to use a "broadcasting" method through directly spreading the cyanobacterial crusts 'diluted' on the soil substrate and created the appropriate conditions for realizing rapid crust development, which is characterized by convenience and easy operation. The results showed that the spreading of cyanobacterial crust organisms onto a soil substrate with the appropriate regulation of the environment is also effective and feasible for the development of cyanobacteria-dominated BSCs. If effective techniques for condition regulation are developed, the current easy method of crust restoration could be widely applied and engineered in the field.

Conclusions

Our study demonstrated that the cultivation and development of cyanobacteria-dominated crusts through a "broadcasting" method is feasible. Soil moisture is a primary factor influencing the growth of cyanobacteria and significantly promoted the development of cyanobacteria-dominated BSCs. In contrast, amendment with NH_4NO_3 had significant negative effects on the growth of cyanobacterial crusts in the current experiment, whereas KH_2PO_4 and $CaCO_3$ had nonsignificant effects on growth. Shading may positively or negatively impact the crust growth depending on whether the temperature and light intensity under shading are lower or higher than the required optimal condition. Five factors impact BSC growth with the following order of importance: NH_4NO_3≈watering frequency>shading> $CaCO_3$≈KH_2PO_4. The specific reason for the negative correlation between EPS and biomass should be studied in further research. In the field, we suggested that maintaining high soil moisture content by reducing runoff and increasing infiltration facilitates the growth and restoration of cyanobacterial crusts, meanwhile fertilization with NH_4NO_3 should be avoided. Our understanding of the factors that influence BSC development will facilitate studies regarding the restoration and cultivation of BSCs.

Acknowledgments

We deeply appreciate the kind assistance of Dr. Natalie Myers, Dr. Ferran Garcia-Pichel, Ruth M. Potrafka, Dr. Yevgeniy Marusenko, and Martha Camacho at Arizona State University during the experimental process. We also wish to thank Dr. Gale W. Jeff for assistance with the language and revision.

Author Contributions

Conceived and designed the experiments: CB SW. Performed the experiments: CB SW. Analyzed the data: CB MZ. Contributed reagents/materials/analysis tools: CB. Wrote the paper: CB SW YY.

References

1. Belnap J (2003) The world at your feet: desert biological soil crusts. Front Ecol Environ 1: 181–189.
2. Meeting B (1991) Biological surface features of semiarid lands and deserts. In: Skujins J, editor. Semiarid and sand deserts: soil resource and Reclamation. New York: Marcel Dekker. 257–293.
3. Tighe M, Haling RE, Flavel RJ, Young IM (2012) Ecological succession, hydrology and carbon acquisition of biological soil crusts measured at the micro-scale. PLoS ONE 7(10): e48565. doi:10.1371/journal.pone.0048565.
4. Angel R, Matthies D, Conrad R (2011) Activation of methanogenesis in arid biological soil crusts despite the presence of oxygen. PLoS ONE 6(5): e20453. doi:10.1371/journal.pone.0020453.
5. Bowker MA, Maestre FT, Escolar C (2009) Biological crusts as a model system for examining the biodiversity ecosystem function relationship in soils. Soil Biol Biochem 42: 1–13.
6. Bu CF, Cai GQ, Zhang XC, Ma L (2008) Review on developmental characteristics and ecological functions of soil crust. Prog Geogr 27: 26–31 (in Chinese with English abstract).
7. Bowker MA, Belnap J, Chaudhary VB, Johnson NC (2008) Revisiting classic water erosion models in drylands: The strong impact of biological soil crusts. Soil Biol Biochem 40: 2309–2316.
8. Zhang YM, Wang HL, Wang XQ, Yang WK, Zhang DY (2006) The microstructure of microbiotic crust and its influence on wind erosion for a sandy soil surface in the Gurbantunggut Desert of Northwestern China. Geoderma 132: 441–449.
9. Zhao YG, Xu MX, Belnap J (2010) Response of biocrusts' photosynthesis to environmental factors: a possible explanation of the spatial distribution of biocrusts in the Hilly Loess Plateau region of China. Acta Ecol Sin 30: 4668–4675 (in Chinese with English abstract).
10. Bu CF, Wu SF, Yang YS, Han FP, Meng J (2013) Interactive effects of moss-dominated biological soil crusts and different types of vegetation on erosion and soil water. Soil Use Manage Under review.
11. Wang XQ, Zhang YM, Zhang WM, Han ZW (2004) Wind tunnel experiment of biological crust effect on wind erodibility of sand surface in gurbantunggut desert, Xinjiang. J Glaciol Geocryol 26: 632–638.
12. Wei JC (2005) Biocarpet engineering using microbiotic crust for controlling sand. Arid Zone Res 22: 287–288 (in Chinese with English abstract).
13. Hu CX, Liu YD, Song LR, Zhang DL (2002) Effect of desert soil algae on the stabilization of fine sands. J Appl Phycol 14: 281–292 (in Chinese with English abstract).
14. Rao BQ, Liu YD, Hu CX, Li DH, Shen YW (2009) The technology of man-made algal crust and its applications in control of desertification. Acta Hydrobiol Sin 33: 756–761 (in Chinese with English abstract).
15. Wang WB, Liu YD, Li DH, Hu CX, Rao BQ (2009) Feasibility of cyanobacteria inoculation for biological soil crusts formation in desert area. Soil Biol Biochem 41: 926–929.
16. Tang DS, Wang WB, Li DH, Hu CX, Liu YD (2009) Effects of artificial algal crust on soil enzyme activities of Hopq desert, China. Acta Hydrobiol Sin 31: 339–344 (in Chinese with English abstract).
17. Xie ZM, Liu YD, Hu CX, Chen LZ, Li DH (2007) Relationships between the biomass of algal crusts in fields and their compressive strength. Soil Biol Biochem 39: 567–572.
18. Bu CF, Wu SF, Zhang KK, Yang YS, Gao GX (2013). Biological Soil Crusts: an Eco-adaptive Biological Conservative Mechanism and Implications for Ecological Restoration. Plant Biosyst DOI:10.1080/11263504.2013.819820.
19. Xie ZM, Liu YD, Chen LZ, Hu CX, Li DH, et al. (2008) The effects of different cultivation conditions on the biomass and exopolysaccharide production by *Microcoleus Vaginatus* Gom. Acta Hydrobiol Sin 32: 272–275 (in Chinese with English abstract).
20. Li XR, He MZ, Zerbe S, Li XJ, Liu LC (2010) Micro-geomorphology determines community structure of biological soil crusts at small scales. Earth Surf Proc Land 35: 932–940.
21. Maestre FT, Martín N, Díez B, Lopez-Poma R, Santos F, et al. (2006) Watering, fertilization, and slurry inoculation promote recovery of biological crust function in degraded doils. Microbial Ecol 52: 365–377.
22. Wu H, Wang HL, Liang SM, Nie HL, Zhang YM (2006) Temporal-spatial dynamics of distribution patterns of microorganism relating to biological soil crusts in the Gurbantunggut Desert. Chinese Sci Bull 51: 124–131.
23. Bu CF, Wu SF, Xie YS, Zhang XC (2013) The Study of Biological Soil Crusts: Hotspots and Prospects. CLEAN: Soil Air Water 41(9): 899–906.
24. Tsujimura S, Nakahara H, Ishida N (2000) Estimation of soil algal biomass in salinized irrigation land: a comparison of culture dilution and chlorophyll a extraction methods. J App Phycol 12: 1–8.
25. Bowker MA, Reed SC, Belnap J, Phillips SL (2002) Temporal variation in community composition, pigmentation, and Fv/Fm of desert cyanobacteria soil crusts. Microbial Ecol 43: 13–25.
26. Lan SB, Wu L, Zhang DL, Hu CX, Liu YD (2011) Ethanol outperforms multiple solvents in the extraction of chlorophyll-*a* from biological soil crusts. Soil Biol Biochem 43: 857–861.
27. Institute of Soil Science, Chinese Academic Sciences (1978) Analysis of physical and chemical properties of soil. Shanghai: Shanghai Science and Technology Press. (in Chinese).
28. Dubois M, Gilles KA, Hamilton JK, Rebers PT, Smith F (1956) Colorimetric method for determination of sugars and related substances. Anal Chem 28: 350–356.
29. Belnap J, Harper KT, Warren SD (1994) Surface disturbance of cryptobiotic soil crusts: Nitrogenase activity, chlorophyll content, and chlorophyll degradation. Arid Soil Res Rehab 8: 1–8.
30. Evans RD, Johansen JR (1999) Microbiotic crusts and ecosystem processes. Crit Rev Plant Sci 18: 183–225.
31. Guo YR, Zhao HL, Zuo XA, Sam D, Zhao XY (2008) Biological soil crust development and its topsoil properties in the process of dune stabilization, Inner Mongolia, China. Environ Geol 54: 653–662.
32. Wu N, Zhang YM, Downing A (2009) Comparative study of nitrogenase activity in different types of biological soil crusts in the Gurbantunggut Desert, Northwestern China. J Arid Environ 73: 828–833.
33. Housman DC, Powers HH, Collins AD, Belnap J (2006). Carbon and nitrogen fixation differ between successional stages of biological soil crusts in the Colorado Plateau and Chihuahuan Desert. J Arid Environ 66: 620–634.
34. Chen YQ, Zhao YG, Ran MY (2011) Influence of 4 nutrients on the development of moss crust. J Northwest A&F Univ (Nat. Sci. Ed.) 39(5): 44–50 (in Chinese with English abstract).
35. Kidron GJ, Vonshak A, Abeliovich A (2009) Microbiotic crusts as biomarkers for surface stability and wetness duration in the Negev Desert. Earth Surf Process Land 34: 1594–1604.
36. Chen LZ, Liu YD, Song LR (2002) The function of exopolysaccharides of *Microcoleus* in the formation of desert soil. Acta Hydrobiol Sin 26(2): 155–159.
37. De Philippis R, Vincenzini M (2003) Outermost polysaccharidic investments of cyanobacteria: Nature, significance and possible applications. Recent Res Devel Microbiol 7: 13–22.
38. Rao BQ, Wang WB, Lan SB, Li DH, Hu CX. (2009) Development characteristics and distribution of microorganisms within 3-year-old artificial algal crusts in Hopq desert. Acta Hydrobiol Sin 33: 937–944 (in Chinese with English abstract).

14

Much beyond Mantel: Bringing Procrustes Association Metric to the Plant and Soil Ecologist's Toolbox

Francy Junio Gonçalves Lisboa[1,4]*, Pedro R. Peres-Neto[2], Guilherme Montandon Chaer[3], Ederson da Conceição Jesus[3], Ruth Joy Mitchell[4], Stephen James Chapman[4], Ricardo Luis Louro Berbara[1]

1 Soil Science Department, Agronomy Institute, Federal Rural University of Rio de Janeiro, Seropédica-RJ, Brazil, **2** Canada Research Chair in Spatial Modelling and Biodiversity; Université du Québec à Montréal, Département des sciences biologiques, Québec, Canada, **3** Embrapa Agrobiologia, Seropédica-RJ, Brazil, **4** The James Hutton Institute, Craigiebuckler, Aberdeen, United Kingdom

Abstract

The correlation of multivariate data is a common task in investigations of soil biology and in ecology in general. Procrustes analysis and the Mantel test are two approaches that often meet this objective and are considered analogous in many situations especially when used as a statistical test to assess the statistical significance between multivariate data tables. Here we call the attention of ecologists to the advantages of a less familiar application of the Procrustean framework, namely the Procrustean association metric (a vector of Procrustean residuals). These residuals represent differences in fit between multivariate data tables regarding homologous observations (e.g., sampling sites) that can be used to estimate local levels of association (e.g., some groups of sites are more similar in their association between biotic and environmental features than other groups of sites). Given that in the Mantel framework, multivariate information is translated into a pairwise distance matrix, we lose the ability to contrast homologous data points across dimensions and data matrices after their fit. In this paper, we attempt to familiarize ecologists with the benefits of using these Procrustean residual differences to further gain insights about the processes underlying the association among multivariate data tables using real and hypothetical examples.

Editor: Andrew R. Dalby, University of Westminster, United Kingdom

Funding: This work was carried out with the aid of a grant from Conselho Nacional de Desenvolvimento Científico e Tecnológico (CNPq - 563304/2010-3 and 562955/2010-0), Fundação de Amparo à Pesquisa de Minas Gerais (FAPEMIG CRA - APQ-00001-11), and the Inter-American Institute for Global Change Research (IAI-CRN II-021). Francy Lisboa greatly acknowledges a research scholarship from CAPES/EMBRAPA (Carbioma) and Dr. Beata Madari. The funders had no role in study design, data collection and analysis, decision to publish, or preparation of the manuscript.

Competing Interests: The authors have declared that no competing interests exist.

* Email: agrolisboa@gmail.com

Introduction

In multidimensional data analysis, ecologists often encounter situations where they need to choose between two or more numerical approaches that are able to tackle the same question of interest. The preference between approaches is based, among other factors, on the familiarity of the user with the method, which in turn depends on the time a particular method has been available in statistical packages and the ease in implementing and interpreting its results. Another relevant factor to consider is "literature-induced use" in which renowned research groups involved in the development, improvement and generation of statistical ecological approaches have a strong influence on the types of statistical approaches other ecologists use.

Determining the strength of the relationships between multivariate datasets is a routine analysis when trying to understand the environmental factors driving the composition and structure of ecological communities. Two approaches, the Mantel test [1] and Procrustes analysis [2], though considered analogous by the literature in the questions they can tackle [3], have not been used to the same extent. Despite the advantages of Procrustes analysis over the Mantel test [3] regarding greater statistical power in detecting significant relationships (i.e., lower type II errors) and the possibility of analyzing further the patterns of association between multivariate matrices (visually and by further statistical analyses), the Procrustean approach remains relatively unused in tackling questions regarding the relationships between data matrices involving plant and soil information or between soil matrices (Fig. 1).

The Mantel test and the Procrustes approach can be both used in many similar situations where the aim is to assess how multivariate data matrices are associated (correlated), though for unknown reasons they have been used in quite different ways in the ecological literature. For example, while the Mantel test has often been applied when testing the relationship between above and below ground data matrices [4], [5], [6], [7], [8], [9], [10], [11], [12], Procrustes analysis has predominantly been used to contrast the results of different ecological ordinations on the same data [13], [14], [15], [16], to compare fingerprinting tools for assessing microbial communities [17], [18], [19] and for deciding between methodological choices [20], [21]. Indeed the Procrustean framework has been rarely used to make inferences about plant and soil relationships [22], [23], [24], [25], [26] and other types of ecological associations between data sets. However there are instances in which the Procrustean and Mantel tests cannot be used interchangeably. Unlike Mantel, the Procrustean approach can be used to compare multiple data matrices. However, when ecologists are interested in correlating distance (or similarity)

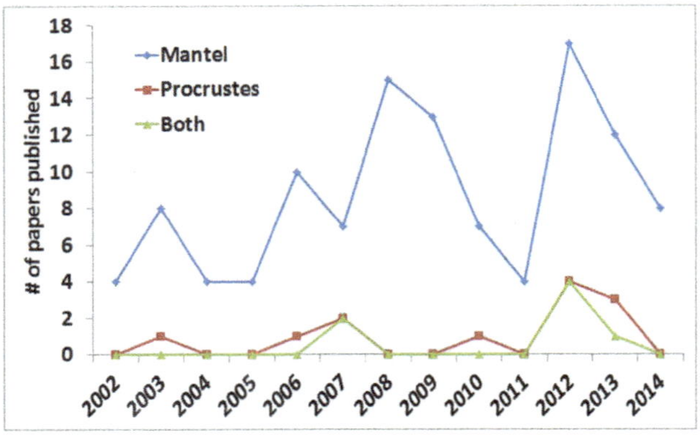

Figure 1. Papers published using Mantel and Procrustes for relating data matrices from soil or plant studies in the ten years since [3] stated the advantages of Procrustes over the Mantel approach. Data obtained using Thompson Reuters database (May, 12, 2014). We searched for papers using uniquely the Mantel approach, uniquely the Procrustes approach and papers using both approaches. The search was based on Procrust* (Procrustean or Procrustes) and PROTEST.

matrices, rather than testing the association among data matrices in their raw form (i.e., not transformed by the property of distance measures), Mantel, rather than Procrustes, is more appropriate. One particular case is the distance-decay of similarity in ecological communities [27] in which one is interested in testing the hypothesis that the similarity in community composition decreases in relation to linear (or log transformed) geographic distance between communities. The differences between raw-based and distance-based approaches have been discussed extensively elsewhere [28], [29].

Despite the relative merits of the Procrustean framework over the Mantel test shown by the relatively well-cited paper by Peres-Neto and Jackson [3], its potential has not yet been tapped. Perhaps the reason for Procrustes analysis not being as popular as the Mantel test among ecologists is the lack of a paper showing that in many situations traditionally investigated by Mantel, the Procrustean analysis can be equally well used. Here, we attempt to familiarize ecologists with the use of Procrustes analysis by using real and hypothetical examples where the Mantel test tends to be preferred. Most importantly, we highlight little explored limits of Procrustes by using its residual vector of association between data tables, hereafter referred as to PAM, in three common statistical approaches: multivariate ordination, variation partitioning and ANOVA.

Procrustes analysis: a foundation for soil and plant ecologists

In ancient Greek mythology there was a character named Procrustes who was a resident of Eleusis Mountain, a known travelers' route. As a "good" host, Procrustes always invited travelers to spend the night at his home; more specifically, he invited them to lie down on his iron bed, which was tailored to fit Procrustes' own body. The guests who did not fit the dimensions of his bed either had their limbs cut off or were stretched until their dimensions approached those of Procrustes's bed. Ironically, none of the guests ever fitted the iron bed because Procrustes secretly had two beds of different sizes [30]. One can easily make a parallel here with ecological data in which data from different sources will almost never easily compare or fit to one another.

Procrustes analysis is based on the search for the best fit between two data tables, hereafter referred to as matrices, where one is kept fixed ("Procrustes' bed" or target matrix), while the other ("Procrustes' guest" or rotated matrix) undergoes a series of transformations (translation, mirror reflection and rotation; [2]) to fit the fixed matrix. Although in this paper we concentrate on fitting two matrices, the extension of Procrustes analysis to multiple matrices is straightforward [3] in which the reference matrix can be either one of the original matrices or their averages (or medians). Hereafter, the target matrix (target) will be referred as to **X**, and the data matrix to be fitted as **Y**. **X** and **Y** are both $n \times p$ matrices, where n is the number of rows and p is the number of columns. The goal of the transformations in **Y** is to minimize the residual sum of squared differences between the corresponding n dimensions between **X** and **Y**; the sum of the squares of these residual differences is termed m^2 (Gower's statistic), representing the optimal fit between the two data matrices, such that the higher the value of m^2, the weaker the relationship between the two data tables is. The significance of m^2 can be estimated through a permutation test (termed PROTEST after [31]; see [3] for further details).

Procrustean association metric (PAM)

The least squares superimposition between the corresponding n observations of **X** and **Y** is one of the main advantages (in addition to the increased statistical power) of the Procrustean framework in contrast to the Mantel test. The Procrustes superimposition generates a $(n \times p)$ matrix of residuals that can be further used to contrast the differences between homologous observations (rows) across matrices in the form of a vector (PAM). Given that within the Mantel approach differences between observations across all dimensions are packed down into a single distance, it cannot be used to assess differences across observations across dimensions. Consistent small and large differences across homologous observations across matrices in regard to other factors of interest can further assist in understanding how **X** and **Y** are related. For example, we could use PAM to assess the degree of observation matching between a plant function trait matrix and a composition matrix and assess whether smaller or greater residual values are a function of the time elapsed since some disturbance event.

PAM is simply a vector of residual differences between the corresponding n observations. For example, assuming that an ecologist wants to correlate two matrices of data **X** and **Y**, both of which are formed by four rows (i.e. sites, plots, observational units), Procrustes analysis will generate four residual differences between the **X** and **Y** configurations. The compilation of these residual differences between homologous rows (observations) across dimensions in the form of a vector – PAM – represents a useful way to represent information on the relationship between two matrices and make it available for further statistical analysis, both parametric and non-parametric; this feature is not offered by the Mantel approach.

The use of the residual vector from Procrustes (PAM) has been quite restricted in the plant and soil ecological literature. To our knowledge the first study was by [32] who assessed the plant-pollinator interaction during three consecutive summers in the southeastern portion of California, USA. These authors employed the PAM to identify which pollinating species exhibited the greatest deviation between two consecutive years. Singh et al. [22] used the PAM in a study on soil microbiology to verify the effect of soil pH on the relationship between arbuscular mycorrhizal fungi (AMF) and plant assemblages. These authors employed the following strategy: 1) Procrustes analysis was applied between the matrices representing the AMF community and that representing the plant community; 2) after detecting a significant relationship ($m_{12} = 0.28$; $P<0.001$), these authors extracted the PAM and used it as a response in a simple regression analysis with the soil pH. No effect of pH on the association between the AMF and plant communities was detected, suggesting that neither the pH nor the identity of the plant species that composed the community affected the AMF community. Other applications can be certainly found (e.g., [24], [25], [26][33]) but its flexibility and general usage remains largely unexplored.

Constructing a practical roadmap for applying PAM

There are few studies in the ecological literature that have used PAM for analyzing relationships between plant and soil datasets. The lack of examples partially explains the low popularity of Procrustes analysis among plant and soil ecologists and ecologists in general as an alternative tool to the more traditional Mantel test. In order to make the possible uses of Procrustean residuals more familiar, we will introduce a number of examples in the form of schematic roadmaps for applying PAM in association with three common statistical approaches: ordination, regression analysis and ANOVA.

Plant and soil ecologists must keep in mind that Procrustes analysis requires that the **X** and **Y** have the same number of rows and columns, though the last dimension is less restricting (see below). Given that the data for both matrices usually originate from the same sites, it is most common in ecology that only the number of columns (descriptors or variables) varies between the two matrices. Therefore, the question arises of how to make the number of columns equal across the two matrices, i.e., how to reduce them to the same dimensionality. Although Procrustes analysis can be performed between matrices having different number of dimensions (i.e., the fit is based on a singular value decomposition (svd) of X^TY, where X and Y are scaled prior to svd and T stands for matrix transpose), traditionally the matrix with the fewer number of columns ("missing columns") is made equal in dimension to the larger matrix by adding columns of zeros in order to keep (Fig. 2a; [2]). Although there are some criticisms related to this practice and alternatives have been suggested [34], the addition of zero columns does not affect the distances between columns among observations and is a convenient device rather than a hurdle [35].

Another convenient way to make **X** and **Y** have the same number of columns is to represent most of the variation in their raw data by matrices formed by the same number of orthogonal axes (Fig. 2b; [3],[35, [36]), i.e., matrices formed by axes derived through ordination methods such as Principal Components Analysis (PCA), Non-metric dimensional scaling, Correspondence Analysis (CA), Principal Coordinate Analysis (PCoA), the choice being dependent on the nature of the data (continuous, presence-absence data, abundance data). Moreover, raw data matrices can be transformed prior to ordination (see [37] for different transformations and their characteristics) or alternatively have pairwise distance matrices calculated from the data matrices that are then orthogonolized via PCoA to extract ordination axes based on the chosen distance measure (e.g., Bray-Curtis, Jaccard, Sorensen, Gower).

Here, for simplicity, we use a PCA in all applications. In cases, where species data (presence/absence or abundance) was used, the data was Hellinger-transformed and PCAs were extracted on species correlation matrix calculated from the transformed data. The Hellinger transformation alleviates the issue of double-zeros in species data matrix transformed into correlation or Euclidean-distance pairwise matrices prior to PCA in which sites sharing no species in common can be found to be more similar than sites sharing a reduced number of species in common (e.g., the horse shoe effect in ordination plots).

The general strategy is as follows:

1) Subject the raw data matrices to an ordination method (here PCA but see above for other strategies);
2) After ordinating **X** and **Y**, use the same number of ordination axes for both matrices (Fig. 2b).

Given that the higher the number of ordination axes used, the higher is the amount of variation explained in **X** and **Y**, it would be interesting run the Procrustean analysis sequentially using matrices made up of an increasing number of ordination axes. It could help ecologists check the consistency of the relationship between **X** and **Y** based on different numbers of ordination axes, which will give more reliability to the results.

The use of PAM in ecological ordination

The first form of PAM shown here is based on ordination methods. Ordination is the graphical representation of the variation of objects (sites), descriptors (species/environmental parameters) or both, in a reduced space formed by orthogonal axes [38].

To illustrate the use of Procrustes analysis associated with ordination we use data derived from Mitchell et al. [39]. This study aimed to compare the plant communities and soil chemistry in their ability to predict changes in the structure of the soil microbial community in three moorland areas established in Northern Scotland called Craggan, Kerrow and Tulchan. The plant community matrices from each area were based on the percent cover. Three matrices for the soil microbial community were obtained for each site: one based on the fatty acid profile of the soil (PLFA analysis), and the other two on the T-RFLP analysis of the communities of fungi and bacteria, respectively. The matrix representing the soil chemistry was based on the concentrations of Na, K, Ca, Mg, Fe, Al, P, total C, total N in addition to pH, loss on ignition and moisture.

There is some consensus that the variation in vegetation can act as a proxy for changes in the soil microbial community, either directly in the case of symbionts, for example, or indirectly via changes in soil chemistry itself. We use Procrustes analysis associated with ordination techniques to verify potential drivers of the soil microbial community and to determine if plant community and soil chemistry are equally related to the microbiological variation. The sequence of analysis was as follows:

1) Ordination analysis: All data matrices (community plant, soil chemistry and soil microbial communities) containing the three chronosequences were subjected to separate PCAs based on correlation matrix. The community plant was Hellinger-transformed prior to PCA. Then, the first six PCA ordination axes from each matrix were retained in order to assemble four PCA matrices representing the variation summarized in the first 3, 4, 5 and 6 PCA axes. Thus, four PCA matrices were obtained from each dataset: plant community, soil chemistry and soil microbial community (PLFA, bacterial and fungal T-RFLP) (Fig. 3a).
2) Procrustes analysis: The PCA matrices of plant community and soil chemistry were used to run Procrustean analyses with the PCA matrices of soil microbial community based on PLFA, and fungal and bacterial T-RFLP datasets.

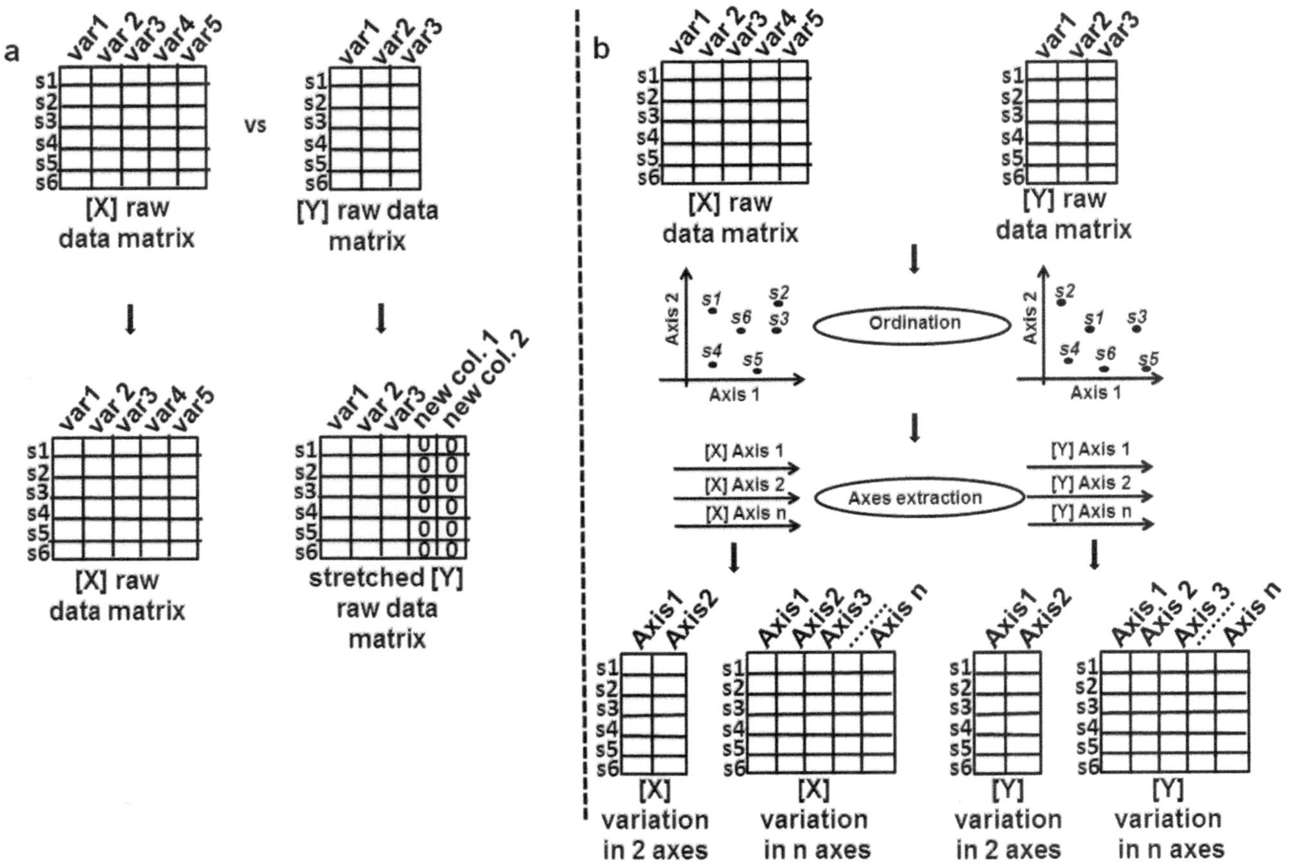

Figure 2. Roadmap for two alternative ways to reach the same dimensionality between matrices, and so relating it by Procrustes analysis. a) Addition of columns containing zeros to the **Y** raw data matrix for matching the **X** raw data matrix dimension; **b)** Application of ordination to raw data matrices to make matrices have equal dimensionality prior to Procrustes analysis.

3) PAM extraction: Since all Procrustean relationships based on PCA matrices with *n* axes were significant, for simplicity, only the PAM obtained from relationships of PCA matrices with 6 axes were used for subsequent analyses. Six PAMs were generated: PAM1 (soil chemistry on PLFA), PAM2 (soil chemistry on bacteria), PAM3 (soil chemistry on fungi), PAM4 (plant on PLFA), PAM5 (plant on bacteria), and PAM6 (plant on fungi) (Fig. 3c).
4) PAM ordination: The PAMs were assembled in a single matrix ("effect matrix") with one PAM per row (Fig. 3c). Therefore, the effect matrix compiled the effects of plant community and soil chemistry on soil microbial community structure derived from the three methods. This effect matrix was submitted to PCA ordination to verify whether the plant community effect on soil microbial community structure differed from the effect of soil chemistry (Fig. 3c).

The results showed that for all chronosequences the plant effect on microbial structure was divergent in relation to the soil chemistry effect, as suggested by the separation along the axis of greatest variation (Fig. 4). Although we cannot apply a proper statistical significance test in one-table based ordination methods (PCA, NMDS, PCoA, etc), visual inferences can be made. For example the Craggan area exhibited a clear distortion between plant community and soil chemistry variation in terms of their effects on the soil microbial community structure depicted by PLFA, bacterial T-RFLP and fungal T-RFLP (Fig. 4a). Also in this area, the response of the microbial community based on PLFA was distant from the response based on molecular data (T-RFLP) (Fig. 4a),

We expected that the effects of soil chemistry on microbial structure were closer to the effect of plant community once the plant community is considered to be a direct and indirect driver for the biotic component of soil [39]. However, these results suggest that plant communities and soil chemistry are acting differently on the soil microbial community structure [24], [40]. They also suggest that the effects of soil chemical properties on the microbial communities may be weakly mediated by above ground alterations [24]. This example shows the usefulness of Procrustes analysis to raise additional evidence in plant and soil ecology studies. (See Text S1 containing the R code used for this example).

The PAM and regression analysis

In regression analysis, 'response' and 'predictor' are common terms. In ecology, predictors can have different natures. Space, time, organic matter and moisture, among other factors, are some examples of predictors. On the other hand the microbial communities are often used as a response variable because they are considered better indicators of a given ecosystem.

Some authors familiar with soil microbial ecology have been using the Mantel test to assess the individual contribution of deterministic and stochastic processes on the soil microbial structure variation [41], [42]. As an example of the utility of the

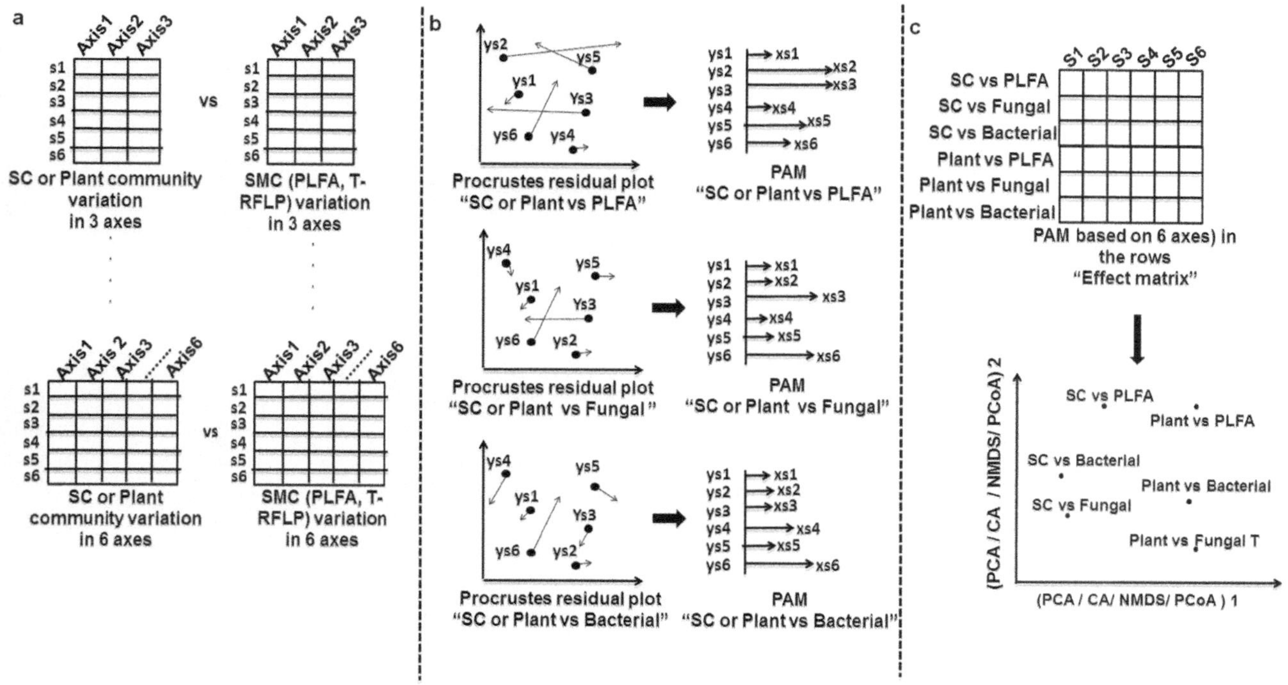

Figure 3. Roadmap for applying the Procrustes association metric (PAM) in the multivariate ordination context using data of [39]. **a)** Assembling matrices with different ordination axes, through Procrustes analysis, soil chemistry (SC) and plant community with soil microbial community (PLFA, and bacterial and fungal T-RFLP); **b)** Extraction of PAM from Procrustean relationships based on matrices with 6 ordination axes; **c)** Assembling of PAM based PCA matrices with 6 axes as rows in a single matrix ("effect matrix"), and using it in an ordination technique (e.g., PCA, PCoA, NMDS) to verify if the different effects diverge.

Procrustes analysis in the context of variation partitioning we can take a hypothetical scenario with four datasets from a given area, corresponding to soil microbial community structure (PLFA), soil microbial functioning (enzyme activities), soil properties and spatial variation. Spatial variation can be represented, for example, by 100 sampling points generated from a 10 m×10 m transects. The matrix of geographical coordinates of the sampling points can be submitted to PCNM (principal coordinates neighbour matrix) analysis generating a matrix of spatial eigenfunctions termed PCNMs [34]. In this scenario, we can assume that the ecologist aims to assess the relative contributions of individual soil properties (deterministic processes) and spatial variation (stochastic event) on the relationship between microbial community structure and soil microbial functioning rather than on these components individually. To use the Procrustean association metric (PAM) in this context, one can use the following steps:

1) Ordinate the two matrices (i.e., the soil microbial community and soil microbial functioning) via PCA (the soil microbial community matrix was Hellinger-transformed) and select a similar number of ordination axes. The multivariate scores of the two matrices across the selected number of axes are subjected to a Procrustes analysis and a PAM was then calculated.
2) Use individual PAMs (based on 2, 3 or more PCA axes) as response variable and soil properties and spatial variation as independent (predictor) variables in a multiple regression framework (Fig. 5b).

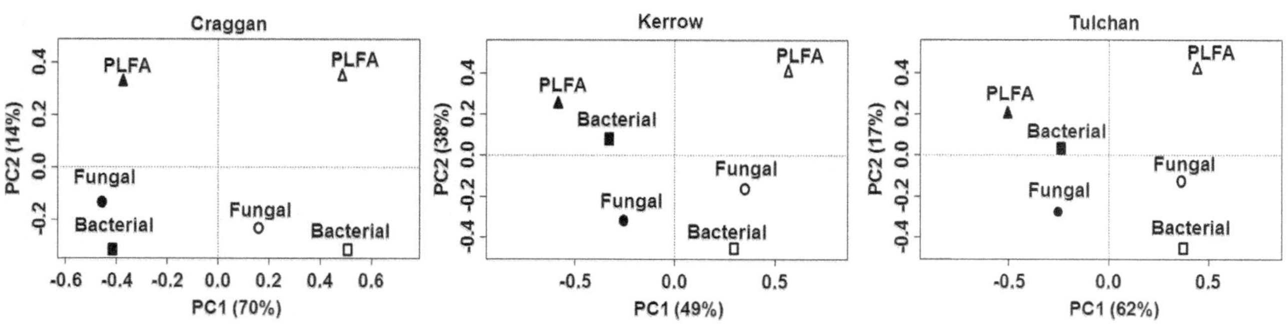

Figure 4. Results from PCA ordination of the Procrustes association metric matrix ("effect matrix") gathering the interactions of soil chemistry and plant community with soil microbial matrices (PLFA, and bacterial and fungal T-RFLP). The filled symbols are the Procrustes relationships between soil chemistry and soil microbial matrices, and the open symbols between plant community and soil microbial matrices. Data from three chronosequences (Craggan, Kerrow and Tulchan) obtained by [39].

3) Finally, the independent contributions of soil properties (independent of space) and unmeasured spatial process and/ or factors (spatial variation independent of soil properties) to the microbial structure can be estimated via variation partitioning [43] and represented by a Venn diagram (Fig. 5c). (See Text S2 containing the R code for this example).

The PAM and Analysis of Variance

Although regression and analysis of variance are ultimately the same analysis in which the response is either continuous (regression) or ascribed to factors (ANOVA), we provide examples for each of them in different sections given that often they are seen as distinct forms of analyses. Evaluation of the effects of land use on soil microbial communities has been a common case-study issue in soil ecology. Some of these studies have been carried out using the Mantel approach to assess how land use type effects soil microbial structure and functioning [44], [45]. However, Mantel does not yield a vector of structure – functioning relationship, that is, a continuous variable, able to be partitioned by categorical variables like land use types. In the following example we show how to use PAM to evaluate the effect of land use type on the relationship between microbial community structure and microbial function in the form of PAM.

In a hypothetical scenario, a researcher is interested in studying whether four different land use types within the Amazon biome are affecting the relationship between microbial structure and microbial functioning. In each of the land uses (original forest fragment, silvipastoral system, improved pasture and unimproved pasture) six plots (10 m×10 m) were established and one composite soil sample (0–10 cm) collected per plot (Fig. 6a). The X dataset (soil microbial structure) was represented by PLFA data, and the Y dataset (microbial functioning) by the abundance of genes associated with microorganisms involved in greenhouse gas emission processes, such as nitrifiers, denitrifiers and methanotrophic organisms. The researcher's hypothesis is that in the original forest (non-altered environment) there is a better matching between microbial structure and microbial function. Thus, in anthropogenically disturbed environments (silvipastoral system, improved pasture, and unimproved pasture) the change in microbial structure relative to the original (forest) is not followed by a change in the microbial functioning to the same magnitude. This hypothesis can be tested using an integration of Procrustes analysis and ANOVA through the following steps:

1) Reduce the datasets **X** (soil microbial structure) and **Y** (soil microbial functioning) to similar dimensions using PCA. Then, run the Procrustean analysis between the PCA matrix of the soil microbial community structure and the PCA matrix of soil microbial functioning and extract the PAM (Fig. 6a).
2) Run an ANOVA with land use type as fixed factor and the PAM as the response variable (Fig. 6b).
3) If the F value of ANOVA is significant, a means test can be performed to compare the mean PAMs of the land use types (Fig. 6c). (See Text S3 containing the R code for this example).

Discussion

In this paper we have attempted to show the advantages of the Procrustean analysis over the Mantel test, in which the former can be used for gaining further information on underlying drivers of data table associations. In particular we have shown how the Procrustean association metric (PAM) constructed of the residuals

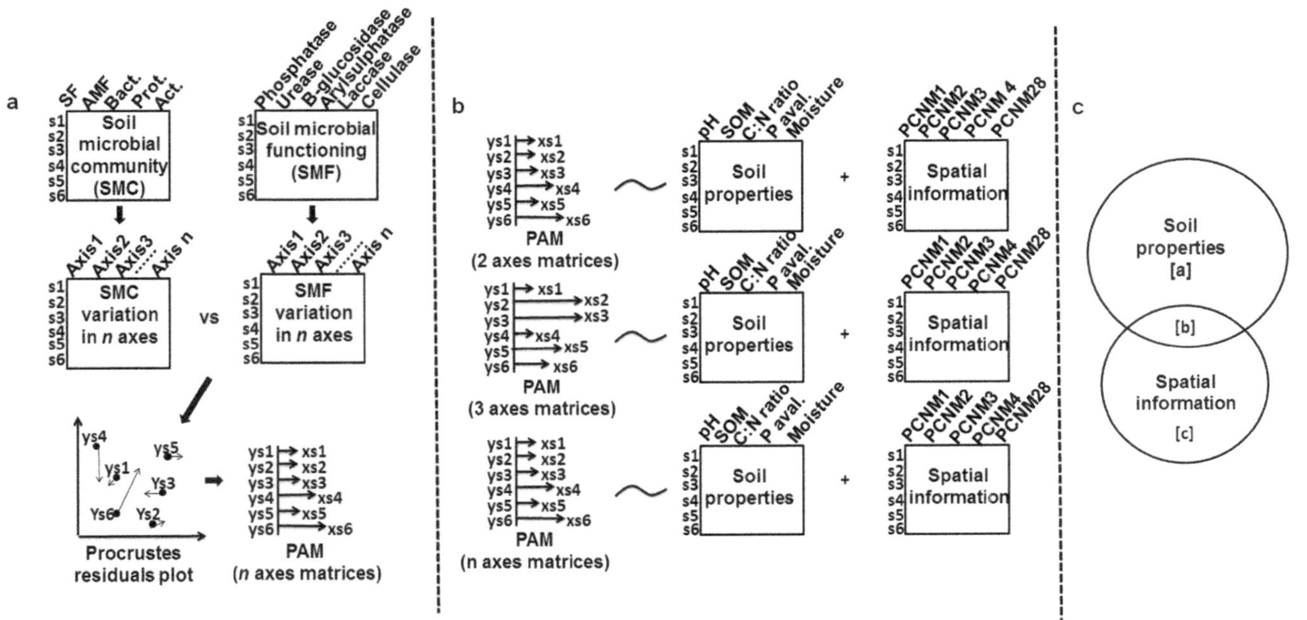

Figure 5. Roadmap for using Procrustes Association Metric (PAM) in a multiple regression analysis framework (variation partitioning). a) Soil microbial community (SMC) and soil microbial functioning (SMF) matrices are submitted to an ordination to reach the same dimensionality, and SMC and SMF matrices formed by 2, 3 and *n* axes related through Procrustes analysis in order to generate PAMs; **b)** PAMs generated were used as response variables in a variation partitioning to verify the individual contribution of soil properties and spatial information (PCNM eigenfunctions) on the SMC-SMF relationship; **c)** Venn diagram depicting the relative contribution of soil properties (niche processes **[a]**) and unmeasured spatial factors (neutral processes **[c]**).

of the vectors after the Procrustes analysis. We concentrated on showing how patterns of concordance between data matrices can be displayed and individual observations contrasted separately using the Procrustean framework, allowing further examination of the common and different association patterns among multiple data matrices. Given that in the Mantel framework, multivariate information is translated into a pairwise distance matrix, we lose the ability to contrast homologous data points across dimensions and data matrices. It is important to notice that it was not our goal to show the statistical advantages of Procrustes over Mantel as done by previous work [3]. Instead, we concentrated on generating different analytical schemes, especially for plant and soil ecologists, to incorporate Procrustes into their statistical toolbox.

What is unique about Procrustean framework? There are at least four characteristics of the approach not shared by others. First, because the approach is correlative rather than regressive, the number of observations (e.g., sites) in the matrices does not have to be greater than the number of columns as in common regression approaches such as RDA and CCA. Second, we can fit as many matrices as we have available; this latter issue is particularly restrictive under a regression approach given the limitation of number of rows versus number of columns. Moreover, all matrices are treated in equal footing as no matrix is treated as response or predictor. Third, the relationships within (only across) matrix columns do not affect the analysis. Fourth, residual values across observations and dimensions can be calculated and explored as shown here. These characteristics should not be necessarily seen as advantages per se over other methods but rather features that are unique and may be useful in many situations. There are certainly other tools that can be used to look at the associations between data sets. RDA and CCA are well-established tools in ecology and are based on regression (asymmetric) methods. Traditionally these approaches may have been thought to be more appropriate for analysis of the examples given in this paper, since they establish relations of cause and effect. However, because these analyses include a regression step, they are limited to situations where the number of rows (sites) in the environmental matrix **X** is higher than the number of columns (variables) [36], [46]. This is not a limitation in Procrustes analysis and moreover, it is not clear how residual variation among homologous observations across dimensions should be explored in the case of RDA and CCA.

At least two other symmetric approaches are similar to the Procrustean approach, namely Co-inertia analysis [47] and symmetric Co-correspondence analysis [48], a form of Co-inertia analysis in which a correspondence analysis is applied to two species matrices prior to the analysis. The main difference resides in the fact that fit is influenced by all variables pairs in Co-inertia analysis (within and between matrices), whereas fit is influenced only by variation between matrices in Procrustean. Co-inertia is always based on ordination within data matrices, whereas in Procrustes either the raw data or their ordination axes can be used. Co-inertia can also take into account row (e.g., sites) and column (e.g., species) weights in the analysis, though the standardization and fit processes in Procrustean analysis could also take these into account [36]. Co-inertia and Procrustean analysis are certainly related in the sense that they both treat matrices as symmetrical during the fitting process, though more studies are necessary to assess in which conditions (e.g., correlation within and across matrices, differences in dimensionality between matrices, outliers within and across matrices) they differ. Finally Dray et al. [36] showed the advantages of merging Co-inertia and Procrustean analysis, where the latter is used as a precursor of the former. In reality, future studies are required to contrast Co-inertia and Procrustean analysis, but in either form of analyses we can produce residual vectors (PAM) that can be further analyzed.

a

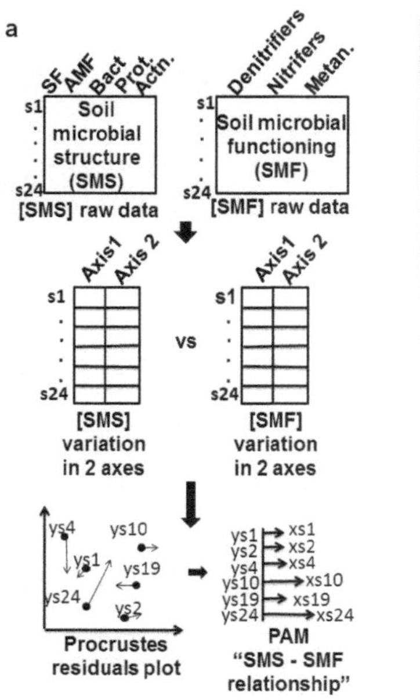

b

Table. One-way ANOVA of the effect of different land use types on the Procrustean relationship between soil microbial structure (SMS) and soil microbial functioning (SMF).

Land use type (fixed factor)	PAM (SMS-SMF relationship)	
	F	*P*
Land use	7.047	0.003**
Residuals	1.345	>0.05

c

Table. Mean effect of different land use types on the Procrustean relationship between soil microbial structure (SMS) and soil microbial functioning (SMF).

Land use types	SMS-SMF relation (PAM in 2 axes)	
	Mean	SE
Original forest	0.021 a	0.003
Silvopastoral system	0.052 b	0.001
Improved pasture	0.074 c	0.003
Unimproved pasture	0.082 c	0.002

Figure 6. Roadmap for using Procrustes association metric (PAM) in an ANOVA context. a) PCA ordination of each SMC and SMF raw data matrices, and then Procrustes correlation from 2 axes-based PCA matrices in order to generate the PAM depicting the SMC-SMF relationship. **b)** Table showing results of a one-way ANOVA for using PAM as response and land use type as fixed factor. **c)** Multiple comparisons test (Tukey, 95%) for means of the Procrustean relationship between soil microbial structure and functioning (PAM in 2 axes) across land use types.

Procrustes can be perhaps best justified when the number of predictors is greater than the number of observations or when **X** and **Y** matrices are equally applicable as explanatory and response variables. In plant-soil ecology, for example, above- and below-ground data matrices can be interchanged as explanatory and response variables. Plant community variation has been shown to be related to variation in below-ground compartments [24]. In addition, soil components such as fertility and the microbial community have been proven to influence aspects of vegetation [49]. Thus, with the literature showing that both types of datasets under analysis can structure each other, the use of Procrustes analysis, as a symmetric canonical analysis method, should be encouraged among plant and soil ecologists and ecologists in general. We hope that this paper has provided enough examples of the potential for using the Procrustes framework as a precursor to further explore ecological data.

Supporting Information

Text S1 R code showing how to use PAM associated to ordination methods (Fig. 4 in the main text). For this example we used data from Mitchell et al. [39].

Text S2 R code showing how to use the PAM in a Regression framework (Fig. 5 in the main text).

Text S3 R code showing how to use the PAM in an ANOVA framework (Fig. 6 in the main text).

Author Contributions

Conceived and designed the experiments: FJGL. Analyzed the data: FJGL. Wrote the paper: FJGL PRPN GMC ECJ SJC RJM RLLB. Contributed with real datasets: SJC RJM.

References

1. Mantel N (1967) The detection of disease clustering and a generalized regression approach. Cancer Res 27: 209–220.
2. Gower JC (1971) Statistical methods of comparing different multivariate analyses on the same data. In: Hodson FR, Kendall DG, Tautu P, editors. Mathematics in the archeological and historical sciences. Edinburgh University Press, Edinburgh. 138–149.
3. Peres-Neto PR, Jackson DA (2001) How well do multivariate data sets match? The advantages of a Procrustean superimposition approach over the Mantel test. Oecologia 129: 169–178.
4. Tuomisto H, Poulsen AD, Ruokolainen K, Moran RC, Quintana C, et al. (2003a) Linking floristic patterns with soil heterogeneity and satellite imagery in Ecuadorian Amazonia. Ecol App 2: 352–371.
5. Tuomisto H, Ruokolainen K, Yli-Halla M (2003b) Dispersal, environment, and the floristic variation of western Amazonian forests. Science 299: 241–244.
6. Tuomisto H, Ruokolainen K, Aguilar M, Sarmiento A (2003c) Floristic patterns along 43-km long transect in an Amazonian rain forest. J Ecol 91: 743–756.
7. Kang S, Mills AL (2004) Soil bacterial community structure changes following disturbance of the overlying plant community. Soil Sci 169: 55–65.
8. Poulsen AD, Tuomisto H, Balsev H (2006) Edaphic and florist variation within a 1-ha plot of lowland Amazonian rain forest. Biotropica 38: 468–478.
9. Fitzsimons MS, Miller RM, Jastrow JD (2008) Scale-dependent niche axes of arbuscular mycorrhizal fungi. Oecologia 158: 117–127.
10. Powers JS, Becknell JM, Irving J, Pères-Aviles D (2009) Diversity and structure of regenerating tropical dry forests in Costa Rica: Geographic patterns and environmental drivers. Forest Ecol Manag 258: 959–970.
11. Castilho-Monroy AP, Bowker MA, Maestre FT, Rodriguez-Echeverria S, Martinez I, et al. (2011) Relationships between biological soil crusts, bacterial diversity and abundance, and ecosystem functioning: insights from a semi - arid Mediterranean environment. J. Veg. Sci 22: 165–174.
12. Pomara LY, Ruokolainen K, Tuomisto H, Young K (2012) Avian composition co-varies with floristic composition and soil nutrient concentration in Amazonian upland forests. Biotropica 44: 545–553.
13. Artz RRE, Chapman SJ, Campbell CD (2006) Substrate utilization profiles of microbial communities in peat are depth dependent and correlate with whole soil FTIR profiles. Soil Biol Biochem, 38: 2958–2962.
14. Trivedi MR, Marecroft MD, Berry PM, Dowson TP (2008) Potential effects of climate change on plant communities in three montane nature reserves in Scotland, UK. Bio Conserv 141: 1665–1675.
15. Jesus EC, Marsh TL, Tiedje JM, Moreira FMS (2009) Changes in land use alter the structure of bacterial communities in Western Amazon soils. ISME J 3: 1004–1011.
16. Merilä P, Lämsa- MM, Stark S, Spetz P, Vierikko K, et al. (2010) Soil organic matter quality as a link between microbial community structure and vegetation composition along a successional gradient in a boreal forest. App Soil Ecol 46: 259–267.
17. Grayston SJ, Campbell CD, Bardgett RD, Mawdsley JL, Clegg CD, et al. (2004) Assessing shifts in microbial community structure across a range of grasslands of differing management intensity using CLPP, PLFA, and community DNA techniques. Appl Soil Ecol 25: 63–84.
18. Singh BK, Munro S, Reid E, Ord B, Potts M, et al. (2006) Investigating microbial community structure in soils by physiological, biochemical and molecular fingerprinting methods. Eur J Soil Sci 57: 72–82.
19. Vinten AJA, Artz RRE, Thomas N, Potts JM, Avery L, et al. (2011) Comparison of microbial community assays for the assessment of stream biofilm ecology. J Microbiol Methods 85: 190–198.
20. Hirst CN, Jackson DA (2007) Reconstructing community relationships: the impact of sampling error, ordination approach, and gradient length. Divers. Distrib 13: 361–371.
21. Poos MS, Jackson DA (2012) Addressing the removal rare species in multivariate bioassessments: The impact of methodological choices. Ecol Indic 18: 82–90.
22. Singh BK, Nunan N, Ridgway KP, Mcnicol J, Young JPW, et al. (2008) Relationship between assemblages of mycorrhizal fungi and bacteria on grass roots. Environ Microbiol 10: 534–542.
23. Burke L, Irwin R (2009) The importance of interannual variation and bottom up nitrogen enrichment for plant – pollinator networks. Oikos 118: 1816–1829.
24. Lisboa FJG, Chaer GM, Jesus EC, Gonçalves FS, Santos FM, et al. (2012) The influence of litter quality on the relationship between vegetation and below-ground compartments: a Procrustean approach. Plant Soil 367 551–562.
25. Landeiro VL, Bini LM, Costa FRC, Franklin E, Nogueira A, et al. (2012) How far can we go in simplifying biomonitoring assessments? An integrated analysis of taxonomy surrogacy, taxonomic sufficiency and numerical resolution in a mega diverse region. Ecol Indic 23: 366–373.
26. Siqueira T, Bini LM, Roque FO, Cottiene K (2012) A metacommunity framework for enhancing the effectiveness of biological monitoring strategies. PLoS One 7: e43626.
27. Nekola JC, White PS (1999) The distance decay of similarity in biogeography and ecology. J Bio 26: 867–878.
28. Legendre P, Borcard D, Peres-Neto PR (2005) Analysing beta diversity: partitioning the spatial variation of community composition data. Ecol Mon 75: 435–450.
29. Tuomisto H, Ruokolainen K (2006) Analysing or explaining beta diversity? Understanding the targets of different methods of analysis. Ecology 87: 2697–2708.
30. Kuehnelt-Leddihn ER (2007) The menace of the herd or Procrustes at large. The Bruce Publishing Company Milwaukee, Auburn. 385 p.
31. Jackson DA (1995) PROTEST: a Procrustean randomization test of community environment concordance. Ecoscience 2: 297–303.
32. Alárcon R, Waser NM, Orlleton J (2008) Year-to-year variation in the topology of a plant - pollinator interaction network. Oikos 117: 1796–1807.
33. Burke L, Irwin R (2009) The importance of interannual variation and bottom up nitrogen enrichment for plant – pollinator networks. Oikos 118: 1816–1829.
34. ten Berge JMF, Kiers HAL, Commandeur JJF (1993) Orthogonal Procrustes rotation for matrices with missing values. B J Math Stat Psyc 46: 119–134.
35. Dijksterhuis GB, Gower JC (1992) The interpretation of Generalized Procrustes Analysis and allied methods. Food Qual Prefer 3: 67–87.
36. Dray S, Chessel D, Thioulouse J (2003) Procrustean Co-inertia analysis for the linking of multivariate datasets. Ecoscience 10: 110–119.
37. Legendre P, Gallagher ED (2001) Ecologically meaningful transformations for ordination of species data. Oecologia 129: 271–280.
38. Legendre P, Legendre L (2012) *Numerical Ecology*, 3rd English edn. Elsevier Science BV, 516 Amsterdam.
39. Mitchell RJ, Hester AJ, Campbell CD, Chapman SJ, Cameron CM (2010) Is vegetation composition or soil chemistry the best predictor of soil microbial community? Plant Soil 333: 417–430.
40. Mitchell RJ, Hester AJ, Campbell CD, Chapman SJ, Cameron CM, et al. (2012) Explaining the variation in the soil microbial community: do vegetation composition and soil chemistry explain the same or different parts of the microbial variation? Plant Soil 351: 355–362.
41. Dumbrell AJ, Nelson M, Helgason T, Dytham C, Fitter AH (2009) Relative roles of niche and neutral processes in structuring a soil microbial community. ISME J doi:10.1038/ismej.2009.122.

42. Zheng YM, Cao P, Fu B, Hughes JM, He JZ (2013) Ecological Drivers of Biogeographic Patterns of Soil Archaeal Community. PloS One 8: e63375.
43. Peres-Neto PR, Legendre P, Dray S, Borcard D (2006) Variation partitioning of species data matrices: estimation and comparison of fractions. Ecology 87: 2603–2613.
44. Chaer GM, Fernandes MF, Myrold DD, Bottomley PJ (2009) Shifts in microbial community composition and physiological profiles across a gradient of induced soil degradation. Soil Sci Soc Am J 73: 1327–1334.
45. Peixoto RS, Chaer GM, Franco N, Reis Junior FB, Mendes IC, et al. (2010) A decade of land use contributes to changes in the chemistry, biochemistry and bacterial community structures of soils in the Cerrado. A Van Leeuw 98: 403–413.
46. Thioulouse J, Simier M, Chessel D (2004) Simultaneous analysis of a sequence of paired ecological tables. Ecology 85: 272–283.
47. Dolédec S, Chessel D (1994) Co-inertia analysis: an alternative method for studying species–environment relationships. Freshwater Biol 31: 277–294.
48. Ter Braak, CJF, Schaffers AP (2004) Co-correspondence analysis: a new ordination method to relate two community compositions. Ecology 85: 834–846.
49. van der Heijden MGA, Klironomos J, Ursic M, Moutoglis P, Streitwolf-Engel R, et al. (1998) Mycorrhizal fungi determines plant biodiversity, ecosystem variability and productivity. Nature 396: 69–72.

15

Analysis of the Diversity of Substrate Utilisation of Soil Bacteria Exposed to Cd and Earthworm Activity Using Generalised Additive Models

Selene Muñiz, Juan Lacarta, María P. Pata, Juan José Jiménez, Enrique Navarro*

Department of Biodiversity & Conservation and Ecosystem Restoration, Pyrenean Institute of Ecology, Spanish National Research Council, Zaragoza, Spain

Abstract

Biolog EcoPlates™ can be used to measure the carbon substrate utilisation patterns of microbial communities. This method results in a community-level physiological profile (CLPP), which yields a very large amount of data that may be difficult to interpret. In this work, we explore a combination of statistical techniques (particularly the use of generalised additive models [GAMs]) to improve the exploitation of CLPP data. The strength of GAMs lies in their ability to address highly non-linear relationships between the response and the set of explanatory variables. We studied the impact of earthworms (*Aporrectodea caliginosa* Savigny 1826) and cadmium (Cd) on the CLPP of soil bacteria. The results indicated that both Cd and earthworms modified the CLPP. GAMs were used to assess time-course changes in the diversity of substrate utilisation (DSU) using the Shannon-Wiener index. GAMs revealed significant differences for all treatments (compared to control -S-). The Cd exposed microbial community presented very high metabolic capacities on a few substrata, resulting in an initial acute decrease of DSU (i.e. intense utilization of a few carbon substrata). After 54 h, and over the next 43 h the increase of the DSU suggest that other taxa, less dominant, reached high numbers in the wells containing sources that are less suitable for the Cd-tolerant taxa. Earthworms were a much more determining factor in explaining time course changes in DSU than Cd. Accordingly, Ew and EwCd soils presented similar trends, regardless the presence of Cd. Moreover, both treatments presented similar number of bacteria and higher than Cd-treated soils. This experimental approach, based on the use of DSU and GAMs allowed for a global and statistically relevant interpretation of the changes in carbon source utilisation, highlighting the key role of earthworms on the protection of microbial communities against the Cd.

Editor: Volker Thiel, Kantonal Hospital St. Gallen, Switzerland

Funding: This work was supported by Spanish Ministry of Economy and Competitiveness (National Research Plan, ref. BFU2010-22053, http://www.mineco.gob.es/portal/site/mineco/). The ARAID Foundation (http://www.araid.es/) is also acknowledged for support provided to J.J. JimÃ©nez. The funders had no role in study design, data collection and analysis, decision to publish, or preparation of the manuscript.

Competing Interests: The authors have declared that no competing interests exist.

* E-mail: enrique.navarro@ipe.csic.es

Introduction

Biolog Ecoplates

Biolog MicroPlates were developed in the late 1980s to assist in the identification of bacterial strains [1]. These 96-well plates contained carbon sources and a tetrazolium violet redox dye that turned purple if inoculated microorganisms utilised these sources. By comparing the obtained carbon substrate utilisation patterns with databases, it was possible to establish a probable identification [2]. Later, microbial ecologists used Biolog plates to investigate patterns at the community level. Therefore, a new plate specifically designed for community analysis and microbial ecological studies was created; this new plate was referred to as the EcoPlate [3]. The EcoPlate contains 31 of the most useful carbon sources for soil community analysis [3], allowing for community-level physiological profiling (CLPP) of heterotrophic bacterial assemblages. This technique has also been widely used to assess the toxicological impacts of different pollutants [4], including different heavy metals [5–9].

The use of EcoPlates results in a CLPP, yielding a very large amount of data that may be difficult to interpret. These difficulties are related to the alteration of the original microbial communities due to soil sampling and pre-treatment as well as bacterial extraction [4,10]. Moreover, the Biolog procedure can be considered a culture method in which the originally inoculated community is altered. The contribution of a certain species population to the colour profile will depend on its culturability in the Biolog wells and on its interactions with other species (differential growth and competition). The tetrazolium dye also introduces some bias in the profile because not all bacteria are able to reduce it [4]. Thus, the Biolog method is more useful for comparing soil microbial communities than for community characterisation [4]. CLPPs provide little insight regarding the function of the community *in situ* unless they are combined with other microbial methods that do not rely on the culturing of the soil microflora [11,12].

The pattern of positive and negative responses as well as substrate oxidation rate and extent are highly reproducible for simple microbial communities, particularly if the inoculum densities are similar [10]. Nonetheless, data corrections are required if inoculum density is not controlled. A kinetic approach must be used to extract data from the plates to avoid problems related to incubation time. However, there is no consensus regarding which statistics are most appropriate, and the interpretation of their meaning remains unclear [4].

Earthworms and cadmium (Cd) on soils

Earthworms are key members of the soil macrofauna in temperate soils [13] that directly or indirectly modulate resource availability (quality, quantity, and distribution) for other organisms. Earthworms are considered ecosystem engineers because they modify, maintain, or create habitats [14] as they build soil biostructures, which consist of aggregates and macropores [15–17]. The soil zone influenced by earthworm burrowing and casting was first termed the drilosphere by Bouché, 1975 [18,19], where microbial biomass is greater than in the surrounding soil [20]. In general, the mechanical and biological activities of earthworms favour organic matter mineralisation and humification and stimulate microbial activity [13].

Cd levels in the environment vary widely, and the average natural abundance of this element in the earth's crust varies from 0.1 to 0.5 mg kg^{-1}. However, human activities, such as mining, metal smelting, urban and industrial emissions, waste incineration, coal combustion, traffic dust, and particularly the use of phosphate fertilisers and sewage sludges, are quoted as the primary reason for the increase in soil Cd content over the last several years in Europe [21]. As heavy metals cannot be degraded, they tend to accumulate in soils. Their disappearance indicates that they have leached into deep soil layers and transferred to subterranean waters or transported to other locations by runoff [22]. The fate of Cd in soils will depend on a balance between adsorption, leaching, and biological uptake. In most soils, more than 99% of Cd is found in the solid phase, and less than 1% is present in the soil solution [21].

Earthworm species belonging to the three main ecological groups, i.e. epigeics, endogeics and anecics, have been shown to increase the mobility and bioavailability of some heavy metals in soils [23,24]. As a result, soil ecosystem function can be altered by heavy metal pollution through changes in microbial communities and earthworms in the drilosphere [25]. These changes in microbial community composition may lead to increased heavy metal tolerance in the bacterial community [26,27]. However, a decrease in microbial diversity also leads to less soil resilience to natural or human-induced perturbations [28]. Thus, studies on the joint effect of earthworm engineering and soil pollution on the bacterial community may help to prevent and remediate human impacts at the ecosystem level.

Experimental approach: the use of generalised additive models (GAMs)

Our goal is to demonstrate that the joint action of earthworm activity and Cd toxicity induce changes in the bacterial community that are measurable by changes in the CLPP. To this end, we have assessed and measured these changes using a combination of descriptive and multivariate statistics together with GAMs [29,30]. GAMs are semiparametric extensions of generalised linear models that allow for the flexible modelling of covariate effects, avoiding the limitations of parametric models.

This methodology is appropriate when the relationship between the variables is expected to be complex and not easily fitted by standard linear or non-linear models, which is the case in most biological studies. One of the most frequent uses of GAMs in ecology is the modelling of species distribution and species abundance [31], related to environmental and geographical variables [32], being plant ecology and fisheries the two fields where the use of GAMs have been generalized in the last decade. Studies of air pollution and health are another domain where these models have been widely used as standard methods since its flexibility in modelling spatial and temporal processes [33]. One of the great advantages of these models is its ability to determine the nature of the relationship between the response and explanatory variables rather than assuming a fixed relationship, allowing for fitting complex relationships in a straightforward manner.

Methods

Soil and earthworm sampling

The earthworm *Aporrectodea caliginosa* is commonly used in soil ecotoxicological studies [34]. This earthworm is present in the natural soils surrounding our facilities, is easy to culture and sensitive enough against different toxicants, and being a key actor of the matter cycling and soil formation, the results obtained with it are environmentally relevant.

A total of 102 individuals were hand-sorted in an experimental field near the Pyrenean Institute of Ecology (Montañana, Zaragoza) in June 2011. Experimental soils were collected in an experimental field near Jaca (northern Spain). No specific permissions were required for these activities since both experimental fields from which the soil and the earthworms were collected belong to the Spanish National Research Council. The activities neither involve vertebrates nor endangered or protected species.

The soil pH was determined from extracts obtained with a 1:2.5 ratio (g soil:g distilled H_2O [pH_w] or in 0.1 M KCl [pH_{KCl}]). Organic matter (OM) content, cation exchange capacity (CEC), and soil texture were measured by calcination, atomic absorption spectrometry, and aerometry, respectively [35]. Field capacity (0.3 bar) and wilting point (15 bar) were determined in two pressure chambers [36]. The pH values at the earthworm sampling site were 7.82 (water) and 7.43 (KCl). The prevailing soil conditions (pH, OM content, and CEC) are provided in Table 1. The soil was

Table 1. Soil pH, organic matter (OM), and cation exchange capacity (CEC) at day 32.

Treatment	pH_w	pH_{KCl}	OM	CEC	[Cd] μg g^{-1}	[Cd] mg L^{-1}
			%	meq 100 g^{-1}	in soils	in pore water
S	8.02	7.89	5.33	17.7	< 0.5	-
Cd	7.8	7.88	5.28	18.3	329	0.78
Ew	7.84	7.91	5.25	19.4	< 0.5	-
EwCd	7.63	7.84	5.16	18.7	311±76	0.49±.002

Soil pH was determined in water (pH_w) and KCl (pH_{KCl}). Prior to Cd analysis in soils, six replicates from Cd treatments were integrated, whereas EwCd replicates were analysed separately (accordingly, standard deviation is available).

loamy with 44.2% sand, 38.1% silt, and 17.7% clay. The bulk density of dried and sieved soil was 1,270.83 kg m^{-3}. The field capacity was 30.3% (w/w), and the wilting point was 14.5% (w/w).

All earthworms remained in a terrarium with the same soil from which they were collected for two weeks (maintenance period). Then, they were moved to another terrarium that contained the experimental soil sieved at < 2 mm. The duration of this preconditioning period was 16 days, and its purpose was to allow the earthworms to empty their guts of the previously ingested soil.

Design and duration of the experiment

We tested the effect of Cd and earthworms on the ability of bacteria to degrade the 31 different sources of carbon included in Biolog EcoPlates. Accordingly, a bifactorial experiment with four treatments was designed (S: soil control; Cd: soil irrigated with Cd; Ew: soil engineered by earthworms; EwCd: soil engineered by earthworms and irrigated with Cd). A detailed schema of the design is provided in Figure 1. Six replicates (pots) per treatment were filled with 1.5 L of sieved soil. The soil surface exposed to air in each pot was approximately 180 cm^2. The earthworms were exposed to a natural light/dark cycle from June to August. The air humidity as well as air and soil temperatures were controlled. The soil moisture was measured by a Tektronix 1502C TDR Cable Tester reflectometer (Bracknell, Berkshire, United Kingdom) [37]. After the preconditioning period, specimens of a similar size and weight were selected and introduced into each pot (three individuals). Each pot was subsequently watered with distilled water to reach field capacity (i.e., 30% w/w).

All pots were watered with 200 mL of distilled water from Days 1–16. From Day 17 onward, the pots were watered with 200 mL of distilled water or Cd solution depending on the treatment (Figure 1). Cd was added to the soil as $CdCl_2 + 2.5\ H_2O$ (Panreac), and the pots were watered with a 343 mg L^{-1} solution. The irrigation was designed to achieve soil concentrations below the median lethal concentration (LC_{50}) to *A. caliginosa, which* has been estimated in 540 µg Cd g-1 for a 56-day exposure [38]. The percolated water was collected daily 24 h after the watering that generated it and before watering the pots again. The soils were exposed to both Cd and earthworms for 16 days (see Figure 1).

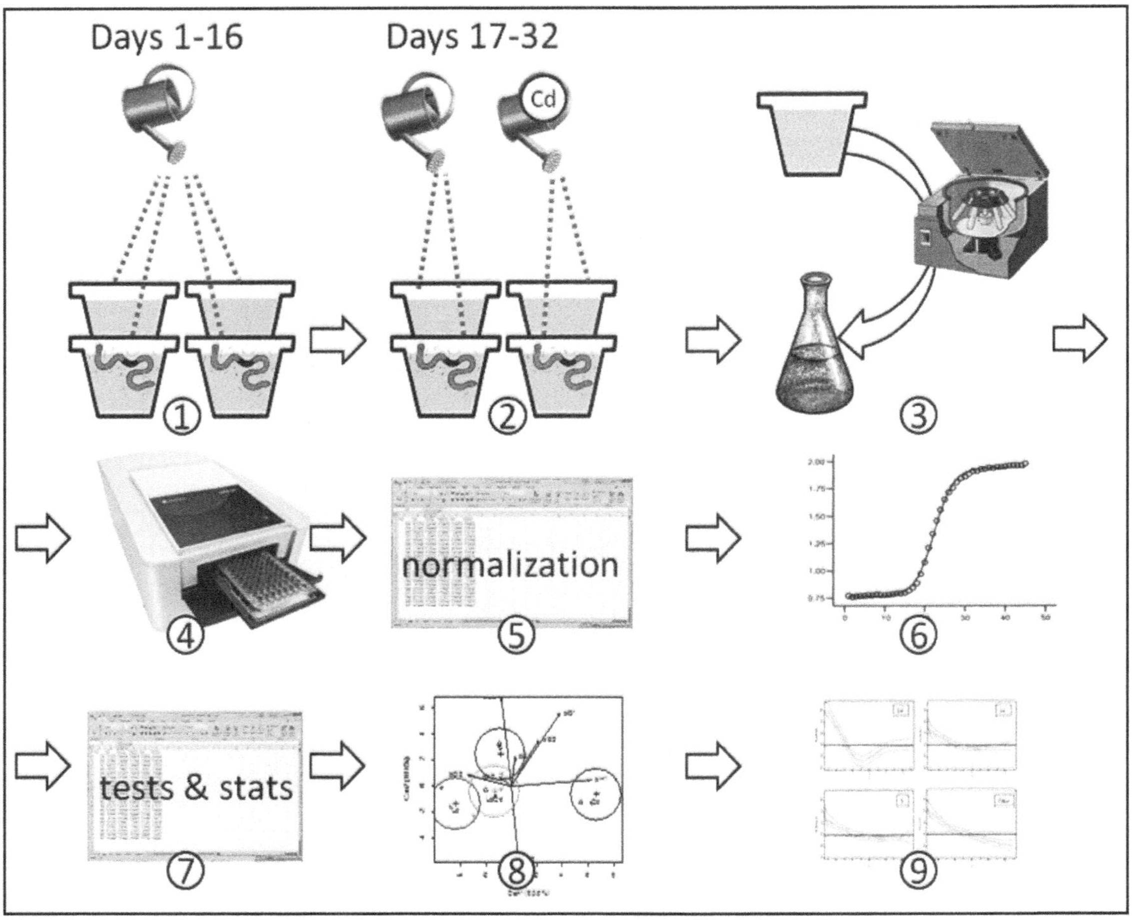

Figure 1. Diagram showing experimental design and data processing. During the first days (1), soil bacteria were exposed to earthworms and irrigated with clean water. Later, soils were irrigated with Cd or clean water depending on the treatment (2). At the end of the experimental period, bacteria were extracted by centrifugation and suspended in liquid media (3). This media was inoculated in Biolog EcoPlates (4). The results of colour development over 167 h were recorded and normalised (5). The results were fitted to a sigmoidal model (6), and their parameters were calculated (slope, time to reach 50% of the maximum colour intensity, and maximum intensity reached). Parameters were compared to assess the effects of the different treatments (7), and the carbon sources that were more informative (accumulating more discriminating power) were selected for multivariate analysis (8). Finally, GAMs were fitted to assess the effect of the different treatments on the use of diverse carbon sources over time, determined by the Shannon-Weaver index (9).

Measurement of heavy metal

Following Hobbelen et al. [39], three soil sample types were obtained on Day 32. First, pore water was collected by placing 40 g samples of fresh soil into 50 mL Falcon tubes and centrifuging for 40 min (2,000 g). The supernatants were filtered with a Büchner funnel and cellulose filters (0.45 µm pore size; Sartorius Stedim Biotech) and fixed with 69% HNO_3 (Hiperpur; Panreac) at a final concentration of 4% (v/v). Next, the soil was air dried. To obtain $CaCl_2$ extracts, a 5:1 mixture (5 g of 0.01 M $CaCl_2$:1 g of soil) was prepared and shaken for 2 h at 200 rpm. The extracts were filtered and fixed. Finally, the soil samples were acid-digested in a microwave (Milestone digester) by adding 4 mL of HNO_3 (PA-ISO 69%) and 1 mL of H_2O_2 (33%) to each 0.1 g of dried soil (20 min at 220°C, final temperature). All samples were stored at 4°C until further analysis. The Cd concentration in the soil and percolated water was determined by inductively coupled plasma-optical emission spectrometry (ICP-OES).

Bacterial community-level physiological profiling (CLPP)

The microbial community population structure and changes in carbon source usage were assessed by live/dead staining and Biolog tests, respectively. Bacterial cells were extracted from soils following Katayama [40] and modified as described in Ikeda [41]. Briefly, 10 g of fresh soil was diluted in 95 mL of Milli-Q water. The samples were homogenised for 15 min, and 10 mL of the soil suspension was pipetted into a 50 mL Falcon tube. The sample was sonicated for 1 min and subsequently centrifuged at 1,000 g for 10 min to obtain 9.5 mL of sample. The remaining 0.5 mL was resuspended in 10 mL and submitted to a new dispersion-centrifuge cycle to obtain additional microbial inocula. This cycle was repeated five times until a 47.5 mL sample was obtained, which was refrigerated at 4°C until analysis.

Bacterial extracts were later filtered using 60 µm cellulose filters to reduce interference during OD measurements (for Biolog). Viable and dead cells (live/dead staining) were stained following the LIVE/DEAD protocol [42]. The cells in 10 randomly selected fields per sample were counted using NIS-ELEMENTS software (Nikon ©). The number of viable cells was calculated as total - dead.

For CLPP, integrated soil samples were used (sampling was performed at different depths). One Biolog™ EcoPlate was inoculated per treatment and incubated at 25°C in a dark chamber. Each plate contained three sets (replicates) of carbon sources, and each plate was filled with samples from one pot. An Anthos 2010 microplate reader (Biochrom) was used to read the absorbance in each well at 590 nm (OD_{590}) at even intervals from 0 to 167 h.

Diversity of substrate utilisation (DSU)

DSU was determined by the Shannon-Weaver index (H), which is one of the most commonly used diversity measures [43,44]. This index considers both the number of species and their abundance. In this case, the H value describes the ability of the bacterial community to degrade more or fewer types of carbon sources, thus being an index of physiological diversity of the bacterial community. Microbial communities that are able to degrade more substrates or/and to degrade them with similar efficiency would have higher values of H. The index was calculated as follows:

$$H = -\sum p_i (\ln p_i)$$

where p_i is the normalised value of OD_{590} (see point 2.6). This equation is commonly used to process Biolog data [45,46].

Statistical analysis

The carbon degradation data from every well were normalised by being divided by the initial OD_{590} at time 0 after averaging the three wells containing the same carbon source. This procedure was performed to reduce the variability due to differences in inocula densities [47,48].

The results of colour development showed sigmoidal relationship between the OD_{590} and time for all carbon sources. Therefore, the curves were fitted to a non-linear model [47,49] with the four-parameter log-logistic function given by the formula:

$$f(x_i,(b,c,d,e)) = c + \frac{d-c}{1+\exp(b(\log(x_i)-e))} + \varepsilon_i$$

Where parameter b is the slope of the curve (S), *c* is the lower asymptote of the sigmoidal curve, *d* is the upper asymptote of the curve representing the maximal degradation increase (M), and parameter *e* is the time to reach 50% of maximum degradation (TM_{50} or ED_{50}). T-tests were performed to evaluate which ratios between parameters from two different treatments were significantly ($p<0.05$) different from 1; p-values were adjusted using Bonferroni correction for multiple test.

For the CLPP analysis, the carbon sources shown to be more informative regarding the effects of heavy metals on bacterial communities were selected. Other sources (such as D-xylose and D-cellobiose) identified in previous studies were also considered [50–52]. Canonical discriminant analysis (CDA) was performed on the selected carbon sources (β-Methyl-D-Glucoside, i-Erythritol, D-Mannitol, α-Cyclodextrin, N-Acetyl-D-Glucosamine, α-D-Lactose, and the two previously mentioned).

Thin-plate regression splines [53] were used as smoothers, with optimal effective degrees of freedom chosen automatically using the generalised cross-validation criteria [30].

All analyses were performed with the free statistical software R [54]. The *drm and compParm* functions of the *drc* package [49] were used to estimate and compare kinetic parameters, the *candisc* package [55] was used for CDA, and the *gam* function of the *mgcv* library [30] was used for GAM modelling.

Results

Bacterial cell density and Cd in soils

Cd-treated soils contained significantly fewer bacterial cells (1.1e7 cells g^{-1} DW) than control soils (paired t-tests, p = 0.03) and a smaller proportion of dead cells (12%). Control soils (S) contained more bacterial cells (4.5e7 cells g^{-1} DW), and earthworm-treated soils had intermediate values (3.0–3.6e7 cells g^{-1} DW). The control and earthworm treatments contained very similar proportions of dead cells (17–20%).

The final Cd concentration in the soils was lower than 0.5 µg g^{-1} $_{dry\ soil}$ in the S and Ew treatments, 328.54 µg g^{-1} $_{dry\ soil}$ in the Cd treatment, and 310.93±75.58 µg g^{-1} $_{dry\ soil}$ (average value, n = 6) in the EwCd treatment. The Cd concentration in the pore water was 0.78 mg L^{-1} in the Cd treatment and 0.49±0.002 mg L^{-1} (n = 2) in the EwCd treatment. The amount of $CaCl_2$-extractable metal was 2.18 µg g^{-1} $_{dry\ soil}$ in the Cd treatment and 2.2±0.55 µg g^{-1} $_{dry\ soil}$ (n = 3) in the EwCd treatment.

Table 2. Comparison of kinetic parameters.

Code	Carbon sources	S/Ew	S/Cd	S/EwCd	Cd/Ew	Cd/EwCd	Ew/EwCd
A2	β-Methyl-D-Glucoside				M:1.7	M:2.08	
A3	D-Galactonic acid α-Lactone	S:0.26			M:1.61	M:1.86	
		(−0.38,0.90)			(1.46,1.75)	(1.68,2.04)	
B1	Pyruvic Acid Methyl Ester						M:0.78
							(0.68,0.87)
B3	D-Galacturonic Acid		M:0.75		M:1.3		
			(0.67,0.83)		(1.22,1.37)		
C2	i-Erythritol					M:1.59	
						(1.37,1.81)	
C3	2-Hydroxy Benzoic Acid						S:0.14
							(−0.27,0.55)
C4	L-Phenylalanine		S:0.23			M:2	
			(−0.12,0.58)			(1.88,2.12)	
D2	D-Mannitol		M:0.68		M:1.71		
			(0.51,0.85)		(1.56,1.85)		
D3	4-Hydroxy Benzoic Acid	T:0.51	M:0.7			M:1.79	
		(0.12,0.90)	(0.55,0.85)			(1.61,1.97)	
			T:0.55				
			(0.15,0.94)				
D4	L-Serine		M:0.7		M:1.53		
			(0.41,0.99)		(1.42,1.63)		
E1	α-Cyclodextrin					M:1.56	
						(1.38,1.74)	
E2	N-Acetyl-D-Glucosamine		M:0.67		M:1.67		
			(0.41,0.93)		(1.44,1.89)		
F1	Glycogen			M:1.59		M:1.7	
				(1.29,1.89)		(1.50,1.90)	
F2	D-Glucosaminic Acid		T:0.68			M:1.61	
			(0.36,0.98)			(1.36,1.86)	
F3	Itaconic Acid					M:0.75	
						(0.65,0.85)	
F4	Glycyl-L-Glutamic Acid				M:1.68	M:1.72	
					(1.49,1.86)	(1.53,1.91)	
G2	Glucose-1-Phosphate	M:1.53			M:1.36		
		(1.33,1.73)			(1.28,1.43)		
G3	α-Ketobutyric Acid		S:0.27			M:1.6	
			(−0.06,0.60)			(1.50,1.70)	
G4	Phenylethyl-amine	S:0.15					
		(−0.18,0.48)					
H1	α-D-Lactose	S:0.35		S:0.31	S:0.29	S:0.29	
		(−0.06,0.76)		(−0.14,0.76)	(−0.04,0.62)	(−0.32,0.66)	
		M:1.51		M:1.59	M:1.25	M:1.32	
		(1.29,1.73)		(1.35,1.83)	(1.10,1.40)	(1.16,1.48)	
H2	D,L-α-Glycerol Phosphate	M:1.84		M:1.64			
		(1.55,2.13)		(1.33,1.95)			
H3	D-Malic Acid	M:1.65	T:1.93	M:1.5	M:1.72		
		(1.46,1.84)	(1.28,2.57)	(1.31,1.69)	(1.54,2.71)		
H4	Putrescine	S:0.19			M:2.18	M:2.2	
		(−0.20,0.58)			(1.64,2.71)	(1.72,2.68)	

Table 2. Continued.

To assess the effect of the different treatments in the degradation of the different carbon sources, the parameters from the 4 parameters log-logistic models, were divided and the resulting ratio compared to 1 using t-tests. The letter indicates the parameter compared (T = TM_{50}, S = slope, M = maximum). Values between brackets are 95% CI. Table shows only those comparisons with significant p-values (adjusted using Bonferroni correction for multiple test), meaning that these ratios significantly ($p<0.05$) different from 1 and thus the parameters compared were significantly different. As example, the first value in the column S/Ew indicates that the slope of the log-logistic model of degradation for A3 in treatment S was four times lower than in treatment Ew.

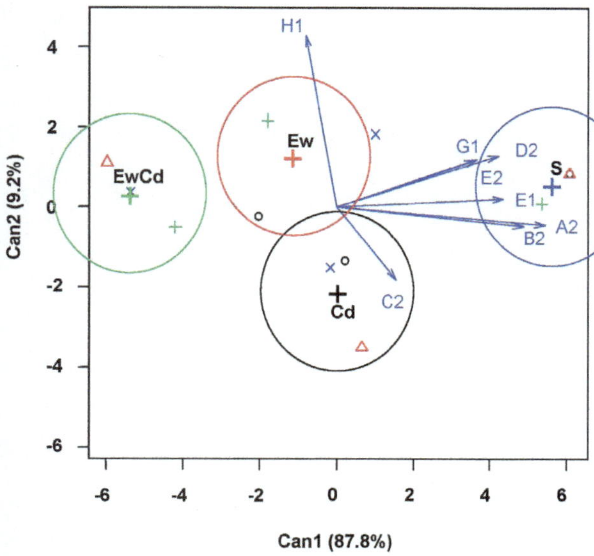

Figure 2. Plot of the two canonical discriminant functions for TM_{50}. TM_{50} refers to the time to reach 50% of the maximum colour intensity. Centroid locations of each treatment are indicated by "+".

Kinetic parameters

The results of colour development revealed sigmoidal relationships between the OD_{590} and time for all carbon sources (see point 6 in Figure 1); thus, the results were adjusted to the four parameters log-logistic model [47]. Three parameters of this model were selected for their usefulness in comparing treatments: the maximum value attained (M), the time at which the colour development is half of its maximum value (TM_{50}), and the slope (S). Those parameters presenting significant differences between treatments are shown in Table 2.

This process is the first step of the experimental approach proposed in this study; the carbon sources exhibiting significant differences between treatments (i.e., discriminant sources) would be potential candidates for the subsequent data analysis (see Figure 1) with the aim of increasing the discriminant power of the procedure.

Treatments provoked differences in the kinetic parameters of 23 of the 31 carbon sources. Specifically, the largest differences in CLPP were promoted by the activity of earthworms in Cd-treated soils, which accounted for the differences in the kinetics of 13 carbon sources. On the other hand, the addition of Cd to earthworm-treated soils only resulted in differences in two carbon sources (Table 2).

Canonical discriminant analysis (CDA)

Figure 2 shows the first two CDA axes based on the TM_{50} values of the sigmoidal models (see section 3.1). The cumulative percentage of the variance explained by the first two canonical

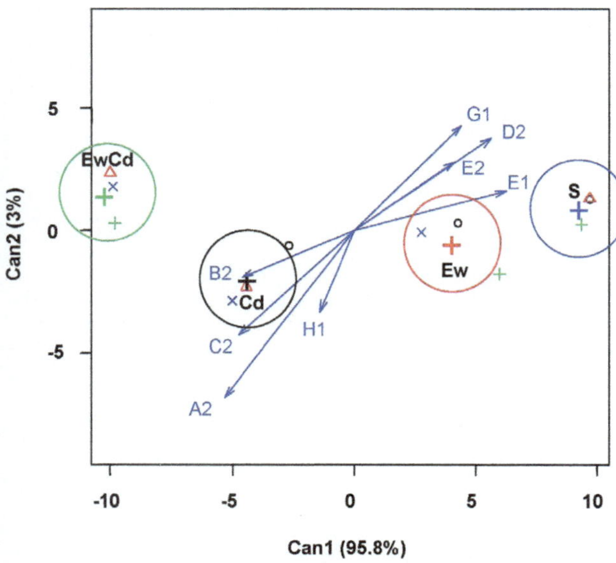

Figure 3. Plot of the two canonical discriminant functions for the maximum colour intensity (M). Centroid locations of each treatment are indicated by "+".

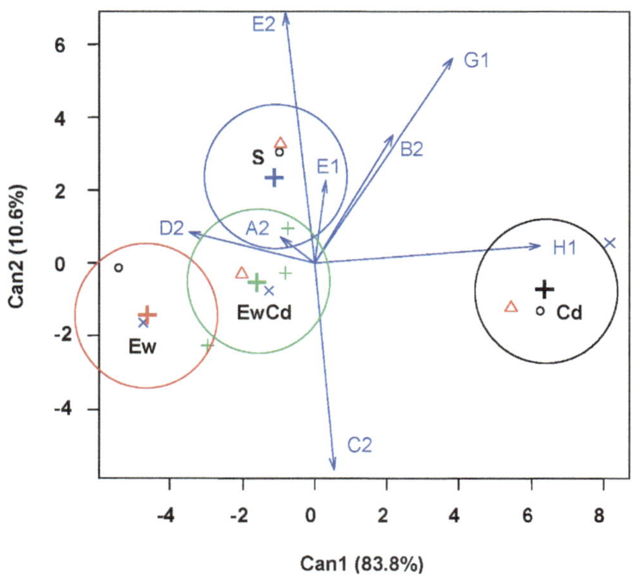

Figure 4. Plot of the two canonical discriminant functions for slope (S). Centroid locations of each treatment are indicated by "+".

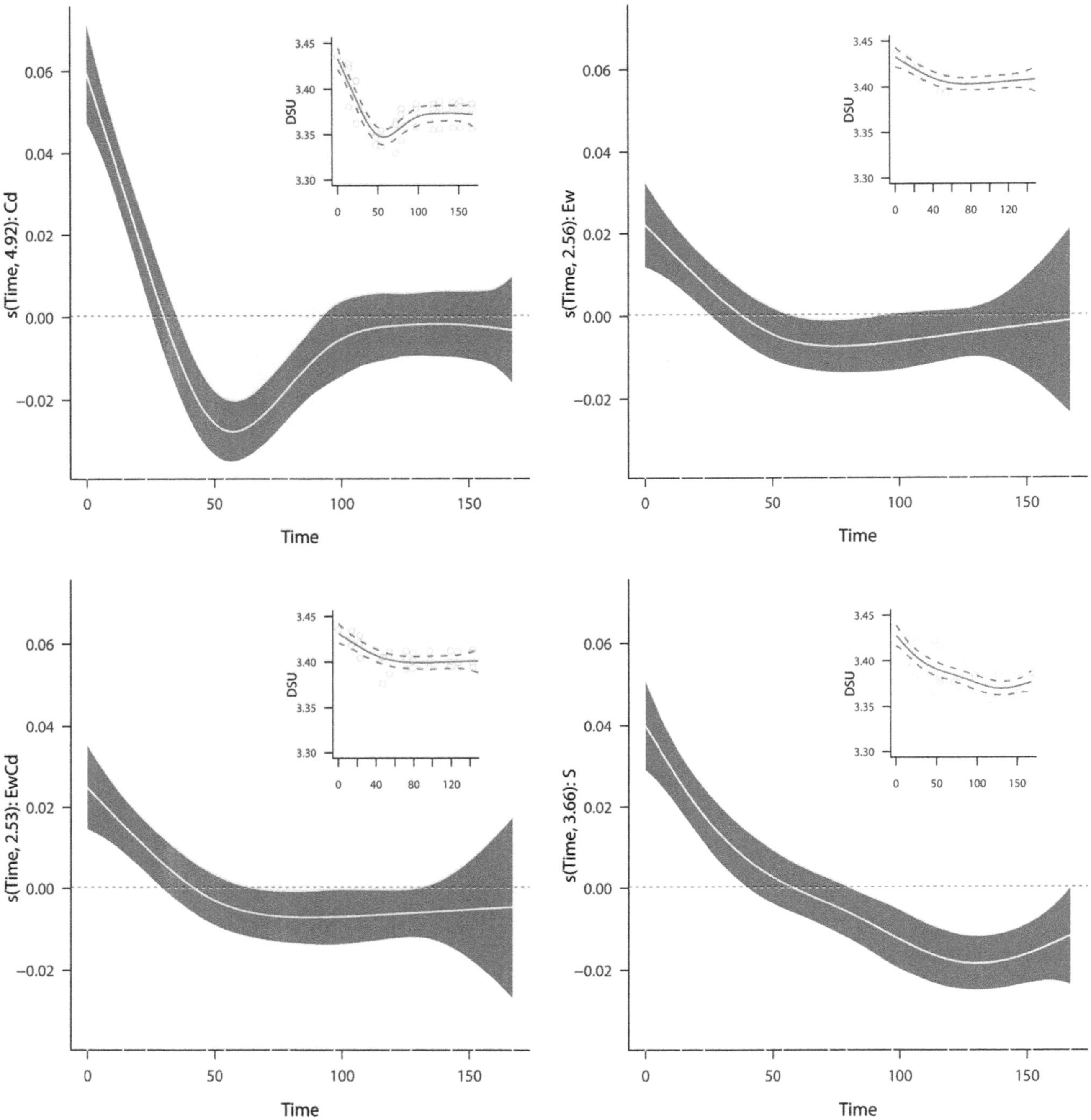

Figure 5. Estimated flexible effects of time-related changes in the diversity of substrate utilisation (DSU). The x axis represents the changes on DSU values over time (hours), obtained from the GAM with the 95% point-wise confidence band (grey coloured) and centred at zero (red dotted line) to allow comparisons. The effective degrees of freedom obtained for each predictor are indicated on the y-axis. The estimated curves in the response scale along with raw data are located in the upper right corner of each plot.

functions is 97%. The first axis accounts for 87% of the variance and is strongly influenced by D-cellobiose. The second function is strongly influenced by D-cellobiose and is also influenced by N-acetyl-D-glucosamine. Figure 3 presents the CDA based on M values. The cumulative percentage of the first two functions is very high (approximately 99%). In this case, the sources with the most influence are also N-acetyl-D-glucosamine and D-cellobiose for both axes. Figure 4 presents the CDA based on slopes. Although the cumulative percentage is also very high (94%), in this case, the standardised coefficients indicate that the weight of the different sources is more similar than that in the previous CDA.

Assessing the time course changes in diversity of substrate utilization (DSU) based on the Shannon-Wiener index using generalised additive models (GAMs)

Results from GAM model shows that treatments exhibit significantly different time course changes in DSU, with a explained deviance of 87%. The effective degrees of freedom, *chi*

Table 3. Estimated non-parametric components of GAM model, with the corresponding effective degrees of freedom, F-statistic and P value.

Smooth effect of variable	edf	F-statistic	P value
f(Time)EwCd	2.525	8.519	< 0.001
f(Time)Cd	4.92	23.868	< 0.001
f(Time)Ew	2.555	6.772	< 0.001
f(Time)S	3.664	18.483	< 0.001

square statistic and p-values of estimated flexible effects from the model are shown in table 3. In Figure 5, the red dotted line represents a hypothetical community with a DSU that was constant as a function of time; in other words, a community exhibiting such a time-course trend would degrade all carbon sources with similar affinity (S) and capacity (M). Thus, this red line can aid in our interpretation of the results. All treatments caused a decrease in DSU during the first 54 h. This decrease was more acute in the bacterial community from the Cd-treated soils. Thereafter, differences were evident between treatments. The communities from soils containing earthworms (Ew and EwCd) presented similar values until the end of the experiment (143 h) regardless the presence of Cd. On the other hand, the community from the Cd-treated soil exhibited an increase in DSU during the next 43 h and did not change thereafter. The community exhibiting the most diverse behaviour was that obtained from control soils (S). In this community, the decline in DSU was constant over time. The model validation did not show any violation of variance homogeneity (i.e. no patterns in residuals) or normality.

Discussion

Kinetic parameter analysis

Cd exposure resulted in the enhanced degradation of eight carbon sources, as shown by increases in S, M, and/or TM_{50} (see column 2 in Table 2). These results indicate the Cd tolerance of soil microbial communities. It has been shown that long-term exposure to Cd (months to years) may cultivate a more tolerant community [26,27], exhibiting enhanced enzyme activities [56]. However, our data indicate that the acclimatisation of the communities to Cd may be achieved after only a few days, which has also been shown in communities exposed to mercury [6]. The low percentage of dead cells in Cd-treated soils also supports this observation.

On the other hand, earthworm activity had a modulating effect on Cd impact. Although earthworm activity allowed for the survival of a higher number of bacterial cells, it also caused reductions in bacterial catabolic activity. As shown in column Cd/EwCd in Table 2, earthworm activity resulted in lower M values for 12 carbon sources. Because EwCd contained three times more active bacteria than Cd, this result indicates a marked reduction in their enzymatic activity. The reduction of bacterial growth and activity by earthworms has also been recently described in the absence of toxicants [57].

Canonical discriminant analysis (CDA)

In all three CDAs (based on TM_{50}, M and S), treatments resulted in different CLPPs, as shown by the centroids. The centroids are the weighted position of each treatment in the multidimensional space created by the eight carbon sources used in the analysis (Figs. 2, 3, and 4); arrows indicate the effect of each source in the discriminatory space. The clear separation between centroids means that these sources gathered a sufficient amount of information to differentiate the effect of each treatment on the CLPP. The CDA illustrates how the CLPP is affected by both Cd and earthworms. Compared with Cd, earthworms had a greater effect on the physiological profile of bacterial communities. The three parameters used (TM_{50}, M, and S) allow for a clear discrimination between the effects of the four treatments; however, these parameters have different sensitivities for each experimental factor.

CDA on TM_{50} (Figure 2). This parameter allowed for discrimination of the control soils and can be used to detect the interactive effects of Cd and earthworms. The EwCd treatment is characterised by faster degradation of D-cellobiose (G1), α-cyclodextrin (E1), and β-methyl-D-glucoside (A2), whereas Cd is characterised by faster degradation of N-acetyl-D-glucosamine (E2) and D-xylose (B2).

CDA on M (Figure 3). Cd provoked clear differences in M values; its presence reduced the capacity for degrading α-D-lactose (H1), β-methyl-D-glucoside (A2), D-xylose (B2), and i-erytritol (C2), whereas the degradation of D-mannitol (D2), α-cyclodextrin (E1), N-acetyl-D-glucosamine (E2), and D-cellobiose (G1) was enhanced.

CDA on S (Figure 4). Slopes also allow for a clear discrimination of Cd impacts. Thus, the Cd treatment appeared clearly separated from the other treatments. The S values of α-D-lactose (H1) and D-cellobiose (G1) were enhanced by Cd.

Sources indicative of specific impacts. The increase in TM_{50} for the degradation of D-cellobiose (code G1 in Biolog plates and in Figure 2) appears to be a promising indicator for earthworm activity. Changes in the slope of α-D-lactose are also indicative of earthworm activity regardless of the presence of Cd. In contrast, an increase in S for 2-hydroxybenzoic acid may be indicative of Cd impacts on bacterial-earthworm consortia.

Time course changes in diversity of substrate utilization (DSU) by GAM analysis

GAMs revealed significant differences for all treatments (compared to control -S-) in the time trend of DSU. However, all treatments shown a common initial decline (the first 54 h) in the DSU; that is generally attributed to changes in the community inoculated [11], which becomes less diverse. The CLPP would be dominated by fast growing species that are positively selected by high nutrient levels [4]. Other mechanisms related to metal tolerance and earthworm's effects would explain the differences shown beginning at 54 h and continuing onward.

The acute initial DSU decrease shown by Cd treated soils, suggests the dominance of a few Cd-tolerant microbial taxa [26,27]. That community presented very high metabolic capacities on a few substrata (as indicated by the results shown in Table 2), and that may explain the acute decrease of DSU (i.e. intense utilization of a few carbon substrata). After 54 h, and over the next 43 h the increase of the DSU suggest that other taxa, less dominant, may reach high numbers in the wells containing sources that are less suitable for the Cd-tolerant taxa.

Earthworms were a much more determining factor in explaining time course changes in DSU than Cd. Accordingly, Ew and EwCd soils presented similar trends and values, regardless the presence of Cd. Moreover, both treatments presented similar number of bacteria (between 3 and 3.6e7 cells g^{-1} DW) and higher than Cd-treated soils (1.1e7). Altogether indicates that earthworm activity provided some protection to the bacterial community against Cd. The bacterial community promoted by earthworms activity during the first 16 days of the experiment, maintained its degrading diversity even after the exposure to a new stressor (Cd). Differently to other studies [58], Cd did not exerted an additional selection pressure. This experimental approach, based on the use of DSU and GAMs allowed for a global and statistically relevant interpretation of the changes in carbon source utilisation, highlighting the key role of earthworms on the protection of microbial communities against the cadmium.

Author Contributions

Conceived and designed the experiments: SM JL JJ EN. Performed the experiments: SM JL. Analyzed the data: SM JL MP EN. Contributed reagents/materials/analysis tools: JJ EN. Wrote the paper: SM EN.

References

1. Garland JL (1997) Analysis and interpretation of community-level physiological profiles in microbial ecology. FEMS Microbiol Ecol 24: 289–300.
2. Klingler JM, Stowe RP, Obenhuber DC, Groves TO, Mishra SK, et al. (1992) Evaluation of the Biolog Automated Microbial Identification System. Appl Environ Microbiol 58: 2089–2092.
3. Insam H (1997) A new set of substrates proposed for community characterization in environmental samples. In: Insam H, Rangger A, editors. Microbial Communities Functional versus structural approaches. Heidelberg: Springer Verlag. Springer. pp. 260–261.
4. Preston-Mafham J, Boddy L, Randerson PF (2002) Analysis of microbial community functional diversity using sole-carbon-source utilisation profiles: a critique. FEMS Microbiol Ecol 42 1–14.
5. Rutgers M, Van't Verlaat IM, Wind B, Posthuma L, Breure AM (1998) Rapid method for assessing pollution-induced community tolerance in contaminated soil. Environ Toxicol Chem 17: 2210–2213.
6. Müller AK, Rasmussen LD, Sorensen JS (2001) Adaptation of the bacterial community to mercury contamination. FEMS Microbiol Lett 204: 49–53.
7. Van Beelen P, Wouterse M, Posthuma L, Rutgers M (2004) Location-specific ecotoxicological risk assessment of metal-polluted soils. Environ Toxicol Chem 23: 2769–2779.
8. Niklińska M, Chodak M, Laskowski R (2005) Characterization of the forest humus microbial community in a heavy metal polluted area. Soil Biol Biochem 37: 2185–2194.
9. Kamitani T, Oba H, Kaneko N (2006) Microbial biomass and tolerance of microbial community on an aged heavy metal polluted floodplain in Japan. Water Air Soil Pollut 172: 185–200.
10. Haack SK, Garchow H, Klug MJ, Forney LJ (1995) Analysis of Factors Affecting the Accuracy, Reproducibility, and Interpretation of Microbial Community Carbon Source Utilization Patterns. Appl Environ Microbiol 61: 1458–1468.
11. Smalla K, Wachtendorf U, Heuer H, Liu WT, Forney L (1998) Analysis of BIOLOG GN substrate utilization patterns by microbial communities. Appl Environ Microbiol 64: 1220–1225.
12. Classen AT, Boyle SI, Haskins KE, Overby ST, Hart SC (2003) Community-level physiological profiles of bacteria and fungi: plate type and incubation temperature influences on contrasting soils. FEMS Microbiol Ecol 44: 319–328.
13. Lavelle P, Decaens T, Aubert M, Barot S, Blouin M, et al. (2006) Soil invertebrates and ecosystem services. Eur J Soil Biol 42: S3–S15.
14. Jones CG, Lawton JH, Shachak M (1994) Organisms as ecosystem engineers. Oikos 69: 373–386.
15. Joschko M, Diestel H, Larink O (1989) Assessment of earthworm burrowing efficiency in compacted soil with a combination of morphological and soil physical measurements. Biol Fertil Soils 8: 191–196.
16. Shipitalo MJ, Protz R (1989) Chemistry and micromorphology of aggregation in earthworm casts. Geoderma 45: 357–374.
17. Marinissen JCY, Dexter AR (1990) Mechanisms of stabilization of earthworm casts and artificial casts. Biol Fertil Soils 9: 163–167.
18. Jégou D, Schrader S, Diestel H, Cluzeau D (2001) Morphological, physical and biochemical characteristics of burrow walls formed by earthworms. Appl Soil Ecol 17: 165–174.
19. Bouché MB (1975) Action de la faune sur les états de la matière organique dans les écosystèmes. In: Gilbertus K, Reisinger O, Mourey A, Cancela da Fonseca JA, editors. Biodegradation et Humification. Sarreguemines: Pierson. pp. 157–168.
20. Zhang H, Schrader S (1993) Earthworm effects on selected physical and chemical properties of soil aggregates. Biol Fertil Soils 15: 229–234.
21. Mc Laughlin MJ, Singh BRE (1999) Cadmium in Soils and Plants. DordrechtThe Netherlands: Kluwer Academic Publishers. 271 p.
22. Violante A, Huang PM, Gadd GME (2008) Biophysico-chemical processes of heavy metals and metalloids in soil environments. New York: John Wiley & Sons. 658 p.
23. Sizmur T, Hodson ME (2009) Do earthworms impact metal mobility and availability in soil?: a review. Environ Pollut 157: 1981–1989.
24. Udovic M, Lestan D (2007) The effect of earthworms on the fractionation and bioavailability of heavy metals before and after soil remediation. Environ Pollut 148: 663–668.
25. Lavelle P, Spain A (2001) Soil Ecology. DordrechtThe Netherlands: Kluwer Academic Publishers. 654 p.
26. Almås ÅR, Bakken LR, Mulder J (2004) Changes in tolerance of soil microbial communities in Zn and Cd contaminated soils. Soil Biol Biochem 36: 805–813.
27. Diaz-Ravina M, Baath E (1996) Development of metal tolerance in soil bacterial communities exposed to experimentally increased metal levels. Appl Environ Microbiol 62: 2970–2977.
28. Degens BP, Schipper LA, Sparling GP, Duncan LC (2001) Is the microbial community in a soil with reduced catabolic diversity less resistant to stress or disturbance? Soil Biol Biochem 33: 1143–1153.
29. Hastie T, Tibshirani R (1990) Exploring the Nature of Covariate Effects in the Proportional Hazards Model. Biometrics 46: 1005–1016.
30. Wood SN (2006) On confidence intervals for generalized additive models based on penalized regression splines. Aust N Z J Stat 48: 445–464.
31. Guisan A, Edwards TC, Hastie T (2002) Generalized linear and generalized additive models in studies of species distributions: setting the scene. Ecol Model 157: 89–100.
32. Yee TW, Mitchell ND (1991) Generalized Additive-Models in Plant Ecology. J Veg Sci 2: 587–602.
33. Dominici F, McDermott A, Hastie TJ (2004) Improved semiparametric time series models of air pollution and mortality. J Amer Statist Assoc 99: 938–948.
34. Römbke J (2006) Tools and techniques for the assessment of ecotoxicological impacts of contaminants in the terrestrial environment. Hum Ecol Risk Assess 12: 84–101.
35. Ministry-of-Agriculture (1975) Orden de 5 de diciembre de 1975, del Ministerio de Agricultura, por la que se aprueban como oficiales los métodos de análisis de suelos y aguas. Boletín Oficial del Estado (B.O.E.) 78: 6458–6491.
36. ASTM Standard D2216 (1998) Standard test method for laboratory determination of water (moisture) content of soil and rock by mass. West Conshohocken, PA: ASTM International. DOI: 10.1520/D2216-98. Available: http://www.astm.org.
37. Moret-Fernández D, Vicente J, Lera F, Latorre B, López MV, et al. (2011) TDR-Lab version 1.2. Users' Guide. Available: http://digital.csic.es/bitstream/10261/35790/3/TDR-Lab%20V.1.2.3_User-Guide.pdf.
38. Khalil MA, Abdel-Lateif HM, Bayoumi BM, van Straalen NM, van Gestel CAM (1996) Effects of metals and metal mixtures on survival and cocoon production of the earthworm Aporrectodea caliginosa. Pedobiologia 40: 548–556.
39. Hobbelen PH, Koolhaas JE, van Gestel CA (2006) Bioaccumulation of heavy metals in the earthworms Lumbricus rubellus and Aporrectodea caliginosa in relation to total and available metal concentrations in field soils. Environ Pollut 144: 639–646.
40. Katayama A, Kai K, Fujie K (1998) Extraction efficiency, size distribution, colony formation and [H-3]-thymidine incorporation of bacteria directly extracted from soil. Soil Sci Plant Nutr 44: 245–252.
41. Ikeda H, Wang X, Inoue K, Someya T (2004) Extraction of bacterial cells from soil by centrifugation. Proceeding of the International Symposium on Identification and Use of Microbial Resources for Sustainable Agriculture. pp. 169.
42. Gregori G, Citterio S, Ghiani A, Labra M, Sgorbati S, et al. (2001) Resolution of viable and membrane-compromised bacteria in freshwater and marine waters based on analytical flow cytometry and nucleic acid double staining. Appl Environ Microbiol 67: 4662–4670.
43. Shannon CE (1948) A Mathematical Theory of Communication. The Bell System Technical Journal 27: 623–656.
44. Tuomisto H (2010) A consistent terminology for quantifying species diversity? Yes, it does exist. Oecologia 164: 853–860.
45. Tam L, Derry AM, Kevan PG, Trevors JT (2001) Functional diversity and community structure of microorganisms in rhizosphere and non-rhizosphere Canadian arctic soils. Biodivers Conserv 10: 1933–1947.
46. Wang Y, Ouyang ZY, Zheng H, Wang XK, Chen FL, et al. (2011) Carbon metabolism of soil microbial communities of restored forests in Southern China. J Soils Sediments 11: 789–799.
47. Hackett CA, Griffiths BS (1997) Statistical analysis of the time-course of Biolog substrate utilization. J Microbiol Methods 30: 63–69.
48. Insam H, Goberna M (2004) Use of Biolog for the Community Level Physiological Profiling (CLPP) of environmental samples. In: Kowalchuk GA, de Bruijn FJ, Head IM, Akkermans AD, van Elsas JD, editors. Molecular Microbial Ecology Manual. Boston: Kluwer Academic Publishers. p. 1780.
49. Ritz C, Streibig JC (2005) Bioassay analysis using R. J Stat Softw 12: 1–22.
50. Jiang W, Wang JJ, Tang JS, Hou F, Lu YT (2010) Soil bacterial functional diversity as influenced by cadmium, phenanthrene and degrade bacteria application. Environ Earth Sci 59: 1717–1722.

51. Akmal M, Xu JM, Li ZJ, Wang HZ, Yao HY (2005) Effects of lead and cadmium nitrate on biomass and substrate utilization pattern of soil microbial communities. Chemosphere 60: 508–514.
52. Yao HY, Xu JM, Huang CY (2003) Substrate utilization pattern, biomass and activity of microbial communities in a sequence of heavy metal-polluted paddy soils. Geoderma 115: 139–148.
53. Wood SN (2003) Thin plate regression splines. J R Stat Soc Series B Stat Methodol 65: 95–114.
54. R_Development_Core_Team (2012) R: A language and environment for statistical computing. Vienna, Austria: R Foundation for Statistical Computing. ISBN 3-900051-07-0. Available: http:www.r-project.org.
55. Friendly M, Fox J (2013) Candisc: Visualizing Generalized Canonical Discriminant and Canonical Correlation Analysis. R package version 0.6–3.
56. Karaca A, Cetin S, Turgay O, Kizilkaya R (2010) Effects of Heavy Metals on Soil Enzyme Activities. In: Sherameti I, Varma A, editors. Soil Heavy Metals. Berlin Heidelberg: Springer Verlag. pp. 237–262.
57. Gomez-Brandon M, Lores M, Dominguez J (2012) Species-Specific Effects of Epigeic Earthworms on Microbial Community Structure during First Stages of Decomposition of Organic Matter. Plos One 7 (2): e31895. doi:10.1371/journal.pone.0031895.
58. Nakatsu CH, Carmosini N, Baldwin B, Beasley F, Kourtev P, et al. (2005) Soil microbial community responses to additions of organic carbon substrates and heavy metals (Pb and Cr). Appl Environ Microbiol 71: 7679–7689.

16

Spatial Ecology of Bacteria at the Microscale in Soil

Xavier Raynaud[1]*, **Naoise Nunan**[2]

1 Sorbonne Universités, UPMC Univ Paris 06, Institute of Ecology and Environmental Sciences - Paris, Paris, France, 2 CNRS, Institute of Ecology and Environmental Sciences - Paris, Campus AgroParisTech, Thiverval-Grignon, France

Abstract

Despite an exceptional number of bacterial cells and species in soils, bacterial diversity seems to have little effect on soil processes, such as respiration or nitrification, that can be affected by interactions between bacterial cells. The aim of this study is to understand how bacterial cells are distributed in soil to better understand the scaling between cell-to-cell interactions and what can be measured in a few milligrams, or more, of soil. Based on the analysis of 744 images of observed bacterial distributions in soil thin sections taken at different depths, we found that the inter-cell distance was, on average 12.46 μm and that these inter-cell distances were shorter near the soil surface (10.38 μm) than at depth (>18 μm), due to changes in cell densities. These images were also used to develop a spatial statistical model, based on Log Gaussian Cox Processes, to analyse the 2D distribution of cells and construct realistic 3D bacterial distributions. Our analyses suggest that despite the very high number of cells and species in soil, bacteria only interact with a few other individuals. For example, at bacterial densities commonly found in bulk soil (10^8 cells g^{-1} soil), the number of neighbours a single bacterium has within an interaction distance of ca. 20 μm is relatively limited (120 cells on average). Making conservative assumptions about the distribution of species, we show that such neighbourhoods contain less than 100 species. This value did not change appreciably as a function of the overall diversity in soil, suggesting that the diversity of soil bacterial communities may be species-saturated. All in all, this work provides precise data on bacterial distributions, a novel way to model them at the micrometer scale as well as some new insights on the degree of interactions between individual bacterial cells in soils.

Editor: Francesco Pappalardo, University of Catania, Italy

Funding: This research was supported by the INSU EC2CO Microbiologie Environnementale and ANR Syscomm Programme MEPSOM. The funders had no role in study design, data collection and analysis, decision to publish, or preparation of the manuscript.

Competing Interests: The authors have declared that no competing interests exist.

* E-mail: xavier.raynaud@upmc.fr

Introduction

The application of novel molecular techniques (such as high throughput sequencing) during the past two decades has uncovered a phenomenal bacterial diversity in soils. For example, a single gram of soil can harbour up to 10^{10} bacterial cells and an estimated species diversity of between $4 \cdot 10^3$ [1] to $5 \cdot 10^4$ species [2]. Several studies have identified major environmental influences on soil bacterial diversity (such as soil pH [3], nitrogen [4], plant communities [5] or land use [6]) and soil bacterial biomass (soil organic carbon [7]), that vary between geographical regions and across biomes. It is intriguing however, that experiments manipulating microbial diversity have found no or only weak links between diversity and many important microbial-driven processes, such as soil carbon mineralization [8–10], nitrite oxidation [8,9] or denitrification [8,9]. This lack of relationship raises the question about the importance of microbial diversity for soil and ecosystem functioning [11] and has even lead some authors to question the value of studying the soil metagenome for understanding soil microbial functioning [12].

The diversity of biological components can affect ecosystem processes through interactions among species, such as when there is competition for resources, mutualism or predation. The extent and intensity of these interactions depend not only on the interacting species but also on their proximity to one another. The role of space in ecosystem function is widely recognised in higher plant and animal ecology: the spatial distribution of species and the spatial organisation of communities regulate the extent to which individuals interact, such as in competition for resources [13–15], mutualism [16] or predation [17,18], which, in turn, affects ecosystem properties [19]. However, compared to the vast amount of studies focusing on microbial diversity in soils, relatively little attention has been paid to spatial aspects of ecology in microbial systems at the scales at which cell-to-cell interactions occur although there have been some attempts to characterize the spatial distribution of diversity and microbial processes at the scale of aggregates [20–22]. As microbial-driven ecosystem processes are sums of the activities of microbial cells, most of which are subject to cell-to-cell interactions, such interactions are likely to have significant effects on overall processes.

In microbial systems, the scale at which individuals interact is related to the distance over which they can effect changes in the concentration of gases or solutes. This may vary depending on the gas or solute and the concentration at which it has an effect on bacterial physiology, however, two notable studies have suggested that the vast majority of interactions occur within 20 μm of bacterial cells [23,24]. Studies on microbial systems (whether they focus on microbial activity or diversity) are generally carried out at scales many orders of magnitude larger than those at which microorganisms interact with other organisms or with their surrounding environment [25]. This disparity of scale is not as prevalent in the study of higher organisms [26] and so the effects of local interactions on ecosystem processes are better understood. In soil microbial ecology, the effects of local interactions are likely to be obscured by relatively large samples that encompass environmental heterogeneity and local interactions [25].

Microbial-scale processes and local spatial organisation are known to be significant regulators of microbial community stability, function and evolution. Spatial separation has been identified as playing a major role in several microbial processes: 1. it is thought to be responsible for the emergence and maintenance of high levels of bacterial diversity observed in structured media [27]; 2. the relative importance of horizontal gene transfer in bacterial evolution is believed to depend on the proximity of bacterial neighbours, with areas of low cell density dominated by clonal reproduction and densely populated areas harbouring communities in which horizontal gene transfer can be significant [28] and 3. it has been shown that the stability of bacterial communities can depend on the distance among constituent members [29]. A common feature of these bacterial community ecology studies is that they are carried out in model or artificial systems rather than *in situ* and, as a result, the pertinence of the processes identified for real communities in their natural habitats can be questioned. An understanding of the importance of space in bacterial ecology requires knowledge of the distribution of bacterial cells in their environment. This paper aims to explore the spatial distribution of soil bacterial cells in soils at the micrometer scale. To this end, we present a method to analyse and model distributions of individual bacterial cells in soil in order to better understand how bacteria interact with one another. Measuring the distribution of bacterial cells in volumes of soil that are relevant to cell-to-cell interactions is not technically possible at present, therefore our analysis was carried out in three steps. We first studied the distributions of bacterial cells measured in 2 dimensional thin sections of soil and then extended these observations to 3 dimensions using point pattern modelling methods. Finally, we used a simple species abundance model to gain some insight into the degree to which different bacterial species may interact with each other.

Materials and Methods

Ethics statement

The Scottish samples on which this study was based were obtained from land belonging to the Scottish Crop Research Institute when one of the authors was a member of staff there and permission was granted. The French samples were given to the authors by Geneviève Grundman of the University of Lyon. However, no new sampling took place for this study. Only previously obtained data were used, most of which has already been published [30–32]. The samples were taken from agricultural fields and did not involve any endangered species. Dataset with bacterial distributions is available from the authors upon request.

Bacterial distribution data

Bacterial distributions in this study consisted of 2D point patterns (the x and y coordinates of individual cells) measured in images of soil thin sections (Fig. 1). We used 752 new or previously measured bacterial distributions taken from 94 soil thin sections sampled at different depths from a Scottish sandy loam (723 images, [30–33]) and 2 soil thin sections of surface soil (20 cm below surface) a French sandy loam (29 images; Table 1). The Scottish samples were taken from topsoil (0–30 cm, sandy loam: 71% sand, 19% silt, 10% clay, pH_{H_2O} 6.2, 1.9% C and 0.07% N) and subsoil (30–80 cm, sandy loam: 72% sand, 17% silt, 11% clay, pH_{H_2O}: 6.5, 0.68% C and 0.02% N) of an arable soil. The French samples were taken from topsoil of an arable soil (0–30, loam: 47.7% sand, 35.3% silt, 17.0% clay, pH_{H_2O}: 7.0, 1.4% C and 0.13% N). Digital images were acquired with a Zeiss Axioplan 2 microscope fitted for epifluorescence and were equivalent to an effective area of 620×460 μm^2 for the Scottish samples and 516×410 μm^2 for the French samples (different cameras were used). Details on the image analysis procedures to extract bacterial coordinates from digital images can be found in [30]. Bacterial distributions that contained 5 cells or less (n = 8) were discarded for the analysis of the spatial distribution of cells, as low cell densities make the analysis of point patterns unreliable. Two thin sections (2×8 microscopic observations) were prepared from a single sample so that one was orthogonal to the other (i.e., XY and XZ). These observations were used to determine whether bacterial distributions at the micro- to millimetre scales were isotropic or not.

Spatial model fitting of bacterial distribution observed in thin sections

The aim of this first section is to describe the spatial structure of the observed distributions of bacteria. Each distribution was compared to two different spatial null models. The first null model, Complete Spatial Randomness (CSR) or Poisson process, assumes that the position of one point in the point pattern is independent of the position of the others. The second null model, Log Gaussian Cox Process (LGCP) is a model for aggregated point patterns where the aggregation is caused by some environmental heterogeneity [34]. As bacteria live in the soil pore network, the soil structure (Fig 1a), that determines the architecture of the pore network [35], is an important environmental heterogeneity affecting the distribution of bacteria. LGCP are processes defined in n-dimensions and are a form of inhomogeneous Poisson process where the intensity is a Gaussian random measure (Fig. 1c). In this study, LGCP with an exponential covariance function were used. These LGCP are determined by three parameters, the mean (μ), variance (σ) and scale (β) of the Gaussian random measure. The three parameters determine the intensity of the point process (the number of points in the point pattern) and the extent of aggregation. The average intensity, λ, of a LGCP is given by:

$$\lambda = e^{\mu + \sigma^2/2} \quad (1)$$

The CSR model is a special case of LGCP where $\sigma \rightarrow 0$. Details on the theory of LGCP are given in the Appendix S1.

All the point patterns used in this study (observations of bacterial distributions and simulations), were characterized by their intensity λ and 2 summary statistics: Ripley's $K(r)$ function and the nearest neighbour distance distribution function, $G(r)$ (see [36] for a mathematical definition of these two functions). The intensity λ is the number of points per unit surface and was estimated by dividing the number of points in the distribution by its surface. Ripley's $K(r)$ summary statistic is related to the number of points in a point pattern that are within distance r of an "average" point. The theoretical expression of Ripley's $K(r)$ is known for the CSR and LGCP models. For 2D point patterns, they are given by the following equations:

$$K_{CSR}(r) = \pi r^2 \quad (2)$$

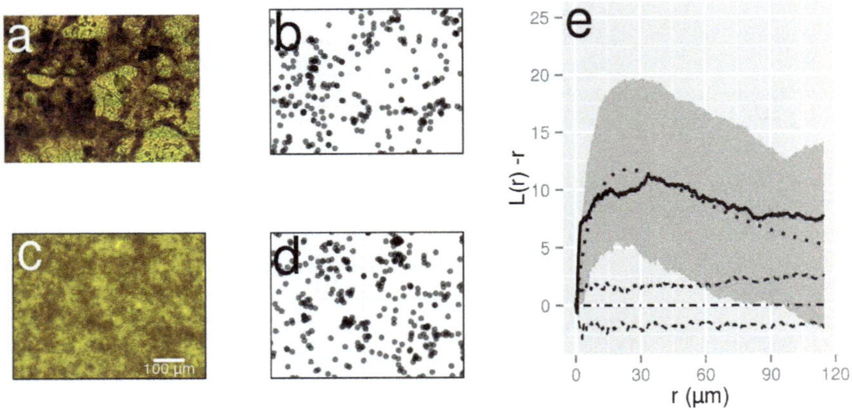

Figure 1. Bacterial habitat, observed and simulated distribution of bacteria in a soil thin section. (a) Bright field image of a soil thin section showing various soil features that characterise the soil microbial habitat and (b) bacterial distribution observed in the same thin section. (c) Random probability field generated using parameters estimated from b ($\mu = -7.64$; $\sigma^2 = 2.0$ and $\beta = 12.95$) and (d) simulated bacterial distribution using c. as random intensity (LGCP model). Colours for the random field were chosen to match those of the thin section image. Darker shades indicate higher probability of bacterial presence. The scale for all panels is identical and is indicated in c. (e) Transformed Ripley functions L(r) - r for the observed bacterial distribution shown in a. (solid line) and envelopes of 99 simulations under CSR and LGCP. The dashed lines indicate the envelope of CSR simulations of the same intensity as a. and the shaded area the envelope of simulations of an LGCP with parameters estimated from the observed point distribution. The dotted line corresponds to the theoretical functions under LGCP and the dashed-dotted line to the theoretical function under CSR. In e., L(r)-r above 0 indicates that the point pattern is more aggregated than a random process.

$$K_{LGCP}(r) = 2\pi \int_0^r s \cdot \exp\left(\sigma^2 e^{-s/\beta}\right) ds \qquad (3)$$

The theoretical expression of *K(r)* for the LGCP model makes it possible to estimate the parameters μ, σ^2 and β for all observed bacterial distributions [37]. The nearest neighbour distance distribution function, *G(r)*, of a point pattern is the distribution function of the distance from an average point to its nearest neighbour. The theoretical expression of *G(r)* for a 2D Poisson process is

$$G_{CSR}(r) = 1 - \exp\left(-\pi^{l} r^2\right) \qquad (4)$$

where λ is the intensity of the point process. There is no theoretical expression for *G(r)* for a LGCP process.

Estimates of the two summary statistics and of the LGCP parameters were all carried out using R 2.15 with packages *spatstat* [38] and *Randomfield* [39]. These packages provide all the necessary functions to manipulate and analyse point patterns. We used the Ripley isotropic correction for estimates of *K(r)* and Kaplan-Mayer estimator for *G(r)* [38].

The goodness of fit between each observed distribution (see example in Fig 1b) and the null models (see example in Fig 1d for the LGCP null model) was carried out as in [40]. To begin with, we simulated 99 point processes of the same intensity under the two null models and calculated the estimated *K(r)* and *G(r)* functions for each simulation. For both the simulations and the observed distributions, we then computed the maximum absolute difference between the estimated summary statistics *K(r)* or *G(r)*

Table 1. General properties of all observed distribution maps of bacteria at different depths in soil.

Site	Depth (cm)	# samples	Cell density(mm^{-2})	Cell density(g$_{soil}^{-1}$)	Average distance to nearest neighbour (µm)
Scotland	0–30 cm	359	668.7±568.0	1.03 10^9±8.74 10^8	12.12±6.06
			(17.7–3572.8)	(2.73 10^7–5.50 10^9)	(0.38–366.17)
Scotland	30–60 cm	261	297.3±289.5	4.57 10^8±4.45 10^8	18.37±11.05
			(17.7–1821.8)	(2.72 10^7–2.80 10^9)	(0.45–366.75)
Scotland	>60 cm	103	121.2±165.5	1.86 10^8±2.54 10^8	28.94±23.98
			(7.1–1003.1)	(1.09 10^7–1.54 10^9)	(0.66–532.36)
France	20 cm	29	3531.28±1809.7	5.43 10^9±2.78 10^9	4.95±0.96
			(155.6–7539.1)	(2.39 10^8–1.16 10^{10})	(0.34–221.12)
	Total	752	576.8±841.0	8.87 10^8±1.29 10^9	12.46±9.38
			(7.1–7539.1)	(1.09 10^7–1.16 10^{10})	(0.34–532.36)

Data for cell numbers, cell densities and average distance to nearest neighbour are given as mean±sd (range). Cell density in g$_{soil}^{-1}$ is calculated assuming a microscope depth of field of 2 µm and a soil density of 1.3 g cm^{-3}.

and its theoretical counterparts to build a two-sided test of significance at $P = 0.01$ which was used to reject the hypothesis that the observed pattern followed the null model. Because there is no theoretical expression of *G(r)* for the LGCP model, the average of the 99 simulations was used as a theoretical counterpart. In the results section, the transformation $L(r) = \sqrt{(K(r)/\pi)}$ was plotted as it is simpler to interpret (e.g., $L(r)-r = 0$ for all r in the case of a 2D CSR process).

Modelling bacterial neighbourhoods in 3 dimensions

As we found that the LGCP model was adequate for modelling bacterial cell distributions in 2D (see Results section), a similar modelling approach, based on LGCP, was used to estimate the number of neighbours a single bacterial cell had within a given distance in 3 dimensions. The expected values of the parameters μ, σ^2, β of an isotropic Gaussian random measure of a 3D LGCP are the same as those of a plane from the same Gaussian field [41,42]. Therefore, we used the 2D estimates of the LGCP parameters to simulate 3D distributions after ascertaining that bacterial distributions were isotropic at the bacterial neighbourhood scale. Simulations of LGCP 3D distributions were carried out following the method given in [34]. Average parameter values corresponding to bacterial densities of ca. 10^8 cells g^{-1} ($\mu = -10.26$, $\sigma^2 = 2.90$, $\beta = 20$), 10^9 cells g^{-1} ($\mu = -7.52$, $\sigma^2 = 1.90$, $\beta = 25$) and 10^{10} cells g^{-1} ($\mu = -4.91$, $\sigma^2 = 1.29$, $\beta = 25$) were used to simulate 39 cubes ($300 \times 300 \times 300$ μm^3) for each bacterial density. The number of cells each bacterium had in its neighbourhood as a function of distance was calculated, and the average across the 39 simulations determined. This was compared with the theoretical number of neighbours for a 3D LGCP, given by the following equation (see Appendix S1 for details):

$$N(r) = e^{\mu + \sigma^2/2} \int_0^r 4\pi s^2 \exp\left(\sigma^2 e^{-s/\beta}\right) ds \qquad (5)$$

where μ, σ^2, and β are the parameters of the LGCP. In our simulations the realised densities were (mean±s.e.) $1.06\ 10^8 \pm 2.48\ 10^6$, $1.04\ 10^9 \pm 2.44\ 10^7$ and $1.02\ 10^{10} \pm 2.47\ 10^8$ cells g^{-1}, respectively.

Diversity in the bacterial neighbourhood

In order to study the number of bacterial species in the bacterial neighbourhood (i.e., the number of species with which a bacterium might interact), 3D bacterial distributions in which bacterial cells were attributed a species identity were simulated. However, the spatial structure of bacterial diversity at the micrometer scale is unknown, so a modelling approach was taken to estimate the number of species a single bacterium interacts with. As the microbial diversity in our soil samples was not known, we used published data of soil microbial diversity [1,2] to simulate the 3D bacterial communities. To do so, a simple species-abundance model, the log-series distribution of species [43], was used to calculate the number of individuals per species from a total number of individuals and species. This species abundance model has been found to fit bacterial species distributions in soils at larger spatial scales [44] and has a very simple mathematical formulation. The Fisher species abundance curve relates the number of species (*S*) to the number of individuals (*N*) as described in the following equation:

$$S = \alpha \ln(1 + N/\alpha) \qquad (6)$$

where α is Fisher's α index of diversity. Values for α were calculated using Eq 6 and data found in [1] ($S = 4000$, $N = 1.5\ 10^{10}$) or [2] (Agricultural soil, Brazil: $S = 3559$, $N = 10^9$; Forest soil, Canada: $S = 15188$, $N = 10^9$; Agricultural soil, Florida, USA: $S = 4477$, $N = 10^9$; Agricultural soil, Illinois, USA $S = 4010$, $N = 10^9$). This gave values of α of 221.86, 233.04, 1107.53, 297.94 and 264.79, respectively. Because the spatial structure of species at such scales is unknown, species identities were assigned at random among individuals in the simulations. In this case, the average number of neighbouring species can be derived from the theoretical number of neighbours (Eq. 5) and the Fisher species abundance curve (Eq. 6) following the equation:

$$\begin{aligned} S(r) &= \alpha \ln(1 + N(r)/\alpha) \\ &= \alpha \ln\left(1 + \left[e^{\mu + \sigma^2/2} \int_0^r 4\pi s^2 \exp\left(\sigma^2 e^{-s/\beta}\right) ds\right] / \alpha\right) \end{aligned} \qquad (7)$$

It should be noted that this type of modelling approach is rather simplistic and was only used to illustrate how interactions between different bacterial species might affect soil functioning (see Discussion).

Results

Bacterial densities in thin sections

The number of bacterial cells in the analysed images ranged from 5 to 1599. This corresponded to densities between 7 and 7539 cells mm^{-2} or ca. $1.09\ 10^7$ to $1.16\ 10^{10}$ cells g^{-1} soil, assuming a microscope depth of field of 2 μm and a soil density of 1.3 g cm^{-3} (Table 1). On average, bacterial densities decreased with depth, with the highest densities found above 30 cm depth and the lowest below 60 cm. However, the variability in cell density was high at all depths (Table 1), indicating that bacterial cells were distributed in a heterogeneous way throughout the soil volume. Due to the decrease in density associated with depth, inter-cell distance also varied with depth, with an average of approximately 10 μm at depths of 0–30 cm (corresponding to high cell densities) and 29 μm at depths below 60 cm (Table 1). It should be noted that inter-cell distances were also highly variable at each depth.

Comparison of observed distributions with the null models

The goodness of fit between the measured distributions and the CSR model revealed clear deviations from CSR ($P<0.05$) in at least 630 of the 744 distributions (Table 2). Clustering was more pronounced in the surface layers of the soil, where bacterial cells were more abundant. Whereas in the surface layers, to a depth of 30 cm, at least 366 of the 387 distributions (94.6%) differed from CSR, only 52 of 98 distributions (53.1%) did so in soil taken from depths below 60 cm (Table 2). An example of the deviation of the Ripley's *K(r)* function of an observed distribution from CSR is given in Fig 1e.

In contrast, the LGCP null model adequately described the observed distributions in, at worst, 80% of the cases (Table 2). Estimates of the LGCP parameters μ, σ^2 and β parameters were in the range [−13.49, −5.81], [0.59, 6.85] and [$2.59\ 10^{-3}$, 161.18], respectively. Complete spatial randomness is the limit of LGCP when σ^2 tends to 0 and therefore the values of σ^2 provide a first indication that the bacterial distributions in these soils ranged from highly aggregated (high σ^2) to near random (low σ^2)

Table 2. Total number of samples, number (proportions %) of samples deviating from CSR and number (proportions %) of samples deviating from the LGCP model for different soil depths for all bacterial distribution having more than 5 bacterial cells in the field of view.

Depth	# samples	samples deviating from CSR		samples deviating from LGCP	
		Ripley's K	G function	Ripley's K	G function
0–30 cm	387	376 (97.1%)	366 (94.6%)	37 (9.6%)	97 (25.1%)
30–60 cm	259	228 (88.0%)	212 (81.8%)	14 (5.4%)	40 (15.4%)
>60 cm	98	68 (69.4%)	52 (53.1%)	4 (4.1%)	11 (11.2%)
Total	744	672 (90.3%)	630 (84.68%)	55 (7.4%)	148 (19.9%)

Deviations are calculated based on a goodness of fit test between summary statistics for each observed distribution (Ripley's K or G function) and the corresponding statistics under the null model (CSR of LGCP). Ripley's K is a summary statistics related to the number of points in a point pattern that are within a certain distance to an average point. The nearest neighbour distance distribution G is the distribution function of the distance from an average point to its nearest neighbour.

distributions. The mean (μ) decreased ($R^2 = 0.25$, $P<0.001$) and variance (σ^2) increased ($R^2 = 0.08$, $P<0.001$) linearly with depth between 30 cm and 80 cm below-ground, but no trends were apparent for depths between 0 and 30 cm. There was no relation between β and depth (data not shown). An example of the concordance between the Ripley's function of an observed distribution and simulations of LGCP distribution is given in Fig 1e. The distributions that did not fit the LGCP null model occurred primarily in topsoil and rhizosphere samples where very high bacterial densities were observed, or in subsoil samples in which individual, isolated colonies were detected.

Isotropy of bacterial distributions

The isotropic nature of the bacterial distributions was tested using the two slides (2×8 images) that were prepared orthogonally. The number of bacterial cells in these slides ranged from 78 to 1008 for one slide and from 31 to 920 for the other. There were no statistical differences in the number of observed cells between the two slides ($P = 0.27$). The estimates of the LGCP parameters for μ, σ^2 and β were, respectively, in the ranges [−10.19, −6.51], [1.44, 3.99] and [8.94, 61.65] for one slide and [−10.17, −6.37], [1.29, 2.60] and [7.21, 60.09] for the other. Here also, there were no statistical differences between these estimates for the two slides ($P>$ 0.16 at least), suggesting that the distribution of bacteria at these micrometer to millimetre scales is isotropic.

Modelling bacterial distributions in 3D

Due to the apparent isotropic nature of the bacterial neighbourhoods at micrometer to millimetre scales, the LGCP parameters obtained in 2D space were used to simulate 39 independent 3D distributions [41]. An example of such a 3D distribution of bacteria (density equivalent to 10^9 cells g^{-1}) is given in Fig. 2a. For each simulation, the number of neighbours each bacterium had as a function of distance was computed (shaded scatterplot in Fig. 2b) and averaged across all bacteria (red line in Fig. 2b). As expected, the average number of neighbours bacteria had as a function of distance was similar to the theoretical number of neighbours derived from the theoretical expression of $K(r)$ for a LGCP (Eq 2, blue line in Fig 2b). The number of neighbours that a single bacterium had within a distance of 20 μm ranged from 7 to 250 for a bacterial density of 10^9 cells g^{-1} (Fig. 2b). The minimum and maximum average number of neighbours obtained across the 39 simulations are given in Fig 3a. Overall, these results suggest that the number of cells in the neighbourhood of a typical bacterium is rather limited. For an average density of 10^9 cells g^{-1} soil, the average number of neighbours around a single cell was ca. 1043 (±250) cells within a distance of 50 μm, decreasing to ca. 120 (±40) cells at 20 μm (Fig 3a). Similarly, for an average density of 10^8 cells g^{-1} soil, the average number of neighbours was ca. 82 (±22) cells within a distance of 50 μm and ca. 12 (±4) cells at 20 μm. Finally, for bacterial densities close to what one would expect in the rhizosphere, (10^{10} cells g^{-1} soil), the average number of neighbours was ca. 5806 (±1000) cells within a distance of 50 μm decreasing to ca. 555 (±100) cells within 20 μm.

Bacterial diversity in the bacterial neighbourhood

In order to estimate the diversity of bacterial species a typical bacterium interacts with, species were distributed at random among individual cells in the 3D simulations of bacterial distributions. Using these simulations, we estimated the number of species in the neighbourhood of each cell as a function of distance (shaded scatterplot in Fig. 2c) and computed its average value (red line Fig. 2c). The average number of species ($S(r)$) in the neighbourhood of a single cell was adequately approximated by the theoretical average number of species around a single bacterium given in Eq. 7 (red and blue lines in Fig 2c).

The average number of species in the neighbourhood of bacteria was estimated from the 39 3D simulations at densities of 10^8, 10^9 and 10^{10} cells g^{-1} soil (Fig 3b). At lower bacterial densities, typical of bulk soil (10^8 cells g^{-1}), the average number of species that a bacterium can be expected to interact with (species within 20 μm, assuming a random distribution of species) was ca. 11 (±4) species. For a density of 10^9 cells g^{-1} soil, the average number of neighbouring species around a single cell increased to 97 (±24) species. Finally, when bacterial densities were set at values expected in the rhizosphere (10^{10} cells g^{-1} soil), the average number of species within an interaction distance of 20 μm was ca. 284 (±30) species. These values were higher when the interaction distance was set at 50 μm (Fig 3b). The diversity indices derived from Roesch et al (2007) [2] were similar to that derived from Torsvik et al (1990) [1] with one exception: bacterial diversity in the forest soil in Roesch et al (2007) [2] was much higher ($\alpha = 1107.53$), resulting in a greater number of species per volume of soil (Fig S1). However, despite the increase in the overall number of species, no increase in the number of neighbouring species within 20 μm was observed for bacterial densities of 10^9 cells g^{-1} soil and lower.

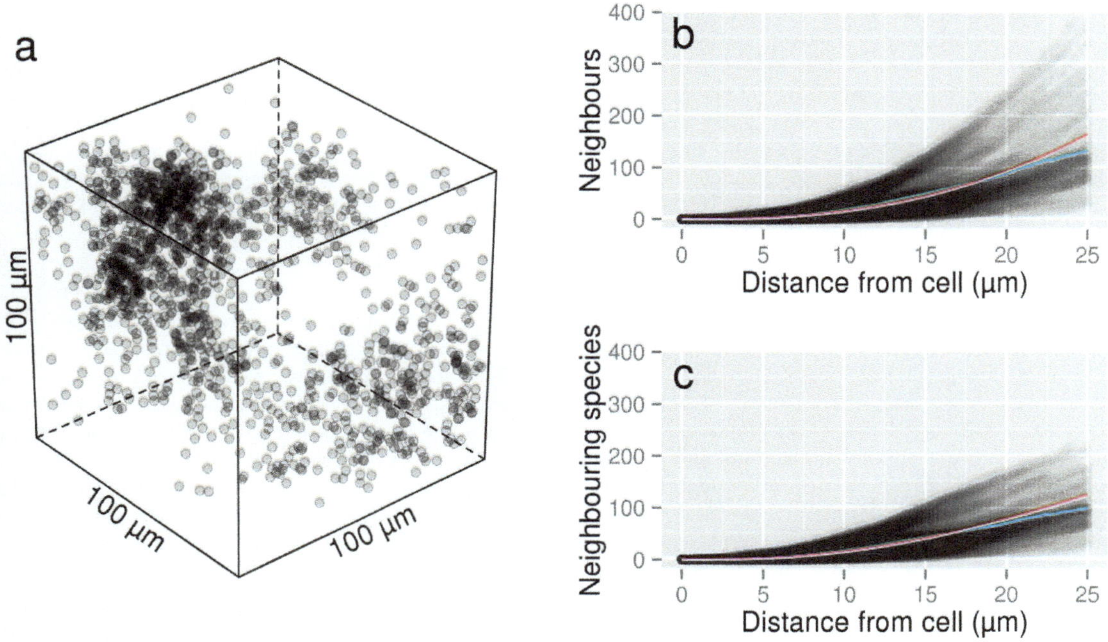

Figure 2. 3D simulation of bacterial distribution. (a) 3D representation of bacterial distribution in a $100 \times 100 \times 100\ \mu m^3$ cube as simulated by LGCP. The model parameters for the simulation were $\mu = -7.52$, $\sigma^2 = 1.90$ and $\beta = 25$. (b) Number of neighbours as a function of distance for each bacterium (scatterplot), average number of neighbours (red line) and theoretical number of neighbours for the LGCP (blue line) and (c) Number of neighbouring species assuming a random distribution of species among individuals (line colours are the same as in b). The number of species considered in this simulation was $S = 450$, corresponding to Fisher's $\alpha = 221.86$ (estimated from [1]).

Discussion

The distribution of bacteria in soils

Unsurprisingly, the bacterial distributions measured in the 744 soil thin sections studied showed a high degree of aggregation at all depths, although aggregation was more frequent in the surface strata than in the subsoil. This corroborates similar observations by [30] on a sample of this dataset. We also found that bacterial distributions were isotropic at the millimetre scale. Although this was based on a limited number of observed distributions and would need to be confirmed on a larger dataset, such an observation is not surprising as the vertical gradients observed in soils generally occur at larger scales [45,46].

Log Gaussian Cox Processes (LGCP) were used to characterise the bacterial distributions and construct 3D distributions from these observations. LGCP are particularly useful for modelling aggregated spatial point patterns where aggregation is due to a stochastic environmental heterogeneity [34]; that is to say, an environmental factor that is continuously and heterogeneously distributed in space. They do not, however, account for the contribution of intrinsic processes, such as birth and death, to the

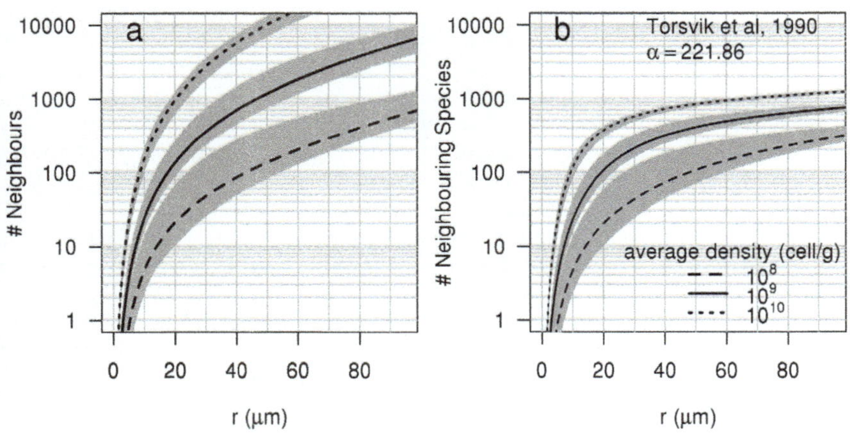

Figure 3. Number of neighbours and neighbouring species. (a) Mean number of neighbours and (b) neighbouring species around an "average" bacterium as a function of distance from the bacterium and as a function of bacterial density. The mean number of neighbours and neighbouring species are derived from Ripley's K(r) function of LGCP with parameters $\mu = -10.26$, $\sigma^2 = 2.90$, $\beta = 20$ for the 10^8 cells g^{-1} soil density, $\mu = -7.52$, $\sigma^2 = 1.90$, $\beta = 25$ for the 10^9 cells g^{-1} soil density and $\mu = -4.91$, $\sigma^2 = 1.29$, $\beta = 25$ for the 10^{10} cells g^{-1} soil density. Grey envelopes surrounding curves represent the maximum and minimum of these numbers calculated from 39 simulations. Note the logarithmic scale on the y axis.

distribution of points in a given point pattern. The distribution of bacteria in soil is clearly a consequence of both extrinsic (environmental conditions, such as pore size and organic matter availability) and intrinsic (reproduction by binary fission) processes. However, it is likely that intrinsic processes such as reproduction are related to the extrinsic processes, as the probability of growth is greater where external environmental conditions are most suitable (i.e. presence of organic substrate, O_2, water…). It is worth noting that the majority of samples that were inadequately described by LGCP were from the rhizosphere or topsoil. There, intrinsic processes (e.g. cell division) would have played a significant role in the generation of these distributions as bacterial growth may be somewhat decoupled from environmental heterogeneity due to the overall high availability of resources. This may possibly be because environmental heterogeneity (with respect to bacterial growth) was reduced due to the input of organic substrate from plants. It can be concluded therefore, that the relative importance of the underlying processes contributing to the generation of the bacterial distributions (intrinsic vs. extrinsic) can change with situation.

Despite the deviation of some of the bacterial distributions from the LGCP model in some specific cases, LGCP were used to simulate bacterial distributions because they have the distinct advantage of being analytically tractable (i.e. the theoretical expression of $K(r)$ is known), making it possible to estimate model parameters from which an observed point pattern could have emerged [37].

The neighbourhood of bacteria in soils

Three properties of microbial communities emerge from our analysis. The first property is that the number of cells within interacting distances in the neighbourhood of a bacterium is, on average, rather limited compared to the number of cells commonly found in a single gram of soil. For densities equivalent to 10^9 cells g^{-1}, the number of neighbours a typical bacterium has within a distance of 20 µm is approximately 120, increasing to 1000 within a distance of 50 µm. The actual number of neighbours within interaction distances may in fact be lower as the calculations carried out here used the euclidian distance between cells rather than the geodesic distance and therefore did not account for pore geometry. For example, two cells on either side of a sand grain will not interact through the sand grain but rather around it.

The second property is that, given the high variability in the number of neighbours around bacterial cells in our dataset (Table 1) and in our simulations (Fig 2 and 3), the density of interactions is highly variable in space, even at very fine scales, with some cells interacting with few others and other cells interacting with many more. For example, some filamentous bacteria or bacteria in colonies or biofilms are likely to have completely different interaction environments to those of isolated individual bacterial cells. Such localized "pockets of interactions" might have important consequences for ecosystem processes and microbial community evolutionary dynamics [47].

The third property to emerge from this analysis, an obvious corollary of low levels of cell-to-cell interactions, is that the number of different species an individual cell interacts with is also limited. The simulations of bacterial distributions in 3D suggest that interspecific interactions in soils are orders of magnitude lower than what is possible in view of the species diversity often measured in soil (4000–50000 species g^{-1} soil; [1,2]). For densities equivalent to 10^9 cells g^{-1} soil, we found that the number of bacterial species within interaction distances (<20 µm) ranged from 1 to 120 (±40) and, even at the highest densities (equivalent to 10^{10} cells g^{-1} soil), never exceeded 1000 species (Fig 3b, Fig S1). Moreover, these data are most likely an overestimate of the actual local diversity, as we assumed that species were distributed randomly among individuals. However, as bacteria reproduce by binary fission, this assumption is almost certainly wrong. The approach taken here thus provides only an upper limit to the number of species in the bacterial neighbourhood. Furthermore, the model also assumes that there is a positive relationship between abundance and diversity as the diversity parameter (α) is constant. This assumption may also be untrue. In reality, species are aggregated and the extent of species aggregation is most probably positively related to growth intensity because bacteria grow by binary fission. This means that diversity may not increase monotonically with abundance when conditions are favourable for growth and it has indeed been found that bacterial diversity in the rhizosphere is lower than in bulk soil [48,49]. A more realistic way to model the spatial distribution of diversity should therefore account for both the environmental determinism of the spatial distribution of cells (as was done here) and cellular reproduction processes in order to yield aggregation in species distribution.

It is noteworthy that the diversity at the microscale (species within 20 µm of a bacterial cell) did not change appreciably as a function of the global diversity (species found in a gram of soil) for bacterial densities of 10^9 cells g^{-1} soil or less (i.e., diversity at 20 µm is similar in Fig 3b and Fig S1, Forest Soil, despite very different α values), suggesting that local communities may be species saturated [50]. It is now widely recognised that local diversity is not solely dependant on local interactions, but that regional processes are also important (in the case of soil bacteria, a gram of soil can be considered a region [26]). Species saturation of local communities can arise from species interactions (community membership is limited by competitive exclusion or local environmental conditions [50]) or from the physical limitations of the environment (if the local environment can only accommodate 100 individuals then there cannot be more than 100 species regardless of the overall diversity [51]). It has been suggested that the lack of relationship between ecosystem processes and diversity in soils is due to functional redundancy within soil microbial communities. Regardless of the underlying cause for the apparent local species saturation, species saturation of local microbial communities may also explain the relative insensitivity of many ecosystem processes (e.g. soil organic matter decomposition, denitrification), as well as the resistance or resilience of these processes to environmental stresses, to experimental changes in microbial diversity [52,53]. If the number of species that the average bacterium interacts with does not change as the overall diversity is changed then the functioning the bacteria are responsible for may not be affected by changes in overall diversity either, as the actual levels of diversity in communities remains the same.

Should this hypothesis be confirmed, there are a number of important consequences for soil microbial ecology. The first is that the relationship between diversity and functioning may only be understood if studied at an appropriate scale. The second is that the relationship between regional (gram of soil) and local diversity (microbial neighbourhood) must be understood if the effects of diversity on soil functioning is to be apprehended.

Conclusions

Our analysis of 744 observations of *in situ* bacterial distributions in soils indicates that bacterial cells are aggregated at the scale of a few micrometers, most likely due to soil structure and the way bacterial cells reproduce. The analysis also suggests that, because cells interact only at very small distances, the number of cells that a typical bacterial cell interact with is relatively limited, as is the number of bacterial species. Such low levels of bacterial interactions could be a reason why several soil microbial processes appear not to be affected during microbial diversity erosion experiments.

Acknowledgments

The authors wish to thank Geneviève Grundmann for a cutting remark that prompted us to initiate this work and C. Chenu and S. Boudsocq for constructive comments on an earlier version of the manuscript.

Supporting Information

Figure S1 Number of neighbouring bacterial species around a typical bacterium as a function of distance at different bacterial densities for the four species diversity levels estimated by Roesch et al (2007). Grey envelopes surrounding curves represent the maximum and minimum of these numbers calculated from 39 simulations with the same parameters values as in Fig. 3. Note the similarity in the number of neighbouring species, for all diversity levels, when bacterial density is 10^9 cells g^{-1} or less. $\alpha = 1107.53$ corresponds to a species richness of 15000 species for 10^9 cells whereas $\alpha = 264.79$ corresponds to a species richness of 4010 species for the same number of cells.

Appendix S1 Notes on the theory of Log Gaussian Cox Processes.

Author Contributions

Conceived and designed the experiments: XR NN. Performed the experiments: XR. Analyzed the data: XR NN. Contributed reagents/materials/analysis tools: NN. Wrote the paper: XR NN.

References

1. Torsvik V, Goksøyr J, Daae FL (1990) High diversity in DNA of soil bacteria. Appl Environ Microbiol 56:782–787.
2. Roesch LFW, Fulthorpe RR, Riva A, Casella G, Hadwin AKM, et al. (2007) Pyrosequencing enumerates and contrasts soil microbial diversity. ISME J 1:283–290. doi:10.1038/ismej.2007.53.
3. Lauber CL, Strickland MS, Bradford MA, Fierer N (2008) The influence of soil properties on the structure of bacterial and fungal communities across land-use types. Soil Biol Biochem 40:2407–2415. doi:10.1016/j.soilbio.2008.05.021.
4. Fierer N, Lauber CL, Ramirez KS, Zaneveld J, Bradford MA, et al. (2012) Comparative metagenomic, phylogenetic and physiological analyses of soil microbial communities across nitrogen gradients. ISME J 6:1007–1017. doi:10.1038/ismej.2011.159.
5. Kowalchuk GA, Buma DS, De Boer W, Klinkhamer PGL, van Veen JA (2002) Effects of above-ground plant species composition and diversity on the diversity of soil-borne microorganisms. A Van Leeuw J Microb 81:509–520. doi:10.1023/A:1020565523615.
6. Ranjard L, Dequiedt S, Chemidlin Prévost-Bouré N, Thioulouse J, Saby NPA, et al. (2013) Turnover of soil bacterial diversity driven by wide-scale environmental heterogeneity. Nat Commun 4:1434. doi:10.1038/ncomms2431.
7. Fierer N, Strickland MS, Liptzin D, Bradford MA, Cleveland CC (2009) Global patterns in belowground communities. Ecol Lett 12:1238–1249. doi:10.1111/j.1461-0248.2009.01360.x.
8. Wertz S, Degrange V, Prosser JI, Poly F, Commeaux C, et al. (2006) Maintenance of soil functioning following erosion of microbial diversity. Environ Microbiol 8:2162–2169. doi:10.1111/j.1462-2920.2006.01098.x.
9. Rousk J, Brookes PC, Glanville HC, Jones DL (2011) Lack of Correlation between Turnover of Low-Molecular-Weight Dissolved Organic Carbon and Differences in Microbial Community Composition or Growth across a Soil pH Gradient. Appl Environ Microbiol 77:2791–2795. doi:10.1128/AEM.02870-10.
10. Langenheder S, Bulling MT, Solan M, Prosser JI (2010) Bacterial biodiversity-ecosystem functioning relations are modified by environmental complexity. PLoS One 5:e10834. doi:10.1371/journal.pone.0010834.
11. Prosser JI (2012) Ecosystem processes and interactions in a morass of diversity. FEMS Microbiol Ecol 81:507–519. doi:10.1111/j.1574-6941.2012.01435.x.
12. Baveye PC (2009) To sequence or not to sequence the whole-soil metagenome? Nat Rev Microbiol 7:756. doi:10.1038/nrmicro2119-c2.
13. Goreaud F, Loreau M, Millier C (2002) Spatial structure and the survival of an inferior competitor: a theoretical model of neighbourhood competition in plants. Ecol Model 158:1–19. doi:10.1016/S0304-3800(02)00058-3.
14. Hodge A (2003) Plant nitrogen capture from organic matter as affected by spatial dispersion, interspecific competition and mycorrhizal colonization. New Phytol 157:303–314. doi:10.1046/j.1469-8137.2003.00662.x.
15. Raynaud X, Jaillard B, Leadley PW (2008) Plants may alter competition by modifying nutrient bioavailability in rhizosphere: a modeling approach. Am Nat 171:44–58. doi:10.1086/523951.
16. Amarasekare P (2004) Spatial dynamics of mutualistic interactions. J Anim Ecol 73:128–142. doi:10.1046/j.0021-8790.2004.00788.x.
17. Holt R (1984) Spatial heterogeneity, indirect interactions and the coexistence of prey species. Am Nat 124:377–406.
18. Jansen V (1995) Regulation of predator-prey systems through spatial interactions: a possible solution to the paradox of enrichment. Oikos 74:384–390.
19. Hooper DU, Chapin III FS, Ewel JJ, Hector A, Inchausti P, et al. (2005) Effects of biodiversity on ecosystem functioning: a consensus of current knowledge. Ecol Monogr 75:3–35. doi:10.1890/04-0922.
20. Dechesne A, Pallud C, Debouzie D, Flandrois JP, Vogel TM, et al. (2003) A novel method for characterizing the microscale 3D spatial distribution of bacteria in soil. Soil Biol Biochem 35:1537–1546. doi:10.1016/S0038-0717(03)00243-8.
21. Stefanic P, Mandic-Mulec I (2009) Social interactions and distribution of *Bacillus subtilis* pherotypes at microscale. J Bacteriol 191: 1756–1764. doi:10.1128/JB.01290-08.
22. Vieublé-Gonod L, Chenu C, Soulas G (2003) Spatial variability of 2,4-dichlorophenoxyacetic acid (2,4-D) mineralisation potential at a millimetre scale in soil. Soil Biol Biochem 35:373–382. doi:10.1016/S0038-0717(02)00287-0.
23. Franklin RB, Mills AL (2008) The importance of microbial distribution in space and spatial scale to microbial ecology. In: Franklin RB, Mills AL, editors. The spatial distribution of microbes in the environment. New York, USA: Kluwer Academic Publishers. pp. 1–30.
24. Gantner S, Schmid M, Dürr C, Schuhegger R, Steidle A, et al. (2006) In situ quantitation of the spatial scale of calling distances and population density-independent N-acylhomoserine lactone-mediated communication by rhizobacteria colonized on plant roots. FEMS Microbiol Ecol 56:188–194. doi:10.1111/j.1574-6941.2005.00037.x.
25. Vos M, Wolf AB, Jennings SJ, Kowalchuk G A (2013) Micro-scale determinants of bacterial diversity in soil. FEMS Microbiol Rev 37:936–954. doi:10.1111/1574-6976.12023.
26. Fierer N, Lennon JT (2011) The generation and maintenance of diversity in microbial communities. Am J Bot 98:439–448. doi:10.3732/ajb.1000498.
27. Dechesne A, Or D, Smets BF (2008) Limited diffusive fluxes of substrate facilitate coexistence of two competing bacterial strains. FEMS Microbiol Ecol 64:1–8. doi:10.1111/j.1574-6941.2008.00446.x.
28. Van Elsas JD, Bailey MJ (2002) The ecology of transfer of mobile genetic elements. FEMS Microbiol Ecol 42:187–197. doi:10.1111/j.1574-6941.2002.tb01008.x.
29. Kim HJ, Boedicker JQ, Choi JW, Ismagilov RF (2008) Defined spatial structure stabilizes a synthetic multispecies bacterial community. Proc Natl Acad Sci USA 105:18188–18193. doi:10.1073/pnas.0807935105.
30. Nunan N, Ritz K, Crabb D, Harris K, Wu KJ, et al. (2001) Quantification of the in situ distribution of soil bacteria by large-scale imaging of thin sections of undisturbed soil. FEMS Microbiol Ecol 36:67–77. doi:10.1111/j.1574-6941.2001.tb00854.x.
31. Nunan N, Wu K, Young IM, Crawford JW, Ritz K (2002) In situ spatial patterns of soil bacterial populations, mapped at multiple scales, in an arable soil. Microb Ecol 44:296–305. doi:10.1007/s00248-002-2021-0.
32. Nunan N, Wu KJ, Young IM, Crawford JW, Ritz K (2003) Spatial distribution of bacterial communities and their relationships with the micro-architecture of soil. FEMS Microbiol Ecol 44:203–215. doi:10.1016/S0168-6496(03)00027-8.
33. Nunan N, Young IM, Crawford JW, Ritz K (2007) Bacterial interactions at the microscale - Linking habitat to function in soil. In: Franklin RB, Mills AL, editors. The spatial distribution of microbes in the environment. New York, USA: Kluwer Academic Publishers. pp. 61–85.
34. Møller J, Waagepetersen RP, Syversveen AR (1998) Log Gaussian Cox processes. Scand J Stat 25: 451–482.
35. Ngom NF, Garnier P, Monga O, Peth S (2011) Extraction of three-dimensional soil pore space from microtomography images using a geometrical approach. Geoderma 163:127–134. doi:10.1016/j.geoderma.2011.04.013.
36. Stoyan D (2006) Fundamentals of Point Process Statistics. In: Baddeley A, Gregori P, Mateu J, Stoica R, Stoyan D, editors. Case studies in spatial point process modeling. NY, USA: Springer. pp. 3–22.

37. Møller J, Waagepetersen RP (2003) Statistical inference and simulation for spatial point processes. Boca Raton, USA: Chapman And Hall/CRC.
38. Baddeley A, Turner R (2005) Spatstat: an R package for analysing spatial point patterns. J Stat Soft 12:1–42.
39. Schlather M (2001) Simulation and analysis of random fields. R News 1:18–20.
40. Andrey P, Kiêu K, Kress C, Lehmann G, Tirichine L, et al. (2010) Statistical analysis of 3D images detects regular spatial distributions of centromeres and chromocenters in animal and plant nuclei. PLoS Comput Biol 6:e1000853. doi:10.1371/journal.pcbi.1000853.
41. Bonami A, Estrade A (2003) Anisotropic Analysis of Some Gaussian Models. J Fourier Anal Appl 9:215–236. doi:10.1007/s00041-003-0012-2.
42. VanMarcke E (1988) Random fields: analysis and synthesis. Massachusetts Institute of Technology. 382p
43. Fisher RA, Corbet A, Williams CB (1943) The relation between the number of species and the number of individuals in a random sample of an animal population. J Anim Ecol 12:42–58.
44. Hill TCJ, Walsh KA, Harris JA, Moffett BF (2003) Using ecological diversity measures with bacterial communities. FEMS Microbiol Ecol 43:1–11. doi:10.1111/j.1574-6941.2003.tb01040.x.
45. Van Groenigen JW, Zwart KB, Harris D, van Kessel C (2005) Vertical gradients of $\delta^{15}N$ and $\delta^{18}O$ in soil atmospheric N_2O–temporal dynamics in a sandy soil. Rapid Commun Mass Spectrom 19:1289–1295. doi:10.1002/rcm.1929.
46. Charley J, West N (1975) Plant-induced soil chemical patterns in some shrub-dominated semi-desert ecosystems of Utah. J Ecol 63:945–963.
47. Kinkel LL, Bakker MG, Schlatter DC (2011) A coevolutionary framework for managing disease-suppressive soils. Annu Rev Phytopathol 49:47–67. doi:10.1146/annurev-phyto-072910-095232.
48. Marilley L, Aragno M (1999) Phylogenetic diversity of bacterial communities differing in degree of proximity of Lolium perenne and Trifolium repens roots. Appl Soil Ecol 13:127–136. doi:10.1016/S0929-1393(99)00028-1.
49. Sanguin H, Remenant B, Dechesne A, Thioulouse J, Vogel TM, et al. (2006) Potential of a 16S rRNA-based taxonomic microarray for analyzing the rhizosphere effects of maize on Agrobacterium spp. and bacterial communities. Appl Environ Microbiol 72:4302–4312. doi:10.1128/AEM.02686-05.
50. Cornell H V, Lawton JH (1992) Species interactions, local and regional processes, and limits to the richness of ecological communities: a theoretical perspective. J Anim Ecol 61:1–12.
51. Loreau M (2000) Are communities saturated? On the relationship between alpha, beta and gamma diversity. Ecol Lett 3:73–76. doi:10.1046/j.1461-0248.2000.00127.x.
52. Wertz S, Degrange V, Prosser JI, Poly F, Commeaux C, et al. (2007) Decline of soil microbial diversity does not influence the resistance and resilience of key soil microbial functional groups following a model disturbance. Environ Microbiol 9:2211–2219. doi:10.1111/j.1462-2920.2007.01335.x.
53. Griffiths BS, Ritz K, Bardgett RD, Cook R, Christensen S, et al. (2000) Ecosystem response of pasture soil communities to fumigation-induced microbial diversity reductions: an examination of the biodiversity-ecosystem function relationship. Oikos 90:279–294. doi:10.1034/j.1600-0706.2000.900208.x.

17

Contributions of Understory and/or Overstory Vegetations to Soil Microbial PLFA and Nematode Diversities in Eucalyptus Monocultures

Jie Zhao[1,3], Songze Wan[1], Chenlu Zhang[2], Zhanfeng Liu[1], Lixia Zhou[1], Shenglei Fu[1]*

1 Key Laboratory of Vegetation Restoration and Management of Degraded Ecosystems, South China Botanical Garden, Chinese Academy of Sciences, Guangzhou, China, 2 State Key Laboratory of Cotton Biology, Key Laboratory of Plant Stress Biology, College of Life Sciences, Henan University, Kaifeng, China, 3 Key Laboratory of Agro-Ecological Processes in Subtropical Region, Institute of Subtropical Agriculture, Chinese Academy of Sciences, Changsha, China

Abstract

Ecological interactions between aboveground and belowground biodiversity have received many attentions in the recent decades. Although soil biodiversity declined with the decrease of plant diversity, many previous studies found plant species identities were more important than plant diversity in controlling soil biodiversity. This study focused on the responses of soil biodiversity to the altering of plant functional groups, namely overstory and understory vegetations, rather than plant diversity gradient. We conducted an experiment by removing overstory and/or understory vegetation to compare their effects on soil microbial phospholipid fatty acid (PLFA) and nematode diversities in eucalyptus monocultures. Our results indicated that both overstory and understory vegetations could affect soil microbial PLFA and nematode diversities, which manifested as the decrease in Shannon–Wiener diversity index (*H'*) and Pielou evenness index (J) and the increase in Simpson dominance index (λ) after vegetation removal. Soil microclimate change explained part of variance of soil biodiversity indices. Both overstory and understory vegetations positively correlated with soil microbial PLFA and nematode diversities. In addition, the alteration of soil biodiversity might be due to a mixing effect of bottom-up control and soil microclimate change after vegetation removal in the studied plantations. Given the studied ecosystem is common in humid subtropical and tropical region of the world, our findings might have great potential to extrapolate to large scales and could be conducive to ecosystem management and service.

Editor: Dafeng Hui, Tennessee State University, United States of America

Funding: This study was financially supported by the Knowledge Innovation Program of the Chinese Academy of Sciences (KSCZ2-EW-Z-6) and Natural Science Fund of China (31210103920). The funders had no role in study design, data collection and analysis, decision to publish, or preparation of the manuscript.

Competing Interests: The authors have declared that no competing interests exist.

* E-mail: sfu@scbg.ac.cn

These authors contributed equally to this work.

Introduction

Biodiversity plays a crucial role in ecosystem function and processes not only because of its importance in production of food, fiber and fuel, but also because of its roles in ground water replenishment, flooding controls, prevention of soil erosion and ecological invasion, influences on restoration succession, and so on [1,2]. Although the relationship between biodiversity and ecosystem functions and processes has been the focuses of quite a few studies over the last two decades, much controversy still exists [3]. The impact of plant diversity on ecosystem functions and processes has been explored mainly in grasslands (e.g. [4,5]); and primary productivity is frequently used to assess the ecosystem stability [4]. In general, ecosystems with high biodiversity level are more stable, sustainable, reliable and predictable [1,4,5,6]. By contrast, several studies reported no significant effect of plant diversity on plant productivity [7,8].

Soil contains the most diverse communities of organisms and soil biodiversity provides a number of ecosystem services [2,9]. Ecological interactions between aboveground and belowground biodiversity have received many attentions in the recent decades [10]. In one hand, plants play a key role in determining soil biota because plants are the main food sources (i.e. leaf litter, dead roots and root exudates) to soil biota, which may exert a bottom-up control on soil organisms. In another hand, soil biota may affect aboveground plant diversity and productivity through influencing nutrient cycling and distribution, and damaging plants by parasites, pathogens and herbivores [11,12].

Plant functional groups (PFG) or plant functional types (PFT) are assemblages of species with similar roles in ecosystem processes by responding in similar ways to multiple environmental factors [13]. They are necessary devices for reducing the complexity of plant community and often uncharted characteristics of species diversity in function and structure when attempting to project the nature and function of species assemblages into future environments [14]. However, due to the diverse aims of different researches, diverse plant functional groups are chosen to simplify the plant community, i.e. legumes, forbs/herbs, C3/C4-plants, overstory vegetation, and understory vegetation.

Vegetation removal has been frequently used to explore how specific plant species or functional groups affect ecosystem processes and functions. Overstory removal (timber harvest) has been a common forest management practice which also provides opportunities for soil ecologists to study relationships between overstory tree species and soil biota. In general, overstory removal suppresses soil biota such as bacteria, fungus, nematodes, and microarthropods in the long run [15,16,17]. However, there are studies which reported overstory removal had slight or non-significant impacts on soil biota, especially in the short term [18,19]. To date, less attention has been paid to understory vegetation in terms of their impacts on soil biota, probably because the biomass and economic value of the understory are less than overstory tree species. Understory vegetation is an important component in many forest ecosystems worldwide [20,21] and drives ecological processes and functions, such as biodiversity, stand productivity, tree-seedling regeneration, forest succession, litter decomposition, soil nutrient cycling and soil water conservation [19,22,23,24]. However, how understory vegetation affects the soil biodiversity has never been studied before.

In the present study, we monitored the soil microbial phospholipid fatty acid (PLFA) and nematode diversity after overstory and/or understory removal in eucalyptus monocultures in South China. The eucalyptus species is *Eucalyptus urophylla*, which is a fast-grow species and has been widely planted for reforestation in South China [19]. The understory vegetation was dominated by *Dicranopteris dichotoma*, a common fern species in humid subtropical and tropical region [20]. Other common understory species included *Miscanthus sinensis* and *Rhodomyrtus tomentosa*. Thus, the diversity of overstory and understory vegetations is not high in the studied ecosystems. Therefore, we did not calculate the plant diversity. In this experiment, overstory and understory vegetations were removed separately or in combination to assess their impacts on soil microbial PLFA profile and nematode community. Both *E. urophylla* and *D. dichotoma* are high in biomass and primary productivity [20,25]. Previous study, conducted at the same site as this study, reported that understory removal showed stronger effects on soil microbial community structure, and abundance of nematode and microarthropod than overstory removal [20]. Therefore, the hypothesis of this study was that the effects of understory removal on soil microbial and nematode diversity were greater than overstory removal. The primary goal of this study was to evaluate the contributions of the understory and/or overstory vegetation to soil microbial and nematode diversity. Our analyses also intend to address the question of which factors (i.e. resource and soil microclimate) affected soil biodiversity after altering of aboveground vegetations.

Materials and Methods

Site Description

This study was conducted at the Heshan Hilly Land Interdisciplinary Experimental Station (112°50′E, 22°34′N), property of the Chinese Academy of Sciences (CAS) in Guangdong Province of China. The climate is subtropical monsoon with a distinct wet and dry Season. The mean annual temperature and precipitation are 21.7°C and 1,700 mm, respectively. The soil is an acrisol [26]. Vegetation at the experimental site consisted of three (replications) 4-year-old *E. urophylla* plantations. The seedlings were planted with a spacing of 3×2 m.

Experimental Design

Our experiment was conducted in three replicated plots of 4-year-old *E. urophylla* monoculture plantations. In September 2009, we established an experiment with a split-plot design. Within each plantation, main plots (15×15-m) were overstory trees treatment (present or removed) and subplots were understory treatment (present or removed). In total, there were four treatments in each plantation, namely, control with intact overstory and understory (CK), understory removal (UR), overstory tree removal (TR) and all-plant removal (PR). The trees were cut with an electrical saw leaving 5 cm stumps aboveground. All woody debris was removed from the plots immediately after vegetation was removed. The shoots of all understory plants were removed manually with the aid of a machete. Trenches were created around each main-plot and subplot to prevent the interactions of the roots from other subplots and to prevent the exchange of nutrients among subplots. Germination of the stumps (TR plots) and understory vegetation (UR and PR subplots) were monthly removed.

Soil Sampling and Analysis

Soils were sampled on 20 September, 1 October and 6 November 2009, 11 March, 15 June and 14 October 2010, corresponding to 10 days before and 7, 37, 160, 256, 376 days after treatments were implemented. Soil cores (2.5 cm in diameter, 5 cm in length) were taken at 0–5 cm and 5–10 cm depths from eight locations selected randomly in each subplot within each plantation. Eight cores of the same depth from each subplot were combined to form one composite sample; there were three replicate samples for each treatment. The surface litter was removed carefully when soil sample was taken.

Soil water content (SWC %, g of water per 100 g dry soil) was measured by oven-drying for 48 h at 105°C. Soil temperature was recorded every 2 h with the DS1922L temperature logger iButtons (Dallas Semiconductor Corp., Dallas, TX) from September 2009 to October 2010. Soil microbial PLFAs were analyzed according to [27]. Briefly, the soil was extracted in a one-phase extraction mixture containing chloroform:methanol:phosphate buffer (1:2:0.8 v/v/v), with the amount of phosphate buffer corrected to account for existing soil water content mixture of chloroform:-methanol:citrate buffer (1:2:0.8 v/v/v). After extraction the phospholipids were separated from neutral lipids and glycolipids on solid phase extraction columns, 0.50 g Si (Supelco, Inc., Bellefonte, Penn). Polar lipids were eluted and were then subjected to mild alkaline methanolysis. Resulting fatty acid methyl esters (FAMEs) were separated, quantified, and identified using capillary gas chromatography (GC). Qualitative and quantitative fatty acid analyses were performed using an Agilent 6890 gas chromatograph (Agilent Technologies, Palo Alto, CA, USA) and the MIDI Sherlock Microbial Identification System (MIDI Inc., Newark, DE, USA). Concentrations of each PLFA were calculated relative to 19:0 internal standard concentrations. The following PLFAs were considered to be of soil microbial origin: i15:0, a15:0, 15:0, i16:0, 16:1ω7, i17:0, a17:0, 17:0, cy17:0, cy19:0 18:2ω6,9, 16:1ω5, 18:1ω7, and 18:1ω9 [28,29,30].

Nematodes were extracted from 50 g of fresh soil using the Baermann funnel method. After fixation in 4% formalin solution, nematodes were counted with a differential interference contrast (DIC) microscope (ECLIPSE 80 i, Nikon), and the first 100 individuals encountered were identified to genus level. All nematodes were identified to genus level when the sample contained fewer than 100 individuals.

Statistical Analysis

Shannon–Wiener diversity index, Pielou evenness index, Margalef richness index, and Simpson dominance index were employed to assess the soil microbial PLFA and nematode diversity.

$$\text{Shannon-Wiener diversity index, } H' = \sum_{i=1}^{S} Pi * \ln Pi \quad (1)$$

$$\text{Pielou evenness index, } J = \frac{H'}{H' \max} \quad (2)$$

$$H' \max = \ln S \quad (3)$$

$$\text{Margalef richness index, } SR = \frac{S-1}{\ln N} \quad (4)$$

$$\text{Simpson dominance index, } \lambda = \sum Pi^2 \quad (5)$$

Where '*Pi*' is the proportion of the individuals of "*i*th" group in the community; '*S*' is the total number of microbial PLFAs or nematode genera in the community; and '*N*' is the amount of microbial PLFAs or number of total nematodes in the community.

Repeated-measure ANOVA was employed to determine the time effect and treatment effect through the whole experimental period with TR and UR as between-subjects factors. To address the relationship between soil microclimates (i.e. soil water content and soil temperature) and soil biodiversity, linear and quadratic regression analysis were performed. Repeated-measure ANOVA and regression model were performed using SPSS software (SPSS Inc., Chicago, IL). Statistical significance was determined at $p<0.05$.

Results

Microbial PLFA Diversity

There was a significant time effect on all the microbial variables, namely, Shannon–Wiener diversity index (H'), Pielou evenness index (J), Margalef richness index (SR), and Simpson dominance index (λ) ($p<0.05$) (Fig. 1). There were trends that understory removal and overstory removal decreased soil microbial PLFA diversity at 0–5 cm soil depth by 0.024 ($p=0.08$) and 0.026 ($p=0.055$) units, respectively (Fig. 1a). Overstoty removal significantly reduced soil microbial PLFA diversity at 5–10 cm soil depth by 0.048 units ($p=0.034$) (Fig. 1b). Repeated measure ANOVA showed that understory removal and overstory removal apparently increased dominance index at 0–5 cm soil depth by 0.007 ($p=0.038$) and 0.008 ($p=0.035$) units, respectively (Fig. 1g); and overstory removal significantly increased dominance index at 5–10 cm soil depth by 0.011 units ($p=0.032$) (Fig. 1h). There was no remarkable effect of vegetation removal on evenness index and richness index of microbial PLFA profile at both 0–5 cm and 5–10 cm soil depths during the study (Fig. 1c, d, e, and f). No interaction effect of understory removal and overstory removal on soil microbial variables was found during the study.

Nematode Diversity

There was a significant time effect on all the nematode variables, namely, Shannon–Wiener diversity index (H'), Pielou evenness index (J), Margalef richness index (SR), and Simpson dominance index (λ) ($p<0.05$) (Fig. 2). There were trends that understory removal decreased soil nematode diversity index by 0.086 units ($p=0.096$) (Fig. 2a) and evenness index by 0.022 units ($p=0.058$) (Fig. 2c) at 0–5 cm soil depth, respectively. Overstory removal tended to increase the soil nematode evenness at 0–5 cm soil depth by 0.022 units ($p=0.062$) (Fig. 2c), and significantly increased the soil nematode evenness at 5–10 cm soil depth by 0.036 units ($p=0.004$) (Fig. 2d). There was a trend that interaction of understory removal and overstory removal tended to affect soil nematode diversity index at 0–5 cm soil depth during the study ($p=0.072$) (Fig. 2a). Neither soil nematode Margalef richness index (SR) nor Simpson dominance index (λ) were significantly affected by understory removal or overstory removal at either 0–5 or 5–10 cm depth (Fig. 2e, f, g, and h).

Correlations between Soil Biodiversity and Soil Microclimates

Regression analysis showed that soil water content and soil biodiversity were not well fitting in either linear or quadratic model (Appendix S1). Coefficients of determination (R^2) of both linear and quadratic regressions between soil water content and biodiversity indices (H', J, SR, and λ) of soil microbial PLFAs and nematodes were very low (Appendix S1). Shannon–Wiener diversity index (H') ($R^2=0.15$, $p=0.006$) and Margalef richness index (SR) ($R^2=0.02$, $p=0.001$) of microbial PLFA profile were quadratic correlated with soil temperature (Fig. 3a and c). Pielou evenness index (J) of nematode community was quadratic correlated with soil temperature ($R^2=0.111$, $p=0.024$) (Fig. 4b).

Discussion

Contributions of Understory or Overstory Vegetations to Soil Biodiversity

Apparently, all plant removal showed strongest suppression on soil microbial and nematode diversity in the present study. In addition, both understory removal and overstory removal reduced soil microbial PLFA and nematode diversities, which was inconsistent with our hypothesis that the effects of understory removal on soil biodiversity might be more powerful than overstory removal. Consistent with our finding, many studies reported that all plant removal had significant negative influence on soil biota in forest and grassland ecosystem [17,31,32]. Our previous study at the same site reported that understory removal had stronger negative effects on abundances and compositions of soil biota (microbes and nematodes) than overstory removal [20]. That was why our hypotheses stated as understory had greater contributions on soil biodiversity than overstory. Contrary to our hypotheses, our analyses clearly showed that the overstory vegetation impacted soil biodiversity too. In addition, overstory removal rather than understory removal significantly reduced amount of total PLFAs in 5–10 cm soil depth during the study (Appendix S2). There are many studies focused on the effects of overstory removal on soil biota because canopy tree harvest is a common forest management practice. For example, fifteen years after harvesting, diversity and structure of soil bacterial and fungal communities remained apparently altered by harvesting disturbances; and three fungi taxa (i.e. ectomycorrhizal fungi, saprobic taxa, and ascomycetes) and actinomycetes were most sensitive to harvesting disturbance in sub-boreal spruce forest in BC, Canada [15]. Both stem-only and whole-tree harvesting decreased microarthropod abundance in a temperate mixed conifer–hardwood stand [33] and in sub-boreal spruce stand [34]. However, these studies usually focused on size of populations and structure of communities of soil biota; there is no mention of how soil biodiversity responses to overstory removal in these studies. Furthermore, most of the studies did not clearly state

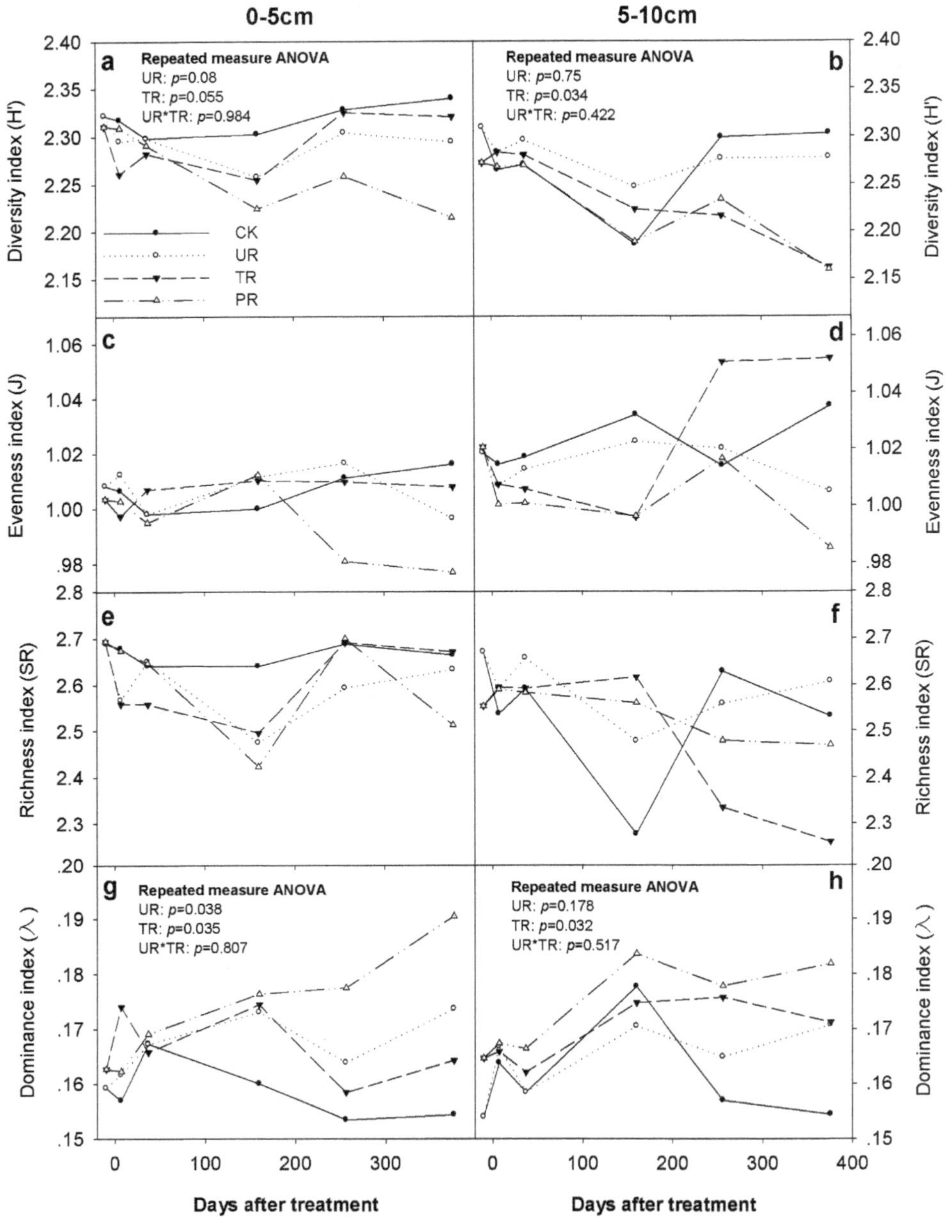

Figure 1. Microbial PLFA diversity as affected by vegetation removal during the study. Shannon–Wiener diversity index (H'), Pielou evenness index (J), Margalef richness index (SR) and Simpson dominance index (λ) under control (CK), understory removal (UR), tree removal (TR), and all-plant removal (PR) in each sampling event at 0–5 cm (a, c, e, g) and 5–10 cm (b, d, f, h) soil depths.

whether understory was remained un-harvesting or not when overstory was harvested. In addition, understory vegetation is important ecological component that plays important roles in driving forest ecosystem functions and processes [20,21]; however, it is usually overlooked in the previous studies. Several studies reported the relationship between understory vegetation and aboveground biodiversity (e.g., birds, reptiles, arthropods) [35,36,37]. Interactions between understory vegetation and soil biodiversity have not been documented yet. This is the first study reported the linkages between understory vegetation and soil microbial PLFA and nematode diversities.

Shannon–Wiener diversity index (H') gives weight to rarer taxa, whereas Simpson dominance index (λ) gives weight to more abundant taxa [38,39]. The suppressive effect of overstory removal on H' of soil microbial PLFA profile might indicate the linkage between overstory vegetation and soil rare microbial taxa. The promotive effect of overstory removal on λ of soil microbial PLFA profile might indicate the linkage between overstory vegetation and soil abundant microbial taxa. Therefore, overstory vegetation might link to both the rare and abundant taxa of soil microbial organisms. Although no significant effect of overstory removal on H' and λ of nematode community, overstory removal increased the Pielou evenness index (J) of nematode community, which implied that the abundance of nematode taxa did not differ from each other as much as their abundance before overstory removal. The likely reason might be that the rare species disappeared and were not collected after overstory removal, which eliminated the weight of the rare taxa to Pielou evenness

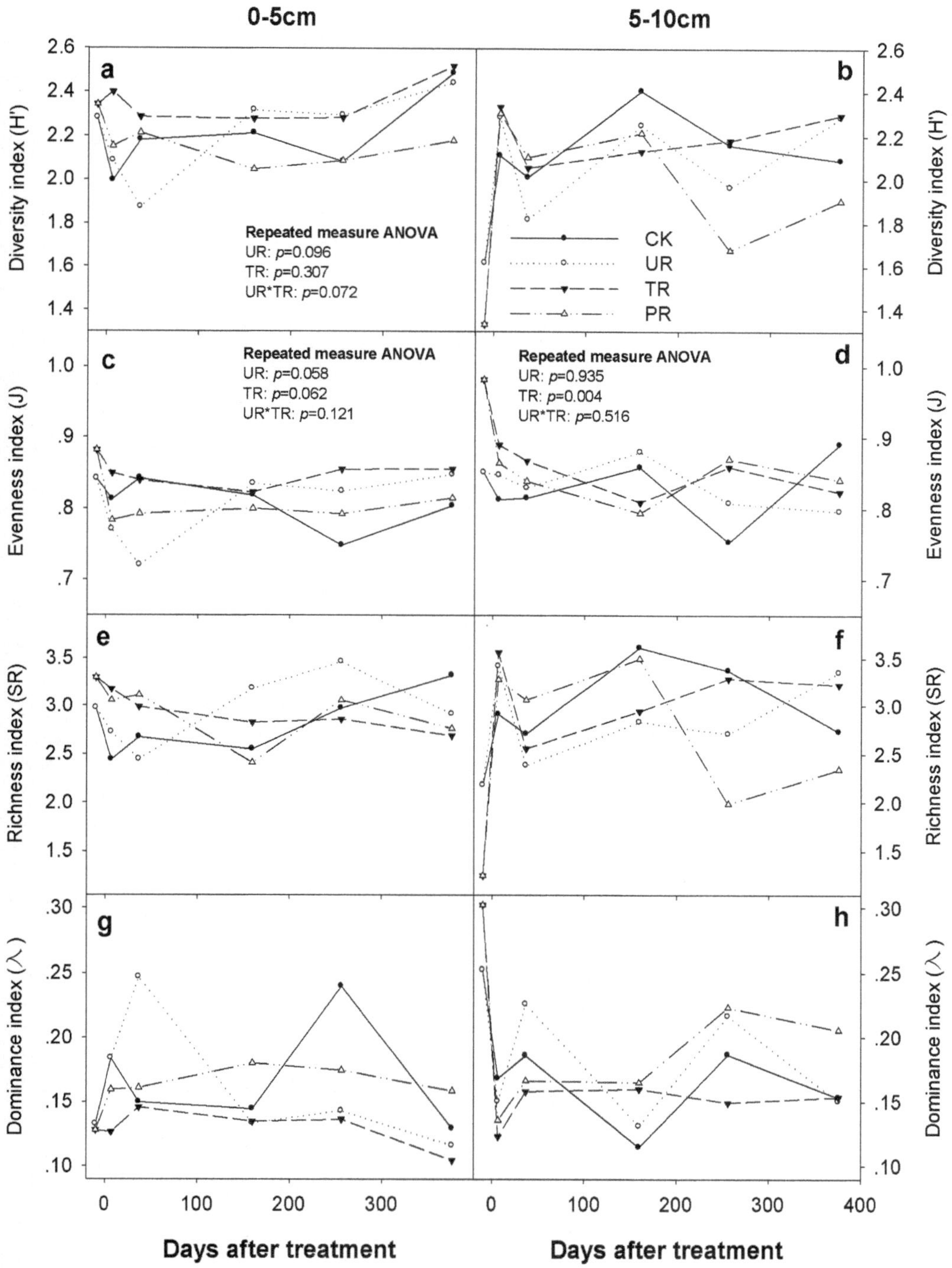

Figure 2. Nematode diversity as affected by vegetation removal during the study. Shannon–Wiener diversity index (*H'*), Pielou evenness index (J), Margalef richness index (SR) and Simpson dominance index (λ) under control (CK), understory removal (UR), tree removal (TR), and all-plant removal (PR) in each sampling event at 0–5 cm (a, c, e, g) and 5–10 cm (b, d, f, h) soil depths.

index. In the similar way, understory vegetation linked to both abundant and rare taxa at 0–5 cm soil depth.

Factor(s) that Affect Soil Biodiversity

Plant diversity may influence soil biodiversity via two main ways. High plant diversity may (1) enhance net primary productivity and (2) lead to high food resources (e.g. litter and root exudates) diversity, and subsequently supports high soil biodiversity [40]. However, many studies found plant species identities were more important than plant diversity in controlling soil biodiversity [41,42,43]; and sometimes the effect of plant identities can overrule the effect of plant diversity [41,44,45].

The bottom-up control might explain the alterations of soil biodiversity after vegetation removal [19,20,32]. *E. urophylla* is the only overstory species and *D. dichotoma* is the dominant understory species in the studied plantations. The productivity of both species are much high. The annual carbon increment of *E. urophylla* can reach 1960 gC m^{-2} yr^{-1} in the developmental stage in the studied

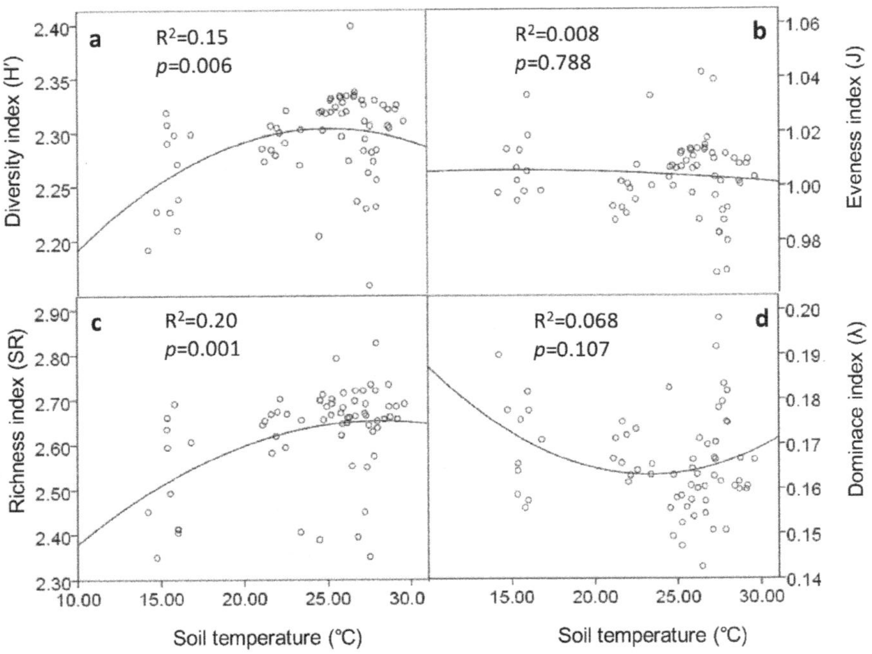

Figure 3. Relationships between indices of microbial PLFAs and soil temperature. Bivariate plots of the microbial PLFA diversity (a), evenness (b), richness (c) and dominance (d) indices versus soil temperature. Points represent results from all six sampling events at 0–5 cm soil depth. Soil temperatures used here were the mean daily temperatures of the six days that soil samplings were carried out.

area [25], while the biomass of *D. dichotoma* may contribute to 22%~48% of total live biomass in *E. urophylla* plantations in the studied area (unpublished data). Both *E. urophylla* and *D. dichotoma* removal resulted in the reduction of carbon input to soil food web [19], which in turn may change soil biodiversity. The bottom-up control opinion was also supported by the differential responses of soil biodiversity at different depths to overstory and understory vegetation. Overstory removal decreased soil biodiversity at both 0–5 cm and 5–10 cm soil depths while understory removal mainly suppressed soil biodiversity at 0–5 cm soil layer. The reason might be that the overstory *E. urophylla* is a deep-rooted species and the dominant understory *D. dichotoma* is a shallow-rooted fern species

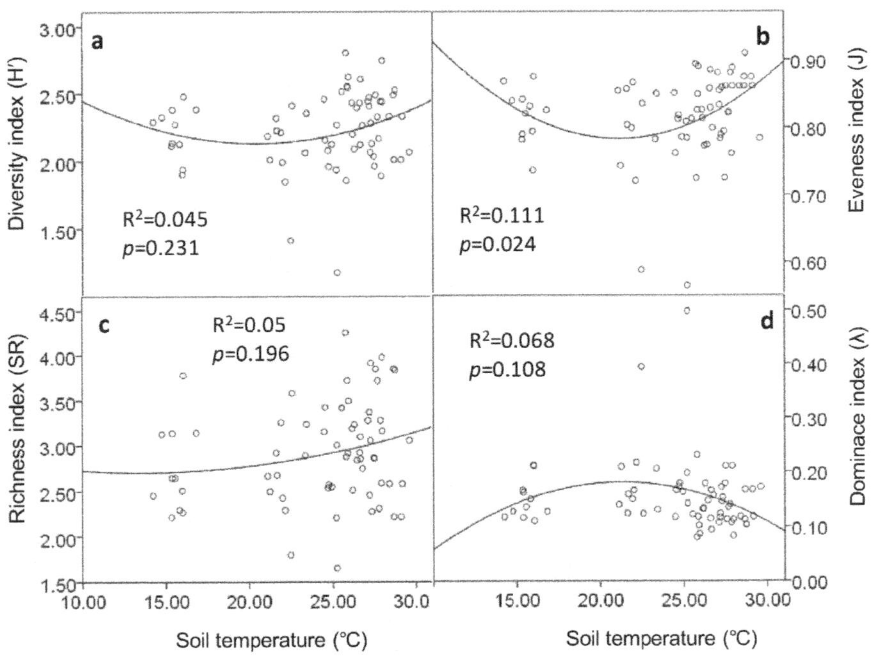

Figure 4. Relationships between indices of nematode diversity and soil temperature. Bivariate plots of the soil nematode diversity (a), evenness (b), richness (c) and dominance (d) indices versus soil temperature. Points represent results from all six sampling events at 0–5 cm soil depth. Soil temperatures used here were the mean daily temperatures of the six days that soil samplings were carried out.

with fibrous root on stolons. The deep-rooted species removal might reduce the availability of soil carbon and nutrients for soil biota at deeper soil depth. Previous studies reported that vegetation removal reduced fine root biomass in eucalyptus plantations in the studied area [19,25]. Unfortunately, none of these studies differentiated either overstory or understory roots from the total plant roots and they did not report the composition of root at different soil depths.

In the present study, soil microclimate (especially soil temperature) explained a certain proportion of variability of soil biodiversity indices. Therefore, soil microclimate change that induced by vegetation removal probably contributed to the alteration of soil biodiversity. Supporting this hypothesis, growth temperature affected the cellular fatty acids composition of sulphate-reducing bacteria that were of marine origin [46]. In addition, our own data revealed that vegetation removal increased soil temperature and decreased soil water content in the later stage of the study, and the effect of understory removal on soil microclimate was greater than overstory removal.

The negative effects of vegetation removal on soil biodiversity also gradually increased over time, which indicated that alterations in bottom-up control and soil microclimate induced by vegetation removal were time-dependent. Firstly, vegetation removal significantly reduced the food resource for soil biota. However, the decomposition of the remained soil residuals (e.g., roots and litter) is time consuming. Therefore, the soil biodiversity declined with the exhaustion of soil residuals. Secondly, the soil microclimate changes that induced by vegetation removal were from March to October (mainly in wet season) in 2010, which was overlapped with the significant declines of soil biodiversity in this study. Thus, the seasonal variation of soil microclimate might be another reason for the time-dependent response of soil biodiversity to vegetation removal.

Conclusions

Here we reported the first study designed to explore the relationship between overstory and understory vegetations and soil biodiversity. Our results showed that both overstory and understory vegetations correlated with soil biodiversity of Eucalyptus plantations in southern China. Interaction between overstory vegetation and soil biodiversity might be due to its high-biomass property that exerts a bottom-up control on soil biodiversity in the studied plantations. Interaction between understory vegetation and soil biodiversity might due to the mixing effect of vegetation identity and soil microclimate change after understory removal that drives soil biodiversity in the studied plantations. Eucalyptus has been extensively planted in many tropical and subtropical regions [25] and *Dicranopteris* is a common fern distributed through humid subtropical and tropical regions of the world [20]. Therefore, our findings might have great potential to extrapolate to large scales, especially to the humid subtropical and tropical region of the world.

Acknowledgments

This study was conducted at Heshan Hilly Land Interdisciplinary Experimental Station, CAS and Heshan National Field Research Station of Forest Ecosystem. We thank Dejun Li for refine the language of the manuscript. We also thank the editor Dafeng Hui and two anonymous reviewers for their comments.

Author Contributions

Conceived and designed the experiments: SF. Performed the experiments: JZ SW CZ ZL LZ. Analyzed the data: JZ SW CZ. Contributed reagents/materials/analysis tools: SF. Wrote the paper: JZ SF.

References

1. Naeem S, Li S (1997) Biodiversity enhances ecosystem reliability. Nature 390: 507–509.
2. Bardgett RD, Usher MB, Hopkins DW (2005) Biological diversity and function in soils. Cambridge, UK: Cambridge University Press.
3. Loreau M, Naeem S, Inchausti P, Bengtsson J, Grime JP, et al. (2001) Biodiversity and ecosystem functioning: current knowledge and future challenges. Science 294: 804–808.
4. Tilman D, Reich PB, Knops J, Wedin D, Mielke T, et al. (2001) Diversity and productivity in a long-term grassland experiment. Science 294: 843–845.
5. Hector A, Schmid B, Beierkuhnlein C, Caldeira MC, Diemer M, et al. (1999) Plant diversity and productivity experiments in European grasslands. Science 286: 1123–1127.
6. McGrady-Steed J, Harris PM, Morin PJ (1997) Biodiversity regulates ecosystem predictability. Nature 390: 162–165.
7. Huston MA (1997) Hidden treatments in ecological experiments: re-evaluating the ecosystem function of biodiversity. Oecologia 110: 449–460.
8. Huston MA, Aarssen LW, Austin MP, Cade BS, Fridley JD, et al. (2000) No consistent effect of plant diversity on productivity. Science 289: 1255.
9. Brussaard L (1997) Biodiversity and ecosystem functioning in soil. Ambio 26: 563–570.
10. Wardle DA, Bardgett RD, Klironomos JN, Setala H, van der Putten WH, et al. (2004) Ecological linkages between aboveground and belowground biota. Science 304: 1629–1633.
11. van der Heijden M, Klironomos J, Ursic M, Moutoglis P, Streitwolf-Engel R, et al. (1998) Mycorrhizal fungal diversity determines plant biodiversity, ecosystem variability and productivity. Nature 396: 69–72.
12. Van Der Heijden MGA, Bardgett RD, Van Straalen NM (2008) The unseen majority: soil microbes as drivers of plant diversity and productivity in terrestrial ecosystems. Ecology Letters 11: 296–310.
13. Lavorel S, McIntyre S, Landsberg J, Forbes TDA (1997) Plant functional classifications: from general groups to specific groups based on response to disturbance. Trends in Ecology & Evolution 12: 474–478.
14. Woodward FI, Cramer W (1996) Plant functional types and climatic change: Introduction. Journal of Vegetation Science 7: 306–308.
15. Hartmann M, Howes CG, VanInsberghe D, Yu H, Bachar D, et al. (2012) Significant and persistent impact of timber harvesting on soil microbial communities in Northern coniferous forests. The ISME Journal.
16. Seastedt T, Crossley Jr D (1981) Microarthropod response following cable logging and clear-cutting in the southern Appalachians. Ecology: 126–135.
17. Forge T, Simard S (2001) Structure of nematode communities in forest soils of southern British Columbia: relationships to nitrogen mineralization and effects of clearcut harvesting and fertilization. Biology and Fertility of Soils 34: 170–178.
18. Panesar TS, Marshall VG, Barclay HJ (2000) The impact of clearcutting and partial harvesting systems on population dynamics of soil nematodes in coastal Douglas-fir forests. Pedobiologia 44: 641–665.
19. Wu J, Liu Z, Wang X, Sun Y, Zhou L, et al. (2011) Effects of understory removal and tree girdling on soil microbial community composition and litter decomposition in two Eucalyptus plantations in South China. Functional Ecology 25: 921–931.
20. Zhao J, Wan S, Li Z, Shao Y, Xu G, et al. (2012) *Dicranopteris*-dominated understory as major driver of intensive forest ecosystem in humid subtropical and tropical region. Soil Biology and Biochemistry 49: 78–87.
21. Nilsson MC, Wardle DA (2005) Understory vegetation as a forest ecosystem driver: evidence from the northern Swedish boreal forest. Frontiers in Ecology and the Environment 3: 421–428.
22. Barrett J, Cochran P, Busse M (1996) Changes in ponderosa pine site productivity following removal of understory vegetation. Soil Science Society of America Journal 60: 1614–1621.

23. Zhao J, Shao Y, Wang X, Neher DA, Xu G, et al. (2013) Sentinel soil invertebrate taxa as bioindicators for forest management practices. Ecological Indicators 24: 236–239.
24. Zhao J, Wan S, Fu S, Wang X, Wang M, et al. (2013) Effects of understory removal and nitrogen fertilization on soil microbial communities in *Eucalyptus* plantations. Forest Ecology and Management 310: 80–86.
25. Chen D, Zhang C, Wu J, Zhou L, Lin Y, et al. (2011) Subtropical plantations are large carbon sinks: Evidence from two monoculture plantations in South China. Agricultural and Forest Meteorology 151: 1214–1225.
26. FAO (2006) World reference base for soil resources 2006. World Soil Resources Report 103 FAO, Rome.
27. Bossio DA, Scow KM (1998) Impacts of carbon and flooding on soil microbial communities: Phospholipid fatty acid profiles and substrate utilization patterns. Microbial Ecology 35: 265–278.
28. Frostegård A, Bååth E (1996) The use of phospholipid fatty acid analysis to estimate bacterial and fungal biomass in soil. Biology and Fertility of Soils 22: 59–65.
29. Joergensen RG, Wichern F (2008) Quantitative assessment of the fungal contribution to microbial tissue in soil. Soil Biology & Biochemistry 40: 2977–2991.
30. Frostegård Å, Tunlid A, Bååth E (2011) Use and misuse of PLFA measurements in soils. Soil Biology and Biochemistry 43: 1621–1625.
31. Wardle DA, Bonner KI, Barker GM, Yeates GW, Nicholson KS, et al. (1999) Plant removals in perennial grassland: vegetation dynamics, decomposers, soil biodiversity, and ecosystem properties. Ecological Monographs 69: 535–568.
32. Zhao J, Wang X, Shao Y, Xu G, Fu S (2011) Effects of vegetation removal on soil properties and decomposer organisms. Soil Biology and Biochemistry 43: 954–960.
33. Bird G, Chatarpaul L (1988) Effect of forest harvest on decomposition and colonization of maple leaf litter by soil microarthropods. Canadian Journal of Soil Science 68: 29–40.
34. Battigelli JP, Spence JR, Langor DW, Berch SM (2004) Short-term impact of forest soil compaction and organic matter removal on soil mesofauna density and oribatid mite diversity. Canadian Journal of Forest Research 34: 1136–1149.
35. Atwell RC, Schulte LA, Palik BJ (2008) Songbird response to experimental retention harvesting in red pine (Pinus resinosa) forests. Forest Ecology and Management 255: 3621–3631.
36. López G, Moro M (1997) Birds of Aleppo pine plantations in south-east Spain in relation to vegetation composition and structure. Journal of Applied Ecology: 1257–1272.
37. Humphrey J, Hawes C, Peace A, Ferris-Kaan R, Jukes M (1999) Relationships between insect diversity and habitat characteristics in plantation forests. Forest Ecology and Management 113: 11–21.
38. Neher DA, Lee Campbell C (1994) Nematode communities and microbial biomass in soils with annual and perennial crops. Applied Soil Ecology 1: 17–28.
39. Zhao J, Zhang W, Wang K, Song T, Du H (2013) Responses of the soil nematode community to management of hybrid napiergrass: The trade-off between positive and negative effects. Applied Soil Ecology. DOI: 10.1016/j.apsoil.2013.10.011.
40. Wardle DA (2006) The influence of biotic interactions on soil biodiversity. Ecology Letters 9: 870–886.
41. De Deyn GB, Raaijmakers CE, Van Ruijven J, Berendse F, Van Der Putten WH (2004) Plant species identity and diversity effects on different trophic levels of nematodes in the soil food web. Oikos 106: 576–586.
42. Viketoft M, Palmborg C, Sohlenius B, Huss-Danell K, Bengtsson J (2005) Plant species effects on soil nematode communities in experimental grasslands. Applied Soil Ecology 30: 90–103.
43. De Deyn GB, Quirk H, Bardgett RD (2011) Plant species richness, identity and productivity differentially influence key groups of microbes in grassland soils of contrasting fertility. Biology Letters 7: 75–78.
44. Wardle DA, Yeates GW, Williamson W, Bonner KI (2003) The response of a three trophic level soil food web to the identity and diversity of plant species and functional groups. Oikos 102: 45–56.
45. Wardle DA, Yeates GW, Barker GM, Bonner KI (2006) The influence of plant litter diversity on decomposer abundance and diversity. Soil Biology and Biochemistry 38: 1052–1062.
46. Könneke M, Widdel F (2003) Effect of growth temperature on cellular fatty acids in sulphate-reducing bacteria. Environmental Microbiology 5: 1064–1070.

18

Effects of Straw Incorporation on Soil Organic Matter and Soil Water-Stable Aggregates Content in Semiarid Regions of Northwest China

Peng Zhang[1,2], Ting Wei[1,2], Zhikuan Jia[1,2]*, Qingfang Han[1,2], Xiaolong Ren[1,2], Yongping Li[3]

1 The Chinese Institute of Water-Saving Agriculture, Northwest A&F University, Yangling, Shaanxi, China, 2 Key Laboratory of Crop Physi-Ecology and Tillage Science in Northwestern Loess Plateau, Ministry of Agriculture, Northwest A&F University, Yangling, Shaanxi, China, 3 Guyuan Institute of Agricultural Sciences, Guyuan, Ningxia, China

Abstract

The soil degradation caused by conventional tillage in rain-fed areas of northwest China is known to reduce the water-use efficiency and crop yield because of reduced soil porosity and the decreased availability of soil water and nutrients. Thus, we investigated the effects of straw incorporation on soil aggregates with different straw incorporation rates in semiarid areas of southern Ningxia for a three-year period (2008–2010). Four treatments were tested: (i) no straw incorporation (CK); (ii) incorporation of maize straw at a low rate of 4 500 kg ha^{-1} (L); (iii) incorporation of maize straw at a medium rate of 9000 kg ha^{-1} (M); (iv) incorporation of maize straw at a high rate of 13 500 kg ha^{-1} (H). The results in the final year of treatments (2010) showed that the mean soil organic carbon storage of the 0–60 cm soil layers were significantly ($P<0.05$) increased with H, M, and L, by 21.40%, 20.38% and 8.21% compared with CK, respectively. Straw incorporation increased >0.25 mm water-stable macroaggregates level, geometric mean diameter, mean weight diameter and the aggregate stability, which were ranked in order of increasing straw incorporation rates: H/M > L > CK. Straw incorporation significantly ($P<0.05$) reduced the fractal dimension in the 0–40 cm soil layers compared with CK. Our results suggest that straw incorporation is an effective practice for improving the soil aggregate structure and stability.

Editor: Raffaella Balestrini, Institute for Plant Protection (IPP), CNR, Italy

Funding: This work was supported by the China Support Program (2012BAD09B03, 2011AA100504, and 2011BAD29B09) for Dryland Farming in the 12th 5-year plan period, the Youth project of National Natural Science Fund (31201156), and the Basic Science Research Fund in Northwest A&F University (QN2013005). The funders had no role in study design, data collection and analysis, decision to publish, or preparation of the manuscript.

Competing Interests: The authors have declared that no competing interests exist.

* E-mail: jiazhk@126.com

These authors contributed equally to this work.

Introduction

Soil infertility [1], soil erosion, and water deficiency [2] are the major factors that limited crop growth in semiarid areas of northwest China. The rates of crop straw use for fuel and forage have declined significantly since the 1980s and crop straw is increasingly burned after the harvest, which leads to high losses of soil organic substances [3,4], and increased emission of CO_2 that pollute the environment [5]. Furthermore, this practice has led to the degradation of the agricultural ecological environment [6].

The soil organic matter (SOM) content is one of the major factors that affects soil properties and functions including a range of physical characteristics such as the water-holding capacity [7], water infiltration [8], and aggregate stability [9]. SOM is considered to be a major binding agent that stabilizes soil aggregates [10,11]. Soil aggregates are the basic units of the soil structure [12], which are composed of primary particles and binding agents that determine the microbial biomass and mineral nutrient reserves [13–15]. These soil properties are also affected by soil organic matter decomposition processes [16,17].

Many studies have shown that crop straw is rich in organic material and soil nutrients, so it is increasingly considered to be an important natural organic fertilizer [18–20]. Straw can be incorporated to soil either directly or indirectly, which can promote the production of a favorable soil environment. Straw also maintains the physicochemical condition of the soil and improves the overall ecological balance of the crop production system [20,21]. Nelson [22] and Wilhelm *et al.* [23] showed that the incorporation of crop residues into soil significantly prevented soil erosion and enhanced the soil quality. Sonnleitner *et al.* [24] found that straw incorporation also improved the aggregate stability and other soil properties compared with farmyard manure. Mulumba and Lal [25] also reported that the addition of crop residues to cultivated soil had positive effects on the soil porosity, available water content, soil aggregation, and bulk density. Bhagat and Verma [26] showed that the incorporation of crop straws for five years significantly increased the crop yield and improved the soil properties.

The soil improvement effect of straw incorporation has been recognized widely [23–25] but information is still limited on the responses of the SOC and water-stable aggregates under different rates of straw incorporation, particularly in the loessal soil in semiarid areas of northwest China. The theory and technique of straw incorporation in this region have also not been reported. Thus, the present study investigated the effects of different crop straw application rates combined with conventional planting on SOC, the >0.25 mm water-stable macroaggregate rate, and various soil properties in the southern Ningxia region of China.

Materials and Methods

Ethics Statements

The study was carried out on the private land, we rent the farmland from the local farmers, and contracts and deeds are signed. No specific permissions were required in this area to run the experiment as the study sites are farming area without any protection zone, and the farming activities won't hurt the local animals. And we only plant the grain crop in the field, so the field studies did not involve endangered or protected species.

Site description

The experiment was conducted between 2008 and 2010 at the Dryland Agricultural Research Station, Pengyang County, Ningxia, China (106°45'N, 35°79'E and 1800 m a.s.l.). The experimental area was in a hilly and gully region of the Loess Plateau, which was characterized by a semiarid, warm temperate, continental monsoon climate. The average annual rainfall was 435 mm, which fell mainly from June to September. The annual mean evaporation was 1050 mm and the annual temperature average was 8.1°C with a frost-free period of 155 days.

Rainfall during the experimental period was measured using an automatic weather station (WS-STD1, England) at the experimental site. Monthly precipitation distributions during the experimental period are shown in Fig. 1. The total precipitation for 2008, 2009, and 2010 was 390.9, 335.2, and 537 mm, while the precipitation during the maize-growing season was 362, 298.2, and 476.1 mm, respectively.

The soil at the experimental site was a loessal soil with a pH of 8.5. In the 0–40 cm soil layer, the organic matter, total N, P, and K were 8.32 g kg^{-1}, 0.61 g kg^{-1}, 0.58 g kg^{-1}, and 5.4 g kg^{-1}, respectively, while the available N, P, and K were 46.25 mg kg^{-1}, 10.41 mg kg^{-1}, and 104.82 mg kg^{-1}. In 2007, the site was planted with maize prior to the experiment

The experimental field was flat and, according to the FAO/ UNESCO Soil Classification [27], the soil was a Calcic Cambisol (sand 14%, silt 26%, and clay 60%) with low fertility. The key physical properties of the soil layers (0–40 cm depth) are shown in Table 1.

Experimental design and field management

The experiment used a randomized block design with three replicates. Each plot was 3 m wide and 6 m long. The experiment included four straw incorporation rate treatments: (i) no straw incorporation (CK); (ii) incorporation of maize straw at a low rate of 4 500 kg ha^{-1} (L); (iii) incorporation of maize straw at a medium rate of 9000 kg ha^{-1} (M); (iv) incorporation of maize straw at a high rate of 13 500 kg ha^{-1} (H).

The maize straws were mixed manually with the top 25 cm of soil in the field. Before mixing with the soil, the maize straws were chopped into 5 cm pieces and then applied to the soil six months before the crop was planted to facilitate decomposition of the straw. The straw was incorporated into the soil layer on 15 October 2007 and after the crop harvests during 2008–2010.

Ten days before sowing, a basis fertilizer containing 102 kg N ha^{-1} and 90 kg P ha^{-1}, was spread evenly over the each plot and plowed into soil layer. Maize (cv. Shendan 16) was sown at a rate of 5.25 seed m^{-2} on 18 April 2008, 15 April 2009 and 20 April 2010 using a holesowing (3 cm in diameter) machine. An additional 102 kg N ha^{-1} was applied as a top dressing in late June. And on 7 October 2008, 5 October 2008, and 10 October 2010. No irrigation was provided during the experimental years. Manual weeding was performed throughout the experiment.

Sampling and measurement

Rainfall data were recorded using a standard weather station located at the experimental site. After the maize harvest in 2008 and 2010, soil samples were collected for the four incorporation treatments. A soil sample was collected from each plot at depths of 0–20 cm, 20–40 cm and 40–60 cm to determine the soil organic matter. A similar soil sample was collected at depths of 0–10 cm, 10–20 cm, 20–30 cm and 30–40 cm to determine the aggregate stability. The soil samples were collected from four points in each plot replicate and mixed to produce a composite sample. Each soil samples was passed through an 8 mm sieve by gently breaking the soil clods, whereas pebbles and stable clods >8 mm were discarded. Soil samples were air-dried for 24 h in the laboratory before analysis.

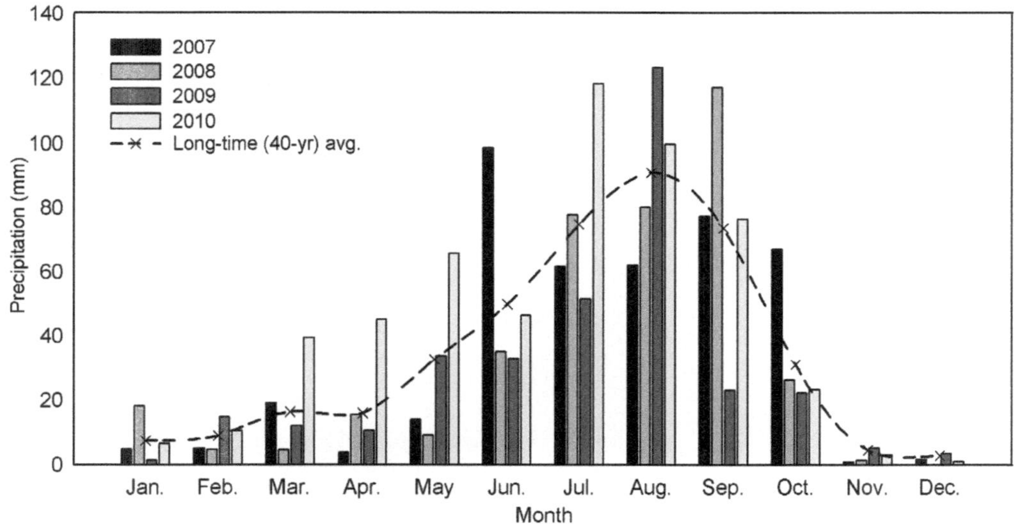

Figure 1. Distribution of mean monthly precipitation at the experimental site during 2007–2010.

Table 1. Physical properties of the tilth soil (0–40 cm depth) in the experimental site.

Depth (cm)	Bulk density (g cm^{-3})	Aggregate size (%)					
		>5 mm	5–2 mm	2–1 mm	1–0.5 mm	0.5–0.25 mm	<0.25 mm
0–10	1.33	0.1	0.27	2.25	4.5	4.47	88.41
10–20	1.33	0.15	0.19	1.38	4.2	4.15	89.93
20–30	1.36	0.1	0.28	1.41	4.02	3.36	90.83
30–40	1.38	0.1	0.22	1.07	3.29	3.8	91.52

Soil organic carbon was determined by the $K_2Cr_2O_7$–H_2SO_4 digestion method, and SOM content was calculated as a portion of SOC which has been described by Wang *et al.* [28].

$$OM_i = O_i \times A \quad (1)$$

$$C_i = \frac{d_i \times \rho_i \times O_i}{10} \quad (2)$$

where OM_i is the soil organic matter content (g kg^{-1}), $O_i$is the soil organic carbon content (g kg^{-1}), C_i is the soil organic carbon storage (Mg C ha^{-1}), $d_i$is the soil depth (cm), ρ_i is the soil bulk density (g cm^{-3}), A is Van Bemmelen coefficient (A = 2) [29].

The size distribution of water-stable aggregates was determined by placing a soil sample on a stack of sieves (5,2, 1, 0.5 and 0.25 mm) fitted with a soil aggregate analyzer (Japan, QD24–DIK–2001). The stacked sieves were immersed in water and moved up and down by 3.5 cm at a frequency of 30 cycles 60s^{-1} for 15 min. The proportions of aggregates that measured >5, 5–2, 2–0.5, 0.5–0.25 and <0.25 mm were calculated [30].

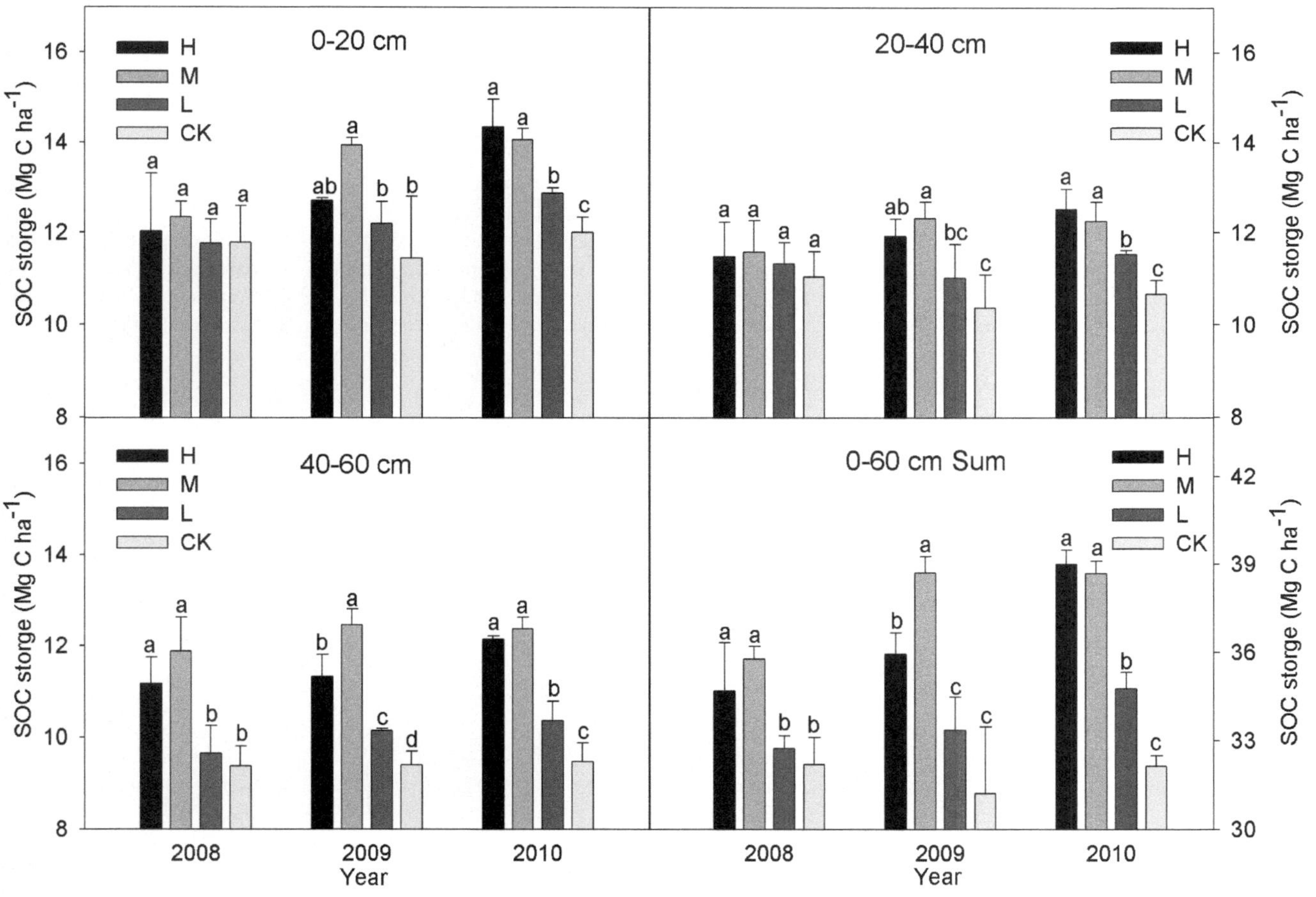

Figure 2. Soil organic carbon storage with different straw incorporation treatments (Mg C ha^{-1}). Note: CK, no straw incorporation; L, incorporation of straw at a low rate of 4 500 kg ha^{-1} maize straw; M, incorporation of straw at a medium rate of 9 000 kg ha^{-1} maize straw; H, incorporation of straw at a high rate of 13 500 kg ha^{-1} maize straw; Sum, the sum value of the 0–60 cm soil layers; Bars with different lower case letters indicate significant differences at $P<0.05$.

Table 2. $WR_{0.25}$ with different straw incorporation treatments (%).

Year	Treatment	Soil depth (cm)				0–40 AVG
		0—10	10—20	20—30	30—40	
2008	H	$13.37a_{1}\pm0.33^{B2}$	$10.57a\pm1.28_{3}^{A}$	$7.72ab\pm0.27^{B}$	$7.33a\pm0.51^{A}$	$9.75a\pm0.31^{B}$
	M	$11.79b\pm1.05^{A}$	$9.28a\pm0.60^{B}$	$7.97a\pm0.65^{A}$	$6.62ab\pm0.85^{B}$	$8.91b\pm0.26^{B}$
	L	$10.10c\pm0.46^{A}$	$7.39b\pm0.58^{A}$	$6.53bc\pm1.08^{A}$	$6.07ab\pm0.81^{A}$	$7.52c\pm0.16^{A}$
	CK	$9.12c\pm0.70^{A}$	$6.93b\pm0.78^{A}$	$5.89c\pm0.57^{A}$	$5.68b\pm0.63^{A}$	$6.91d\pm0.44^{A}$
2009	H	$14.64a\pm0.49^{AB}$	$11.46a\pm0.17^{A}$	$9.51a\pm0.26^{AB}$	$7.85a\pm1.00^{A}$	$10.86a\pm0.19^{AB}$
	M	$14.93a\pm0.89^{A}$	$11.22a\pm0.14^{AB}$	$9.34a\pm0.32^{A}$	$7.06ab\pm1.46^{AB}$	$10.64a\pm0.27^{AB}$
	L	$10.39b\pm0.26^{A}$	$9.48b\pm1.49^{A}$	$7.02b\pm1.18^{A}$	$6.48ab\pm1.10^{A}$	$8.34b\pm0.32^{A}$
	CK	$9.75b\pm0.26^{A}$	$7.18c\pm0.73^{A}$	$6.40b\pm1.00^{A}$	$5.65b\pm0.35^{A}$	$7.25c\pm0.10^{A}$
2010	H	$15.81a\pm1.20^{A}$	$12.76a\pm0.44^{A}$	$11.50a\pm0.20^{A}$	$8.57a\pm0.05^{A}$	$12.16a\pm0.59^{A}$
	M	$16.72a\pm1.36^{A}$	$12.81a\pm0.05^{A}$	$11.30a\pm1.10^{A}$	$8.81a\pm0.44^{A}$	$12.41a\pm0.72^{A}$
	L	$11.00b\pm0.78^{A}$	$9.06b\pm0.81^{A}$	$8.03b\pm2.14^{A}$	$6.96b\pm0.81^{A}$	$8.76b\pm0.66^{A}$
	CK	$10.23b\pm0.46^{A}$	$7.79b\pm0.23^{A}$	$6.07b\pm0.46^{A}$	$5.71c\pm0.23^{A}$	$7.45c\pm0.25^{A}$

Note: CK, no straw incorporation; L, incorporation of straw at a low rate of 4 500 kg ha^{-1} maize straw; M, incorporation of straw at a medium rate of 9 000 kg ha^{-1} maize straw; H, incorporation of straw at a high rate of 13 500 kg ha^{-1} maize straw; AVG, the mean value of the 0–40 cm soil layers.
[1]Values followed by the same lowercase letter in the same line are not significantly different according to Duncan's multiple range test ($P<0.05$) between the four straw incorporation treatments in the same year.
[2]Values followed by the same uppercase letter in the same line are not significantly different according to Duncan's multiple range test ($P<0.05$) between the different years of the same straw incorporation treatment.
[3]Means ± standard deviations.

And the proportions of aggregates were used to calculate the water-stable macroaggregates content with a diameter of >0.25 mm [31], mean weight diameter (MWD) [32,33], geometric mean diameter (GMD) [34], and the soil aggregate stability (WSAR) [35]. These parameters were calculated as follows:

$$macroaggregates(>0.25mm)=\sum_{i=1}^{n}W_i \tag{3}$$

$$MWD=\sum_{i=1}^{n}\overline{X_i}W_i \tag{4}$$

$$GMD=\exp\left(\frac{\sum_{i=1}^{n}W_i\log\overline{X_i}}{\sum_{i=1}^{n}W_i}\right) \tag{5}$$

$$WSAR=WSA/A\times100\% \tag{6}$$

where $macroaggregates(>0.25mm)$ is the volume of soil particles with a diameter of >0.25 mm, $W_i$is weight of the aggregates in that size range as a fraction of total dry weight of the sample analysed, and n is number of sieves, $\overline{X_i}$ is the mean diameter of aggregates over each sieve size, WSA is the mass of $macroaggregates(>0.25mm)$, and A is the mass of the soil aggregates with a diameter of >0.25 mm.

Fractal dimension D were then obtained to describe the characteristics of soil aggregate size distribution. As suggested by Tyler and Wheatcraft [36] and Zhang *et al.*, [37], the volume of soil with particle diameter $>d_i(d_i>d_{i+1},i=1,2,\ldots)$ is defined as:

$$V(\delta>d_i)=A\left[1-(d_i/k)^{3-D}\right], \tag{7}$$

where δ is yard measure, k and A are the constants representing size and shape, respectively, and D is the fractal dimension. For a given soil, d_i represents the average value of soil particles between d_i and d_{i+1}.Generally, variations of particle density ρ among different soil particles could be ignored. And hence ρ is a constant. Therefore, another expression (Eq. (8)) is derived from Eq. (7):

$$W(\delta>d_i)=V(\delta>d_i)\rho=\rho A\left[1-(d_i/k)^{3-D}\right], \tag{8}$$

where $W(\delta>d_i)$ is the cumulative mass of particles with sizes $\delta>d_i$, and W_0 is the total mass of any sizes of soil particles. The fractal equation, reflecting the relationship between the mass distributions of soil particles and average particle diameter, can be obtained as follows:

$$\frac{W(\delta>d_i)}{W_0}=1-\left(\frac{\overline{d_i}}{d_{\max}}\right)^{3-D}\ or\ \frac{W(\delta>d_i)}{W_0}=\left(\frac{\overline{d_i}}{d_{\max}}\right)^{3-D} \tag{9}$$

Then after regression analysis between $\log(\overline{d_i}/d_{\max})$ and $\log((W(\delta<\overline{d_i}))/W_0)$, the fractal dimension D can be calculated.

Statistical analysis

Statistical analysis was carried out by using the SPSS 13.0 (Statistical Package for the Social Sciences) package. The effects of treatments on the measured parameters were evaluated using a one-way ANOVA. Duncan's new multiple range test was used to calculate the least significant difference (LSD) between means

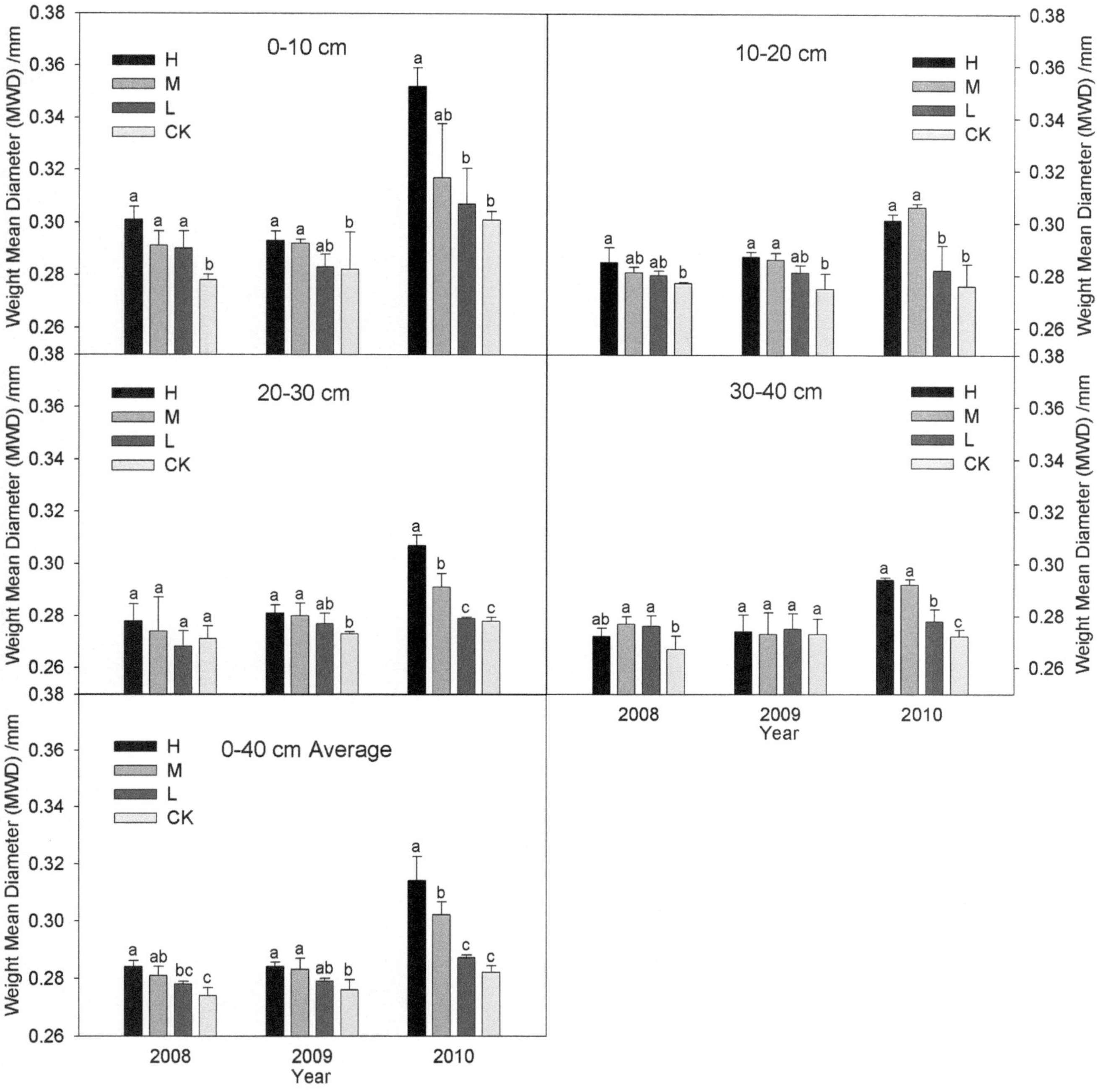

Figure 3. MWD values with wet sieving under the different straw incorporation treatments. Note: MWD, weight mean diameter; CK, no straw incorporation; L, incorporation of straw at a low rate of 4 500 kg ha^{-1} maize straw; M, incorporation of straw at a medium rate of 9 000 kg ha^{-1} maize straw; H, incorporation of straw at a high rate of 13 500 kg ha^{-1} maize straw; Average, the mean value of the 0–40 cm soil layers. Bars with different lower case letters indicate significant differences at $P<0.05$.

when F-values were significant. In all cases, differences were deemed to be significant if $P<0.05$.

Results

Soil organic carbon storage (SOC)

The effect of straw incorporation on SOC storage is shown in Fig. 2, where the soil organic carbon storage increased with the straw incorporation. The sum of SOC storage in 0–60 cm layers with the three incorporation treatments were higher than CK, i.e., 7.71% ($P<0.05$), 11.14% ($P<0.05$) and 1.70% in 2008, 15.15% ($P<0.05$), 24.00% ($P<0.05$) and 6.86% in 2009, and 21.40% ($P<0.05$), 20.38% ($P<0.05$) and 8.21% ($P<0.05$) in 2010, respectively. The SOC storage increased with the number of years of incorporation, i.e., the SOC storage (0–60 cm depth) in 2010 had increased by 6.19–12.48% compared with 2008, and decreased with the soil layer depth, i.e., by 3.75–25.68% in 2008, 11.85–21.70% in 2009 and 13.51–26.64% in 2010. The SOC storage of H and M was slightly higher than CK in 2008, although the difference was significant in 40–60 cm layer only. In 2009, compared with CK, H and M significantly increased the SOC storage by 11.01% and 21.74% ($P<0.05$) in 0–20 cm layer,

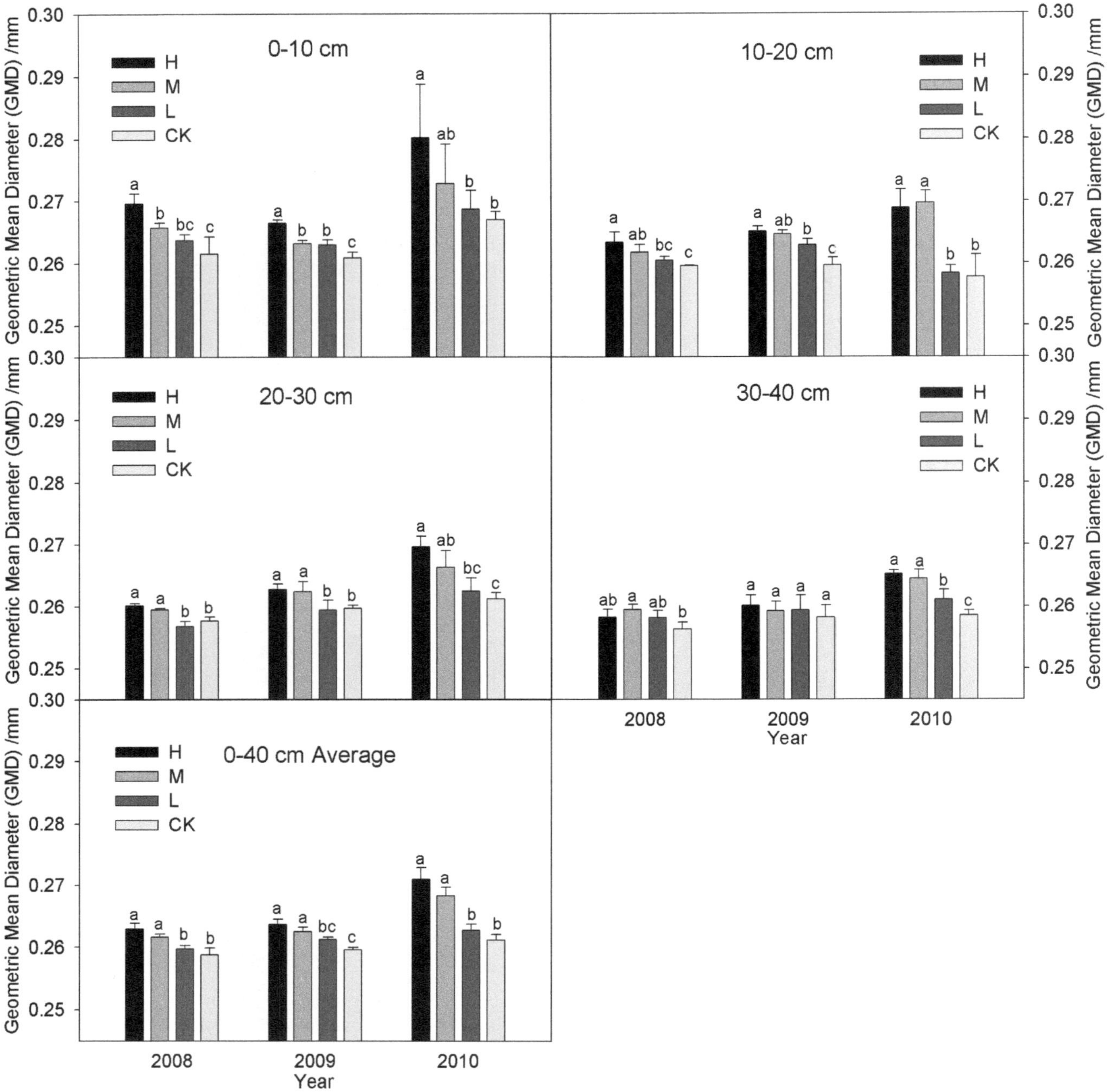

Figure 4. GMD values with wet sieving under the different straw incorporation treatments. Note: GMD, geometric mean diameter; CK, no straw incorporation; L, incorporation of straw at a low rate of 4 500 kg ha^{-1} maize straw; M, incorporation of straw at a medium rate of 9 000 kg ha^{-1} maize straw; H, incorporation of straw at a high rate of 13 500 kg ha^{-1} maize straw; Average, the mean value of the 0–40 cm soil layers. Bars with different lower case letters indicate significant differences at $P<0.05$.

14.94% and 18.81% ($P<0.05$) in 20–40 cm layer, and 20.40% and 32.47% ($P<0.05$) in 40–60 cm layer, respectively, and L was significantly increased by 7.89% ($P<0.05$) in 40–60 cm layer only. There was a significant difference between straw incorporation treatments and CK for each of the three soil layers in 2010, but there were no significant differences between H and M.

Water-stable macroaggregates (>0.25 mm)

The effects of straw incorporation on >0.25 mm macroaggregates are shown in Table 2. The mean >0.25 mm macroaggregates contents of the 0–40 cm layers of incorporation treatments were significantly ($P<0.05$) higher than CK during 2008–2010, i.e., 8.91–41.14%, 15.13–49.94%, and 17.64–66.58%, respectively. The >0.25 mm macroaggregates content also decreased with the soil layer depth, and increased with the number of years of incorporation,, i.e., the H, M, L and CK in 2010 was 24.72% ($P<0.05$), 39.28% ($P<0.05$), 16.49% and 7.81% higher than that in 2008. Compared with CK, the >0.25 mm macroaggregates content of three straw incorporation treatments in the 0–10 cm layer was increased by 10.75–46.60% in 2008, 6.56–53.13% in 2009, and 7.53–63.44% in 2010, respectively, while the differences between L and CK were not significant ($P>0.05$) during 2008–2010. The

Table 3. Effects on the soil aggregate stability rate with different straw incorporation treatments (%).

Year	Treatments	Soil Depth (cm)				0–40 AVG
		0–10	10–20	20–30	30–40	
2008	H	21.86a[1]±0.89[A2]	16.29a±0.41[3][A]	10.83a±0.54[B]	9.19a±0.65[A]	14.60a±0.33[A]
	M	19.62ab±1.09[A]	15.31b±0.62[A]	11.16a±0.82[A]	8.47b±0.21[A]	13.70b±0.62[A]
	L	17.83bc±1.56[A]	12.88c±0.24[A]	9.70b±0.64[A]	8.00bc±0.42[A]	11.83c±0.18[A]
	CK	16.45c±2.42[A]	12.52c±0.58[A]	9.39b±0.80[A]	7.53c±0.37[A]	11.79c±0.45[A]
2009	H	20.32a±0.27[A]	16.14a±0.54[A]	13.15a±0.34[AB]	9.74a±0.99[A]	14.84a±0.23[A]
	M	20.02a±0.46[A]	15.41a±0.50[A]	12.61a±0.41[A]	8.66ab±1.28[A]	14.18b±0.22[A]
	L	15.76b±0.49[A]	14.45a±1.78[A]	10.16b±1.36[A]	8.28ab±1.01[A]	12.16c±0.35[A]
	CK	15.47b±0.30[A]	11.81b±0.83[A]	9.59b±1.10[A]	7.50b±0.41[A]	11.09d±0.05[A]
2010	H	20.19a±1.93[A]	16.48a±1.73[A]	15.24a±0.12[A]	10.45a±0.11[A]	15.59a±0.26[A]
	M	20.87a±0.68[A]	15.78a±1.32[A]	13.99a±1.23[A]	10.08a±0.43[A]	15.18a±0.64[A]
	L	15.72b±0.72[A]	13.39a±1.81[A]	12.53b±0.46[A]	8.66b±0.77[A]	12.58b±1.00[A]
	CK	15.11b±0.12[A]	12.79a±1.70[A]	12.00b±0.76[A]	7.52c±0.26[A]	11.85b±1.32[A]

Note: CK, no straw incorporation; L, incorporation of straw at a low rate of 4 500 kg ha^{-1} maize straw; M, incorporation of straw at a medium rate of 9 000 kg ha^{-1} maize straw; H, incorporation of straw at a high rate of 13 500 kg ha^{-1} maize straw; AVG, the mean value of the 0–40 cm soil layers.
[1]Values followed by the same lowercase letter in the same line are not significantly different according to Duncan's multiple range test ($P<0.05$) between the four straw incorporation treatments in the same year.
[2]Values followed by the same uppercase letter in the same line are not significantly different according to Duncan's multiple range test ($P<0.05$) between the different years of the same straw incorporation treatment.
[3]Means ± standard deviations.

>0.25 mm macroaggregates content showed the same trends in the 10–20 and 20–30 cm layers, and the H and M treatments had the significant difference ($P<0.05$) with CK. The H, M and L increased by 29.05% ($P<0.05$), 16.55% and 6.87% in 2008, 38.94% ($P<0.05$), 24.96% and 14.69% in 2009, and 50.09% ($P<0.05$), 54.29% ($P<0.05$), 21.89% ($P<0.05$) in 2010, respectively. A linear correlation was found between the >0.25 mm macroaggregates and soil organic carbon ($R^2>0.64$, $P<0.01$).

Mean weight diameter (*MWD*) and Geometric mean diameter (*GMD*)

Fig. 3 & 4 show that the *MWD* and *GMD* values with the three incorporation treatments increased significantly throughout the three-year study. The average *MWD* and *GMD* values under the incorporation treatments were higher in the 0–40 cm layers than CK during 2008–2010, i.e., 1.46–3.65% and 0.39–1.54% in 2008, 1.09–2.90% and 0.62–1.55% in 2009, 1.77–11.35% and 0.77–3.83% in 2010, respectively. There was no significant difference ($P>0.05$) between L and CK during the study period. The *MWD* and *GMD* values decreased with the soil layer depth and increased with the number of years of incorporation, i.e., the H, M, L and CK in 2010 was 10.56% ($P<0.05$) and 3.04% ($P<0.05$), 7.47% ($P<0.05$) and 2.29% ($P<0.05$), 3.24% and 1.15%, and 2.92% and 0.77% higher than that in 2008. The *MWD* and *GMD* values of treatments in the 0–10 and 10–20 cm layers were ranked in the order: H > M > L > CK. The H and M levels were significantly higher than CK while H and M levels were similar during 2008–2010 (Fig. 3–4). The *MWD* and *GMD* exhibited the same trends in 20–30 cm layers and there were no significant ($P>0.05$) differences among incorporation treatments during 2008–2009, the *MWD* values with the H were significantly ($P<0.05$) higher than the other three treatments during 2010, i.e., by 5.50%, 10.04%, and 10.43%, respectively. The *GMD* did not differ significantly among the four treatments, with the exceptions of H and L, and H and CK in 2010. The *MWD* and *GMD* values in the 30–40 cm layers under the incorporation treatments were increased by 1.87%, 3.75% ($P<0.05$), 3.37% ($P<0.05$) and 0.76%, 1.76% ($P<0.05$), 0.78% in 2008, 0.37%, 0.01%, 0.73% and 0.73%, 0.45%, 0.45% in 2009, 8.09% ($P<0.05$), 7.35% ($P<0.05$), 2.21% ($P<0.05$) and 2.32% ($P<0.05$), 1.93% ($P<0.05$), 0.77% ($P<0.05$) in 2010, compared with CK, respectively. There was no significant difference ($P>0.05$) between H and M during the study period.

Soil aggregate stability (*WSAR*)

The straw incorporation significantly affected the *WSAR* after three years (Table 3). The *WSAR* values in the 0–40 cm layers of incorporation treatments was significantly higher than CK during 2008–2010, i.e., 0.34–19.80%, 9.65–33.79%, and 6.16–35.08%, respectively. The *WSAR* also decreased with the soil layer depth, while it increased with the number of years of incorporation. The *WSAR* in the 0–10 cm layers increased significantly with the amount of straw and the *WSAR* was highest with H throughout 2008–2010, i.e., the *WSAR* significantly ($P<0.05$) increased by 28.35%, 28.93% and 25.78% compared with L, respectively, and by 33.62%, 31.35% and 32.89% ($P<0.05$) compared with CK, respectively. Compared with L and CK, the WSAR values with M increased significantly by 20.08–24.67% and 26.87–29.78% during 2008–2010, respectively, but there were no significant ($P>0.05$) differences between H and M, and L and CK. With the four treatments, the *WSAR* values in the 10–20 and 20–30 cm layers were lower than the 0–10 cm layers, which were ranked in the order: H > M > L > CK. There were no significant differences ($P>0.05$) among the four treatments in the 30–40 cm layers throughout the three years, and the *WSAR* maintained at 7.50–10.45%.

Fractal dimension (D)

Fig. 5 shows that the fractal dimensions with the three incorporation treatments decreased significantly ($P<0.05$) after three years of straw incorporation (the R^2 of fitting curve is:

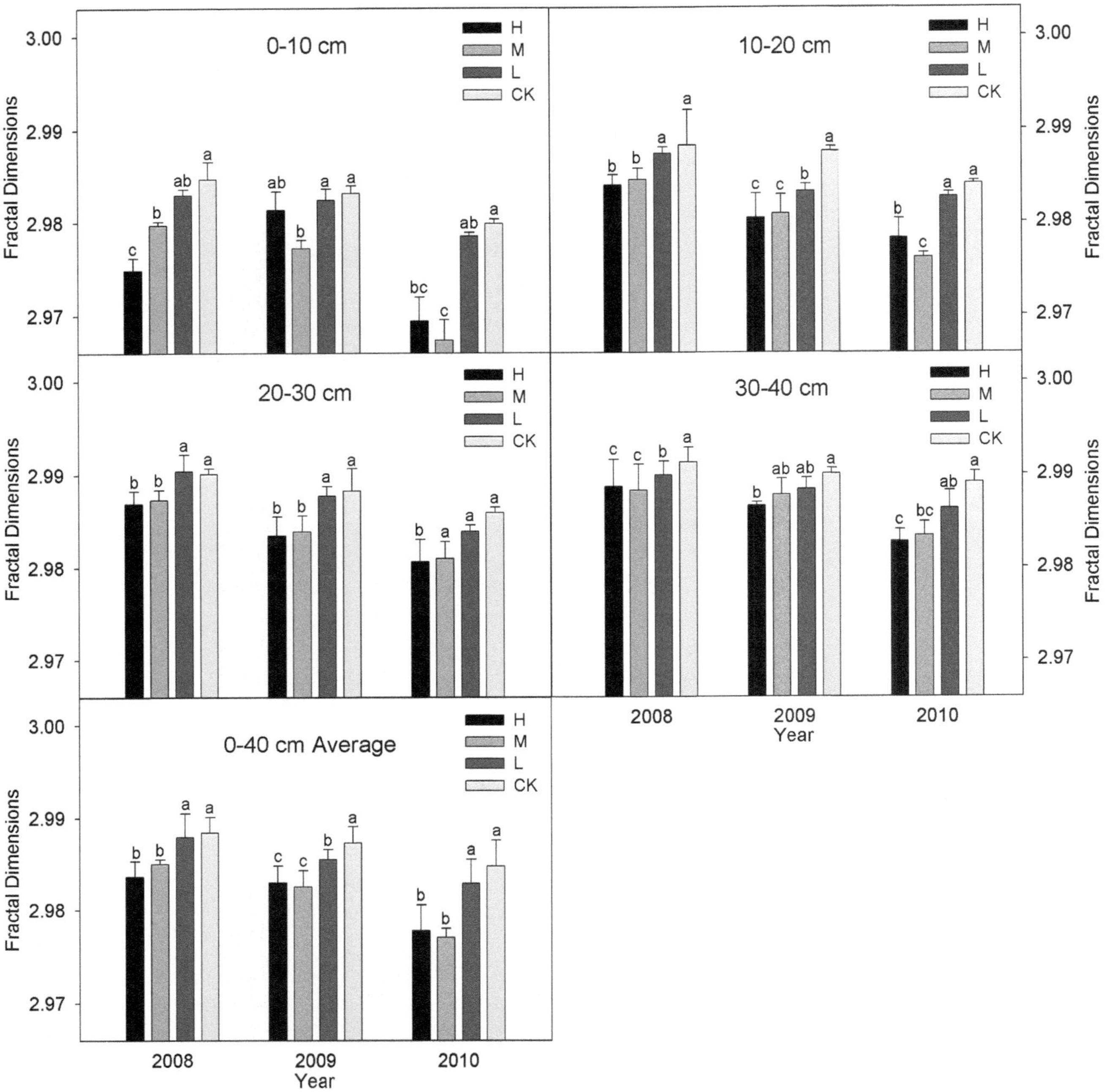

Figure 5. Fractal dimensions (D) of soil aggregates with different straw incorporation treatments. Note: CK, no straw incorporation; L, incorporation of straw at a low rate of 4 500 kg ha^{-1} maize straw; M, incorporation of straw at a medium rate of 9 000 kg ha^{-1} maize straw; H, incorporation of straw at a high rate of 13 500 kg ha^{-1} maize straw; Average, the mean value of the 0–60 cm soil layers; Bars with different lower case letters indicate significant differences at $P<0.05$; $0.92<R^2<0.99$.

0.92~0.99). The fractal dimensions (0–40 cm layer) with the incorporation treatments were lower than those of CK during 2008–2010, i.e., 0.16% ($P<0.05$), 0.11% ($P<0.05$) and 0.02% in 2008, 0.14% ($P<0.05$), 0.16% ($P<0.05$) and 0.06% ($P<0.05$) in 2009, and 0.24% ($P<0.05$), 0.26% ($P<0.05$) and 0.06% in 2010, respectively. The fractal dimensions also increased with the soil layer depth and decreased with the number years of incorporation, i.e., the H, M, L and CK in 2010 was 0.20% ($P<0.05$), 0.27% ($P<0.05$), 0.17% ($P<0.05$) and 0.12% higher than that in 2008. The fractal dimensions in the 0–10 cm layers with the four treatments were ranked in the order: H<M<L<CK in 2008. The fractal dimensions with H and M were significantly ($P<0.05$) lower than CK, i.e., 0.33%, 0.17% and 0.06%, respectively. However, the ranking was M<H<L<CK during 2009–2010, i.e., 0.06%, 0.20% ($P<0.05$), 0.02% in 2009, 0.35% ($P<0.05$), 0.42% ($P<0.05$) and 0.04% in 2010, respectively. While the differences between L and CK were not significant ($P>0.05$) during 2008–2010. Compared with L and CK, the fractal dimensions in the 10–20 cm layers under H was significantly ($P<0.05$) reduced the fractal dimensions by 0.11% and 0.14% in 2008, and 0.10% and 0.24% in 2009,, while M significantly ($P<0.05$) reduced by 0.09% and 0.12% in 2008, and 0.08% and 0.23% in 2009, respectively.

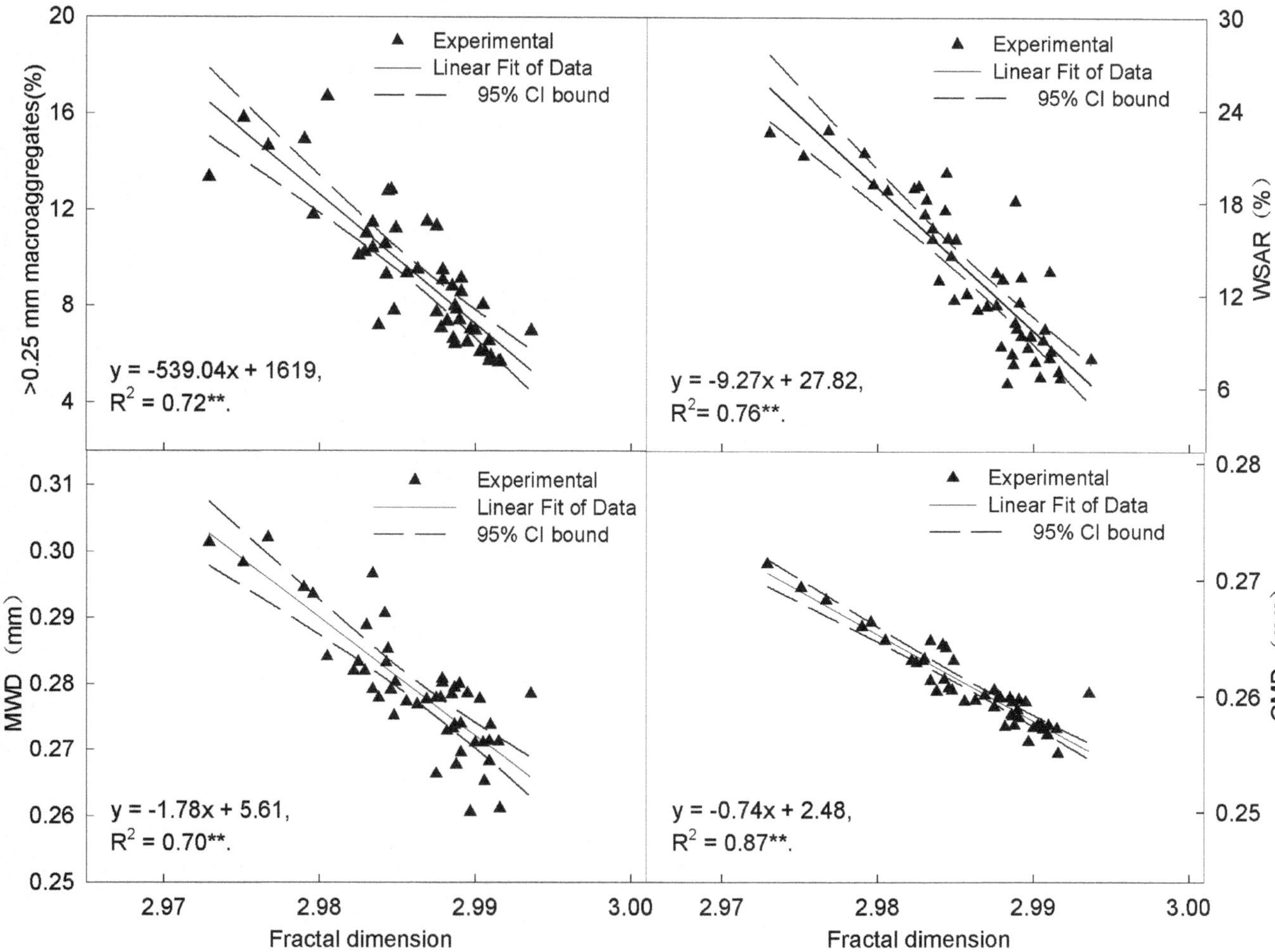

Figure 6. Correlations between D and >0.25 mm macroaggregates, MWD, GMD, WSAR of the soil aggregates. Note: MWD, weight mean diameter of soil aggregates; GMD, geometric mean diameter of soil aggregates; WSAR, soil aggregate stablility rate; D, Fractal dimension of soil aggregates. 95% CI, 95% confidence interval.** $P<0.01$.

The fractal dimension was lowest with M during 2010, i.e., it was decreased by 0.22% ($P<0.05$) compared with L, and by 0.27% ($P<0.05$) compared with CK. The differences among the four treatments decreased gradually in the 20–40 cm layer and the fractal dimension remained at 2.987~2.991.

Data fitting detected a linear correlation between the indexes of soil aggregates and the fractal dimension (Fig. 6).

Discussion

Results of this study demonstrated that application of crop straws had positive effects on the soil physico-chemical properties. On the other hand, there are Loess plateau regions at northwest China with intensive cultivation systems and poor soil management strategies. Therefore, crop residual management is very important for preserving natural ecosystems [38]. Another problem for this area is conventional tillage that reduces the soil water storage and destroys the aggregates and soil structure [39] which prompts deterioration of crop yield.

The results indicated that the SOC storage increased significantly in the 0–60 cm layers after three years of straw incorporation (Fig. 2). Rasmussen and Collins [40] reported that the soil organic matter content was strongly related to the amount of residues added and only weakly related to the type of residue applied. The H and M treatments were significantly different from the L throughout the three years and the soil organic matter level increased as the straw incorporated and decomposed [20], which effectively mitigated the loss of soil organic carbon from in the agroecosystem caused by intensive cropping [41]. The SOC storage decreased gradually in all the treatments with the soil layer depth because the degree of straw incorporation was lower in the deeper layers compared with the surface layers (0–20 cm) of the soil [42]. This was because the amount of straw incorporated in the topsoil was greater than that in the deeper layers [40,43].

Soil aggregates are the basic units of the soil structure and they are composed of primary particles and binding agents [12]. They are also necessary soil conditions for high crop yields [44]. Conventional tillage disturbs the soil and increase the effects of drying–rewetting and freezing–thawing, which increases the susceptibility of the macroaggregate (>0.25 mm) to disruption [45–47]. Pinheiro *et al.* [48] showed that soil exposure with tillage and the lack of residue inputs led to a decline in aggregation and organic carbon, both of which made the soil susceptible to erosion. Our study showed that the straw incorporation of straws determined significantly more and larger soil aggregates than CK, thus indicating an improvement of soil physical quality. This may have been attributable to the significant increase in the SOC storage (an average increase by 8.21–21.40% in 2010), the lower soil bulk density (an average decreased by 1.80–4.13% in 2010, data no shown) [49], and increased soil porosity (an average increased by 1.70–3.90% in 2010, data no shown) [50] after straw incorporation. It also stimulated the activity of soil microorganisms [51] and an abundance of polyose metabolites were produced during the straw decay process [52]. These soil physical and chemical conditions may have accelerated the incorporated SOM decomposing process and increased soil aggregation [24].

Soil aggregate structural stability is widely recognized as a key indicator of soil quality, which is closely related to a number of soil properties, processes, and functions, e.g., the quantity and composition of SOM [48], infiltration capacity [53], soil biotic activity [54] and the resistance to erosion [55,56]. Wei *et al.* [57] showed that the addition of crop residues was the most effective measure for increasing the rhizosphere aggregate stability. Sonnleitner *et al.* [24] and Karami *et al.* [41] also found that straw application improved the aggregate stability and other soil properties. In our study, the soil aggregate stability of the straw incorporation treatments were significantly higher than CK in 2010, and it was decreased with the soil layer depth. These results agreed with studies by Tripathy and Singh [42] and Karami *et al.* [41]. Our results also indicated that straw incorporation was positively related to the physical protection of organic matter [20] and an increased aggregate quantity [58], but it also improved the soil aggregate stability [20] and reduced soil degeneration [22,23].

Many studies have shown that soil is a porous medium with fractal characteristics [59–60]. Thus, fractal theory can be used to describe the complex characteristic of soil structure [61]. Castrignanò and Stelluti [62] reported that a higher fractal dimension indicated the heavier texture of a soil and its inferior permeation properties. This showed that fractal theory is an effective method to describing the soil aggregate distribution [37] and changed with different levels of straw incorporation [31]. The fractal dimensions of the 0–40 cm layers with the four treatments after three years were ranked in the order: H<M<L<CK and the three straw incorporation treatments were significantly different from CK. These results agreed with Zhang *et al.* [37] and Zhang *et al.* [31]. The low values of D indicated a size distribution dominated by a large number of macroaggregates (>0.25 mm) [59,63]. This improvement in the fractal dimension may have been accelerated by the incorporation of straw, which improved the soil structure, increased the SOM content and microbial activity [64], and significantly increased the mount and size of soil aggregates [37]. Our results indicated there were significant improvements in the soil macroaggregates and the aggregate structure after straw incorporation [20].

Conclusion

The incorporation of different amount of straw significant increased the *SOC* storage, >0.25 mm macroaggregates, *MWD* and *GMD* in a semiarid soil. The *SOC* storage, >0.25 mm macroaggregates, *MWD* and *GMD* also increased with higher straw incorporation rates. The fractal dimension decreased with increasing straw incorporation rates. Therefore, the incorporation of straw into the soil in semiarid areas is an effective practice for improving the soil aggregate content and stability.

Acknowledgments

We are grateful to Mr Li Yongping and Liu Shixin for managing the field experiments and professional English editor Jackson who is from UK, and kind help with the language of this manuscript.

Author Contributions

Conceived and designed the experiments: ZJ QH. Performed the experiments: PZ TW. Analyzed the data: PZ. Contributed reagents/materials/analysis tools: XR YL. Wrote the paper: PZ.

References

1. Hu KL, Li BG, Lin QM, Li GT, Chen DL (1999) Spatial variabi lity of soil nutrient in wheat field. Transactions of the Chinese Society of Agricultural Engineering 15:33–38. (in Chinese with English abstract)
2. Mupangwa W, Twomlow S, Walker S (2008) The influence of conservation tillage methods on soil water regimes in semi-arid southern Zimbabwe. Phys Chem Earth, Parts A/B/C 33: 762–767.
3. Biederbeck V, Campbell C, Bowren K, Schnitzer M, McIver R (1980) Effect of burning cereal straw on soil properties and grain yields in Saskatchewan. Soil Sci Soc Am J 44: 103–111.
4. Wuest SB, Caesar-TonThat T, Wright SF, Williams JD (2005) Organic matter addition, N, and residue burning effects on infiltration, biological, and physical properties of an intensively tilled silt-loam soil. Soil Tillage Res 84: 154–167.
5. Duan F, Liu X, Yu T, Cachier H (2004) Identification and estimate of biomass burning contribution to the urban aerosol organic carbon concentrations in Beijing. Atmos Environ 38: 1275–1282.
6. Mandal KG, Misra AK, Hati KM, Bandyopadhyay KK, Ghosh PK, et al. (2004) Rice residue-management options and effects on soil properties and crop productivity. J Food Agric Environ 2: 224–231.
7. Carter MR (2002) Soil quality for sustainable land management. Agron J 94: 38–47.
8. Hillel D (2004) Encyclopedia of Soils in the Environment: London, Elsevier, v. 4, p. 295–303.
9. Six J, Bossuyt H, Degryze S, Denef K (2004) A history of research on the link between (micro) aggregates, soil biota, and soil organic matter dynamics. Soil Tillage Res 79: 7–31.
10. Tisdall J, Oades J (1982) Organic matter and water-stable aggregates in soils. J Soil Sci 33:141–163. doi:10.1111/j.1365-2389.1982.tb01755.x
11. Haynes R, Beare M (1997) Influence of six crop species on aggregate stability and some labile organic matter fractions. Soil Biol Biochem 29: 1647–1653.
12. Scanlon BR, Andraski BJ, Bilskie J (2002) Methods of Soil Analysis: Part 4 Physical Methods: Soil Science Society of America. 643–668 p.
13. Hernández-Hernández R, López-Hernández D (2002) Microbial biomass, mineral nitrogen and carbon content in savanna soil aggregates under conventional and no-tillage. Soil Biol Biochem 34: 1563–1570.
14. Villar M, Petrikova V, Dıaz-Ravina M, Carballas T (2004) Changes in soil microbial biomass and aggregate stability following burning and soil rehabilitation. Geoderma 122: 73–82.
15. Ashagrie Y, Zech W, Guggenberger G (2005) Transformation of a *Podocarpus falcatu* dominated natural forest into a monoculture *Eucalyptus globulus* plantation at Munesa, Ethiopia: soil organic C, N and S dynamics in primary particle and aggregate-size fractions. Agric Ecosyst Environ 106: 89–98.
16. Jastrow J (1996) Soil aggregate formation and the accrual of particulate and mineral-associated organic matter. Soil Biol Biochem 28: 665–676.
17. Chevallier T, Blanchart E, Albrecht A, Feller C (2004) The physical protection of soil organic carbon in aggregates: a mechanism of carbon storage in a Vertisol under pasture and market gardening (Martinique, West Indies). Agric Ecosyst Environ 103: 375–387.
18. Duiker S, Lal R (1999) Crop residue and tillage effects on carbon sequestration in a Luvisol in central Ohio. Soil Tillage Res 52: 73–81.
19. Saroa G, Lal R (2003) Soil restorative effects of mulching on aggregation and carbon sequestration in a Miamian soil in central Ohio. Land Degrad Dev 14: 481–493.
20. Tan DS, Jin JY, H SW, Li ST, He P (2007) Effect of long-term application of K fertilizer and wheat straw to soil on crop yield and soil K under different planting systems. Agric Sci in China 6: 200–207.
21. Li Q, Chen X, Zhang F, Römheld V (2002) Study on balance of phosphorus and potassium in winter wheat and summer maize rotation system. Plant Nutr Fert Sci 8: 152–156. (in Chinese with English abstract)
22. Nelson RG (2002) Resource assessment and removal analysis for corn stover and wheat straw in the Eastern and Midwestern United States—rainfall and wind-induced soil erosion methodology. Biomass Bioenergy 22: 349–363.
23. Wilhelm W, Johnson JMF, Karlen DL, Lightle DT (2007) Corn stover to sustain soil organic carbon further constrains biomass supply. Agron J 99: 1665–1667.
24. Sonnleitner R, Lorbeer E, Schinner F (2003) Effects of straw, vegetable oil and whey on physical and microbiological properties of a chernozem. App Soil Ecol 22: 195–204.
25. Mulumba LN, Lal R (2008) Mulching effects on selected soil physical properties. Soil Tillage Res 98: 106–111.
26. Bhagat R, Verma T (1991) Impact of rice straw management on soil physical properties and wheat yield. Soil Sci 152: 108–115.

27. FAO/UNESCO (1993) 'World soil resources. An explanatory note on the FAO World Soil Resource Map at 1 : 25,000,000 scales.' (FAO: Rome)
28. Wang S, Tian H, Liu J, Pan S (2003) Pattern and change of soil organic carbon storage in China: 1960s–1980s. Tellus B 55: 416–427.
29. Pribyl DW (2010) A critical review of the conventional SOC to SOM conversion factor. Geoderma 156: 75–83.
30. Oades J, Waters A (1991) Aggregate hierarchy in soils. Soil Res 29: 815–828.
31. Zhang P, Jia ZK, Wang W, Lu WT, Gao F, et al. (2012) Effects of straw returning on characteristics of soil aggregates in semi-arid areas in southern Ningxia of China. Sci Agric Sin 45:1513-1520. (in Chinese with English abstract)
32. Van Bavel C (1949) Mean weight diameter of soil aggregates as a statistical index of aggregation. Soil Sci Soc Am Proc. pp. 20–23.
33. Youker R, McGuinness J (1957) A short method of obtaining mean weight-diameter values of aggregate analyses of soils. Soil Sci 83: 291–294.
34. Mazurak AP (1950) Effect of gaseous phase on water-stable synthetic aggregates. Soil Sci 69: 135–148.
35. Hou XQ, Li R, Jia ZK, Han QF, Wang W, et al. (2012) Effects of rotational tillage practices on soil properties, winter wheat yields and water-use efficiency in semi-arid areas of north-west China. Field Crops Res 129: 7–13.
36. Tyler SW, Wheatcraft SW (1992) Fractal scaling of soil particle-size distributions: analysis and limitations. Soil Sci Soc Am J 56: 362–369.
37. Zhang Z, Wei C, Xie D, Gao M, Zeng X (2008) Effects of land use patterns on soil aggregate stability in Sichuan Basin, China. Particuology 6: 157–166.
38. Zhu H, Wu J, Huang D, Zhu Q, Liu S, et al. (2010) Improving fertility and productivity of a highly-weathered upland soil in subtropical China by incorporating rice straw. Plant Soil 331: 427–437.
39. Mele PM, Crowley DE (2008) Application of self-organizing maps for assessing soil biological quality. Agric Ecosyst Environ 126: 139–152.
40. Rasmussen PE, Collins HP (1991) Long-term impacts of tillage, fertilizer, and crop residue on soil organic matter in temperate semiarid regions. Adv Agron 45: 93–134.
41. Karami A, Homaee M, Afzalinia S, Ruhipour H, Basirat S (2012) Organic resource management: Impacts on soil aggregate stability and other soil physico-chemical properties. Agric Ecosyst Environ 148: 22–28.
42. Tripathy R, Singh AK (2004) Effect of water and nitrogen management on aggregate size and carbon enrichment of soil in rice-wheat cropping system. J Plant Nutr Soil Sci 167: 216–228.
43. Prasad R, Power J (1991) Crop residue management. Adv Soil Sci 15: 205–251.
44. Limon-Ortega A, Govaerts B, Sayre KD (2009) Crop rotation, wheat straw management, and chicken manure effects on soil quality. Agron J 101: 600–606.
45. Beare M, Hendrix P, Cabrera M, Coleman D (1994) Aggregate-protected and unprotected organic matter pools in conventional-and no-tillage soils. Soil Sci Soc Am J 58: 787–795.
46. Paustian K, Levine E, Post WM, Ryzhova IM (1997) The use of models to integrate information and understanding of soil C at the regional scale. Geoderma 79: 227–260.
47. Mikha MM, Rice CW (2004) Tillage and manure effects on soil and aggregate-associated carbon and nitrogen. Soil Sci Soc Am J 68: 809–816.
48. Pinheiro E, Pereira M, Anjos L (2004) Aggregate distribution and soil organic matter under different tillage systems for vegetable crops in a Red Latosol from Brazil. Soil Tillage Res 77: 79–84.
49. Lal R (2000) Physical management of soils of the tropics: priorities for the 21st century. Soil Sci 165: 191–207.
50. Pagliai M, Vignozzi N, Pellegrini S (2004) Soil structure and the effect of management practices. Soil Tillage Res 79: 131–143.
51. Kasteel R, Garnier P, Vachier P, Coquet Y (2007) Dye tracer infiltration in the plough layer after straw incorporation. Geoderma 137: 360–369.
52. Pascual JA, García C, Hernandez T (1999) Comparison of fresh and composted organic waste in their efficacy for the improvement of arid soil quality. Bioresour Technol 68: 255–264.
53. Abu-Sharar T, Salameh A (1995) Reductions in hydraulic conductivity and infiltration rate in relation to aggregate stability and irrigation water turbidity. Agric Water Manage 29: 53–62.
54. Roldán A, Salinas-García J, Alguacil M, Caravaca F (2005) Changes in soil enzyme activity, fertility, aggregation and C sequestration mediated by conservation tillage practices and water regime in a maize field. Appl Soil Ecol 30: 11–20.
55. Barthes B, Roose E (2002) Aggregate stability as an indicator of soil susceptibility to runoff and erosion; validation at several levels. Catena 47: 133–149.
56. Ramos M, Nacci S, Pla I (2003) Effect of raindrop impact and its relationship with aggregate stability to different disaggregation forces. Catena 53: 365–376.
57. Wei CF, Gao M, Shao JG, Xie DT, Pan GX (2006) Soil aggregate and its response to land management practices. China Particuology 4: 211–219.
58. Tarafdar JC, Meena SC, Kathju S (2001) Influence of straw size on activity and biomass of soil microorganisms during decomposition. Eur J Soil Biol 37: 157–160.
59. Perfect E, Kay B (1991) Fractal theory applied to soil aggregation. Soil Sci Soc Am J 55: 1552–1558.
60. Rieu M, Sposito G (1991) Fractal fragmentation, soil porosity, and soil water properties: II. Applications. Soil Sci Soc Am J 55: 1239–1244.
61. Bird N, Bartoli F, Dexter A (2005) Water retention models for fractal soil structures. Eur J Soil Sci 47: 1–6.
62. Castrignanò A, Stelluti M (1999) Fractal geometry and geostatistics for describing the field variability of soil aggregation. J Agric Eng Res 73: 13–18.
63. Ding Q, Ding W (2007) Comparing stress wavelets with fragment fractals for soil structure quantification. Soil Tillage Res 93: 316–323.
64. Liu JF, Hong W, Wu CZ (2002) Fractal features of soil clusters under some precious hardwood stands in the central subtropical region, China. Acta Ecol. Sin. 22, 197. (in Chinese with English abstract).

19

Cross-Species Extrapolation of Prediction Model for Lead Transfer from Soil to Corn Grain under Stress of Exogenous Lead

Zhaojun Li[1]*, Hua Yang[1,2], Yupeng Li[1,3], Jian Long[2], Yongchao Liang[1]

1 Institute of Agricultural Resources and Regional Planning, Chinese Academy of Agricultural Sciences, Key Laboratory of Plant Nutrition and Fertilizer, Ministry of Agriculture, Beijing, China, **2** Guizhou Key Laboratory of Mountain Environment, Guizhou Normal University, Guiyang, China, **3** College of City and Environmental Sciences, Shanxi Normal University, Linfen, China

Abstract

There has been increasing concern in recent years regarding lead (Pb) transfer in the soil-plant system. In this study the transfer of Pb (exogenous salts) was investigated from a wide range of Chinese soils to corn grain (Zhengdan 958). Prediction models were developed with combination of the Pb bioconcentration factor (BCF) of Zhengdan 958, and soil pH, organic matter (OM) content, and cation exchange capacity (CEC) through multiple stepwise regressions. Moreover, these prediction models from Zhengdan 958 were applied to other non-model corn species through cross-species extrapolation approach. The results showed that the soil pH and OM were the major factors that controlled Pb transfer from soil to corn grain. The lower pH and OM could improve the bioaccumulation of Pb in corn grain. No significant differences were found between two prediction models derived from the different exogenous Pb contents. When the prediction models were applied to other non-model corn species, the ratio ranges between the predicted BCF values and the measured BCF values were within an interval of 2-fold and close to the solid line of 1:1 relationship. Moreover, the prediction model i.e. Log[BCF] = −0.098 pH-0.150 log[OM] −1.894 at the treatment of high Pb can effectively reduce the measured BCF intra-species variability for all non-model corn species. These suggested that this prediction model derived from the high Pb content was more adaptable to be applied to other non-model corn species to predict the Pb bioconcentration in corn grain and assess the ecological risk of Pb in different agricultural soils.

Editor: Xianglin Shi, University of Kentucky, United States of America

Funding: This work was jointly supported by the Special Fund of Public Industry in China (Agriculture, project no. 200903015, http://www.moa.gov.cn/) and the Natural Science Foundation of Institute of Agricultural Resources and Regional Planning, Chinese Academy of Agricultural Sciences (2013-9, http://www.iarrp.cn/SoisWeb/home/index.html). The funders had no role in study design, data collection and analysis, decision to publish, or preparation of the manuscript.

Competing Interests: The authors have declared that no competing interests exist.

* E-mail: lizhaojun@caas.cn

Introduction

Lead (Pb) is one of the most toxic heavy metals for people in the world [1–3]. In children, Pb has been known to cause decreases in intelligence quotient (IQ) scores, retardation of physical growth, and hearing problems. In individuals of all ages, it has been known to cause anemia brain damage, impaired function of the peripheral nervous system, high blood pressure, reproductive abnormalities, development defects, abnormal vitamin D metabolism, and in some situations death of people [4–5]. Soil-to-plant transfer of heavy metals is the major exposure pathway for humans with regards to soil contamination [6]. Therefore, the accumulation of heavy metals in plants in agricultural soils, especially in edible part of plants, is great concern regarding its posing potential threats [7–8]. The efficiency of the plants for absorbing metals is mainly evaluated by plant uptake or the metal transfer factors from soil to plant [9]. Evaluating the metal transfer factors from soil to plant especially to edible parts can further provide the information about accumulation of contaminants into the food chain [10].

The Pb transfer factor from soil to plant mainly depends on its availability in soils. Soil properties can determine the availability of Pb to crops, including soil pH, organic matter (OM), cation exchange capacity (CEC), redox potential, clay minerals, Fe and Mn oxides, and calcium carbonate content [11–13]. In addition, it has been reported that soil pH is a key factor in controlling the cation mobility and regulating the solubility of heavy metals in soils [14–15]. The heavy metals tended to be more available in acid soil [16]. The low soil pH is favorable for the transference of Pb from soil to plant [17–19]. As well as, soil OM has also been found to play an important role in influence Pb adsorption processes in soils [20–21]. Pb in soils with low OM content was difficult to be adsorbed but was easy to be uptaken by plants [22]. The importance for assessing the ecological risk of heavy metals in complex soil types has urged a lot of prediction models to be developed to predict heavy metals transfer from soil to plant [23]. The heavy metal transfer factor from soil to plant also depends on plant species. It has been observed that significant differences exist in heavy metal uptake among cultivars [24]. Due to the expenses and efforts involved, these models were limited to specific contaminated sites or single species [25–27]. In addition, most of these models were developed to describe the transfer of heavy metals from soil to plant leaves, vegetables, earthworms by using the log-transformed concentrations [28–29]. A few attempts were only made to describe heavy metals uptake by rice, wheat and barley grain directly from soil with combination of total concentrations of heavy metal and soil basic properties [30–31]. Little is reported about the prediction models developed to describe the transfer of heavy metals, especially Pb, from soils to corn grain. Intake of heavy metals from dietary sources may represent a significant exposure pathway for human populations. Prediction of heavy metals including Pb concentrations in edible part of crops based on their concentration in soil and other environmental factors are urgently required for human risk assessment and derivation of soil environmental benchmark croplands.

The species sensitivity distributions (SSD) have been proved to be successful and were increasingly used in ecological risk assessment procedures [32–33]. However, the toxicology data from the soils with different soil properties need to be normalized using prediction models to eliminate the influence of soil factors [34]. It can improve the accuracy of the sensitivity distribution of species and environmental quality standard values [35]. These models can also provide great opportunities to carrying out ecological risk assessments and establishing soil quality criteria for heavy metals in croplands. Due to the fact that significant differences existed among cultivars for heavy metal uptake, it is ideal that separate species-specific models should be developed for each single species. However, this is unrealistic due to the expenses and efforts involved. Thus the cross-species extrapolation of toxicity data models was often applied in some cases of pollutants risk assessment practice. For example, Deleebeeck et al. [36–37] have applied biotic bigand model (BLM) from *O. mykiss* to other fish species and applied BLM from D. *magna* to other cladocera species. Schlekat et al. [38] applied chronic nickel BLMs developed for the cladocera such as Daphnia magna and Ceriodaphnia dubia to predict chronic toxicity of nickel to three other invertebrates including snail *(Lymnaea stagnalis)*, insect (*Chironomus tentans*), and rotifer (*Brachionus calyciflorus*). Yang et al. [16] applied corn grain Cd BCF predication models developed for Zhengdan 958 to predict other non-model corn species and wheat grain Cd BCF. However, little effort has been put on establishing models to describe the relationship between Pb accumulation in corn grain and soil properties. The objectives of this study are: 1) to establish the prediction model for Pb accumulation in corn grain on seventeen soils with greatly different properties; 2) to assess the feasibility of applying these models to other non-model corn species, and 3) to investigate their accuracy in predicting accumulation of Pb in non-model corn species.

Materials and Methods

None of these 17 soil samples were collected from national parks or other protected areas. It is confirmed that no specific permissions were required for the soils sampling activities in the 17 locations in China. It is also can be confirmed that the field studies did not involve endangered or protected species. No tested corn species are under first- or second-class state protection, and they are not listed in the Inventory of Rare and Endangered Plants of China (http://zrbhq.forestry.gov.cn/portal/zrbh/s/3053/

Table 2. Soil Environmental Quality Standards of China (GB15618-1995) and the added contents of exogenous Pb (mg.kg^{-1}).

pH	<6.5	6.5~7.5	>7.5
Grade Two Standard	250	300	350
Low Pb	125	150	175
High Pb	250	300	350

Table 1. Selected soil properties of the 17 soils used in the lead bioconcentration factor test.

Soil NO.	Location*	pH	OM (g.kg^{-1})	CEC (cmol.kg^{-1})	TN (g.kg^{-1})	TP (mg.kg^{-1})	TK (g.kg^{-1})	Background Pb (mg.kg^{-1})
S1	Hunan	4.90	15.52	10.85	1.14	0.47	15.26	23.12
S2	Chongqing	5.74	17.48	21.34	1.00	0.55	22.61	29.72
S3	Liaoning	5.74	25.84	12.19	1.00	0.73	23.94	27.94
S4	Yunnan	5.92	34.26	11.10	2.01	0.81	4.77	24.58
S5	Jiangxi	6.01	11.69	8.70	0.51	0.52	9.96	33.48
S6	Anhui	6.25	20.04	19.08	0.99	0.35	15.41	29.34
S7	Heilongjiang	6.27	35.69	28.59	1.74	0.48	24.70	33.99
S8	Jilin	6.82	32.85	31.11	1.75	0.35	24.58	31.01
S9	Jiangsu	6.93	47.69	26.20	2.44	0.69	21.03	36.81
S10	Shaanxi	7.90	16.49	22.37	1.36	0.98	24.37	37.30
S11	Hebei	7.98	8.57	8.12	0.68	0.53	24.22	31.96
S12	Henan	8.07	17.79	16.01	1.07	0.75	19.86	35.89
S13	Xinjiang	8.12	19.43	25.25	1.32	0.78	25.49	31.73
S14	Shanxi	8.24	23.17	16.80	1.13	0.95	23.70	37.97
S15	Tianjin	8.29	22.02	24.67	1.42	0.92	24.63	34.18
S16	Gansu	8.37	19.27	11.23	1.05	0.74	23.62	35.22
S17	Shandong	8.65	11.84	13.09	0.93	0.97	21.37	34.55

*Soil sample locations are listed in the order of increasing pH.

content-457748.html), or the Key Protected Inventory of Wild Plants of China (http://zrbhq.forestry.gov.cn/uploadfile/zrbh/2010-10/file/2010-10-14-bb296addeaa047798d6b6c476aaa1da9.doc). These corn species were used for only scientific research as permitted by the Ministry of Agriculture of China.

Soil Samples

A set of 17 soils covering a wide range of soil properties was sampled from typical locations in China. In each sample location, the soils were collected from the top 20 cm of the soil profile. For analysis of their phys-chemical characteristics, the soils were air dried and sorted to pass a 2-mm sieve. Soil pH was measured in deionized water (soil: solution ratio, 1:5) [39]. CEC was determined by the unbuffered silver-thiourea method [40]. OM was measured by dry combustion [41]. The total phosphorus in soils (TP) was measured by colorimetric method [42]. The total nitrogen in soils was determined by the Kjeldahl's method [43]. The background Pb content in the soils was determined by aqua regia (1:3 fresh mixture of concentrated HNO_3 and HCl) digestion [44]. The selected properties of the 17 soil samples were shown in Table 1.

Experimental Design

Bioaccumulation factors for Pb by corn grain from seventeen soils. According to Grade Two Standard for Pb in the Soil Environmental Quality Standards of China (GB15618-1995), the tested levels of Pb added in soils were shown in Table 2. To obtain the tested levels, a certain amount of exogenous Pb i.e. $Pb(NO_3)_2$ was added to the 8 kg air-dried soils. All the soils were thoroughly mixed and placed into pots ($\phi \times h$, 23 cm×26.5 cm). Then the soils were moistened with deionized water to the 60% of field moisture capacity. Each tested level of Pb was replicated three times. Then the soils containing Pb at different levels were covered with plastic film with six small holes for air ventilation. All the pots were kept for 3 months for Pb aging at a greenhouse at the temperature of 25±3°C during the daytime and 20±3°C at night with a natural light photoperiod. During the period of aging, water contents of each soil sample was maintained by adding deionized water every 3 days. After 3 months' aging, corn seeds of Zhengdan 958 were sown in each soil containing Pb. During the experimental period, soil moisture content was maintained at approximately 60% of water holding capacity through weighing method and adding deionized water to the pots. The corn with grain were sampled and harvested till the corn maturity.

Bioaccumulation factors for Pb of eight corn species from two soils. Two soils with typical properties sampled from Jiangxi and Shanxi provinces were selected for this experiment. The tested levels of Pb and exogenous Pb addition were the same as those described in the experiment of bioaccumulation factors for Pb by corn from different soils (Table 2). Three replications were conducted on each tested level of Pb. After three months' aging of soils, the corn seeds, such as Jingketian 183, Chuandan 30, Liaodan 565, Tunyu 88, Zhongdan 808, and Nongda 84,were surface sterilized with 1% (v/v) NaOCl, rinsed, and soaked in distilled water for 24 h at 33°C in dark. Then these seeds were sown in each soil containing Pb. All these pots were placed in the greenhouse at the temperature of 25±3°C during the daytime and 20±3°C at night with a natural light photoperiod.

Soil Sampling and Analysis

After three months' aging, soils in the pots of the above two experiments were sampled, air-dried, passed through a <0.20-mm sieve, and then thoroughly mixed. 0.5 g of the soil sample was weighed and added to digestion tubes containing 9 ml HNO_3 and 3 ml HF.

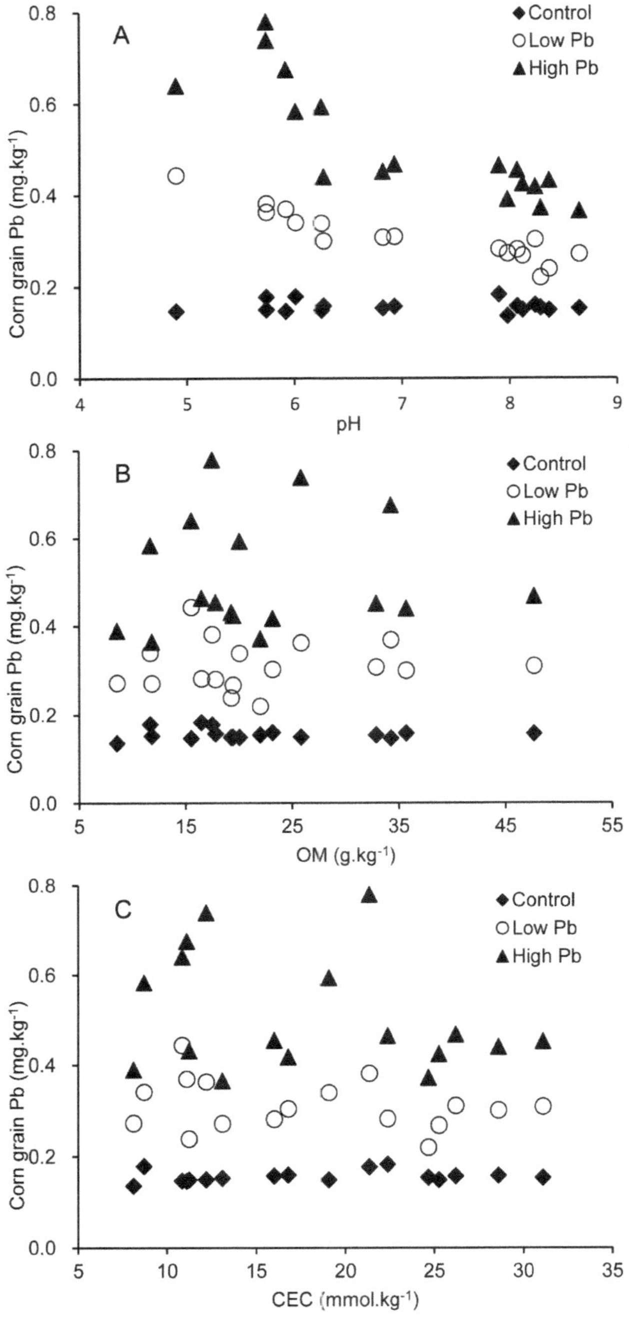

Figure 1. Relationships between Pb in corn grain and soil properties.

The corn grains were harvested and oven-dried at 105°C for 30 minutes and then kept at 70°C until the weight of the grains was stable. 0.5 g of corn grain samples oven-dried were weighed and digested in 6 ml concentrated HNO_3 (70% w/v) and 3 ml H_2O_2 in a CEM Mars X microwave oven (CEM Mars) at a pressure of 3.1 MPa. The Pb concentration in the digestion solution from soil and grain was determined by inductively coupled plasma-mass spectrometry (ICP-MS, Agilent 7500a, Agilent Technologies Co. Ltd., USA).

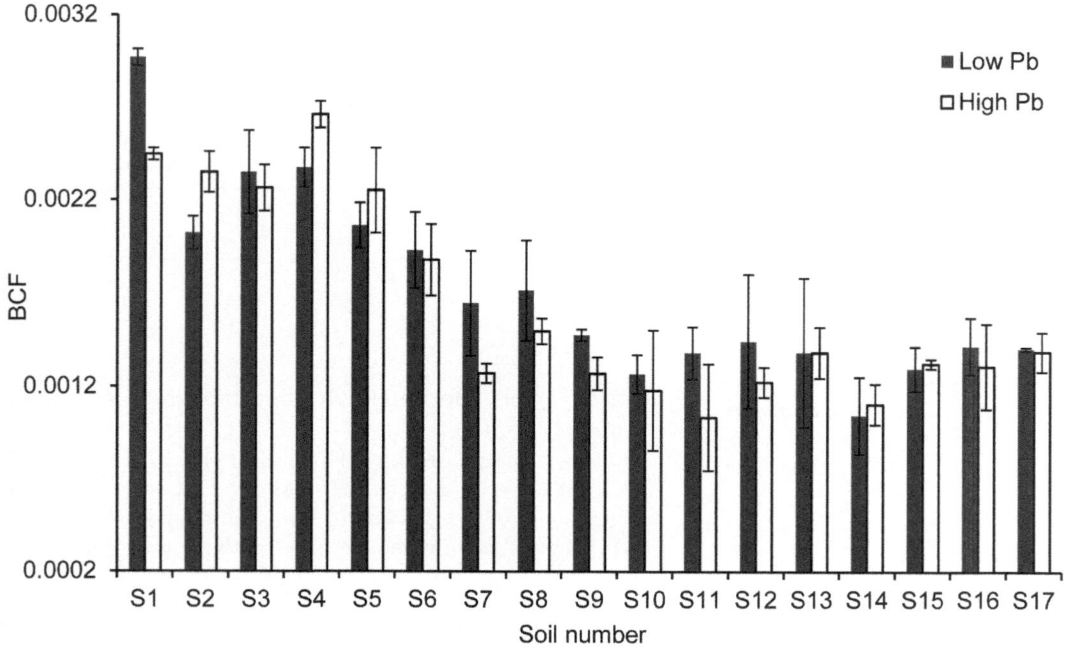

Figure 2. Effects of different soil types on the bioconcentration factor under stress of exogenous Pb.

Data Analysis

The bioconcentration factor (BCF) was calculated as the ratio of the content of Pb in the corn grain to that in the soil (equation 1) [45–46].

$$BCF = C_{grain}/C_{soil} \quad (1)$$

Where C_{grain} is the Pb concentration in the corn grain and C_{soil} is Pb concentration in the soil.

SPSS 16.0 for Windows® 107 (SPSS Inc, Chicago, IL, USA) was used for the regression analysis and statistical analysis of the significant differences. Origin 8.0 (OriginLab Co., Northampton, MA, USA) was employed for figure rendering.

Model

The prediction models for Pb transfer were established through multiple stepwise regression of the Pb BCF from Zhengdan 958 in a wide range of soil properties. The basis of these models is on the basis of equation 2 [38].

$$Log[BCF] = a \times pH + b \times \log[OM] + c \times \log[CEC] + k \quad (2)$$

Where Log [BCF], Log [OM], and Log [CEC] were the logarithm base 10 of the BCF values, the content of organic matter ($g.kg^{-1}$) in soils, and the cation exchange capacity ($cmol.kg^{-1}$) of soils, respectively? The slope of soil property parameters such as a, b, and c indicate the impacts of soil properties on Pb accumulation in corn grain. The intercept k is the intrinsic sensitivity that characterizes the ability of corn species to absorbing Pb.

Cross-species Extrapolation

In the process of cross-species extrapolation, an interim alternative was to assume that interactions between Pb accumulation in corn grain and soil pH, OM, CEC were the same among related corn species. In other words, the models stability constants including a, b, c were assumed the same among related species, and the only difference between related species was assumed to be their intrinsic sensitivity (k) [38]. The variation of intrinsic sensitivities within a species among plants reflects residual variation [34]. Under the condition of the squared error with the lowest value between predicted BCF value and the measured BCF value calculated as $\sum_{i=1}^{n} (measured\,BCF_i - predicited\,BCF_i)^2$, the intercept ($k$) for different models corresponding to various species were obtained through Excel Solver for linear optimization [47].

The accuracy of the model predictions was evaluated by comparing the measured BCF of the other non-model corn species with the predicted BCF from the model. The predictions for other non-model corn species were calculated with the prediction model derived from corn species i.e. Zhengdan 958.

Table 3. Prediction models for the different Pb levels.

Model No.	Lead levels	Prediction models	R^2	*P*
Model 1	Low Pb	Log[BCF] = −0.098 pH-0.150 log[OM] −1. 894	0.9085	<0.001
Model 2	High Pb	Log[BCF] = −0.108 pH-0.178 log[OM] −1. 806	0.8603	<0.001

Table 4. Intrinsic sensitivity (*k*) for non-model species fitted by models from Zhengdan 958.

Prediction Model	Intrinsic sensitivities (*k*) Jingketian 183	Chuandan 30	Liaodan 565	Tunyu 88	Zhongdan 808	Nongda 84
Model 1	−1.823	−1.881	−1.924	−2.295	−2.288	−2.152
Model 2	−1.681	−1.785	−1.846	−1.857	−1.794	−1.717

Analysis of the Reduction of Intra-species Variability

The Pb BCF values of non-model corn species were normalized to the specific soil conditions through the obtained model for Zhengdan 958 and the intra-species variability was computed by equation 3 [47].

$$f=\sqrt{\frac{\sum_{i=1}^{n}(BCFsi-\overline{BCFs})^2}{(n-1)\times(\overline{BCFs})^2}} \quad (3)$$

Where the BCF of i-th conditions for specific corn species was normalized to specific soil conditions. *BCFsi* is the value that Pb BCF specific corn species were normalized to i-th soil conditions. $\overline{BCFs}$ is the mean of n BCFs. *n* is the number of different conditions for specific corn species, *f* is the intra-species variability.

The BCFs that Pb for different corn species was normalized by the prediction models to a set soil condition should be equal. Therefore, the decreases in intra-species variability indicate that normalization processing eliminated soil properties to a certain extent.

Results

Major Factors Affecting Pb Accumulation in Corn Grain in Different Soils

The relationships between Pb concentration in corn grain at different Pb levels and soil pH, OM and CEC were shown in Figure 1. At the treatment of low concentration Pb, the Pb concentration in corn grain decreased with the increasing pH of soil. The similar trend was also found at the treatment of high concentration Pb. There was no significant difference was found among Pb concentration in corn grain with the change of pH at the treatment of control. The significantly negatively direct correlations were observed between Pb accumulation in corn grain and soil pH at low concentration Pb, and high concentration Pb, respectively. However, no significantly direct correlations between Pb concentration in corn grain and soil OM, CEC were found.

Effects of Different Soil Types on Pb Transfer from Soil to Corn Grain

As shown in Figure 2, the BCF values from Zhengdan 958 under stresses of exogenous Pb significantly differed in 17 soils with a wide range of soil properties in China. The variation tendency of BCF values from Zhengdan 958 was generally consistent at different Pb levels of treatments. These BCF values decreased with

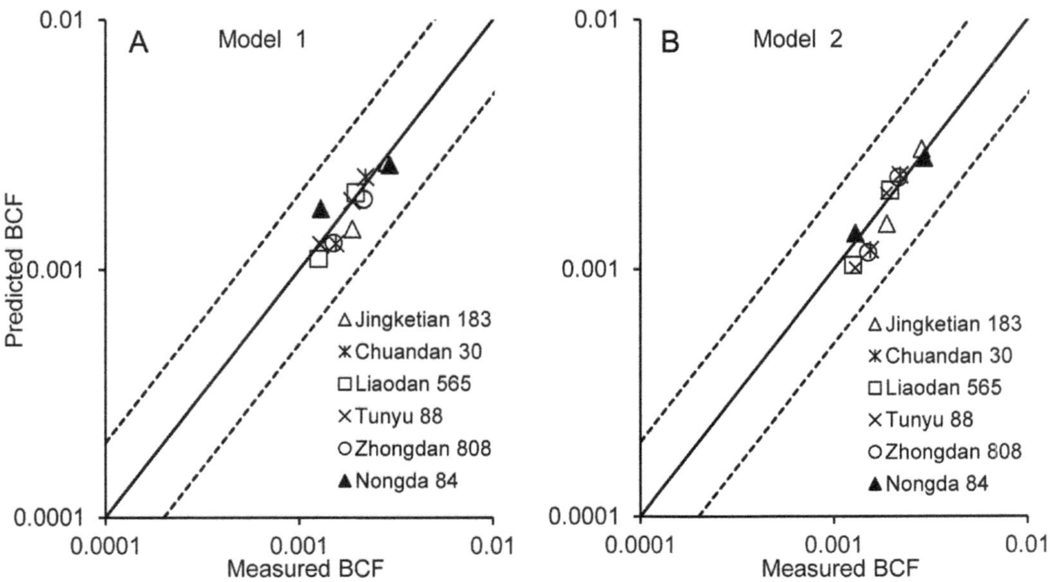

Figure 3. Relativity between measured and predicted BCF values for Pb in non-model corn grains. The predicted BCF values were estimated by model 1 and model 2 in Table 3; the solid line represents a 1:1 relationship; the interval between two dashed lines indicates a 2-fold prediction interval between the predicted and measured values.

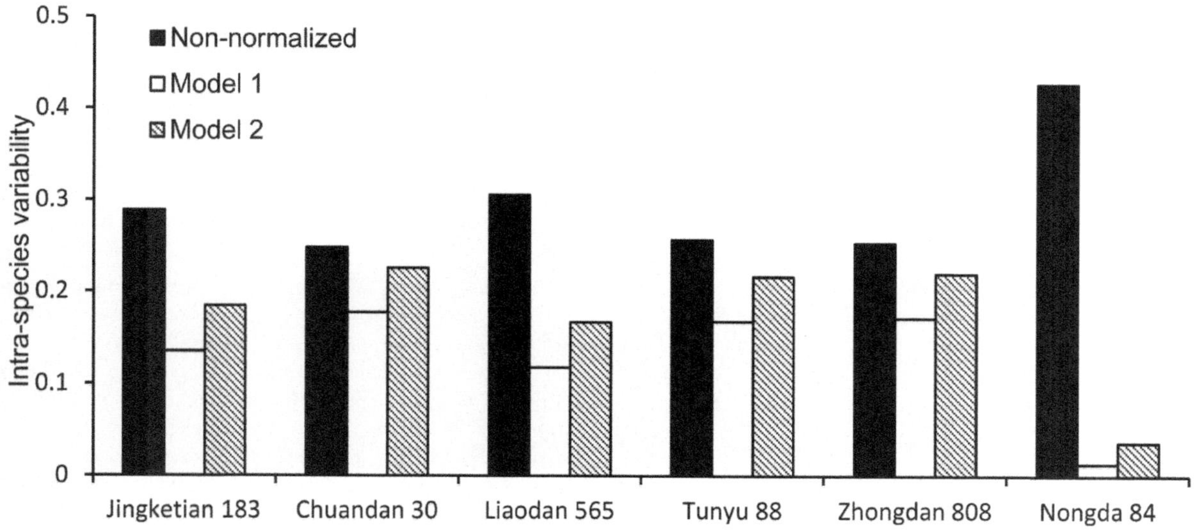

Figure 4. Intra-species variability of Pb BCF. The data from all non-model corn species were normalized with the models listed in Table 3.

the increases of soil pH. The BCF values detected from S1 to S9 (pH from 4.90 to 6.93) were significantly higher than those from S10 to S17 (pH from 7.90 to 8.80). At the treatment of low Pb, the maximum BCF (0.0030) among the 17 soils was observed in S1, and the minimum value for BCF (0.0010) was found in S14. The maximum value was 3.0 folds the minimum. At the same time, the highest BCF (0.0024) values were once again found in S1, and the lowest BCF (0.0010) was found in S11 at the treatment of high Pb, The maximum was 2.40 folds the minimum. These results indicate that Pb under acidic soil conditions is more highly bioavailable and leads to increased absorption by the plants.

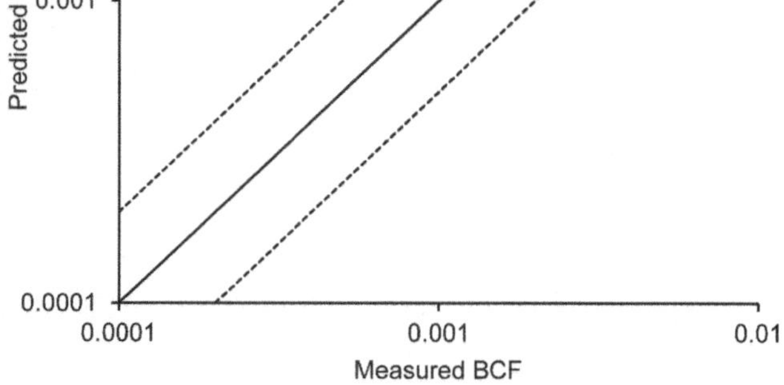

Figure 5. Relativity between measured and predicted BCF values for Pb in corn grains. The measured BCF were obtained from references. The predicted BCF values in Figure 5 were estimated by model 1. The solid line represents a 1:1 relationship; the dashed lines indicate a 2-fold prediction interval between the predicted and measured values.

Prediction Models

The prediction models established by stepwise regression were shown in Table 3. The result showed that pH and OM were the major factors affecting Pb accumulation in corn grain. A significantly negative correlation was observed between the Log[BCF] and soil pH and Log[OM] at the different treatments including low Pb, and high Pb, with the R^2 values of 0.8227, and 0.8955 ($P<0.001$), respectively. There was no significant difference between the two equations at the different treatment of Pb levels.

Cross-species Extrapolation

The intercept (*k*) can indicate the sensitivity of corn species to Pb accumulation. The intercept (*k*) for different corn species deduced by two different models were shown in Table 4. In terms of the same corn cultivar, no significant differences were observed among the *k* values for all tested non-model corn species deduced by the same model including model 1, and 2. However, the variation of *k* values among corn species deduced by the model 2 was stronger than that by the model 1. For example, the *k* values of corn species were different with the range from −1.888 to −1.736 deduced by model 1 and with range from −2.295 to −1.823 deduced by the model 2.

On the basis of the *k* values, the BCF values for non-model corn species were predicted by different models developed for Zhengdan958. The relationship between the predicted BCF values and the measured BCF values for non-model corn species was shown in Figure 3. The ratio between the predicted BCF values and the measured BCF values was within 2-fold interval and close to the solid line of 1:1 relationship. It indicates that the two models from Zhengdan 958 can be applied to predict the Pb BCF for non-model corn species.

Reduction of Intra-species Variability

Figure 4 shows the intra-species variability of Pb BCF for non-model species which were normalized with the prediction models listed in Table 3. The two prediction models can reduce the measured BCF intra-species variability for all non-model corn species. The intra-species variability of Pb BCF values fitted by model 1 was found to be significantly lower than those by model 1. It indicates that model 1 is more effective in reducing the uncertainty caused by the soil property differences than model 2.

Table 5. Soil properties and BCF values for Pb from reference.

Cultivation soil	cultivar species	pH	OM (g.kg-1)	CEC (cmol.kg^{-1})	Pb (g.kg^{-1})	BCF	Reference
Soil	Zhengdan 958	6.3	17.6	22.6	30.62	0.0020	[61]
Soil	Zhengdan 958	5.9	38.6	–	30.72	0.0024	[62]

This novel result also indicate that model 1 can be more reasonable and to be applied to other non-model corn species to establish the soil Pb SSD curves and to provide an ecological benchmark in China in the future.

Discussion

The bioavailable concentration of heavy metals is a key factor affecting its uptake and accumulation in plants [48]. In the present study, the results showed that corn grain taken up Pb from soil was mainly governed by soil pH and OM content. The Pb accumulation in the corn grain increased with the decreasing soil pH. This mainly attributed to pH played an important role in Pb uptake by plants from soil, this process through diverse reactions such as adsorption, ionic exchange, redox reactions, and precipitation-dissolution. These reactions are mainly affected by soil pH because of pH affecting Pb speciation and the charge of soil surface groups [49–50]. Christensen [51] and Naidu R et al [52] have found that the pH also influence on the solution activity of Pb and the distribution of Pb between the soil phase and the solution phase. The increased sorption of Pb at high pH values could reduce the Pb speciation in soils. In the present study, the BCF values observed in acidic soils (S1 to S9) with low pH values were significantly higher than those in alkaline soils (S10 to S17) with high pH values (Figure 2). These new results suggest that Pb is much more bioavailable under acidic conditions and is easily accumulated in corn grain. In the present paper, it is newly found that the Pb accumulation in corn grain was negatively related with OM. This is alike to the Cu accumulation in plant shoots [53], however it is different from Cd accumulation in corn grain [17]. This may be attributed to organic matter action as a primary sorbent of Pb because of its high sorption affinity [54–56].

In the present paper, no significant differences were observed between model 1 and model 2 at the different exogenous Pb concentration. In addition, there were also no significantly differences among Pb BCF values of corn grain at different exogenous Pb concentration, either. This was consistent with the findings of Rezvani and Zaefarian [57].

More importantly, we confirmed that the Pb BCF prediction models from Zhengdan 958 could be used to predict Pb BCF of other non-model corn species according to the soils pH and OM through cross-species extrapolation approach. Further, it was more accurate to apply model 2 to predict the BCF values of other corn species than model 1. Although many prediction models were developed to describe the relationship between plants uptake heavy metals and soil properties in terrestrial ecosystems, these models can't be extrapolated to non-model plant species, being only for single species [58–59]. Only a few studies were carried out to develop the prediction models for one species in aquatic ecosystems and then they are applied to other non-model species [36,38]. Few prediction models were developed for Pb accumulation in corn grain [60].

In order to further validate the feasibility of the prediction modes developed in the present study, the measured Pb BCF values of corn grains from soils were selected from the references [61,62], which were shown in Table 5. Although the corn species was also the same as that used to develop the model 1, the properties of the soils used in these references including OM, pH, and so on were different from those in the present paper. Therefore it is suitable that these data were used to test the prediction accuracy of model 1. The relativity between measured and predicted BCF values for Pb is shown in Figure 5. The result showed that the ratios between the predicted BCF values and the measured BCF values were also within 2 folds interval and close to 1:1 relationship. This indicates that the model 1 developed from Zhengdan 958 in the present study is feasible. It could be applied to other non-model corn species.

Conclusions

The soil pH and OM were the most major factors that can control Pb uptake in corn grain. The Pb accumulation in corn grain was negatively related with pH and Log [OM], respectively. The Pb BCF in corn grain can be well predicted by the model 1, which could be applied to other non-model corn species.

Acknowledgments

We are indebted to Prof. Lifang Hong for her help in soils collection, and Prof. Yibin Ma and Dr. Xiaoqing Wang for their help in the data processing and model derivation.

Author Contributions

Conceived and designed the experiments: ZL Y. Liang. Performed the experiments: Y. Li. Analyzed the data: HY. Contributed reagents/materials/analysis tools: Y. Li JL. Wrote the paper: ZL HY.

References

1. Cecchi M, Dumat C, Alric A, Felix-Faure B, Pradere P, et al (2008) Multi-metal contamination of a calcic cambisol by fallout from a lead-recycling plant. Geoderma 144: 287–298.
2. Grover P, Rekhadevi P, Danadevi K, Vuyyuri S, Mahboob M, et al (2010) Genotoxicity evaluation in workers occupationally exposed to lead. Int J Hyg Environ Health 213: 99–106.
3. Shahid M, Pinelli E, Pourrut B, Silvestre J, Dumat C (2011) Lead-induced genotoxicity to *Viciafaba L.* roots in relation with metal cell uptake and initial speciation. Ecotoxicol Environ Saf 74: 78–84.
4. Hrudey SE, Chen W, Rousseaux CG (1995) Bioavailability in environmental risk assessment. Lewis Publ, Boca Raton, FL.
5. USEPA (1992) Common chemicals found at Superfund sites. EPA540/R-94/044. U S Gov Print Office, Washington, DC.
6. Khan S, Cao Q, Zheng YM, Huang YZ, Zhu YG (2008) Health risks of heavy metals in contaminated soils and food crops irrigated with wastewater in Beijing, China. Environmental Pollution 152: 686–692.
7. Li ZJ, Xu JM, Tang CX, Wu JJ, Muhammad A, et al. (2006) Application of 16S rDNA-PCR amplification and DGGE fingerprinting for detection of shift in microbial community diversity in Cu-, Zn-, and Cd-contaminated paddy soils. Chemosphere 26: 1374–1380.
8. Chien CC, Kuo YM, Chen CC, Hung CW, Yeh CH, et al. (2008) Microbial diversity of soil bacteria in agricultural field contaminated with heavy metals. Journal of Environmental Sciences 20: 359–363.

9. Rattan RK, Datta SP, Chhonkar PK, Suribabu K, Singh AK (2005) Long-term impact of irrigation with sewage effluents on heavy metal content in soils, crops and groundwater-a case study. Agriculture Ecosystems and Environment 109: 310–322.
10. Chopin EIB, Alloway BJ (2007) Distribution and mobility of trace elements in soils and vegetation around the mining and smelting areas of Tharsis, Ríotinto and Huelva, Iberian Pyrite Belt, SW Spain. Water, Air, and Soil Pollution 182: 245–261.
11. Jung MC (2008) Heavy metal concentration in soils and factors affecting metal uptake by plants in the vicinity of a Korean Cu-W mine. Sensors 8: 2413–2423.
12. Wang J, Zhang CB, Jin ZX (2009) The distribution and phytoavailability of heavy metal fractions in rhizosphere soils of paulowniu fortune (seem) Hems near a Pb/Zn smelter in Guangdong, PR China. Geodema 148: 299–306.
13. Chapman EEV, Dave G, Murimboh JD (2013) A review of metal (Pb and Zn) sensitive and pH tolerant bioassay organisms for risk screening of metal-contaminated acidic soils. Environmental Pollution 179: 326–342.
14. Rodrıguez JA, Nanos N, Grau JM, Gil L, Lopez-Arias M (2008) Multiscale analysis of heavy metal contents in Spanish agricultural topsoils. Chemosphere 70: 1085–1096.
15. Bang J, Hesterberg D (2004) Dissolution of trace element contaminants from two coastal plain soils as affected by pH. Journal of Environmental Quality 33: 891–901.
16. Yang H, Li ZJ, Lu L, Long J, Liang YC (2013) Cross-species extrapolation of prediction models for cadmium transfer from soil to corn grain. PLoS ONE 8(12): e80855.doi: 10.1371/journal.pone.0080855.
17. Wang AS, Angle JS, Chaney RL, Delorme TA, Reeves RD (2006) Soil pH effects on uptake of Cd and Zn by Thlaspi caerulescens. Plant and Soil 281: 325–337.
18. Kirkham MB (2006) Cadmiumin plantson polluted soils: Effect of soil factors, hyperaccumulation and amendments. Geoderma 137: 19–32.
19. Zeng F, Ali S, Zhang H, Quyang Y, Qiu B, et al. (2001) The influence of pH and organic matter content in paddy soil on heavy metal availability and their uptake by rice plants. Environmental Pollution 159: 84–91.
20. Hettiarachchi GM, Ryan JA, Chaney RL, La Fleur CM (2003) Sorption and desorption of cadmium by different fractions of biosolids-amended soils. Journal of Environmental Quality 32: 1684–1693.
21. Antoniadis V, Robinson JS, Alloway BJ (2008) Effects of short-term pH fluctuations on cadmium, nickel, lead, and zinc availability to ryegrass in a sewage sludge-amended field. Chemosphere 71: 759–764.
22. Zeng FR, Shafaqat Ali, Zhang HT, Ouyang YN, Qiu BY, et al. (2011)The influence of pH and organic matter content in paddy soil on heavy metal availability and their uptake by rice plants. Environmental Pollution 159: 84–91.
23. Ge Y, Murray P, Hendershot WH (2000) Trace metal speciation and bioavailability in urban soils. Environ Pollut 107: 137–144.
24. Chen WP, Li LQ, Chang AC, Wu LS, Chaney RL, et al. (2009) Characterizing the solid-solution partitioning coefficient and plant uptake factor of As, Cd, and Pb in California croplands. Agriculture, Ecosystems and Environment 129: 212–220.
25. Zarcinas BA, Ishak CF, McLaughlin MJ, Cozens G (2004) Heavy metals in soils and crops in southeast Asia. 1. Peninsular Malaysia. Environmental Geochemistry and Health 26: 343–357.
26. Simmons RW, Noble AD, Pongsakul P, Sukreeyapongse O, Chinabut N (2008) Analysis of field-moist Cd contaminated paddy soils during rice grain fill allows reliable prediction of grain Cd levels. Plant and Soil, 302: 125–137.
27. Kuo S, Huang B, Bembenek R (2004) The availability to lettuce of zinc and cadmium in a zinc fertilizer. Soil Science, 169: 363–373.
28. Hough RL, Breward N, Young SD, Crout NMJ, Tye AM, et al. (2004) Assessing potential risk of heavy metal exposure from consumption of home-produced vegetables by urban populations. Environmental Health Perspectives, 112: 215–221.
29. Nahmani J, Hodson ME, Black S (2007) A review of studies performed to assess metal uptake by earthworms. Environmental Pollution, 145: 402–424.
30. Zhang HZ, Luo YM, Song J, Zhang HB, Xia JQ, et al. (2011) Predicting As, Cd and Pb uptake by rice and vegetables using field data from China. Journal of Environmental Sciences 23: 70–78.
31. Adamsa ML, Zhao FJ, McGratha SP, Nicholsonb FA, Chambersb BJ (2004) Predicting cadmium concentrations in wheat and barley grain using soil properties. Journal of Environmental Quality 33: 532–541.
32. Zhao XJ, Ma RC, Li ZJ (2013) Non-target organism species sensitivity distributions to oxytetracycline in water. Research Journal of Chemistry and Environment 17: 74–81.
33. Larras F, Bouchez A, Rimet FR, Montuelle B (2012) Using bioassays and species sensitivity distributions to assess herbicide toxicity towards benthic diatoms. PloS ONE 8: 1–9.
34. Van Sprang PA, Verdonck FAM, Van Assche F, Regoli L, De Schamphelaere KAC (2009) Environmental risk assessment of zinc in European freshwaters: a critical appraisal. Sci Tot Environ 407: 5373–5391.
35. Allard P, Fairbrother A, Hope BK, Hull RN, Johnson MS, et al. (2010) Recommendations for the development and application of wildlife toxicity reference values. Intergrated Environmental Assessment and Management 6: 28–37.
36. Deleebeeck NME, De Schamphelaere KAC, Janssen CR (2007a) A bioavailability model predicting the toxicity of nickel to rainbow trout (Oncorhynchus mykiss) and fathead minnow (Pimephales promelas) in synthetic and natural waters. Ecotoxicol Environ Saf 67: 1–13.
37. Deleebeeck NME, Muyssen BTA, De Laender F, Janssen CR, De Schamphelaere KAC (2007b) Comparison of nickel toxicity to cladocerans in soft versus hard surface waters. Aquat Toxicol 84: 223–35.
38. Schlekat CE, Genderen EV, Schamphelaere KACD, Antunes PMC, Rogevich EC, et al. (2010). Cross-species extrapolation of chronic nickel biotic ligand models. Science of the Total Environment 408: 6148–6157.
39. Sparks DL, Page AL, Helmke PA, Loeppert RH, Soltanpour PN, et al. (1996) Methods of soil analysis. Part 3 - chemical methods. Soil Science Society of America, Inc, Madison. WI.
40. Dohrmann R (2006) Cation exchange capacity methodology II: A modified silver-thiourea method. Applied Clay Science 34: 38–46.
41. Ball DF (1964) Loss-on-ignition as an estimate of organic matter and organic carbon in noncalcareous soils. Journal of Soil Science 15: 84–92.
42. Bray RH, Kurtz LT (1945) Determination of total, organic, and available forms of phosphorus in soils. Soil Science 59: 39–46.
43. Bremner JM (1996) Nitrogen total Methods of soil analysis. Part 3: Chemical methods, American Society of Agronomy-Soil Science Society of America, 1996, PP 1085–1121.
44. Zarcinas BA, McLaughlin MJ, Smart MK (1996) The effect of acid digestion technique on the performance of nebulisation systems used in inductively coupled plasma spectrometry. Commun Soil Sci Plant Anal 27: 1331–1354.
45. Baker AJM, Reeves RD, Hajar ASM (1994) Heavy metal accumulation and tolerance in British populations of metallophyte Thlaspi caerulescens. New Phytol 127: 61–68.
46. Raskin I, Kumar PBAN, Dushenkov S, Salt DE (1994) Bioconcentrationof heavy metals by plants. Curr Opin Biotechnol 5: 285–290.
47. Wang XQ (2012) Ecological Thresholds for Copper and Nickel in Chinese Agricultural Soils, Beijing, China Mining University 1–50.
48. Sarwar N, Saifullah, Malhi SS, Zia MH, Naeem A, et al. (2010) Role of mineral nutrition in minimizing cadmium accumulation by plants. J Sci Food Agric 90: 925–937.
49. Martınez-Villegas N, Flores-Velez LM, Domınguez O (2004) Sorption of lead in soil as a function of pH:a study case in Mexico. Chemosphere 57: 1537–1542.
50. Smical AI, Hotea V, Oros V, Juhasz J, Pop I (2008) Studies on transfer and bioaccumulation of heavy metals from soil into lettuce. Environmental Engineering and Management Journal 7: 609–615.
51. Christensen TH (1984) Cadmium soil sorption at low concentrations: I. Effect of time, cadmium load, pH, and calcium. Water. Air. Soil Pollut 21: 105–114.
52. Naidu R, Bolan NS, Kookana RS, Tiller KG (1994) Ionic-strength and pH effects on the sorption of cadmium and the surface charge of soils. European Journal of Soil Science 45: 419–429.
53. Moreno JL, Garcia C, Hernandez T, Pascual JA (1996) Transference of heavy metals from a calcareous soil amended with sewage sludge compost to barley plants. Bioresource Technology 55: 251–258.
54. Ge Y, Hendershot W (2005) Modeling sorption of Cd, Hg and Pb in soils by the NICA (non-ideal competitive adsorption)-Donnan model. Soil Sediment Contam 14: 53–69.
55. Sauve S, Manna S, Turmel MC, Roy AG, Courchesne F (2003) Solid solution partitioning of Cd, Cu, Ni, Pb, and Zn in the organic horizons of a forest soil. Environ Sci Technol 37: 5191–5196.
56. Puschenreiter M, Horak O, Friesl Wand Hartl W (2005) Low-cost agricultural measures to reduce heavy metal transfer into the food chain – a review. Plant Soil Environ 51: 1–11.
57. Rezvani M, Zaefarian F (2011) Bioaccumulation and translocation factors of cadmium and lead in Aeluropus littoralis. Australian Journal Agricultural Engineering 2: 114–119.
58. Gandoisa L, Probstb A, Dumata C (2010) Modelling trace metal extractability and solubility in French forest soils by using soil properties. European Journal of Soil Science 61: 271–286.
59. Wang XP, Shan XQ, Zhang SZ, Wen B (2004) A model for evaluation of the phytoavailability of trace elements to vegetables under the field conditions. Chemosphere 55: 811–822.
60. Rainbow PS (2007) Trace metal bioaccumulation: Models, metabolic availability and toxicity. Environment international 33: 576–582.
61. Cao Y, Liu Y, Wang GJ, Han Y, Wang SB (2009). Differences in Lead and Cadmium Concentrations among Plant Tissues of Maize Cultivars under the Combined Stress of Lead and Cadmium. Journal of Maize Sciences 17: 80–85.
62. Li YF, Liu L, Chen X, Hu LL, Tang JJ (2010) Plant Growth,Lead Uptake and Partitioning of Maize (*Zea mays L.*) Under Simulated Mild/Moderate Lead Pollution Stress. Journal of Agro-Environment Science 29: 2260–2267.

Land-Use History and Contemporary Management Inform an Ecological Reference Model for Longleaf Pine Woodland Understory Plant Communities

Lars A. Brudvig[1]*, John L. Orrock[2], Ellen I. Damschen[2], Cathy D. Collins[3], Philip G. Hahn[2], W. Brett Mattingly[2¤], Joseph W. Veldman[2], Joan L. Walker[4]

1 Department of Plant Biology, Michigan State University, East Lansing, Michigan, United States of America, **2** Department of Zoology, University of Wisconsin, Madison, Wisconsin, United States of America, **3** Department of Biology, Colby College, Waterville, Maine, United States of America, **4** U.S. Forest Service Southern Research Station, Clemson, South Carolina, United States of America

Abstract

Ecological restoration is frequently guided by reference conditions describing a successfully restored ecosystem; however, the causes and magnitude of ecosystem degradation vary, making simple knowledge of reference conditions insufficient for prioritizing and guiding restoration. Ecological reference models provide further guidance by quantifying reference conditions, as well as conditions at degraded states that deviate from reference conditions. Many reference models remain qualitative, however, limiting their utility. We quantified and evaluated a reference model for southeastern U.S. longleaf pine woodland understory plant communities. We used regression trees to classify 232 longleaf pine woodland sites at three locations along the Atlantic coastal plain based on relationships between understory plant community composition, soils (which broadly structure these communities), and factors associated with understory degradation, including fire frequency, agricultural history, and tree basal area. To understand the spatial generality of this model, we classified all sites together and for each of three study locations separately. Both the regional and location-specific models produced quantifiable degradation gradients–i.e., progressive deviation from conditions at 38 reference sites, based on understory species composition, diversity and total cover, litter depth, and other attributes. Regionally, fire suppression was the most important degrading factor, followed by agricultural history, but at individual locations, agricultural history or tree basal area was most important. At one location, the influence of a degrading factor depended on soil attributes. We suggest that our regional model can help prioritize longleaf pine woodland restoration across our study region; however, due to substantial landscape-to-landscape variation, local management decisions should take into account additional factors (e.g., soil attributes). Our study demonstrates the utility of quantifying degraded states and provides a series of hypotheses for future experimental restoration work. More broadly, our work provides a framework for developing and evaluating reference models that incorporate multiple, interactive anthropogenic drivers of ecosystem degradation.

Editor: Ben Bond-Lamberty, DOE Pacific Northwest National Laboratory, United States of America

Funding: This study was funded by the Strategic Environmental Research and Development Program (Project RC-1695) and by funds provided to the Department of Agriculture, Forest Service, Savannah River, under Interagency Agreement DE-AI09-00SR22188 with the Department of Energy, Aiken, SC. The funders had no role in study design, data collection and analysis, decision to publish, or preparation of the manuscript.

Competing Interests: The authors have declared that no competing interests exist.

* E-mail: brudvig@msu.edu

¤ Current address: Department of Biology, Eastern Connecticut State University, Willimantic, Connecticut, United States of America

Introduction

Ecological restoration efforts guided by a target range of reference conditions [1] often fail to achieve these targets [2]. In part, this may be because knowledge of reference conditions, while useful, is by itself insufficient for guiding or prioritizing restoration due to variation in degraded states (unrestored conditions that deviate from reference conditions). Human-modified landscapes support a range of degraded states, resulting from land-use legacies and variation in contemporary management [3,4]. Thus, an important early step in the restoration of human-modified landscapes is the formalization of ecological reference models, which describe both reference conditions and the spectrum of degraded states that are common for a given ecosystem [3,5]. Reference models have been formulated for many ecosystems [6–8]; however, these models are frequently qualitative and models of ecosystem degradation have rarely been quantitatively developed or evaluated [4].

Data-driven ecological reference models that incorporate both the consequences (e.g., altered species compositions) and causes of degradation (e.g., altered disturbance regimes) can promote a better understanding of degraded landscapes, help to prioritize restoration and management activities, and contribute toward the goal of tailoring restoration strategies to specific degraded states [3–5]. For example, in fire-maintained ecosystems, qualitative reference models simply predict increasing degradation with fire suppression [6–8], but do not detail the nature of this relationship, such as the rate at which degradation increases with fire suppression or whether thresholds exist where degradation increases abruptly with fire suppression. In contrast, a quantitative reference model could describe degradation based on thresholds in fire frequency, allowing for sites to be classified along this axis, and restoration planning and approaches could be tailored accordingly [4].

We suggest that a quantitative ecological reference model should have three features to be both ecologically relevant and useful to land managers. First, it would classify sites based on relevant ecological communities, which in many cases will be plants – the basis for many management and restoration decisions [9,10]. Second, suspected drivers of ecosystem degradation would be incorporated into the classification by linking site conditions to factors associated with degradation [4,5]. Together, these two steps would describe the range of degraded states and quantify how, in terms of degrading factors, they differ from each other and from reference conditions. Third, to make such a model applicable to restoration management, it would use data routinely available to land managers across sites spanning landscapes, the scale at which restoration planning and many management efforts typically operate [4,5,11]. We define landscape as "an area that is spatially heterogeneous in at least one factor of interest" [12] – in our case, degradation of sites across a location for which restoration or management might be coordinated. In this study, we incorporate these three features to develop a quantitative ecological reference model for longleaf pine (*Pinus palustris*) woodlands at three locations (i.e., landscapes) in the southeastern United States.

Fire-maintained longleaf pine woodlands support species-diverse understory plant communities that have been widely degraded by human land uses and are an active target for restoration [8]. The starting point for our work is a previously published qualitative reference model [8], which describes a degradation gradient in longleaf pine understory plant diversity and composition caused by past agricultural land use, altered fire regimes, and silvicultural activities (see also [13]). Agricultural legacies can persist for decades following abandonment, leading to reduced understory diversity and modified community composition on post-agricultural sites [14–16]. Fire suppression leads to increased tree abundance, canopy cover, and an accumulation of leaf litter and duff (i.e., forest floor), each of which may reduce understory diversity and modify understory community composition [17–20]. Historically, lightning and human-ignited surface fires burned longleaf pine woodlands as frequently as every 1–6 years, but today fire suppression is widespread and, where fires do occur, they are implemented through prescribed burning [21]. Overstory trees reduce understory plant diversity through competition with understory plants for light and water [18,22]; tree density is also altered by silvicultural management, including tree planting and harvesting [8,23]. Based on these consequences of human land use, we predict longleaf pine understory communities to be most degraded at sites with a history of agriculture, contemporary fire suppression, and a dense overstory (Figure 1), which are also determinants of degradation in many other ecosystems [24,25]. Guided by this past work, we focus on fire history, agricultural legacies, and overstory density as likely degrading factors for longleaf pine understory communities; however, we know of no efforts to quantify a reference model based on these factors.

The goals of our study were to quantify and then evaluate a reference model for longleaf pine woodlands, based on a previously described qualitative model [8] and associated literature (references above). To achieve these goals, we pursued three specific objectives: 1) classify longleaf pine understory plant communities based on a set of previously identified degrading factors (agricultural history, fire frequency, overstory tree basal area; references above), 2) evaluate the spatial generality of this model by evaluating the roles of degrading factors across sites that vary in soil conditions and by comparing models for three different landscape-scale locations and a region-scale model (three locations combined), and 3) evaluate the resulting regional and location-specific reference models by comparing degraded states to a set of reference sites.

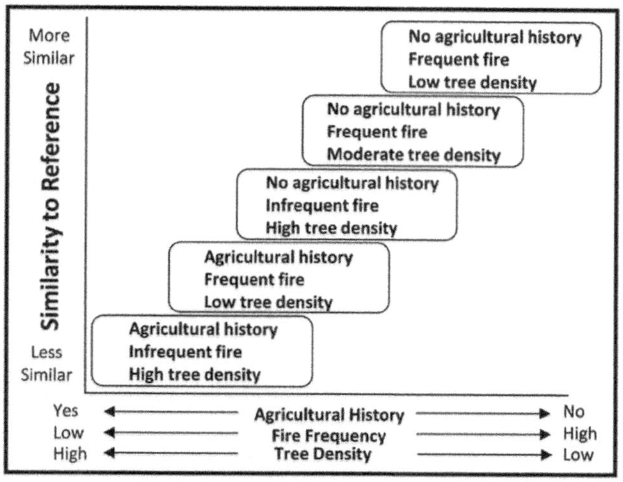

Figure 1. Conceptual model of degradation for understory plant communities in longleaf pine woodlands. Understory degradation (deviation in community composition from reference site conditions) is predicted to increase with occurrence of agricultural history, increasing overstory density, and declining fire frequency. Note: not all combinations of model components are presented in this figure and, while depicted as a linear process in this conceptual diagram, nonlinearities may exist during restoration from degraded states. Model is based on [8].

We pursued our first objective through regression tree analysis and data from 232 longleaf pine woodland sites, which were selected to span a range of biophysical conditions across our three study locations in the southeastern U.S. (Figure 2). This analysis groups sites with similar plant communities based on data-defined levels and combinations of agricultural history, fire frequency, and overstory tree basal area. We also included soil attributes in this classification because of their importance for determining plant community composition in this system [26] and to assist with our second objective. We recognize that additional degrading factors might be identified (e.g., invasive species); however, we selected this set for consideration based on clear linkages with understory degradation (references above), applicability to our study landscapes (e.g., invasive species were in low abundance at all of our study sites), and the likelihood that these data would be easily obtained by land managers, facilitating application of the model to land management and restoration planning. We pursued our second objective in two ways. First, the inclusion of soil variables in our models provides insight into whether degrading factors (e.g., fire frequency) operate generally to determine degraded states, or differently for some types of longleaf pine woodlands compared to others (e.g., those that occur on Entisols vs. Ultisols). Second, by including three locations in our study, we can evaluate the generality of a regional model (spanning all sites) relative to models for three separate landscapes (Table 1). This is important for understanding the degree to which we can generalize our results across a region or to which landscape-to-landscape variation might preclude broad application of the regional model. Finally, to address our third objective, we compared the degraded states resulting from the regional and location-specific models to a set of 38 reference sites based on a suite of biophysical characteristics that are relevant to restoration and land management. In doing so, we evaluated whether each model produces a quantifiable gradient in degradation. Given the geographic distribution of our study locations, our results may be most applicable to Atlantic Coast longleaf pine woodlands, but the degrading factors we study – agricultural legacies, altered fire regimes, and altered overstory tree abundance – are broadly relevant across the longleaf pine ecosystem [8,13].

Methods

Ethics Statement

Prior to conducting field data collection, we obtained approvals from the Fish and Wildlife Branch and Forestry Branch at Fort Stewart, the USDA Forest Service-Savannah River at Savannah River Site (a National Environmental Research Park), and the Endangered Species Branch, Forestry Branch, and the Cultural Resources Program at Fort Bragg. All data collection occurred on publicly owned land. We observed and surveyed, but made no collections of protected species.

Study System

This study was carried out in fire-dependent longleaf pine woodlands on sandy soils of the southeastern United States (Figure 2). Frequently burned longleaf pine woodlands are characterized by a relatively sparse, longleaf pine-dominated overstory and a species-diverse understory of graminoids, forbs, and shrubs [21,26]. Longleaf pine ecosystems, of which woodlands are a component, span from Texas to Virginia (Figure 2), but today less than 3% of historic area remains due to agricultural conversion, urbanization, and fire suppression [21]. As such, longleaf pine ecosystems are of high priority for conservation and restoration [8,21].

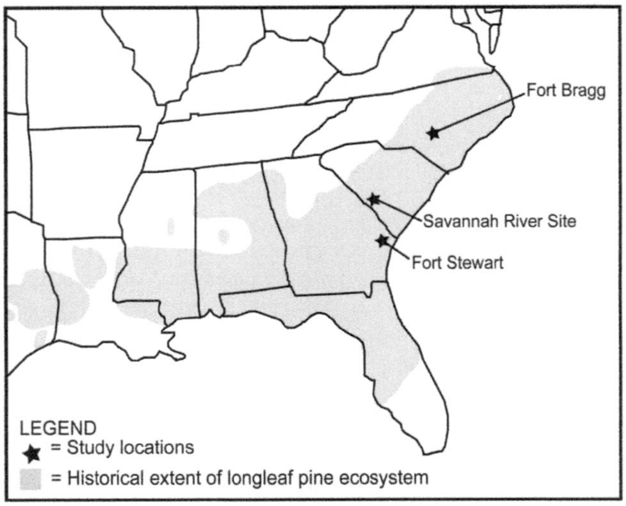

Figure 2. Geographic range of the longleaf pine ecosystem and map of study locations. Study locations (Fort Bragg [NC], 31°11′ N, 79°15′ W; Fort Stewart [GA], 31°56′ N, 81°36′ W; and Savannah River Site [SC], 33°20′ N, 81°40′ W) were in three different physiographic regions (sandhills, southern coastal plain, and Atlantic coastal plain, respectively [26]) allowing for creation and evaluation of an ecological reference model across a range of ecological settings. This model was based on data from 232 sites, which varied in their levels of degradation, with results subsequently compared to data from 38 reference sites.

We sampled longleaf pine woodlands at three locations within the historical range of the longleaf pine ecosystem: Fort Bragg (North Carolina), Fort Stewart (Georgia), and Savannah River Site (South Carolina) (Figure 2). Like much of the longleaf pine region [21], the landscapes at these study locations were historically fragmented by agriculture. Following federal government acquisition (Fort Bragg: 1919, Fort Stewart: 1940, Savannah River Site: 1951) agriculture was abandoned, resulting in contemporary landscapes supporting mosaics of longleaf pine woodlands with and without agricultural land-use histories. Study locations are managed with prescribed fire, but substantial variation in recent fire history exists across the sites, resulting in a range of fire histories from frequently burned to fire suppressed (Table 1).

Site Selection

We selected a set of study sites that characterized the range in variation of degraded conditions (fire history, overstory density, agricultural history) at each location. Sites were each ≥1 ha, supported overstory longleaf pines, and lacked firebreaks, drainages, or other features causing abrupt transitions in understory vegetation. Sites varied in overstory tree density, recent fire history (1991–2009), and agricultural land-use history (Table 1); stand age from the 165 of 232 non-reference sites with available data was 62±1.2 years (mean±1SE). We classified each site as having a "forest" or "agricultural" land-use history based on its status in historical aerial photographs (Fort Stewart, Savannah River Site) or maps (Fort Bragg) from the year of federal acquisition. We obtained GIS data from prescribed fire managers at each location to reconstruct fire history (prescribed and wild) between 1991 and 2009 for each site. Sites were considered burned in a given year if they occurred within the boundaries of a fire management unit that was burned in that year. We determined overstory basal area for each site during vegetation sampling.

We quantified reference conditions by sampling a set of reference sites at each location (Fort Bragg n = 15, Fort Stewart n = 14, Savannah River Site n = 9). These sites had been previously identified by regional botanical experts at the Carolina Vegetation Survey (CVS) "to document the composition and status of the natural vegetation of the Carolinas" (http://cvs.bio.unc.edu/). Reference sites had no known history of cultivation, were generally well maintained by prescribed fire, and were located within the boundaries of the respective study locations.

All study sites were located on soils that are primary substrates for longleaf pine communities [26], with three soil orders characterizing the majority of our sites (97%): 212 of our 270 sites (78.5%) were located on Ultisols, 38 sites (14.0%) were on Entisols, 13 sites (4.8%) were on Spodosols, and 7 sites (2.6%) were on Inceptisols.

Data Collection

Between 20 August and 13 November 2009, we surveyed the 270 study sites: 99 at Fort Bragg; 82 at Fort Stewart; 89 at Savannah River Site. We used one randomly located and oriented 20 m×50 m plot at each site. This plot design was a modified version of the CVS protocol [27], which is broadly employed throughout and beyond our region to characterize forest, savanna, and grassland plant communities. We identified and assigned a percent cover value to all understory plant species (herbaceous species and woody species <2.5 cm diameter at 1.4 m height) rooted within or overhanging each of eight 1 m×1 m subplots located within a 20 m×20 m portion of each plot. Taxonomy follows Radford and colleagues [28], except for the genus

Table 1. Attributes across study sites and for each of the three study locations.

Variable	All Sites	Fort Bragg	Fort Stewart	Savannah River Site
Number of sites	232	84	68	80
Canopy cover (%)	51.7±2.8	30.8±1.9	62.7±4.8	64.5±3.5
Total basal area (m^2/ha)	18.8±1.0	19.4±1.5	15.8±1.9	20.7±1.8
Pinus basal area (m^2/ha)	17.0±1.0	17.6±1.7	14.2±1.7	18.8±1.6
Non-Pinus basal area (m^2/ha)	1.8±0.4	1.8±0.8	1.7±0.6	1.8±0.8
Years since fire	3.4±0.8	1.0±0.2	2.3±0.7	6.8±2.1
Number of fires (1991–2009)	4.6±0.3	5.8±0.3	5.0±0.6	3.0±0.5
Soil water holding capacity (%)	39.5±0.8	41.2±1.4	40.8±1.6	36.8±1.0
Soil organic matter	1.8±0.1	2.0±0.2	2.0±0.2	1.5±0.1
Species richness/m^2	5.1±0.4	4.7±0.6	6.7±0.8	4.2±0.6
Species evenness/m^2	0.7±0.02	0.6±0.03	0.7±0.02	0.7±0.02
Vegetation cover (%)	23.2±2.3	13.4±2.0	38.3±4.7	20.5±3.0
Bare ground (%)	8.2±1.5	11.4±2.7	10.2±3.2	3.1±1.2
Down woody debris (%)	5.2±0.7	2.1±0.4	5.2±1.2	8.4±1.5
Litter depth (cm)	2.2±0.3	1.0±0.1	2.7±0.4	3.1±0.4
Duff depth (cm)	0.9±0.2	0.2±0.03	0.6±0.2	1.9±0.3

Values exclude data from reference sites and are mean ±95% confidence interval.

Dichanthelium, which follows Weakley [29], and the genera *Lyonia* and *Persea*, which follow Wunderlin and Hansen [30]. Within each subplot, we estimated the percent cover of green vegetation, bare ground, and down woody debris (logs, sticks, pine cones, and bark). We measured the depth of leaf litter and duff in the center of each subplot. To characterize overstory conditions within each 20 m×50 m plot, we recorded canopy cover with a spherical densiometer held at 1.4 m above six points spaced at 10 m intervals along the plot center line, and identified and measured the diameter of all trees ≥2.5 cm diameter at 1.4 m, within the plot.

To characterize site-level soil conditions, we collected soil cores (2.5 cm diameter by 15 cm deep) at 10 m intervals along the center line of each 20 m×50 m plot. Samples were composited by site and analyzed by Brookside Laboratories, Inc. (New Knoxville, OH) for soil organic matter content (SOM) [31], which is an indicator of soil degradation in post-agricultural forests [3,24]. We also analyzed soil samples for water holding capacity as the proportionate difference between saturated wet weight and oven-dried weight, following the method of [32], as described by [15]. Soil moisture availability correlates with longleaf pine understory diversity and productivity [33].

Statistical Methods

Following data collection, we constructed a site-by-species matrix for subsequent plant community analyses, using the mean cover for each species across the eight subplots at each site. To develop ecological reference models, we used multivariate classification and regression tree analysis to classify non-reference sites based on a combination of plant community composition and environmental data [34]. This analysis creates a dichotomous tree with splits based on environmental data that minimize compositional dissimilarity within groups of sites (i.e., 'classes'). In our case, we used environmental data that corresponded to the degrading factors in Figure 1: agricultural history, tree basal area (total basal area, *Pinus* spp. basal area, non-*Pinus* spp. basal area), and fire frequency (number of burns 1991–2009, time since last fire). We investigated *Pinus* and non-*Pinus* overstory separately because these groups respond differently to fire frequency (*Pinus* is more fire tolerant; [35,36]) and because land managers often remove hardwoods during restoration [8,20]. We also included soil order (Entisol, Inceptisol, Spodosol, Ultisol) and soil water holding capacity in the classification, which are two aspects of soils that broadly structure longleaf pine communities [26]. As we explain above, the inclusion of these soil variables in our models provides an opportunity to evaluate the generality of degrading factors across longleaf pine communities that occur on different substrates. We conducted classifications for all 232 non-reference sites together to accomplish our first objective ('All Sites' analysis) and for each of the three locations separately to address our second objective ('Fort Bragg', 'Fort Stewart', and 'Savannah River Site' analyses). To determine the number of final classes generated by each analysis, we conducted 500 cross-validations of the model and selected the most frequently occurring tree size using the 1-SE rule [37]. For ease of interpretation, we numbered the resulting classes to align with Figure 1, so that Class 1 was most degraded. Classes defined by soil attributes do not align with our conceptual framework in Figure 1 and we do not attempt to attribute degradation to soil conditions; however, we attempt to integrate soil-defined classes into this alignment based on divergence in biophysical factors from reference conditions (see below).

To accomplish our third objective, we evaluated the ecological relevance of the classes resulting from our classification analyses by comparing biophysical attributes of classes to reference sites. We calculated means per site (±95% CI) of understory richness and evenness, canopy cover, basal area (total, *Pinus*, non-*Pinus*), years since last fire, number of fires between 1991 and 2009, percent cover of vegetation, bare ground, and down woody debris, litter and duff depth, SOM, and soil water holding capacity. To visualize plant community composition among classes and to compare these classes to reference sites, we used Canonical Analysis of Principal Coordinates (CAP), with Bray-Curtis similarity as the distance measure following square-root transformation of the raw species abundance data [38]. CAP is a constrained ordination analysis that characterizes multivariate differences among groups (i.e., classes, reference sites) [38]. We note that, while many of the attributes we investigated were unrelated to our classification analyses, some attributes were components of the classifications (e.g., community composition). The purpose of these analyses was to quantitatively compare among classes and reference sites, rather than to formally test research hypotheses. To identify species that distinguished classes and reference sites from one another, we used Indicator Species Analysis (ISA) [39]. ISA results are presented in Tables S1, S2, S3 and S4. We ran separate analyses for each of the four classifications (All Sites, Fort Bragg, Fort Stewart, Savannah River Site). We used PRIMER-E version 6 for CAP analyses and calculation of species richness and Pielou's evenness [40], SAS version 9.2 for calculation of confidence intervals [41], R version 3.0 for multivariate regression trees [42], and PC-ORD version 5.31 for ISA [43].

Results

All Sites Classification

The classification of All Sites from the combined three locations resulted in five splits and six classes (Figure 3a). Sites first split according to fire frequency. Infrequently burned sites (≤4 burns since 1991) were further classified based on soil moisture: Class 1 was characterized by infrequent fire and lower soil moisture (<45.08%), whereas Class 2 was characterized by infrequent fire and higher soil moisture (≥45.08%). Frequently burned sites were further classified based on land-use history, with Class 3 characterized by frequent fire and agricultural history. Among sites with forested history, the model made two splits, with the first based on overstory density and the second based on soil moisture. Class 4 was characterized by frequent fire, forested history, high basal area (≥9.965 m^2/ha), and lower soil moisture (<42.12%), class 5 by frequent fire, forested history, high basal area, and higher soil moisture (≥42.12%), and class 6 by frequent fire, forested history, and low basal area (<9.965 m^2/ha).

Location-specific Classifications

Location-specific classifications illustrated a number of differences from the All Sites classification, in terms of the identity and ordering of degrading factors, as well as the role of soils (Figure 3). The Fort Bragg classification contained a single split and two groups defined by agricultural land-use history. The Fort Stewart classification contained three splits and four groups, which illustrated a prominent influence of soils. The first split was between Entisol/Ultisol sites and Inceptisol/Spodosol sites. Inceptisol/Spodosol sites (Class 1) formed a unique soil-vegetation association in the model, which was not related to degradation. The only degrading factor in this model, non-*Pinus* basal area, distinguished among sites on Entisol and Ultisol soils (low basal area sites were further split by soil moisture). The Savannah River Site classification contained splits based on all three degrading factors, but with differing order of importance relative to the All Sites classification, as basal area was the most important factor, followed by agricultural land-use history.

Degradation Classes in Relation to Reference Conditions

The All Sites classification revealed three main delineations in community composition among classes: sites with forested history (Classes 4–6, reference sites), sites with agricultural history or sites with infrequent fire and high soil moisture (Classes 2, 3), and sites with infrequent fire and low soil moisture (Class 1) (Figure 4a). Species richness declined with degradation, whereas species evenness was generally greater at infrequently burned sites (Figure 4b,c). Among ground cover attributes, vegetation cover was greatest at sites with high soil moisture and at reference sites (Figure 4e), bare ground was greatest at two of the frequently burned classes with forested history, and little difference in down woody debris was apparent (Figure 4e). Forest floor accumulation was related to fire suppression, with greater depths of litter and duff in Classes 1 and 2 (Figure 4f). Soil organic matter was markedly higher on sites with high soil moisture and at reference sites, but not strongly associated with degradation (Figure 4d; Table 2).

Models resulting from the location-specific classifications produced classes that were compositionally distinct, with the exception of Classes 1 and 3 at Fort Stewart (Figures S1, S2, and S3). Reference sites ranged from compositionally unique at Fort Bragg (Figure S1) to comparable to less degraded classes on at least one CAP axis at Fort Stewart and Savannah River Site (Figures S2,S3). Species richness declined with degradation in each location-specific model, whereas many other attributes illustrated location-specific patterns (Figures S1, S2, and S3, Table S5). Metrics used to classify sites (e.g., years since last fire, basal area; Table S5) largely mirrored each classification; however, the ways that these variables changed with degradation varied with the different classification rules at each location. Several variables were associated with degradation at one or more sites, including soil organic matter, vegetation cover, and forest floor depth at Savannah River Site, vegetation cover at Fort Bragg, and forest floor depth at Fort Stewart. Other metrics, such as species evenness and soil variables, illustrated location-specific patterns.

Discussion

An Ecological Reference Model for Longleaf Pine Woodlands

Ecological reference models are commonly employed during restoration, but are frequently qualitative, which limits their utility for prioritizing sites for restoration and guiding restoration efforts on the ground [4]. Using data from 232 sites, we quantified an ecological reference model for longleaf pine woodland understory plant communities spanning a broad geographic region of the Southeastern United States (Figure 3a). Three degrading factors – agricultural legacies, recent fire history, and overstory tree abundance – as well as soil moisture holding capacity, delineated classes of plant communities in this regional model. Location-specific models, however, illustrated substantial variation relative to this regional model and each other (Figure 3). We suggest that our regional model can help prioritize longleaf pine woodlands for restoration across our study region; however, due to substantial landscape-to-landscape variation, location-specific reference models can help guide local management decisions.

Our regional model provides general support for the core components of a qualitative model of longleaf pine degradation (Figure 1); however, by quantifying this conceptual model, we illustrate the relative contributions of degrading factors, as well as threshold values that determine differences between degraded classes. In this regional model, fire frequency was the most important factor (i.e., the first split in the classification), followed by agricultural legacies (Figure 3a). Overstory basal area, while a significant factor in the regional model, was the least important. Based on similarity in community composition to reference sites, Classes 1–3 (infrequently burned and post-agricultural sites) were the most degraded, whereas historically forested sites were less degraded, being both compositionally similar to each other and to reference sites (Figure 4a). Our regional model further quantifies 4

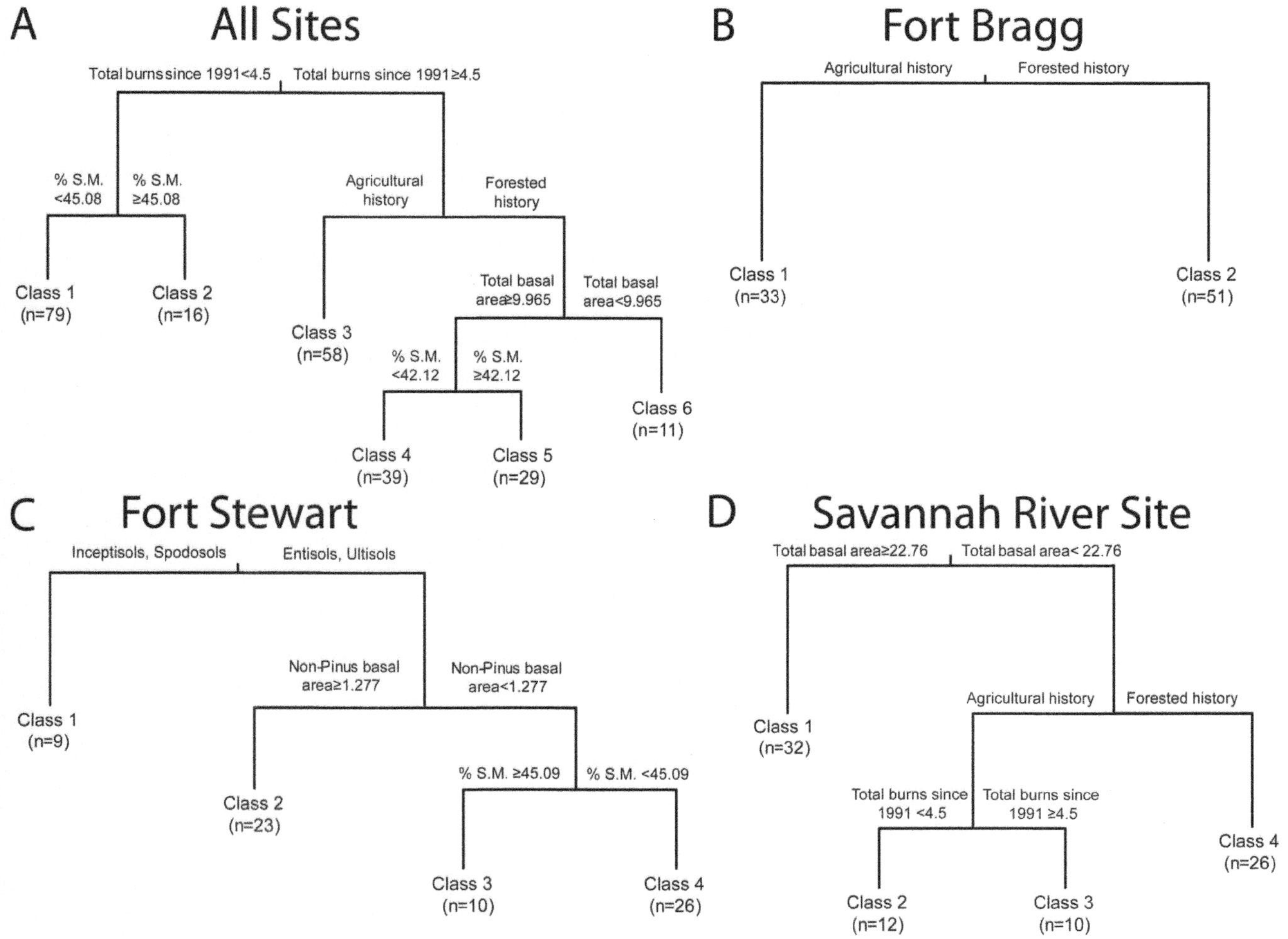

Figure 3. Results of multivariate classification and regression tree analysis. A) All study sites, B-D) separate study locations. In each analysis, sites are classified based on plant community composition and environmental data and classes are ordered to align with the conceptual model in Figure 1 (Class 1 = most degraded), with the exception of Class 1 at Fort Stewart, which represents a soil-vegetation association not related to degradation. Branch length at each split is scaled to the variance explained by the corresponding environmental factor. The number of study sites in each class is presented below each class label.

vs. 5 fires since 1991 and ~10 m^2/ha as breakpoints between degraded classes related to burning and tree basal area, respectively. These results reinforce the importance of frequent prescribed fires for management of longleaf pine woodlands [8,13], because of its positive influence on understory plant communities [15,17]. Our results further highlight the role of agricultural legacies in the degradation of longleaf pine plant communities [14–16]. The pronounced deviation of fire suppressed and post-agricultural plant communities from reference conditions (Figure 4) suggests that sites supporting these degraded conditions may require the greatest efforts to restore.

The resulting classes from our regional and location-specific analyses captured relevant gradients of degradation (Figures 4, S1, S2, and S3). As degradation increased, levels of biotic and abiotic variables became more dissimilar from values at reference sites. This pattern was consistent across the regional and location-specific models for several variables including understory richness, cover, and composition, forest floor accumulation, and tree basal area, whereas other variables including soil organic matter showed a degradation gradient in one or more models. Importantly, a number of these variables were not used in the construction of our ecological reference models (e.g., species diversity metrics, ground cover components, forest floor depth, and soil attributes), yet in many cases these variables corresponded with the degradation gradient.

The explicit consideration of land-use legacies has received increasing attention in conservation and restoration [3,10,24,44,45]. Our results illustrate how multiple drivers of ecosystem degradation, including land-use legacies, can be quantitatively incorporated into ecological reference models. This may be of particular importance when legacies interact with contemporary management to influence population viability and community composition. For example, past agricultural activities can mediate the effects of present-day fire management on plant community diversity, such that frequent burning may lead to increased richness, but only in sites with an agricultural land-use history [15]. Similarly, past land use can modify levels and patterns of soil nutrients, with ensuing effects on plant populations [46]. Future research will be necessary before specific restoration strategies might be tailored to the various degraded states in our regional and location-specific models (see below) and our work shows that consideration of both contemporary and historical causes of degradation will be important during this process.

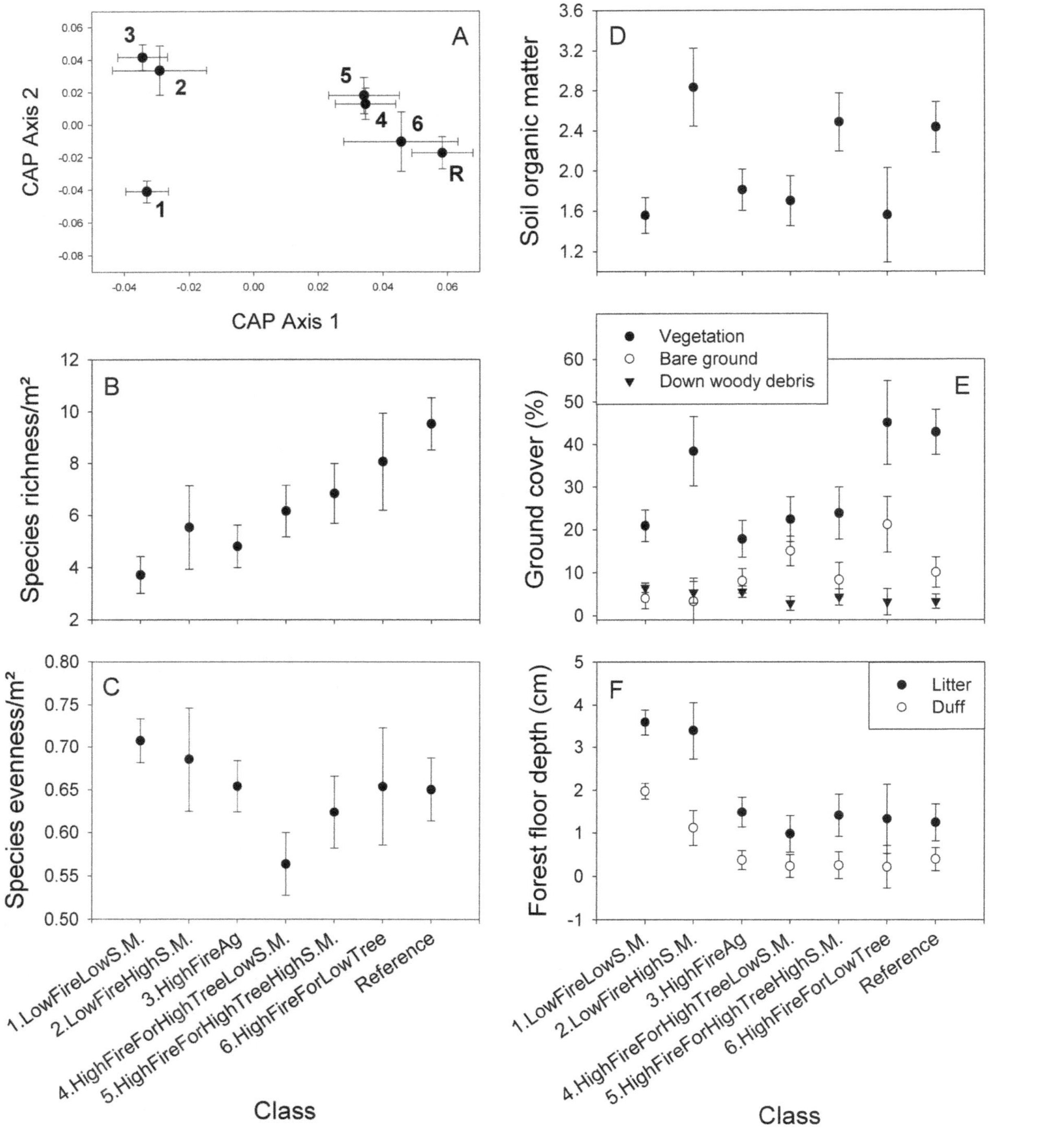

Figure 4. Comparison of Classes (1–6) from the All Sites classification and regression tree analyses to reference conditions. A) Understory community composition, B) understory species richness, C) understory species evenness, D) soil organic matter content, E) ground cover variables, and F) forest floor depth. All values are means ±95% confidence intervals.

Generality of the Ecological Reference Model

Our location-specific models illustrated notable variation, both relative to each other and to the regional All Sites model (Figure 3). Only one location-specific model (Savannah River Site) contained all three degrading factors, but with the reverse order of importance compared to the regional model. Conversely, Fort Bragg supported a simple two class model based only on agricultural land-use history, whereas Fort Stewart's model illustrated a prominent role of soils, in addition to hardwood abundance, for structuring understory plant communities (Figure 3). This variation may reflect differences in underlying environmental factors or differences in land-use and management histories among our three study locations (Table 1). For example, the presence or absence of fire frequency as a variable in location-specific models may be explained in part by variation in prescribed fire management among locations (Table 1). At Fort Bragg, where

Table 2. Attributes of degradation classes (1–6) for the All Sites classification and reference sites.

Class	Canopy cover (%)	Total basal area (m^2/ha)	Pinus basal area (m^2/ha)	Non-Pinus basal area (m^2/ha)	Years since fire	# Fires 1991–2009	Soil water holding capacity (%)
1. Low fire/Low SM[1]	62.4±4.2	20.3±1.6	18.2±1.6	2.0±0.7	7.6±1.2	2.3±0.4	37.0±1.3
2. Low fire/High SM	63.0±9.4	18.3±3.5	16.2±3.5	2.2±1.7	3.1±2.6	2.7±0.8	48.9±3.0
3. Ag/High fire	51.8±4.9	20.5±1.8	18.5±1.9	2.0±0.9	1.0±1.3	5.9±0.4	38.9±1.55
4. Forest/High Fire/High BA[2]/Low SM	34.8±6.0	16.2±2.2	14.8±2.6	1.4±1.1	0.8±1.6	6.4±0.5	37.2±1.9
5. Forest/High Fire/High BA/High SM	45.7±7.0	19.6±2.6	18.2±2.6	1.3±1.2	1.0±1.9	6.4±0.6	46.3±2.2
6. Forest/High fire/Low BA	35.1±11.3	7.3±4.2	7.0±4.3	0.4±2.0	1.5±3.1	5.7±1.0	37.9±3.6
Reference	44.4±6.1	15.3±2.4	14.0±2.3	1.3±1.1	1.0±1.7	6.0±0.5	46.6±2.3

Values are mean ±95% confidence interval.
[1]Soil moisture holding capacity.
[2]Basal area.

fire frequency was not selected as a model component, prescribed fire is highly regimented, resulting in frequent fires and little variation in time since fire among sites (Figure 3). Conversely, substantial variation in fire frequency exists among sites at the Savannah River Site (Table 1), where fire frequency was selected as a model component (Figure 3). Furthermore, our study spanned three physiographic regions of the southeastern U.S. coastal plain [26], capturing variation in important ecological factors such as dominant species (Tables S2, S3, and S4) and soil conditions (Table 1) [26,33]. Longleaf pine communities are broadly structured by soils [26] and this influence of soils was most prominent at Fort Stewart – a location that supports longleaf pine woodlands underlain by a variety of soil orders, including Spodosols and Inceptisols, which formed a unique class in the Fort Stewart model. Fort Stewart also illustrated the lone example of how the influence of a degrading factor depended on soil conditions, as non-*Pinus* basal area was important to plant communities on Entisol and Ultisol soils, but not Inceptisol and Spodosol soils. Importantly, at least one hypothesized degrading factors (Figure 1) was a significant model component at each location, illustrating that these factors structured understory degradation (i.e., departure from reference conditions) at both regional and landscape scales in this study. The differences in reference model details between our three study locations (Figure 3) suggests that, while our regional reference model may provide broad-scale insight into patterns of longleaf pine degradation, locations with available resources should consider collecting data to parameterize location-specific reference models.

Implications for Conservation and Restoration

We suggest that our regional ecological reference model provides a way for managers to broadly infer the degradation status of longleaf pine understory communities in our study region. This model, as well as the location-specific models, employs relatively easily measured data, which serve as proxies for ecological characteristics of interest to land managers (Figure 4, S1, S2, and S3). Agricultural and fire history and, perhaps, overstory basal area, may be available to land managers as GIS data. Thus, our regional and location-specific reference models may be mapped at large spatial scales to assist management and conservation decisions. Longleaf pine understory communities are notable for high levels of species diversity from local (e.g., 1 m×1 m) to regional scales [13,26]. Our findings provide guidance over much of this range in scales, spanning sites, landscapes, and portions of a region (the Atlantic coastal plain). Future work might explore additional reference models to inform small-scale (e.g., within site) restoration decisions.

Our approach to classifying communities based on degraded conditions should be broadly applicable to other ecosystems around the world modified by human land use and altered fire regimes, including fragmented woodlands in Australia [45], fire-suppressed savannas in Brazil [47], and forests in the western United States [6,48], among others. During such application, our framework for developing reference models should be modified to include the relevant suspected drivers of degradation for an ecosystem of interest. For example, the presence of invasive species is an important consideration during longleaf pine woodland restoration [8] and, while invasive species were not abundant at our study sites, their inclusion as a model variable might be important at other longleaf pine sites or in other ecosystems.

The remaining challenge is to determine how to best restore longleaf pine understory communities once patterns of degradation have been assessed. Our regional and location-specific ecological reference models suggest some strategies for restoration, but these should be interpreted with caution. For example, transitioning sites between Classes 4/5 and 6 in our regional model might simply entail mechanical thinning of overstory trees, a strategy that can increase plant diversity in some contexts [49] but has limited effects in others [20,50]. Other transitions might require multiple restoration strategies. For example, restoration of fire regime at post-agricultural sites (e.g., transitioning between Classes 2 and 3 at Savannah River Site) might need to be coupled with introduction of propagules – particularly those of dispersal-limited plant groups, such as passive and ant dispersed species like *Tephrosia virginiana* and *Aristida stricta/beyrichiana* [51] (Tables S1, S2, S3, and S4). Future experimental work will be necessary to evaluate these hypotheses suggested by our models. Further, reinstating processes may lead to unexpected outcomes for some degraded states. For example, reintroducing fire to long-unburned sites may produce novel fire behavior, leading to unexpected outcomes, such as mortality of longleaf pine overstory trees [52]. Moreover, while our conceptual reference model (Figure 1) depicts a simplistic set of linear transitions among degraded states, it remains an open question as to whether a simple linear or a non-linear, such as alternative stable states [53] or state and transition (e.g. [54]), model of restoration will be most appropriate during restoration of longleaf pine woodlands. Finally, more work is needed to understand how to best tailor combinations and sequences of restoration strategies (e.g., seed addition, prescribed fire, overstory thinning) to the variety of degraded conditions illustrated by our reference models for longleaf pine woodlands.

Supporting Information

Figure S1 Comparison of Classes (1–2) from the Fort Bragg classification and regression tree analyses to reference conditions. A) understory community composition, B) understory species richness, C) understory species evenness, D) soil organic matter content, E) ground cover variables, and F) forest floor depth. All values are means ±95% confidence intervals.

Figure S2 Comparison of Classes (1–4) from the Fort Stewart classification and regression tree analyses to reference conditions. A) understory community composition, B) understory species richness, C) understory species evenness, D) soil organic matter content, E) ground cover variables, and F) forest floor depth. All values are means ±95% confidence intervals.

Figure S3 Comparison of Classes (1–4) from the Savannah River Site classification and regression tree analyses to reference conditions. A) understory community composition, B) understory species richness, C) understory species evenness, D) soil organic matter content, E) ground cover variables, and F) forest floor depth. All values are means ±95% confidence intervals.

Table S1 Species with the 10 highest indicator values (from Indicator Species Analysis) for each site class in the All Sites classification. Species identified as indicators of individual site classes ($p<0.05$) are noted by *. P-values are listed for significant indicator species; where not provided p-values were ≥0.05. The number of significant indicator species decreased strongly with degradation; Classes 1–6 contained 2, 18, 0, 3, 0, 40 significant indicator species, respectively. Thirteen species were indicative of reference sites.

Table S2 Species with the 10 highest indicator values (from Indicator Species Analysis) for each site class in the Fort Bragg classification. Species identified as indicators of individual site classes are noted by *. Indicator species were primarily present at reference sites (n = 59), with 0 and 4 in Classes 1 and 2, respectively.

Table S3 Species with the 10 highest indicator values (from Indicator Species Analysis) for each site class in the Fort Stewart classification. Species identified as indicators of individual site classes are noted by *. 7, 1, 9, and 10 species were significant indicators of classes 1–4, respectively, and 19 species were indicative of reference sites.

Table S4 Species with the 10 highest indicator values (from Indicator Species Analysis) for each site class in the Savannah River Site classification. Species identified as indicators of individual site classes are noted by *. There were 0, 3, 14, and 4 significant indicators of classes 1–4, respectively, and 26 species indicative of reference sites.

Table S5 Attributes of classes resulting from the location-specific classifications, compared to references sites. Values are mean ±95% confidence interval.

Acknowledgments

For logistical support at field locations, we thank David Heins and the Endangered Species Branch at Fort Bragg; the Fish and Wildlife Branch at Fort Stewart; and John Blake, Chris Hobson, and Ed Olson at SRS. For assistance with data collection, we thank Lauren Bizzari, Cory Christopher, Amanda Powell, and Rick Ranalli.

Author Contributions

Conceived and designed the experiments: EID JLO JLW LAB. Performed the experiments: CDC EID JLO LAB WBM. Analyzed the data: JLO LAB. Contributed reagents/materials/analysis tools: EID JLO LAB. Wrote the paper: CDC EID JLO JLW JWV LAB PGH WBM.

References

1. White PS, Walker JL (1997) Approximating nature's variation: Selecting and using reference information in restoration ecology. Restor Ecol 5: 338–349.
2. Rey Benayas JM, Newton AC, Diaz A, Bullock JM (2009) Enhancement of biodiversity and ecosystem services by ecological restoration: a meta-analysis. Science 325: 1121–1124.
3. Foster D, Swanson F, Aber H, Burke I, Brokaw N, et al. (2003) The importance of land-use legacies to ecology and conservation. Bioscience 53: 77–88.
4. Hobbs RJ (2007) Setting effective and realistic restoration goals: Key directions for research. Restor Ecol 15: 354–357.
5. Hobbs RJ, Harris JA (2001) Restoration ecology: repairing the Earth's ecosystems in the new millennium. Restor Ecol 9: 239–246.
6. Allen CD, Savage M, Falk DA, Suckling KF, Swetnam TW, et al. (2002) Ecological restoration of Southwestern ponderosa pine ecosystems: A broad perspective. Ecol Appl 12: 1418–1433.
7. Asbjornsen H, Brudvig LA, Mabry CM, Evans CW, Karnitz HM (2005) Defining reference information for restoring ecologically rare tallgrass oak savannas in the Midwestern United States. J For 103: 345–350.

8. Walker JL, Silletti AM (2006). Restoring the ground layer of longleaf pine ecosystems. In: Jose S, Jokela EJ, Miller DL, editors. The longleaf pine ecosystem: Ecology, silviculture, and restoration. New York: Springer. 297–325.
9. Bestelmeyer BT, Miller JR, Wiens JA (2003) Applying species diversity theory to landscape management. Ecol Appl 13: 1750–1761.
10. Brudvig LA (2011) The restoration of biodiversity: Where has research been and where does it need to go? Am J Bot 98: 549–558.
11. Palik BJ, Goebel PC, Kirkman LK, West L (2000). Using landscape hierarchies to guide restoration of disturbed ecosystems. Ecol Appl 10: 189–202.
12. Turner MG, Gardner RH, O'Neill RV (2001) Landscape ecology in theory and practice: pattern and process. New York: Springer-Verlang. 401 p.
13. Mitchell RJ, Duncan SL (2009) Range of variability in southern coastal plain forests: Its historical, contemporary, and future role in sustaining biodiversity. Ecol Soc 14(1): 17.
14. Hedman CW, Grace SL, King SE (2000) Vegetation composition and structure of southern coastal plain pine forests: an ecological comparison. For Ecol Manage 134: 233–247.
15. Brudvig LA, Damschen EI (2011) Land-use history, historical connectivity, and land management interact to determine longleaf pine woodland understory richness and composition. Ecography 34: 257–266.
16. Brudvig LA, Grman E, Habeck CW, Orrock JL, Ledvina JA (2013) Strong legacy of agricultural land use on soils and understory plant communities in longleaf pine woodlands. For Ecol Manage 310: 944–955.
17. Brockway DG, Lewis CE (1997) Long-term effects of dormant season prescribed fire on plant community diversity, structure and productivity in a longleaf pine wiregrass ecosystem. For Ecol Manage 96: 167–183.
18. Harrington TB, Edwards MB (1999) Understory vegetation, resource availability, and litterfall responses to pine thinning and woody vegetation control in longleaf pine plantations. Can J For Res 29: 1055–1064.
19. Hiers JK, O'Brien JJ, Will RE, Mitchell RJ (2007) Forest floor depth mediates understory vigor in xeric *Pinus palustris* ecosystems. Ecol Appl 17: 806–814.
20. Kirkman LK, Barnett A, Williams BW, Hiers JK, Pokswinski SM, et al. (2013) A dynamic reference model: a framework for assessing biodiversity restoration goals in a fire-dependent ecosystem. Ecol Appl 23: 1574–1587.
21. Frost C (2006) History and future of the longleaf pine ecosystem. In: Jose S, Jokela EJ, Miller DL, editors. The longleaf pine ecosystem: Ecology, silviculture, and restoration. New York: Springer. 9–42.
22. Harrington TB, Dagley CM, Edwards MB (2003) Above- and belowground competition from longleaf pine plantations limits performance of reintroduced herbaceous species. Forest Sci 49: 681–695.
23. Van Lear DH, Carroll WD, Kapeluck PR, Johnson R (2005) History and restoration of the longleaf pine-grassland ecosystem: Implications for species at risk. For Ecol Manage 211: 150–165.
24. Flinn KM, Vellend M (2005) Recover of forest plant communities in agricultural landscapes. Front Ecol Environ 3: 243–250.
25. Pyke DA, Brooks ML, D'Antonio C (2010) Fire as a restoration tool: A decision framework for predicting the control or enhancement of plants using fire. Restor Ecol 18: 274–284.
26. Peet RK (2006) Classification of longleaf pine woodlands. In: Jose S, Jokela EJ, Miller DL, editors. The longleaf pine ecosystem: Ecology, silviculture, and restoration. New York: Springer. 51–93.
27. Peet RK, Wentworth TR, White PS (1998) A flexible, multipurpose method for recording vegetation composition and structure. Castanea 63: 262–274.
28. Radford AE, Ahles HE, Bell CR (1964) Manual of the vascular flora of the Carolinas. Chapel Hill: University of North Carolina Press. 1183 p.
29. Weakley AS (2008) Flora of the Carolinas, Virginia, and Georgia, and surrounding areas. Chapel Hill: University of North Carolina Press. 924 p.
30. Wunderlin RP, Hansen BF (2003) Guide to the vascular plants of Florida, Second edition. Gainesville: University Press of Florida. 787 p.
31. Schulte EE, Hopkins BG (1996) Estimation of soil organic matter by weight Loss-On-Ignition. In: Magdoff FR, Tabatabai MA, Hanlon Jr EA, editors. Soil organic matter: Analysis and interpretation. Special publication No. 46. Madison: Soil Science Socity of America. 21–32.
32. Salter PJ, Williams JB (1967) The influence of texture on the moisture characteristics of soil. IV. A method of estimating the available-water capacities of profiles in the field. J Soil Sci 18: 174–181.
33. Kirkman LK, Mitchell RJ, Helton RC, Drew MD (2001) Productivity and species richness across and environmental gradient in a fire-dependent ecosystem. Am J Bot 88: 2119–2128.
34. De'ath G (2002) Multivariate regression trees: A new technique for modeling species-environment. Ecology 83: 1105–1117.
35. Glitzenstein JS, Platt WJ, Streng DR (1995) Effects of fire regime and habitat on tree dynamics in north Florida longleaf pine savannas. Ecol Monogr 65: 441–476.
36. Veldman JW, Mattingly WB, Brudvig LA (2013) Understory plant communities and the functional distinction between savanna trees, forest trees, and pines. Ecology 94: 424–434.
37. De'ath G, Fabricius KE (2000) Classification and regression trees: a powerful yet simple technique for ecological data analysis. Ecology 81: 3178–3192.
38. Anderson MJ, Willis TJ (2003) Canonical analysis of principal coordinates: a useful method of constrained ordination for ecology. Ecology 84: 511–525.
39. Dufrene M, Legendre P (1997) Species assemblages and indicator species: the need for a flexible asymmetrical approach. Ecol Monogr 67: 345–366.
40. Clark KR, Gorley RN (2006) PRIMER v6: User Manual/Tutorial. Plymouth: PRIMER-E.
41. SAS Institute (2008) SAS version 9.2. Cary: SAS Institute.
42. R Development Core Team (2005) R: A Language and Environment for Statistical Computing, version 3.0. Vienna: R Foundation for Statistical Computing.
43. McCune B, Mefford MJ (2006) PC-ORD: Multivariate Analysis of Ecological Data, version 5.31. Gleneden Beach: MjM Software.
44. Honnay O, Bossuyt B, Verheyen K, Butaye J, Jacquemyn H, et al. (2002) Ecological perspectives for the restoration of plant communities in European temperate forests. Biodivers Conserv 11: 213–242.
45. Lunt ID, Spooner PG (2005) Using historical ecology to understand patterns of biodiversity in fragmented agricultural landscapes. J Biogeogr 32: 1859–1873.
46. Fraterrigo JM, Turner MG, Pearson SM (2006) Previous land use alters plant allocation and growth in forest herbs. J Ecol 94: 548–557.
47. Roitman I, Felfili JM, Rezende AV (2008) Tree dynamics of a fire-protected cerrado sensu stricto surrounded by forest plantations, over a 13-year period (1991–2004) in Bahia, Brazil. Plant Ecol 197: 255–267.
48. Fulé PZ, Covington WW, Moore MM (1997) Determining reference conditions for ecosystem management of southwestern ponderosa pine forests. Ecol Appl 7: 895–908.
49. Platt WJ, Carr SM, Reilly M, Fahr J (2006) Pine savanna overstorey influences on ground-cover biodiversity. Appl Veg Sci 9: 37–50.
50. Provencher L, Herring BJ, Gordon DR, Rodgers HL, Galley KEM, et al. (2001) Effects of Hardwood Reduction Techniques on Longleaf Pine Sandhill Vegetation in Northwest Florida. Restor Ecol 9: 13–27.
51. Kirkman LK, Coffey KL, Mitchell RJ, Moser EB (2004). Ground cover recovery patterns and life-history traits: implications for restoration obstacles and opportunities in a species-rich savanna. J Ecol 92: 409–421.
52. Varner III JM, Gordon DR, Putz FE, Hiers JK (2005) Restoring fire to long-unburned *Pinus palustris* ecosystems: Novel fire effects and consequences for long-unburned ecosystems. Restor Ecol 13: 536–544.
53. Suding KN, Gross KL, Houseman GR (2004) Alternative stable states and positive feedbacks in restoration ecology. Trends Ecol Evol 19: 46–53.
54. Yates CJ, Hobbs RJ (1997) Woodland restoration in the Western Australian wheatbelt: A conceptual framework using a state and transition model. Restor Ecol 5: 28–35.

Effects of Soil Moisture on the Temperature Sensitivity of Soil Heterotrophic Respiration: A Laboratory Incubation Study

Weiping Zhou[1,2], Dafeng Hui[3], Weijun Shen[1]*

1 Key laboratory of Vegetation Restoration and Management of Degraded Ecosystems, South China Botanical Garden, Chinese Academy of Sciences, Guangzhou, China, 2 Graduate University of the Chinese Academy of Sciences, Beijing, PR China, 3 Department of Biological Sciences, Tennessee State University, Nashville, Tennessee, United States of America

Abstract

The temperature sensitivity (Q_{10}) of soil heterotrophic respiration (R_h) is an important ecological model parameter and may vary with temperature and moisture. While Q_{10} generally decreases with increasing temperature, the moisture effects on Q_{10} have been controversial. To address this, we conducted a 90-day laboratory incubation experiment using a subtropical forest soil with a full factorial combination of five moisture levels (20%, 40%, 60%, 80%, and 100% water holding capacity - WHC) and five temperature levels (10, 17, 24, 31, and 38°C). Under each moisture treatment, R_h was measured several times for each temperature treatment to derive Q_{10} based on the exponential relationships between R_h and temperature. Microbial biomass carbon (MBC), microbial community structure and soil nutrients were also measured several times to detect their potential contributions to the moisture-induced Q_{10} variation. We found that Q_{10} was significantly lower at lower moisture levels (60%, 40% and 20% WHC) than at higher moisture level (80% WHC) during the early stage of the incubation, but became significantly higher at 20%WHC than at 60% WHC and not significantly different from the other three moisture levels during the late stage of incubation. In contrast, soil R_h had the highest value at 60% WHC and the lowest at 20% WHC throughout the whole incubation period. Variations of Q_{10} were significantly associated with MBC during the early stages of incubation, but with the fungi-to-bacteria ratio during the later stages, suggesting that changes in microbial biomass and community structure are related to the moisture-induced Q_{10} changes. This study implies that global warming's impacts on soil CO_2 emission may depend upon soil moisture conditions. With the same temperature rise, wetter soils may emit more CO_2 into the atmosphere via heterotrophic respiration.

Editor: Shuijin Hu, North Carolina State University, United States of America

Funding: This study was financially supported by the Natural Science Foundation of China (31130011 and 31290222), and the Natural Science Foundation of Guangdong Province, China (S2012020011084). The funders had no role in study design, data collection and analysis, decision to publish, or preparation of the manuscript.

Competing Interests: Co-author DH is an academic editor of PLOS ONE. This does not alter our adherence to all the PLOS ONE policies on sharing data and materials.

* E-mail: shenweij@scbg.ac.cn

Introduction

Temperature sensitivity of soil respiration, usually termed as Q_{10}, is defined as the increase of soil respiration rate by a 10°C rise in temperature [1]. Q_{10} has been considered an important model parameter in predicting terrestrial ecosystem carbon cycle and feedback to climate warming [2]. In the past several decades, Q_{10} has been investigated extensively, particularly through field-observed soil respiration and environmental factor data [3,4]. It has been found that Q_{10} is not a constant of 2, but varies with vegetation and edaphic conditions such as temperature, moisture, and substrate availability [2]. As global temperature continues to rise [5], it is of paramount importance to understand how Q_{10} is influenced by these factors individually and interactively. Since under field conditions, effects of soil temperature and moisture on Q_{10} are often confounded with each other and with other factors, laboratory incubation has the advantage of deriving the primary and interactive effects of the environmental factors on Q_{10}.

Many studies have demonstrated that Q_{10} can be influenced by a variety of biological and environmental factors [1,6,7]. Soil temperature itself has been found to have a negative correlation with Q_{10}. For example, at lower temperature regions (e.g., tundra), Q_{10} tends to be higher than the estimates at warmer temperature regions (e.g., warm desert) [8]. A manipulated warming experiment also demonstrates that Q_{10} is significantly lower at high temperature treatments than at the low temperature control [1]. Thus, the temperature effects on Q_{10} have been generally consistent; i.e., Q_{10} decreases with increasing temperature. However, the effects of other factors such as soil moisture on Q_{10} have been less certain and deserve more research.

Soil moisture plays a critical role in soil respiration and may have a significant impact on Q_{10} [9–11]. The basic concepts and mechanisms of soil moisture on soil respiration have been discussed by many researchers [12–14]. The optimum soil moisture for soil respiration is frequently found at intermediate levels, above or below which soil respiration decreases [15]. At the optimum soil moisture, the macropore spaces are filled with adequate amounts of air and water which can facilitate the diffusion of both oxygen and soluble substrates [16]. In very wet soils oxygen limitation occurs, and in very dry soils the movement of soluble substrates via water films is restricted. Although the mechanistic understanding on the effects of soil moisture on R_h has been largely advanced, its influence on the Q_{10} of R_h is still inconclusive. For example, Wang et al. [17] reported that Q_{10} increased with soil moisture until reaching a threshold, and then declined in six temperate forests of China. Carlyle and Than [18] showed that soil moisture limited the Q_{10} of soil respiration beneath a *pinus radiata* stand in south-eastern Australia. But Reichstein et al. [19] found that Q_{10} was insensitive to the drying of a spruce forest soil. The inconsistency of soil moisture effects on Q_{10} is probably due to the confounding influences of different environmental factors under field conditions. One recent incubation study showed that soil moisture indeed influenced Q_{10} and the moisture-Q_{10} relationship differed between soils obtained at different topographic positions [20], but the underlying mechanisms remained unclear.

Effects of soil moisture on Q_{10} may be ascribed to changes in microbial biomass and community structure, and the physical and chemical properties of the soil [7,21]. Changes in soil moisture can affect the composition and function of soil microbial community due to differences in drought tolerance among taxonomic and functional groups of microorganisms [22]. For example, fungi can survive drought stress better than bacteria due to their ability to grow at lower matric potentials [23,24]. Soil moisture can also affect the quantity of soil microbial biomass carbon (MBC) and dissolved organic carbon (DOC) [25,26]. Despite a general understanding of the above processes, whether soil moisture effects on Q_{10} can be related to its influences on soil properties such as MBC, DOC, nutrient availability, and microbial community structure is still in active debate.

In this study, we investigated soil moisture effects on Q_{10} by incubating a subtropical forest soil under five temperature levels and five moisture levels over 90 days. Soil R_h and other properties such as MBC and DOC, nitrogen and phosphorous contents, and microbial community phospholipid fatty acids (PLFAs) were measured several times during the incubation period. Our objectives were first to analyze how changes in soil moisture influenced Q_{10}, and second to explore whether the moisture effects on Q_{10} could be related to its impacts on the soil microbial and chemical properties measured.

Materials and Methods

Ethics Statement

Soils were sampled from a study site that is maintained by the South China Botanical Garden, Chinese Academy of Sciences. All necessary permits were obtained for the described study. This study did not involve endangered or protected species.

Site Description

Incubation soils used in this study were collected from an evergreen broadleaved forest stand at the Heshan Hilly Land Interdisciplinary Experimental Station (22°34′N, 112°50′E) in Guangdong Province of China. The region has a subtropical humid monsoon climate with apparent dry and wet seasons. The wet season starts in April and ends in October, and the dry season begins in November and lasts through March of the following year. The mean annual precipitation and temperature are 1700 mm and 21.7°C, respectively. The forest stand is 29 years old and mainly dominated by native tree species (*Schima superba* and *Michelia macclurei*) with an average height of 15 m and an average diameter at breast height (DBH) of 30 cm. The soil is categorized as Oxisols based on the US Soil Classification System [27,28], with a bulk density of 1.4 g cm^{-3}, total organic carbon (TOC) of 2.80%, total nitrogen (TN) of 0.15%, and total phosphorous (TP) of 0.02% at the depth of 0–20 cm.

Incubation Experimental Design

In the field, four sampling areas, with a distance of at least 10 m between each, were selected to collect the incubation soils. In each area, five sampling sites (20×10 cm^2) were randomly selected and sampled to the depth of 20 cm. These five random samples were homogenized to form a composite sample. Before sampling, the uppermost layer of litter with visible un-decomposed materials was excluded. We had four composite samples as four experimental replicates, each one weighing about 50 kg in fresh weight. All soil samples were transported to the laboratory and passed through a 2 mm sieve with apparent plant roots and stones being removed.

To investigate soil moisture effects on the temperature sensitivity (Q_{10}) of soil R_h, we used five soil moisture levels: 20%, 40%, 60%, 80%, and 100% water holding capacity (WHC). For each moisture level, soils were incubated under five temperature levels: 10, 17, 24, 31, and 38°C. A full factorial combination of the two factors and five levels for each factor produced 25 experimental treatments. Each treatment had four replicates from the four composite samples. Each replicate further had 6 duplicates, with one duplicate for measuring R_h and the other 5 for destructive sampling. As a result, we had 600 incubation soil samples in total with 120 (= 5 moisture levels × 4 replicates × 6 duplicates) in each of the five static temperature incubators (RXZ-600B, Southeast Instrument Co., Ltd., Ningbo, China). The temperature and relative humidity deviations of the incubators are ±1.5°C and ±7%, respectively. Each air-dried incubation soil sample (equivalent to 50 g of oven-dried soil) was added to each triangle flask and its soil water content was adjusted to the corresponding soil moisture level by adding deionized water. The flasks were covered by rubber stoppers with small holes to reduce water loss via evaporation and maintain gas exchange. In order to maintain constant soil moisture levels, water loss was checked and corrected weekly by weighing each flask and adding water as necessary. At most, 1 ml of water (equivalent to 3.3% of changes in 100% WHC) was added every week to the flask at the temperature level of 38°C.

Measurements of Soil R_h, Microbial and Chemical Properties

The incubation experiment lasted for 90 days. Soil R_h rates were measured using the Li-6262 Infrared Gas Analyzer (Li-Cor Inc., Lincoln, NE) on days 1, 2, 3, 4, 6, 7, 13, 18, 27, 34, 41, 53, 62, 74, and 90. Electrical fans blowing air into each incubator for 30 min every four days were used to maintain an aerobic incubation environment. Before R_h measuring, each triangle flask was ventilated for 3 minutes to minimize gas accumulation in the headspace. After ventilation, another type of rubber stoppers with two plastic tubes for gas inlet and outlet was used to seal the flask and the tubes were connected to Li-6262 for measuring headspace CO_2 concentration. The CO_2 concentration in the headspace was recorded every second for 2 minutes and R_h rate was calculated using the linear portion of the response curve of CO_2 concentration versus time [29]. At each moisture level and each measurement day, Q_{10} was calculated by fitting an exponential function to the measured R_h against the 5 temperature levels:

$$R_h = ae^{bT} \qquad (1)$$

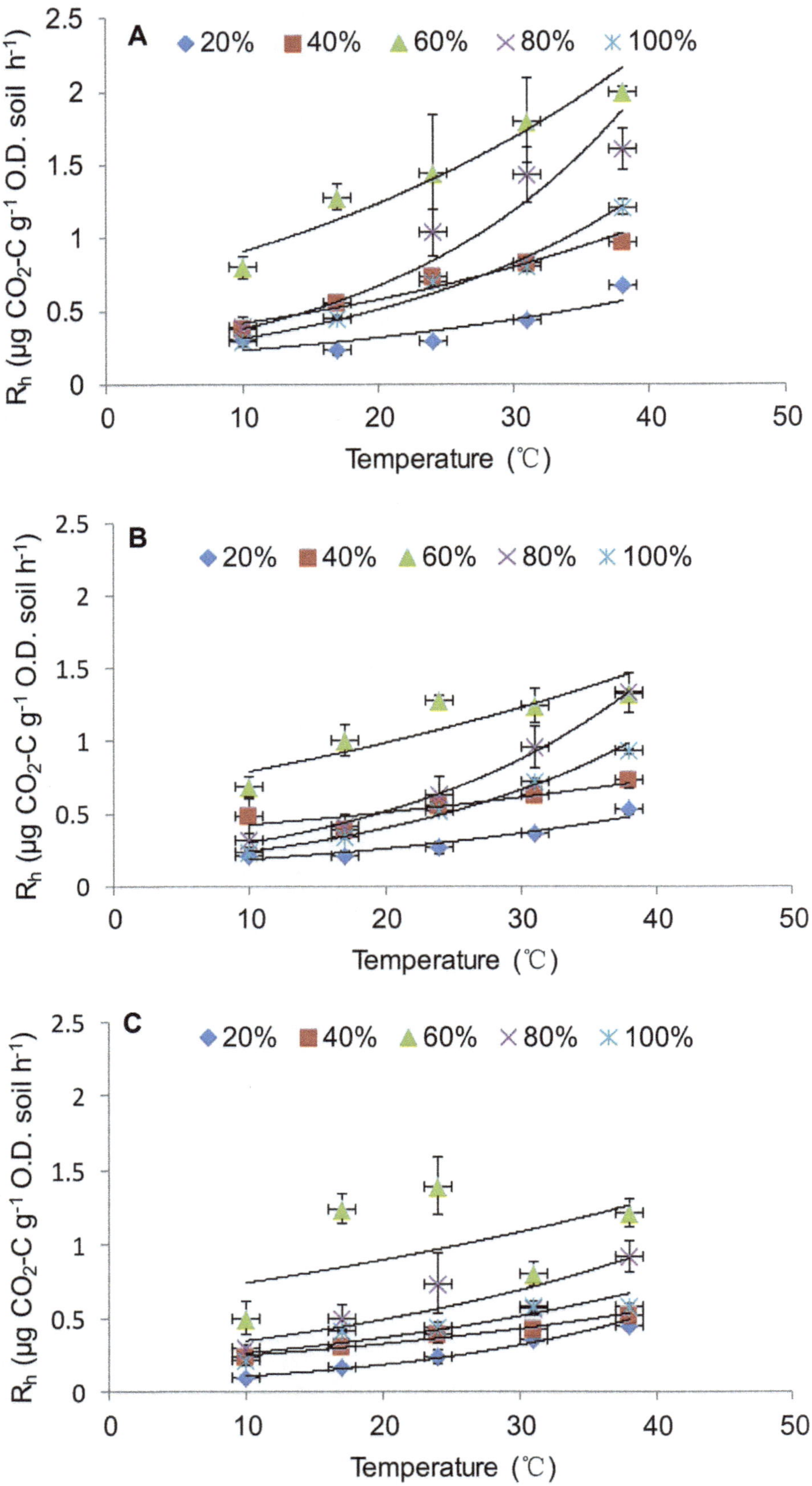

Figure 1. Responses of soil heterotrophic respiration (R_h) to changes in soil temperature after 7 (A), 30 (B), and 90 (C) days of incubation. Each data point is the mean of four replicates under each soil moisture treatment. Error bars represent standard errors (n = 4).

Table 1. Repeated-measures ANOVA for the temperature and time effects on R_h under different soil moisture treatments.

Moisture (^{a} %WHC)	Temperature (°C)		Time (day)		Temperature×Time	
	$F_{(4, 15)}$	P	$F_{(14, 210)}$	P	$F_{(56, 210)}$	P
20%	36.96	<0.01	26.59	<0.01	5.46	<0.01
40%	35.77	<0.01	50.03	<0.01	3.88	<0.01
60%	10.90	<0.01	45.26	<0.01	6.63	<0.01
80%	12.24	<0.01	52.44	<0.01	6.47	<0.01
100%	139.15	<0.01	159.63	<0.01	17.60	<0.01

[a]%WHC: percent of water holding capacity.

where R_h is the measured soil heterotrophic respiration rate (μg g^{-1} oven dried soil h^{-1}), T is the incubation temperature (°C), a and b are the fitted model parameters. Q_{10} is calculated individually for each of the four replicates using the following equation:

$$Q_{10} = e^{10b} \quad (2)$$

As intrinsic Q_{10} is based on the kinetic reactions of molecular structure changes with temperature [2], the Q_{10} of this study is the apparent temperature sensitivity.

For measuring MBC, DOC, total organic carbon (TOC), total nitrogen (TN), total phosphorous (TP), inorganic N (NH_4^+ and NO_3^-), and PLFAs, three flasks (or replicates) of each treatment were harvested on days 7, 30, and 90. Part of the soil in each flask was collected and stored at −20°C for later analysis of PLFAs. The remaining soil was used to analyze microbial biomass and chemical properties. It is noted here that TOC, TN, and TP were only analyzed for the samples harvested on days 7 and 90.

Soil MBC was measured using the modified fumigation-extraction method [30]. MBC was calculated as the difference in extractable C concentrations between the fumigated and un-fumigated samples divided by a K_{EC} factor of 0.38 [30]. The extractable C concentrations of un-fumigated samples were the soil DOC [31]. Soil NH_4^+ and NO_3^- were measured by the method proposed in Dorich and Nelson [32]. Soil TOC was measured using the potassium bichromate-concentrated sulphuric acid heating method. Soil TN and TP were measured using the Kjeldahl resolution Auto Flow Injection method. Microbial community PLFAs were analyzed according to Bossio and Scow [33]. Concentrations of each PLFA were calculated relative to the 19:0 internal standard concentrations. 15:0, i15:0, a15:0, i16:0, 16:1w7c, 17:0, i17:0, a17:0, cy17:0, 18:1w7c, 19:0 cyclow8c were selected as bacterial biomarkers and 18:2w6, 9c were selected as fungal biomarkers [34,35].

Statistical Analysis

Repeated-measures ANOVA was employed to determine the effects of sampling time and soil moisture on Q_{10} and R_h on the 15 measuring days. One-way ANOVA was used to analyze soil moisture effects on Q_{10} for days 7, 30, and 90; Tukey's HSD multiple comparison method was used to test Q_{10} differences among soil moisture levels. Regression analysis was used to derive the relationships between R_h and temperature, and between Q_{10} (or R_h) and incubation time. Pearson correlation analysis was applied to detect the potential contributions of soil microbial and chemical properties on the Q_{10} variations with moisture. All these statistical analyses were performed using SPSS software 16.0 (SPSS Inc., Chicago, IL).

Table 2. Regression equations of heterotrophic respiration (R_h) with temperature under different moisture treatments.

Soil moisture (^{a} %WHC)	Day 7	Day 30	Day 90
20	^{b} $R_h = 0.1684e^{0.0318T}$	$R_h = 0.1333e^{0.0335T}$	$R_h = 0.062e^{0.0546T}$
	^{c} $R^2 = 0.72$*	$R^2 = 0.92$**	$R^2 = 0.98$**
40	$R_h = 0.3014e^{0.0324T}$	$R_h = 0.3526e^{0.0183T}$	$R_h = 0.1896e^{0.0273T}$
	$R^2 = 0.95$**	$R^2 = 0.72$*	$R^2 = 0.97$**
60	$R_h = 0.6639e^{0.031T}$	$R_h = 0.6382e^{0.0218T}$	$R_h = 0.6148e^{0.019T}$
	$R^2 = 0.92$**	$R^2 = 0.78$*	$R^2 = 0.25$
80	$R_h = 0.2136e^{0.057T}$	$R_h = 0.1774e^{0.0532T}$	$R_h = 0.2484e^{0.0342T}$
	$R^2 = 0.92$**	$R^2 = 0.99$**	$R^2 = 0.77$*
100	$R_h = 0.1926e^{0.0485T}$	$R_h = 0.1469e^{0.0501T}$	$R_h = 0.19e^{0.0333T}$
	$R^2 = 0.98$**	$R^2 = 0.99$**	$R^2 = 0.81$*

[a]%WHC: percent of water holding capacity.
[b]R_h represents soil heterotrophic respiration rate and T represents temperature.
[c]R^2 is the coefficient of determination; * and ** indicate significance at P≤0.05 and P≤0.01, respectively.

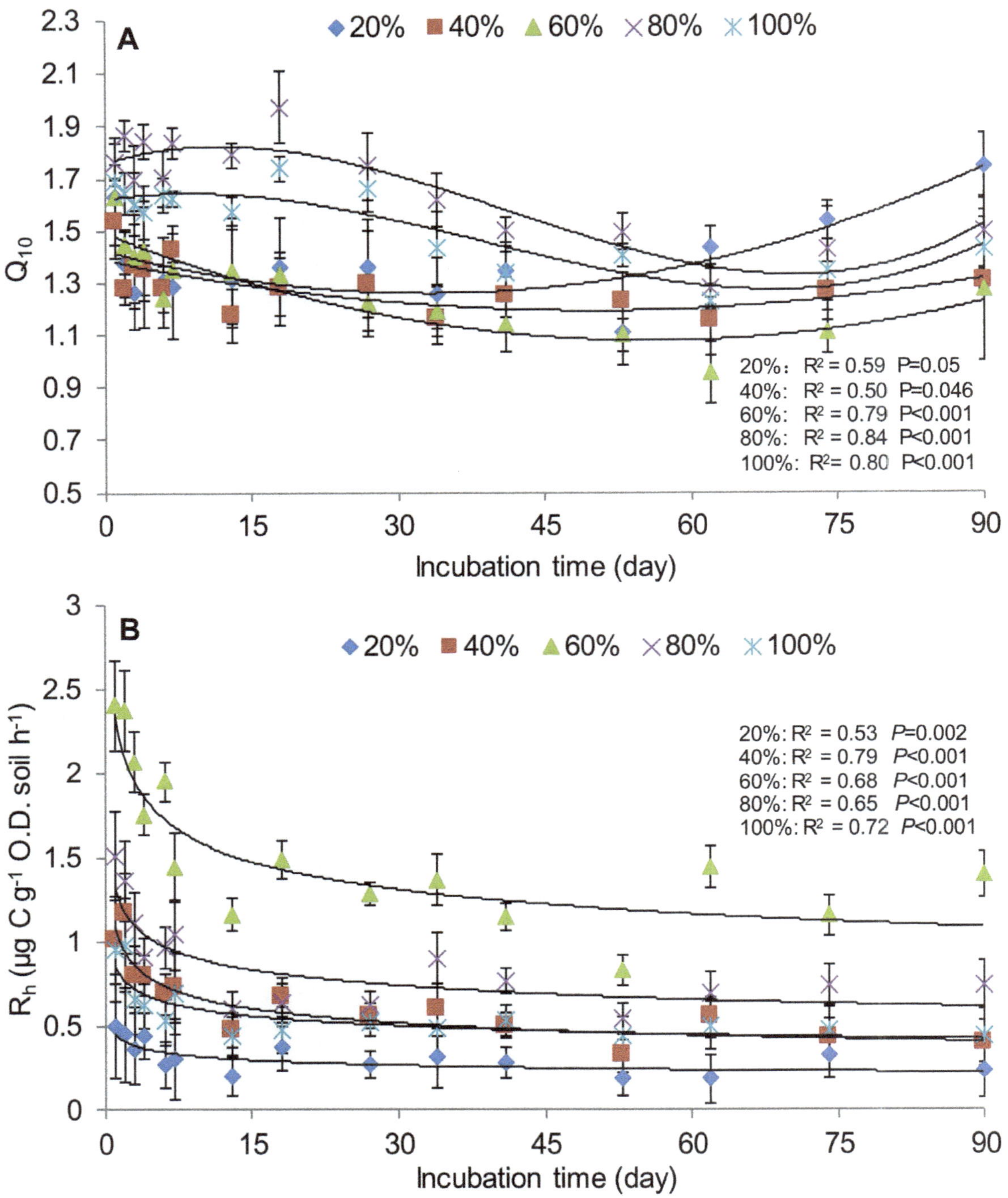

Figure 2. Changes of Q_{10} and R_h with incubation time at different soil moisture treatments. The Q_{10} regression functions are quadratic for 20%, 40% and 60% WHC and cubic for 80% and 100% WHC. Error bars (n = 4) represent standard deviations. R^2 is the coefficient of determination. P is the significance level.

Results

Relationships between Soil R_h and Temperature under Different Soil Moisture Treatments

Soil R_h varied significantly with temperature and incubation time under all the 5 moisture treatments (Table 1). R_h responses to temperature changes on days 7, 30, and 90 were displayed to represent the early, middle, and late incubation stages (Fig. 1). Throughout the whole incubation period, soil R_h was highest at 60% WHC, lowest at 20% WHC, and intermediate at the other moisture levels (Fig. 1). Among the three measurement days, soil R_h had the highest values on day 7, especially at 60% and 80% WHC, compared to those on days 30 and 90. Soil R_h declined with incubation time, declining in smaller magnitudes at 20% WHC and 100% WHC.

The temperature response of soil R_h could be well fitted using the exponential model for each soil moisture treatment and measurement day (Fig. 1). Model parameters for the three representative days are presented in Table 2. All models were significant with the coefficient of determination (R^2) ranging from

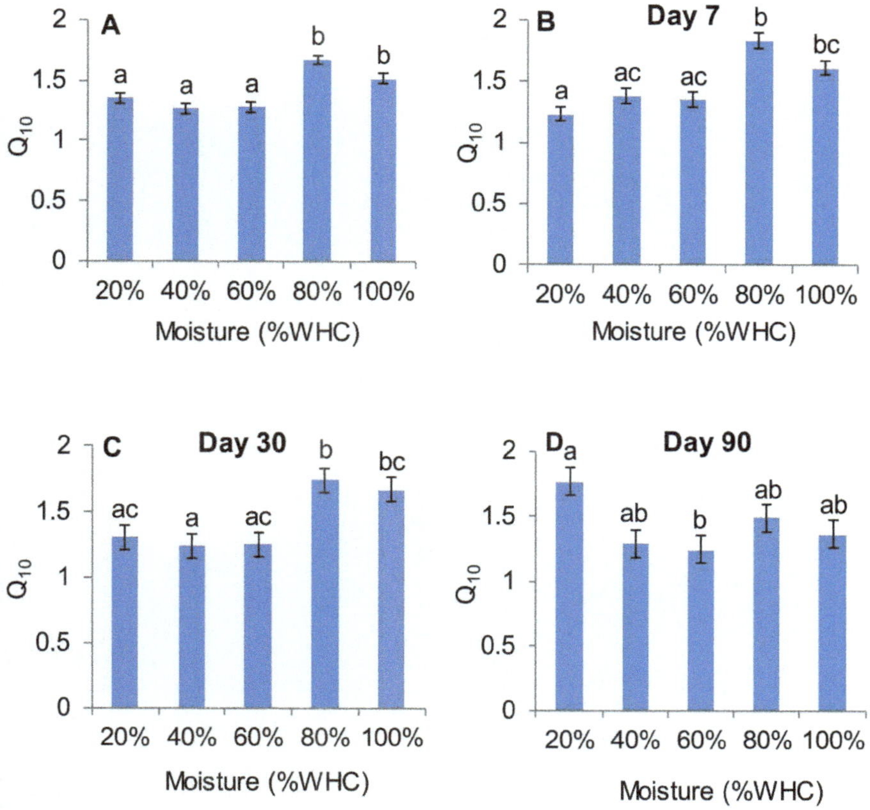

Figure 3. Comparisons of mean Q_{10} under different soil moisture treatments. The mean Q_{10} in panel (A) is calculated from 4 replicates across 15 measurement times (n = 60), and the mean Q_{10}s in panel (B), (C) and (D) are calculated from 4 replicates on day 7, day 30 and day 90 (n = 4), respectively. Error bars represent standard errors. Lower case letters represent significant difference at $P \leq 0.05$.

0.72 to 0.98, except for ($R^2 = 0.25$) at 60% WHC on day 90. Estimated basal soil respiration (i.e. parameter a) ranged from 0.06 to 0.66 and was higher at 60% WHC than at the other soil moisture treatments. The exponent (i.e. parameter b) ranged from 0.02 to 0.05 and was also lowest (0.019) at 60% WHC on day 90.

Changes of Q_{10} with Incubation Time under Different Soil Moisture Treatments

Under all the 5 moisture treatments, Q_{10} values ranged from 0.95 to 1.97 and varied markedly with incubation time. Q_{10} was higher in the beginning of the incubation, and declined with time to around day 60, then increased slightly to the end of the experiment (Fig. 2A). Among different soil moisture treatments, Q_{10} at 80% and 100% WHC was higher than at the other soil moisture treatments. Q_{10} at 60% WHC had the lowest value, especially after day 30. Q_{10} at 20% WHC was among the lowest in the beginning of the incubation, but increased with incubation time and had the highest values at the end of the experiment (Fig. 2A).

To better show the Q_{10} variation pattern with incubation time, polynomial regression models were used to fit Q_{10} with incubation time. Findings indicate that a quadratic regression model could fit Q_{10} well at 20%, 40% and 60% WHC with $R^2 \geq 0.50$ (Fig. 2A), while a cubic regression model should be applied at 80% and 100% WHC ($R^2 \geq 0.80$; Fig. 2A). Further inspection revealed the days on which the minimum and maximum Q_{10} appeared. At 20%, 40%, and 60% WHC, the lowest Q_{10} appeared on days 53, 62, and 62 with their values being 1.11, 1.15, and 0.95, respectively. At 80% and 100% WHC, Q_{10} showed the highest values of 1.94 and 1.74 on day 18, and the lowest values of 1.28 and 1.22 on day 62, respectively.

In contrast to the Q_{10} dynamics, R_h was always higher at 60% WHC, lower at 20% WHC and somewhere in between at the other three moisture levels (80%, 40%, and 100% WHC) (Fig. 2B). It is noted here that the R_h data shown in Fig. 2B are only those under the incubation temperature of 24°C, because under the other temperature treatments the general variation patterns of R_h with moisture and incubation time were similar. In our incubation experiment, the declination of R_h with incubation time could be best fitted using a power law decay function, with the coefficient of determination ranging from 0.53 to 0.78 (Fig. 2B).

Effects of Soil Moisture on Q_{10}

Repeated-measures ANOVA showed that Q_{10} was significantly influenced by soil moisture ($F_{(4, 15)} = 18.41$, $P<0.01$), incubation time ($F_{(14, 210)} = 15.41$, $P<0.01$) and the interaction of the two ($F_{(56, 210)} = 4.44$, $P<0.01$). Averaged over the 15 measurement times, Q_{10} was significantly lower at 60%, 40% and 20% WHC compared to those at 80% and 100% WHC (Fig. 3A). Q_{10} at 80% WHC had the highest value but was not significantly different from that at 100% WHC (Fig. 3A).

Since the interactive effect of soil moisture and incubation time was significant, we further compared Q_{10} values among soil moisture treatments on three typical measurement days (Fig. 3B–D). On days 7 and 30, Q_{10} at 80% WHC was not significantly different from that at 100% WHC but significantly higher than those at the three lower moisture levels, which is similar to the all-day average results shown in Fig. 3A. On day 90, Q_{10} at 20%

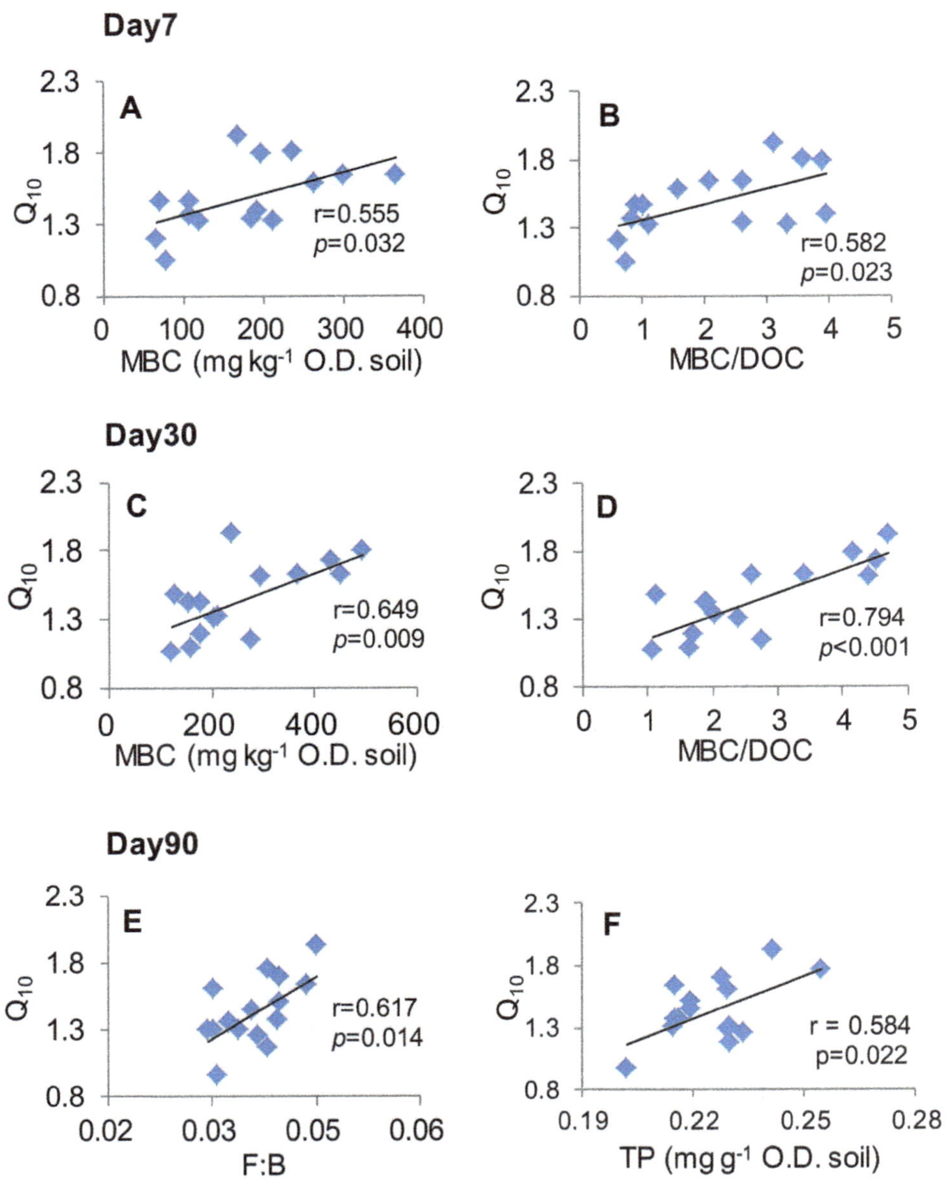

Figure 4. Correlations between Q_{10} and soil microbial and chemical properties. MBC: microbial biomass carbon, DOC: dissolved organic carbon, TP: total phosphorus, and F:B: ratio of fungi to bacteria. r is the correlation coefficient. P is the significance level.

WHC was significantly higher than at 60% WHC, but there was no significant difference among the other three soil moisture treatments.

Correlations between Q_{10} and Soil Properties

Pearson correlation analysis showed that on days 7 and 30, Q_{10} was positively correlated with MBC and the ratio of MBC to DOC (Fig. 4A–D). On day 90, no significant correlation was found between Q_{10} and MBC or MBC/DOC; instead, Q_{10} positively correlated with the F:B ratio (Fig. 4E) and TP (Fig. 4F). We did not find significant correlations between Q_{10} and other soil chemical properties (inorganic nitrogen, DOC, TOC, and TN), which was therefore not presented.

Discussion

Moisture Effects on R_h Responses to Temperature

Similar to many previous results [2,36,37], soil R_h increased with temperature exponentially in our study; however response curves varied among different soil moisture treatments (Fig. 1, Table 2). Some previous studies showed that soil respiration may be decoupled from temperature under certain soil moisture levels resulting in soil respiration that is unaffected by temperature under water stress [38]. For example, Yu et al. [39] found that temperature was the determinant factor and R_h increased with it exponentially only when soil moisture was not limited. However, our results showed that soil R_h could still increase with temperature, even at 20% WHC, though at a relatively slow rate (Table 2). The discrepancy between our results and theirs may be because the studied soils differ in chemical, physical and microbial properties.

Regardless of temperature variations, soil R_h at 60% WHC tended to be higher than at both lower and higher soil moisture treatments. This result is also consistent with some other studies that reported higher R_h at intermediate moisture content [40–42]. The decrease of soil R_h at lower soil moisture has been attributed to soluble substrate limitation, whereas at higher soil moisture level, especially at saturated soil moisture, R_h was mainly limited by oxygen [12,36]. The decrease of soil R_h over incubation time was probably caused by the depletion of labile substrate [43,44]. However, the persistently high R_h at 60% WHC during the 90-day incubation could be due to the persistently high microbial activity, because this was the optimal moisture level and high microbial activity might override the influence of substrate limitation.

Variations of Q_{10} with Incubation Time

The variation of Q_{10} with time has been found not to be uniform [45]. A laboratory incubation study found that Q_{10} increased with incubation time, which was ascribed to substrate quality change from labile to recalcitrant [46]. In some long-term warming experiments, Q_{10} was found to decline over time [1,45]. In our relatively short-term incubation experiment, we found that Q_{10} declined with incubation time initially but increased during later incubation stages, and quadratic or cubic regression models were fitted to quantify the changes of Q_{10} at different soil moisture treatments (Fig. 2A). Over the 90-day incubation period, mean Q_{10} values was mostly <2.0, which is lower than the conventional estimates (2.0–2.6) probably due to less confounding factors involved in our incubation experiment [47]. The changes of Q_{10} might be related to the changes of soil R_h, as many laboratory studies have shown that soil R_h decreases with incubation time [48]. The underlying mechanisms were ascribed to substrate depletion [43,44]: the longer the incubation time, the more time microbes had to consume the labile carbon, leaving less to remain in the soil. In the absence of labile carbon, microbial mediated soil R_h tends to have lower Q_{10} [49,50]. Similar variation patterns of Q_{10} with incubation time have been observed by Tuomi et al. [51] and Hamdi et al. [52], in which quadratic and cubic functions were also used to describe the relationships between Q_{10} and incubation time.

The increases of Q_{10} at the later stage might be related to soil substrate quality changes (Fig. 2A). As the labile carbon decreased, recalcitrant carbon could be decomposed. It has been previously reported that Q_{10} tends to be higher in this situation [53,54]. In this study, it was not clear what caused the higher Q_{10} at 20% WHC at the later stage, but NH_4^+-N and TP were also higher at 20% WHC, which may be related to the higher Q_{10}.

Moisture Effects on Q_{10}

We found that soil moisture had a significant effect on Q_{10}, which aligns with the findings from several previous studies [17,20,55]. Our results showed that at the intermediate soil moisture level (i.e. 60% WHC), Q_{10} was lower than at the other soil moisture levels. While there was no significant difference of Q_{10} among 60%, 40% and 20% WHC, Q_{10} at 60% WHC was significantly lower than at 80% and 100% WHC (Fig. 3). Previous studies have shown that drying can decrease Q_{10} of soil respiration and total ecosystem respiration [2,56], and this may be largely due to substrate limitation caused by the limited diffusion of solutes in thin soil water films [57,58].

We further tested which soil properties would influence Q_{10} at different incubation days and found that, at the early and middle incubation stages, Q_{10} had a significant positive correlation with MBC and the ratio of MBC to DOC (Fig. 4). The higher MBC and MBC to DOC ratio were particularly associated with higher soil moisture levels, under which labile substrate might be more available to microbes due to less water limitation. However, the Arrhenius equation shows that reactants with lower activation energies (i.e. more reactive and less recalcitrant) should have lower temperature sensitivity [2]. Our incubation results indicated that Q_{10} might not only be determined by substrate availability, but also by microbial properties such microbial biomass.

At the late stage of incubation, Q_{10} was significantly related to F:B and TP. The tight correlation of Q_{10} with F:B ratio was quite interesting. Both fungi and bacteria are important decomposers, but their structures and chemical compositions are very different. Fungi have hyphae that allow them to move, colonize and degrade surface litters, and fungal cell walls are the polymers of melanin and of chitin, much more resistant to degradation [59,60]. At the late stage, labile substrate diminishing may favor fungi communities which can degrade more recalcitrant substrate. As suggested by the carbon quality hypothesis [2,61], soils with more fungi or higher F:B ratio would have larger Q_{10}, as demonstrated here. Bradford et al. [62] also reported a shift in microbial community structure could alter the Q_{10} of R_h. The positive correlation between Q_{10} and TP suggested that P availability might also influence Q_{10}. For example, the Q_{10} value for the 20% WHC was higher than those for the other moisture levels (Fig. 2A) and TP was correspondingly higher, probably due to the lower rate of consumption by microbes at lower moisture levels. A field study also showed that summer drought caused a 22–64% reduction of microbial phosphorus [63], indicating lower microbial consumption of P under water stress. Furthermore, forest soils in subtropical China are often phosphorous limited [64]. The phosphorous saved by the lower rate of consumption might therefore contribute to the higher Q_{10} at 20%WHC during the late stage of the incubation.

Conclusions

By incubating a subtropical forest soil under five temperature levels and five moisture levels and measuring soil R_h and microbial and chemical properties throughout the incubation, we found that: 1) soil moisture significantly influenced Q_{10}, with Q_{10} being higher at higher soil moisture levels than at the lower moisture levels during the early stage of the incubation;2) soil heterotrophic respiration was highest at intermediate moisture and lowest when the soil was very dry; 3) Q_{10} mostly declined with incubation time and could be best described by quadratic or cubic functions; and 4) moisture-induced Q_{10} changes were associated with soil microbial biomass at the early stage of incubation, but to the ratio of fungi-to-bacteria at the late stage. These results imply that the response of soil R_h to future global warming may be shaped by changes in precipitation patterns. In dry conditions, global warming may stimulate less soil CO_2 emission, but in wet conditions, relatively more soil CO_2 may be emitted. Considering that more soil organic carbon has often been accumulated in the wet areas, with the same temperature rise high Q_{10} would mean more soil CO_2 emission to the atmosphere from these areas in the future.

Acknowledgments

We thank Ms. Chunqing Long for her assistance in field sampling and laboratory analysis, and Mr. Brett Seybert for his critical review and editing of the manuscript. Two anonymous reviewers provided insightful comments on the early versions of the manuscript.

Author Contributions

Conceived and designed the experiments: WS DH. Performed the experiments: WZ. Analyzed the data: WZ DH WS. Wrote the paper: WZ DH WS.

References

1. Luo Y, Wan S, Hui D, Wallace LL (2001) Acclimatization of soil respiration to warming in a tall grass prairie. Nature 413: 622–625.
2. Davidson EA, Janssens IA (2006) Temperature sensitivity of soil carbon decomposition and feedbacks to climate change. Nature 440: 165–173.
3. Davidson EA, Belk E, Boone RD (1998) Soil water content and temperature as independent or confounded factors controlling soil respiration in a temperate mixed hardwood forest. Global Change Biol 4: 217–227.
4. Jenkinson DS, Adams DE, Wild A (1991) Model estimates of CO_2 emissions from soil in response to global warming. Nature 351: 304–306.
5. Solomon S, Qin D, Manning M, Chen Z, Marquis M, et al. (2007) IPCC Climate Change 2007: The physical science basis. Contribution of working group I to the fourth assessment report of the intergovernmental panel on climate change, Cambridge University Press, Cambridge, United Kingdom and New York, NY, USA.
6. Cook FJ, Orchard VA (2008) Relationships between soil respiration and soil moisture. Soil Biol Biochem 40: 1013–1018.
7. Richardson J, Chatterjee A, Jenerette GD (2012) Optimum temperatures for soil respiration along a semi-arid elevation gradient in southern California. Soil Biol Biochem 46: 89–95.
8. Zhou T, Shi P, Hui D, Luo Y (2009) Spatial patterns in temperature sensitivity of soil respiration in China: Estimation with inverse modeling. Sci China Ser C 52: 982–989.
9. Raich JW, Schlesinger WH (1992) The global carbon dioxide flux in soil respiration and its relationship to vegetation and climate. Tellus B 44: 81–99.
10. Parton W, Silver WL, Burke IC, Grassens L, Harmon ME, et al. (2007) Global-scale similarities in nitrogen release patterns during long-term decomposition. Science 315: 361–364.
11. Hui D, Luo Y (2004) Evaluation of soil CO_2 production and transport in Duke Forest using a process-based modeling approach. Global Biogeochem Cycle 18: GB4029.
12. Orchard VA, Cook FJ (1983) Relationship between soil respiration and soil moisture. Soil Biol Biochem 15: 447–453.
13. Shen W, Jenerette GD, Hui D, Phillips RP, Ren H (2008) Effects of changing precipitation regimes on dryland soil respiration and C pool dynamics at rainfall event, seasonal and interannual scales. J Geophys Res-Biogeosci 113: G03024.
14. Shen W, Reynolds JF, Hui D (2009) Responses of dryland soil respiration and soil carbon pool size to abrupt vs. gradual and individual vs. combined changes in soil temperature, precipitation, and atmospheric [CO_2]: a simulation analysis. Global Change Biol 15: 2274–2294.
15. Oberbauer SF, Gillespie CT, Cheng W, Gebauer R, Serra AS, et al. (1992) Environmental effects on CO_2 efflux from riparian tundra in the northern foothills of the Brooks Range, Alaska, USA. Oecologia 92: 568–577.
16. Linn DM, Doran JW (1984) Aerobic and anaerobic microbial populations in no-till and plowed soils. Soil Sci Soc Amer J 48: 794–799.
17. Wang C, Yang J, Zhang Q (2006) Soil respiration in six temperate forests in China. Global Change Biol 12: 2103–2114.
18. Carlyle JC, Than UB (1988) Abiotic controls of soil respiration beneath an eighteen-year old *pinus* radiata stand in south-eastern Australia. J Eco 76: 654–662.
19. Reichstein M, Subke JA, Angeli AC, Tenhunen JD (2005) Does the temperature sensitivity of decomposition of soil organic matter depend upon water content, soil horizon, or incubation time? Global Change Biol 11: 1754–1767.
20. Craine JM, Gelderman TM (2011) Soil moisture controls on temperature sensitivity of soil organic carbon decomposition for a mesic grassland. Soil Biol Biochem 43: 455–457.
21. Moyano FE, Manzoni S, Chenu C (2013) Responses of soil heterotrophic respiration to moisture availability: An exploration of processes and models. Soil Biol Biochem 59: 72–85.
22. Gray SB, Classen AT, Kardol P, Yermakov Z, Michael Mille R (2011) Multiple climate change factors interact to alter soil microbial community structure in an old-field ecosystem. Soil Sci Soc Amer J 75: 2217–2226.
23. Schnürer J, Clarholm M, Boström S, Rosswall T (1986) Effects of moisture on soil microorganisms and nematodes: A field experiment. Microbial Ecol 12: 217–230.
24. Nakas JP, Klein DA (1979) Decomposition of microbial cell components in a semi-arid grassland soil. Appl Environ Microbiol 38: 454–460.
25. Christ MJ, David MB (1996a) Temperature and moisture effects on the production of dissolved organic carbon in a Spodosol. Soil Biol Biochem 28: 1191–1199.
26. Lundquist EJ, Jackson LE, Scow KM (1999) Wet–dry cycles affect dissolved organic carbon in two California agricultural soils. Soil Biol Biochem 31: 1031–1038.
27. Li Z, Peng S, Rae D, Zhou G (2001) Litter decomposition and nitrogen mineralization of soils in subtropical plantation forests of southern China, with special attention to comparisons between legumes and non-legumes. Plant Soil 229: 105–116.
28. Chen D, Zhang Y, Lin Y, Chen H, Fu S (2009) Stand level estimation of root respiration for two subtropical plantations based on *in situ* measurement of specific root respiration. Forest Ecol Manage 257: 2088–2097.
29. Yuste JC, Baldocchi DD, Gershenson A, Goldstein A, Misson L, et al. (2007) Microbial soil respiration and its dependency on carbon inputs, soil temperature and moisture. Global Change Biol 13: 2018–2035.
30. Vance ED, Brookes PC, Jenkinson DS (1987) An extraction method for measuring soil microbial biomass C. Soil Biol Biochem 19: 703–707.
31. Joergensen RG, Mueller T (1996) The fumigation-extraction method to estimate soil microbial biomass: Calibration of the k_{EC} value. Soil Biol Biochem 28: 33–37.
32. Dorich RA, Nelson DW (1983) Direct colorimetric measurement of ammonium in potassium chloride extracts of soils. Soil Sci Soc Amer J 47: 833–836.
33. Bossio DA, Scow KM (1998) Impacts of carbon and flooding on soil microbial communities: phospholipid fatty acid profiles and substrate utilization patterns. Microbial Ecol 35: 265–278.
34. Zak DR, Ringlberg DB, Pregitzer KS, Randlett DL, White DC, et al. (1996) Soil microbial communities beneath populus grandidentata grown under elevated atmospheric CO_2. Ecol Appl 1996: 257–262.
35. Frostegård A, Bååth E (1996) The use of phospholipid fatty acid analysis to estimate bacterial and fungal biomass in soil. Biol Fert Soils 22: 59–65.
36. Reichstein M, Rey A, Freibauer A, Tenhunen J, Valentini R, et al. (2003) Modeling temporal and large-scale spatial variability of soil respiration from soil water availability, temperature and vegetation productivity indices. Global Biogeochem Cycle 17: 1104–1118.
37. Deng Q, Hui D, Zhang D, Zhou G, Liu J, et al. (2012) Effects of precipitation increase on soil respiration: a three-year field experiment in subtropical forests in China. PLoS ONE 7: e41493.
38. Jassal RS, Black TA, Novak MD, Gaumont-Guay D, Nesic Z (2008) Effect of soil water stress on soil respiration and its temperature sensitivity in an 18-year-old temperate Douglas-fir stand. Global Change Biol 14: 1305–1318.
39. Yu X, Zha T, Pang Z, Wu B, Wang X, et al. (2011) Response of soil respiration to soil temperature and moisture in a 50-year-old oriental arborvitae plantation in China. PLoS ONE 6: e28397.
40. Suseela V, Conant RT, Wallenstein MD, Dukes JS (2012) Effects of soil moisture on the temperature sensitivity of heterotrophic respiration vary seasonally in an old-field climate change experiment. Global Change Biol 18: 336–348.
41. Xu L, Baldocchi DD, Tang J (2004) How soil moisture, rain pulses, and growth alter the response of ecosystem respiration to temperature. Global Biogeochem Cycle 18: GB4002.
42. Williams MA (2007) Response of microbial communities to water stress in irrigated and drought-prone tallgrass prairie soils. Soil Biol Biochem 39: 2750–2757.
43. Fang CM, Smith P, Moncrieff JB, Smith JU (2005) Similar response of labile and resistant soil organic matter pools to changes in temperature. Nature 433: 57–59.
44. Kirschbaum MUF (2006) The temperature dependence of organic-matter decomposition - still a topic of debate. Soil Biol Biochem 38: 2510–2518.
45. Billings SA, Ballantyne F (2013) How interactions between microbial resource demands, soil organic matter stoichiometry, and substrate reactivity determine the direction and magnitude of soil respiratory responses to warming. Global Change Biol 19: 90–102.
46. Wang G, Zhou Y, Xu X, Ruan H, Wang J (2013) Temperature sensitivity of soil organic carbon mineralization along an elevation gradient in the Wuyi Mountains, China. PLoS ONE 8: e53914.
47. Mahecha MD, Reichstein M, Carvalhais N, Lasslop G, Lange H, et al. (2010) Global convergence in the temperature sensitivity of respiration at ecosystem level. Science 329: 838–840.
48. Song M, Jiang J, Cao G, Xu X (2010) Effects of temperature, glucose and inorganic nitrogen inputs on carbon mineralization in a Tibetan alpine meadow soil. Eur J Soil Biol 46: 375–380.
49. Arevalo CBM, Bhatti JS, Chang SX, Jassal RS, Sidders D (2010) Soil respiration in four different land use systems in north central Alberta, Canada. J Geophys Res-Biogeosci 115: G01003.
50. Yuste Curiel J, Ma S, Baldocchi DD (2010) Plant-soil interactions and acclimation to temperature of microbial-mediated soil respiration may affect predictions of soil CO_2 efflux. Biogeochemistry 98: 127–138.
51. Tuomi M, Vanhala P, Karhu K, Fritze H, Liski J (2008) Heterotrophic soil respiration–Comparison of different models describing its temperature dependence. Ecol Model 211: 182–190.
52. Hamdi S, Moyano F, Sall S, Bernoux M, Chevallier T (2013) Synthesis analysis of the temperature sensitivity of soil respiration from laboratory studies in relation to incubation methods and soil conditions. Soil Biol Biochem 58: 115–126.

53. Conant RT, Steinweg JM, Haddix ML, Paul EA, Plante AF, et al. (2008) Experimental warming shows that decomposition temperature sensitivity increases with soil organic matter recalcitrance. Ecology 89: 2384–2391.
54. Craine JM, Fierer N, McLauchlan KK (2010) Widespread coupling between the rate and temperature sensitivity of organic matter decay. Nat Geosci 3: 854–857.
55. Moyano FE, Vasilyeva N, Bouckaert L, Cook F, Craine J, et al. (2012) The moisture response of soil heterotrophic respiration: interaction with soil properties. Biogeosciences 9: 1173–1182.
56. Reichstein M, Tenhunen JD, Roupsard O, Ourcival JM, Rambal S, et al. (2002) Ecosystem respiration in two Mediterranean evergreen Holm Oak forests: drought effects and decomposition dynamics. Funct Ecol 16: 27–39.
57. Rey A, Pegoraro E, Tedeschi V, De Parri I, Jarvis PG, et al. (2002) Annual variation in soil respiration and its components in a coppice oak forest in Central Italy. Global Change Biol 8: 851–866.
58. Curiel Yuste J, Janssens IA, Carrara A, Ceulemans R (2004) Annual Q_{10} of soil respiration reflects plant phenological patterns as well as temperature sensitivity. Global Change Biol 10: 161–169.
59. Holland EA, Coleman DC (1987) Litter placement effects on microbial and organic matter dynamics in an agroecosystem. Ecology 68: 425–433.
60. Guggenberger G, Frey SD, Six J, Paustian K, Elliott ET (1999) Bacterial and fungal cell-wall residues in conventional and no-tillage agroecosystems. Soil Sci Soc Amer J 63: 1188–1198.
61. Bosatta EÅ, Göran I. (1999) Soil organic matter quality interpreted thermodynamically. Soil Biol Biochem 31: 1889–1891.
62. Bradford MA, Davies CA, Frey SD, Maddox TR, Melillo JM, et al. (2008) Thermal adaptation of soil microbial respiration to elevated temperature. Eco Lett 11: 1316–1327.
63. He Z, Wu J, O'Donnell AG, Syers JK (1997) Seasonal responses in microbial biomass carbon, phosphorus and sulphur in soils under pasture. Biol and Fert Soils 24: 421–428.
64. Huang W, Liu J, Wang Y, Zhou G, Han T, et al. (2013) Increasing phosphorus limitation along three successional forests in southern China. Plant Soil 364: 181–191.

22

Linking Stoichiometric Homeostasis of Microorganisms with Soil Phosphorus Dynamics in Wetlands Subjected to Microcosm Warming

Hang Wang[1], HongYi Li[1], ZhiJian Zhang[1]*, Jeffrey D. Muehlbauer[2], Qiang He[3], XinHua Xu[1], ChunLei Yue[4], DaQian Jiang[5]

1 College of Natural Resource and Environmental Sciences, China Academy of West Development, Zhejiang University, Hangzhou, Zhejiang Province, China, **2** Curriculum for the Environment & Ecology, University of North Carolina at Chapel Hill, Chapel Hill, North Carolina, United States of America, **3** Department of Civil & Environmental Engineering, University of Tennessee, Knoxville, Tennessee, United States of America, **4** Institute of Ecology, Zhejiang Forestry Academy, Hangzhou, China, **5** Department of Earth and Environmental Engineering, Henry Krumb School of Mines, Columbia University, New York, New York, United States of America

Abstract

Soil biogeochemical processes and the ecological stability of wetland ecosystems under global warming scenarios have gained increasing attention worldwide. Changes in the capacity of microorganisms to maintain stoichiometric homeostasis, or relatively stable internal concentrations of elements, may serve as an indicator of alterations to soil biogeochemical processes and their associated ecological feedbacks. In this study, an outdoor computerized microcosm was set up to simulate a warmed (+5°C) climate scenario, using novel, minute-scale temperature manipulation technology. The principle of stoichiometric homeostasis was adopted to illustrate phosphorus (P) biogeochemical cycling coupled with carbon (C) dynamics within the soil-microorganism complex. We hypothesized that enhancing the flux of P from soil to water under warming scenarios is tightly coupled with a decrease in homeostatic regulation ability in wetland ecosystems. Results indicate that experimental warming impaired the ability of stoichiometric homeostasis (*H*) to regulate biogeochemical processes, enhancing the ecological role of wetland soil as an ecological source for both P and C. The potential P flux from soil to water ranged from 0.11 to 34.51 mg m^{-2} d^{-1} in the control and 0.07 to 61.26 mg m^{-2} d^{-1} in the warmed treatment. The synergistic function of C-P acquisition is an important mechanism underlying C:P stoichiometric balance for soil microorganisms under warming. For both treatment groups, strongly significant ($p<0.001$) relationships fitting a negative allometric power model with a fractional exponent were found between n-$H_{C:P}$ (the specialized homeostatic regulation ability as a ratio of soil highly labile organic carbon to dissolved reactive phosphorus in porewater) and potential P flux. Although many factors may affect soil P dynamics, the n-$H_{C:P}$ term fundamentally reflects the stoichiometric balance or interactions between the energy landscape (i.e., C) and flow of resources (e.g., N and P), and can be a useful ecological tool for assessing potential P flux in ecosystems.

Editor: Roeland M. H. Merks, Centrum Wiskunde & Informatica (CWI) & Netherlands Institute for Systems Biology, Netherlands

Funding: The authors wish to thank the National Natural Science Foundation of China (41373073) and Zhejiang Science and Technology Program (2011F20025) for providing the financial support for this project. The funders had no role in study design, data collection and analysis, decision to publish, or preparation of the manuscript.

Competing Interests: The authors have declared that no competing interests exist.

* E-mail: zhangzhijian@zju.edu.cn

Introduction

The principle of stoichiometric homeostasis states that organisms maintain relatively stable levels of biologically-relevant elements (e.g., carbon, nitrogen, and phosphorous) over time. This concept is based on ecological stoichiometry and is primarily applied to trophic interactions [1]. By modulating organism responses to key environmental drivers (e.g., nutrient fertilization), stoichiometric homeostasis is also a major mechanism responsible for the structure, function, and stability of ecosystems. Therefore, deviations from stoichiometric homeostasis can serve as indicators of environmental fluctuations that may be useful in ecological evaluation. The homeostatic regulation coefficient (*H*), which may be accessed through a regression model (equation 1, in the Materials and Methods) is a measure of homoeostatic regulation ability and is commonly used as an indicator of these ecological changes [2].

The average global surface temperature has increased by 0.74°C since 1850 and is likely to increase an additional 1.1–6.4°C by the end of this century [3,4]. Temperature is one of the primary determinants affecting the metabolic rates of organisms, from cells to global-scale ecosystems, based on the metabolic theory of ecology [5,6]. Even a slight increase in temperature can alter energy flow and resource cycles in ecosystems and may affect environmental quality [3,7]. For instance, global soil respiration increased by 0.1 Pg yr^{-1} between 1989 and 2008 due to a high temperature anomaly [8]. Understanding the response of stoichiometric homeostasis to global warming and its associated ecological feedback to the biosphere is gaining an increasing degree of attention worldwide.

Studies of stoichiometric homeostasis have strongly emphasized physiological variation in elemental composition in macroscopic aquatic species [9]. To a lesser extent, some studies have also quantified the degree of stoichiometric homeostasis in terrestrial plants [2,10,11]. Stoichiometric homeostasis investigations in soil microorganisms subjected to environmental changes, however, are rare. Biogeochemical processes at the soil-water interface play a fundamental role in overall soil development, and they are the primary driving force behind key ecosystem functions such as community productivity and water quality [12,13]. Although stoichiometric homeostasis reflects the net outcome of many underlying physiological and biochemical adjustments as organisms respond to their surroundings [14], to our knowledge, few studies have linked any change in stoichiometric homeostasis to biogeochemical processes in a soil ecosystem affected by global warming.

Wetlands are one of the most productive and biologically diverse ecosystem types on the planet, where the chemical and, in particular, nutrient composition of shallow groundwater and surface water can be altered by a range of biogeochemical processes [15]. Many of the elements and nutrients that cycle through wetlands have ecosystem and global-scale effects; for example, phosphorous (P), is the key element affecting eutrophication in aquatic ecosystems [16], limiting coastal ecosystem processes [17], and in regulating allometry in soil food webs [18]. However, global warming may impact the inherent biogeochemical balance of P and other major elements. Diffusive models predict that elemental fluxes of P and nitrogen (N) to receiving waters are more than 2 times higher during summer (i.e., when temperatures are warmer) than in other seasons [19,20]. Further, an 18-month microcosm investigation using a minute and seasonal scale temperature manipulation verified that warming induced substantial mobilization of P from wetland sediment to water [21,22]. In other studies, excessive P loading under high temperature accelerated the risk of eutrophication in receiving waters [15,20,23] and altered carbon (C) assimilation and N accumulation in aquatic plants [24]. These studies suggest that warming or increasing temperature disturbs P equilibrium between sediment and water. However, the link between stoichiometric homeostasis and P dynamics in wetland soil subjected to global warming is poorly understood, which impedes the development of ecological regulatory strategies for coping with global warming scenarios. Additionally, wetlands play a vital role in the global C cycle and may respond strongly to climate change [25]. This is important because the flow of C energy and material also controls the biogeochemical cycling of many other elements in Earth's ecosystems [5,26,27,28]. Thus, integrating C:P stoichiometry is vital to understanding the connection between soil P dynamics and the regulation of stoichiometric homeostasis in wetland ecosystems.

The objective of our study was to shed new light on P dynamics in response to global warming, from the perspective of C:P stoichiometric homeostasis in six subtropical wetlands located in the Yangtze River delta in southeast China. We simulated global warming using a novel, in situ deployment of a microcosm device that mimics a warming scenario of 5°C above ambient temperature. The link between potential P flux from soil to water and the homeostatic regulation ability of soil microorganisms was also examined. We hypothesized that enhancing P flux from soil to water under warming scenarios is tightly coupled with a decrease in homeostatic regulation ability in wetland ecosystems.

Materials and Methods

This work is unrelated to any ethics issues. No specific permit was required for the described field study because the sampling locations were not located in protected areas or private land. The experimental field studies did not involve endangered or protected species.

Study sites

The study sites (120°41′31″E, 30°53′55″N to 120°33′32″E, 30°01′58″N) were located in the southern region of the Taihu Lake Basin and the NingShao Plain within the Yangtze River delta in southeast China. The climate in this area is subtropical monsoon with an annual average rainfall of 1350 mm and an annual average temperature of 26°C in summer and 4°C in winter.

The Supporting Information contains details about the six sites related to geographical position, hydrological parameters, and dominant vegetation during sampling (Table S1), as well as the physico-chemical properties of the six collected soils (Table S2). In brief, YaTang riverine wetland (YT) is in an advanced state of eutrophication and is classified as eutrophic, with the highest P and C total soil content among the six study sites. Soils sampled from XiaZhuhu wetland (XZ), XiXi National Wetland Park (XX), BaoYang riverine wetland (BY), and JinHu wetland (JH) are in a meso-eutrophic state, while the lowest soil C and P content among the sites was found in ShiQiu multipond wetland (SQ). Because these selected study sites represented the range of typical soil C and P conditions across subtropical wetland ecosystems in southeast China, we did not design any additional C or P treatments, which would have required artificial manipulations of the soil concentrations of these elements.

Microcosm configuration and sampling regime

A custom-built, novel microcosm (Fig. S1) simulating climate warming was developed under both present-day ambient temperature conditions (Control) and simulated warming conditions of 5°C above ambient temperature (Warmed). Details on the configuration of this microcosm system and its operation were either described previously [21] or can be found in Text S1. Compared to fixed-temperature laboratory incubations, this novel microcosm offers higher temporal resolution temperature control (on a minute scale), which simulates more realistic warming conditions.

Details on the establishment of wetland columns and water-soil samplings during the microcosm investigation are provided in Text S2. Briefly, each wetland column consisted of a prefabricated PVC pipe assembly (45.0 cm in height and 10.0 cm in internal diameter) that was designed to hold 20 cm of fresh soil and 20 cm of corresponding overlying water. Columns were installed in May 2008, and three replicates were placed inside each of the two incubation boxes. In this study, three samplings of water (overlying water and porewater) and 0–5 cm of topsoil were carried out in July and November 2010 and March 2011, which were used to illustrate the link between soil microorganism stoichiometric homeostasis and soil P dynamics in wetlands subjected to experimental warming.

Water and soil analysis

All water and soil samples were frozen at −15°C prior to analysis. After thawing, water was filtered through a 0.45-μm filter. Phosphorous in filtered samples was measured using a continuous flow analyzer (Autoanalyzer III, Bran+Luebbe, Germany) with a spectrophotometer set at 880 nm (Murphy and Riley, 1962). This is the absorbance of dissolved reactive phosphorus (DRP), which represents the bioavailable P fraction for microorganisms and/or algae in surface water. Porewater samples for dissolved organic carbon (DOC) analysis were first acidified (10% HCl) and purged with inert gas to remove any inorganic carbon, then analyzed using a Shimadzu TOC 5000 analyzer (Shimadzu Scientific Instruments, Columbia, Maryland, USA).

After thawing, soil samples were analyzed for C and P in microbial biomass (Table 1) using the fumigation-extraction method described by Wu et al. [29] and Brookes et al. [30]. In brief, soil C and P in fumigated and non-fumigated soil samples was extracted using solutions of 0.5 mol L^{-1} K_2SO_4 and 0.5 mol L^{-1} $NaHCO_3$. The differences in extractable C and P between fumigated and non-fumigated soil were assumed to be released from lysed soil microbes, i.e., microbial biomass C and P (MB-C, MB-P). By utilizing the susceptibility of organic C to $KMnO_4$ oxidation [31], three fractions of labile organic C in these wetland columns; namely, highly labile organic C (HLOC), mid-labile organic C (MLOC) and labile organic C (LOC), were determined using 33 mmol L^{-1}, 167 mmol L^{-1}, and 333 mmol L^{-1} of $KMnO_4$, respectively.

Calculation of homeostatic regulation ability of soil microorganisms

The homeostatic regulation ability of soil microorganisms was estimated by calculating the homeostatic regulation coefficient (*H*) according to the following equation [1]:

$$y = c \cdot x^{(1/H)} \qquad (1)$$

where y is the microbial C or P concentration (mg g^{-1} or mg kg^{-1} of dry soil weight) or the C:P molar ratio of soil microorganisms; x is the organic C concentration (including HLOC, MLOC, and LOC) in soil, DRP or DOC concentration in porewater, or the C:P molar ratio among the soil organic C and porewater DRP; and c is a constant. H is defined as the homeostatic regulation ability of soil microorganisms; as the value of H increases, the ecosystem trends toward a higher degree of ecological stability and resilience. For example, higher values of H_N and H_P in grasses receiving field N and P fertilizations were significantly associated with higher species dominance and stability [2]. According to ecological allometry [5,32], H may be classified as an equilibrium constant with respect to functions of organism metabolic and stoichiometric processes.

The first derivative of regression equation (1), reflecting the response rate of y to x, is given as follows:

$$\frac{dy}{dx} = c \cdot (1/H) \cdot \left(x^{(1/H-1)}\right) = \left(c \cdot x^{1/H}\right) \cdot (1/x) \cdot (1/H) \qquad (2)$$

By substituting in variables from equation (1), equation (2) can be transformed as:

$$\frac{dy}{dx} = y \cdot (1/\mathrm{x}) \cdot (1/\mathrm{H}) = (\mathrm{y}/\mathrm{x}) \cdot (1/H) \qquad (3)$$

Here, the value of (y/x) is defined as the specific variable for $1/H$, which may be transformed into (x/y) responding to H. Therefore, we define n-H_i as the specialized homeostatic regulation ability for each of the tested wetland soil columns:

$$n - H_i = (x_i/y_i) \cdot H_i \qquad (4)$$

where x_i is the measured value of the organic C concentration (including HLOC, MLOC, and LOC) in soil, DRP or DOC concentration in porewater, or the C:P molar ratio among the soil organic C and porewater DRP for each soil sample; y_i is the microbial C or P concentration (mg g^{-1} or mg kg^{-1} of dry soil weight) or the C:P molar ratio of soil microorganisms for each soil sample; i is the type of parameter (form of HLOC, MLOC, LOC, DRP, DOC, or ratio of C:P for soil or soil microbes); and H refers to the homeostatic regulation coefficient calculated by equation (1) according to the respective for each of the two treatments (Control and Warmed) in terms of the i parameter.

Potential P flux (F_P, mg m^{-2} d^{-1}) was used to evaluate the potential P transfer out of porewater under experimental warming according to the following equation:

$$F_p = (C_{in} - C_{out}) \times V/(S \times T) \qquad (5)$$

where $C_{in}C_{in}$ is the DRP concentration in the porewater (mg L^{-1}), C_{out} is the P concentration in the overlying water (mg L^{-1}), V is the volume of overlying water (mL), S is the area of the wetland column (m^2) and T is the interval between two sampling dates (d). Due to the relatively static hydrological conditions in the wetlands (Table S1), DRP was the most important form for P exchange between soil and water in the study. Soil microorganisms are also more sensitive to DRP than to TP due to its bioavailability. Therefore, values of F_P (and n-H_i) were evaluated by the form of DRP rather than TP. The potential P flux defined here provides a good indicator for assessing P concentration gradients between overlying water and porewater, with high values suggesting a high risk of P transfer from soil to overlying water.

In this study, aquatic macrophytes (e.g., *Phragmites communis*) were temporarily excluded from soil columns, but small benthic organisms (e.g., *Margarya melanoide*) and plankton (e.g., *Spirogyra*) were preserved. Data collected from water and soil samples were expressed as the mean plus standard error. Paired Student's t-tests were used to compare the effects of warming on soil microorganism and C fractions, P concentrations in porewater and overlying water, and potential P flux. Two-way analysis of variance (repeated ANOVA) tests were carried out using SPSS software (version 15.0) to examine the effects of experimental warming, sampling time, and their interaction on soil biochemical traits and potential P flux. Statistical test results were considered significant at the $p<0.05$ level.

Results

Soil carbon and phosphorus potential flux in response to experimental warming

After >2.5 y of incubation (in Nov 2010), experimental warming increased soil microbial biomass, soil labile C, and porewater DOC (Table 1); however, warming decreased concentrations of total soil C at all wetlands except site SQ. Meanwhile, although fluctuations in DRP concentrations were found in the overlying water, DRP in porewater increased in the warmed treatment by 52% to 137% compared to the ambient treatment, except in the SQ wetland (Table 2). On average, potential P flux in the six wetlands ranged from 0.11 to 34.51 mg m^{-2} d^{-1} in the control and 0.07 to 61.26 mg m^{-2} d^{-1} in the warmed treatment (Table 2). Two-way analysis of variance indicated that experimental warming significantly increased potential P flux, except in the SQ wetland. There were also significant differences in soil C pools and potential P flux between Warmed and Control treatments (Table 3). A significant interaction between treatment and season (sampling date) was found for soil C pools and potential

Table 1. Concentrations of soil microbial biomass (MB; MB-C: microbial carbon; MB-P: microbial phosphorus), soil carbon (TOC: total organic carbon; HLOC: highly labile organic carbon; MLOC: mid-labile organic carbon; and LOC: labile organic carbon), and carbon in porewater (DOC: dissolved organic carbon) in wetland columns in the microcosm experiment (Control: ambient temperature; Warmed: ambient temperature +5°C).

Treatment		Soil microbial biomass				Soil carbon				DOC in porewater (mg kg^{-1})
		Total mass (mg g^{-1})	MB-C (mg g^{-1})	MB-P (mg kg^{-1})	Microbial C:P stoichiometric ratio	TOC (g kg^{-1})	HLOC (mg g^{-1})	MLOC (mg g^{-1})	LOC (mg g^{-1})	
JH	Control	0.402±0.023	0.386±0.022	15.0±2.3	68.6:1	25.6±1.4	2.28±0.09	2.75±0.66	3.75±0.71	13.2±2.3
	Warmed	0.493±0.010	0.477±0.008	16.1±2.1	79.0:1	22.3±2.1	2.31±0.37	3.16±0.70	5.48±1.31	16.6±3.1
p		0.064	**0.050**	0.568		**0.028**	0.899	0.498	0.113	0.182
XZ	Control	1.12±0.02	1.08±0.02	25.9±2.4	111:1	54.8±5.9	5.78±0.31	8.55±0.29	9.93±0.43	17.3±0.9
	Warmed	1.47±0.03	1.42±0.04	33.4±2.3	113:1	49.2±2.6	6.35±0.06	8.67±0.13	9.93±0.32	21.6±2.2
p		**<0.001**	**0.009**	**0.005**		0.054	0.055	0.575	0.990	**0.038**
YT	Control	1.42±0.01	1.37±0.01	36.4±2.9	100:1	133±10	9.13±0.23	13.0±1.0	18.9±4.2	20.2±3.3
	Warmed	1.73±0.06	1.66±0.04	46.1±5.7	96.0:1	117±3	10.9±0.1	18.6±0.9	20.5±4.3	22.7±9.9
p		**0.001**	**0.035**	**0.045**		**0.024**	**<0.001**	**0.002**	0.681	0.701
XX	Control	1.01±0.03	0.964±0.024	33.4±4.4	77.0:1	35.6±1.7	2.14±0.07	3.12±0.23	3.70±0.43	16.2±3.0
	Warmed	1.31±0.08	1.25±0.11	42.0±2.3	79.4:1	31.8±1.3	3.39±0.06	3.65±0.19	4.32±0.83	24.3±8.7
p		**0.003**	0.067	**0.028**		**0.011**	**<0.001**	**0.026**	**0.039**	0.208
BY	Control	1.07±0.06	1.04±0.06	34.5±2.4	80.4:1	37.6±1.9	2.90±0.12	3.98±0.05	4.50±0.09	18.6±1.5
	Warmed	1.13±0.07	1.10±0.06	36.6±2.6	80.1:1	32.4±2.1	3.59±0.36	5.91±0.16	7.06±0.33	21.8±1.0
p		0.299	0.317	0.435		**0.043**	**0.033**	**<0.001**	**<0.001**	**0.027**
SQ	Control	0.183±0.018	0.179±0.017	4.25±0.87	112:1	14.2±2.3	0.556±0.025	0.889±0.673	1.73±0.33	8.14±0.96
	Warmed	0.469±0.011	0.456±0.013	13.5±2.7	90.1:1	15.3±1.7	2.61±0.05	3.74±0.15	4.23±0.16	12.3±1.2
p		**<0.001**	**<0.000**	**0.005**		0.443	**<0.001**	**0.002**	**<0.001**	**0.006**

p-values refer to significant differences between two treatments. Sampling date is November 2010.

Table 2. Phosphorus (P) concentrations in overlying water and porewater, and the potential flux of dissolved reactive P in wetland columns in the microcosm experiment (Control: ambient temperature; Warmed: ambient temperature +5°C).

Wetland column	Treatments	Overlying water (mg L^{-1})	Porewater (mg L^{-1})	Potential P flux (mg m^{-2} d^{-1})
	July 2010			
JH	Control	0.066±0.038	0.063±0.005	0.21±0.07
	Warmed	0.017±0.009	0.095±0.001	0.32±0.09
p		**0.022**	0.004	**0.018**
XZ	Control	0.070±0.026	0.266±0.003	0.90±0.03
	Warmed	0.024±0.024	0.612±0.063	2.50±0.38
p		**0.012**	**0.039**	**0.003**
YT	Control	1.507±1.22	2.334±0.703	15.30±3.56
	Warmed	0.909±0.51	4.537±1.510	35.26±4.61
p		0.304	**0.002**	**<0.000**
XX	Control	0.106±0.076	0.330±0.015	2.14±0.17
	Warmed	0.038±0.031	0.570±0.033	3.65±0.10
p		0.091	**0.023**	**0.004**
BY	Control	0.063±0.061	0.104±0.001	0.61±0.02
	Warmed	0.051±0.079	0.231±0.001	0.98±0.24
p		0.783	**<0.000**	**0.017**
SQ	Control	0.016±0.008	0.088±0.015	0.11±0.02
	Warmed	0.012±0.010	0.021±0.009	0.08±0.01
p		0.431	0.025	0.412
	March 2011			
JH	Control	0.018±0.000	0.182±0.102	0.63±0.37
	Warmed	0.016±0.000	0.354±0.048	1.07±0.23
p		0.467	**0.016**	**0.004**
XZ	Control	0.029±0.014	0.720±0.286	2.10±0.13
	Warmed	0.018±0.000	1.704±0.189	4.36±0.78
p		0.187	**0.013**	**0.006**
YT	Control	0.042±0.008	5.83±1.23	34.51±10.36
	Warmed	0.047±0.012	8.34±0.41	61.26±6.12
p		0.572	**0.011**	**<0.000**
XX	Control	0.016±0.006	1.17±0.18	3.38±0.71
	Warmed	0.013±0.004	2.53±0.34	11.72±2.10
p		0.320	**0.004**	**<0.000**
BY	Control	0.016±0.000	0.390±0.067	1.06±0.12
	Warmed	0.022±0.000	0.653±0.182	2.38±0.53
p		**0.012**	**0.039**	**0.034**
SQ	Control	0.012±0.001	0.068±0.019	0.16±0.09
	Warmed	0.016±0.000	0.044±0.040	0.07±0.03
p		0.490	0.193	0.387

p-values refer to significant differences between two treatments.

P flux as well, although there were no significant differences in these variables by season alone.

Microorganism stoichiometric homeostasis in response to experimental warming

The relationships between soil microorganism biomass and three forms of soil carbon concentrations (HLOC, MLOC, and LOC) sampled in Mar. 2011 were rigorously and significantly described by the stoichiometric homeostasis model (Fig. 1, $p<0.01$). The values of H in the warmed treatment were lower than those in the control, with H values of 1.22 and 1.73 for HLOC, 1.48 and 1.75 for MLOC, and 1.09 and 2.20 for LOC in warmed and control treatments, respectively (Fig. 1). Results from the other two samplings (data not shown) were fundamentally similar to the Mar. 2011 sampling. There was no consistent pattern in the homeostasis model fit between DOC and microorganism C among the three samplings (Table 4). However, strongly significant and positive correlations between DRP in porewater and soil microorganism biomass P fit a power model for

Table 3. Results of two-way analysis of variance (repeated ANOVA) showing the *p*-values for soil microorganisms and soil carbon and potential phosphorus (P) flux under experimental warming and sampling time. MB-C: microbial carbon, MB-P: microbial phosphorus, TOC: total organic carbon, HLOC: highly labile organic carbon, MLOC: mid-labile organic carbon, and LOC: labile organic carbon.

Factor	Soil microorganisms			Soil carbon				Potential P flux ($mg\ m^{-2}\ d^{-1}$)
	Total ($mg\ g^{-1}$)	MB-C ($mg\ g^{-1}$)	MB-P ($mg\ kg^{-1}$)	TOC ($g\ kg^{-1}$)	HLOC ($mg\ kg^{-1}$)	MLOC ($mg\ kg^{-1}$)	LOC ($mg\ kg^{-1}$)	
Treatment	**0.007**	**0.005**	**0.008**	**0.013**	**0.011**	**0.038**	**0.007**	**0.033**
Time	0.285	0.321	0.308	0.079	0.563	0.439	0.407	0.185
Treatment×Time	**0.017**	**0.016**	**0.009**	**0.006**	**0.019**	0.076	**0.016**	**0.047**

p-values smaller than 0.05 are in bold to indicate statistical significance.

these three consecutive samplings (Fig. 2), with lower values of H under the warmed treatment (1.79 to 2.49) compared to the control (3.08 to 4.24). Although no significant difference between these two treatments was found for the C:P stoichiometric ratios of soil microorganisms (paired *t*-test $p = 0.607$, Table 1), strongly significant correlations existed between soil C:P ratios of HLOC to DRP and soil microorganism C:P for these three samplings under the two treatments (Fig. 3). More importantly, the C:P ratio-related H values (i.e., $H_{C:P}$) obtained from the warmed group were consistently lower than those from the control group.

No significant relationship was found between porewater ratios of DOC:DRP and soil microbial C:P (Table 4). Relationships were observed for other comparisons of soil LOC to porewater DRP and soil MLOC to porewater DRP vs soil microorganism C:P ratios (Table 4); however, among the three samplings, corresponding $H_{C:P}$ values showed no consistent trends between the two treatment groups. This is in contrast to soil HLOC to porewater DRP results, in which a consistent increase in $H_{C:P}$ was observed from control to warmed treatments across all three samplings (Fig. 3). Specific n-$H_{C:P}$ values for each wetland column and soil treatment (Warmed or Control) were subsequently calculated using the $H_{C:P}$ values derived from these HLOC:DRP regressions, using C:P ratios (HLOC:DRP) in porewater as the 'x_i' variable and soil microbial C:P ratios as the 'y_i' variable in equation (4). Again, strongly significant ($p<0.001$) linkages that fit a power model between n-$H_{C:P}$ and potential P flux (i.e., allometric relationships [5]) were found for both treatment groups (Fig. 4). Notably, these two fitted curves strongly overlapped, and the regression equation for the two datasets combined ($y = 5.261 \times^{-1.7509}$, $R^2 = 0.5893$) was almost identical to the equations derived from each dataset individually. The data for potential P flux and n-$H_{C:P}$ values were concentrated in ranges less than 20 $mg\ m^{-2}\ d^{-1}$ and 3.0, respectively (Fig. 4).

Discussion

Does homeostasis relate to the ecological role of soil in C and P cycling?

The tight linkages in this study between different pools of soil labile C and soil microbial C (Fig. 1), and porewater DRP and soil microbial P (Fig. 2) fit the stoichiometric homeostasis model. This supports the view that stoichiometric homeostasis is an important mechanism underpinning the dynamics of C and P resource availability at the soil-microbe interphase in wetland ecosystems. It confirms that the composition of C and P in microbial assemblages is closely related to resource availability in ecosystems [33,34], as soil microorganism C and P increases were related to the increases in soil C and P availability observed in the warmed treatment in this study (Table 1 & 2). Based on the principle of homeostatic regulation ability [1] and its prior application to a grassland ecosystem [2], decreased H values for soil labile C and DRP in warmed columns indicate that experimental warming impaired the ability of soil organisms to regulate C and P biogeochemical turnover in the studied wetland ecosystems. This loss of homoeostatic regulation ability is consistent with the observed decrease in soil TOC (Table 1) and the increase in potential P flux (Table 2). Comparison of the data for the decrease in soil TOC and the increase in both soil labile C and microorganism C (Table 1) further indicates that losses of total organic C in warmed soils were mainly derived from non-active C pools; i.e., the recalcitrant organic C. Recalcitrant organic fractions generally represent a large proportion of the total C pool and have relatively slow turnover rates [35,36], and generation of recalcitrant organic matter is regarded as a vital process for C sequestration [36]. Excluding P that is assimilated into microbial biomass (Table 1), warming subjects the remaining P pool to increases in enzymatic biodegradation of organic phosphorus [21], P desorption [20,37], and inorganic P solubilization [38], which could enhance the strength of potential P flux (Table 2) as well as increase DRP in porewater. The ecological role of wetland soil as a source for C and P, coupled with lower H values observed in the warming treatment compared to the control, suggests that H may yield valuable insight into the ecological role of soil in C and P biogeochemical turnover and its response to global warming.

Does homeostasis relate to stoichiometric balance?

Although experimental warming did affect the soil microbial C:P stoichiometric ratios for particular samples to some degree (Table 1), no overall significant difference was found between the two (Warmed and Control) groups. However, C:P stoichiometric ratios in the soil microbial community did seem to vary according to geographical (biotic and abiotic) features [39] across the six wetlands (Table 1 and Fig. 3). Because macronutrients are coupled with various biochemical and cellular constituents, the ratios of these elements remained relatively constant compared to the potentially decoupled resource stoichiometry changes in the surrounding environment that were induced by warming [1,39].

In contrast, tight and significant associations between ratios of soil C:P and microorganism C:P were obtained with the fitted curves of the power model (Fig. 3), which seems contradictory to the results described above. This may be explained by soil microorganisms regulating the equilibrium between the elemental

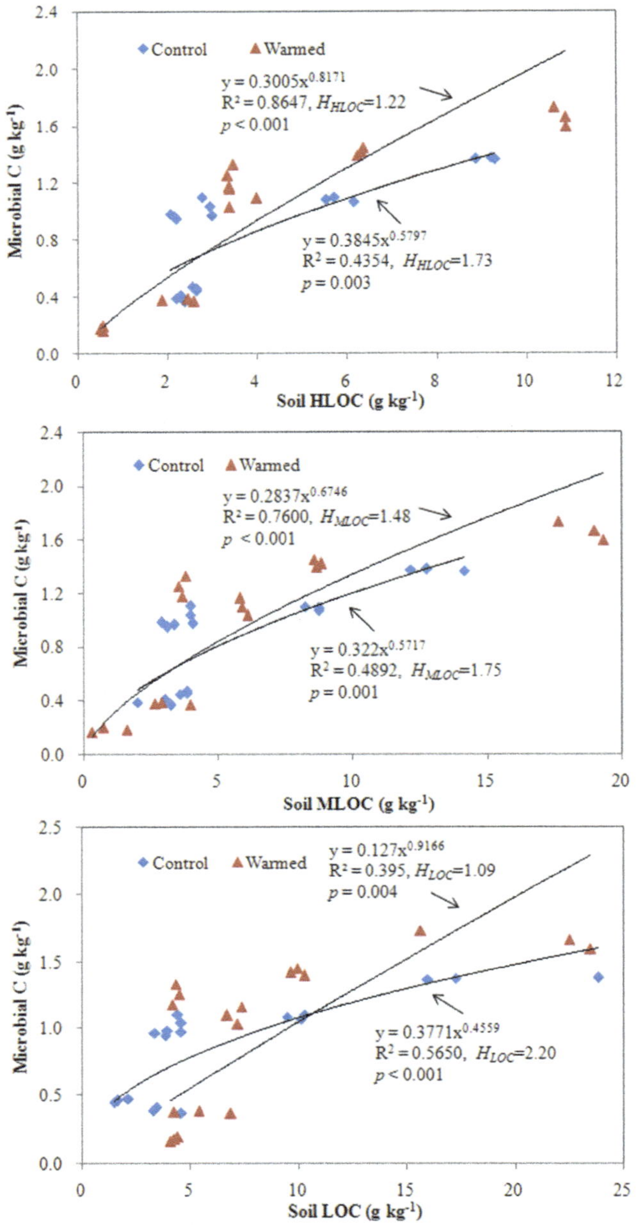

Figure 1. Relationships between soil microbial carbon (MB-C) and soil organic carbon (in forms of HLOC: highly labile organic carbon, MLOC: mid-labile organic carbon, and LOC: labile organic carbon). Control and Warmed represent treatments of ambient temperature and ambient temperature +5°C, respectively. Sampling date is March 2011. Coefficients fitting the data to the equation for stoichiometric homeostasis are given along with an estimate of H_C.

composition of their biomass and the efficiencies of microbial nutrient assimilation and growth through ecoenzymatic interactions [26]. Further, simultaneous N and P enrichment (by photosynthetic biota) produces strongly positive, synergistic responses in ecosystems [16]. This synergistic effect can aid in maintaining "Redfield-like" ratios in soil microorganisms [39], and may also have served to regulate the stoichiometric balance of soil microbial C and P acquisition for the warmed wetland soils in this study. This is also in line with general ecological theory on the elemental stoichiometries of microbial biomass and environmental availability relative to microbial nutrient assimilation and growth [1,32]. For example, increases in N concentration in biomass are usually accompanied by increases in P uptake for vascular plants [11]. This suggests that the synergistic function of C-P acquisition is an important mechanism underlying the C:P stoichiometric balance for soil microorganisms in wetland ecosystems subjected to warming.

Field N and P fertilization in another study showed that higher values of $H_{N:P}$ in grasses were significantly associated with higher species dominance and stability [2]. In this study, the homeostatic regulation coefficient for C:P stoichiometry in the soil microorganism community vs soil HLOC:DRP ratios (i.e., $H_{C:P}$) was lower under the warmed treatment than the ambient treatment (Fig. 3). In ecoenzymatic stoichiometry theory [26,32], the relative availability of hydrolyzed P via ecoenzymes (such as phosphatase) in soils relative to C in the microbial community composition was higher under experimental warming than in the control treatments. Ecosystem function and stability may have been impacted in the warmed soil columns in this study, with a negative allometric relationship to soil microorganism $H_{C:P}$. Our previous research has shown that a significant increase in soil microorganism biomass measured by total phospholipid fatty acids (PLFAs) occurred under warming [21], which is line with the data in Table 1; however, bacterial abundance in the soil tended to decrease along with an increase in the ratio of fungi:bacteria [21,40,41]. Furthermore, relatively high abundances of fungi in soil ecosystems increase the secretion of P-hydrolyzing enzymes in soils [21,42,43]. Meanwhile, functional gene array data verified that soil P utilization genes (such as polyphosphate kinase and exopolyphosphate) were also enhanced by warming [27], leading to increases in soil microbial P pools (Table 1) and speeding up nutrient cycling processes.

In a 13 d laboratory incubation test of YT wetland sediment (Text S3), lower dissolved oxygen (DO) concentrations in overlying water (Fig. S2) and higher ferrous iron (Fe^{2+}) concentrations in sediment (Fig. S3) were found for sediment sampled from the warmed column compared to the control treatment. This indicates that the redox-enhanced sediment under the warmed treatment could increase dissimilatory reduction of P-bound metal oxides and the decomposition of organic matter, liberating P from sediment aggregates. This may explain the relatively high levels of P in porewater and potential P flux under warming (Table 2). In addition to stimulating net primary production, warming could significantly increase both soil respiration and the associated C-degrading microbial genes [27,44], increasing the C flux from terrestrial ecosystems to the atmosphere in the global C cycle [8].

Results from our previous investigation [21], coupled with the data reported here, lend further support to the use of $H_{C:P}$, H_C, and H_P as ecological 'indicators' (Fig. 3, 1, & 2, respectively). In other words, $H_{C:P}$ values in the soil microorganism community in this study were positively correlated with C-P-related ecological function and stability. Because C:P stoichiometric homeostasis is associated with synergistic C-P acquisitions (as discussed above), a decreased $H_{C:P}$ value under warming conditions (Fig. 4) means that C biodegradation and P mineralization exceeded microorganism utilization of these elements, leading to increases in both C loss and P export. This has a "double-negative" effect on climate change and water quality.

We did not find a consistent relationship between $H_{C:P}$ values in warmed vs. control treatments in terms of LOC:DRP and MLOC:DRP ratios (Table 4). Soil C losses in this study were traced to the recalcitrant fraction, while pools of labile C were not depleted (Table 1). Carbon availability from HLOC for the soil microorganism assemblage and metabolism is generally higher

Table 4. Homeostatic regulation coefficient (*H*) in soils collected from wetland columns in the microcosm experiment (Control: ambient temperature; Warmed: ambient temperature +5°C).

Variable	Dependent variable	Sampling data	Control group			Warmed group		
			H	R^2	*p*	*H*	R^2	*p*
Porewater DOC	Microbial biomass C	Jul 2010	−5.00	0.084	0.203	4.26	0.085	0.219
		Nov 2010	2.76	0.274	**0.004**	3.00	0.257	**0.008**
		Mar 2011	1.71	0.386	**0.000**	−20.0	0.019	0.452
Porewater DOC:DRP	Microbial C:P	Jul 2010	500	0.014	0.672	100	0.005	0.953
		Nov 2010	250	0.000	–	71.4	0.002	0.976
		Mar 2011	55.6	0.007	0.872	43.4	0.016	0.553
Soil LOC: porewater DRP	Microbial C:P	Jul 2010	6.76	0.303	**0.004**	9.43	0.463	**0.000**
		Nov 2010	12.9	0.275	**0.003**	13.7	0.237	**0.010**
		Mar 2011	12.6	0.154	**0.013**	7.46	0.545	**0.000**
Soil MLOC: porewater DRP	Microbial C:P	Jul 2010	8.26	0.340	**0.003**	6.90	0.391	**0.000**
		Nov 2010	16.1	0.268	**0.008**	22.2	0.071	0.237
		Mar 2011	12.8	0.213	**0.013**	5.18	0.461	**0.000**

p-values refer to significant differences between two treatments. DOC and DRP refer to dissolved organic carbon and dissolved reactive phosphorus in porewater while MLOC and LOC refer to mid-labile organic carbon and labile organic carbon in soil, respectively.

than for the rest of the C sources [31]; therefore the index of HLOC is relatively sensitive to C dynamics for microorganism acquisition. Further mechanistic investigations are needed to establish direct links between the destabilization of soil organics and soil microorganism community dynamics in order to better assess the relationship between the availability of different forms of C to C:P stoichiometric dynamics and their ultimate effects on $H_{C:P}$.

Does homeostasis link to soil phosphorus flux?

Efforts to predict P flux or P dynamics from the soil biosphere to aquatic ecosystems are becoming increasingly important in light of concerns related to eutrophication, climate change, and other anthropogenic impacts [19,20,45,46]. Statistics-based modeling integrates various parameters, such as soil physicochemical properties, regional meteorology, and anthropogenic activities [19,20,23], and can be used to predict P flux under IPCC climate scenarios [4]. However, these models are insufficiently related to the ecological forcing of microorganisms in the soil micro-ecosystem, even though such microorganisms essentially mediate and drive Earth's biogeochemical cycles [13,27]. We found that potential P flux (Table 2) was significantly related to n-$H_{C:P}$ according to a negative allometric model (Fig. 4), verifying the hypothesis that a n-$H_{C:P}$ index may be useful as an ecological tool to predict P flux. Although C:P stoichiometric dynamics alone can provide some insight into P dynamics in an ecosystem [5,39], the n-$H_{C:P}$ term (as outlined in equation 4) comprehensively incorporates both soil C and P availability and soil microorganism nutrient acquisition under the principle of stoichiometric homeostasis [1]. The n-$H_{C:P}$ response mirrors mechanisms related to the metabolic theory of ecology [5,6] in that the soil micro-environment of wetland ecosystems responds positively (albeit via a negative allometric relationship) to temperature increases.

According to the derived models of potential P flux to n-$H_{C:P}$ (Fig. 4), lower n-$H_{C:P}$ values in this study indicate higher P flux, which is fundamentally consistent with the ecological interpretation of H_C (Fig. 1), H_P (Fig. 2), and $H_{C:P}$ (Fig. 3) between the two treatments. Although regression models differed, the ecological role of n-$H_{C:P}$ in this study is essentially identical in its nature to the linear and positive relationships between community *H* and ecological production and stability in a Mongolian grassland [2]. The consistency of these results suggests that although diverse factors may affect P dynamics, n-$H_{C:P}$ may be a useful ecological tool for assessing potential P flux. Notably, the mathematic formulations between n-$H_{C:P}$ and potential P flux were identical between the two experimental treatments (Fig. 4), suggesting that the feasibility of n-$H_{C:P}$ to quantitatively assess potential P flux might be mainly determined by the features of soil C-P stoichiometric homeostasis itself.

It should be noted that both n-$H_{C:P}$ and potential P flux are calculated using one of the same parameters; namely, soil porewater DRP, and are therefore inherently interrelated to some extent. However, n-$H_{C:P}$ (Eq 4) is computed using a multifactor equation linking soil C:P ratio (in form of soil HLOC: soil porewater DRP), soil microorganism biomass C:P ratio, and the associated homeostatic coefficient ($H_{C:P}$). Similarly, changes in potential P flux (Eq 5) were jointly determined by differences in the concentration of DRP in soil porewater vs. overlying water, by the quantity of overlying water and the area of topsoil in the wetland column, and by the overlying water replacement time interval. Moreover, values of these two indices were distinctly influenced by experimental warming (Table 2, Fig. 3). Therefore, the strong, negative allometric relationship observed between these indices (Fig. 4) is indeed an indication of the potential ecological management role of n-$H_{C:P}$ in assessing potential P flux, and is not be expected outright based only on co-variation arising from their common incorporation of a soil porewater DRP parameter.

The stoichiometric parameter $S_{C/P}$ (or $S_{C/N}$) has been defined as a scalar for the relative availability of hydrolyzed P (or N) in relation to microbial community composition [32,47]. Utilizing this $S_{C/P}$ (or $S_{C/N}$) metric, an allometric biogeochemical equilibrium model was developed to predict microbial growth efficiency (GE) from elemental C:N and C:P ratios in biomass ($B_{C/N}$ and $B_{C/P}$, respectively) and environmental substrate sources

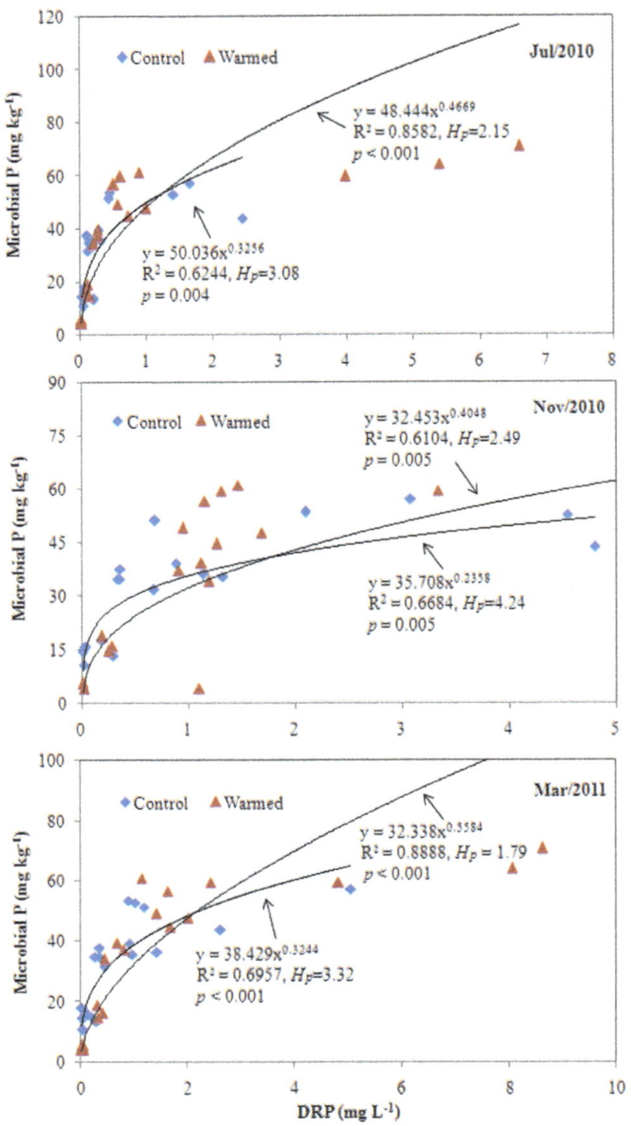

Figure 2. Relationships between soil microbial phosphorus (MB-P) content and dissolved reactive phosphorus (DRP) in porewater. Control and Warmed represent treatments of ambient temperature and ambient temperature +5°C, respectively. Coefficients fitting the data to the equation for stoichiometric homeostasis are given along with an estimate of H_P.

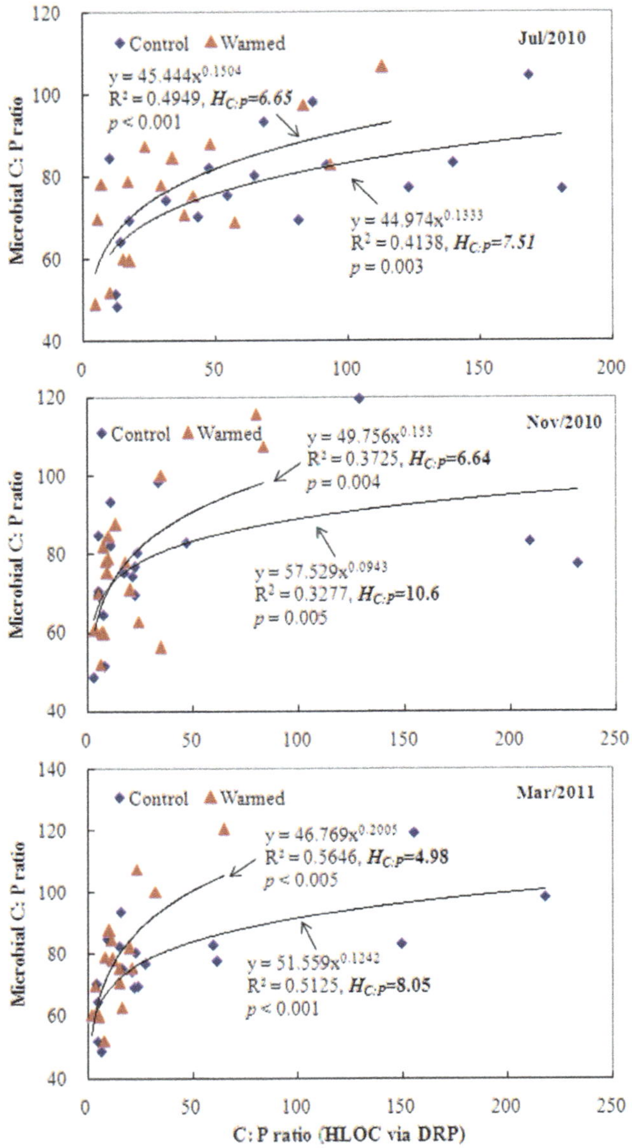

Figure 3. Relationships between the molar ratios of carbon to phosphorus (C:P) in soil microbial biomass and the C:P ratios of soil highly labile organic carbon (HLOC) to dissolved reactive phosphorus (DRP) in porewater. Control and Warmed represent treatments of ambient temperature and ambient temperature +5°C, respectively. Coefficients fitting the data to the equation for stoichiometric homeostasis are given along with an estimate of $H_{C:P}$.

($L_{C/N}$, $L_{C/P}$), integrated with ratios of ecoenzymatic activities (EEAs) that mediate C, N, and P acquisition ($EEA_{C/N}$, $EEA_{C/P}$) [32]. Although the intended prediction outcomes of these approaches differ, the nature of the relationship between microbial n-$H_{C:P}$ and sediment potential P flux is fundamentally similar to the use of the $S_{C/P}$ indictor, because both scalars reflect how stoichiometric balance/interaction is regulated by environmental signals. Further, they can also both be viewed in terms of an energy landscape (i.e., C) that directs the availability and flow of resources (e.g., N and P). Therefore, our experimental data and the measurement of soil n-$H_{C:P}$ might be practical for comparative investigations of soil P dynamics across a variety of wetland ecosystems. This n-$H_{C:P}$ method offers a new alternative for the measurement and understanding of soil P biogeochemical cycling, comparable to current methodologies that utilize biochemistry [37,42], molecular biology [21,27,43,46], isotopic tracing [48], and mechanistic models [19,23,45]. To the best of our knowledge, this is the first study of integrated C-P stoichiometric homeostasis to clearly demonstrate the linkage between soil n-$H_{C:P}$ and potential P flux in a freshwater wetland ecosystem.

Undoubtedly, aquatic plants are another primary pool for P biogeochemical cycling, because 12%–85% of P in water may be immobilized by wetland plants [48,49]. However, addressing the influence of aquatic plants on n-$H_{C:P}$ and other parameters was beyond the purpose of this study, and thus they were temporarily excluded from the soil columns during the course of the experiment. The involvement of aquatic plants in influencing soil

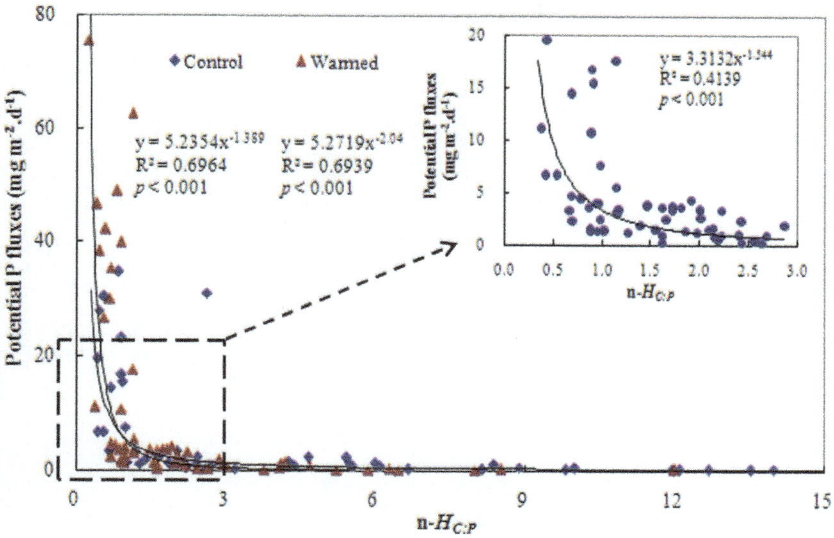

Figure 4. Relationship between potential phosphorus (P) flux and the specialized ability of stoichiometric homeostasis (n-$H_{C:P}$) as a molar ratio of soil highly labial organic carbon (HLOC) to dissolved reactive phosphorus (DRP) in each of the wetland soil columns. Control and Warmed represent treatments of ambient temperature and ambient temperature +5°C, respectively. The data for P flux and n-$H_{C:P}$ less than 20 mg m^{-2} d^{-1} and 3.0, respectively, are plotted in the figure inset.

C:P stoichiometric homeostasis and associated variations in n-$H_{C:P}$ thus requires further investigation.

Although soil respiration at the global scale responds positively to air temperature, in the long term warming may also mobilize older stored carbon, potentially resulting in higher carbon inputs to the soil biosphere [8]. If so, labile P forms would be 'returned' to the soil-microorganism complex under warming according to principles of ecoenzymatic stoichiometry [26] and mechanisms of synergistic C-P acquisition. Such an outcome would be somewhat inconsistent with the near-term decrease in soil C stock (Table 1) and increase in potential P flux (Table 2) we observed in the current microcosm warming experiment. Thus, it will also be necessary to further explore the linkage between n-$H_{C:P}$ and P flux under long-term warming.

In summary, microcosm warming impaired the stoichiometric homeostatic ability (H) of soil microorganisms to regulate P and C biogeochemical processes. This resulted in a "double negative" effect for wetland ecosystem services: increasing their potential for P export and enhancing their role as a recalcitrant organic C source. Moreover, the specialized homeostatic regulation ability (n-$H_{C:P}$, as a function of the C:P ratio in the form of HLOC to DRP in soil porewater relative to the C:P ratio in soil microbial biomass) of soil microorganisms was inversely linked to potential P flux at the soil-water interface. Based on our results, we advocate further use of homeostatic regulation indices (H) as novel ecological tools for assessing potential P flux and other biogeochemical dynamics in the face of global climate change.

Supporting Information

Figure S1 The design of the experimental wetland microcosm system setup by using independently monitored water bath jackets under the current climate condition (Left: Ambient temperature, Control) and the warming climate condition (Right: Ambient temperature +5°C, Warmed treatment).

Figure S2 Dynamics of dissolved oxygen (DO) concentration in overlying water during 12 d of sediment incubation. Sediment samples were collected in July 2010 from YaTang riverine wetland (YT) under control (ambient temperature) and warmed (ambient temperature +5°C) treatments in microcosm experiment. Error bars show ± SD. The differences between control and warmed treatments were tested by Student's t-test for each sampling point, indicated by * $p<0.05$, ** $p<0.01$.

Figure S3 Dynamics of ferric iron (Fe^{3+}) and ferrous iron (Fe^{2+}) concentration in sediment measured during a 13-d laboratory incubation for YaTang riverine wetland (YT) sediment samples under control (ambient temperature) and warmed (ambient temperature +5°C) treatments. Error bars show ± SD. The differences between control and warmed treatments were tested by Student's t-test for each sampling point, indicated by * $p<0.05$, ** $p<0.01$.

Table S1 Details of the six selected wetland sites used in the study.

Table S2 Select physico-chemical parameters of 20-cm depth soil samples collected from the JinHu wetland (JH), XiaZhuhu wetland (XZ), YaTang riverine wetland (YT), XiXi national wetland park (XX), BaoYang riverine wetland (BY), and ShiQiuyang multipond wetland (SQ) in May 2008.

Text S1 Microcosm configuration.

Text S2 Wetland column preparation and field sampling.

Text S3 Laboratory incubation for sediment oxygen demand and reducing capability.

Author Contributions

Conceived and designed the experiments: ZZ JDM XX QH. Performed the experiments: HW CY. Analyzed the data: HW HL DJ. Wrote the paper: HW JDM ZZ. Contributed the experimental materials: ZZ.

References

1. Sterner RW, Elser JJ (2002) Ecological Stoichiometry: The Biology of Elements From Molecules to the Biosphere. Princeton University Press, Princeton.
2. Yu Q, Chen QS, Elser JJ, He NP, Wu HH, et al. (2010) Linking stoichiometric homoeostasis with ecosystem structure, functioning and stability. Ecology Letters 13: 1390–1399.
3. Hansen J, Sato M, Ruedy R, Lo K, Lea DW, et al. (2006) Global temperature change. Proceedings of the National Academy of Sciences of the United States of America 103: 14288–14293.
4. IPCC (2001) Climate Change 2001:The Scientific Basis. Intergovernmental Panel on Climate Change. Cambridge University Press, Cambridge.
5. Allen AP, Gillooly JF (2009) Towards an integration of ecological stoichiometry and the metabolic theory of ecology to better understand nutrient cycling. Ecology Letters 12: 369–384.
6. Gillooly JF, Brown JH, West GB, Savage VM, Charnov EL (2001) Effects of size and temperature on metabolic rate. Science 293: 2248–2251.
7. Englund G, Ohlund G, Hein CL, Diehl S (2011) Temperature dependence of the functional response. Ecology Letters 14: 914–921.
8. Bond-Lamberty B, Thomson A (2010) Temperature-associated increases in the global soil respiration record. Nature 464: 579-U132.
9. Rhee GY (1978) Effects of N: P atomic ratios and nitrate limitation on algal growth, cell compostion and nitrate uptake. Limnology and Oceanography 23: 10–25.
10. Elser JJ, Fagan WF, Kerkhoff AJ, Swenson NG, Enquist BJ (2010) Biological stoichiometry of plant production: metabolism, scaling and ecological response to global change. New Phytologist 186: 593–608.
11. Yu QA, Elser JJ, He NP, Wu HH, Chen QS, et al. (2011) Stoichiometric homeostasis of vascular plants in the Inner Mongolia grassland. Oecologia 166: 1–10.
12. Totsche KU, Rennert T, Gerzabek MH, Kogel-Knabner I, Smalla K, et al. (2010) Biogeochemical interfaces in soil: The interdisciplinary challenge for soil science. Journal of Plant Nutrition and Soil Science 173: 88–99.
13. Falkowski PG, Fenchel T, Delong EF (2008) The microbial engines that drive Earth's biogeochemical cycles. Science 320: 1034–1039.
14. Hessen DO, Agren GI, Anderson TR, Elser JJ, De Ruiter PC (2004) Carbon, sequestration in ecosystems: The role of stoichiometry. Ecology 85: 1179–1192.
15. Verhoeven JTA, Arheimer B, Yin CQ, Hefting MM (2006) Regional and global concerns over wetlands and water quality. Trends in Ecology & Evolution 21: 96–103.
16. Elser JJ, Bracken MES, Cleland EE, Gruner DS, Harpole WS, et al. (2007) Global analysis of nitrogen and phosphorus limitation of primary producers in freshwater, marine and terrestrial ecosystems. Ecology Letters 10: 1135–1142.
17. Sundareshwar PV, Morris JT, Koepfler EK, Fornwalt B (2003) Phosphorus limitation of coastal ecosystem processes. Science 299: 563–565.
18. Mulder C, Elser JJ (2009) Soil acidity, ecological stoichiometry and allometric scaling in grassland food webs. Global Change Biology 15: 2730–2738.
19. Serpa D, Falcao M, Duarte P, da Fonseca LC, Vale C (2007) Evaluation of ammonium and phosphate release from intertidal and subtidal sediments of a shallow coastal lagoon (Ria Formosa-Portugal): A modelling approach. Biogeochemistry 82: 291–304.
20. Nicholls KH (1999) Effects of temperature and other factors on summer phosphorus in the inner Bay of Quinte, Lake Ontario: implications for climate warming. Journal of Great Lakes Research 25: 250–262.
21. Zhang ZJ, Wang ZD, Holden J, Xu XH, Wang H, et al. (2012) The release of phosphorus from sediment into water in subtropical wetlands: a warming microcosm experiment. Hydrological Processes 26: 15–26.
22. Wang H, Holden J, Spera K, Xu XH, Wang ZD, et al. (2013) Phosphorus fluxes at the sediment-water interface in subtropical wetlands subjected to experimental warming: A microcosm study. Chemosphere 90: 1794–1804.
23. Jeppesen E, Kronvang B, Meerhoff M, Sondergaard M, Hansen KM, et al. (2009) Climate change effects on runoff, catchment phosphorus loading and lake ecological state, and potential Adaptations. Journal of Environmental Quality 38: 1930–1941.
24. Cheng WG, Sakai H, Matsushima M, Yagi K, Hasegawa T (2010) Response of the floating aquatic fern Azolla filiculoides to elevated CO(2), temperature, and phosphorus levels. Hydrobiologia 656: 5–14.
25. Hilbert DW, Roulet N, Moore T (2000) Modelling and analysis of peatlands as dynamical systems. Journal of Ecology 88: 230–242.
26. Sinsabaugh RL, Hill BH, Shah JJF (2009) Ecoenzymatic stoichiometry of microbial organic nutrient acquisition in soil and sediment. Nature 462: 795-U117.
27. Zhou JZ, Xue K, Xie JP, Deng Y, Wu LY, et al. (2012) Microbial mediation of carbon-cycle feedbacks to climate warming. Nature Climate Change 2: 106–110.
28. Bradford MA, Fierer N, Reynolds JF (2008) Soil carbon stocks in experimental mesocosms are dependent on the rate of labile carbon, nitrogen and phosphorus inputs to soils. Functional Ecology 22: 964–974.
29. Wu J, Joergensen RG, Pommerening B, Chaussod R, Brookes PC (1990) Measurement of soil microbial biomass C by fumigation extraction - an automated procedure. Soil Biology & Biochemistry 22: 1167–1169.
30. Brookes PC, Powlson DS, Jenkinson DS (1982) Measurement of microbial biomass phosphorus in soil. Soil Biology & Biochemistry 14: 319–329.
31. Logninow W, Wisniewski W, Strony WM (1987) Fractionation of organic carbon based on susceptibility to oxidation Polish. Journal of Soil Science 20.
32. Sinsabaugh RL, Follstad Shah JJF (2012) Ecoenzymatic stoichiometry and ecological theory. In: Futuyma DJ, editor. Annual Review of Ecology, Evolution, and Systematics, Vol 43. Palo Alto: Annual Reviews. pp. 313–343.
33. Small GE, Pringle CM (2010) Deviation from strict homeostasis across multiple trophic levels in an invertebrate consumer assemblage exposed to high chronic phosphorus enrichment in a Neotropical stream. Oecologia 162: 581–590.
34. Marichal R, Mathieu J, Couteaux MM, Mora P, Roy J, et al. (2011) Earthworm and microbe response to litter and soils of tropical forest plantations with contrasting C:N:P stoichiometric ratios. Soil Biology & Biochemistry 43: 1528–1535.
35. Olk DC, Gregorich EG (2006) Overview of the symposium proceedings, "Meaningful pools in determining soil carbon and nitrogen dynamics". Soil Science Society of America Journal 70: 967–974.
36. Schmidt MWI, Torn MS, Abiven S, Dittmar T, Guggenberger G, et al. (2011) Persistence of soil organic matter as an ecosystem property. Nature 478: 49–56.
37. Liikanen A, Murtoniemi T, Tanskanen H, Vaisanen T, Martikainen PJ (2002) Effects of temperature and oxygen availability on greenhouse gas and nutrient dynamics in sediment of a eutrophic mid-boreal lake. Biogeochemistry 59: 269–286.
38. Rinu K, Pandey A (2010) Temperature-dependent phosphate solubilization by cold- and pH-tolerant species of Aspergillus isolated from Himalayan soil. Mycoscience 51: 263–271.
39. Cleveland CC, Liptzin D (2007) C: N: P stoichiometry in soil: is there a "Redfield ratio" for the microbial biomass? Biogeochemistry 85: 235–252.
40. Zhang W, Parker KM, Luo Y, Wan S, Wallace LL, et al. (2005) Soil microbial responses to experimental warming and clipping in a tallgrass prairie. Global Change Biology 11: 266–277.
41. Feng XJ, Simpson MJ (2009) Temperature and substrate controls on microbial phospholipid fatty acid composition during incubation of grassland soils contrasting in organic matter quality. Soil Biology & Biochemistry 41: 804–812.
42. Newman S, McCormick PV, Backus JG (2003) Phosphatase activity as an early warning indicator of wetland eutrophication: problems and prospects. Journal of Applied Phycology 15: 45–59.
43. Allison VJ, Condron LM, Peltzer DA, Richardson SJ, Turner BL (2007) Changes in enzyme activities and soil microbial community composition along carbon and nutrient gradients at the Franz Josef chronosequence, New Zealand. Soil Biology & Biochemistry 39: 1770–1781.
44. Wang H, He ZL, Lu ZM, Zhou JZ, Van Nostrand JD, et al. (2012) Genetic linkage of soil carbon pools and microbial functions in subtropical freshwater wetlands in response to experimental warming. Applied and Environmental Microbiology 78: 7652–7661.
45. Alvarez-Cobelas M, Sanchez-Carrillo S, Angeler DG, Sanchez-Andres R (2009) Phosphorus export from catchments: a global view. Journal of the North American Benthological Society 28: 805–820.
46. He ZL, Deng Y, Van Nostrand JD, Tu QC, Xu MY, et al. (2010) GeoChip 3.0 as a high-throughput tool for analyzing microbial community composition, structure and functional activity. Isme Journal 4: 1167–1179.
47. Sinsabaugh RL, Van Horn DJ, Shah JJF, Findlay S (2010) Ecoenzymatic Stoichiometry in Relation to Productivity for Freshwater Biofilm and Plankton Communities. Microbial Ecology 60: 885–893.
48. Noe GB, Scinto LJ, Taylor J, Childers DL, Jones RD (2003) Phosphorus cycling and partitioning in an oligotrophic Everglades wetland ecosystem: a radioisotope tracing study. Freshwater Biology 48: 1993–2008.
49. White JR, Reddy KR, Majer-Newman J (2006) Hydrologic and vegetation effects on water column phosphorus in wetland mesocosms. Soil Science Society of America Journal 70: 1242–1251.

23

Bioavailable Soil Phosphorus Decreases with Increasing Elevation in a Subarctic Tundra Landscape

Andrea G. Vincent[1,2]*, Maja K. Sundqvist[2], David A. Wardle[2], Reiner Giesler[1,3]

1 Department of Ecology and Environmental Sciences, Umeå University, Umeå, Sweden, 2 Department of Forest Ecology and Management, Swedish University of Agricultural Sciences, Umeå, Sweden, 3 Climate Impacts Research Centre, Department of Ecology and Environmental Sciences, Umeå University, Abisko, Sweden

Abstract

Phosphorus (P) is an important macronutrient in arctic and subarctic tundra and its bioavailability is regulated by the mineralization of organic P. Temperature is likely to be an important control on P bioavailability, although effects may differ across contrasting plant communities with different soil properties. We used an elevational gradient in northern Sweden that included both heath and meadow vegetation types at all elevations to study the effects of temperature, soil P sorption capacity and oxalate-extractable aluminium (Al_{ox}) and iron (Fe_{ox}) on the concentration of different soil P fractions. We hypothesized that the concentration of labile P fractions would decrease with increasing elevation (and thus declining temperature), but would be lower in meadow than in heath, given that N to P ratios in meadow foliage are higher. As expected, labile P in the form of Resin-P declined sharply with elevation for both vegetation types. Meadow soils did not have lower concentrations of Resin-P than heath soils, but they did have 2-fold and 1.5-fold higher concentrations of NaOH-extractable organic P and Residual P, respectively. Further, meadow soils had 3-fold higher concentrations of Al_{ox} + Fe_{ox} and a 20% higher P sorption index than did heath soils. Additionally, Resin-P expressed as a proportion of total soil P for the meadow was on average half that in the heath. Declining Resin-P concentrations with elevation were best explained by an associated 2.5–3.0°C decline in temperature. In contrast, the lower P availability in meadow relative to heath soils may be associated with impaired organic P mineralization, as indicated by a higher accumulation of organic P and P sorption capacity. Our results indicate that predicted temperature increases in the arctic over the next century may influence P availability and biogeochemistry, with consequences for key ecosystem processes limited by P, such as primary productivity.

Editor: Xiujun Wang, University of Maryland, United States of America

Funding: This project was funded by a Wallenberg Scholars award to DAW (http://www.wallenberg.com/kaw/) and a grant from the Centre for Environmental Research in Umeå (CMF) to DAW and RG (grant number 0632292; http://www8.umu.se/cmf/). The funders had no role in study design, data collection and analysis, decision to publish, or preparation of the manuscript.

Competing Interests: The authors have declared that no competing interests exist.

* E-mail: vincent.andrea@gmail.com

Introduction

Phosphorus (P) is an important macronutrient in subarctic tundra, where it is often co-limiting with nitrogen (N), and is sometimes the main limiting nutrient for plant growth [1–3]. The main source of plant-available, inorganic P (hereafter 'available P') in subarctic tundra is the biological mineralization of organic P [1,4,5]. Most soil P in the surface of tundra soils is organic [1,4,6] and is dominated by highly labile compounds [7]. Temperature is one of the main controls of organic matter decomposition in the arctic [8–10], meaning it is likely to constrain organic P mineralization and the supply of available P for plants. As such, warming experiments often show increases in P mineralization and/or plant P availability [9,11,12]. Likewise, an increase in foliar and litter P concentrations, together with a decrease in foliar and litter N to P ratios (indicative of greater relative P availability), have been observed with decreasing elevation (thus increasing temperature) in subarctic tundra [13]. The predicted annual average air temperature increases of 3–5°C in the subarctic during this century [14–16] could therefore influence the availability of P. Current knowledge on the distribution of different P forms in tundra landscapes and how they may be affected by temperature is limited, despite this information being crucial for understanding future temperature effects on bioavailable P.

Elevational gradients are powerful tools for studying how temperature and associated climatic factors, which shift with elevation, influence ecosystems properties and processes [17–21]. As such, an increasing number of studies in a wide range of ecosystems have used elevational gradients to study how temperature affects ecological processes [22–24] including in the subarctic [13,25–28]. Elevational gradients also provide excellent opportunities for exploring the impacts of temperature on the availability of soil P over larger spatial scales and timeframes than what is possible through conventional experiments [19]. There are, to our knowledge, no studies available on the responses of different P pools to elevation in subarctic or arctic tundra. These landscapes are biogeochemically heterogeneous as a consequence of spatial variation in topography and plant community structure [29,30]. This results in high spatial variation in P availability [6,25] and in the concentrations of Al and Fe [6,31], which influence P availability [32,33]. The Fennoscandian tundra therefore consists of a mosaic of highly contrasting vegetation types. Specifically, heath vegetation occurs on soils with low pH and N availability and is dominated by slow-growing dwarf-shrubs, while meadow vegetation grows on soils that are more N-rich and is dominated by faster-growing herbaceous species [25,29,30]. The biogeochemistry of P differs between the two vegetation types [6] and soil phosphate concentrations and plant foliar N to P ratios suggest that the relative importance of P versus N limitation is greater in meadow than in heath [6,13,25]. As such, obtaining a representative picture of how P biogeochemistry varies in the subarctic requires explicit recognition of both vegetation types.

In this study, we used a well-established elevational gradient [13,25,26,34] in which both heath and meadow vegetation types occur at all elevations, to study the effects of elevation-associated variation in temperature on P availability and biogeochemistry in a subarctic ecosystem. We also used this gradient to examine whether previously reported changes in foliar and litter P contents and N to P ratios with elevation are matched by shifts in the concentration of P fractions of different lability. To determine P fractions we used the Hedley fractionation method [35–38], an approach widely used to determine landscape-level variation in P availability and dynamics [6,39,40–42]. Additionally, we examined if other known drivers of P availability, such as soil P sorption capacity and Al and Fe concentration, influence the distribution of P fractions across the gradient for both the heath and meadow vegetation. Specifically, we tested the hypotheses that (1) The concentrations of labile P fractions decline with elevation (and therefore temperature) regardless of vegetation type, (2) Across all elevations, meadow soils have consistently lower concentrations of labile P than heath soils, together with higher concentrations of Al and Fe and higher soil P sorption capacity. By addressing these hypotheses we aim to better understand how temperature changes, such as those that are expected through climate warming, may affect P availability across two dominant vegetation types in subarctic tundra ecosystems.

Materials and Methods

Ethics statement

This study was carried out across an elevational gradient ranging from 500 to 1000 m above sea level (a.s.l.) along the north-east facing slope of Mount Suorooaivi (1193 m), located approximately 20 km south-east of Abisko, northern Sweden (68°21′ N, 18°49′ E), as described in [25]. Figure 1 shows a map with the location of the study site and the elevational gradient. No part of this gradient is located within national reserves and the land is public and not government protected. We confirm that all national and international rules were observed during the field work. This investigation did not involve measurements on animals or humans. The soils collected for this research were sampled at very small spatial scales and thus had negligible effects on ecosystem functioning. We have no commercial interests or conflicts of interest in performing this work.

Study site

The mean annual precipitation in the area, measured at the Abisko Scientific Research Station, was 310 mm for the period 1913–2000, with the highest mean monthly precipitation in July (51 mm) and the lowest in April (12 mm) [43]. The treeline in this area is formed by *Betula pubescens* spp. *czerepanovii* (mountain birch) and is located at 500–650 m a.s.l. at the study site. Two types of vegetation, heath and meadow, grow in a mosaic across the study area and co-occur on all elevations, with the meadow generally found in shallow depressions. The heath is dominated by ericaceous dwarf-shrubs such as *Vaccinium vitis-idaea*, *V. uliginosum* spp. *uliginosum*, *Empetrum hermaphroditum* and *Betula nana*. The meadow vegetation is dominated by graminoids such as *Deschampsia flexuosa* and *Anthoxanthum alpinum*, herbs such as *Saussurea alpina*, *Viola biflora* and *Solidago virgaurea*, and sedges, notably *Carex bigelowii* [25]. The bedrock is comprised of salic igneous rocks and quartic and phyllitic hard schists. The soils are podzols at lower elevations and cryosols at higher elevations. For more details on the study system see Table S1 and [13,25].

In the summer of 2007, four replicate plots (2×2 m) were established in each of the two vegetation types in each of six elevations (every 100 m ranging from 500 to 1000 m) rendering a total of 48 plots as described by [13,25]. To minimize pseudoreplication within each elevation, the average distance between each plot and the next nearest plot was c. 15 m (with the mean distance between the two most distant plots in each vegetation type being c. 100 m). Because the microtopography, hydrology, and soil fertility of these communities is highly spatially heterogeneous over short distances (i.e. in the order of a few metres) [29], it is expected that the 15 m distance among plots is enough to ensure adequate independence among them [13,25]. Plots at the 500 m elevation site were located in open birch forest, plots at 600 m were situated immediately above the forest line, and plots from 700 m to 1000 m were devoid of trees [25]. Monthly mean air temperatures during August 2009 at 500 m, 700 m and 1000 m at the study site were 12.1°C, 11.8°C, and 9.9°C, respectively. The daily mean temperature across the elevational gradient during the growing season of 2009 is given in Figure S1; similar data for the previous year (2008) is also given in Figure S2 [25]. Elevational gradients of this type serve as useful natural experiments to inform on the effects of temperature on ecological properties when other potentially co-varying factors can be kept constant [17,18]; as such, all plots in our study have the same aspect (north-east facing slope), parent material, and slopes of 4–18°.

Soil sampling

Humus soils were sampled on August 4, 2009, in a 1×1 m subplot inside each 2×2 m plot. For each plot, a minimum of four 4.5 cm diameter cores were sampled to the full humus depth to ensure a total sample volume of approximately 0.3 L humus, and humus depth was recorded. The humus depth (mean ± standard error) across all heath plots was 5.4±0.2 cm (for each elevation mean depths are 6.3, 6.1, 5.0, 4.0, 5.8, and 5.4 cm starting at 500 m.a.s.l. in ascending order), and for meadow plots it was 3.0±0.6 cm (starting at the 500 m.a.s.l. site and in ascending order, 7.2, 2.2, 1.7, 1.1, 3.2, and 2.8 cm). Within each plot, the cores were sieved (2 mm mesh) in the field to homogenize the samples, and combined to yield a single bulked sample per plot. Samples were sealed in polyethylene bags and transported to the laboratory on the same day as sampling. From each sample, a subsample was immediately stored at 2°C (<48 h) and the remaining portion was frozen at −20°C.

Hedley fractionation

In order to characterize soil P composition, we performed a five-step sequential extraction [38] with some modifications [44,45], outlined in Figure 2. This method was chosen because it provides a direct estimate of the lability of different operationally-defined P pools, and is the most commonly used method to investigate differences in P availability and dynamics in natural soils [35–37,46]. In Step 1, 2 g (dry weight) of humus soil were combined with 180 mL deionized water and one 9×62 mm anion exchange membrane (hereafter called 'resin') (55164 2S, BDH Laboratory Supplies, Poole, England) [47] in a 250 mL centrifuge bottle and shaken overnight (16 h, 150 rpm). The following day, the resins were removed and the sample centrifuged (14 000 *g*, 15 min, 10°C), after which the supernatant was discarded and the

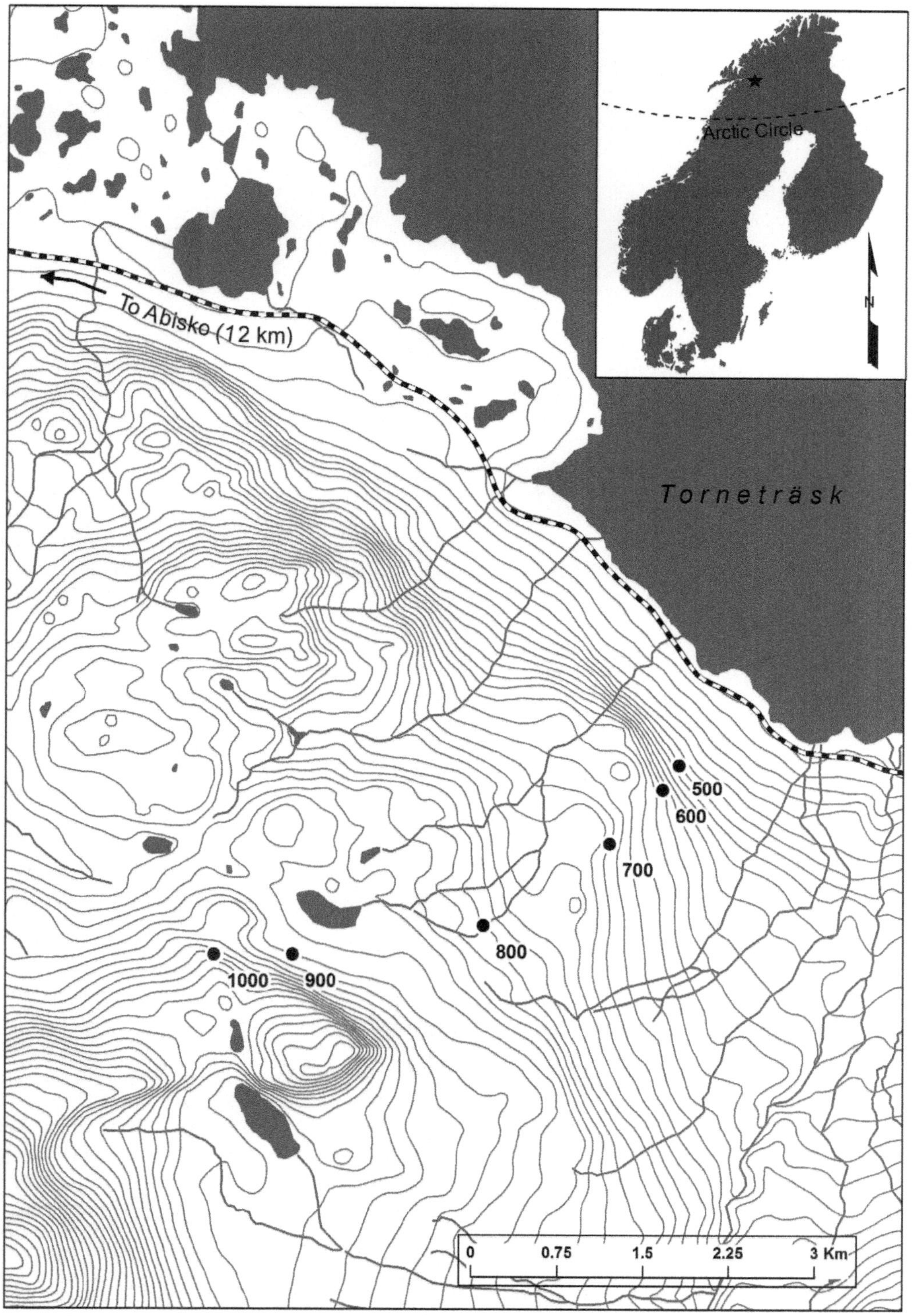

Figure 1. Location of the study elevational gradient. Filled black circles indicate each of the six study sites, ranging in elevation from 500 to 1000 m above sea level. The nearest town is Abisko, and its position in Sweden is indicated by a star in the map inset.

remaining soil used in Step 2. The resin was transferred to a bottle and eluated on a shaker (1 h, 150 rpm) with 40 mL NaCl. The eluate was immediately stored at −20°C until further analysis of this resin-extractable P fraction, hereafter referred to as 'Resin-P'. In Step 2, the soil remaining from Step 1 was combined with 180 mL 0.5 M $NaHCO_3$, set on a shaker (16 h, 150 rpm), centrifuged (14 000 *g*, 15 min, 10°C), and 40 mL of the supernatant removed and stored at −20°C for determination of organic and inorganic $NaHCO_3$-extractable P (hereafter referred to as 'Bic-P_o' and 'Bic-P_i', respectively). In Step 3, the soil remaining from Step 2 was extracted with 0.2 M NaOH following the same procedure as in Step 2 and stored at −20°C for further analysis of organic and inorganic NaOH-extractable P (hereafter referred to as 'NaOH-P_o' and 'NaOH-P_i', respectively). In Step 4, the soil remaining from Step 3 was combined with 1.0 M HCl, following the exact same extraction procedure as in Steps 2 and 3 and the supernatant stored at −20°C for further analysis of HCl-extractable P (hereafter, 'HCl-P'). In the final and fifth step, the soil remaining from Step 4 was washed with 180 mL of deionised water by shaking for 1 h, centrifuging, and discarding the supernatant, after which the soil was set to air dry at room temperature. This residual air-dried soil was ground in a ball mill and a 200 mg subsample was combined with 4 mL of concentrated nitric acid (HNO_3) and 1 mL of hydrogen peroxide (H_2O_2) (soil to solution ratio 1:40) and digested in a microwave (Mars XPress, CEM, Germany). This constitutes the 'Residual-P' fraction.

Total labile P consists of resin-extractable and bicarbonate-extractable P [48]. Resin-P is well correlated with P uptake by plants [49,50], has a rapid turnover and high bioavailability, and consists of P in a form that can exchange freely between the solid phase of the soil and the soil solution [35,51]. Bicarbonate-P_i (Bic-P_i) has similar sources to resin-P, turns over fast and is also bioavailable in the short-term [35], while bicarbonate-P_o (Bic-P_o) is easily mineralizable and supplies plant-available P. Hydroxide-extractable P_i (NaOH-P_i) and P_o (NaOH-P_o) are associated with Al and Fe phosphates, have lower plant availability, and longer turnover times. HCl-extractable P represents calcium-bound P_i and is often taken to represent P associated with primary minerals; it is also considered to be more stable [48]. By using extractants of different strength, this method is considered to quantify pools according to their 'lability' to plants [48].

Sorption index

In order to estimate the relative P sorption capacity of each soil sample from each plot, we used a single point P sorption method [52]. For each sample, 2 g (dry weight) of soil were weighed into each of two 60 mL bottles respectively, and 40 mL of 100 mM KCl was added to each bottle. For one of the two bottles, 1.6 mg P g soil^{-1} (dry weight basis, equivalent to 50 mmol kg^{-1} soil) of phosphate (KH_2PO_4) was added, which yielded two suspensions per soil sample: one with added P (spiked) and one without added P (unspiked). The suspensions were shaken for 24 h, filtered (Munktell 00H filter paper, pore size approx. 1 µm; Grycksbo, Sweden) and the amount of phosphate in each of the two filtrates was determined colorimetrically. The amount of sorbed P was estimated as the difference between the phosphate concentration in spiked and unspiked samples. We calculated the P sorption capacity using the equation

$$Sorption = \frac{x}{\log c}$$

where x is the quantity of P sorbed onto soil constituents and c is the equilibrium P concentration in the soil +P solution [52]. Thus, a high index indicates a high P sorption capacity.

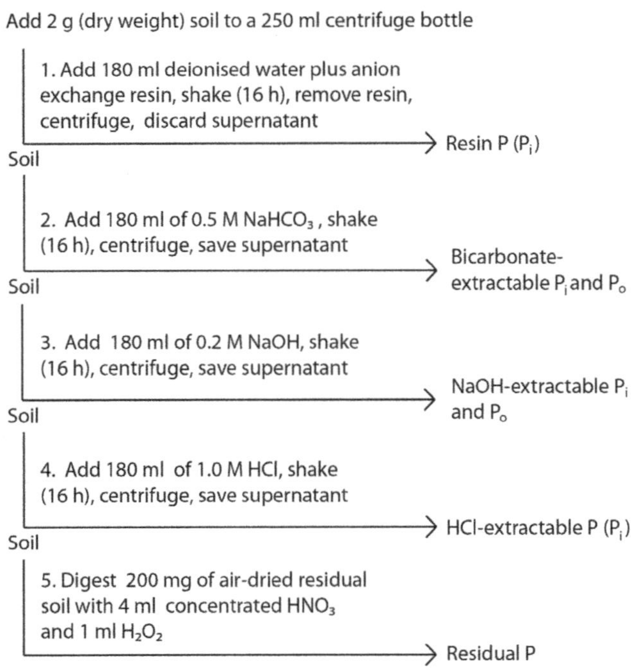

Figure 2. Flow chart showing the five steps involved in the sequential phosphorus fractionation used in this study.

Determination of P in extracts

All Hedley fractionation and sorption index extracts were analysed for molybdate-reactive P using a flow injection analyzer (FIA) (FIAstarTM 5000 Analyzer, FOSS Analytical AB, Höganäs, Sweden). The $NaHCO_3$ and NaOH solutions were diluted by a factor of 5–10 and amended with sulphuric acid (20 µl concentrated H_2SO_4 to 5 mL diluted extract) in order to precipitate organic material, after which they were centrifuged and the supernatant analyzed by FIA. Since the $NaHCO_3$ and NaOH extracts were colored, the measured P concentration was corrected by subtracting the effect of the color in the analysis. Total P in the $NaHCO_3$ and NaOH extracts was determined after digestion with acidified potassium persulfate ($K_2S_2O_8$), and inorganic P was analysed as above. The concentration of organic P was calculated as the difference between total and inorganic P. Phosphorus in the digests was analysed as above. Total soil P was calculated as the sum of all P fractions measured in the Hedley fractionation [6,39,44,53] and we refer to it hereafter as 'Total P'. Since all of our samples are from organic soils, we express the concentrations of all P fractions as mg kg^{-1} soil dry weight, which would be equivalent to expressing them on a per unit organic matter basis.

Aluminium and iron

Aluminium (Al) and iron (Fe) concentrations in the soil samples were determined following extraction by 0.2 M acid oxalate ($C_2H_8N_2O_4$) adjusted to a pH of 3 [54]. A subsample of 0.5 g dry weight of soil from each sample was combined with acid oxalate solution in a 1:30 soil to solution ratio and shaken on an orbital shaker for 4 h in the dark. Extracts were subsequently filtered (00H, Munktell Filter AB, Grycksbo, Sweden) and stored at 5°C (~2 months) following analysis of oxalate extractable Al and Fe (hereafter referred to as Al_{ox} and Fe_{ox}, respectively) determined by inductively-coupled plasma optical-emission spectroscopy (ICP-OES) (Perkin Elmer). The oxalate extractant is assumed to release exchangeable Al and Fe and dissolve non-crystalline and poorly crystalline oxides of Al and Fe (i.e. organic and amorphous Al and Fe forms), which are major P sorbents in non-calcareous soils [55]. Concentrations of Al_{ox}, Fe_{ox}, and all P fractions are expressed as mg kg^{-1} soil dry weight on the basis of oven-dried soils (105°C, 24 h).

Statistical analysis

We used multivariate ANOVA (MANOVA) for the entire dataset given the large number of response variables measured per soil sample. We followed a significant MANOVA with a two-way ANOVA to test for the effect of elevation and vegetation type and

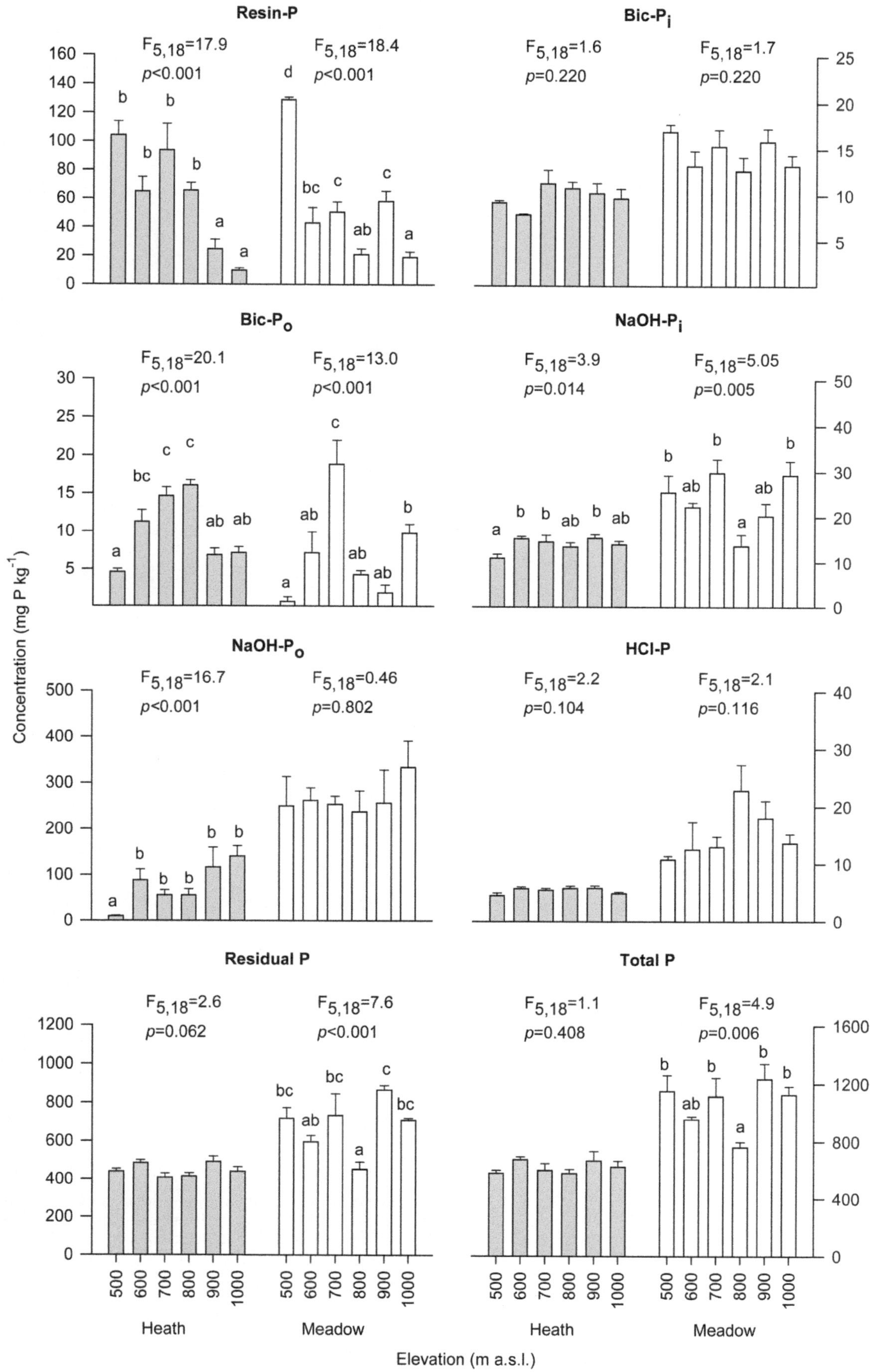

Resin-P
F5,18=17.9
p<0.001
F5,18=18.4
p<0.001
Bic-Pi
F5,18=1.6
p=0.220
F5,18=1.7
p=0.220
Bic-Po
F5,18=20.1
p<0.001
F5,18=13.0
p<0.001
NaOH-Pi
F5,18=3.9
p=0.014
F5,18=5.05
p=0.005
NaOH-Po
F5,18=16.7
p<0.001
F5,18=0.46
p=0.802
HCl-P
F5,18=2.2
p=0.104
F5,18=2.1
p=0.116
Residual P
F5,18=2.6
p=0.062
F5,18=7.6
p<0.001
Total P
F5,18=1.1
p=0.408
F5,18=4.9
p=0.006
Concentration (mg P kg-1)
Heath
Meadow
Elevation (m a.s.l.)

Figure 3. Concentration of phosphorus fractions in humus soils in contrasting vegetation types across an elevational gradient. Soils were collected in subarctic heath and meadow vegetation along an elevational gradient (500–1000 m) in Abisko, Sweden. Panels represent phosphorus fractions extractable with: anion-exchange resins (Resin-P); $NaHCO_3$ (inorganic fraction - Bic-P_i; and organic fraction - Bic-P_o); NaOH (inorganic fraction - NaOH-P_i; and organic fraction - NaOH-P_o); HCl (HCl-P fraction); non-extractable P (Residual P fraction); and Total P (arithmetic sum of all P fraction). Bars represent mean concentration (+1 SE) for four plots; for each P fraction and within each vegetation type, F and p values (with d.f.) are from a one-way ANOVA testing for the effect of elevation within each vegetation type, and bars topped with the same letter do not differ at $p = 0.05$ (Tukey's h.s.d.). Note the difference in y-axis scales.

their interaction on each response variable [13,25,26]. We chose ANOVA because it is the most powerful way of detecting significant responses to the underlying gradient even when these responses are not unidirectional or simple [56]. To further explore the effects of elevation within vegetation types, one-way ANOVA testing for the effect of elevation was performed separately for both heath and meadow. Where significant effects of elevation were found, data were further analyzed for differences among means using Tukey's honestly significant difference (h.s.d.) at $p = 0.05$. Tukey's h.s.d. was chosen because it reduces Type I error when multiple comparisons are being performed [57,58].

In order to account for potential effects of co-variation of soil P sorption capacity with elevation on labile P concentrations, we divided the concentration of each of the three labile components (i.e. Resin-P, Bic-P_i, and Bic-P_o) for each soil sample by the sorption index for that sample. We then performed separate one-way ANOVA testing for the effect of elevation on the transformed data, followed by Tukey's h.s.d. at $p = 0.05$ after a significant ANOVA result. We used Pearson's correlation to test for the relationship between Al_{ox} and Fe_{ox}; linear regression was used to test for the relationship between soil P fractions with $Al_{ox} + Fe_{ox}$, sorption index and temperature. When required, data were transformed to conform to the assumptions of parametric tests. For statistical analyses we used SPSS PASW Statistics 18.0 and R version 3.3.0 (www.r-project.org).

Results

Effects of elevation

Overall, Residual-P was the most abundant P fraction in humus across all elevations and vegetation types (59−76% of total soil P), followed by NaOH-P_o (1.6–31%), and Resin-P (1.5–18%) (Figure 3, Table S2). Total labile P (i.e. the sum of Resin-P, Bic-P_i and Bic-P_o) represented 4–20% of total soil P (Table S2). The concentrations of the remaining fractions ranged from 0.8–3.0% of total soil P (Table S2). The most unidirectional effect of elevation was found for Resin-P, for which the highest concentrations were recorded at the lowest elevation and the lowest concentrations at the highest elevation, for both vegetation types (Figure 3, Table 1). Resin-P concentrations at the highest elevation were 7–fold and 11–fold lower than at the lowest elevation for heath and meadow, respectively. Total labile P trends mirrored those of Resin P, concentrations at the highest elevation (1000 m) in heath were less than one fifth of those recorded at the 500 m (lowest elevation) and 700 m sites. In meadow, Total labile P concentrations at the highest elevation were less than one third of those recorded at the lowest elevation. While elevation also had a significant effect on all other soil properties except Bic-P_i and HCl-P (which was almost significant) (Table 1, Table 2), there were no simple unidirectional trends with elevation for any of the other P fractions measured (Figure 3) or for $Al_{ox} + Fe_{ox}$ or the sorption index (Figure 4). After concentrations were divided by the sorption index, the unidirectional elevational trend remained for Resin-P and the non-unidirectional effect remained for Bic-P_o; further, a significant effect of elevation emerged for Bic-P_i, through it being highest at the lowest elevation for the meadow (Figure S3). When all elevations and vegetation types were combined, Resin-P was significantly and negatively related with both $Al_{ox} + Fe_{ox}$ and the Sorption Index, although the relationship was weak (Table 3). Significant positive relationships were observed between most other P fractions and $Al_{ox} + Fe_{ox}$ and Sorption Index, the strongest being for NaOH-P_o and HCl-P (Table 3). Finally, Resin-P and Total labile P were significantly positively related with temperature in both the heath and meadow vegetation (Table S3), as was the case for pH in the heath. NaOH-P_o was negatively correlated with temperature in the heath (Table S3).

Effects of vegetation

The concentrations of all P fractions except Resin and Total labile P were significantly different between meadow and heath,

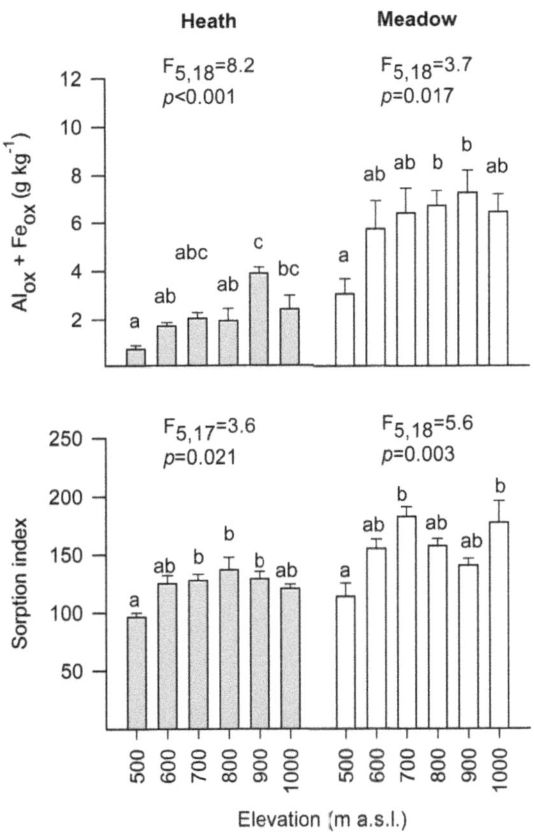

Figure 4. Concentration of aluminum and iron, and phosphorus sorption index for humus soils across an elevational gradient. Soils were collected in subarctic heath and meadow vegetation types along an elevational gradient (500–1000 m) in Abisko, Sweden. The top panel represents the concentration of the sum of oxalate-extractable Al and Fe ($Al_{ox} + Fe_{ox}$), the bottom panel represents the soil phosphorus sorption index. Bars represent mean values (+1 SE) for four plots; within each vegetation type, bars toped with the same letter do not differ at $p = 0.05$ (Tukey's h.s.d.) after a significant ANOVA (ANOVA results in Table 2).

Table 1. Effect of vegetation type and elevation as determined by multivariate analysis of variance (MANOVA) (F- values, with *p* in parenthesis) and two-way ANOVA on the concentration (mg kg^{-1}) of different phosphorus (P) fractions in humus soils along an elevational gradient in Abisko, Sweden.

	ANOVA results		
Variables	**Vegetation type (V)**	**Elevation (E)**	**V x E interaction**
Multivariate analyses			
MANOVA	5.602 (<0.001)	23.34 (<0.001)	2.89 (<0.001)
Univariate analysis			
Resin P[a]	0.140 (0.710)	27.4 (<0.001)	8.8 (<0.001)
Bic-extractable P_i	47.5 (<0.001)	1.8 (0.147)	1.4 (0.255)
Bic-extractable P_o	12.7 (0.001)	21.5 (<0.001)	7.6 (<0.001)
Total labile P[b]	0.94 (0.339)	27.6 (<0.001)	7.6 (<0.001)
NaOH-extractable P_i[a]	89.3 (<0.001)	4.6 (0.002)	5.1 (0.001)
NaOH-extractable P_o[a]	147.8 (<0.001)	11.5 (<0.001)	8.2 (<0.001)
HCl-extractable P	59.0 (<0.001)	2.4 (0.058)	1.8 (0.142)
Residual P[a]	104.7 (<0.001)	7.9 (<0.001)	4.7 (0.002)
Total P[a,c]	160.5 (<0.001)	4.0 (0.005)	2.7 (0.038)

For all variables, degrees of freedom for V = 1,36, E = 5,36, V*E = 5, 36.
Bic = Bicarbonate; P_i = inorganic P; P_o = organic P.
[a]Data were log transformed prior to analysis.
[b]Sum of Resin P, Bic-extractable P_i and Bic-extractable P_o.
[c]Sum of all sequentially extracted P fractions.

and of these all were highest in the meadow except for Bic-P_o (Table 1, Figure 3). The NaOH-P_o fraction had the largest difference in concentration between the two vegetation types (Figure 3); it was on average three–fold higher in meadow than in heath and represented ~25% of total P in the meadow but only ~13% in the heath. The concentrations of Al_{ox} and Fe_{ox} were highly correlated with each other ($R^2 = 0.951$, $p<0.001$), and their sum (Al_{ox} + Fe_{ox}) was on average three times higher in meadow than in heath (Figure 4). The sorption index was also significantly higher in meadow than in heath, on average by 20% (Figure 4, Table 2). Additionally, sorption index was significantly positively related to Al_{ox} + Fe_{ox} (Figure 5). After concentrations were divided by the sorption index, Bic-P_i and Bic-P_o remained highest in the meadow and heath, respectively, and Resin-P was also highest in the heath (Figure S3). There was an interactive effect between vegetation type and elevation for all P fractions except Bic-P_i and HCl-P, meaning that soil P composition responds differently to changes in elevation depending on vegetation type (Table 1, Figure 3).

Discussion

Effects of elevation on the concentration of labile P fractions

The concentrations of labile P fractions were hypothesized to decline with increasing elevation and associated declines in temperature, regardless of vegetation type. We consider Resin-P, Bic-P_i and Bic-P_o to represent the most labile P fractions [35]; Resin-P is well correlated with plant P uptake [49,59], and is considered to be the most highly bioavailable fraction [35,51]. Further, Bic-P_i is considered to derive from similar sources to Resin-P, while Bic-P_o to be easily mineralized [35,48]. These three fractions are discussed separately given the differences in their concentration and dynamics observed in this study. Partially in line with our predictions, Resin-P concentration declined with elevation in both heath and meadow. However, for this P form there was also a strong interactive effect of vegetation × elevation, meaning that this pattern of decline differed between the two vegetation types and that elevation (and thus temperature) effects on P availability therefore depend on vegetation type. Concen-

Table 2. Effects of vegetation type and elevation as determined by ANOVA (F- values, with *p* values in parenthesis) on phosphorus sorption index and the concentration of oxalate-extractable Al and Fe (Al_{ox} + Fe_{ox}) in humus soils along an elevational gradient in Abisko, Sweden.

	ANOVA results		
Variables	**Vegetation type (V)**	**Elevation (E)**	**V x E interaction**
Al_{ox} + Fe_{ox} (mg kg^{-1})[a]	107.9 (<0.001)	8.8 (<0.001)	0.6 (0.666)
Sorption index	36.3 (<0.001)	7.2 (<0.001)	2.4 (0.060)

Degrees of freedom (d.f.) for V = 1,36, E = 5,36, V*E = 5, 36 for Al_{ox} + Fe_{ox}; and d.f. for V = 1,35, E = 5,35, V*E = 5, 35 for Sorption index.
[a]Data were square-root transformed prior to analysis.

Table 3. Linear regressions between the sum of oxalate-extractable aluminium and iron (Al_{ox}+ Fe_{ox}) and the Sorption Index with phosphorus (P) fractions in humus soils along an elevational gradient in Abisko, Sweden.

P fraction	Al_{ox}+Fe_{ox}		Sorption Index	
	R^2 (*p*-value)	Direction	R^2 (*p*-value)	Direction
Resin P	**0.165 (0.002)**	**Negative**	**0.136 (0.006)**	**Negative**
Bic-P_i	**0.264 (<0.001)**	**Positive**	**0.124 (0.009)**	**Positive**
Bic-Po	0.002 (0.305)	N/A	0.078 (0.032)	**N/A**
Total labile P[a]	**0.138 (0.005)**	**Negative**	**0.070 (0.039)**	**N/A**
NaOH-P_i	**0.241 (<0.001)**	**Positive**	**0.385 (<0.001)**	**Positive**
NaOH-P_o	**0.467 (<0.001)**	**Positive**	**0.356 (<0.001)**	**Positive**
HCl P	**0.654 (<0.001)**	**Positive**	**0.223 (<0.001)**	**Positive**
Residual P	**0.352 (<0.001)**	**Positive**	**0.118 (0.010)**	**Positive**
Total P[b]	**0.419 (<0.001)**	**Positive**	**0.218 (<0.001)**	**Positive**

Values are for heath and meadow vegetation data combined. Degrees of freedom for all P pools are 1,46 for Al_{ox}+Fe_{ox}; and 1,45 for the Sorption Index due to one missing value.
[a]Sum of Resin P, Bic-extractable P_i and Bic-extractable P_o.
[b]Sum of all sequentially extracted P fractions.

trations of Bic-P_i and P_o did not show any unidirectional trends with elevation, but given that they occurred in much lower concentrations, the overall trend is still one of declining bioavailable P with increasing elevation. In some organic soils, the concentration of bioavailable inorganic P has been shown to be influenced by Al and/or Fe concentration and soil P sorption capacity [60–62]. While Al_{ox} + Fe_{ox} concentrations and the sorption index in our study sites differed across elevations (Figure 4) and were both weakly negatively correlated with Total labile P (Table 3), the elevational trends in Resin-P remained even after correcting for the sorption index (Figure S3), suggesting that they are largely explained by factors other than sorption.

A number of factors could explain the observed decrease in Resin-P with elevation. Organic P (as NaOH-extractable and Residual P) is the dominant form of P in these humus soils and enzymatic hydrolysis of organic P is a likely driver for the release of bioavailable inorganic P, as has been shown for Alaskan tundra

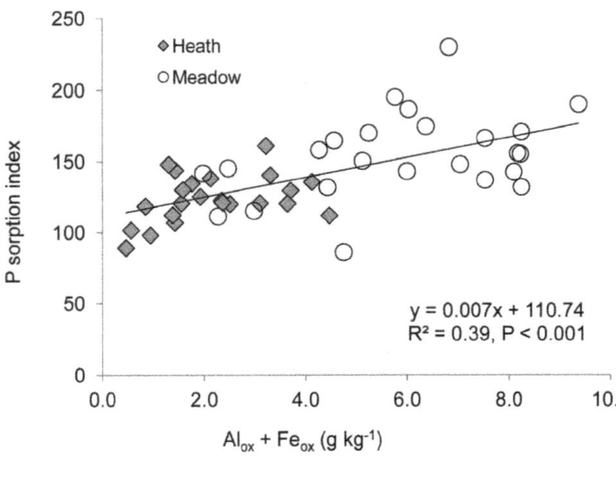

Figure 5. Relationship between phosphorus sorption capacity and metal concentration in humus soils across an elevational gradient. Sorption index versus the sum of oxalate-extractable Al and Fe (Al_{ox} + Fe_{ox}) in subarctic heath and meadow vegetation types across an elevational gradient in Abisko, Sweden.

[1]. Temperature is the strongest driver of soil enzyme activity in the subarctic [63] and warming experiments show that even a relatively small increase in temperature (1.2–1.7°C) can cause a large increase in organic P mineralization in alpine ecosystems [64]. The average air temperature difference across our elevational gradient during the vegetation growing season is about 2.5–3.0°C (Figures S1 and S2), which would be sufficient to explain the differences in labile P we observed. This suggests that the decline in temperature associated with increasing elevation is an important driver of the elevational decline in Resin-P observed and is consistent with our hypothesis. However, further information on organic P mineralization processes, for example using mineralization and soil enzyme studies together with ^{31}P nuclear magnetic resonance (NMR) spectroscopy to characterize soil organic P [65,66], would be useful to increase our mechanistic understanding of the trends observed here. Our results are in line with the few studies that have measured labile inorganic P concentrations at different elevations in subarctic tundra, as they report lower concentrations of 1.0 M KCl-extractable P at high elevations (1150 and 1000 m) than at low elevations (450 and 500 m) [25,67]. Our results are also consistent with [13] who showed a decrease in foliar P concentrations and an increase in foliar N to P ratios with elevation independent of vegetation type. Decreasing organic P mineralization with increasing elevation should lead to the accumulation of soil organic P, which was the case for NaOH-P_o in heath (Figure 3). However, a greater accumulation of soil organic P with elevation may have been negated to some extent by declining primary productivity, as is observed with increasing elevation both in the vicinity of the study area [28] and globally [19].

Temperature can affect P availability directly via effects on microbial mineralization [68] and soil process rates [69] but also indirectly by influencing factors that affect soil processes such as plant [70] and microbial [71] community composition. Temperature variation across the elevational gradient studied here has previously been shown to be related to a range of variables including soil pH, total N and ammonium concentration, C to N ratio, vegetation density, plant and microbial community composition, and fungal to bacterial ratios [25]. As such, many of the soil and vegetation properties that vary along the elevational gradient are likely to represent indirect temperature controls on P availability. Our interpretation is supported by many other studies that have used elevational gradients to understand how temperature affects ecological properties and processes [17,22–25,72–75].

To our knowledge, this is the first time that a Hedley P fractionation analysis has been carried out along a subarctic elevational gradient [6] and the two most comprehensive reviews on P fractionation lack data for these ecosystems [35,36]. These data are necessary to constrain soil P pools in terrestrial biogeochemical models, which are invaluable to understand the processes controlling P cycling and the role of P in driving terrestrial plant productivity [36]. The dominance of organic and residual P fractions (~87% of total soil P on average) and the concentrations of Resin-P measured in these soils are comparable with previous findings for other high latitude tundra ecosystems [1,4,6]. Additionally, the highest Resin-P concentrations measured (i.e. at the lowest elevations) were around two thirds of those reported for a Swedish boreal forest humus with low P sorption capacity [39]. While the concentration of Total labile P (sum of Resin P and Bicarbonate-extractable P) is considered to be low in tundra relative to other ecosystems [1,4–6], our results show that this may not necessarily hold when a wide range of elevations is considered. The 3–5 fold variation in Total labile P concentrations that we found along this elevational gradient encompasses the whole range of Total labile P concentrations recently reported in a world synthesis of Hedley P fractionation studies in natural ecosystems that spans 11 soil orders [36]. Our results further suggest that because small changes in elevation (and thus temperature) were associated with large changes in available P, increases in temperature according to current climate change predictions [14–16] may have a significant impact on future P availability in arctic tundra.

Vegetation differences and distribution of different P fractions

We hypothesized that the concentration of labile P fractions (Resin-P, Bic-P_i and Bic-P_o) would be lower in soils under meadow than heath vegetation, concomitant with higher Al and Fe concentrations and soil P sorption capacity. Our results partially support this, as overall Bic-P_o concentration was significantly lower, and sorption index and Al_{ox} + Fe_{ox} concentration were significantly higher, in meadow than in heath soils. Nevertheless, the concentration of Bic-P_i showed the opposite pattern and the concentration of Resin-P, the largest labile P pool, was not significantly different between vegetation types. Because higher Al and/or Fe has been shown to be positively correlated with P retention in organic soils [60,76–78], we expected that it would also lead to lower concentrations of Resin-P, which has a high propensity for sorption. However, Al_{ox} and Fe_{ox} concentrations in our soils were much lower than those of other organic soils for which negative correlations between labile inorganic P and Al and/or Fe have been reported [60,61,76]. No correlations have been reported for other organic soils that have Al_{ox} and Fe_{ox} concentrations more within the range of what we measured here [6,40]. Taken together, our results suggest that the Al and Fe concentrations in meadow soils are insufficient to exert a strong control on the concentration of labile inorganic P.

The higher Al_{ox} + Fe_{ox} concentrations and sorption index values in the meadow soils had little apparent effect on the concentration of Resin-P, but they may have resulted in higher sorption of organic P. This is suggested by the fact that concentrations of NaOH-P_o and Residual P (which in our soils is mostly organic) were 2.0– and 1.5– fold higher in meadow than in heath. Some organic P compounds have a high affinity for Al and Fe oxides [79–81], and soil organic P is often strongly positively related with Al and Fe concentration in organic soils, as seen in both this (Table 3) and other studies [40,45,61,77,82,83]. Soil organic P is generally considered to be less prone to sorption than are labile inorganic P forms such as Resin-P [84], but in some humus soils organic P is correlated with Al_{ox} and Fe_{ox} concentration while labile inorganic P is not [40], which is consistent with our findings. A higher organic P sorption capacity in meadow soils could lower P availability indirectly by protecting organic P from microbial mineralization. This is supported by our results showing that Resin-P concentration expressed as a percentage of total soil P was approximately half in meadow than in heath soils (Table S2), and is consistent with findings of relatively higher P limitation in meadow than in heath vegetation [6,13,25].

Conclusions

A 500 m decrease in elevation accompanied by a 2.5–3.0°C increase in temperature resulted in approximately 10–fold higher Resin-P concentration in subarctic soils under contrasting tundra vegetation types. However, the specific way in which Resin-P concentration changed with elevation (and thus temperature) differed with vegetation type. In meadow soils, the higher concentrations of oxalate-extractable Al and Fe, higher P sorption capacity, higher accumulation of organic P and proportionally lower concentrations of Resin-P than in heath soils is consistent with previous reports of higher relative P limitation in meadow vegetation. Our results suggest that predicted temperature increases of 3–5°C for the arctic in the next century could increase the concentrations of labile P in soils, but that the specific pattern of this increase is likely to depend on vegetation type. This is supported by previous work showing significant relationships between temperature and soil and vegetation properties known to influence P availability. Our results also emphasize the need for a better mechanistic understanding of P dynamics in arctic environments. Specifically, knowledge of organic P forms and of the processes that affect their mineralization is crucial for furthering our understanding of how temperature and vegetation affect P availability. New methods which apply 2-D ^{1}H-^{31}P NMR spectroscopy on soils [65] used in combination with enzyme assays and tracer experiments with ^{32}P and ^{33}P radioisotopes would be a natural follow-up of this study. This knowledge is important because future increases in soil P availability may affect key ecosystem processes such as primary production in these highly nutrient-limited tundra ecosystems.

Supporting Information

Figure S1 Temperature along the elevational gradient in 2009. Daily mean temperature (°C) in July and August 2009 at 500 m, 700 m and 1000 m, along the elevational study gradient.

Figure S2 Temperature along the elevational gradient in 2008. Daily mean temperature (°C) from 28 June to 31 August 2008 measured at 400, 700 and 1000 m along the elevational study gradient.

Figure S3 Concentration of soil phosphorus fractions divided by phosphorus sorption index along the elevational gradient. Panels represent phosphorus (P) fractions extractable with: anion-exchange resins (Resin-P) and $NaHCO_3$ (inorganic fraction – Bic-P_i; and organic fraction – Bic-P_o). Bars represent mean values (+1 SE) for four plots; for each P fraction and within each vegetation type, F and *p* values (with d.f.) are from a one-way ANOVA testing for the effect of elevation within each vegetation type, and bars topped with the same letter do not differ at $p = 0.05$ (Tukey's h.s.d.). Note the difference in y-axis scales.

Table S1 Selected properties of humus soils in contrasting vegetation types along the elevational gradient. Values represent the mean (±1 SE) of four plots, data from [13,25].

Table S2 Concentrations of different phosphorus fractions across an elevational gradient. Concentrations are expressed as proportions (%) of total soil phosphorus (P) in humus soils in contrasting vegetation types (heath and meadow). P_o = organic P, P_i = inorganic P. Values represent the means (±1 SE) of four replicate plots.

Table S3 Linear regressions between soil phosphorus fractions and air temperature. Temperature is the August 2009 mean, and phosphorus (P) is measured on humus soils collected under each of two vegetation types along an elevational gradient.

Acknowledgments

We are grateful to the staff at the Climate Impacts Research Centre (CIRC) for help in the field, to Keith Larson for producing Figure 1, and to Leo Condron, Jonathan De Long, and three anonymous reviewers for helpful comments on previous versions of this manuscript.

Author Contributions

Conceived and designed the experiments: AGV MKS DAW RG. Performed the experiments: AGV MKS DAW RG. Analyzed the data: AGV MKS RG. Contributed reagents/materials/analysis tools: RG. Wrote the paper: AGV MKS DAW RG. Wrote the grant application that funded the work: DAW RG.

References

1. Chapin FS, Barsdate RJ, Barel D (1978) Phosphorus cycling in Alaskan coastal tundra - hypothesis for the regulation of nutrient cycling. Oikos 31: 189–199.
2. Nadelhoffer KJ, Johnson L, Laundre J, Giblin AE, Shaver GR (2002) Fine root production and nutrient content in wet and moist arctic tundras as influenced by chronic fertilization. Plant and Soil 242: 107–113.
3. Shaver GR, Chapin FS (1995) Long-term responses to factorial NPK fertilizer treatment by Alaskan wet and moist tundra wet sedge species. Ecography 18: 259–275.
4. Giblin AE, Nadelhoffer KJ, Shaver GR, Laundre JA, McKerrow AJ (1991) Biogeochemical diversity along a riverside toposequence in Arctic Alaska. Ecological Monographs 61: 415–435.
5. Weintraub MN (2011) Biological phosphorus cycling in arctic and alpine soils. In: Bünemann E, Oberson A, Frossard E, editors. Phosphorus in action. Berlin Heidelberg: Springer-Verlag. pp. 295–316.
6. Giesler R, Esberg C, Lagerström A, Graae B (2012) Phosphorus availability and microbial respiration across different tundra vegetation types. Biogeochemistry 108: 429–445.
7. Turner BL, Baxter R, Mahieu N, Sjögersten S, Whitton BA (2004) Phosphorus compounds in subarctic Fennoscandian soils at the mountain birch (*Betula pubescens*) - tundra ecotone. Soil Biology and Biochemistry 36: 815–823.
8. Rustad LE, Campbell JL, Marion GM, Norby RJ, Mitchell MJ, et al. (2001) A meta-analysis of the response of soil respiration, net nitrogen mineralization, and aboveground plant growth to experimental ecosystem warming. Oecologia 126: 543–562.
9. Schmidt IK, Jonasson S, Shaver GR, Michelsen A, Nordin A (2002) Mineralization and distribution of nutrients in plants and microbes in four arctic ecosystems: responses to warming. Plant and Soil 242: 93–106.
10. Stark S (2007) Nutrient cycling in the tundra. In: Marschner P, Rengel Z, editors. Nutrient cycling in terrestrial ecosystems. Heidelberg: Springer-Verlag. pp. 309–330.
11. Chapin FS, Shaver GR, Giblin AE, Nadelhoffer KJ, Laundre JA (1995) Responses of arctic tundra to experimental and observed changes in climate. Ecology 76: 694–711.
12. Jonasson S, Michelsen A, Schmidt IK, Nielsen EV (1999) Responses in microbes and plants to changed temperature, nutrient, and light regimes in the arctic. Ecology 80: 1828–1843.
13. Sundqvist MK, Giesler R, Wardle DA (2011b) Within- and across-species responses of plant traits and litter decomposition to elevation across contrasting vegetation types in subarctic tundra. PLoS ONE 6(10): e27056. doi:10.1371/journal.pone.0027056.
14. ACIA (2005) Arctic Climate Impact Assessment. Cambridge, U.K: Cambridge University Press.
15. IPCC (2007a) Climate Change 2007: Impacts, Adaptation and Vulnerability. Contribution of Working Group II to the Fourth Assessment Report of the Intergovernmental Panel on Climate Change. Cambridge, UK: Cambridge University Press.
16. IPCC (2007b) Climate Change 2007: The Physical Science Basis. Contribution of Working Group I to the Fourth Assessment Report of the Intergovernmental Panel on Climate Change. Cambridge, UK and New York, USA: Cambridge University Press.
17. Fukami T, Wardle DA (2005) Long-term ecological dynamics: reciprocal insights from natural and anthropogenic gradients. Proceedings of the Royal Society B-Biological Sciences 272: 2105–2115.
18. Körner C (2007) The use of 'altitude' in ecological research. Trends in Ecology and Evolution 22: 569–574.
19. Sundqvist MK, Sanders NJ, Wardle DA (2013) Community and ecosystem responses to elevational gradients: processes, mechanisms and insights for global change. Annual Review of Ecology, Evolution and Systematics (in press).
20. Vitousek PM, Matson PA, Turner DR (1988) Elevational and age gradients in Hawaiian montane rainforest: foliar and soil nutrients. Oecologia 77: 565–570.
21. Vitousek PM, Turner DR, Parton WJ, Sanford RL (1994) Litter decomposition on the Mauna Loa environmental matrix, Hawai'i: Patterns, mechanisms and models. Ecology 75: 418–429.
22. Bragazza L, Parisod J, Buttler A, Bardgett RD (2013) Biogeochemical plant-soil microbe feedback in response to climate warming in peatlands. Nature Climate Change 3: 273–277.
23. Salinas N, Malhi Y, Meir P, Silman M, Cuesta RR, et al. (2011) The sensitivity of tropical leaf litter decomposition to temperature: results from a large-scale leaf translocation experiment along an elevation gradient in Peruvian forests. New Phytologist 189: 967–977.
24. Girardin CAJ, Aragao LEOC, Malhi Y, Huaraca Huasco W, Metcalfe DB, et al. (2013) Fine root dynamics along an elevational gradient in tropical Amazonian and Andean forests. Global Biogeochemical Cycles 27: 252–264.
25. Sundqvist MK, Giesler R, Graae BJ, Wallander H, Fogelberg E, et al. (2011a) Interactive effects of vegetation type and elevation on aboveground and belowground properties in a subarctic tundra. Oikos 120: 128–142.
26. Sundqvist MK, Wardle DA, Olofsson E, Giesler R, Gundale MJ (2012) Chemical properties of plant litter in response to elevation: subarctic vegetation challenges phenolic allocation theories. Functional Ecology 26: 1090–1099.
27. Karlsson J, Jonsson A, Jansson M (2005) Productivity of high-latitude lakes: climate effect inferred from altitude gradient. Global Change Biology 11: 710–715.
28. Jansson M, Hickler T, Jonsson J, Karlsson J (2008) Links between terrestrial primary production and bacterial production and respiration in lakes in a climate gradient in subarctic Sweden. Ecosystems 11: 367–376.
29. Björk RG, Klemedtsson L, Molau U, Harndorf J, Odman A, et al. (2007) Linkages between N turnover and plant community structure in a tundra landscape. Plant and Soil 294: 247–261.
30. Eskelinen A, Stark S, Männistö M (2009) Links between plant community composition, soil organic matter quality and microbial communities in contrasting tundra habitats. Oecologia 161: 113–123.
31. Litaor MI (1992) Aluminium mobility along a geochemical catena in an alpine watershed, Front Range, Colorado. Catena 19: 1–16.
32. Guzman G, Alcantara E, Barron V, Torrent J (1994) Phytoavailability of phosphate adsorbed on ferrihydrite, hematite, and goethite. Plant and Soil 159: 219–225.
33. Brady NC, Weil RR (1999) The nature and properties of soils. New Jersey: Prentice-Hall.
34. Milbau A, Shevtsova A, Osler N, Mooshammer M, Graae BJ (2013) Plant community type and small-scale disturbances, but not altitude, influence the invasability in subarctic ecosystems. New Phytologist 197: 1002–1011.
35. Cross AF, Schlesinger WH (1995) A literature review and evaluation of the Hedley fractionation: Applications to the biogeochemical cycle of phosphorus in natural ecosystems. Geoderma 64: 197–214.
36. Yang X, Post WM (2011) Phosphorus transformations as a function of pedogenesis: A synthesis of soil phosphorus data using Hedley fractionation method. Biogeosciences 8: 2907–2916.
37. Johnson AH, Frizano J, Vann DR (2003) Biogeochemical implications of labile phosphorus in forest soils determined by the Hedley fractionation procedure. Oecologia 135: 487–499.

38. Hedley MJ, Stewart JWB, Chauhan BS (1982) Changes in inorganic and organic phosphorus fractions induced by cultivation practices and by laboratory incubations. Soil Science Society of America Journal 46: 970–976.
39. Lagerström A, Esberg C, Wardle DA, Giesler R (2009) Soil phosphorus and microbial response to a long-term wildfire chronosequence in northern Sweden. Biogeochemistry 95: 199–213.
40. Achat DL, Bakker MR, Augusto L, Derrien D, Gallegos N, et al. (2013) Phosphorus status of soils from contrasting forested ecosystems in southwestern Siberia: effects of microbiological and physicochemical properties. Biogeosciences 10: 733–752.
41. Dieter D, Elsenbeer H, Turner BL (2010) Phosphorus fractionation in lowland tropical rainforest soils in central Panama. Catena 82: 118–125.
42. Kitayama K, Majalap-Lee N, Aiba S (2000) Soil phosphorus fractionation and phosphorus-use efficiencies of tropical rainforests along altitudinal gradients of Mount Kinabalu, Borneo. Oecologia 123: 342–349.
43. Kohler J, Brandt O, Johansson M, Callaghan T (2006) A long-term Arctic snow depth record from Abisko, northern Sweden, 1913-2004. Polar Research 25: 91–113.
44. Binkley D, Giardina C, Bashkin MA (2000) Soil phosphorus pools and supply under the influence of *Eucalyptus saligna* and nitrogen-fixing *Albizia facaltaria*. Forest Ecology and Management 128: 241–247.
45. Giesler R, Satoh F, Ilstedt U, Nordgren A (2004) Microbially available phosphorus in boreal forests: Effects of aluminum and iron accumulation in the humus layer. Ecosystems 7: 208–217.
46. Condron LM, Newman S (2011) Revisiting the fundamentals of phosphorus fractionation of sediments and soils. Journal of Soils and Sediments 11: 830–840.
47. Saggar S, Hedley MJ, White RE (1990) A simplified resin membrane technique for extracting phosphorus from soils. Fertilizer Research 24: 173–180.
48. Tiessen H, Moir JO (1993) Characterization of available P by sequential extraction. In: Carter MR, editor. Soil Sampling and Methods of Analysis. Ann Arbor, Michigan: Lewis Publishers. pp. 75–86.
49. Sibbesen E (1978) An investigation of anion exchange resin method for soil phosphorus extraction. Plant and Soil 50: 305–321.
50. Tran TS (1992) A comparison of four resin extractions and 32P isotopic exchange for teh assessment of plant available P. Canadian Journal of Soil Science 72: 281–294.
51. Frossard E, Condron LM, Oberson A, Sinaj S, Fardeau JC (2000) Processes governing phosphorus availability in temperate soils. Journal of Environmental Quality 29: 15–23.
52. Bache BW, Williams EG (1971) A phosphate sorption index for soils Journal of Soil Science 22: 289–301.
53. Hedley MJ, Kirk GJD, Santos MB (1994) Phosphorus efficiency and the forms of soil phosphorus utilized by upland rice cultivars. Plant and Soil 158: 53–62.
54. Buurman P, van Lagen B, Velthorst EJ (1996) Manual for soil and water analyses. Leiden: Blackhuys.
55. McKeague JA, Day JH (1966) Dithionite- and oxalate-extractable Fe and Al as aids in differentiating various classes of soils. Canadian Journal of Soil Science 46: 13–22.
56. Wardle DA, Bardgett RD, Walker LR, Peltzer DA, Lagerstrom A (2008) The response of plant diversity to ecosystem retrogression: evidence from contrasting long-term chronosequences. Oikos 117: 93–103.
57. Zar JH (2010) Biostatistical Analysis: Prentice-Hall/Pearson. 944 p.
58. Quinn GP, Keough MJ (2002) Experimental Design and Data Analysis for Biologists. New York: Cambridge University Press.
59. Tran TS (1992) A comparison of four resin extractions and 32P isotopic exchange for the assessment of plant available P. Canadian Journal of Soil Science 72: 281–294.
60. Dell'Olio LA, Maguire RO, Osmond DL (2008) Influence of Mehlich-3 extractable aluminum on phosphorus retention in organic soils. Soil Science 173: 119–129.
61. Giesler R, Petersson T, Hogberg P (2002) Phosphorus limitation in boreal forests: Effects of aluminum and iron accumulation in the humus layer. Ecosystems 5: 300–314.
62. Nieminen M, Jarva M (1996) Phosphorus adsorption by peat from drained mires in southern Finland. Scandinavian Journal of Forest Research 11: 321–326.
63. Wallenstein MD, McMahon SK, Schimel JP (2009) Seasonal variation in enzyme activities and temperature sensitivities in Arctic tundra soils. Global Change Biology 15: 1631–1639.
64. Rui YC, Wang YF, Chen CR, Zhou XQ, Wang SP, et al. (2012) Warming and grazing increase mineralization of organic P in an alpine meadow ecosystem of Qinghai-Tibet Plateau, China. Plant and Soil 357: 73–87.
65. Vestergren J, Vincent AG, Jansson M, Persson P, Istedt U, et al. (2012) High-Resolution Characterization of Organic Phosphorus in Soil Extracts Using 2D H-1-P-31 NMR Correlation Spectroscopy. Environmental Science & Technology 46: 3950–3956.
66. Vincent AG, Vestergren J, Gröbner G, Persson P, Schleucher J, et al. (2013) Soil organic phosphorus transformations in a boreal chronosequence. Plant and Soil 367: 149–162.
67. Jonasson S, Havstrom M, Jensen M, Callaghan TV (1993) In-situ mineralization of nitrogen and phosphorus of arctic soils after perturbations simulating climate change. Oecologia 95: 179–186.
68. Schimel JP, Bilbrough C, Welker JM (2004) Increased snow depth affects microbial activity and nitrogen mineralization in two Arctic tundra communities. Soil Biology & Biochemistry 36: 217–227.
69. Jenny H (1994) Climate as a soil-forming factor. In: Jenny H, editor. Factors of soil formation: A system of quantitative pedology. New York: Dover Publications. pp. 104–196.
70. Kardol P, Cregger MA, Campany CE, Classen AT (2010) Soil ecosystem functioning under climate change: plant species and community effects. Ecology 91: 767–781.
71. Strickland MS, Lauber C, Fierer N, Bradford MA (2009) Testing the functional significance of microbial community composition. Ecology 90: 441–451.
72. Sveinbjörnsson B, Davis J, Abadie W, Butler A (1995) Soil carbon and nitrogen mineralization in the Chugach Mountains of South-Central Alaska, USA. Arctic Antarctic and Alpine Research 27: 29–37.
73. Kitayama K, Aiba S-I, Majalap-Lee N, Ohsawa M (1998) Soil nitrogen mineralization rates of rainforests in a matrix of elevations and geological substrates on Mount Kinabalu, Borneo. Ecological Research 13: 301–312.
74. Wang S, Ruan H, Wang B (2009) Effects of soil microarthropods on plant litter decomposition across an elevation gradient in the Wuyi Mountains. Soil Biology & Biochemistry 41: 891–897.
75. Hoch G, Körner C (2012) Global patterns of mobile carbon stores in trees at the high-elevation tree line. Global Ecology and Biogeography 21: 861–871.
76. Giesler R, Andersson T, Lovgren L, Persson P (2005) Phosphate sorption in aluminum- and iron-rich humus soils. Soil Science Society of America Journal 69: 77–86.
77. Kang J, Hesterberg D, Osmond DL (2009) Soil organic matter effects on phosphorus sorption: a path analysis. Soil Science Society of America Journal 73: 360–366.
78. Richardson CJ (1985) Mechanisms controlling phosphorus retention capacity in freshwater wetlands. Science 228: 1424–1427.
79. Celi L, Lamacchia S, Marsan FA, Barberis E (1999) Interaction of inositol hexaphosphate on clays: adsorption and charging phenomena. Soil Science 164: 574–585.
80. McBridge M, Kung K (1989) Complexation of glyphosate and related ligands with iron (III). Soil Science Society of America Journal 53: 1668–1673.
81. Ognalaga M, Frossard E, Thomas F (1994) Glucose-1-phosphate and myo-inositol hexaphosphate adsorption mechanisms on goethite. Soil Science Society of America Journal 58: 332–337.
82. Gerke J (2010) Humic (organic matter)-Al(Fe)-phosphate complexes: An underestimated phosphate form in foils and source of plant-available phosphate. Soil Science 175: 417–425.
83. Vincent AG, Schleucher J, Grobner G, Vestergren J, Persson P, et al. (2012) Changes in organic phosphorus composition in boreal forest humus soils: the role of iron and aluminium. Biogeochemistry 108: 485–499.
84. Celi L, Barberis E (2005) Abiotic stabilization of organic phosphorus in the environment. In: Turner BL, Frossard E, Baldwin DS, editors. Organic Phosphorus in the Environment. Wallingford, UK.: CAB International, pp. 113–132.

Evaluation of Dissolved Organic Carbon as a Soil Quality Indicator in National Monitoring Schemes

David L. Jones[1]*, Paul Simfukwe[1], Paul W. Hill[1], Robert T. E. Mills[2], Bridget A. Emmett[3]

1 School of the Environment, Natural Resources & Geography, Bangor University, Bangor, Gwynedd, United Kingdom, **2** Ecole Polytechnique Fédérale de Lausanne (EPFL), School of Architecture, Civil and Environmental Engineering (ENAC), Laboratory of Ecological Systems (ECOS), Lausanne, Switzerland, **3** Centre for Ecology and Hydrology, Environment Centre Wales, Bangor, Gwynedd, United Kingdom

Abstract

Background: Monitoring the properties of dissolved organic carbon (DOC) in soil water is frequently used to evaluate changes in soil quality and to explain shifts in freshwater ecosystem functioning.

Methods: Using >700 individual soils (0–15 cm) collected from a 209,331 km^2 area we evaluated the relationship between soil classification (7 major soil types) or vegetation cover (8 dominant classes, e.g. cropland, grassland, forest) and the absorbance properties (254 and 400 nm), DOC quantity and quality (SUVA, total soluble phenolics) of soil water.

Results: Overall, a good correlation ($r^2 = 0.58$) was apparent between soil water absorbance and DOC concentration across the diverse range of soil types tested. In contrast, both DOC and the absorbance properties of soil water provided a poor predictor of SUVA or soluble phenolics which we used as a measure of humic substance concentration. Significant overlap in the measured ranges for UV absorbance, DOC, phenolic content and especially SUVA of soil water were apparent between the 8 vegetation and 7 soil classes. A number of significant differences, however, were apparent within these populations with total soluble phenolics giving the greatest statistical separation between both soil and vegetation groups.

Conclusions: We conclude that the quality of DOC rather than its quantity provides a more useful measure of soil quality in large scale surveys.

Editor: Dafeng Hui, Tennessee State University, United States of America

Funding: This work was funded by the United Kingdom Department for Environment, Food and Rural Affairs. The funders had no role in study design, data collection and analysis, decision to publish, or preparation of the manuscript.

Competing Interests: The authors have declared that no competing interests exist.

* E-mail: d.jones@bangor.ac.uk

Introduction

Measurement of dissolved organic matter (DOM) concentrations have provided a sensitive indicator for assessing large scale geographical and temporal (e.g. decadal) changes in water quality [1]. In contrast, while DOM is frequently measured in soil studies, it has rarely been used for evaluating changes in soil quality in large scale (e.g. national) monitoring programmes [2]. To some extent, the DOM concentration of freshwaters should reflect that of the surrounding catchment area suggesting that an assessment of soil DOM might provide a sensitive soil quality indicator. This would also match with studies investigating the progressive increase in DOM within rivers, a widespread trend observed in both Europe and North America [3–4]. An investigation of the spatial or temporal changes in soil DOM within different soil types could therefore prove useful for explaining these trends in catchments containing a complex mixture of soils. One potential drawback, however, is that determination of DOM in soil can be very time consuming and expensive when large numbers of samples need to be processed. One potential solution is to use the spectral properties of soil water which provides a rapid proxy for estimating dissolved organic C (DOC) concentration [5–10]. For a single geographical location, strong relationships are often apparent between DOC concentration and UV absorbance [11–13]. However, these relationships are often non-linear and the mathematical function describing the relationship can vary with season or changes in hydrological flow patterns [14–16]. Further, whilst differences in the quality and quantity of DOC are frequently reported for contrasting soil and vegetation types, whether these dissimilarities remain valid over large geographical scales remains uncertain. In this context, we undertook a national large scale survey of the quality and quantity of DOM across Great Britain to evaluate its potential for evaluating changes to soil quality in national soil quality monitoring programmes. Our first aim was to critically evaluate the relationship between direct and indirect methods for DOC quantification. Secondly, we aimed to evaluate whether different soil classes and vegetation cover types possessed unique DOM signatures in terms of both quantity and quality and whether these might be useful for explaining changes in DOM occurring at a national scale.

Methods

Soil Sampling

To encompass all the major soil and land use types, a total of 702 soil samples were collected throughout Great Britain in June-

July, 2007 (area 209,331 km^2; ca. 300 km^2 $sample^{-1}$) as part of the Centre for Ecology and Hydrology Countryside Survey (CS) [17]. Samples were selected, based on a stratified random sample of 1 km squares at gridpoints on a 15 km grid using the Institute of Terrestrial Ecology (ITE) Land Classification as the basis of the stratification (Fig. S1). At each grid intersection, a 1 km^2 sample area was selected. Within the 1 km^2 sample area, a 5×5 m^2 plot was randomly located and replicate 15 cm long ×4 cm diameter soil cores were collected. Topsoils (0–15 cm) were only sampled to reflect past and current (1978-present) standard practice in UK national monitoring [18–20]. This depth was originally selected by the UK Government's Department for Environment, Food and Rural Affairs as it was hypothesized that it would show the greatest change in response to environmental perturbation (including changes in land use, climate, agronomic management or atmospheric deposition). The soil horizons sampled included H, O and A horizons with *a*, *e*, *i*, *h*, *g*, *k* and *p* sub-designations [21]. Across all land use categories, the dominant eight soil groupings (% of total) were: Brown soils (31%), Podzolic soils (15%), Surface water (SW) gley soils (18%), Peat soils (13%), Groundwater (GW) gley soils (12%), Lithomorphic soils (8%), and Pelosol soils (3%) [22]. The FAO World Reference Base Classification equivalent categories for these soil groups are presented in Table S1 and their major chemical, physical and biological properties presented in Table 1. Vegetation cover at each sampling point was classified into eight aggregated vegetation classes (AVC) with the following groupings (% of total): Cropland (15%), Tall grass and herbs (4%), Fertile grassland (19%), Infertile grassland (21%), Lowland woodland (3%), Upland woodland (8%), Moorland grassland mosaic (11%), and Heathland and bog (19%) [17]. Aggregate vegetation classes were derived by cluster analysis of the mean DECORANA scores for 100 smaller classes obtained by TWINSPAN analysis of plant species data in each sample plots [37]. Further descriptions of the vegetation types can be found in Table S2.

Soil Water Collection and Analysis

Soil water was obtained by adding artificial rainwater to the intact soil columns and collecting the leachate as detailed in [23–24]. The absorbance of the soil leachate water was measured at 254 and 400 nm on a Synergy 96 well plate spectrophotometer (BioTek Instruments Inc., Winooski, VT) using Falcon flat-bottom UV well plates (BD Biosciences, Franklin Lakes, NJ). DOC concentrations were measured with a TOC-V analyser (Shimadzu Corp., Kyoto, Japan). Total dissolved phenolics and tannins were assayed colorimetrically using the Folin-Ciocalteu reagent (F9252; Sigma-Aldrich Inc.) according to [25] using gallic acid as a standard. Specific ultraviolet absorbance (SUVA) was calculated by dividing the absorbance at 254 nm (cm^{-1}) by the DOC concentration (mg l^{-1}).

Background Soil Analysis

Background soil characteristics were measured on replicate cores matching those used in the analysis above. After collection from the field, the soil was extruded from the core, roots and stones removed and the soil homogenised and dried (105 °C). Total soil C and N were analysed on an Elementar Vario-EL elemental analyser (Elementar Analysensysteme GmbH, Hanau, Germany) using the UKAS accredited method SOP3102 [23]. Bulk density was calculated as mass/volume after the removal of stones (>2 mm) and accounting for their volume [23]. Soil organic matter was determined by loss-on-ignition (LOI) by first drying soil (10 g) at 105°C and then measuring the mass loss after further heating at 375 °C for 16 h. Available soil P was measured using

Table 1. Properties of the major soil groups.

Soil group	Soil pH	Soil organic matter (%)	C-to-N ratio	Olsen P (mg kg^{-1})	Bulk density (g cm^{-3})	Exchangeable Ca (mmol kg^{-1})	Exchangeable Al (mmol kg^{-1})	Soil respiration (mg C cm^{-3} h^{-1})	Soil respiration (g C g SOC^{-1} h^{-1})
Pelosol	6.8±1.4^{a}	7.3±3.3^{cd}	10.7±0.8^{de}	28±22^{abc}	1.10±0.21^{a}	62±40^{a}	1.3±4.8a^{b}	10.2±26.4^{bc}	4.7±3.4^{a}
Brown	6.5±1.2^{a}	10.0±12.5^{d}	12.2±4.1^{e}	32±27^{abc}	1.04±0.33^{a}	39±27^{a}	2.1±4.8^{b}	8.6±11.0^{c}	7.3±5.6^{a}
Groundwater gley	6.6±1.2^{a}	10.9±13.5^{d}	13.4±4.4^{cde}	32±34^{abc}	0.98±0.36^{a}	42±31^{a}	2.3±5.4^{ab}	13.2±24.1^{bc}	8.1±6.4^{a}
Surface water gley	5.9±1.3^{a}	24.9±30.0^{bc}	14.2±5.5^{cd}	25±28^{bc}	0.78±0.43^{b}	54±49^{a}	6.3±14.6^{ab}	14.9±19.2^{bc}	6.6±5.0^{ab}
Podzolic	4.9±0.9^{b}	28.8±27.8^{b}	17.9±7.0^{b}	21±20^{c}	0.57±0.33^{c}	20±29^{b}	7.1±8.9a^{b}	17.9±32.9^{bc}	5.1±3.9^{b}
Lithomorphic	6.0±1.4^{a}	35.1±32.8^{b}	16.2±6.4^{bc}	41±69^{ab}	0.56±0.38^{c}	42±39^{a}	3.4±5.6^{ab}	20.8±33.0^{b}	6.4±5.6^{ab}
Peat	4.6±0.6^{b}	75.9±27.9^{a}	24.1±8.5^{a}	44±51^{a}	0.20±0.21^{d}	46±50^{a}	9.6±38.0^{a}	43.1±37.1^{a}	4.5±3.8^{b}
ANOVA	***	***	***	**	***	***	**	***	***

Values are expressed on a dry weight basis and represent mean ± SD.
** and *** indicate significant differences between soil groups at the $P<0.01$ and $P<0.001$ levels respectively. Superscript letters indicate significant between soil groups at the $P<0.05$ level.

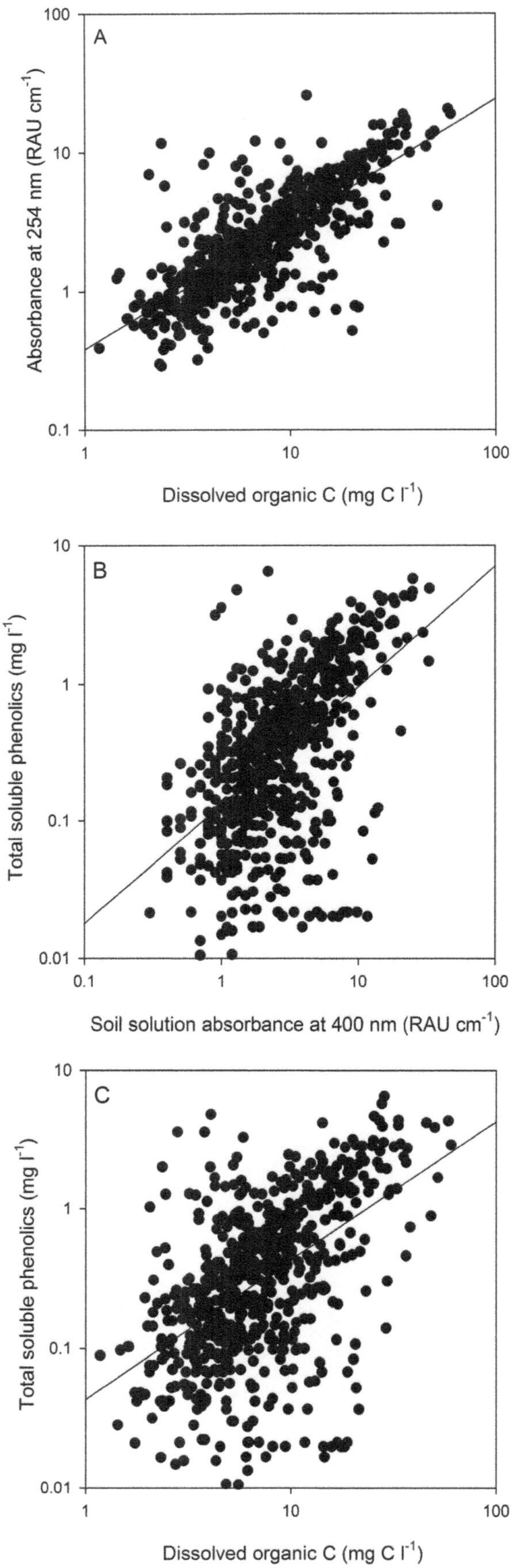

Figure 1. Relationship between the absorbance of soil water from 702 individual soils measured in the UV (254 nm) range and DOC concentration (Panel A), the relationship between total soluble phenolics and either solution absorbance at 400 nm (Panel B) or DOC concentration (Panel C). The lines are linear regression for all the data in the plots (Panel A, $r^2 = 0.579$; $y = 0.387 + 0.301x$; Panel B, $r^2 = 0.432$; $y = 0.12 + 0.139x$; Panel C, $r^2 = 0.375$; $y = 0.025 + 0.068x$).

the Olsen method whereby 5 g of soil was extracted with 100 ml of 0.5 M sodium bicarbonate (pH 8.5). The P in the extract was determined colorimetrically using the molybdate blue method (880 nm) using a continuous flow analyser [23]. Soil pH was measured by equilibrating 10 g of field-moist soil with 25 ml of deionised water. Exchangeable Ca and Al were determined by shaking 5 g of soil with 25 ml of 1.0 M NH_4Cl (250 rev min^{-1}, 60 min). Subsequently, the extracts were centrifuged (5000 *g*, 10 min) and the supernatant recovered for analysis. Ca in the extracts was determined by atomic absorption spectrometry on a Perkin Elmer Analyst 400 Atomic Absorption Spectrometer (PerkinElmer, Waltham, MA, USA). The extracts were diluted with $LaCl_3$ (0.5% w/v) prior to Ca determination. Al concentration in the extracts was determined using the modified catechol violet method [35]. The absorbance of the solution was measured at 580 nm using a PowerWave XS scanning microplate spectrophotometer (BioTek Instrument, Winooski, VT). Basal soil respiration was determined on one replicate core. The cores were wet to field capacity as described previously, placed in a sealed chamber (1250 cm^3 head space). The soils were then incubated at 10°C (average UK air temperature) for 1 h (at which linearity was known to be established following testing on selected cores which covered the range of soil types sampled). Subsequently, the head space gas was analysed for CO_2 concentration using a Clarus 500 Gas Chromatograph (PerkinElmer). SR was determined as the change in CO_2 concentration before and after incubation corrected for soil dry weight and soil organic matter content.

Statistical Analysis

Linear and stepwise regression analysis was undertaken using Minitab v16 (Minitab Inc., State College, PA). When the solution DOC, absorbance, soluble phenolic and SUVA values were grouped according to soil and vegetation type the data failed normality testing (Shapiro-Wilk). Consequently, the data were log transformed, normality verified and an ANOVA performed with Tukey-pairwise comparisons ($P < 0.05$ cut-off) using Minitab v16.

Results

Soil Water Properties

The relationship between the absorbance of soil water in the UV (254 nm) and visible (400 nm) range across 702 individual sites showed a strong linear correlation ($r^2 = 0.931$; $P < 0.001$; Fig. S2). A strong positive correlation was also observed between the DOC concentration in soil water and both absorbance at 254 nm ($r^2 = 0.58$; $P < 0.001$; Fig. 1) and 400 nm ($r^2 = 0.47$; $P < 0.001$). The prediction of DOC concentration using both absorbance values in a stepwise regression model did not result in a significantly better fit ($r^2 = 0.61$). In contrast, the absorbance characteristics of soil water proved to be a less reliable predictor of total soluble phenolics (254 nm $r^2 = 0.41$; $P < 0.001$; 400 nm $r^2 = 0.43$; $P < 0.001$; Fig. 1). Similarly, DOC concentration either with or without inclusion of spectral properties into the regression model, proved to be the least reliable predictor of total soluble phenolics ($r^2 = 0.38$; $P < 0.001$). SUVA proved to be the worst predictor of

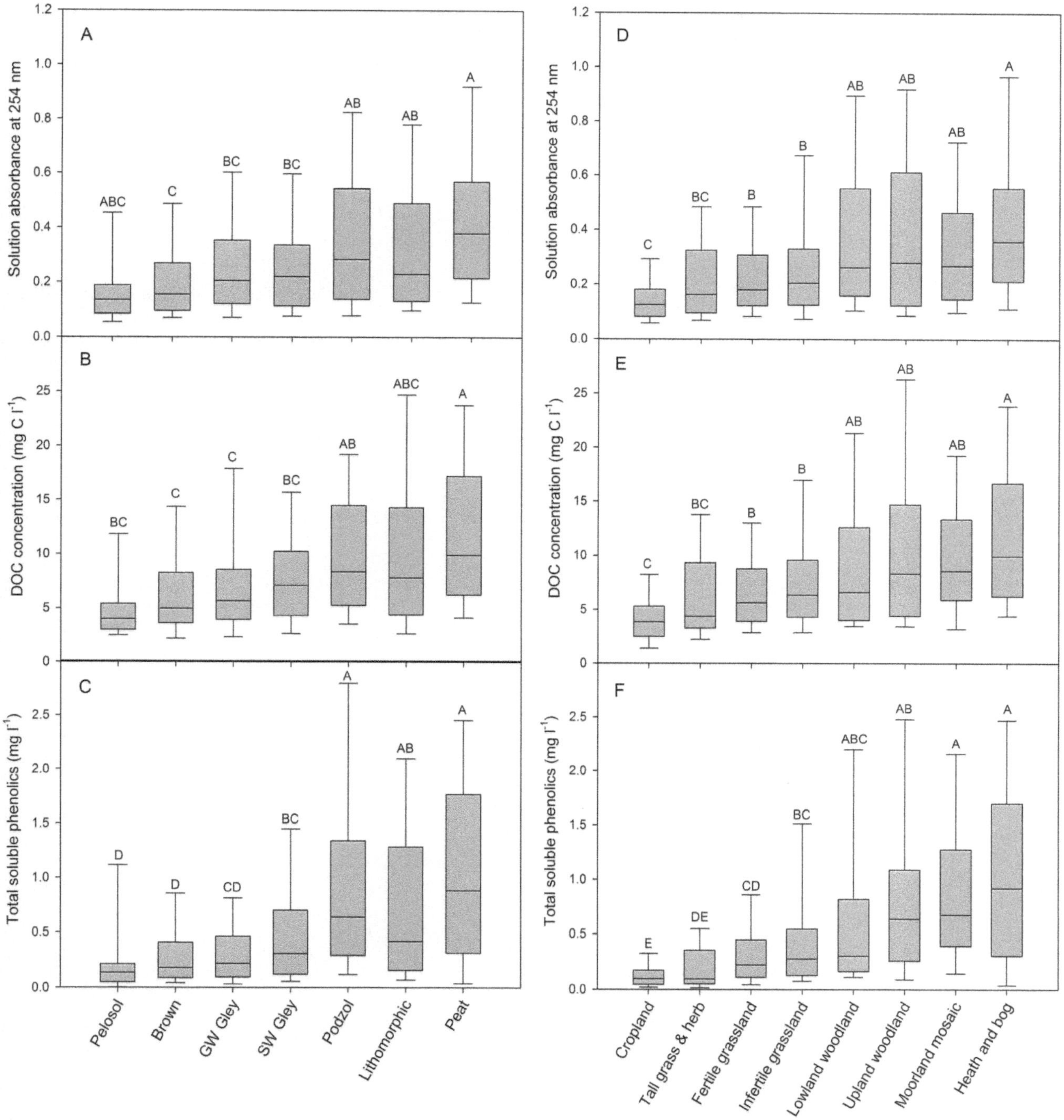

Figure 2. Box plots showing the influence of soil class (panels A–C) and vegetation cover type (panels D–F) on the absorbance of soil solutions at 254 nm, their DOC and total soluble phenolic concentration from 702 individual soils sampled as part of a nationwide soil quality assessment. The boundary of the box closest to zero indicates the 25th percentile, the line within the box marks the median, and the boundary of the box farthest from zero indicates the 75th percentile. Whiskers (error bars) above and below the box indicate the 90th and 10th percentiles respectively. Different letters indicate significant difference between individual groups ($P<0.05$). Soils are ranked in organic matter (OM) content from left (low OM) to right (high OM) while vegetation is broadly ranked according to agricultural productivity.

total soluble phenolics with either linear ($r^2 = 0.13$) or non-linear models (data not presented).

Is Soil or Vegetation a Better Predictor of DOC Quantity and Quality?

The quantity and quality of DOC as a function of different soil class and vegetation types is shown in Figure 2. Overall, there was a large similarity in the measured ranges of DOC, absorbance at 254 nm and soluble phenolics between the different groupings for both soil and vegetation. Despite this, ANOVA revealed significant differences between groups (all $P<0.001$) with response gradients apparent when the groups were ranked according to either organic matter content (soils) or an intensification/altitude scale (vegetation)(Fig. 2). Of the parameters measured, soluble phenolic content gave the greatest separation between groups. In contrast, SUVA revealed no significant differences between groupings for both soil ($P = 0.483$) and vegetation ($P = 0.819$; Fig. S3).

Discussion

Method of Soil Sampling

The sampling undertaken here formed part of a nationwide monitoring programme which is carried out each decade to assess changes in soil and water quality, landscape features and aquatic and terrestrial biodiversity [19,26]. Whilst the sampling depth (0–15 cm) is highly suited to agricultural soils which show a high degree of vertical homogeneity (due to cultivation), we acknowledge that the sampling regime may be less suited to highly stratified soils [27]. In these horizonated soils, differential amounts of O, E, A and possibly B horizons could be sampled by coring from 0–15 cm, depending on their thickness. This has led to many studies, particularly in forests, where soil quality is measured on organic and mineral horizons separately [27–28]. In our case, the historical legacy is such that the sampling protocols will not be changed for the foreseeable future (on scientific, socio-political and economic grounds) and therefore our results should be viewed in the context of this.

DOC and Soil Quality Assessment

Overall, our results show that in large geographical scale soil quality assessments across a broad range of vegetation and soil types, soil water DOC concentrations can be rapidly and cheaply estimated from their spectral properties. This is particularly relevant as DOC in itself can be used to predict the movement of both organic and inorganic pollutants in soil and is often seen as a pollutant itself when entering freshwaters [9,29–30]. Disappointingly, however, we found that all soil and vegetation types had a wide variability in their DOC concentrations. This indicates that the quantity of DOC may not represent a sensitive indicator for monitoring the stability of ecosystems when faced with anthropogenic perturbation (e.g. land use change). This is in contrast to [24] who found different rates of N mineralization (NH_4^++NO_3^-) in the same soils alongside other soil quality indicators as shown in Table 1. While SUVA is frequently used for assessing the quality of DOC in freshwaters [30], our results indicated that it was incapable of separating between land uses and soil classes over a wide geographical range and is therefore probably unsuited to large scale soil quality assessments. In contrast, total soluble phenols gave the best separation between ecosystem types. This was a surprising result as we had assumed that SUVA and soluble phenols would be highly correlated. This disagreement could be due to interference in both the determination of total phenolics (e.g. SO_2, DOC; [31–32]) or SUVA (e.g. pH, Fe^{3+}; [12]). However, more likely it is due to the natural variability in the total phenol content of humic substances originating from different soil classes and vegetation cover types [33–34]. As expected, the quality and quantity of DOC were dependent on both vegetation and soil type. This is to some extent expected considering that certain soil types favour certain vegetation covers, however, it does indicate that 2-way stratification by soil and vegetation type may prove more useful for revealing unique DOC signatures.

With respect to the measurement of DOC quality, we acknowledge that recent methodological advancements are increasing our potential to characterise the many thousands of compounds that comprise DOC [36]. It is highly likely that these new analytical approaches will reveal compounds that are specific to different functional soil or plant types (e.g. keystone compounds) improving our capacity to use DOC as a soil quality indicator. In addition, this analysis may also directly support the interpretation of temporal changes in other soil quality indicators (e.g. soil organic matter quality and quantity, soil biodiversity).

Supporting Information

Figure S1 Map of the UK showing the individual soil sampling locations used in the study. The total land area is 209,331 km2.

Figure S2 Relationship between the absorbance of soil water from 702 individual soils measured in either the UV (254 nm) or visible (400 nm) range. The line is a linear regression for all the data in the plot ($r^2 = 0.931$; $y = -0.01+0.123x$).

Figure S3 Box plots showing the influence of soil type (panel A) and vegetation cover (panel B) on the specific UV absorbance (SUVA) values from 702 individual soils sampled as part of a nationwide soil quality assessment. The boundary of the box closest to zero indicates the 25th percentile, the line within the box marks the median, and the boundary of the box farthest from zero indicates the 75th percentile. Whiskers (error bars) above and below the box indicate the 90th and 10th percentiles respectively. No significant differences were apparent between treatments.

Table S1 Comparable classification of the UK soil groups with those in the FAO World Reference Base Classification (WRB, 2006).

Table S2 Summary descriptions of the eight aggregate vegetation classes represented. Table adapted from Smart et al. (2003).

Acknowledgments

We would like to thank Ed Rowe at the Centre for Ecology and Hydrology for help with the experiments and data handling. Countryside Survey (CS) is funded by a consortium of UK government funded bodies led by the Natural Environment Research Council (NERC) and the Department for Environment, Food and Rural Affairs (Defra), which includes the Centre for Ecology and Hydrology, Countryside Council for Wales, Forestry Commission, Natural England, the Northern Ireland Environment Agency, the Scottish Government, Scottish Natural Heritage, and the Welsh Assembly Government. The completion of CS was only possible with the support and advice of many dedicated individuals from these and other organizations. The authors would like to thank the land owners, farmers, and other land managers who gave permission for the field surveyors to collect data and samples from their land; without such cooperation scientific field studies like CS would not be possible. Additionally, the authors would especially like to acknowledge the major contribution made by the team of laboratory staff who processed all the soils and undertook the chemical analyses.

Author Contributions

Conceived and designed the experiments: DLJ BAE. Performed the experiments: PS RTEM. Analyzed the data: PS PWH DLJ RTEM. Wrote the paper: DLJ PWH PS BAE.

References

1. Camarero L, Rogora M, Mosello T, Anderson NJ, Barbieri A, et al. (2009) Regionalisation of chemical variability in European mountain lakes. Freshwater Biol 54: 2452–2469.
2. Filep T, Rekasi M, Kincses I (2008) Effect of soil quality on dissolved organic matter (DOM) content of soils. Cereal Res Commun 36: 667–670.
3. Evans CD, Monteith DT, Cooper DM (2005) Long-term increases in surface water dissolved organic carbon: Observations, possible causes and environmental impacts. Environ Pollut 137: 55–7.
4. Clutterbuck B, Yallop AR (2010) Land management as a factor controlling dissolved organic carbon release from upland peat soils. 2 Changes in DOC productivity over four decades. Sci Total Environ 408: 6179–6191.
5. Mattson JS, Smith CA, Jones TT, Gerchako SM, Epstein BD (1974) Continuous monitoring of dissolved organic-matter by UV-visible photometry. Limnol Oceanogr 19: 530–535.
6. Dittman JA, Shanley JB, Driscoll CT, Aiken GR, Chalmers AT, et al. (2009) Ultraviolet absorbance as a proxy for total dissolved mercury in streams. Environ Pollut 157: 1953–1956.
7. Brandstetter A, Sletten RS, Mentler A, Wenzel WW (1996) Estimating dissolved organic carbon in natural waters by UV absorbance (254 nm). J Soil Sci Plant Nutr 159: 605–607.
8. Jaffrain J, Gerard F, Meyer M, Ranger J (2007) Assessing the quality of dissolved organic matter in forest soils using ultraviolet absorption spectrophotometry. Soil Sci Soc Am J 71: 1851–1858.
9. Amery F, Degryse F, Cheyns K, De Troyer I, Mertens J, et al. (2008) The UV-absorbance of dissolved organic matter predicts the fivefold variation in its affinity for mobilizing Cu in an agricultural soil horizon. Eur J Soil Sci 59: 1087–1095.
10. Hansson K, Kleja DB, Kalbitz K, Larsson H (2010) Amounts of carbon mineralised and leached as DOC during decomposition of Norway spruce needles and fine roots. Soil Biol Biochem 42: 178–185.
11. Dilling J, Kaiser K (2002) Estimation of the hydrophobic fraction of dissolved organic matter in water samples using UV photometry. Water Res 36: 5037–5044.
12. Weishaar JL, Aiken GR, Bergamaschi BA, Fram MS, Fujii R, et al. (2003) Evaluation of specific ultraviolet absorbance as an indicator of the chemical composition and reactivity of dissolved organic carbon. Environ Sci Technol 37: 4702–4708.
13. Tipping E, Corbishley HT, Koprivnjak JF, Lapworth DJ, Miller MP, et al. (2009) Quantification of natural DOM from UV absorption at two wavelengths. Environ Chem 6: 472–476.
14. Wheeler JR (1977) Dissolved organic-carbon - spectral relationships in coastal waters. Limnol Oceanogr 22: 573–575.
15. Edwards AC, Cresser MS (1987) Relationships between ultraviolet absorbency and total organic-carbon in 2 upland catchments. Water Res 21: 49–56.
16. Izbicki JA, Pimentel IM, Johnson R, Aiken GR, Leenheer J (2007) Concentration, UV-spectroscopic characteristics and fractionation of DOC in stormflow from an urban stream, Southern California, USA. Environ Chem 4: 35–48.
17. Emmett BA, Reynolds B, Chamberlain PM, Rowe E, Spurgeon D, et al. (2010) Countryside Survey: Soils Report from 2007. CS Technical Report No. 9/07, Centre for Ecology and Hydrology, Wallingford, UK.
18. Bellamy PH, Loveland PJ, Bradley RI, Lark RM, Kirk GJD (2005) Carbon losses from all soils across England and Wales 1978–2003. Nature 437: 245–248.
19. Norton LR, Maskell LC, Smart SS, Dunbar MJ, Emmett BA, et al. (2012) Measuring stock and change in the GB countryside for policy - Key findings and developments from the Countryside Survey 2007 field survey. J Environ Manage 113: 117–127.
20. Reynolds B, Chamberlain PM, Poskitt J, Woods C, Scott WA, et al. (2013) Countryside Survey: National "soil change" 1978-2007 for topsoils in Great Britain-acidity, carbon, and total nitrogen status. Vadose Zone J 12: vzj2012.0114
21. FAO (2006) Guidelines for Soil Description. Fourth Edition.Food and Agriculture Organization of the United Nations, Rome, Italy.
22. Avery BW (1990) Soils of the British Isles. Cambridge University Press, Cambridge.
23. Emmett BA, Frogbrook ZL, Chamberlain PM, Griffiths R, Pickup R, et al. (2008) Countryside Survey. Soils Manual. NERC/Centre for Ecology & Hydrology, (CS Technical Report No. 3/07, CEH Project Number: C03259), 180 pp.
24. Rowe EC, Emmett BA, Frogbrook ZL, Robinson DA, Hughes S (2012) Nitrogen deposition and climate effects on soil nitrogen availability: Influences of habitat type and vegetation. Sci Total Environ 434: 62–70.
25. Velioglu YS, Mazza G, Gao L, Oomah BD (1998) Antioxidant activity and total phenolics in selected fruits, vegetables and grain products. J Agric Food Chem 46: 4113–4117.
26. Smart SM, Clarke RT, van de Poll HM, Robertson EJ, Shield ER, et al. (2003) National-scale vegetation change across Britain; an analysis of sample-based surveillance data from the Countryside Surveys of 1990 and 1998. J Environ Manage 67: 239–254.
27. Lawrence GB, Fernandez IJ, Richter DD, Ross DS, Hazlett PW, et al. (2013) Measuring environmental change in forest ecosystems by repeated soil sampling: a North American perspective. J Environ Qual 42: 623–639.
28. O'Neill KP, Amacher MC, Palmer CJ (2005) Developing a national indicator of soil quality on U.S forestlands: Methods and initial results. Environ Monit Assess 107: 59–80.
29. Steinberg CEW, Hoss S, Kloas W, Lutz I, Meinelt T, et al. (2003) Hormonelike effects of humic substances on fish, amphibians, and invertebrates. Environ Toxicol 19: 409–411.
30. Zhang Q, Kuang WF, Liu LY, Li KX, Wong KH, et al. (2009) Trihalomethane, haloacetonitrile, and chloral hydrate formation potentials of organic carbon fractions from sub-tropical forest soils. J Hazard Mater 172: 880–887.
31. Matilainen A, Gjessing ET, Lahtinen T, Hed L, Bhatnagar A, et al. (2011) An overview of the methods used in the characterisation of natural organic matter (NOM) in relation to drinking water treatment. Chemosphere 83: 1431–1442.
32. Ohno T, First PR (1998) Assessment of the Folin and Ciocalteu's method for determining soil phenolic carbon. J Environ Qual 27: 776–782.
33. Sharma OK, Krishnan PS (1966). Colorimetric estimation of humic acid with phenol reagent of Folin and Ciocalteu. Anal Biochem 14: 11–17.
34. Thorn KA, Folan DW, MacCarthy P (1989) Characterization of the International Humic Substances Society standard and reference fulvic and humic acids by solution state carbon-13 (^{13}C) and hydrogen-1 (^{1}H) nuclear magnetic resonance spectrometry. U.S. Geological Survey, Water-Resources Investigations Report 89-4196, Denver, CO, USA.
35. Dougan WK, Wilson AL (1974) The absorptiometric determination of aluminium in water. A comparison of some chromogenic reagents and the development of an improved method. Analyst 99: 413–30.
36. Osborne DM, Podgorski DC, Bronk DA, Roberts Q, Sipler RE, et al. (2013) Molecular-level characterization of reactive and refractory dissolved natural organic nitrogen compounds by atmospheric pressure photoionization coupled to Fourier transform ion cyclotron resonance mass spectrometry. Rapid Comm. Mass Spectrom. 27: 851–858.
37. Bunce RGH, Barr CJ, Gillespie MK, Howard DC, Scott RA, et al. (1999) Vegetation of the British Countryside—the Countryside Vegetation System. Department of the Environment, London.

Dynamics of Soil Bacterial Communities in Response to Repeated Application of Manure Containing Sulfadiazine

Guo-Chun Ding[1], Viviane Radl[2], Brigitte Schloter-Hai[2], Sven Jechalke[1], Holger Heuer[1], Kornelia Smalla[1]*, Michael Schloter[2]

1 Julius Kühn-Institut - Federal Research Centre for Cultivated Plants (JKI), Institute for Epidemiology and Pathogen Diagnostics, Braunschweig, Germany, **2** Helmholtz Zentrum München, German Research Center for Environmental Health, Research Unit for Environmental Genomics, Neuherberg, Germany

Abstract

Large amounts of manure have been applied to arable soils as fertilizer worldwide. Manure is often contaminated with veterinary antibiotics which enter the soil together with antibiotic resistant bacteria. However, little information is available regarding the main responders of bacterial communities in soil affected by repeated inputs of antibiotics via manure. In this study, a microcosm experiment was performed with two concentrations of the antibiotic sulfadiazine (SDZ) which were applied together with manure at three different time points over a period of 133 days. Samples were taken 3 and 60 days after each manure application. The effects of SDZ on soil bacterial communities were explored by barcoded pyrosequencing of 16S rRNA gene fragments amplified from total community DNA. Samples with high concentration of SDZ were analyzed on day 193 only. Repeated inputs of SDZ, especially at a high concentration, caused pronounced changes in bacterial community compositions. By comparison with the initial soil, we could observe an increase of the disturbance and a decrease of the stability of soil bacterial communities as a result of SDZ manure application compared to the manure treatment without SDZ. The number of taxa significantly affected by the presence of SDZ increased with the times of manure application and was highest during the treatment with high SDZ-concentration. Numerous taxa, known to harbor also human pathogens, such as *Devosia*, *Shinella*, *Stenotrophomonas*, *Clostridium*, *Peptostreptococcus*, *Leifsonia*, *Gemmatimonas*, were enriched in the soil when SDZ was present while the abundance of bacteria which typically contribute to high soil quality belonging to the genera *Pseudomonas* and *Lysobacter*, *Hydrogenophaga*, and *Adhaeribacter* decreased in response to the repeated application of manure and SDZ.

Editor: Gabriele Berg, Graz University of Technology (TU Graz), Austria

Funding: This work was funded by the Deutsche Forschungsgemeinschaft (DFG) in the framework of the Research Unit FOR 566 "Veterinary medicines in soil: Basic research for risk analysis" (SM59/5-3). The funders had no role in study design, data collection and analysis, decision to publish, or preparation of the manuscript.

Competing Interests: The authors have declared that no competing interests exist.

* E-mail: kornelia.smalla@jki.bund.de

Introduction

The use of animal manure for fertilization of agricultural soils has a long tradition in many parts of the world and is generally assumed to be ecologically more friendly and sustainable than mineral fertilizer. In particular organic and bio-dynamic farms depend on manure or compost as sources for fertilization. However, industrial husbandries with typically large numbers of animals sharing limited space depend on the prophylactic and therapeutic use of antibiotics. In addition, in several parts of the world antibiotics are still being used as growth promoters. Although antibiotics as growth promoters were banned in many European countries, considerable amounts of antibiotics such as tetracyclines, β-lactams and sulfonamides are still used in animal husbandries [1]. Depending on their physicochemical properties, many antibiotics such as sulfonamides are to a large extent excreted via urine or feces and are not or only to a low extent degraded during manure storage [2,3]. Thus, spreading manure on agricultural soils does not only introduce nutrients required for maintaining the soil fertility but also antibiotics, their metabolites and antibiotic resistant bacteria. Indeed, antibiotics have been detected in the environment due to the use of manure for soil fertilization or direct deposition via dung and urine of animals grazing on pastures [4,5].

The rapid sequestration of most antibiotics in soils, as e.g. observed for the sulfonamide antibiotic SDZ [6], leads usually to low concentrations of bioavailable SDZ, which are far below the minimal inhibitory concentrations after a single application of manure. Most data published indicated only short-term effects on the microbial community after a single application of manure spiked with antibiotics followed by a fast regeneration of the community structure [7,8] and its function [9–11]. However, agricultural management implies in most cases a repeated application of manure mainly during the vegetation period to keep the level of nutrients needed for best possible plant growth. Surprisingly, little is known of the effects of repeated application of manure containing antibiotics on microbial communities in soil which might cause cumulative effects. Despite a rapid dissipation and sequestration of SDZ in soil [3,6,12] an increased abundance of *sul1* and *sul2* resistance genes and their transferability was observed in soils treated with manure containing SDZ compared to control manure under microcosm, mesocosm and field conditions [13–15], suggesting that manure containing antibiotics enhanced the spreading of the antibiotic resistance genes in soils. In particular the repeated application of manure in combination with antibiotics may set the ground for an increased abundance of resistant bacteria as recently reported by Heuer *et al.* [16] and thus might stimulate the spreading of antibiotic resistance genes and mobile genetic elements in agricultural ecosystems. Antibiotic resistance genes localized on mobile genetic elements can be captured by human and veterinary pathogens and thus pose a threat to the treatment of bacterial diseases [17–19].

In the present microcosm experiment, the effect of repeated application of manure containing antibiotics on the soil bacterial community was investigated by 454 sequencing of 16S rRNA gene fragments amplified from total community (TC)-DNA. We have chosen the sulfonamide SDZ as a model compound as it is still frequently used in pig husbandry [4] and highly persistent in manure [20]. Manure unspiked and spiked with SDZ in two concentrations was applied three times, and the bacterial community composition was monitored over a period of 193 days. We hypothesized that repeated application of manure spiked with SDZ to soil would increase the degree of disturbance and reduce the resilience of soil bacterial communities compared to a single application of manure spiked with SDZ or soils treated with manure not containing antibiotics. Resilience was measured as the stability of bacterial diversity after changes in soil properties ([21]; here manure addition). To differentiate between short and long-term effects, the soil was sampled 3 and 60 days after each manure application, respectively.

Materials and Methods

Experimental design

The experimental design is described in detail by Heuer *et al.* [16]. In short, topsoil samples (Ap horizon) of a silt loam soil (Orthic Luvisol) with no history of manure application were used. Pots were filled with 100 g of 2 mm-sieved soils and treated with 4 g of manure from healthy pigs (approximately 30 m^3/ha) spiked with SDZ or not to achieve a SDZ concentration in soil of 10 mg/kg, 100 mg/kg, or no SDZ (Sigma-Aldrich, Germany), which corresponded to samples S10, S100, and S0, respectively. Soil samples without manure application served as controls (U). To avoid effects based on different manure composition at the different time points of application, manure was taken before the experiment and frozen in aliquots at −20°C. For each time point of application, an aliquot of the manure was thawn and the corresponding amount of SDZ was directly applied before mixing the manure into the soil. The SDZ concentration S10 was chosen in accordance to maximal inputs of sulfonamides detected in agricultural soils [3,22]. Manure loads were applied on days 0, 63 and 133. The loosely covered pots were incubated at 15°C in the dark. Water was added to the microcosms twice a week to compensate for weight losses and to maintain a soil moisture of about 55% of the soil's water-holding capacity. For each treatment and time point 5 replicates were prepared and treated individually for the subsequent analysis. Samples were taken every 3rd and 60th day after manure application (days 3, 60, 66, 123, 136 and 193) and kept at −80°C until further analysis. No specific permissions were required for collecting the manure and soil and the study did not involve endangered or protected species.

DNA extraction

Extraction of total community DNA from soil samples was performed according to Griffith *et al.* [23]. Briefly, 0.4 g soil per replicate were added to a lysing matrix tube (MP Biomedicals, Germany) and submitted to phenol:chloroform extraction starting with homogenization for 30 seconds at 5.5 m S-1 (Precellys, PeqLab, Germany). Extracted DNA was finally resolved in 50 µl DNase-free water. Quality and quantity of DNA extracts were checked using a spectrophotometer (Nanodrop, PeqLab, Germany). For the extraction of total community DNA from manure, 10 ml manure were centrifuged at about 6500 *g* for 10 min, the pellet was homogenized and DNA was extracted from 0.5 g manure pellet using the FastDNA SPIN kit for soil (MP Biomedicals, Heidelberg, Germany), followed by a purification step using the Geneclean spin kit (MP Biomedicals, Heidelberg, Germany), according to the manufacturer's instructions.

Preparation of the 16S rRNA gene amplicons, pyrosequencing and data processing

To generate the amplicon library for *Bacteria*, 16S primers 926-F (5′-AAACTYAAAKGAATTGACGG-3′), *Escherichia coli* position 907–926 and 630-R (5′-CAKAAAGGAGGTGATCC-3′), *E. coli* position 1528–1544 [24], were selected. Fusion primers were designed according to the guidelines of ROCHE (www.my454.com), by extending the specific primers with a 10 base multiplex identifier (MID), a 4 base key and the respective sequencing primers A or B for bidirectional sequencing. The optimal PCR conditions were determined by gradient PCRs with annealing temperatures ranging from 50°C to 60°C. Amplicons were finally generated using 50 ng of extracted DNA, an annealing temperature of 50°C and 22 PCR cycles. The PCR products were purified with AMPure Beads (Agencourt, Beckman Coulter, Krefeld, Germany) according to the Amplicon Library Preparation Method Manual (www.my454.com) and pooled in equimolar amounts.

Sequencing of the 16S rRNA genes was performed on a second-generation pyrosequencer (454 GS FLX Titanium, ROCHE, Germany) following the manufacturer's protocol for amplicon sequencing (www.my454.com). The number of replicates per treatment included in the final sequencing run can be found in Table S1 in File S1.

The automatic amplicon pipeline of the GS Run Processor (ROCHE) was used to perform an initial quality filtering of the pyrosequencing raw reads in order to remove failed reads, low quality reads and adaptor sequences. Sequence files were submitted to the NCBI Sequence Read Archive (www.ncbi.nlm.nih.gov/sra/) and are available with the study accession number SRP038712.

Sequence analysis

The two data sets acquired by the forward and reverse primers were analyzed separately to avoid the systematical bias of individual primers. The analyses were mainly performed according to Ding *et al.* [25]. Only those sequences with a length above 200 bp after removing the barcode, primer and unpaired regions were subjected to further analysis. The unpaired regions were truncated based on a standalone BlastN analysis against a bacterial database described by Pruesse *et al.* [26]. Sequences were grouped into operational taxonomic units (OTUs>97% sequence identity) using software package mothur (v1.14.0). The classification of sequences was performed using software RDP MultiClassifier at >80% confidence [27]. A taxonomic OTU report with each row representing one OTU containing taxonomic positions (from phylum to genus) and the number of sequences for each sample was constructed based on the OTU assignment and on the classification.

Analyses based on the taxonomic OTU report were done with R (version 2.14). To allow the comparison of samples with different number of reads, all algorithms selected rely on relative

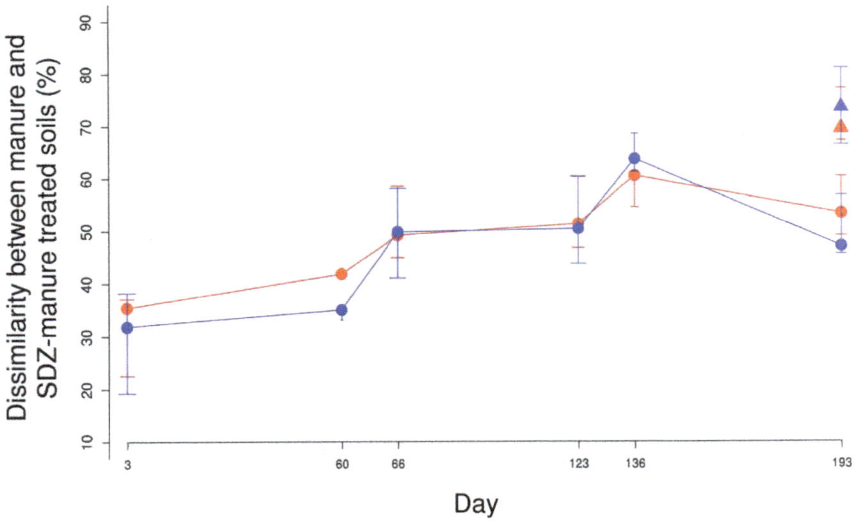

Figure 1. Dissimilarity between soils treated with manure (S0) and manure spiked with SDZ (S10 indicated by circles; S100 indicated by triangles) at different sampling times. Red symbols: results based on the data set acquired by the forward primer; blue symbols: results based on the data set acquired by the reverse primer. Error bars indicate the first and third quartiles.

abundance of taxonomic groups. To check whether the forward and reverse primers revealed a similar taxonomic composition or not, the percent of Bray-Curtis similarity of taxonomic composition between two data sets was calculated for each sample based on the percent of classified taxa at phylum, class, order, family, and genus level using the R add-on package "vegan". To evaluate the influence of repeated application of manure containing SDZ on the bacterial community compositions, the pairwise Pearson dissimilarities between S0 and S10 or between S0 and S100 for day 193 were calculated. To compare the microbial community composition between samples, non-metric multidimensional scaling (NMDS) analysis was carried out based on the Pearson dissimilarity matrix using R package MASS. The 3-dimensional plots of NMDS analysis were created with the R package "rgl" in conjunction with the Linux command "import". Tukey's honest significance tests under a generalized linear model via a logistic function for binomial data with the package multcomp [28] was applied to identify the discriminative taxa between treatments using a Bonferroni adjusted p value<0.05.

Results

Effects of the repeated application of manure containing SDZ on the composition of soil bacterial communities

In this study, bacterial 16S rRNA gene fragments amplified from TC-DNA extracted from 50 composite soil samples were sequenced using forward (926F) and reverse (630R) primers. A total of 132,278 sequences for 50 samples with 790–5685 high quality sequences per sample were obtained with the forward primer, and 162,446 sequences for 49 samples with 1105–7053 high quality sequences per sample with the reverse primer.

To study the effect of repeated applications of manure containing SDZ on the soil bacterial community composition, 16S rRNA gene pyrosequencing data sets acquired from TC-DNA of soils treated with SDZ manure or control manure were compared. Sequences obtained with the forward and reverse primers were treated independently and grouped into OTUs (>97% sequence identity). Both datasets revealed similar changes of bacterial community composition after manure application (Figure 1). In general, the Pearson dissimilarities of the bacterial community compositions between the S0 and S10 soils increased with repeated manure application (Figure 1). Shortly after each manure application the dissimilarity between the bacterial community composition of S0 and S10 soils was strongly increased. Sixty days after the first and second manure applications the dissimilarity remained at the same level as shortly after manure application, while 60 days after the third application a decrease in dissimilarity between S0 and S10 soil was observed. Compared with the S10 soils on day 193, the S100 soils were less similar to S0 soils in the bacterial community composition, indicating that a higher concentration of SDZ caused more pronounced changes of the soil bacterial communities (Figure 1). Non-metric multidimensional scaling based on the OTU reports acquired for the forward (Figure S1 in File S1) and reverse (Figure 2) data set confirmed the findings presented in Figure 1 as it showed that the bacterial community composition of S10 soils increasingly deviated from the S0 soils with time, and this difference was even more pronounced for S100 soils. Furthermore, the faint color symbols in Figure 2 representing the bacterial community composition shortly after manure application strongly deviated from the community composition of untreated soil suggesting a pronounced disturbance of the system. In contrast to S10 and S100 soils, the bacterial community composition of S0 soils (indicated by cubes) 60 days after the third manure application still grouped in the vicinity of the untreated soils (Figures 2 and S1) suggesting a stronger resilience of the soil bacterial community towards manure when no antibiotic was present.

The difference between bacterial communities of the initial soil (day 3 untreated) and of the treated soil was calculated based on Pearson dissimilarity (Figure 3). A pronounced increase in difference was observed for soils shortly after each manure application (S0, S10). The dissimilarity increased with the repeated applications of manure containing SDZ for samples collected 60 days after each manure application. The highest deviation was observed for soil treated three times with manure containing the high concentration (100 mg/kg) of SDZ. At the same sampling time, the least difference was found for the untreated control soils.

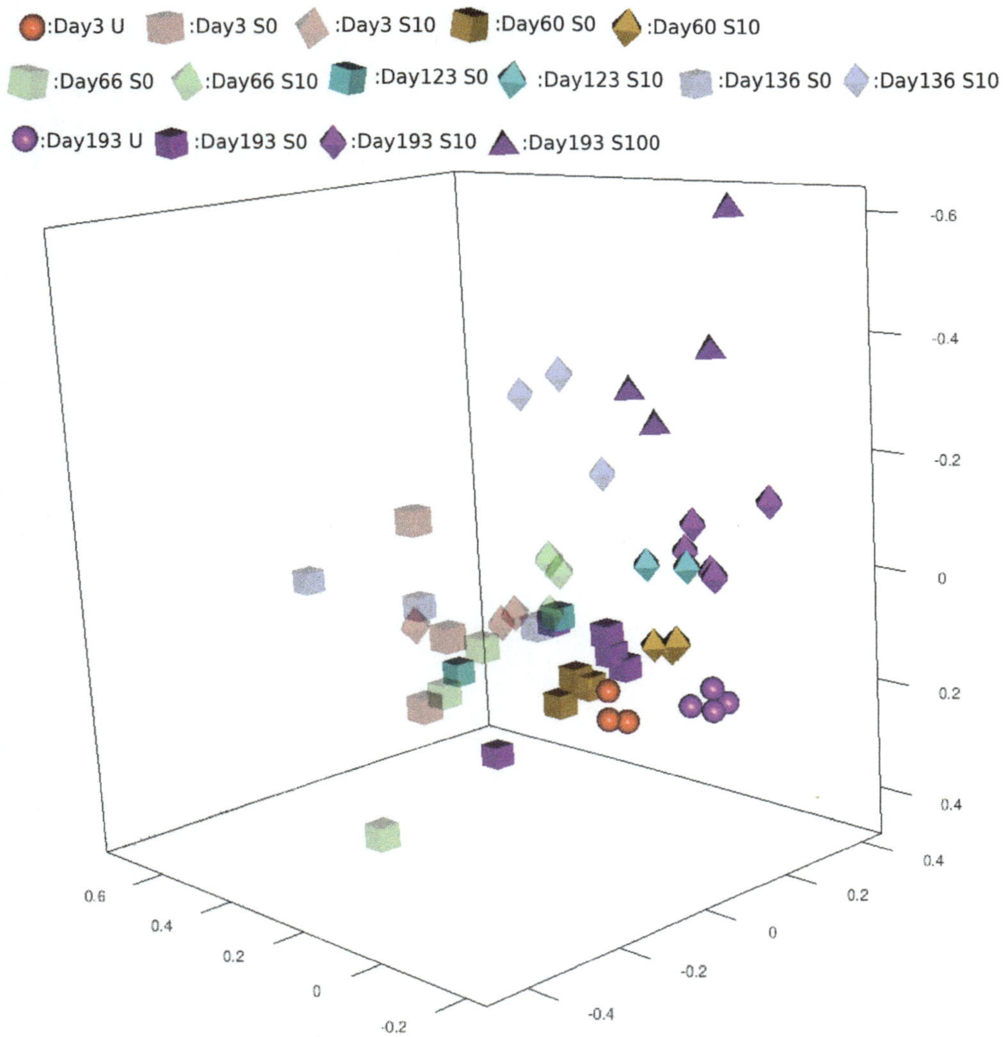

Figure 2. Non metric multidimensional scaling based on OTU reports acquired by the reverse primer sequencing. Balls: untreated soils; cubes: S0; octahedron: S10; pyramid: S100.

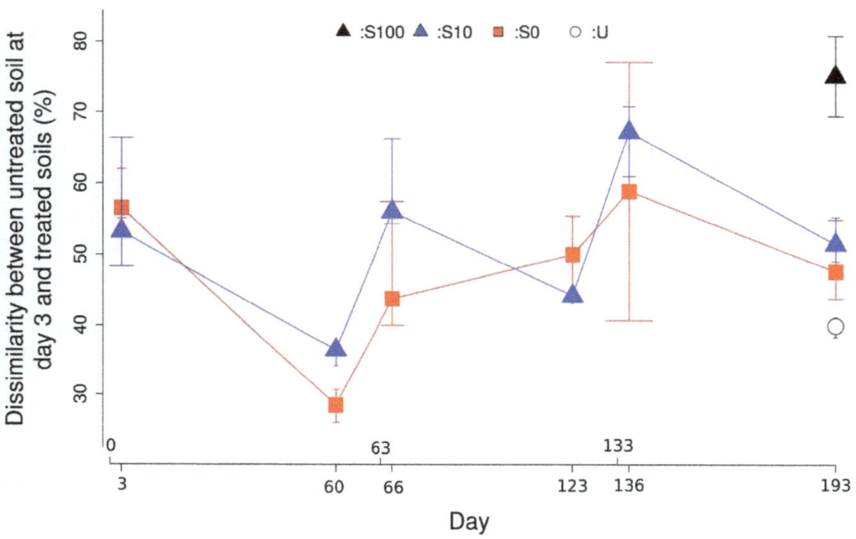

Figure 3. Dissimilarity between soils treated with manure (S0, S10 and S100) or not (U) and the untreated initial soils (Day3 U) at different sampling times. Dissimilarity was calculated from the data acquired by the reverse primer. Error bars indicate the first and third quartiles.

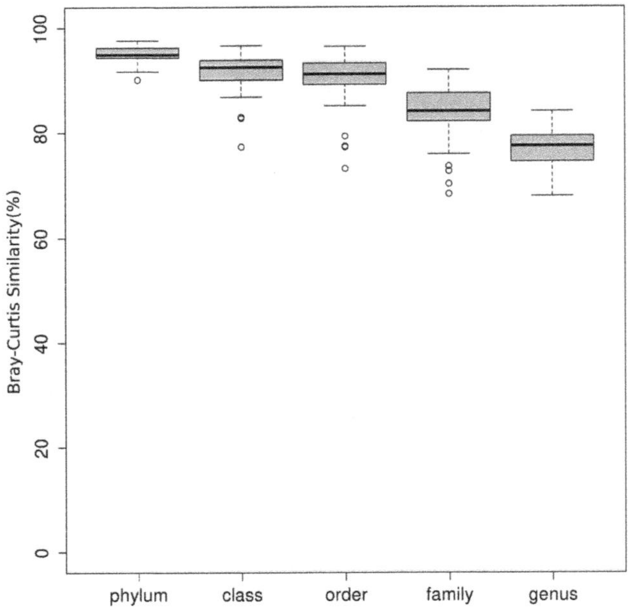

Figure 4. Box plot showing the similarity of taxonomic composition at different ranks between data acquired by the forward and reverse primers.

Taxa with significantly altered relative abundance in response to repeated application of manure spiked with SDZ or unspiked

About 86% of the sequences obtained with the forward primer could be assigned to 21 phyla while 90% of the reverse primer derived sequences were affiliated to 19 phyla. The percent Bray-Curtis similarities of the taxonomic composition between the data sets acquired with the forward and reverse primers, respectively, were calculated for each sample based on the percent of the classified taxa at different taxonomic ranks (from phylum to genus level). Although from the phylum to the genus level the similarity of the taxonomic composition decreased as shown in Figure 4, the overall taxonomic composition of both data sets was very similar.

To identify the taxa which significantly (Bonferroni adjusted p value$<$0.05) responded to the repeated application of manure, comparisons were made between soils treated with S0 or not (U). All taxa with significantly different relative abundance (Bonferroni adjusted p value$<$0.05) were considered. In soil samples taken three days after the addition of manure short-term responders were identified (summarized in Table S2 for the reverse primer derived sequence set and in Table S3 in File S1 for the sequence set obtained with the forward primer). Based on these data (Table S2 and S3) we could mainly identify genera belonging to the *Proteobacteria* such as *Delftia* (*Betaproteobacteria*), *Pseudomonas*, *Pseudoxanthomonas*, *Stenotrophomonas* (*Gammaproteobacteria*), *Bacillus*, *Lactobacillus*, *Streptocococcus*, *Clostridiales* (*Firmicutes*) and *Arthrobacter* (*Actinobacteria*) which were significantly increased in relative abundance in S0 soils compared to the untreated soil as short time responders to the manure application. The increase in the relative abundance of *Firmicutes* (Table S4 in File S1) is likely related to the high abundance of these organisms in the manure (data not shown). In turn the relative abundance of different genera of the *Acidobacteria* and *Planctomycetaceae* but also *Gemmatimonadetes*, the genus *Methylibium* (*Betaproteobacteria*) decreased in S0 soils compared to the untreated soils (Table S2 and S3 in File S1). An increased relative abundance of the genera *Devosia*, *Clostridium*, *Peptostreptococcus* and *Adheribacter* in response to the manure application was observed by comparing U and S0 soil on day 193 (Table S5 and S6 in File S1). In contrast, *Myxococcales* seemed to be negatively affected by manure. On day 193 the relative abundance of *Acidobacteria* was still significantly decreased in all manure treatments compared to untreated soil with the exception of GP10 which showed a higher relative abundance in S10. Some taxa, such as the family *Planctomycetaceae*, were not affected by manure application.

With repeated application of manure the number of taxa with significantly altered relative abundance in S10 compared to S0 soil increased (summarized in Table 1 for the reverse primer and in Table S7 in File S1 for the forward primer). An increased abundance in S10 soils was observed in particular after repeated manure application for the genera *Devosia*, *Sphingobium*, Gp10 and *Gemmatimonas*. Shortly after manure application a significantly increased abundance was observed in S10 compared to the S0 soils for the phylum *Chloroflexi*, the class of the *Deltaproteobacteria* and the orders of *Clostridales* and *Actinomycetales*. Acidobacterial genera Gp4, Gp6 and Gp10 with one exception showed always higher relative abundances in S10 but only in some cases these differences were found to be significant (Table 1). The relative abundances of *Pseudomonas*, *Lysobacter* and *Adheribacter* were lower in all S10 compared to S0 soils indicating that the presence of SDZ exerted a negative effect on these taxa.

Samples taken 60 days after the first, second and third manure applications mainly revealed the same taxa with increased or decreased relative abundance in S10 compared to S0 as in the samples taken shortly after manure application (shown in Table 2 for the reverse primer and in Table S8 in File S1 for the forward primer). With repeated application of manure (S0, S10) the number of taxa with significantly different relative abundance increased from 3 (60 days after the first manure application) to 11 (60 days after the third manure application). Although not significant at all time points the genera *Devosia*, *Gemmatimonas* and *Acidobacteria* Gp6 and Gp10 showed an increased relative abundance in S10 compared to S0 soils. In S100 soils analyzed only for the samples taken 60 days after the third manure application (day 193) the relative abundances of *Devosia* and *Gemmatimonas* were increased compared to their abundance in S10 and S0 soils clearly indicating that bacteria belonging to these genera take profit from high concentrations of SDZ. A higher relative abundance in S100 soils was also observed for *Anerolineaceae* (*Chloroflexi*), *Microbacteriaceae* and the genus *Clostridium*. Most striking was the selection of *Stenotrophomonas* in S100 soils as sequences of this genus were not detected in S0 and in very low relative abundance in S10 soils on day 193. Also acidobacterial genera were identified as responders to SDZ. However, the different genera of *Acidobacteria* responded differently to SDZ manure. The S10 soils had a higher abundance of *Acidobacteria* belonging to the genera Gp10 and Gp6 (except day 60) compared to S0. On day 193, the relative abundance of Gp4, Gp6 and Gp10 was lower in S100 than in S10 soil. The negative effect of SDZ on *Pseudomonas*, *Lysobacter* and *Adheribacter* was also detected in S100 but there was no significant difference in their relative abundance between S10 and S100.

Discussion

The importance of the soil microbiome for various ecosystem services such as nutrient cycling, soil fertility, degradation of pollutants, and plant growth promotion is well recognized [29]. However, antibiotics introduced into agricultural fields via manure might alter the ability of soil microbes to fulfill crucial ecosystem services, changing the diversity and activity of key functional

Table 1. Taxa with significantly different relative abundance between manured soils with or without SDZ treatments shortly after manure addition as revealed by barcoded pyrosequencing using the reverse primer (Bonferroni adjusted p value<0.05).

Phylum	Class	Order	Family	Genus	Day 3		Day 66		Day136	
					S0	S10	S0	S10	S0	S10
Proteobacteria	*Alphaproteobacteria*	*Rhizobiales*	*Hyphomicrobiaceae*	*Devosia*	0.2±0	0.3±0	0.5±0	**1.5±0**	1.1±0	**2.4±0**
		Sphingomonadales	*Sphingomonadaceae*	*Sphingobium*	0.1±0	0.1±0	0.1±0	0.3±0	0.1±0	**0.4±0**
	Betaproteobacteria	*Burkholderiales*	*Comamonadaceae*		**3.8±3**	1.6±0	3.3±1	2.1±0	2.6±1	3±1
			Oxalobacteraceae	*Herminiimonas*	0.1±0	0.1±0	0.5±0	0±0	**3±4**	0±0
	Deltaproteobacteria				1.1±0	1.2±0	1.4±0	1.3±0	3.1±1	**4.4±2**
	Gammaproteobacteria	*Pseudomonadales*	*Pseudomonadaceae*	*Pseudomonas*	**0.8±1**	0.2±0	**2.6±4**	0.2±0	0.6±1	0±0
		Xanthomonadales	*Xanthomonadaceae*	*Lysobacter*	**2.1±1**	1.2±0	**4.1±1**	0.8±1	**1.7±1**	0.2±0
				Stenotrophomonas	**1.6±2**	0.1±0	0±0	0±0	0.1±0	0.3±0
Acidobacteria	*Acidobacteria_Gp10*			*Gp10*	1±0	1.5±0	1.7±0	**3±1**	1±0	**2.1±0**
	Acidobacteria_Gp4			*Gp4*	3.3±0	**4.7±1**	4.8±1	5.3±0	4.2±2	4.1±2
	Acidobacteria_Gp6			*Gp6*	8.7±1	10.5±2	12.1±2	**16.1±1**	8.7±4	9.4±2
Firmicutes	*Bacilli*	*Lactobacillales*			5.4±1	5.3±2	3.9±1	**6.3±2**	2.8±1	3.8±1
	Clostridia	*Clostridiales*			5.9±2	6.4±1	6±1	8.2±1	7.5±1	**10.5±1**
Actinobacteria	*Actinobacteria*	*Actinomycetales*			7.1±1	5.8±2	2.9±1	3.3±1	4.7±5	**5.9±6**
Bacteroidetes	*Sphingobacteria*	*Sphingobacteriales*	*Cytophagaceae*	*Adhaeribacter*	**1.5±0**	0.7±0	**2.3±1**	0.6±0	**1.5±0**	0.4±0
Chloroflexi					3.3±1	3.6±0	1.9±1	3.8±1	4.5±0	**6.2±1**
Gemmatimonadetes	*Gemmatimonadetes*	*Gemmatimonadales*	*Gemmatimonadaceae*	*Gemmatimonas*	0.2±0	0.3±0	0.3±0	0.5±0	0.4±0	**1±1**

Bold numbers indicate taxa with significantly higher relative abundance in SDZ treated or non treated soils. Soils were collected three days after each treatment.

Table 2. Taxa with significantly different relative abundance between manured soils with or without SDZ treatments as revealed by barcoded pyrosequencing using the reverse primer (Bonferroni adjusted p value<0.05).

Phylum	Class	Order	Family	Genus	Day 60		Day 123		Day 193		
					S0	S10	S0	S10	S0	S10	S100
Proteobacteria	*Alphaproteobacteria*	*Rhizobiales*	*Hyphomicrobiaceae*	*Devosia*	0.4±0	**1.5±0**	1.8±1	2.2±0	1.5±0b	2.6±1a	3.5±1a
			Rhizobiaceae		0.4±0	0.3±0	0.3±0	0.5±0	0.6±0a	0.4±0ab	0.4±0b
		Rhodospirillales	*Rhodospirillaceae*	*Magnetospirillum*	0.4±0	0.4±0	0.3±0	0.6±0	0.8±0ab	0.9±0a	0.5±0b
		Sphingomonadales	*Sphingomonadaceae*		4±1	3.4±1	4±1	3.4±1	5.4±1a	3.6±0b	3.4±1b
	Betaproteobacteria	*Burkholderiales*	*Alcaligenaceae*		0.1±0	0.1±0	0.1±0	0.1±0	0.1±0b	0.1±0b	0.8±0a
			Burkholderiales_incertae_sedis	*Methylibium*	1.8±0	1.3±0	1.2±0	1.3±0	2±1a	1.8±1a	0.9±0b
			Comamonadaceae	*Hydrogenophaga*	0.1±0	0.1±0	0.6±0	0.2±0	0.5±0a	0.1±0b	0±0b
			Oxalobacteraceae	*Herminiimonas*	0±0	0±0	**1.2±0**	0±0	0.5±0a	0±0ab	0±0b
		Methylophilales	*Methylophilaceae*		0.1±0	0.1±0	0.3±0	0.1±0	0.3±0a	0.1±0ab	0±0b
		Rhodocyclales	*Rhodocyclaceae*	*Shinella*	0±0	0±0	0±0	0.1±0	0±0b	0.2±0b	1.2±1a
	Gammaproteobacteria	*Pseudomonadales*	*Pseudomonadaceae*	*Pseudomonas*	0.5±1	0.1±0	0.1±0	0±0	1.2±2a	0.2±0b	0±0b
		Xanthomonadales	*Xanthomonadaceae*	*Lysobacter*	**1.6±0**	0.7±0	**1.7±1**	0.6±0	0.9±0a	0.2±0b	0.1±0b
				Pseudoxanthomonas	0.7±1	0.4±0	**1.9±1**	0.4±0	0.7±0b	0.4±0b	2±2a
				Rhodanobacter	0±0	0±0	0.3±0	0.3±0	0.3±0b	0.9±1a	0.2±0b
				Stenotrophomonas	0±0	0±0	0.1±0	0.3±0	0±0b	0.2±0b	3.5±3a
	Deltaproteobacteria	*Myxococcales*	*Haliangiaceae*	*Haliangium*	0.5±0	0.1±0	**0.8±0**	0.1±0	0.8±0a	0.1±0b	0.1±0b
			Nannocystaceae		0.3±0	0±0	0.2±0	0±0	0.3±0a	0.2±0ab	0±0b
			Polyangiaceae		0.2±0	0.3±0	0.2±0	0.3±0	0.3±0b	0.2±0b	0.8±1a
Acidobacteria	*Acidobacteria_Gp10*			*Gp10*	2.8±0	4.3±1	2±0	**5.4±1**	1.6±0c	4.4±0a	2.5±1b
	Acidobacteria_Gp4			*Gp4*	7.9±1	8.6±1	4.4±0	5±1	7.5±2a	5.8±2a	4.5±1b
	Acidobacteria_Gp6			*Gp6*	18.2±2	**22.1±3**	14.2±2	**17.8±0**	13.6±3b	15.8±3a	11.1±2c
Firmicutes	*Bacilli*	*Lactobacillales*			0.1±0	0.1±0	0±0	0.4±0	0.1±0b	0.2±0ab	0.5±0a
	Clostridia	*Clostridiales*	*Clostridiaceae*	*Clostridium*	0.5±0	0.4±0	1.2±1	1.1±0	1.4±0b	1.7±0b	3.7±1a
	Clostridia		*Peptostreptococcaceae*	*Peptostreptococcus*	0.1±0	0.1±0	0.1±0	0.5±0	0.3±0b	0.4±0ab	0.8±0a
Bacteroidetes	*Sphingobacteria*	*Sphingobacteriales*	*Cytophagaceae*	*Adhaeribacter*	1.7±0	0.9±0	1.5±0	0.7±0	2.1±1a	0.6±0b	0.4±0b
Actinobacteria	*Actinobacteria*	*Actinomycetales*	*Microbacteriaceae*	*Leifsonia*	0±0	0.1±0	0±0	0.1±0	0±0b	0.2±0b	0.4±0a
Chloroflexi	*Anaerolineae*	*Anaerolineales*	*Anaerolineaceae*		1±0	1.1±0	1.9±0	1.9±0	1.5±0b	1.8±1ab	2.2±1a
Gemmatimonadetes	*Gemmatimonadetes*	*Gemmatimonadales*	*Gemmatimonadaceae*	*Gemmatimonas*	1.1±1	1.6±0	0.5±0	**1.7±2**	0.7±0c	1.6±1b	2.5±1a
Planctomycetes	*Planctomycetacia*	*Planctomycetales*	*Planctomycetaceae*		2.7±0	2.6±1	2.5±0	2.4±0	2.6±1a	2.5±0a	1.7±0b
Verrucomicrobia	*Subdivision3*			*Subdivision3_genera_incertae_sedis*	0.4±0	0.3±0	0.6±0	0.5±0	0.4±0a	0.5±0a	0.1±0b

Bold numbers indicate taxa with significantly higher relative abundance in SDZ treated or non treated soils; letters a, b and c indicate taxa with significant difference in relative abundance between S0, S10 and S100 treated soils collected at day 193. Soils were collected 60 days after each manure application.

groups by enhancing antibiotic resistant populations while decreasing the abundance of sensitive populations in soils [16,30]. In the present study, the effects of repeated application of SDZ manure on soil bacterial communities were explored by barcoded pyrosequencing of 16S rRNA gene amplicons from TC-DNA. More pronounced changes of bacterial communities were observed for soils that were repeatedly treated with SDZ manure. The strongest effects were observed for soils treated three times with manure containing a high concentration of SDZ (Figures 2 and 4). For the first time, taxonomic groups affected by the presence of SDZ were systematically identified, and numerous taxa were found with significantly altered relative abundance (Tables 1 and 2). These findings largely extended our understanding of the influence of SDZ introduced via manure on soil bacterial communities.

Although the bioavailable fraction of SDZ in soil rapidly decreased after each manure application [16], cumulative inputs of SDZ via manure still affected bacterial communities. Also in the study by Byrne-Bailey *et al.* [31], the proportion of sulfachloropyridazine resistant bacteria increased in soils after treatment with manure containing antibiotics. The repeated application of SDZ manure to grassland soil was found to affect nitrite oxidizing bacteria [32]. The influence of SDZ on soil bacterial communities is more likely to be detected when additional nutrients (root exudates, manure) are added to stimulate bacterial growth [33,34] as SDZ affects only growing bacteria and bacterial activities are generally low in oligotrophic environments such as soils.

The amendment of soil with SDZ manure resulted in a significantly increased relative abundance of numerous Gram-negative and Gram-positive taxa such as *Devosia*, *Shinella*, *Stenotrophomonas*, *Clostridium*, *Peptostreptococcus*, *Leifsonia*, *Gemmatimonas*, suggesting that the presence of SDZ provided a selective advantage for these taxa. Mainly the increase in bacteria affiliated to *Clostridia* detected shortly after the addition of manure likely resulted from manure-derived bacteria. In our study more than 60% of sequences in manure could be affiliated to *Clostridium* (data not shown), which confirms data from previous studies where the dominance of *Clostridium* in manure has been reported [35–37]. A steep decrease of *Clostridia* was observed 60 days after the application. Only in S100 soils the relative abundance of *Clostridia* was found to be significantly higher than in the other soils on day 193, suggesting that the high concentration of SDZ slowed down the decline of *Clostridium* populations. Whether the enhanced abundance of *Clostridium* in S100 soils was due to the selection of SDZ resistant populations deserves further research.

Although the resistance to SDZ was not reported yet for most of these taxa, it is likely that some of the populations enriched in the present study might be resistant to SDZ either due to the presence of resistance genes or due to intrinsic resistance. Isolates belonging to *Stenotrophomonas* and *Clostridium* were previously reported to be resistant to other sulfonamide antibiotics [38,39]. Based on the same TC-DNA, Heuer *et al.* [16] found an accumulation of *sul* gene carrying populations in the soil repeatedly treated with SDZ manure. *Sul* genes located on mobile genetic elements such as class 1 integrons or plasmids might have spread among different taxa [40,41]. The presence of antibiotics, even at very low concentrations could accelerate the genetic exchanges between bacteria [42,43]. The enhanced transferability of resistant genes was recently also reported by Jechalke *et al.* [15] in a field experiment using manure from SDZ-treated pigs. However, we cannot exclude that some of the taxa were intrinsically resistant to, or even able to mineralize SDZ as a SDZ degrader identified as *Microbacterium lacus* was recently isolated from the same Merzenhausen soil [44]. Interestingly, in the present study an OTU affiliated to *Leifsonia*, which is phylogenetically closely related to *Microbacterium* was found to be enriched in S100 soils.

In this study, the consensus sequences of an enriched OTU in soils treated with manure shared high similarity with the 16S rRNA gene of *S. maltophilia* which is also known as a human pathogen [38]. However, these enriched bacteria are not necessarily human pathogens as *Stenotrophomonas* was also found in different environments such as rhizosphere or soils [45,46]. Nevertheless, the enrichments of *Shinella*, *Stenotrophomonas*, *Clostridium* and *Peptostreptococcus* in soil which was continually treated with manure containing a high concentration of SDZ still urge for further investigation due to potential implications for public health. The enrichment of these bacteria, which are phylogenetically closely related to human pathogens, may improve the chance of transferring antibiotic resistance genes to human pathogens, since horizontal gene transfer is more prevalent between closely related organisms than between those distantly related (reviewed by Boto [47]). Soil particles carrying viable bacteria can be transported over long distances and might contribute to the spreading of antibiotic resistant bacteria over wide geographic ranges [48]. The ecological role of taxa such as *Devosia*, *Leifsonia* and *Gemmatimonas* has not been fully understood and thus the effect of their changed abundance on soil functions remains to be explored. Few studies suggested that some members of *Devosia* [49,50] and *Gemmatimonas* [51] might be associated with nitrogen cycling.

The relative abundances of several other taxa such as *Pseudomonas*, *Lysobacter*, *Hydrogenophaga*, *Haliangium*, and *Adhaeribacter* were found to be significantly lower in the soils treated by SDZ manure. Several members of *Lysobacter* or *Pseudomonas* are known as biological control agents against soilborne phytopathogens such as *Rhizoctonia solani*, *Thielaviopsis basicola* [52,53]. Several studies suggested that these beneficial bacteria belonging to *Pseudomonas* or *Lysobacter* probably play an important role for soil suppressiveness and plant growth [45,54]. In the present study, the consensus sequences of an OTU with decreased relative abundance in soil treated with SDZ manure shared 100% similarities with 16S rRNA gene of the strain of *P. brassicacearum* (NCBI accession number: NR_074834) which is a beneficial root-associated bacterium [55]. The decline of *Pseudomonas* and *Lysobacter* in arable soils might increase the susceptibility towards fungal pathogens. However, in general the application of manure to soils is one strategy to suppress soilborne diseases. Our data showed that the relative abundance of *Pseudomonas* was much higher in soils treated with manure than in untreated soils on days 3 and 193. The presence of SDZ in manure likely reduces the beneficial effects of manure on improving soil health as the abundance of *Pseudomonas* was reduced. However, resilience of these beneficial populations might still occur in the rhizosphere as decreased concentrations of SDZ were recently observed in the rhizosphere of maize and grass compared to bulk soil [6]. Little is known about the ecological roles of *Hydrogenophage*, *Haliangium*, *Adhaeribacter* in soil. *Hydrogenophaga* spp. were reported as main biphenyl degraders in the rhizosphere of horseradish (*Armoracia rusticana*) contaminated with polychlorinated biphenyls [56].

The present study is the first to explore the influence of veterinary antibiotics entering soil via manure on the diversity and abundance of *Acidobacteria*. The results revealed that Gp4 was negatively affected by repeated application of manure containing the high concentration of SDZ. Low concentrations of SDZ in manure seemed to favor Gp6 and Gp10, while high concentrations of SDZ could deprive the effects. *Acidobacteria* are known to be adapted to an oligotrophic lifestyle. Indeed, in the present study the relative abundance of *Acidobacteria* was significantly higher in the untreated soil than in the soil amended with manure. Due to their high abundance and ubiquitous distribution in soils, *Acidobacteria* might play an important role in terrestrial ecosystems. In combination with the analysis of three genomes of *acidobacterial* isolates, Ward *et al.* [57] suggested that *Acidobacteria* might significantly contribute to the terrestrial carbon cycle.

Along with nutrients, a large amount of manure bacteria was also introduced into soil which could have also contributed to the dramatically altered bacterial community compositions observed in the present study shortly after each manure application. However, the majority of these introduced manure bacteria might not be very well adapted to soil, allowing the resilience of the indigenous soil bacteria community. In the present study we showed that repeated application of manure spiked with SDZ to soil increased the degree of disturbance and reduced the resilience of soil bacterial communities compared to a single application of manure spiked with SDZ or soils treated with manure without antibiotics. In a recent study by Poulsen *et al.* [58], the taxonomic composition of bacterial communities based on the abundance of phyla and proteobacterial classes were highly similar between soils with amendments of different organic wastes such as manure. In the present study, we observed clear effects of manure on the soil bacterial community composition as the NMDS analysis revealed separated groups for S0 and untreated soils on day 193 (Figure 3). In contrast to several subgroups of the *Acidobacteria*, *Myxococcales* (*Deltaproteobacteria*), numerous taxa belonging to different phyla (*Proteobacteria*, *Firmicutes* and *Bacteriodetes*) were observed with significantly higher relative abundance in the soils amended with manure than in the untreated control 60 days after the third application. Very likely the added nutrients favored taxa with a more copiotrophic lifestyle. In contrast to the field experiment analyzed by Poulsen *et al.* [58], the conditions in the present experiment were more controlled and the time period after the last manure amendment might have been shorter although this was not specified in the Poulsen study.

In summary, repeated input of SDZ into the Merzenhausen soil via manure, especially at a high concentration, caused pronounced changes in soil bacterial communities. The presence of SDZ provided a selective advantage for species affiliated to the genera *Devosia, Shinella, Stenotrophomonas, Clostridium, Peptostreptococcus, Leifsonia, Gemmatimonas*, while suppressing *Pseudomonas, Lysobacter, Hydrogenophage, Haliangium, Adhaeribacter*. We could show that SDZ-containing manure caused a more pronounced disturbance of the soil bacterial community compared to control manure and reduced resilience at least for the time period studied. However, whether the findings obtained can be generalized to other soil types with different soil properties remains an open question which needs further attention. Furthermore, mainly the long-term consequences of repeated manure application for soil ecosystem functions and human health demand further investigations.

Supporting Information

File S1 Supporting Files. Table S1. The number of replicates per treatment included in the final sequencing run using the reverse primer. The numbers in brackets indicate the number of samples sequenced using the forward primer. Table S2. Taxa with significantly different relative abundance between soils on day 3 using the reverse primer (Bonferroni adjusted p value<0.05). Table S3. Taxa with significantly different relative abundance between soils on day 3 using the forward primer (Bonferroni adjusted p value<0.05). Table S4. Relative abundance of detected phyla based on the average of the forward and reverse dataset. Table S5. Taxa with significantly different relative abundance between soils on day 193 using the reverse primer (Bonferroni adjusted p value<0.05). Table S6. Taxa with significantly different relative abundance between soils on day 193 using the forward primer (Bonferroni adjusted p value<0.05). Table S7. Taxa with significantly different relative abundance between manured soils with or without SDZ treatments as revealed by barcoded pyrosequencing using the forward primer (Bonferroni adjusted p value<0.05). Table S8. Taxa with significantly different relative abundance between manured soils with or without SDZ treatments as revealed by barcoded pyrosequencing using the forward primer (Bonferroni adjusted p value<0.05). Figure S1. Non metric multidimensional scaling based on OTU reports acquired by the forward primer sequencing. Balls: untreated soils; cubes: S0; octahedron: S10; pyramid: S100.

Author Contributions

Conceived and designed the experiments: HH KS MS. Performed the experiments: VR BSH. Analyzed the data: GCD. Contributed reagents/materials/analysis tools: GCD VR KS MS. Wrote the paper: GCD BSH HH KS MS SJ.

References

1. Kools SAE, Moltmann JF, Knacker T (2008) Estimating the use of veterinary medicines in the European union. Regulatory Toxicology and Pharmacology 50: 59–65.
2. Lamshöft M, Sukul P, Zühlke S, Spiteller M (2007) Metabolism of C-14-labelled and non-labelled sulfadiazine after administration to pigs. Anal Bioanal Chem 388: 1733–1745.
3. Heuer H, Focks A, Lamshöft M, Smalla K, Matthies M, et al. (2008) Fate of sulfadiazine administered to pigs and its quantitative effect on the dynamics of bacterial resistance genes in manure and manured soil. Soil Biol Biochem 40: 1892–1900.
4. Sarmah AK, Meyer MT, Boxall ABA (2006) A global perspective on the use, sales, exposure pathways, occurrence, fate and effects of veterinary antibiotics (VAs) in the environment. Chemosphere 65: 725–759.
5. Sittig S, Kasteel R, Groeneweg J, Vereecken H (2012) Long-term sorption and sequestration dynamics of the antibiotic sulfadiazine: a batch study. J Environ Qual 41: 1497–1506.
6. Rosendahl I, Siemens J, Groeneweg J, Linzbach E, Laabs V, et al. (2011) Dissipation and sequestration of the veterinary antibiotic sulfadiazine and its metabolites under field conditions. Environ Sci Technol 45: 5216–5222.
7. Hammesfahr U, Heuer H, Manzke B, Smalla K, Thiele-Bruhn S (2008) Impact of the antibiotic sulfadiazine and pig manure on the microbial community structure in agricultural soils. Soil Biol Biochem 40: 1583–1591.
8. Hammesfahr U, Bierl R, Thiele-Bruhn S (2011) Combined effects of the antibiotic sulfadiazine and liquid manure on the soil microbial-community structure and functions. J of Plant Nutr Soil Sci 174: 614–623.
9. Schauss K, Focks A, Heuer H, Kotzerke A, Schmitt H, et al. (2009) Analysis, fate and effects of the antibiotic sulfadiazine in soil ecosystems. Trac-Trends Anal Chem 28: 612–618.
10. Kotzerke A, Sharma S, Schauss K, Heuer H, Thiele-Bruhn S, et al. (2008) Alterations in soil microbial activity and N-transformation processes due to sulfadiazine loads in pig-manure. Environ Pollut 153: 315–322.
11. Kotzerke A, Hammesfahr U, Kleineidam K, Lamshoft M, Thiele-Bruhn S, et al. (2011) Influence of difloxacin-contaminated manure on microbial community structure and function in soils. Biol Fertil Soils 47: 177–186.
12. De Gelder L, Ponciano JM, Abdo Z, Joyce P, Forney LJ, et al. (2004) Combining mathematical models and statistical methods to understand and predict the dynamics of antibiotic-sensitive mutants in a population of resistant bacteria during experimental evolution. Genetics 168: 1131–1144.
13. Heuer H, Smalla K (2007) Manure and sulfadiazine synergistically increased bacterial antibiotic resistance in soil over at least two months. Environ Microbiol 9: 657–666.
14. Kopmann C, Jechalke S, Rosendahl I, Groeneweg J, Krögerrecklenfort E, et al. (2013) Abundance and transferability of antibiotic resistance as related to the fate of sulfadiazine in maize rhizosphere and bulk soil. FEMS Microbiol Ecol 83: 125–134.

15. Jechalke S, Kopmann C, Rosendahl I, Groeneweg J, Weichelt V, et al. (2013) Increased abundance and transferability of resistance genes after field application of manure from sulfadiazine-treated pigs. Appl Environ Microbiol 79: 1704–1711.
16. Heuer H, Solehati Q, Zimmerling U, Kleineidam K, Schloter M, et al. (2011) Accumulation of sulfonamide resistance genes in arable soils due to repeated application of manure containing sulfadiazine. Appl Environ Microbiol 77: 2527–2530.
17. Heuer H, Schmitt H, Smalla K (2011) Antibiotic resistance gene spread due to manure application on agricultural fields. Curr Opin Microbiol 14: 236–243.
18. Gaze WH, Krone SM, Larsson DGJ, Li X, Robinson JA, et al. (2013) Influence of humans on evolution and mobilization of environmental antibiotic resistome. Emerg Infect Dis 19: Article 871.
19. Forsberg KJ, Reyes A, Wang B, Selleck EM, Sommer MOA, et al. (2012) The shared antibiotic resistome of soil bacteria and human pathogens. Science 337: 1107–1111.
20. Sukul P, Lamshoft M, Zuhlke S, Spiteller M (2008) Sorption and desorption of sulfadiazine in soil and soil-manure systems. Chemosphere 73: 1344–1350.
21. Griffiths BS, Philippot L (2013) Insights into the resistance and resilience of the soil microbial community. FEMS Microbiol Rev 37: 112–129.
22. Lamshöft M, Sukul P, Zühlke S, Spiteller M (2010) Behaviour of C-14-sulfadiazine and C-14-difloxacin during manure storage. Sci Total Environ 408: 1563–1568.
23. Griffiths RI, Whiteley AS, O'Donnell AG, Bailey MJ (2000) Rapid method for coextraction of DNA and RNA from natural environments for analysis of ribosomal DNA- and rRNA-based microbial community composition. Appl Environ Microbiol 66: 5488–5491.
24. Timmers RA, Rothballer M, Strik D, Engel M, Schulz S, et al. (2012) Microbial community structure elucidates performance of *Glyceria maxima* plant microbial fuel cell. Appl Microbiol Biotechnol 94: 537–548.
25. Ding G-C, Heuer H, Smalla K (2012) Dynamics of bacterial communities in two unpolluted soils after spiking with phenanthrene: soil type specific and common responders. Front Microbiol 3: doi: 10.3389/fmicb.2012.00290.
26. Pruesse E, Quast C, Knittel K, Fuchs BM, Ludwig W, et al. (2007) SILVA: a comprehensive online resource for quality checked and aligned ribosomal RNA sequence data compatible with ARB. Nucleic Acids Res 35: 7188–7196.
27. Wang Q, Garrity GM, Tiedje JM, Cole JR (2007) Naïve bayesian classifier for rapid assignment of rRNA sequences into the new bacterial taxonomy. Appl Environ Microbiol 73: 5261–5267.
28. Hothorn T, Bretz F, Westfall P (2008) Simultaneous inference in general parametric models. Biometrical J 50: 346–363.
29. Torsvik V, Øvreås L (2002) Microbial diversity and function in soil: from genes to ecosystems. Curr Opin Microbiol 5: 240–245.
30. Zhu YG, Johnson TA, Su JQ, Qiao M, Guo GX, et al. (2013) Diverse and abundant antibiotic resistance genes in Chinese swine farms. Proc Natl Acad Sci U S A 110: 3435–3440.
31. Byrne-Bailey KG, Gaze WH, Kay P, Boxall ABA, Hawkey PM, et al. (2009) Prevalence of sulfonamide resistance genes in bacterial isolates from manured agricultural soils and pig slurry in the United Kingdom. Antimicrob Agents Chemother 53: 696–702.
32. Ollivier J, Schacht D, Kindler R, Groeneweg J, Engel M, et al. (2013) Effects of repeated application of sulfadiazine-contaminated pig manure on the abundance and diversity of ammonia and nitrite oxidizers in the root-rhizosphere complex of pasture plants under field conditions. Front Microbiol 4: Article 22.
33. Brandt KK, Sjøholm OR, Krogh KA, Halling-Sørensen B, Nybroe O (2009) Increased pollution-induced bacterial community tolerance to sulfadiazine in soil hotspots amended with artificial root exudates. Environ Sci Technol 43: 2963–2968.
34. Zielezny Y, Groeneweg J, Vereecken H, Tappe W (2006) Impact of sulfadiazine and chlorotetracycline on soil bacterial community structure and respiratory activity. Soil Biol Biochem 38: 2372–2380.
35. Doornbos RF, van Loon LC, Bakker P (2012) Impact of root exudates and plant defense signaling on bacterial communities in the rhizosphere. A review. Agron Sustain Dev 32: 227–243.
36. Peu P, Brugere H, Pourcher AM, Kerouredan M, Godon JJ, et al. (2006) Dynamics of a pig slurry microbial community during anaerobic storage and management. Appl Environ Microbiol 72: 3578–3585.
37. Yi J, Zheng R, Li FG, Chao Z, Deng CY, et al. (2012) Temporal and spatial distribution of *Bacillus* and *Clostridium histolyticum* in swine manure composting by fluorescent in situ hybridization (FISH). Appl Microbiol Biotechnol 93: 2625–2632.
38. Brooke JS (2012) Stenotrophomonas maltophilia: an emerging global opportunistic pathogen. Clin Microbiol Rev 25: 2–41.
39. Neikirk RL, Krieg NR (1964) Sulfonamide resistance of *Clostridium sporogenes*. J Bacteriol 87: 1526–1527.
40. Heuer H, Kopmann C, Binh CTT, Top EM, Smalla K (2009) Spreading antibiotic resistance through spread manure: characteristics of a novel plasmid type with low %G plus C content. Environ Microbiol 11: 937–949.
41. Heuer H, Smalla K (2012) Plasmids foster diversification and adaptation of bacterial populations in soil. FEMS Microbiol Rev 36: 1083–1104.
42. Davies J, Spiegelman GB, Yim G (2006) The world of subinhibitory antibiotic concentrations. Curr Opin Microbiol 9: 445–453.
43. Martinez JL (2009) Environmental pollution by antibiotics and by antibiotic resistance determinants. Environ Pollut 157: 2893–2902.
44. Tappe W, Herbst M, Hofmann D, Koeppchen S, Kummer S, et al. (2013) Degradation of sulfadiazine by *Microbacterium lacus* strain SDZm4, isolated from lysimeters previously manured with slurry from sulfadiazine-medicated pigs. Appl Environ Microbiol 79: 2572–2577.
45. Hayward AC, Fegan N, Fegan M, Stirling GR (2010) *Stenotrophomonas* and *Lysobacter*: ubiquitous plant-associated gamma-proteobacteria of developing significance in applied microbiology. J Appl Microbiol 108: 756–770.
46. Ryan RP, Monchy S, Cardinale M, Taghavi S, Crossman L, et al. (2009) The versatility and adaptation of bacteria from the genus *Stenotrophomonas*. Nat Rev Microbiol 7: 514–525.
47. Boto L (2010) Horizontal gene transfer in evolution: facts and challenges. Proc Biol Sci 277: 819–827.
48. Prospero JM, Blades E, Mathison G, Naidu R (2005) Interhemispheric transport of viable fungi and bacteria from Africa to the Caribbean with soil dust. Aerobiologia 21: 1–19.
49. Vanparys B, Heylen K, Lebbe L, De Vos P (2005) *Devosia limi* sp. nov., isolated from a nitrifying inoculum. Int J Syst Evol Microbiol 55: 1997–2000.
50. Rivas R, Velazquez E, Willems A, Vizcaino N, Subba-Rao NS, et al. (2002) A new species of *Devosia* that forms a unique nitrogen-fixing root-nodule symbiosis with the aquatic legume *Neptunia natans* (L.f.) druce. Appl Environ Microbiol 68: 5217–5222.
51. Yang Y, Wang X, Shi JF, Li JJ (2012) The influence of the discharging sewage on microbial diversity in sludge from Dongting Lake. World J Microb Biot 28: 421–430.
52. Sullivan RF, Holtman MA, Zylstra GJ, White JF, Kobayashi DY (2003) Taxonomic positioning of two biological control agents for plant diseases as *Lysobacter enzymogenes* based on phylogenetic analysis of 16S rDNA, fatty acid composition and phenotypic characteristics. J Appl Microbiol 94: 1079–1086.
53. Weller DM (2007) Pseudomonas biocontrol agents of soilborne pathogens: looking back over 30 years. Phytopathology 97: 250–256.
54. Mendes R, Kruijt M, de Bruijn I, Dekkers E, van der Voort M, et al. (2011) Deciphering the rhizosphere microbiome for disease-suppressive bacteria. Science 332: 1097–1100.
55. Ortet P, Barakat M, Lalaouna D, Fochesato S, Barbe V, et al. (2011) Complete genome sequence of a beneficial plant root-associated bacterium, *Pseudomonas brassicacearum*. J Bacteriol 193: 3146.
56. Uhlik O, Jecna K, Mackova M, Vlcek C, Hroudova M, et al. (2009) Biphenyl-metabolizing bacteria in the rhizosphere of horseradish and bulk soil contaminated by polychlorinated biphenyls as revealed by stable isotope probing. Appl Environ Microbiol 75: 6471–6477.
57. Ward NL, Challacombe JF, Janssen PH, Henrissat B, Coutinho PM, et al. (2009) Three genomes from the phylum *Acidobacteria* provide insight into the lifestyles of these microorganisms in soils. Appl Environ Microbiol 75: 2046–2056.
58. Poulsen PHB, Abu Al-Soud W, Bergmark L, Magid J, Hansen LH, et al. (2013) Effects of fertilization with urban and agricultural organic wastes in a field trial - prokaryotic diversity investigated by pyrosequencing. Soil Biol Biochem 57: 784–793.

Permissions

All chapters in this book were first published in PLOS ONE, by The Public Library of Science; hereby published with permission under the Creative Commons Attribution License or equivalent. Every chapter published in this book has been scrutinized by our experts. Their significance has been extensively debated. The topics covered herein carry significant findings which will fuel the growth of the discipline. They may even be implemented as practical applications or may be referred to as a beginning point for another development.

The contributors of this book come from diverse backgrounds, making this book a truly international effort. This book will bring forth new frontiers with its revolutionizing research information and detailed analysis of the nascent developments around the world.

We would like to thank all the contributing authors for lending their expertise to make the book truly unique. They have played a crucial role in the development of this book. Without their invaluable contributions this book wouldn't have been possible. They have made vital efforts to compile up to date information on the varied aspects of this subject to make this book a valuable addition to the collection of many professionals and students.

This book was conceptualized with the vision of imparting up-to-date information and advanced data in this field. To ensure the same, a matchless editorial board was set up. Every individual on the board went through rigorous rounds of assessment to prove their worth. After which they invested a large part of their time researching and compiling the most relevant data for our readers.

The editorial board has been involved in producing this book since its inception. They have spent rigorous hours researching and exploring the diverse topics which have resulted in the successful publishing of this book. They have passed on their knowledge of decades through this book. To expedite this challenging task, the publisher supported the team at every step. A small team of assistant editors was also appointed to further simplify the editing procedure and attain best results for the readers.

Apart from the editorial board, the designing team has also invested a significant amount of their time in understanding the subject and creating the most relevant covers. They scrutinized every image to scout for the most suitable representation of the subject and create an appropriate cover for the book.

The publishing team has been an ardent support to the editorial, designing and production team. Their endless efforts to recruit the best for this project, has resulted in the accomplishment of this book. They are a veteran in the field of academics and their pool of knowledge is as vast as their experience in printing. Their expertise and guidance has proved useful at every step. Their uncompromising quality standards have made this book an exceptional effort. Their encouragement from time to time has been an inspiration for everyone.

The publisher and the editorial board hope that this book will prove to be a valuable piece of knowledge for researchers, students, practitioners and scholars across the globe.

List of Contributors

Yunbin Qin, Zhongbao Xin, Xinxiao Yu and Yuling Xiao
Institute of Soil and Water Conservation, Beijing Forestry University, Beijing, China

Xiao-Tao Lü
State Key Laboratory of Forest and Soil Ecology, Institute of Applied Ecology, Chinese Academy of Sciences, Shenyang, China

Hai-Wei Wei
State Key Laboratory of Forest and Soil Ecology, Institute of Applied Ecology, Chinese Academy of Sciences, Shenyang, China
Graduate University of Chinese Academy of Sciences, Beijing, China

Xing-Guo Han
State Key Laboratory of Forest and Soil Ecology, Institute of Applied Ecology, Chinese Academy of Sciences, Shenyang, China
State Key Laboratory of Vegetation and Environmental Change, Institute of Botany, Chinese Academy of Sciences, Beijing, China

Fu-Mei Lü
State Key Laboratory of Vegetation and Environmental Change, Institute of Botany, Chinese Academy of Sciences, Beijing, China

Guang-Lei Gao, Yan-Feng Bao, Yun-Dong Liu, Li Wan and Ji-Feng Deng
College of Soil & Water Conservation, Beijing Forestry University, Beijing, China

Guo-Dong Ding, Bin Wu, Yu-Qing Zhang, Shu-Gao Qin and Yuan-Yuan Zhao
College of Soil & Water Conservation, Beijing Forestry University, Beijing, China
Yanchi Research Station, Yanchi, China

Atanu Mukherjee and Rattan Lal
Carbon Management and Sequestration Center, School of Environment and Natural Resources, The Ohio State University, Columbus, Ohio, United States of America

Max Brunke
Science Team, National Ecological Observatory Network (NEON), Boulder, Colorado, United States of America

Henry Loescher and Hongyan Luo
Science Team, National Ecological Observatory Network (NEON), Boulder, Colorado, United States of America
Institute of Arctic and Alpine Research (INSTAAR), University of Colorado, Boulder, Colorado, United States of America

Edward Ayres
Science Team, National Ecological Observatory Network (NEON), Boulder, Colorado, United States of America
Institute of Arctic and Alpine Research (INSTAAR), University of Colorado, Boulder, Colorado, United States of America
Natural Resource Ecology Laboratory, Colorado State University, Fort Collins, Colorado, United States of America

Paul Duffy
Neptune and Company Inc., Lakewood, Colorado, United States of America

Belén Carbonetto, Nicolás Rascovan and Martin P. Vázquez
Instituto de Agrobiotecnología de Rosario (INDEAR), Predio CCT Rosario, Santa Fe, Argentina

Roberto Álvarez
Facultad de Agronomía, Universidad de Buenos Aires, Buenos Aires, Argentina

Alejandro Mentaberry
Departamento de Fisiología y Biología Molecular y Celular, Facultad de Ciencias Exactas y Naturales, Universidad de Buenos Aires, Buenos Aires, Argentina

Xingliang Xu, Qingkang Li, Jingyuan Wang and Leiming Zhang
Key Laboratory of Ecosystem Network Observation and Modeling, Institute of Geographic Sciences and Natural Resources Research, Chinese Academy of Sciences, Beijing, China

Qianru Li and Yue Sun
Key Laboratory of Ecosystem Network Observation and Modeling, Institute of Geographic Sciences and Natural Resources Research, Chinese Academy of Sciences, Beijing, China
Graduate School of the Chinese Academy of Sciences, Beijing, China

Shengni Tian
College of Life Science, Anhui Agriculture University, Hefei, Anhui Province, China

Lin Zhi
College of Life Science, Anhui Agriculture University, Hefei, Anhui Province, China
Xikou Junior Middle School, Xikou City, Zhejiang Province, China

Thomas Püschel
Faculty of Life Sciences, University of Manchester, Michael Smith Building, Oxford Road, Manchester, United Kingdom

Hugo A. Benítez
Faculty of Life Sciences, University of Manchester, Michael Smith Building, Oxford Road, Manchester, United Kingdom
Instituto de Alta Investigación, Universidad de Tarapacá, Casilla, Arica, Chile

Darija Lemic, Maja Čačija, Antonela Kozina and Renata Bažok
University of Zagreb, Faculty of Agriculture, Department for Agricultural Zoology, Svetošimunska, Zagreb, Croatia

Dylan P. Smith and Kabir G. Peay
Department of Biology, Stanford University, Stanford, California, United States of America

Xiaohua Wei
Department of Earth and Environmental Sciences, University of British Columbia, Kelowna, British Columbia, Canada

Juan A. Blanco
Departmento de Ciencias del Medio Natural, Universidad Pública de Navarra, Pamplona, Navarra, Spain

Naomi Kodama, Seiichiro Yonemura and Satoshi Kaneda
National Institute for Agro-Environmental Sciences (NIAES), 3-1-3 Kannondai, Tsukuba, Ibaraki, Japan

Toshifumi Kimura, Mizue Ohashi and Hidetoshi Ikeno
School of Human Science and Environment, University of Hyogo, 1-1-12 Shinzaike-honcho, Himeji, Hyogo, Japan

Xing Liu, Wenming Zhang and Huizhen Qiu
College of Resources and Environmental Sciences, Gansu Agricultural University, Lanzhou, Gansu Province, China
Gansu Provincial Key Laboratory of Aridland Crop Science, Gansu Agricultural University, Lanzhou, Gansu Province, China

Junlian Zhang
College of Agronomy, Gansu Agricultural University, Lanzhou, Gansu Province, China
Gansu Key Laboratory of Crop Genetic and Germplasm Enhancement, Gansu Agricultural University, Lanzhou, China

Tianyu Gu and Shixue Yin
College of Environmental Science and Engineering, Yangzhou University, Yangzhou, Jaingsu Province, China

Qirong Shen
Jiangsu Provincial Key Lab of Organic Solid Waste Utilization, Nanjing Agricultural University, Nanjing, China

Chongfeng Bu
Institute of Soil and Water Conservation, Northwest A & F University, Yangling, Shaanxi, China
Institute of Soil and Water Conservation, Chinese Academy of Sciences and Ministry of Water Resources, Yangling, Shaanxi, China

Yongsheng Yang
Institute of Soil and Water Conservation, Chinese Academy of Sciences and Ministry of Water Resources, Yangling, Shaanxi, China

Shufang Wu
Institute of Soil and Water Conservation, Northwest A & F University, Yangling, Shaanxi, China
College of Water Resources and Architectural Engineering, Northwest A & F University, Yangling, Shaanxi, China

Mingguo Zheng
Institute of Geographic Sciences and Natural Resources Research, Chinese Academy of Sciences, Beijing, China

Ricardo Luis Louro Berbara
Soil Science Department, Agronomy Institute, Federal Rural University of Rio de Janeiro, Seropédica-RJ, Brazil

Pedro R. Peres-Neto
Canada Research Chair in Spatial Modelling and Biodiversity; Université du Québecá Montréal, Département des sciences biologiques, Québec, Canada

Guilherme Montandon Chaer and Ederson da Conceição Jesus
Embrapa Agrobiologia, Seropédica-RJ, Brazil

Francy Junio Gonçalves Lisboa
Soil Science Department, Agronomy Institute, Federal Rural University of Rio de Janeiro, Seropédica-RJ, Brazil
The James Hutton Institute, Craigiebuckler, Aberdeen, United Kingdom

Ruth Joy Mitchell, Stephen James Chapman, Ruth Joy Mitchell and Stephen James Chapman
The James Hutton Institute, Craigiebuckler, Aberdeen, United Kingdom

Selene Muñiz, Juan Lacarta, María P. Pata, Juan José Jiménez and Enrique Navarro
Department of Biodiversity & Conservation and Ecosystem Restoration, Pyrenean Institute of Ecology, Spanish National Research Council, Zaragoza, Spain

Xavier Raynaud
Sorbonne Universités, UPMC Univ Paris 06, Institute of Ecology and Environmental Sciences - Paris, Paris, France

Naoise Nunan
CNRS, Institute of Ecology and Environmental Sciences - Paris, Campus AgroParisTech, Thiverval-Grignon, France

Songze Wan, Zhanfeng Liu, Lixia Zhou and Shenglei Fu
Key Laboratory of Vegetation Restoration and Management of Degraded Ecosystems, South China Botanical Garden, Chinese Academy of Sciences, Guangzhou, China

Chenlu Zhang
State Key Laboratory of Cotton Biology, Key Laboratory of Plant Stress Biology, College of Life Sciences, Henan University, Kaifeng, China

Jie Zhao
Key Laboratory of Vegetation Restoration and Management of Degraded Ecosystems, South China Botanical Garden, Chinese Academy of Sciences, Guangzhou, China
Key Laboratory of Agro-Ecological Processes in Subtropical Region, Institute of Subtropical Agriculture, Chinese Academy of Sciences, Changsha, China

Peng Zhang, Ting Wei, Zhikuan Jia, Qingfang Han and Xiaolong Ren
The Chinese Institute of Water-Saving Agriculture, Northwest A&F University, Yangling, Shaanxi, China
Key Laboratory of Crop Physi-Ecology and Tillage Science in Northwestern Loess Plateau, Ministry of Agriculture, Northwest A&F University, Yangling, Shaanxi, China

Yongping Li
Guyuan Institute of Agricultural Sciences, Guyuan, Ningxia, China

Zhaojun Li and Yongchao Liang
Institute of Agricultural Resources and Regional Planning, Chinese Academy of Agricultural Sciences, Key Laboratory of Plant Nutrition and Fertilizer, Ministry of Agriculture, Beijing, China

Hua Yang
Institute of Agricultural Resources and Regional Planning, Chinese Academy of Agricultural Sciences, Key Laboratory of Plant Nutrition and Fertilizer, Ministry of Agriculture, Beijing, China
Guizhou Key Laboratory of Mountain Environment, Guizhou Normal University, Guiyang, China

Jian Long
Guizhou Key Laboratory of Mountain Environment, Guizhou Normal University, Guiyang, China

Yupeng Li
Institute of Agricultural Resources and Regional Planning, Chinese Academy of Agricultural Sciences, Key Laboratory of Plant Nutrition and Fertilizer, Ministry of Agriculture, Beijing, China
College of City and Environmental Sciences, Shanxi Normal University, Linfen, China

Lars A. Brudvig
Department of Plant Biology, Michigan State University, East Lansing, Michigan, United States of America

Philip G. Hahn, W. Brett Mattingly, Joseph W. Veldman, John L. Orrock and Ellen I. Damschen
Department of Zoology, University of Wisconsin, Madison, Wisconsin, United States of America

Cathy D. Collins
Department of Biology, Colby College, Waterville, Maine, United States of America

Joan L. Walker
U.S. Forest Service Southern Research Station, Clemson, South Carolina, United States of America

Weijun Shen
Key laboratory of Vegetation Restoration and Management of Degraded Ecosystems, South China Botanical Garden, Chinese Academy of Sciences, Guangzhou, China

Weiping Zhou
Key laboratory of Vegetation Restoration and Management of Degraded Ecosystems, South China Botanical Garden, Chinese Academy of Sciences, Guangzhou, China
Graduate University of the Chinese Academy of Sciences, Beijing, PR China

Dafeng Hui
Department of Biological Sciences, Tennessee State University, Nashville, Tennessee, United States of America

Hang Wang, HongYi Li, ZhiJian Zhang and XinHua Xu
College of Natural Resource and Environmental Sciences, China Academy of West Development, Zhejiang University, Hangzhou, Zhejiang Province, China

Jeffrey D. Muehlbauer
Curriculum for the Environment & Ecology, University of North Carolina at Chapel Hill, Chapel Hill, North Carolina, United States of America

Qiang He
Department of Civil & Environmental Engineering, University of Tennessee, Knoxville, Tennessee, United States of America

ChunLei Yue
Institute of Ecology, Zhejiang Forestry Academy, Hangzhou, China

DaQian Jiang
Department of Earth and Environmental Engineering, Henry Krumb School of Mines, Columbia University, New York, New York, United States of America

Andrea G. Vincent
Department of Ecology and Environmental Sciences, Umeå University, Umeå, Sweden
Department of Forest Ecology and Management, Swedish University of Agricultural Sciences, Umeå, Sweden

Maja K. Sundqvist and David A. Wardle
Department of Forest Ecology and Management, Swedish University of Agricultural Sciences, Umeå, Sweden

Reiner Giesler
Department of Ecology and Environmental Sciences, Umeå University, Umeå, Sweden
Climate Impacts Research Centre, Department of Ecology and Environmental Sciences, Umeå University, Abisko, Sweden

David L. Jones, Paul Simfukwe and Paul W. Hill
School of the Environment, Natural Resources & Geography, Bangor University, Bangor, Gwynedd, United Kingdom

Robert T. E. Mills
Ecole Polytechnique Fédérale de Lausanne (EPFL), School of Architecture, Civil and Environmental Engineering (ENAC), Laboratory of Ecological Systems (ECOS), Lausanne, Switzerland

Bridget A. Emmett
Centre for Ecology and Hydrology, Environment Centre Wales, Bangor, Gwynedd, United Kingdom

Guo-Chun Ding, Sven Jechalke, Holger Heuer and Kornelia Smalla
Julius Kühn-Institut - Federal Research Centre for Cultivated Plants (JKI), Institute for Epidemiology and Pathogen Diagnostics, Braunschweig, Germany

Viviane Radl, Brigitte Schloter-Hai and Michael Schloter
Helmholtz Zentrum München, German Research Center for Environmental Health, Research Unit for Environmental Genomics, Neuherberg, Germany

Index